Communications
in Computer and Information Science 678

Commenced Publication in 2007
Founding and Former Series Editors:
Alfredo Cuzzocrea, Dominik Ślęzak, and Xiaokang Yang

Vladimir M. Vishnevskiy · Konstantin E. Samouylov
Dmitry V. Kozyrev (Eds.)

Distributed Computer and Communication Networks

19th International Conference, DCCN 2016
Moscow, Russia, November 21–25, 2016
Revised Selected Papers

 Springer

Editors

Vladimir M. Vishnevskiy
V.A. Trapeznikov Institute of Control
  Sciences
Russian Academy of Sciences
Moscow
Russia

Konstantin E. Samouylov
RUDN University
Moscow
Russia

Dmitry V. Kozyrev
V.A. Trapeznikov Institute of Control
  Sciences
Russian Academy of Sciences
Moscow
Russia

and

RUDN University
Moscow
Russia

ISSN 1865-0929   ISSN 1865-0937   (electronic)
Communications in Computer and Information Science
ISBN 978-3-319-51916-6   ISBN 978-3-319-51917-3   (eBook)
DOI 10.1007/978-3-319-51917-3

Library of Congress Control Number: 2016963656

Printed on acid-free paper

This Springer imprint is published by Springer Nature
The registered company is Springer International Publishing AG
The registered company address is: Gewerbestrasse 11, 6330 Cham, Switzerland

Preface

This volume contains a collection of revised selected full-text papers presented at the 19th International Conference on Distributed Computer and Communication Networks (DCCN 2016), held in Moscow, Russia, November 21–25, 2016.

The conference is a continuation of traditional international conferences of the DCCN series, which took place in Bulgaria (Sofia, 1995, 2005, 2006, 2008, 2009, 2014), Israel (Tel Aviv, 1996, 1997, 1999, 2001), and Russia (Moscow, 1998, 2000, 2003, 2007, 2010, 2011, 2013, 2015) in the past 19 years. The main idea of the conference is to provide a platform and forum for researchers and developers from academia and industry from various countries working in the area of theory and applications of distributed computer and communication networks, mathematical modeling, methods of control and optimization of distributed systems, by offering them a unique opportunity to share their views, discuss prospective developments, and pursue collaborations in this area. The content of this volume is related to the following subjects:

1. Computer and communication networks architecture optimization
2. Control in computer and communication networks
3. Performance and QoS/QoE evaluation in wireless networks
4. Analytical modeling and simulation of next-generation communications systems
5. Queuing theory and reliability theory applications in computer networks
6. Wireless 4G/5G networks, cm- and mm-wave radio technologies
7. RFID technology and its application in intellectual transportation networks
8. Internet of Things, wearables, and applications of distributed information systems
9. Probabilistic and statistical models in information systems
10. Mathematical modeling of high-tech systems
11. Mathematical modeling and control problems
12. Distributed and cloud computing systems, big data analytics

The DCCN 2016 conference gathered 208 submissions from authors from 20 different countries. From these, 141 high-quality papers in English were accepted and presented during the conference, 56 of which were recommended by session chairs and selected by the Program Committee for the Springer proceedings.

All the papers selected for the proceedings are given in the form presented by the authors. These papers are of interest to everyone working in the field of computer and communication networks.

We thank all the authors for their interest in DCCN, the members of the Program Committee for their contributions, and the reviewers for their peer-reviewing efforts.

November 2016
Vladimir M. Vishnevskiy
Konstantin E. Samouylov

Organization

DCCN 2016 was jointly organized by the Russian Academy of Sciences (RAS), the V. A. Trapeznikov Institute of Control Sciences of RAS (ICS RAS), the Peoples' Friendship University of Russia (RUDN), the National Research Tomsk State University, and the Institute of Information and Communication Technologies of Bulgarian Academy of Sciences (IICT BAS).

Steering Committee

General Chairs

S.N. Vasilyev	ICS RAS, Russia
V.M. Filippov	RUDN University, Russia
V.M. Vishnevskiy	ICS RAS, Russia
K.E. Samouylov	RUDN University, Russia

Program Committee

G. Adam	Joint Institute for Nuclear Research, Romania
A.M. Andronov	Transport and Telecommunication Institute, Latvia
E.A. Ayrjan	Joint Institute for Nuclear Research, Armenia
L.I. Abrosimov	Moscow Power Engineering Institute, Russia
Mo Adda	University of Portsmouth, UK
T.I. Aliev	ITMO University, Russia
S.D. Andreev	Tampere University of Technology, Finland
G. Araniti	University Mediterranea of Reggio Calabria, Italy
Bijan Saha	Joint Institute for Nuclear Research, Bangladesh
J. Busa	Technical University of Košice (TUKE), Slovakia
H. Chaouchi	Institut Télécom SudParis, France
T. Czachorski	Institute of Informatics of the Polish Academy of Sciences, Poland
B.N. Chetverushkin	Keldysh Institute of Applied Mathematics of RAS, Russia
O. Chuluunbaatar	National University of Mongolia, Mongolia
A.N. Dudin	Belarusian State University, Belarus
D. Fiems	Ghent University, Belgium
V.P. Gerdt	Joint Institute for Nuclear Research, Russia
A. Gelman	IEEE Communications Society, USA
D. Grace	York University, UK
A.A. Grusho	Federal Research Center "Computer Science and Control" of RAS, Russia
M. Hnatich	Pavol Jozef Šafárik University in Košice (UPJŠ), Slovakia
J. Hošek	Brno University of Technology, Czech Republic
J. Kolodziej	Cracow University of Technology, Poland

V.Y. Korolev	Lomonosov Moscow State University, Russia
B. Khoromskij	Max Planck Institute for Mathematics in the Sciences, Germany
C. Kim	Sangji University, Korea
G. Kotsis	Johannes Kepler University Linz, Austria
A. Krishnamoorthy	Cochin University of Science and Technology, India
A.E. Kucheryavy	Bonch-Bruevich St. Petersburg State University of Telecommunications, Russia
E.A. Kucheryavy	Tampere University of Technology, Finland
L. Lakatos	Budapest University, Hungary
R. Lazarov	Texas A&M University, USA
E. Levner	Holon Institute of Technology, Israel
B.Y. Lemeshko	Novosibirsk State Technical University, Russia
S.D. Margenov	Institute of Information and Communication Technologies of Bulgarian Academy of Sciences, Bulgaria
O. Martikainen	Service Innovation Research Institute, Finland
L. Militano	University Mediterranea of Reggio Calabria, Italy
E.V. Morozov	Institute of Applied Mathematical Research of the Karelian Research Centre RAS, Russia
G.K. Mishkoy	Academy of Sciences of Moldova, Moldavia
A.A. Nazarov	Tomsk State University, Russia
I. Novak	Brno University of Technology, Czech Republic
D.A. Novikov	ICS RAS, Russia
Y.N. Orlov	Keldysh Institute of Applied Mathematics of RAS, Russia
M. Pagano	Pisa University, Italy
I.V. Puzynin	Joint Institute for Nuclear Research, Russia
Y.P. Rybakov	RUDN University, Russia
V.V. Rykov	Gubkin Russian State University of Oil and Gas, Russia
Z. Saffer	Budapest University of Technology and Economics, Hungary
L.A. Sevastianov	RUDN University, Russia
S.Ya. Shorgin	Federal Research Center "Computer Science and Control" of RAS, Russia
A.L. Skubachevskii	RUDN University, Russia
P. Stanchev	Kettering University, USA
A.M. Turlikov	St. Petersburg State University of Aerospace Instrumentation, Russia
D. Udumyan	University of Miami, USA
S.I. Vinitsky	Joint Institute for Nuclear Research, Russia
J.P. Zaychenko	Kyiv Polytechnic Institute, Ukraine

Executive Committee

D.V. Kozyrev (Chair)	RUDN University and ICS RAS, Russia
S.P. Moiseeva	Tomsk State University, Russia
T. Atanasova	IICT BAS, Bulgaria

Y.V. Gaidamaka   RUDN University, Russia
D.S. Kulyabov   RUDN University, Russia
A.V. Demidova   RUDN University, Russia
S.N. Kupriyakhina   ICS RAS, Russia

Organizers and Partners

Organizers

Russian Academy of Sciences
RUDN University
V.A. Trapeznikov Institute of Control Sciences of RAS (ICS RAS)
National Research Tomsk State University (NR TSU)
Institute of Information and Communication Technologies of Bulgarian Academy of
Sciences (IICT-BAS)
Research and Development Company "Information and Networking Technologies"

Support

Information support was provided by the Moscow department of the IEEE
Communication Society. Financial support was provided by the Russian Foundation
for Basic Research.

Contents

Mathematical Modeling and Computation

Computer and Communication Networks

Enhanced C-RAN Architecture Supporting SDN and NFV Functionalities for D2D Communications

Antonino Orsino[1]([✉]), Giuseppe Araniti[1], Li Wang[2], and Antonio Iera[1]

[1] ARTS Laboratory, DIIES Department, University Mediterranea of Reggio Calabria, Loc. Feo di Vito, Via Graziella 2, 89100 Reggio Calabria, Italy
{antonino.orsino,araniti,antonio.iera}@unirc.it
[2] Beijing University of Posts and Telecommunications,
10 Xitucheng Rd, Haidian, Beijing, China
liwang@bupt.edu.cn

Abstract. Future Fifth Generation (5G) cellular systems will be characterized by ultra-dense areas, where users are gradually asking for new multimedia applications and hungry-bandwidth services. Therefore, a promising solution to boost and optimize this future wireless heterogeneous networks is represented by the Cloud Radio Access Network (C-RAN) with the joint use of Software Defined Networking (SDN) and Network Function Virtualization (NFV). In such a scenario, low power base stations and device-to-device communications (D2D), involved into traditional cellular network, represented a possible solution to offload the heavy traffic of macrocells, while guaranteeing user experience as well. Nevertheless, the high centralization and the limited-capacity backhauls makes it difficult to perform centralized control plane functions on a large network scale. To address this issue, we investigate the integration of two enabling technologies for C-RAN (i.e., SDN and NFV) in the current 5G heterogeneous wireless architecture in order to exploit properly proximity-based transmissions among devices. Then, in order to validate the applicability of our proposed architecture, we consider the case of D2D pair handover where we show that our solution is able to decrease the number of signaling messages needed to handoff the D2D pair from a source to a target base station and, at the same time, the time execution for the entire handover process.

Keywords: Wireless network virtualization · SDN · NFV · C-RAN · D2D · Handover

1 Introduction

Future 5G networks are expected to deal with different multi-connectivity and multi-technology tiers all of them deployed within the same ares of interest. For this reason, such kind of environment are one of the foreseeable mission-critical

© Springer International Publishing AG 2016
V.M. Vishnevskiy et al. (Eds.): DCCN 2016, CCIS 678, pp. 3–12, 2016.
DOI: 10.1007/978-3-319-51917-3_1

hybrid networks connecting machines and humans to provide various public services through highly reliable, ultra-low latency and broadband communications [1,2]. The benefits expected to be introduced by future 5G systems, indeed, are widely and will change completely the perception that the end-users have with the surrounding environment. Nevertheless, jointly with these enhancements, the avalanche of data traffic and new multimedia services, e.g., massive sensor deployment and vehicular to anything communication, will lead to strict requirements concerning latencies, signaling overhead, energy consumption, and data rates [3].

To overcome these issues, D2D communications have gained momentum among the research community as a possible enabling technology for 5G future systems. However, Proximity Services (ProSe) standardization is still on going and only few works in literature proposed solutions and architectures in order to manage efficiently short-range transmissions in current and future cellular networks [4,5]. For this reason, it is expected that the integration of D2D communications within the cellular infrastructure may be facilitated by SDN, NVF, and C-RAN.

For instance, in [6] the integration of D2D communications exploiting the paradigms of SDN and NVF is addressed. Specifically, the authors consider how to manage properly a pool of radio resources belonging to multiple Infrastructure Providers (InPs) through short-range transmissions in virtual wireless networks with the aim of maximizing the network-wide welfare. An SDN-controlled optical mobile fronthaul (MFH) architecture, instead, is proposed in [7] for bidirectional coordinated multipoint (CoMP) and low latency inter-cell D2D connectivity in a 5G mobile scenario. In particular, the SDN controller *OpenFlow* is exploited in order to control dynamically CoMP and inter-cell D2D features by monitoring the behavior of both optical and electrical SDN switching elements.

Although a *centralized* SDN-NFV solution works well for the standard cellular infrastructure (i.e., core and access network), this approach results not completely suitable when focusing on proximity-based communications. To this end, a possible solution is represented by a distributed approach where part of the features typically located in a *global* SDN-controller are implemented in *local* SDN-controllers deployed, in a *distributed* way [8], inside the radio access network (i.e., the local controller usually is implemented within the base stations/access points). The concept of hierarchical SDN has been also addressed in [9], where the authors propose a hierarchical D2D communications architecture with a centralized SDN controller communicating with the cloud head (CH) to reduce the number of requested LTE communication links, thereby improving the overall system energy consumption. However, in this last work the word *hierarchical* is referred only to the direct links used between an SDN controller and a cloud head (i.e., similar to the cluster head of a grouped users). In our work instead, we aim to efficiently manage D2D communications by using jointly a multi-layer SDN infrastructure and an enhanced C-RAN architecture.

Beside SDN and NFV, indeed, the C-RANs, differently from the conventional RANs, decouple the baseband processing unit (BBU) from the remote radio head (RRH) allowing centralized operation of BBUs and scalable deployment of light-weight RRHs as small cells. To the best of our knowledge, only few papers in literature deal with the C-RAN architecture for proximity-based communications. In particular, in [10] the authors propose a D2D service selection framework in C-RAN networks. The optimal solution is achieved theoretically by using queue theory and convex optimization. Further, an energy efficient resource allocation algorithm through joint channel selection and power allocation design is presented in [11]. The approach proposed by the authors includes a hybrid structure that exploits the C-RAN architecture (i.e., distributed RRHs and centralized BBU pool). Nevertheless, the aim of our work is not to exploit the existing C-RAN concepts like has been made in the cited works, but to propose enhancements in the current C-RAN infrastructure concerning new modules and procedures for proximity services is future 5G scenarios.

Therefore, the aim of this work is to propose a new hybrid architecture where hierarchical SDN will be used in conjunction with the C-RAN to improve not only the users' experience (i.e., QoE), but also to decrease drastically the amount of signaling messages performed by the cellular radio access and core network. The scenario considered deals with the usage of low power base stations (i.e., femtocells, microcells, picocells) and proximity connections (i.e., D2D) in a heterogeneous scenario in order to offload the heavy traffic among the macro BSs, and from these ones toward the core network infrastructure. The proposed solution will also guarantee an improved quality of service (QoS) in terms of spectrum utilization, energy efficiency and mobility management. The motivation behind the utilization of a hierarchical SDN architecture is that, normally, the high centralization as well as the limited-capacity backhauls makes it difficult to perform centralized control plane functions on a large network scale. Thus, in our work distributed local SDN controllers (in help to the global SDN controller) are dynamically set up and configured in a programmable way, such that all the transmissions paths and the system, generally, can be well managed. Finally, with a practical example we show that the usage of local SDN controllers, jointly with the concepts of NFV and new D2D-aware C-RAN architecture, will allow D2D users' handover to be managed efficiently by avoiding signaling overload and useless network procedures.

The rest of the paper is summarized as follows. In Sect. 2 the enhanced Software Defined Proximity Services Networking (eSDN-ProSe) architecture is described with particular emphasis on the SDN and C-RAN integration over ProSe network deployment. A possible application of the proposed architecture is provided in Sect. 3 whereas conclusive remarks are illustrated in Sect. 4.

2 Enhanced Software Defined Proximity Services Networking (eSDN-ProSe) Architecture

The basic theory and key technology of the wireless network design, management and scheduling, based on the cloud architecture and multi-resource virtualization, is a great driving force to solve the problems and challenges faced by the heterogeneous ultra-dense future 5G scenarios. The C-RAN architecture has the characteristics of communication computing integration, which can meet the needs of high performance computing in large-scale complex networks. On the other hand, network virtualization and SDN paradigm can integrate network resources and achieve centralized control thus improving the overall usage of the radio resource and managing efficiently uncontrolled and unplanned environment driven by the mobility of the users and network entities (e.g., mobile femtocells). Therefore, designing a network architecture with the combined use of C-RAN, NFV, and SDN functionalities is an effective way to achieve an efficient management of the overall system.

2.1   Hierarchical Enhanced-SDN Architecture (eSDN)

The idea behind the proposal of a hierarchical SDN architecture, is not only to expand the network connections and the way the users grant the access, but also offload the macro layer traffic and improve the radio resource management and issues related to mobility. In fact, in such a solution part of the *global* controller functionalities are moved to *local* controllers deployed within the access nodes (i.e., the base station) and, in a limited way, also on the different devices and users. Thereby, the huge pressure on the macro base stations in ultra dense mobility-aware environments can be relieved. In addition, the single node access failure and backhaul link delay problems, typical of such scenarios, are widespread. To overcome these issues, we propose a hierarchical SDN architecture with NFV functionalities, where D2D communications, that represent a feasible solution for traffic offloading, can be managed "locally" with deployed SDN controllers (either global or local) within base stations (or generally within the access nodes).

Our vision is, indeed, to enhance the current LTE cellular network infrastructure with a novel SDN architecture based on OpenFlow (or other available SDN controller software), and virtualized networks where different service providers (SPs) can dynamically share the physical substrate of wireless networks operated by mobile network operators (MNOs). In addition, given the significant amount of modifications needed to integrate "natively" D2D communications in future 5G systems (i.e., both control plane and data plane of radio access networks and core networks), SDN and NVF can provide a versatile framework for the integration of new communications schemes (i.e., D2D) in legacy cellular systems (i.e., LTE, LTE-A).

To cope with this purpose, in a D2D-based network environment an SDN controller (either deployed globally or locally) will be responsible for detecting the user traffic so that the potential D2D users can be paired if it is feasible. In addition, this new architecture entity will introduce new useful functions (e.g., traffic earmarking for gateways, new radio resource management schemes for access points) and new signaling protocols between network entities under the legacy wireless mobile network framework. Indeed, if two networks devices will be identified as potential D2D partners, the network controller further decides whether to perform D2D transmission or to utilize an access point to relay the information. Moreover, since it is a virtual wireless network, it will also consider which access point is responsible for relaying the traffic if the access point relay mode is selected.

Nevertheless, even if SDN can assure a considerable management of proximity-based transmissions, in such a situation one of the key questions is how to make the best use of the precious radio resources. One solution to overcome this issue is to decouple the control and data planes and to have the control logic located inside a controller (i.e., globally or locally distributed) via the software defined networking paradigm. The benefit is that networks controlled by the SDN architecture are also mapped into virtual infrastructures and elements in order to make it available "everywhere" and "anywhere". Then the virtual entities are aggregated and sliced into different virtual networks by a virtual resource manager or hypervisor.

In conclusion, the improvements of the proposed architecture are that network coverage and network access during instances of very high demand are cost effectively increased at the additional cost of an overlay network and SDN infrastructure, thereby increasing revenue. Furthermore, in the rare instances when there is a lack of the network infrastructure (i.e., due to a disaster or tragic event), communications are still guaranteed by D2D-based transmissions among the devices.

2.2 C-RAN Deployment for the eSDN

The virtualization and softwarisation paradigms are gaining ground in the mobile networking ecosystem, particularly in conjunction with C-RAN. In this field, great effort has gone into virtualizing radio access technology as this enables the virtualization of edge functions of the core network without incurring additional hardware costs. Concerning the enhancement already introduced by the current C-RAN architecture in the current cellular an wireless architectures, our main idea is to propose new C-RAN modules and procedures for proximity services that can cooperate in an efficient way with the NFV and SDN paradigms. The reason, is that works present in literature only deal with the exploitation on the C-RAN architecture to improve the performance of D2D communications without focusing, indeed, if this architecture (at the present stage) is effectively suitable or not for managing short-range transmissions.

Looking at the common C-RAN architecture, the main idea is the replacement of self-contained base stations at each radio mast with shared- and cloud-based processing and distributed radio elements. In particular, the related main components/entities are: (i) the BBUs that represents the pool of processing resources useful to provide enhanced functionalities (i.e., signal processing and cells coordination capabilities) to all the network base stations within a given area of interest; (ii) the Fronthaul layer consisting in all the transmission links exploited to carry the baseband information that have to be transmitted in the RAN; (iii) the RRHs entities identified as the antenna equipment to whom the end-user connect toward the RAN.

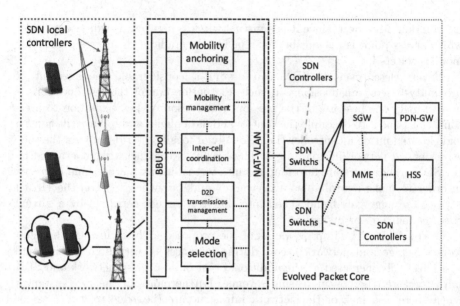

Fig. 1. Proposed C-RAN D2D-aware Architecture

However, the exploitation of the C-RAN architecture in current LTE systems poses limitations and challenges regarding the overall network efficiency procedures. Indeed, the core network of current 4G networks is experiencing a fast growing of signaling messages given by the usage of new transmission paradigms and technologies. Although a portion of this new signaling is required for new services and new devices types, over 50% of the signaling is related to mobility and paging, due to the greater node density. Focusing of an expected avalanche of short-range transmissions (i.e., D2D connections) performed not only by users (i.e., through smartphone or tablet) but also by a multitude of sensors and "smart" devices referred to the Internet of Things (IoT) paradigm, is clear the need of a more signaling-effective architecture in order to face these issues [12].

To this end, in this paper, jointly with the hierarchical enhanced-SDN architecture already mentioned in Sect. 2.1, we improve the current C-RAN paradigm in order to manage in an efficient way the D2D transmissions performed by the users/devices without overload strongly the already stressed LTE core network. In particular, starting from the C-RAN baseline already proposed in [13], we propose to major enhanced to be added in the C-RAN core in addition to the existing ones. As is shown in Fig. 1, the first module is identified as *D2D transmission management* whereas the second one is named as *mode selection*.

The idea at the basis of these new features is to virtualize possible D2D transmitters together with the cells in order to exploit the VLAN and NAT functionalities already proposed in [13]. In particular, the D2D transmission management is responsible to identify the possible D2D forwarder/transmitter and perform all the procedures to establish the D2D links among the users. The transmission mode selection entity, instead, selects the interface on which the users download/upload a given data content. Of course, the selection may be performed based on given metrics of interest, the application considered, or the scenario taken into consideration.

The need of the mode selection module is also justified by upcoming new wireless technology that are the hard core of the future 5G systems. In particular, users that in the past 3GPP Release 12 could exploit direct connection over licensed (e.g., LTE) and unlicensed (e.g., WiFi) bands, now have the chance to exploit also higher frequencies dealing with the new paradigm of the milliter-wave (mmWave) communications. In such a case, users will have the possibility to switch among four possible mode of transmissions when using D2D communications: (i) below 6 GHZ licensed bands (e.g., LTE), (ii) below 6 GHz unlicensed bands (e.g., WiFi), (iii) above 6 GHz licensed bands (e.g., 28 GHz mmWave), and (iv) above 6 GHz unlicensed bands (e.g., 60 GHz mmWave).

3   Applicability of the eSDN-ProSe Architecture: The Case of D2D Pair Handover

As a possible example of the applicability of the proposed eSDN-ProSe architecture, in this section we analyze the case where a pair of D2D users (with an active connection) are moving across the overlapping zone of two eNodeBs and have to perform a handover procedure from, e.g., a source eNodeB and a target eNodeB. In such a case, following the 3rd Generation Partnership Project (3GPP) LTE standard handover procedures[1], the handoff of the entire D2D pair results in an expensive process in terms of signaling messages overhead and resource allocation (and reservation) between the two eNodeBs [14] and the core networks. To overcome these issues, exploiting our proposed eSDN architecture, we are able to reduce sensibly the number of messages that have to be exchanged between the source and target eNodeB and toward the Evolved Packet Core (EPC) network entity. In particular, all the messages exchanged (e.g., the handover request message with its acknowledgement) through the X2 protocol between the source and target eNodeB could be avoided and demanded to the new evolved EPC. Further, the part related to the new path request and the bearer setting, typical of the legacy 3GPP X2 handover protocol, thanks to the C-RAN and NFV functionalities could be performed directly within the core network, without overload with signaling messages the links that connect the EPC with the radio access base station. Finally, for the lack of clarification, in Fig. 2 is shown the possible message flow diagram of the new D2D handover procedure proposed.

In details, the evolved EPC (i.e., enhanced with SDN, NFV, and C-RAN functionalities) is responsible of gathering the measurements reports from the D2D pair and, if needed, starting the handover requests and the RRC reconfiguration. In such a case, the source and target eNodeB have to deal only with

---

[1] Please, refer to the 3GPP specification 23.009 – Available at: http://www.3gpp.org/ DynaReport/23009.htm.

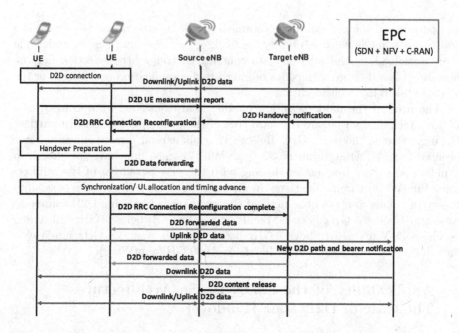

Fig. 2. Possible example of D2D handover procedure exploiting the eSDN architecture

the RCC synchronization and pre-allocation (from the target eNodeB) of the spectrum resources in order to host the upcoming D2D pair. Once that all the procedures regarding the resource allocation and RRC configuration are completed, the EPC, by exploiting the SDN and NFV functionalities, establish a new data flow for the D2D pair and inform with a bearer notification either the source and target eNodeB of the new D2D data path. Finally, the source eNodeB releases the resource previously allocated for the D2D users and these ones can continue their D2D transmission within the target eNodeB (i.e., their D2D connection is never interrupted). It is worth noticing, that the EPC is able to decide whether or not the handover has to be performed by scheduling the periodic channel measurements sent by the D2D pair. In addition, the exploitation of the new D2D-aware C-RAN functionalities described in Sect. 2.2, allows the EPC to start the handover procedure and inform both the involved eNodeB that the RRC configuration and pre-resource allocation have to be started.

In conclusion, this novel approach is able to reduce substantially the signaling messages overload when considering the handover of a D2D pair between two eNodeB of the same operator. In particular, the amount of signaling needed to offload the D2D pair and perform all the handover procedure is reduced up to 9 messages instead of the legacy LTE X2 handover protocol that requests, at least, 22 messages[2].

---

[2] For the lack of space the messages diagram of the legacy LTE handover has been omitted, but the reader can have a look at the 3GPP specification 23.009 (available at http://www.3gpp.org/DynaReport/23009.htm) for a detailed description.

4 Conclusions

In this paper we have presented a D2D-aware hierarchical SDN architecture where NFV and C-RAN functionalities are used to improve the users' Quality of Service (QoS) in an ultra-dense heterogenous scenario where the network traffic is expected to offloaded, not only through small cell (e.g., picocell or femtocell) disseminated within the given area of interest, but also through direct links among users (i.e., by using D2D transmissions). After describing the hierarchical SDN and NFV infrastructure (i.e., named enhanced SDN – eSDN), a new D2D-aware C-RAN architecture has been proposed with the aim of managing efficiently the connection establishment of the D2D links among users in proximity. For the best of our knowledge, this is still an aspect that has not been well investigated in literature and the present work is one of the first that is providing a novel solution to this important topic. Finally, in the last section of the paper we have provided a practical example about how this new D2D-aware eSDN architecture could be exploited by focusing on the case of handover procedure of a D2D pair between a source and target eNodeB. In such a case, we have shown that our solution is able not only to decrease drastically the number of signaling messages needed to perform all the D2D handover procedures, but also achieve good results in terms of time needed to perform the entire handover process thus allowing to avoid the ping pong effect and the unnecessary handover requests on the network operator-side.

References

1. Ometov, A., Masek, P., Malina, L., Florea, R., Hosek, J., Andreev, S., Hajny, J., Niutanen, J., Koucheryavy, Y.: Feasibility characterization of cryptographic primitives for constrained (wearable) IoT devices. In: Proceedings of IEEE International Conference on Pervasive Computing and Communication Workshops (PerCom Workshops), pp. 1–6. IEEE (2016)
2. Fadda, M., Popescu, V., Murroni, M., Angueira, P., Morgade, J.: On the feasibility of unlicensed communications in the tv white space: field measurements in the uhf band. Int. J. Digit. Multimedia Broadcast. **2015** (2015)
3. Ometov, A., Andreev, S., Turlikov, A., Koucheryavy, Y.: Characterizing the effect of packet losses in current WLAN Deployments. In: Proceedings of 13th International Conference on ITS Telecommunications (ITST), pp. 331–336. IEEE (2013)
4. Andreev, S., Moltchanov, D., Galinina, O., Pyattaev, A., Ometov, A., Koucheryavy, Y.: Network-assisted device-to-device connectivity: contemporary vision and open challenges. In: Proceedings of 21th European Wireless Conference European Wireless, pp. 1–8. VDE (2015)
5. Ometov, A., Zhidanov, K., Bezzateev, S., Florea, R., Andreev, S., Koucheryavy, Y.: Securing network-assisted direct communication: the case of unreliable cellular connectivity. In: Proceedings of Trustcom/BigDataSE/ISPA, vol. 1, pp. 826–833. IEEE (2015)
6. Cai, Y., Yu, F.R., Liang, C., Sun, B., Yan, Q.: Software defined device-to-device (D2D) communications in virtual wireless networks with imperfect network state information (NSI). IEEE Trans. Veh. Technol. **PP**(99), 1 (2015)

7. Cvijetic, N., Tanaka, A., Kanonakis, K., Wang, T.: SDN-controlled topology-reconfigurable optical mobile fronthaul architecture for bidirectional CoMP and low latency inter-cell D2D in the 5G mobile era. Opt. Express **22**(17), 20809–20815 (2014)

8. Zhang, H., Zhou, M., Song, L., Zhang, S.: Demo: software-defined device to device communication in multiple cells. In: Proceedings of the 16th ACM International Symposium on Mobile Ad Hoc Networking and Computing. MobiHoc 2015, pp. 401–402. ACM, New York (2015)

9. Usman, M., Gebremariam, A.A., Raza, U., Granelli, F.: A software-defined device-to-device communication architecture for public safety applications in 5G networks. IEEE Access **3**, 1649–1654 (2015)

10. ul Hassan, S.T., Ashraf, M.I., Katz, M.D.: Mobile Cloud based architecture for Device-to-Device (D2D) communication underlying cellular network. In: 2013 IFIP Wireless Days (WD), pp. 1–3, November 2013

11. Zhou, Z., Dong, M., Ota, K., Wang, G., Yang, L.: Energy-efficient resource allocation for D2D communications underlaying cloud-RAN based LTE-a networks. IEEE Internet Things J. **PP**(99), 1 (2015)

12. Pyattaev, A., Johnsson, K., Andreev, S., Koucheryavy, Y.: Communication challenges in high-density deployments of wearable wireless devices. IEEE Wirel. Commun. **22**(1), 12–18 (2015)

13. Dawson, A.W., Marina, M.K., Garcia, F.J.: On the benefits of RAN virtualisation in C-RAN based mobile networks. In: 2014 Third European Workshop on Software Defined Networks (EWSDN), pp. 103–108, September 2014

14. Orsino, A., Gapeyenko, M., Militano, L., Moltchanov, D., Andreev, S., Koucheryavy, Y., Araniti, G.: Assisted handover based on device-to-device communications in 3GPP. LTE systems. In: 2015 IEEE Globecom Workshops (GC Wkshps), PP. 1–6, December 2015

On Internet of Things Programming Models

Dmitry Namiot[1](✉) and Manfred Sneps-Sneppe[2]

[1] Faculty of Computational Mathematics and Cybernetics,
Lomonosov Moscow State University, GSP-1, 1-52,
Leninskiye Gory, Moscow 119991, Russia
dnamiot@gmail.com
[2] Institute of Mathematics and Computer Science,
University of Latvia, Raina Bulvaris 29, Riga 1459, Latvia

Abstract. In this paper, we present the review of existing and proposed programming models for Internet of Things (IoT) applications. The requests by the economy and the development of computer technologies (e.g., cloud-based models) have led to an increase in large-scale projects in the IoT area. The large-scale IoT systems should be able to integrate diverse types of IoT devices and support big data analytics. And, of course, they should be developed and updated at a reasonable cost and within a reasonable time. Due to the complexity, scale, and diversity of IoT systems, programming for IoT applications is a great challenge. And this challenge requires programming models and development systems at all stages of development and for all aspects of IoT development. The first target for this review is a set of existing and future educational programs in information and communication technologies at universities, which, obviously, must somehow respond to the demands of the development of IoT systems.

Keywords: Internet of Things · Smart Cities · Streaming · Sensor fusion · Programming · Education

1 Introduction

The Internet of Things (IoT) world is becoming an important direction for technology development. In general, the IoT promotes a heightened level of awareness about our world. IoT plays a basic role in many other things. For example, IoT is a base for Smart Cities, etc.

IoT ecosystem is currently presented by multiple (sometimes - competing) technologies and platforms. IoT platforms (at least, nowadays) are varied across the vertical and horizontal segments of the markets. Of course, it complicates and delays the development and deployment, makes the support of IoT systems more expensive than it should be, etc. So, IoT standards are highly demanded [1].

In the same time, it is a very competitive area. We cannot expect that a general solution will be agreed upon by all players. Standards proposals in IoT (and M2M) come from formal standards development organizations (e.g., the European

© Springer International Publishing AG 2016
V.M. Vishnevskiy et al. (Eds.): DCCN 2016, CCIS 678, pp. 13–24, 2016.
DOI: 10.1007/978-3-319-51917-3_2

Telecommunications Institute - ETSI) or non-formal groups (the Institute of Electrical and Electronics Engineers - IEEE). Standards can target the connectivity for a particular set of devices (e.g., Bluetooth Low Energy) or provide common application interfaces up to developers (e.g., oneM2M) [2]. In this paper, we would like to discuss the common elements of IoT programming models and perform this review from the perspective of educational programs.

The rest of the paper is organized as follows. In Sect. 2, we discuss programming systems for IoT. In Sect. 3, we discuss data models, data persistence, data processing and educational programs for IoT. In Sect. 4, we discuss cloud computing for IoT.

2   IoT Programming Models

The choice of programming languages for IoT platforms does not depend on a hardware platform. Also, new hardware platforms make programming embedded (nowadays - cyber-physical) systems easier. Even more, the diversity in hardware platforms enhances the interest to the platform independent on hardware.

Standards for the IoT could be classified as downward-facing standards that establish connectivity with devices and upward-facing standards that provide common application interfaces up to end users and application developers.

By our opinion, confirmed by the practical experience and academic papers, the key moment for software development in telecom and related areas (IoT is among them) is time to market indicator [3]. The main question to any software standard is the generalization. Shall the standard follow to the "all or nothing" model and covers all the areas of the life cycle? In software standards, the excessive generalization (unification) could be the biggest source of the problems. Actually, all the standards should make its implementation by the most convenient way for developers. Because only the developers are finally responsible for the putting new services in place.

It is especially true for such areas as IoT or Smart Cities. The services here are not finalized (and it is very probably that they could not be finalized at all). This means that we will constantly try (test) new services and to refuse from the old ones. Naturally, this process needs to be fast and inexpensive. As the next step, it means that the most of IoT application could be described as mashups [4]. Mashups use data from several data sources. On programming level, it should stimulate the interest in scripting languages and to the systems for fast prototyping. In the modern software architecture world, we can mention also micro-services approach [5]. Of course, these directions should have an appropriate reflection in educational programs.

Another direction, which is very close to mashups, actually, is so-called Data-as-a-Service (DaaS) approach [6]. In its technical aspects, DaaS is an information provision and distribution model in which data files (including text, images, sounds, and videos) are made available to customers over a network. The key moment is the separation for data and proceedings. It lets delivery data (e.g., in some open format, like JSON), rather than some API with the predefined model for data processing.

The next significant visible trend is the growing interest in the dynamic languages. And the perfect example here is JavaScript. We can mention the following reasons for JavaScript in IoT applications [7]. At the first hand, there is a big army of web developers. So, the entry level for programming is low. Most of the Internet applications already use JavaScript. JavaScript nowadays covers both server-side and client-side programming. It could be useful to use the same language across the whole project. So, it makes sense to extend the same standard platform to the Internet of Things, communicating to a larger set of devices using the same language.

JavaScript has matured as a language and international standards cover its extensions. JavaScript has a range of already existing libraries, plugins, etc. And what is also important, there is a huge Open Source community behind JavaScript.

Technically, this language has got a great support for event-driven apps. The nature of IoT project is mostly associated with asynchronous communications. And event-driven models are the most suitable solution. In JavaScript, it is very easy to implement models, where an application can receive and respond to events, then wait for a callback from each event that notifies us once it is complete. It lets respond to events as they happen, performing many tasks simultaneously as they come in.

Also, the recent development shows more and more direct involvement JavaScript into data processing. Actually, the winning data format (JSON) has its origin in JavaScript.

As a recent example of JavaScript in IoT, we can mention the developments from Samsung. Samsung Electronics recently opened the development of IoT.js, a web-based Internet of Things (IoT) platform that connects lightweight devices. Examples of lightweight devices include micro-controllers or devices with only a few kilobytes of RAM available [8]. The idea is to make all devices interoperable in the IoT space by enabling more devices to be interoperable, from complex and sophisticated devices such as home appliances, mobile devices, and televisions, to lightweight and small devices such as lamps, thermometers, switches, and sensors. The IoT.js platform is comprised of a lightweight version of the JavaScript engine, and a lightweight version of node.js. We think that JavaScript for IoT world should be in educational programs. This is very important because until now, this language is often seen as simple web pages scripting. But it's not for a long time already.

By the same reason, any attempt to replace JavaScript with the similar idea of portability could be also interested in IoT programming. In this connection, we should mention Dart programming language from Google [9].

3   Data Persistence and Processing

It terms of the data processing in IoT applications, we should pay attention in educational programs to the following two moments: sensor fusion and streaming. There are different data mining and data science approaches which are applicable to IoT. And of course, they should be a subject of the separate courses for statistics, machine learning, etc. For example, in many cases, IoT (Smart City) measurements are time series. Of course, it should be a subject of a separate course among other data-mining techniques [10].

But one moment is important, in our opinion, and should be discussed separately for IoT applications. It is sensor fusion. Sensor fusion is combining of sensory data or data derived from disparate sources such that the resulting information has less uncertainty than would be possible when these sources were used individually [11]. It is illustrated in Fig. 1.

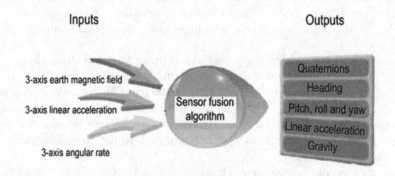

Fig. 1. Sensor fusion [12]

There are many ways of fusing sensors into one stream. Each sensor has its own strengths and weaknesses. The idea of sensor fusion is to take readings from each sensor and provide a more useful result which combines the strengths of each. Actually, such a fusion is the main idea for all IoT and Smart City projects, related to some measurements. The next big issue for IoT data processing is streaming. By our opinion, it is a key technology in data acquisition and proceeding for Smart Cities and IoT.

There are many tasks in IoT with the requirements for real-time (or near real-time) processing. In this case, the common architecture is associated with some messaging bus. And it is very important to present the software architectures associated with streaming. At the first hand, it is so-called Lambda Architecture [13]. Originally, the Lambda Architecture is an approach to building stream processing applications on top of MapReduce and Storm or similar systems (Fig. 2). Currently, we should link it to Spark and Spark streaming too [14].

The main idea behind this schema is the fact that an immutable sequence of source data is captured and fed into a batch system and a stream processing system in parallel. Of course, the negative impact of this decision is the need to implement business logic twice, once in the batch system and once in the stream processing system.

The Lambda Architecture targets applications built around complex asynchronous transformations that need to run with low latency. One proposed

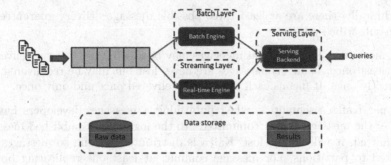

Fig. 2. Lambda architecture [15]

approach to fixing this is to have a language or framework that abstracts over both the real-time and batch framework [16].

Another solution here is so-called Kappa architecture [17]. The Kappa architecture simplifies the Lambda architecture by removing the batch layer and replacing it with a streaming layer (Fig. 3).

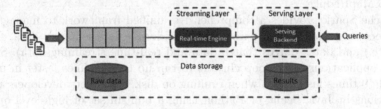

Fig. 3. Kappa architecture [15]

With Kappa, everything in the system is a stream. All batch operations become a subset of streaming operations. Data source (raw data) is persisted and views are derived. Of course, a state can always be recomputed where the initial record is never changed. This feature lets us support replay functionality. Computations and results can evolve by replaying the historical data from a stream. With Kappa, only a single analytics engine is required. It means that code is considerably reduced. Also, maintenance and upgrades are cheaper.

The hearth for such implementations is a scalable, distributed messaging system with events ordering and at-least-once delivery guarantees. At this moment, it is almost always Kafka system [18].

Apache Kafka is a distributed publish-subscribe messaging system. It is designed to provide high throughput persistent scalable messaging. Kafka allows parallel data loads into Hadoop. Its features include the use of compression to optimize performance and mirroring to improve availability, scalability. Kafka is optimized for multiple-cluster scenarios. In general, publish-subscribe architecture is the most suitable approach for mostly asynchronous measurements in IoT.

Technically, there are at least three possible message delivery guarantees in publish-subscribe systems:

(1) At most once. It means that messages may be lost but are never redelivered.
(2) At least once. It means messages are never lost but may be redelivered.
(3) Exactly once. It means each message is delivered once and only once.

As per Kafka's semantics, when publishing a message, developers have a notion of the message being "committed" to the log. Once a published message is committed, it will not be lost. Kafka is distributed system, so messages are replicated to partitions. For message commit, at least one replicating broker should be alive.

Kafka guarantees at-least-once delivery by default. It also allows the user to implement at most once delivery by disabling retries on the producer and committing its offset prior to processing a batch of messages. Exactly-once delivery requires co-operation with the destination storage system (it is some sort of two-phase commit).

In connection with Kafka, we should highlight Apache Spark [19]. Apache Spark is an open-source cluster computing framework for big data processing. It has emerged as the next generation big data processing engine, overtaking Hadoop MapReduce.

Apache Spark provides a comprehensive, unified framework to manage big data processing requirements with a variety of diverse data sets (text data, graph data, etc.) and data sources (batch data and real-time streaming data). Spark enables applications in Hadoop clusters to run up to 100 times faster in memory and 10 times faster even when running on disk. Spark lets developers write applications in Java, Scala, or Python using a built-in set of high-level operators. In addition to MapReduce operations, Apache Spark supports SQL queries, streaming data, graph data processing, and machine learning. Developers can use these capabilities stand-alone or combine them to run in a single data pipeline use case.

Another model is the recently introduced Kafka Streams. Kafka models a stream as a log, that is, a never-ending sequence of key/value pairs. Kafka Streams is a library for building streaming applications, specifically applications that transform input Kafka topics into output Kafka topics (or calls to external services, or updates to databases, or whatever). It lets you do this with concise code in a way that is distributed and fault-tolerant [20].

The above-mentioned models describe the modern view of the building IoT systems from the position of data architecture. A review for some IoT and/or Smart Cities related program (of course, we target technology-related education only) is presented in [21], for example.

Time-series databases historically play the important role for IoT applications. Technically, most of the applications (especially, in M2M area) collect and proceed some measurements. And time-series are the natural way of saving measurements.

One important element of IoT programming is associated with meta-data. In the most cases, the public APIs IoT systems are dealing with are based on

REST model. It is true for data persistence interfaces too. REST architecture proposes the uniform interface. In REST model, all resources present the same interface to clients. And it is one of the reasons for REST popularity. Alternatively, the Service Oriented Architecture (SOA) approach may offer personalized interfaces for the different resources. The whole SOA model is based on the idea that different services have different interfaces. In SOA, we need to provide the definition for used interfaces. The definition of the services (Web Service Definition Language - WSDL) is a key part of SOA. Any WSDL definition of a Web Service defines operations in terms of their underlying input and output messages. Unlike this, REST is based on the self-described messages. WSDL defines the form of the data that accompany the messages in SOA. REST does not provide this information. In other words, SOA has got a rich set of metadata, where REST model does not have meta-data at all.

Metadata support lets discover information about interfaces programmatically. It is a key moment. With the program-based discovery, we can automate the programming. And automation is a very important issue for Internet of Things due to high diversity in hardware (e.g., sensors, actuators, etc.). So, in our opinion, adding some standard form of metadata for REST API is very important for Internet of Things programming.

The educational program should include the following parts (elements):

sensing,
network connectivity,
IoT security
data integration, data processing, and applications.

In the first part, we present some overview for the modern sensors. Network connectivity section should discuss IoT networks, such as Bluetooth, Bluetooth Low Energy, ZigBee, Wi-Fi, WiMAX, LTE. There are several key dimensions for IoT protocols: their communication range, application duty cycle, data rate, and battery consumption. Also, we should talk here about data protocols, such as CoAP, MQTT, HTTP (HTTP/2).

Data integration elements should include IoT middleware, data storage options, principles of processing for unstructured data. This topic should cover data architectures for IoT systems too.

Data processing part includes real-time processing engines and algorithms, as well as stream analytics.

Applications-related section should include gateways, user applications as well as top-level architectures, such as edge processing and fog computing. Also, we will discuss here such things as localization: localization algorithms, indoor and mobile localization. In this section, we place also context-aware applications and utilizing sensors to gain greater visibility and real-time situational awareness.

In general, we followed the following schema (Fig. 4) in our educational program.

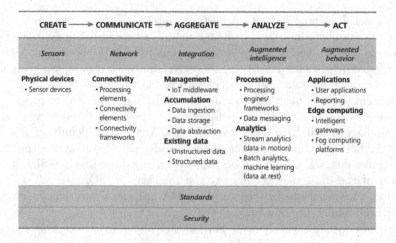

Fig. 4. Sensor fusion [22]

4   Cloud Computing and Related Areas for IoT

In this section, we would like to discuss the cloud computing models for IoT. Cloud architecture provides new opportunities in aggregating IoT data (e.g. data from sensors) and exploiting the aggregates for larger coverage and relevancy. In the same time, cloud models affect privacy and security.

There are several important moments. Firstly, we would like to highlight the important role of Amazon S3 for all tasks related to media data persistence. Almost all existing projects use Amazon Simple Storage Service (S3) for media data. The classical model is when Amazon S3 stores media objects and a separate relational database (it could be some NoSQL database, e.g., key-value store) keeps keys for objects.

The OpenStack project [23] produces the open standard cloud computing platform for both public and private clouds. OpenStack has a modular architecture with various code names for its components. OpenStack Object Storage (Swift) is a scalable redundant storage system [24]. With Swift, objects and files are written to multiple disk drives spread throughout servers in the data center, with the OpenStack software responsible for ensuring data replication and integrity across the cluster. It lets scale storage clusters scale horizontally simply by adding new servers. Swift is responsible for replication its content.

Another important moment for IoT programming and cloud is so-called MBaaS (Mobile Backend As A Service). It is a model for providing the web and mobile app developers with a way to link their applications to backend cloud storage [25]. The key moments are simplicity for the developers and time to market factor. Actually, the additional services are the key idea behind MBaaS. MBaaS provides public application program interfaces (APIs) and custom software development kits (SDKs) for mobile and web developers. Also, MBaaS provides such features as user management, push notifications, and integration with social networking services. The key moment here is the simplicity for the developers.

Usually, MBaaS API (SDK) hides data persistence details from the developers. So, in the most cases, for the developers, it looks like some unrestricted key-value data store.

Also, in this connection, we can mention such IoT frameworks as oneM2M or FIWARE. They are pretending to be IoT standards. But standards in IoT (M2M) do not provide dedicated data persistence solutions. They rely on the existing cloud solutions.

Our vision for the future of cloud computing in IoT is based on the conception of fog computing. Fog computing is an extension of classic cloud computing to the edge of the network. It has been designed to support IoT applications, characterized by latency constraints and a requirement for mobility and geo-distribution [26]. It is illustrated in Fig. 5.

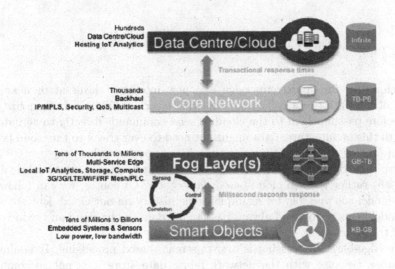

Fig. 5. Fog computing [27]

Fog computing architecture is not just about aggregation or concatenation of various physical data. It is about real-time distributed intelligence.

The common model is a bit more complex. It will include also extreme edge computing and software defined networks. It is illustrated in Fig. 6.

As per Cisco, extreme edge part includes data collecting elements: vehicles, ships, railways, roadways, factory floors. And of course, nowadays data could be processed on the collecting devices too. In general, the most time-sensitive should be analyzed on the node closest to the data collector (collectors). It is the main idea behind the fog computing. In this paradigm, we could analyze data even on the data collector itself. Especially, if our collector has the same computing power as fog's node.

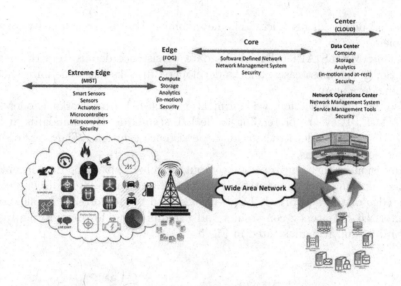

Fig. 6. The common architecture computing [28]

Technically, for or extreme edge is a new intelligent layer at or near the source of the data (data collector). And this layer can filter and normalize the data before passing them to the cloud or send commands directly to actuators.

With this architecture, data might not need to ever travel to the cloud layer. We can process data in real-time at the edge of our network.

With this architecture, we can store data on our devices (network fabric data store [29]) rather than in cloud-based data-center. Of course, we can follow to mixed model too and log, for example, some history on our cloud. But anyway, this model helps as to avoid always passing big data to our cloud and reduce the cost of transportation.

The edge-based processing is always stream-based processing. It is almost mandatory, because with the network fabric data store it is not so common to save big volumes of data. This emphasizes again the importance of stream processing, which we discussed in the section devoted to training. Stream processing is a key element for IoT programming.

Some of the authors describe this model as a shift in the way business is organized. The infrastructure based model is being replaced by the service-based model [30]. It corresponds to the common trend in which business is being virtualized and digitalized.

In this connection, we should mention the importance of another element in Fig. 5 - Software Defined Networks [31]. With Software Defined Networks (SDN) we can virtualize and digitalize network's hardware. Firstly, SDN separate (decouple) the data plane and control plane. SDN establish open interfaces between them. Secondly, SDN proposes a centralized control plane, thus having a global view of the network.

The ability to program the network, provided by SDN, can be compared with the mobile applications running on a mobile Operating System (Android or IoS) [31]. Similar to mobile applications, network-based applications can use resource and services (public interfaces) provided by SDN.

For IoT applications, SDN model brings yet another set of public interfaces. As per reviews, the specific SDN capabilities which will be useful for IoT application, are dynamic load management, service chaining, and bandwidth management.

Dynamic load management enables to monitor and orchestrate bandwidth changes automatically depends on the overall load of the network. It helps providers to support data peaks from IoT devices. Service chaining enables to sequence application-specific processing procedures to a given clients job [32]. SDN will ease the provisioning and service management processes for IoT devices. Service chaining allows to integrate unchanged network service software that is unaware of its operating environment. Many IoT devices (e.g., sensors) send data (measurements) periodically. Bandwidth management allow scheduling when and how much traffic an IoT application will need at a given time [33].

References

1. Chen, Y.-K.: Challenges and opportunities of internet of things. In: Design Automation Conference (ASP-DAC), pp. 383–388. IEEE Press, New York (2012)
2. Namiot, D., Sneps-Sneppe, M.: On IoT programming. Int. J. Open Inf. Technol. 2(10), 25–28 (2014)
3. Namiot, D., Sneps-Sneppe, M.: On software standards for smart cities: API or DPI. In: ITU Kaleidoscope Academic Conference: Living in a Converged World-Impossible Without Standards? pp. 169–174. IEEE Press, New York (2014)
4. Im, J., Seonghoon, K., Daeyoung, K.: IoT mashup as a service: cloud-based mashup service for the internet of things. In: 2013 IEEE International Conference on Services Computing (SCC), pp. 462–469. IEEE Press, New York (2013)
5. Namiot, D., Sneps-Sneppe, M.: On micro-services architecture. Int. J. Open Inf. Technol. 2(9), 24–27 (2014)
6. Bahrami, M., Singhal, M.: The role of cloud computing architecture in big data. In: Pedrycz, W., Chen, S.-M. (eds.) Information Granularity, Big Data, and Computational Intelligence, pp. 275–295. Springer, Heidelberg (2015)
7. Raggett, D.: The internet of things: W3C plans for developing standards for open markets of services for the IoT. Ubiquity 10, 3–6 (2015)
8. Samsung Iot.js. https://news.samsung.com/global/samsung-electronics-opens-development-of-iot-js-an-iot-platform-that-expands-interoperability-to-lightweight-devices. Accessed May 2016
9. Who uses Dart. https://www.dartlang.org/community/who-uses-dart.html. Accessed May 2016
10. Aggarwal, C.C.: Managing and Mining Sensor Data. Springer Science & Business Media, New York (2013)
11. Wang, M., et al.: City data fusion: sensor data fusion in the internet of things. arXiv preprint arXiv:1506.09118 (2015)
12. Introduction to sensor fusion. http://projects.mbientlab.com/introduction-to-sensor-fusion/. Accessed May 2016

13. Marz, N., Warren, J.: Big Data: Principles and Best Practices of Scalable Realtime Data Systems. Manning Publications Co., Greenwich (2015)
14. Ranjan, R.: Streaming big data processing in datacenter clouds. IEEE Cloud Comput. **1**, 78–83 (2014)
15. Applying the Kappa architecture in the telco industry. https://www.oreilly.com/ideas/applying-the-kappa-architecture-in-the-telco-industry. Accessed May 2016
16. Villari, M., et al.: AllJoyn Lambda: an architecture for the management of smart environments in IoT. In: Smart Computing Workshops (SMARTCOMP Workshops), pp. 9–14. IEEE Press, New York (2014)
17. Erb, B., Kargl, F.: A conceptual model for event-sourced graph computing. In: Proceedings of the 9th ACM International Conference on Distributed Event-Based Systems, pp. 352–355. ACM, New York (2015)
18. Garg, N.: Apache Kafka. Packt Publishing Ltd., Birmingham (2013)
19. Shanahan, J.G., Laing, D.: Large scale distributed data science using apache spark. In: Proceedings of the 21th ACM SIGKDD International Conference on Knowledge Discovery and Data Mining, pp. 2323–2324. ACM, New York (2015)
20. Kafka Streams. http://www.confluent.io/blog/introducing-kafka-streams-stream-processing-made-simple. Accessed May 2016
21. Namiot, D.: On internet of things and smart cities educational courses. Int. J. Open Inf. Technol. **4**(5), 26–38 (2016)
22. Inside the Internet of Things (IoT). http://dupress.deloitte.com/dup-us-en/focus/internet-of-things/iot-primer-iot-technologies-applications.html/. Accessed Aug 2016
23. OpenStack. https://www.openstack.org/. Accessed Aug 2016
24. Jackson, K., Bunch, C., Sigler, E.: OpenStack Cloud Computing Cookbook. Packt Publishing Ltd., Birmingham (2015)
25. Sneps-Sneppe, M., Namiot, D.: On mobile cloud for smart city applications. arXiv preprint arXiv:1605.02886 (2016)
26. Bonomi, F., Milito, R., Zhu, J., Addepalli, S.: Fog computing and its role in the internet of things. In: Proceedings of the 1st edn. of the MCC workshop on Mobile Cloud Computing, pp. 13–16. ACM, New York (2012)
27. Byers, C.C., Wetterwald, P.: Fog computing distributing data and intelligence for resiliency and scale necessary for IoT. Ubiquity **11**, 1–12 (2015)
28. Edge Computing - Where data comes alive! https://vividcomm.com/2016/04/08/edge-computing-where-data-comes-alive/. Accessed Sept 2016
29. Greenberg, A., et al.: VL2: a scalable and flexible data center network. ACM SIGCOMM Comput. Commun. Rev. **39**(4), 51–62 (2009)
30. di Costanzo, A., de Assuncao, M.D., Buyya, R.: Harnessing cloud technologies for a virtualized distributed computing infrastructure. IEEE Internet Comput. **13**(5), 24–33 (2009)
31. Caraguay, V., Leonardo, A., et al.: SDN: evolution and opportunities in the development IoT applications. Int. J. Distrib. Sens. Netw., 1–10 (2014)
32. Blendin, J., et al.: Software-defined network service chaining. In: 2014 Third European Workshop on Software Defined Networks, pp. 109–114. IEEE, New York (2014)
33. Kim, H., Feamster, N.: Improving network management with software defined networking. IEEE Commun. Mag. **51**(2), 114–119 (2013)

A Trial of Yoking-Proof Protocol in RFID-based Smart-Home Environment

Anton Prudanov[1], Sergey Tkachev[1], Nikolay Golos[1], Pavel Masek[3],
Jiri Hosek[3], Radek Fujdiak[3], Krystof Zeman[3], Aleksandr Ometov[2],
Sergey Bezzateev[1], Natalia Voloshina[1], Sergey Andreev[2(✉)], and Jiri Misurec[3]

[1] Saint Petersburg University of Aerospace Instrumentation,
Bolshaya Morskaya St. 67, 190000 St. Petersburg, Russia
[2] Tampere University of Technology, Korkeakoulunkatu 1, 33720 Tampere, Finland
aleksandr.ometov@tut.fi, serge.andreev@gmail.com
[3] Brno University of Technology, Technicka 12, 616 00 Brno, Czech Republic

Abstract. Owing to significant progress in the Internet of Things (IoT) within both academia and industry, this breakthrough technology is increasingly penetrating our everyday lives. However, the levels of user adoption and business revenue are still lagging behind the original expectations. The reasons include strong security and privacy concerns behind the IoT, which become critically important in the smart home environment. Our envisioned smart home scenario comprises a variety of sensors, actuators, and end-user devices interacting and sharing data securely. Correspondingly, we aim at investigating and verifying in practice the Yoking-proof protocol, which is a multi-factor authentication solution for smart home systems with an emphasis on data confidentiality and mutual authentication. Our international team conducted a large trial featuring the Yoking-proof protocol, RFID technology, as well as various sensors and user terminals. This paper outlines the essentials of this trial, reports on our practical experience, and summarizes the main lessons learned.

Keywords: Authentication · IoT · RFID · Smart-Home · Yoking-proof protocol

1 Introduction

The rapid proliferation of smart devices prepares to invade many areas of modern life: from connected home appliances and furniture to wearables and other personal systems, known altogether as the Internet of Things (IoT) [1]. What appeared a decade ago as a science fiction, is steadily becoming real today by bringing along over 24 billion of networked smart objects by 2020 [2]. On the other hand, despite all the enthusiasm around the IoT, a significant number of implementation-related challenges still remain to be solved [3].

© Springer International Publishing AG 2016
V.M. Vishnevskiy et al. (Eds.): DCCN 2016, CCIS 678, pp. 25–34, 2016.
DOI: 10.1007/978-3-319-51917-3_3

The smart home (or, home automation) sector belongs to the most ambitious business drivers across the entire IoT domain. However, in order to boost the end-user adoption of smart home applications and services built over different communication technologies [4] – and thus reach the anticipated market benefits – the resultant environment has to be easy to use, secure, and trustworthy [5]. It is believed by many that such a *success story* environment should be orchestrated by a single device, often named the Smart-Home Gateway (SH-GW). Following the recent developments in the field, it is evident that the SH-GW (deployed typically by telecommunication operators) will play a crucial role in the customer-centric IoT ecosystem [6]. To this end, the SH-GWs evolve by aiming to provide support for: (i) multiple connected smart devices, (ii) different types of user services, (iii) content access via different communication platforms, (iv) unrestricted mobility for end-users, and (v) enhanced security and reliability.

As known from other information and communication domains, sufficient levels of security and privacy are the key ingredient to the success of any user technology, and the smart home business sector is no exception [7]. Hence, despite the fact that SH-GW-controlled residential networks do offer a range of value-added services, this area is highly vulnerable to numerous security threats [8]. In particular, smart home sensing devices with relatively low computational capabilities are often connected to the digital residential networks via a central node [9]; they could thus become subject to several types of attacks, such as eavesdropping and replay, among others. Remember that home networks are intended to handle private user information and may provide critical services, including healthcare and protection of property. Therefore, attacks on such systems may lead to violating privacy and ultimately threatening the very life of residents, and appropriate security measures must therefore be considered carefully.

As the question of holistic security in the IoT ecosystem is very complex, in this work we focus on a particular problem of authentication, which we consider to be one of the critical *information-security* elements of the IoT. The main security risk in the systems without a straightforward authentication is vulnerability to the person-in-the-middle attacks that are particularly crucial for applications with stringent security requirements, such as those related to healthcare. Governments as well as industry leaders engage with trusted but overly complex authorities that provide authentication and authorization services while attempting to protect from such vulnerabilities [10]. However, said systems might be cumbersome to use and prohibitively expensive, especially when implemented in small-scale scenarios, including the residential environment.

Following the above reasoning, we propose a lightweight authentication protocol enabling simultaneous identification of newly-added IoT devices (e.g., sensors, actuators, meters, etc.) towards the central communication unit in a residential network, which is based on Yoking-proof protocol [11]. Summarizing the main requirements in the context of smart home scenarios, the target authentication scheme should satisfy the following security objectives [12]: (i) data confidentiality; (ii) mutual authentication; and (iii) forward security. To make an authentication protocol more secure, it is common to utilize Multi-Factor Authentication (MFA) [13]. The main feature of the MFA protocol is the involvement of two or more different devices with their own secrets. In order to increase efficiency of the MFA protocol, not only different devices but also various communication channels are usually utilized.

The main goal of our work is therefore to offer a universal communication scheme for the MFA that would be independent of the utilized communication technology between the sensors and the SH-GW. In our practical demonstrator, the MFA is implemented and verified in real-world smart home environment using smartphones and other wireless handheld devices. Further, we demonstrate the applicability of Yoking-proof approach for the MFA protocol construction. The rest of this paper is organized as follows. Section 2 is devoted to a discussion on the existing types of Yoking-proof protocol with respect to the MFA. Then, in Sect. 3, a detailed description of the considered smart home scenario is provided, together with an introduction of all the needed communication phases. Finally, the lessons learned during the trial implementation of our prototype are summarized in Sect. 4.

2   Overview of Yoking-Proof Protocol Principles

The Yoking-proof protocol comprises a set of elements, known as tags. These tags are controlled by two nodes – a reader and a verifier. At the time when the reader is capable of simultaneous identification of the tags and the verifier can provide a proof that two or more tags have been scanned simultaneously – the tags are sharing the secret keys with the verifier, but the reader's knowledge of this procedure is limited [14].

Each tag has its own *ID*, status value c, and the corresponding secret information – *Key*. The final result (if positive) is computed by the reader and constitutes the proof of simultaneous presence of all the tags involved. There are several types of Yoking-proof protocols to be considered:

1. *Online protocols* utilizing the so-called grouping-proof [15] and its modification, the existence-proof approach [16].
2. *Offline protocols* based on the chaining-proof approach [17].
3. *Broadcast style protocols* that are not studied here due to their higher complexity as a result of involving the Public Key Infrastructure (PKI) and asymmetric cryptography (Elliptic curves, ECC), which may not be fully operational on the target embedded devices [17].

An example of the *online protocol* is presented in Fig. 1. Here, Key_1 and Key_2 are the secret keys of the IoT devices 1 and 2, correspondingly (these keys are known to the reader). ID_1 and ID_2 represent the identifiers of the IoT devices 1 and 2. Further, SN_1 and SN_2 are special data junks generated by the IoT devices in the course of operation, i.e., at the time of a request from the reader reception. Note that SN_1 generated by device 1 at step 2 is not equal to SN_1 generated by the same device at step 5, see Fig. 1. The main advantage of this Yoking-proof protocol type is in the capabilities of the verifier – the protocol structure becomes more elaborate and provides a secure verification (compared to the offline protocol).

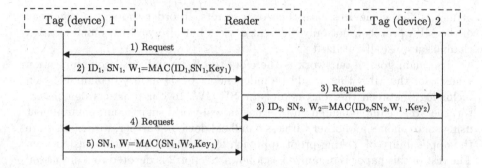

Fig. 1. Communication flow of the online Yoking-proof protocol.

Another alternative is based on the chaining-proof approach, which is named the *offline protocol*. The description of the respective communication flow is provided in Fig. 2. Here, *random(*)* is a pseudo random number generator initialized with the value *. For each device and for each instant of time, the result of this generator is assumed to vary even for the same parameter *. It is worth mentioning that the main difference compared to the online protocols is represented by the fact that the verifier is used only as an arbiter (thus, the absence of the verifier does not collapse the operation of this protocol type).

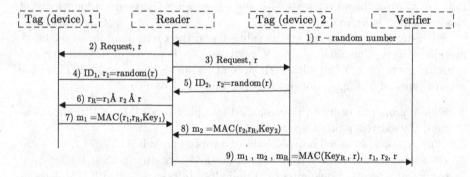

Fig. 2. An example of the offline Yoking-proof protocol.

3   Yoking-proof Protocol Prototype in Smart Home Scenario

In this section, our prototype implementation is outlined. We discuss the proposed communication scheme and then describe our Yoking-proof protocol trial demonstrating authentication of the Radio Frequency Identification (RFID)-equipped temperature/humidity sensors.

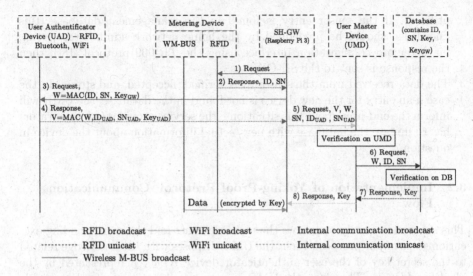

Fig. 3. Proposed communication scheme for the smart home scenario.

3.1   Trial Construction Conditions

For the purposes of this demonstration, we propose a novel communication scheme, see Fig. 1. Following the introductory information above, we reiterate that various types of smart IoT devices are flooding the market today. Conventionally, these devices (sensors, wearables, smart objects, etc.) are deployed in households and controlled via the SH-GW, which is the entity responsible for handling privacy of the collected information. First, we begin by describing the initial conditions for our test scenario. The SH-GW realized on a Raspberry Pi 3 interconnected with the RFID reader[1] represents the executive entity in a home network. The complete prototype installation is introduced in Fig. 3 and the implemented communication flow is described below:

1. When a new IoT device is joining the network, the SH-GW sends the request (using the RFID technology) to the incoming device. As a response, the identification and serial number of this new device are sent back to the SH-GW. Based on the Yoking-proof protocol functionality, the obtained data is sent to the User Authenticator Device (UAD) in the form of a Message Authentication Code (MAC) function $W = MAC(ID, SN, Key_{GW})$.
2. Next, the UAD computes a response $V = MAC(W, ID_{UAD}, SN_{UAD}, Key_{UAD})$. The request from the SH-GW, containing V, W, SN, ID_{UAD}, and SN_{UAD}, is then sent to the User Master Device (UMD). Further, the UMD verifies whether the new device is allowed to join the home network by sending the request to the database (in our implementation, the SQLite has been

---

[1] **See Cottonwood: USB Long Range UHF RFID reader, 2016** http://store. linksprite.com/cottonwood-usb-long-range-uhf-rfid-reader-iso18000-6c-epc-g2/.

utilized). If the sensor entry is found in the management database (this knowledge should be predefined by the home network administrator, e.g., via a remote configuration interface using the TR-069 protocol [18]), then the response is sent to the SH-GW.

3. The data received from the sensor(s) is verified, accepted, and stored. In the case if an entry for the new device is not found in the database, the UMD will inform the end-user about this situation – the service provider (administrator) has to update the database with new/actual information about the device in question.

3.2 Implementation of Yoking-Proof Protocol: Communication Flow

This subsection briefly overviews the main elements and the corresponding execution steps of our protocol. First, the UMD disposes with the knowledge related to the secret key of the user authenticator device Key_{UAD} represented by the identifier ID_{UAD}. Therefore, the UMD has the possibility to verify the variable value V received from the UAD compared with (i) the parameter W obtained from the SH-GW and (ii) the value SN_{UAD} received from the UAD.

The implemented MAC employs the symmetric key cryptographic systems computed as a hash function (i.e., MD5, SHA1, SHA256, etc.) of concatenated messages received from the identifier (time stamp, secret key, etc.), see below:

$$V = MAC(W, ID_{UAD}, SN_{UAD}, Key_{UAD}) \tag{1}$$

$$\Downarrow$$

$$hash\,(W \| ID_{UAD} \| SN_{UAD} \| Key_{UAD}) \tag{2}$$

Similarly to the classical Yoking-proof protocol, we use SN_{UAD} to protect our system against a replay attack at the side of the UAD. In case of a successful verification on the UMD, the final step of the verification process may be completed by utilizing the secret key of the Gateway (Key_{GW}) stored in the database. We use SN to prevent from the replay attack on the Metering Device (MD) as well. If the verification at the UMD fails, our proposed protocol returns to the step 3, see Fig. 3. In case of a successful verification at the database side, the SH-GW receives the secret key to execute the authentication process and the data processing with such newly added MD (our IoT device). In case the verification process fails during the search phase in the database, the system returns to the phase 1.

3.3 Real-World Demonstration

Further, we have tested the proposed communication scheme in a real-world smart home scenario, see Fig. 4. As the SH-GW entity, the Raspberry Pi 3 platform was utilized and the main programming language for our test implementation has been Python. The developed script enables the connection between the SH-GW and the RFID reader utilizing the USB interface (AT commands were used to send specific requests to the reader controller). When the initialization process is completed, the reader starts to continuously scan for new devices (equipped with the RFID tags). When such a tag appears, its ID is concatenated with the SH-GW's ID and the secret key. Based on this resultant string, the Python script calculates the MAC using the SHA-256 hash function. At step 3, see Fig. 3, the socket-based connection (data is transmitted in the JSON format) towards the UAD is created. After that, the SH-GW sends the MAC and waits for the response from the UAD.

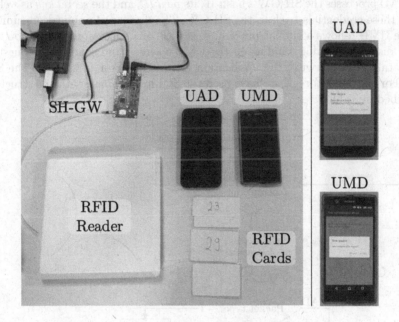

Fig. 4. Prototyped smart home scenario.

When a reply from the UAD is received (step 4), the SH-GW creates a client request (on the UMD, step 5) with the corresponding content and waits for the MD's secret key (stored in the database, see Fig. 3). During the final stage (after obtaining the secret key), the SH-GW starts a data transfer with the authenticated MD (sensor). As it is indicated in Fig. 4, two Android-based smartphones were used as the UAD and the UMD, respectively. In order to develop the required software running on these devices, Java programming language has been utilized. The key component of the created application is represented by the socket-based server operating in a separate thread. The requested data is provided as a JSON-object and contains the packet type that determines the application mode – UAD or UMD, respectively.

When the UAD receives an authentication request, a pop-up window with the query is displayed as to whether the authentication procedure should be accepted

Table 1. Created table structure in utilized SQLite database.

Keys	
recID	int, AI, PK
tagID	string
tagKey	string

or declined (it contains the tag's ID received from the SH-GW). This procedure represents the proof of the user involvement, When the user accepts the request, the UAD processes the SH-GW's hash W, its *own ID*, and the *secret key* as well as sends these credentials back to the SH-GW. The developed application running on the UMD follows a similar procedure. In case of a successful verification, the UMD displays a dialog window on the UMD's screen and sends the request to the database. Otherwise, the application returns an error message to the SH-GW. For the data storage, we have used the SQLite database with the structure described in Table 1.

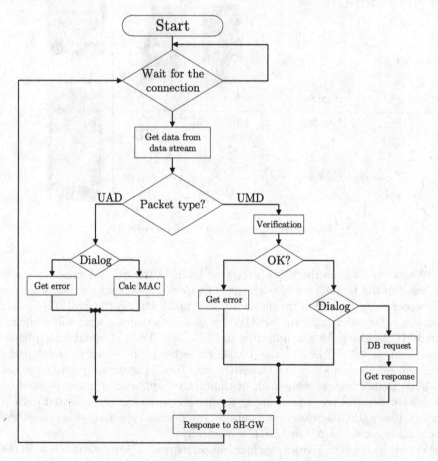

Fig. 5. Proposed logic for UAD and UMD.

Another part of the database module is represented by the PHP interface allowing for the data selection. When the UMD is requesting a PHP webpage, the verification is performed. In case of a successful request, the UMD connects to the database and selects the required information with an *SQL SELECT* request. Otherwise, the UMD displays an error message on the screen, see Fig. 5.

Our applications created in Python (Raspberry Pi 3) and Java (Android smartphones running Android 4.4.2) were tested in a smart home scenario, where the sensing device was represented by an electricity meter. As a communication interface for the multi-factor authentication, the RFID technology was used. For the user data transmission, the Wireless M-BUS protocol was utilized as it is a preferred solution for smart metering systems across modern Europe. The functionality of the implemented logic was evaluated during a long-term trial that mimicked user habits when adding new IoT device(s)[2].

4   Lessons Learned and Conclusions

The full-scale smart home deployments are still cumbersome mostly due to the users' security concerns that often outweigh any potential benefits. Despite the fact that academia as well as industry invested considerable efforts into the development of security and privacy protecting technologies for smart home applications, many of them fail to provide with the required ease of use and cost-efficiency in real residential environments.

Therefore, in this paper, we considered Yoking-proof protocol as a promising light-weight tool for authentication of various IoT devices. Specifically, our goal was to demonstrate the potential of Yoking-proof protocol while implementing it as part of a practical smart home demonstrator. With our trial, we uncovered the way how such technology can be easily implemented and operated over the off-the-shelf components. Our unique prototype enables secure authentication of new IoT devices joining the home network, which increases the user's perceived trust of different smart home solutions and thus promises to boost the expected revenues of this industry.

References

1. Atzori, L., Iera, A., Morabito, G.: The Internet of Things: a survey. Comput. Netw. **54**(15), 2787–2805 (2010)
2. Cisco, Cisco Visual Networking Index, Global mobile data traffic forecast update, 2015–2020, White paper, 2016
3. Perera, C., Liu, C.H., Jayawardena, S.: The emerging Internet of Things marketplace from an industrial perspective: a survey. IEEE Trans. Emerg. Top. Comput. **3**(4), 585–598 (2015)

---

[2] **See a showcase video capturing the functionality of our utilized testbed:** https://www.youtube.com/watch?v=zDP2MpO0YHY.

4. Vishnevsky, V.M., Larionov, A.: Design concepts of an application platform for traffic law enforcement and vehicles registration comprising RFID technology. In: Proceedings of IEEE International Conference on RFID-Technologies and Applications (RFID-TA), pp. 148–153. IEEE (2012)

5. Vishnevsky, V., Kozyrev, D., Rykov, V.: New generation of safety systems for automobile traffic control using RFID technology and broadband wireless communication. In: Vishnevsky, V., Kozyrev, D., Larionov, A. (eds.) DCCN 2013. CCIS, vol. 279, pp. 145–153. Springer, Heidelberg (2014). doi:10.1007/978-3-319-05209-0_13

6. Miori, V., Russo, D.: Home automation devices belong to the IoT world. ERCIM News **101**, 22–23 (2015)

7. Vishnevsky, V.M., Larionov, A., Ivanov, R.: Architecture of application platform for RFID-enabled traffic law enforcement system. In: Proceedings of 7th International Workshop on Communication Technologies for Vehicles (Nets4Cars-Fall), pp. 45–49. IEEE (2014)

8. Komninos, N., Philippou, E., Pitsillides, A.: Survey in smart grid and smart home security: issues, challenges and countermeasures. IEEE Commun. Surv. Tutorials **16**(4), 1933–1954 (2014)

9. Hosek, J., Masek, P., Kovac, D., Ries, M., Kröpfl, F.: IP home gateway as universal multi-purpose enabler for smart home services. e & i Elektrotechnik und Informationstechnik **131**(4–5), 123–128 (2014)

10. He, D., Zeadally, S.: An analysis of RFID authentication schemes for Internet of Things in healthcare environment using elliptic curve cryptography. IEEE Internet Things J. **2**(1), 72–83 (2015)

11. Yu, Y.-C., Hou, T.-W., Chiang, T.-C.: Low cost RFID real lightweight binding proof protocol for medication errors and patient safety. J. Med. Syst. **36**(2), 823–828 (2012)

12. Ning, H., Liu, H., Yang, L.T.: Aggregated-proof based hierarchical authentication scheme for the Internet of Things. IEEE Trans. Parallel Distrib. Syst. **26**(3), 657–667 (2015)

13. Ramsey, B.W., Temple, M.A., Mullins, B.E.: PHY foundation for multi-factor ZigBee node authentication. In: Proceedings of IEEE Global Communications Conference (GLOBECOM), pp. 795–800. IEEE (2012)

14. Juels, A.: 'Yoking-proofs' for RFID tags. In: Proceedings of the Second IEEE Annual Conference on Pervasive Computing and Communications Workshops, pp. 138–143. IEEE (2004)

15. Saito, J., Sakurai, K.: Grouping proof for RFID tags. In: Proceedings of 19th International Conference on Advanced Information Networking and Applications (AINA 2005) Volume 1 (AINA papers), vol. 2, pp. 621–624. IEEE (2005)

16. Nuamcherm, T., Kovintavewat, P., Tantibundhit, C., Ketprom, U., Mitrpant, C.: An improved proof for RFID tags. In: Proceedings of 5th International Conference on Electrical Engineering/Electronics, Computer, Telecommunications and Information Technology (ECTI-CON), vol. 2, pp. 737–740. IEEE (2008)

17. Chen, C.-L., Wu, C.-Y.: An RFID system yoking-proof protocol conforming to EPCglobal C1G2 standards. Secur. Commun. Netw. **7**(12), 2527–2541 (2014)

18. Masek, P., Hosek, J., Zeman, K., Stusek, M., Kovac, D., Cika, P., Masek, J., Andreev, S., Kröpfl, F.: Implementation of true IoT vision: survey on enabling protocols and hands-on experience. Int. J. Distrib. Sens. Netw. **2016** (2016)

Analysis and Simulation of UHF RFID Vehicle Identification System

Vladimir Vishnevskiy$^{(\boxtimes)}$, Andrey Larionov, and Roman Ivanov

V.A. Trapeznikov Institute of Control Sciences of Russian Academy of Sciences,
Profsoyuznaya 65, 117997 Moscow, Russia
vishn@inbox.ru, larioandr@gmail.com, iromcorp@gmail.com
http://www.ipu.ru

Abstract. In this paper a model of UHF RFID Vehicle Identification System based on EPC Class1 Gen2 is described. The model takes into account the influence of protocol settings, antenna and tranceiver parameters and signal propagation along roads on the system performance. It is shown that the two-ray pathloss model and Rayleigh distribution for BER computation allows to simulate RFID system operation adequately. The estimated protocol settings providing reliable vehicle identification at speed up to 220 kmph are given.

Keywords: UHF RFID · EPC Class 1 Gen. 2 · RFID link budget · Propagation models · Vehicle identification · Wireless network simulation · Passive RFID · Traffic law enforcement

1 Introduction

Vehicle identification is one of the primary RFID applications. First RFID-based tolling system appeared in the USA in 1991 [10]. RFID usages in entrance control systems, tolling and others are devoted a significant amount of researches [2,5, 7–9,21]. RFID can be applied to solve other problems, such as performance enhancement of the traffic law enforcement systems making use of cameras to identify vehicle licence plates [19,20].

RFID system consists of a Reader, or an Interrogator, and Tags. Such systems can be active, if the tag has inner power source and can independently initiate message exchange with the reader, or passive, if the tag has no inner power source. Passive RFID [3,6] generally provides less reading distance as the tag is required to be close enough to the reader to power up its chip and transmit the identifier, but such tags are significantly cheaper then the active ones.

In 2014 in Kazan city (Russia, Tatarstan) an expirement was carried out. During the experiment the performance of the EPC Class1 Gen2 [4] passive

I. Roman—This research has been financially supported by the Russian Science Foundation and the Department of Science and Technology (India) via grant 16-49-02021 for the joint research project by the V.A. Trapeznikov Institute of Control Science and the CMS College Kottayam.

© Springer International Publishing AG 2016
V.M. Vishnevskiy et al. (Eds.): DCCN 2016, CCIS 678, pp. 35–46, 2016.
DOI: 10.1007/978-3-319-51917-3_4

RFID system to identify vehicles in urban area was studied. Four identification points were deployed and 1000 buses were equiped with tags placed in license plates and under the windshields. The experiment was running for several winter months and showed that system provided 0.92 buses recognition rate [20].

To analyze the efficiency of RFID application for massive identification of fast moving vehicles achieving 220 kmph speed it is neccessary to study all factors affecting system performance. The RFID system performance is affected by the factors which are as follows: significant contribution is made by antenna systems and transeivers [3,6,13–16,18], propagation conditions [1,7] and protocol settings [12,17]. Thus, it is necessary to take into account both protocol specifics and signal propagation properties to build an adequate model.

In this paper a detailed model enabling precise UHF RFID vehicle identification system performance evaluation is presented. Peculiarities of wave propagation along roads, characteristics and mounting parameters of the antennas are considered. The model also allows to analyze the influence of different protocol settings on the system performance.

The paper is organized as follows. In Sect. 2 the principles of UHF RFID vehicle identification system and fundamental specifics of UHF RFID EPC Class1 Gen2 protocol are described. In Sect. 3 the model of RFID channel is given, link budget is calculated and computation of the bit error probability (BER) is performed. Section 4 provides the parameters estimation and performance evaluation using system simulation modeling. Section 5 concludes the paper.

2   Operation of UHF RFID Vehicle Identification System

UHF RFID Vehicle Identification System consists of a reader, an antenna connected to the reader and disposed over a lane at the 5 m altitude and a passive tag, embedded into a licence plate (see Fig. 1).

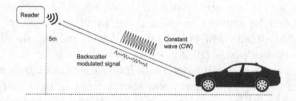

Fig. 1. The scheme of UHF RFID vehicle identification system

A reader radiates a constant or sine wave (CW). As a tag has no energy source it makes use of CW to power up and respond to the reader via backscatter modulation. Thus, the connection between a reader and a tag is possible if and only if (a) the tag receives enough power to operate (approximately −18 dBm) and (b) the power of the backscattered signal is enough for successful decoding at the reader side. Standard tags provide read distance up to 8–15 m from the reader that has transmitting power of 31.5 dBm (1.4 W) and 8dBi antenna gain

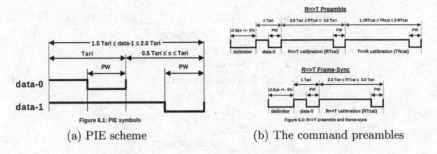

(a) PIE scheme   (b) The command preambles

Fig. 2. Reader command encoding [4]

in a lab environment. However this distance is overestimated and can shorten in a real environment affected by multipath fading.

Reader sends commands to a tag using DSB-ASK, SSB-ASK or PR-ASK modulation. These commands are encoded with PIE (Pulse Interval Encoding); according to PIE, the duration of the data-0 and data-1 symbols differ in 1.5–2 times (see Fig. 2a). Before transmitting a command the reader sends either a preamble or a frame-sync (see Fig. 2b). Due to PIE any command transmission duration significantly depends on the content of the message. Data-0 symbol duration is refered to as Tari and variates from 6.25 to 25 μs.

Tag takes advantage of ASK or/and PSK modulation. It makes use of FM0 encoding scheme or Miller code with 2, 4 or 8 symbols per bit. The symbol rate is defined as $BLF = \frac{DR}{TRcal}$, where DR is equal to 8 or 64/3. Each tag response follows a preamble whose length depends on the chosen coding scheme and can be around 10–12 bits or 18–22 bits in extended mode. All tag operation settings except a modulation are transmitted by the reader in the session beginning.

The communication protocol between a reader and a tag is based on Slotted ALOHA. All operating time is divided into rounds (see Fig. 3a). Each round starts with the QUERY command transmitted by the reader. A QUERY command follows a preamble containing tag settings and carries Q parameter value defining time slots number in the round as 2^Q. At the beginning of all following slots the reader transmits QREP (QueryRep) command that has significantly less size then QUERY.

After receiving QUERY the tag assigns slot counter a random number in $[0, 2^Q - 1]$ range. After receiving QREP a tag decrements the counter. When it reaches 0 a tag transmits a random 16-bit word in RN16 response. The reader receives the word and transmits it back to the tag in ACK command. Then the tag should transmit a response with its parameters (PC/XPC), memory content (EPC bank) and CRC (Cyclic Redundancy Check) as a response to ACK (see Fig. 3b). The EPC size may vary from tag to tag, we will consider it be equal to 96 bit. If two of more tags have chosen the same slot, then the collision arises during RN16 transmission and the reader either would not able to receive any response or would transmit ACK with a wrong word (see Fig. 3c). In both cases the tag will fail to transmit its identifier in the round.

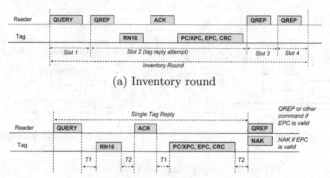

(a) Inventory round

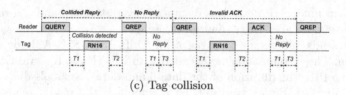

(b) Tag answer transmission in the first slot

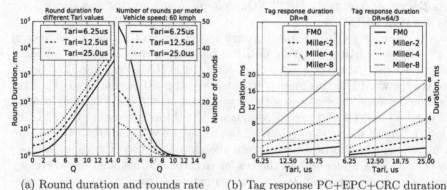

(c) Tag collision

Fig. 3. The structure and duration of rounds and slots

Thus, the successful vehicle identification probability P_s can be evaluated using the following formula:

$$P_s = 1 - (1 - (1 - P_c)(1 - P_e^{RN16})(1 - P_e^{EPC}))^{N_{rounds}}, \qquad (1)$$

where N_{rounds} is a number of rounds in which the tag participates, P_c is the collision probability (depends on a tags and slots number), P_e^{RN16} is RN16 transmission error probability, P_e^{EPC} is the probability of PC+EPC+CRC response transmission error. A number of rounds (see Fig. 4a) depends on the vehicle speed and the duration of each round which, in turn, depends on the duration of commands and responses (see Fig. 4b).

(a) Round duration and rounds rate   (b) Tag response PC+EPC+CRC duration

The duration of commands and answers may vary in a wide range (see Table 1) depending on protocol settings. The slot duration differs when the slot is free, a collision accures or an identifier transmission takes place. The slot duration also depends on the commands and responses duration. The number of slots in a round is 2^Q, where $Q = \overline{0, 15}$. Thus, the duration of a round may vary from a millisecond to several seconds depending on the settings chosen, see Fig. 4a and b.

Table 1. Computation of commands and responses duration depending on the protocol parameters. Tari, TRcal, QUERY and PC+EPC+CRC columns are measured in μs.

DR	Tari	TRcal	BLF, kHz	M	QUERY	PC+EPC+CRC
64/3	6.25	33.38	639.2	FM0	252.13	211.20
64/3	6.25	33.38	639.2	Miller-8	270.88	1739.67
64/3	12.5	66.75	319.6	FM0	491.75	422.4
64/3	25.0	133.5	159.8	FM0	971.0	844.8
64/3	25.0	225.0	94.81	FM0	1062.5	1423.83
8	6.25	33.38	239.7	FM0	245.8	563.2
8	25.0	225.0	33.56	Miller-8	1112.5	31275.0

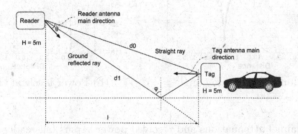

Fig. 4. Straight and ground reflected rays geometry

Tags can move with very large speeds when mounted on cars. This implies that the large round duration may result in a tag identifier transmission failure since the tag won't have enough time to repeat the attempt if any error occurs. On the other hand, decreasing commands and responses duration may lead to higher error rate, and shortening the slots number per round leads to higher collision probability. It will be shown that decreasing the slots number along with slow and reliable coding schemes utilization (Miller-4, Miller-8) allows to enhance the system performance. In the next chapter we are going to study the propagation properties and their influence on the responses delivery error rates.

3   The Model of RFID Channel

The vehicle identification probability in an arbitrary inventory round is defined by the probability of successful tag identifier transmission. At the same time this probability depends on the probability of successful transmission of all messages between the reader and the tag. The command is successfully received by the tag only if the tag has beed powered on since the round beginning, i.e. it has been extracting enough power from the CW. The reader should also receive a tag response of sufficient power for its decoding. Furthermore, while transmitting any message the interference with other tags responses is not allowed. Thus, there are three reasons of tag identification failure: collision, insufficient power for a tag to operate and a weak tag response that cannot be separated from the noise. The collision probability depends on a number of slots in a round, a number of tags and protocol settings. This section analyzes the last two reasons.

To perform this analysis it is neccessary to consider the whole propagation path, attenuations and losses arising out on this path and evaluate the probability of successful message transmission via evaluation of the bit error probability.

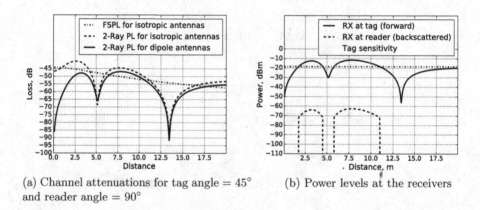

(a) Channel attenuations for tag angle $= 45°$ and reader angle $= 90°$

(b) Power levels at the receivers

Fig. 5. Channel attenuations and received power versus tag-reader distance

To define the power on the reader and tag receivers let us investigate the link budget [14,16], which allows, among others, to find areas of sufficient power levels for a tag [13,18]. Let P_r^{TX} be the reader power and G_r be a reader antenna gain. Then $P_r^{TX}G_r = EIRP$, equivalent isotropically radiated power. A signal experiences path loss P_L during propagation over the radio channel. To compute P_L free-space path loss (FSPL) or multiray path loss models can be used. Due to significant influence of ground reflected ray on the path loss (see Figs. 4 and 5a) the multiray propagation model will be used further. It is defined as [16]:

$$P_L = \left(\frac{\lambda}{4\pi}\right)^2 \left| \frac{\Gamma_0(\theta)}{d_0}e^{-j\frac{\lambda}{2\pi}d_0} + \sum_{i=1}^{N}\frac{\Gamma_i(\theta)R_i(\phi)}{d_i}e^{-j\frac{\lambda}{2\pi}d_i}\right|^2 \tag{2}$$

where λ is a wavelength, d_0 is a direct ray path length, d_i is a length of the i-th reflected ray, $\Gamma_0(\theta) = \Gamma_0^{(r)}(\theta)\Gamma_0^{(t)}(\theta)$ is an attenuation coefficient of the direct ray corresponding to the radiation pattern of the reader and the tag, $\Gamma_i(\theta) = \Gamma_i^{(r)}(\theta)\Gamma_i^{(t)}(\theta)$ is a proper attenuation coefficient of the i-th reflected ray, $R_i(\phi)$ is a ground reflection coefficient of the i-th ray, N is a number of reflected rays.

Radiation patterns significantly affects path loss. For instance, Fig. 5a illustrates the path loss computation for an isotropic and dipole antennas. The radiation patterns are chosen according to antenna specifications. In this paper a dipole radiation pattern is used.

The ground reflection coefficient is defined by the following formula [7]:

$$R = \frac{\sin\phi - \sqrt{C}}{\sin\phi + \sqrt{C}} \tag{3}$$

where ϕ is a grazing angle, $C = \eta - \cos^2\phi$ for horizontally polarized component and $C = \frac{\eta - \cos^2\phi}{\eta^2}$, for vertical one; $\eta = \epsilon_r(f) - j60\lambda\sigma(f)$, where $\epsilon_r(f)$ is a relative permittivity of a ground surface at the frequency f, $\sigma(f)$ is a conductivity.

Unfortunately, it is difficult to take into account all summands due to chaotic nature of the signal propagation. For this reason, in the given paper two-ray propagation model considering ground reflected ray is concerned (see Fig. 4).

The received power at a tag antenna is

$$P_t^{RX} = EIRP \cdot P_L. \tag{4}$$

As a rule, readers take advantage of circular polarization to identify a tag in any space orientation, and a tag uses linear polarisation. Thus, if a tag receives a message from a reader (in contrast with backscatter modulation) the energy that tag obtains equals to $P_t^{RX}p$, where $p = 1/2$ is the polarization loss. If this value is found to be low and insufficient to power up the tag, the tag chip switches off and the round is lost. If the tag needs to send an answer via backscatter modulation, power P_t^{RX} should be multiplied by the modulation loss $K = \alpha|\rho_o - \rho_c|$, where ρ_o and ρ_c is reflection coefficient of a tag antenna in a reflection and absorption states.

Since backward propagation is computed in the same way, we obtain Friis' formula for RFID channel:

$$P_r^{RX} = P_r^{TX}G_r^2 P_L^{(fwd)} P_L^{(back)} G_t^2 K, \tag{5}$$

where $P_L^{(fwd)}$ is the path loss in forward channel (from the reader) and $P_L^{(back)}$ is the path loss in backscattered channel (from the tag). For the sake of simplicity, if the polarization loss is considered to be equal for both directions, $P_L^{(fwd)}$ equals $P_L^{(back)}$.

To compute the successful pocket receiving probability it is neccessary to find a signal-noise ratio $SNR = \frac{P_r^{RX}}{N_r}$, where N_r is a noise power at the receiver antenna. Since the reader should radiate CW all time including the time of tag response reception and since the RX and TX channels are not isolated completely, a noise contained in CW leaks into RX channel. The attenuation of RX-TX isolation may vary around 25–40 dB and under this condition the noise produced by this leakage is much greater then the thermal noise (about $-115\,\mathrm{dBm}$) and it is a dominant contribution into the N_r value [16].

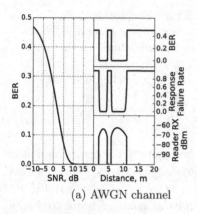

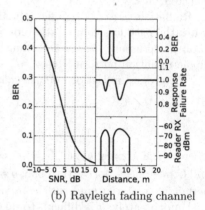

(a) AWGN channel   (b) Rayleigh fading channel

Fig. 6. BER and a tag answer transmission probability (PC+EPC+CRC) versus the distance between the reader and the tag.

Besides direct and ground reflected signal replicas, there are other dublicates resulted from the vehicles or road objects reflections in the received signal. This spontaneous reflections changes rapidly leading to fast multipath fading. Thus, usage of AWGN channel as a model (being often used for RFID-system analysis [12]) is not adequate and gives too optimistic results. The dependency of BER from tag-reader distance is illustrated in Fig. 6a (for the sake of convenience, the area where the reader receives the tag answers is illustrated on the bottom plot). As we can see on the tag response failure rate plot there exist areas where a tag transmits its response too reliably under the condition of collision absence that disagrees with the experiment results.

To describe the multipath fading correctly let us make use of Rayleigh distribution (see Fig. 6b). Thus, BER can be calculated as [1]:

$$BER = \frac{1}{2} - \frac{1}{\sqrt{1 + \frac{2}{M \cdot SNR}}} + \frac{2}{\pi} \frac{\arctan(\sqrt{1 + \frac{2}{M \cdot SNR}})}{\sqrt{1 + \frac{2}{M \cdot SNR}}}. \tag{6}$$

This model provides higher BER which provides the response error rate closer to the experimentally measured results.

Let us take a note that in open areas without obstacles AWGN channel can be used for BER computation.

In next section the results of system simulation based on the results of this and previous sections are given.

4 Analysis of UHF RFID Vehicle Identification System

To evaluate the performance of vehicle identification system a simulation model have been developed [11]. In this model tags mounted on vehicles are simulated, tag activation time and the probability that the response being received by the reader are computed using models defined in the previous section. After multiple simulation runs the successful vehicle identification probabilities and mean read number of each tag are computed by Monte-Carlo method. The simulation model is defined by the parameters as follows:

- reader and tags antennas radiation patterns, polarizations, gains, altitudes and mounting angles
- reader transmission power, tag sensitivity, modulation loss
- propagation loss and bit error probability models
- protocol settings: M – tag encoding, Tari – a data-0 symbol duration, RTcal, TRcal, DR, Q – an exponent of the slots number in a round
- vehicle speed, vehicles number on the simulated road area.

The model was developed in Python3 programming as IPython notebook; source code is available [11] on GitHub under GNU GPLv3 license.

The equipment parameters of all computations have been given from those had been used in the experiment in Kazan city:

- reader transmission power: 31.5 dBm
- reader antenna: dipole radiation pattern, 8dBi gain with circular polarization, 5m altitude above the lane, 30° vertical angle
- tag antennas: dipole radiation pattern, 2dBi gain, oriented along the road
- tag sensitivity: −18.5 dBm

Two-ray propagation model and formula 6 based on Rayleigh distribution for BER computation have been used for the reasons mentioned above. An interrogator read distance (a maximum distance between the reader and vehicles to be identified along the road) has been restricted by 20 m that results in the number of tags in reading area not exceeding 4 without side tags.

First of all the parameters providing tag reading under conditions corresponding to the Kazan experiment (vehicle speed was 60 kmph, up to three tags in reading range) with probability over 0.92 were discovered. The probability turned out to be achieved under major settings, some of them are given in Table 2. It should be mentioned that the best results were achieved under small values of Q parameter – the reason is a small collision probability due to the small tags number on the one hand, and the increase in rounds number the tag takes part in on the other hand.

Further the configurations allowing to identify the vehicles with probability above 0.95 when vehicles move at 180 kmph speed and 6 tags are in a reading area have been found. The results are given in Table 3. As it can be seen, the general trend is in rising the command transmission rate (Tari has minimum $6.25\mu s$ value for almost all configurations) and increasing the number of symbols per

Table 2. Some configurations for stable identification of the vehicle moving at 60 kpmh speed and 3 tags presence in the reading area.

M	Tari	DR	Q	$P_{success}$
FM0	6.25	8	2	0.93
Miller-2	6.25	8	2	1.0
Miller-4	6.25	64/3	3	1.0
FM0	12.5	64/3	2	0.93
Miller-2	12.5	8	2	0.99
Miller-2	18.75	8	2	0.95
Miller-4	25.0	8	4	0.95

Table 3. Configurations for stable identification of the vehicle moving at 180 kpmh speed and 6 tags presence in the reading area.

M	Tari	DR	Q	$P_{success}$
Miller-2	6.25	64/3	3	0.96
Miller-2	6.25	64/3	4	0.95
Miller-4	6.25	64/3	2	0.99
Miller-4	6.25	64/3	3	0.99
Miller-4	6.25	64/3	4	0.99
Miller-8	6.25	64/3	2	0.99
Miller-8	6.25	64/3	3	1.0
Miller-8	6.25	64/3	4	1.0
Miller-8	12.5	64/3	3	0.96
Miller-8	12.5	64/3	4	0.95

bit in tag responses (only two configurations have used Miller-2 encoding, the rest of them has used Miler-4 or Miler-8). At that time it was quite unexpected that the rise of Q value doesn't lead to identification probability increase. It is resulted from the fact that the drop of Q value allows to exponentially decrease the round duration and if the tag is able to transmit the response it almost surely achieves the reader when more resistant encoding is used. That is the increase of a number of transmission attempts with more reliable encoding has greater effect then the decrease of collision probability.

For some configurations the dependencies of identification probabilities versus speed and reader antenna angle were studied. The results are given in Figs. 7a and b. They show that it is possible to reliably identify the vehicles moving with speed up to 220 kmph and with presence of up to 6 vehicles in a reading area under the usage of $Q = 3$, $Tari = 6.25$, angle of slope to vertical equals to $\pi/6$, $DR = 64/3$ and Miller-8 encoding.

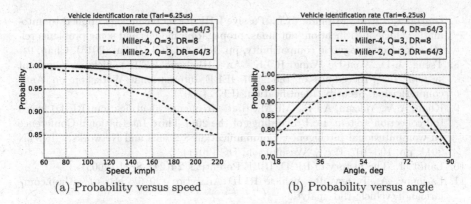

(a) Probability versus speed   (b) Probability versus angle

Fig. 7. Vehicle identification probability dependence on vehicles speed and antenna angle of slope to vertical

5   Conclusion

In the paper a model of UHF RFID Vehicle Identification System based on EPC Class1 Gen2 is described. The model considers the influence of protocol settings, antenna and tranceiver parameters, signal propagation along roads on the vehicle identification probability. The results computed by the given model agreed with the results of the vehicle identification system experiment carried out in Kazan city. The protocol settings estimation for reliable identification of vehicles moving at 60–220 kmph speed under the presence of several tags in a reading area was performed. The results presented in this paper show that the passive UHF RFID-based vehicle identification can be successfully used in modern traffic law enforcement systems.

References

1. Lazaro, A., Girbau, D., Villarino, R.: Effects of interferences in UHF RFID systems. Prog. Electromagn. Res. **98**, 425–443 (2009)
2. Blythe, P.: RFID for road tolling, road-use pricing and vehicle access control. In: IEEE Colloquium on RFID Technology, pp. 8–16 (1999)
3. Dobkin, D.M.: The RF in RFID: Passive UHF RFID in Practice. Elsevier Inc., New York (2008)
4. EPCGlobal: EPC Radio-frequency Identity Protocols Generation-2 UHF RFID. Specification for RFID Air Interface. Protocol for Communications at 860 MHz–960 MHz. Version 2.0.1 Ratified. EPCglobal Gen2 Specification, GS1 EPCglobal Inc. (2015)
5. Al-Naima, F.M., Al-Any, H.: Vehicle location system based on RFID. In: 2011 Proceedings of the Developments in E-systems Engineering (DeSE). IEEE (2011)
6. Finkenzeller, K.: RFID Handbook: Fundamentals and Applications in Contactless Smart Cards and Identification, 2nd edn. Wiley, New York (2003)

7. Gonzalez, S.R.M., Miranda, R.L.: Passive UHF RFID technology applied to automatic vehicle identification: antennas, propagation models and some problems relative to electromagnetic compatibility, pp. 188–220. IGI Global (2013). Chap. 9

8. Tseng, J.-D., Wen-De Wang, R.J.K.: An UHF band RFID vehicle management system. In: Proceedings of the 2007 IEEE International Workshop on Anti-counterfeiting, Security, Identification. IEEE (2007)

9. Khan, A.A., Yakzan, A.I.E., Ali, M.: Radio frequency identification (RFID) based toll collection system. In: Proceedings of the 2011 Third International Conference on Computational Intelligence, Communication Systems and Networks, CICSYN 2011, pp. 103–107. IEEE, Washington, DC (2011)

10. Landt, J.: The history of RFID. IEEE Potentials **24**(24), 8–11 (2005)

11. Larionov, A., Ivanov, R.: Vehicle RFID Analytics (2016). https://github.com/larioandr/vehicle-rfid-analytic

12. Mohaisen, M., Yoon, H., Chang, K.: Radio transmission performance of EPCglobal Gen-2 RFID system. CoRR abs/0911.0542 (2009)

13. Nikitin, P.V., Rao, K.V.S.: Performance limitations of passive UHF RFID systems. In: IEEE Antennas and Propagation Society International Symposium, pp. 1011–1014. IEEE (2006)

14. Nikitin, P., Rao, K., Lam, E., Pinc, P.: UHF RFID tag characterization: overview and state-of-the-art. In: AMTA. Seattle, WA (2012)

15. Nikitin, P.V., Rao, K.V.S.: Measurement of backscattering from RFID tags. In: Proceedings of Antennas Measurement Techniques Association Symposium (2005)

16. Nikitin, P.V., Rao, K.V.S.: Antennas and propagation in UHF RFID systems. In: Proceedings of the IEEE International Conference on RFID, pp. 277–288. IEEE (2008)

17. Nikitin, P.V., Rao, K.V.S.: Effect of Gen2 protocol parameters on RFID tag performance. In: Proceedings of the IEEE International Conference on RFID. IEEE (2009)

18. Rao, K.V.S., Nikitin, P.V., Lam, S.F.: Antenna design for UHF RFID tags: a review and a practical application. IEEE Trans. Antennas Propag. **53**(12), 3870–3876 (2005)

19. Vladimir M. Vishnevsky, A.L.: Design concepts of an application platform for traffic law enforcement and vehicles registration comprising RFID technology. In: 2012 IEEE International Conference on RFID-Technologies and Applications (RFID-TA). IEEE (2012)

20. Vishnevsky, V.M., Andrey Larionov, R.I.: Architecture of application platform for RFID-enabled traffic law enforcement system. In: Proceedings of the 2014 7th International Workshop on Communication Technologies for Vehicles (Nets4Cars-Fall), IEEE (2014)

21. Yoon, W.J., Chung, S.H., Lee, S.J.: Implementation and performance evaluation of an active RFID system for fast tag collection. Comput. Commun. **31**(17), 4107–4116 (2008)

Modeling and Performance Comparison of Caching Strategies for Popular Contents in Internet

Natalia M. Markovich[1]([✉]), Vladimir Khrenov[1], and Udo R. Krieger[2]

[1] V.A. Trapeznikov Institute of Control Sciences, Russian Academy of Sciences, Profsoyuznaya Str. 65, 117997 Moscow, Russia
markovic@ipu.rssi.ru
[2] Fakultät WIAI, Otto-Friedrich-Universität, An der Weberei 5, 96047 Bamberg, Germany
udo.krieger@ieee.org

Abstract. The paper is devoted to caching of popular multimedia and Web contents in Internet. We study the Cluster Caching Rule (CCR) recently proposed by the authors. It is based on the idea to store only popular contents arising in clusters of related popularity processes. Such clusters defined as consecutive exceedances of popularity indices over a high threshold are caused by dependence in the inter-request times of the objects and, hence, their related popularity processes. We compare CCR with the well-known Time-To-Live (TTL) and Least-Recently-Used (LRU) caching schemes. We model the request process for objects as a mixture of Poisson and Markov processes with a heavy-tailed noise. We focus on the hit probability as a main characteristic of a caching rule and introduce cache effectiveness as a new metric. Then the dependence of the hit probability on the cache size is studied by simulation.

Keywords: Caching · Cluster Caching Rule · TTL · LRU · Hit/miss probability · Popularity process · Clusters of exceedances · Inter-request times

1 Introduction

Nowadays, caching of contents is intensively applied in the Internet to provide multimedia or Web objects on demand to the users with a minimal delay. The idea stems from computer systems where frequently demanded files have to be cached in a short memory to accelerate the exchange between the processor and the operative memory. In telecommunication systems this concept is used to keep the requested content in a cache, e.g. at an edge router in fog computing (cf. [19–21]), or a hierarchy of caches (cf. [3,4]). Numerous problems arising from the randomness of the inter-request time (IRT) sequences concern the optimal cache size, cache utilization and occupancy, and the replacement of objects within a cache to provide the fast availability of the requested content. The latter item is characterized by the hit/miss probability, i.e. the probability to find/miss a requested content in the cache.

© Springer International Publishing AG 2016
V.M. Vishnevskiy et al. (Eds.): DCCN 2016, CCIS 678, pp. 47–56, 2016.
DOI: 10.1007/978-3-319-51917-3_5

Among these cache replacement rules the Least-Recently-Used (LRU) (cf. [1]), the Least-Frequently-Used (LFU) (cf. [2]) and the Time-to-Live (TTL) policy (cf. [3–5]) are the most popular schemes. Usually, the Independent Reference Model (IRM) that summarizes a number of assumptions is used to simplify the formulation of the hit/miss probability, the cache utilization and occupancy problems. According to IRM it is assumed that the inter-request times are independent and exponentially distributed (i.e. the request process is a Poisson renewal process), and that the popularity of contents or Web objects and content sizes are constant. The IRM implies a time and space locality regarding the object popularity. It should be noted that normally a non-Poisson renewal process model cannot capture the superposition of request processes that arise in cache networks (cf. [3]).

Not much work has been done when the IRM model is not appropriate. Then the IRT sequence may be correlated, heavy-tailed and non-stationary. Our first objective is to show how one can handle the caching problem in this case and what is the impact of such conditions on the effectiveness and utilization of a cache. Correlated IRTs are particularly realistic if some content has become very popular and many users are interested in it. Therefore, such correlations generate clusters of peaks of the popularity index. Following [6] we determine the cluster as a conglomerate of consecutive exceedances of the popularity process over a threshold between two consecutive non-exceedances. A cluster structure of the popularity process is shown in Fig. 1.

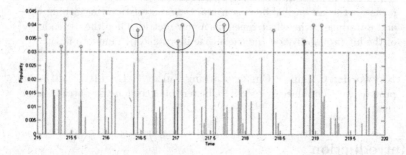

Fig. 1. The sequence of content popularity against the time including some indicated clusters of exceedances over a given threshold.

We focus on the Cluster Caching Rule (CCR) policy proposed in [7] and studied in [8]. Dealing with a single cache we propose here an *effectiveness of a cache* as new caching metric. It is defined as total popularity of objects placed in the cache at time t. The second objective is given by the analysis and comparison of the CCR, TTL and LRU rules by a simulation study. Both the CCR and TTL rule use *timers* as tuning knobs for individual objects to stay in the cache, but they apply different arguments. We propose to select the TTL timers depending on the popularity of the cached objects.

The paper is organized as follows. In Sect. 2 related work is discussed. In Sect. 3 we propose the effectiveness of a cache as characteristic metric. In Sect. 4 we modify the TTL rule regarding the specific TTL timers which depend on the popularity indices. Moreover, we compare the hit probabilities of the CCR, LRU and TTL rules depending on the cache size and the TTL timer selection by simulation. The results are summarized in the Conclusion.

2   Related Work

Cache replacement schemes can be split into capacity-driven and TTL-based policies (cf. [9]). The hit (or miss) probability determines the long-term frequency to find (or not to find) a requested object in the cache. The LRU and LFU policies belong to the capacity-driven group since objects are evicted from the cache by arrivals of those objects not yet stored. According to LRU a new requested object is placed into the cache and the least recently requested object is evicted from the cache. In case the requested object is found in the cache, it is put on the first position while the residual cache contents is shifted upwards. According to the TTL policy objects are evicted according to individual timers, i.e. life times to be in cache (cf. [3]). It was found that LFU is better than LRU (cf. [10]). Thus, modifications of LRU were proposed like persistent-access-caching (PAC) to improve its miss probability (cf. [11]).

The CCR policy [7] is related to a popularity oriented, threshold-driven policy. It allows to cache only those contents corresponding to related popularity clusters, i.e. those objects are cached whose popularity index exceeds a sufficiently high threshold u. The hit probability is then determined as the probability to enter the cluster and the time of an object to stay in the cache is determined by the duration of consecutive clusters containing that object and the corresponding inter-cluster times (see Fig. 2). The CCR scheme provides some kind of congestion control that allows to drive cache utilization. The threshold u determines the popularity level which is exceeded and impacts on the cluster sizes of the popularity process. CCR is in a way similar to LFU where only popular objects may be placed in the cache. Caching only frequently referenced objects has also been developed as central processing unit (CPU) approach in [1].

Regarding the stochastic analysis of caching rules for *correlated request processes with heavy tails* not much research has been done yet. Poisson arrival processes were considered in [12–14] with light- and heavy-tailed request rates λ_i, i.e. $\lambda_i \sim c\exp(-\xi i^\beta)$ for $i = 1, 2, \ldots$ with $c, \xi, \beta > 0$ and $\lambda_i \sim c/i^\alpha$ for $i = 1, 2, \ldots$ with $\alpha > 1$, $c > 0$, respectively. The miss probability of the LRU policy was shown to decrease following a power law or exponentially, respectively, for heavy- and light-tailed λ_i as the cache size C tends to infinity. It was derived that the correlation does not impact the miss probability for unlimited cache size. Markov arrival processes (MAPs) were also used to model correlated requests (cf. [3]), since they are self-contained regarding superposition. Regarding the LRU strategy and moderate cache sizes, non-stationary and dependent request processes and the average miss probability were considered as input and metric in [15].

Cache utilization determines an important metric and raises several issues. To optimize cache utilization based on TTL policies, it was proposed in [16] to maximize the sum of the utilities of all objects regarding the TTL timers. Therein, each content item is associated with a utility metric that is a function of the corresponding content hit probability. The latter approach assumes a Poisson renewal process as request model. In [7] the utilization with regard to the CCR strategy has been determined by the ratio of the cluster and the cache sizes where the cluster implies a set of consecutive exceedances of the popularity index over a sufficiently high threshold. Then the average cache utilization was considered both for fixed and random object sizes.

3   Effectiveness of the Cluster Caching Rule

The analysis of real traces has shown that about 70% of contents in caches is requested only once. It translates into an even higher miss ratio of 0.88 (cf. [1]). The LRU and TTL cache policies do not prevent to place unpopular contents in the cache. To prevent caching of a large portion of rarely requested objects, we propose to maximize the *effectiveness* of a cache. It is reflected by the new metric

$$e(t) = \sum_{i=1}^{C} p_i(t) \mathbb{I}\{\text{ith object } o_{j_i} \text{ from the catalog is in the cache at time t}\}.$$

We assume that all objects $\{o_j \mid j \in M\}, M = \{1, \ldots, N\}$ in the catalog have equal size s and $\widehat{C} = C \cdot s$ is the cache size. N denotes the size of the catalog. $p_i(t)$ is the popularity of the ith object o_{j_i} in the cache at epoch t. $e(t)$ indicates the total popularity of all those objects $\{o_{j_1}, \ldots o_{j_C}\}$ stored in the cache at time t. It holds $j_C \leq C$ since the cache may not be full. According to the CCR policy, the ith object o_{j_i} may be placed in the cache if its popularity $p_i(t)$ at time t exceeds a given threshold u.

As the cache load is provided by clusters of highly popular objects, their indices $p_i(t)$ may belong only to one cluster. This means that the number of cached objects is limited by the cluster size $T_2(u)$ or more exactly by the maximal cluster size. The notion of the cluster size of a stationary process $\{X_t\}_{t \geq 1}$

$$T_2(u) = \min\{j \geq 1 : L_{1,j} > u, X_{j+1} \leq u | X_1 \leq u\},$$

where $M_{1,j} = \max\{X_2, \ldots, X_j\}$, $M_{1,1} = -\infty$, $L_{1,j} = \min\{X_2, \ldots, X_j\}$, $L_{1,1} = +\infty$ is mentioned in [6,7] following [17]. Regarding the CCR policy, we then get the effectiveness

$$e_u(t) = \sum_{i=1}^{C} p_i(t) \mathbb{P}\{p_i(t) > u | i\text{th object is in the latest cluster at time t}\}$$

$$\leq \sum_{i=1}^{j} p_i(t) \mathbb{P}\{T_2(u) = j\} \tag{1}$$

where $j \leq C$ is the observed cluster size. In case $j > C$ we can load the rest of those objects in the next cache of a cache hierarchy or increase u to decrease the cluster size.

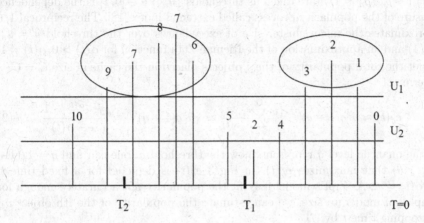

Fig. 2. Illustration of the CCR caching mechanism and the popularity clustering for different threshold values U_1 and U_2 over time.

Example 1. Figure 2 illustrates the dynamics of CCR caching and the calculation of the effectiveness. At time T_1 the cache contains objects with numbers 1, 2 and 3 because their popularity exceeds the threshold U_1. If U_2 were the threshold, the objects with numbers $0 - 4$ would be cached. Let us consider the threshold U_1. The next cluster begins at the object with number 6. The popularity of the object 2 decreases and it falls between two clusters. Nevertheless, at time T_1 it remains in the cache. In the second cluster the object with number 7 occurs twice. At time T_2 we have the objects with numbers $6 - 8$ in the cache. The objects $1 - 3$ are evicted from the cache. Hence, the effectiveness at time T_1 is calculated as the sum of the popularity of objects $1 - 3$ and at time T_2 by means of the objects $6 - 8$.

The probability $\mathbb{P}\{T_2(u) = j\}$ in (1) does not take into account possible repetitions of the same objects in the popularity clusters. Therefore, it provides an upper bound of the real effectiveness.

The effectiveness metric $e_u(t)$ in (1) is driven by u. We can find such u that provides a maximal value $e_u(t)$ for a fixed time t. To this end, let us assume that the objects' popularity is determined by Zipf's law, i.e. $p_i \sim \chi/i^\alpha$, where $\chi > 0$ is a constant. $\alpha > 0$ is the tail index. It shows the heaviness of the tail of the popularity distribution. As the popularity index may change over time, we can take $p_i(t) \sim \chi/i^{\alpha(t)}$.

Regarding a sequence of increasing thresholds $\{u_n\}_{n \geq 1}$, the probability of $T_2(u_n)$ derived in [6] satisfies for each $\varepsilon > 0$ and some n_ε and $j_0(n_\varepsilon)$ the following expression

$$| \, \mathbb{P}\{T_2(x_{\rho_n}) = j\}/(\theta^2 q_n (1 - q_n)^{(j-1)\theta}) - 1 \, | < \varepsilon$$

for all $n > n_\varepsilon$ and sufficiently large j, i.e. $j > j_0(n_\varepsilon)$. Here high quantiles $\{x_{\rho_n}\}$ of the common popularity process of all objects in the catalog w.r.t. the levels $q_n = 1 - \rho_n$, $\rho_n \sim 1/n$ are taken as thresholds $\{u_n\}$. $\theta \in [0, 1]$ is the dependence measure of the popularity process called extremal index [18]. The reciprocal $1/\theta$ approximates the mean cluster size of exceedances over the threshold $u = u_n$. By (1) and an approximation of the Rieman Zeta function for $\alpha(t) > 0, \alpha(t) \neq 1$, we get the total popularity of the j objects placed in the cache of size $\widehat{C} = C \cdot s$ in terms of

$$e_q(t) \approx \theta^2 q (1 - q)^{(j-1)\theta} \sum_{i=1}^{j} \frac{\chi}{i^{\alpha(t)}} \approx \chi \theta^2 q (1 - q)^{(j-1)\theta} \frac{j^{1-\alpha(t)} - 1}{1 - \alpha(t)}. \qquad (2)$$

As the quantile level q represents now the threshold u, one can find $q = 1/(1 + (j - 1)\theta)$ that maximizes $e_q(t)$. In Fig. 3 $e_q(t)$ is depicted for a fixed time t, i.e. $\alpha(t) = \alpha$. As Zipf's model may fit the popularity not accurately enough for samples of moderate size, we can estimate the popularity of the ith object o_{j_i} at stopping time t by [7]

$$p_i(t) = J_{i,t}/N_t. \qquad (3)$$

Here $J_{i,t}$ and N_t denote the number of requests for the ith object o_{j_i} and for all objects $o_j, j \in M$ in the catalog at time t, respectively, that progress in time. The cluster size probability can be evaluated as ratio of the number of requests R_t with popularity exceedances over u to the total number of requests N_t at time t. Here, R_t contains only exceedances corresponding to different objects falling in the clusters. Then we get from (1) the empirical effectiveness

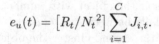

$$e_u(t) = \left[R_t/N_t^2 \right] \sum_{i=1}^{C} J_{i,t}.$$

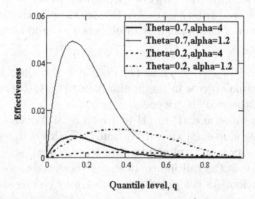

Fig. 3. Effectiveness (2) with $C = j = 10$ for the CCR policy against the quantile level q of the extremal index $\theta \in \{0.2, 0.7\}$ and the tail index $\alpha \in \{1.2, 4\}$, where $q \in \{0.137, 0.357\}$ corresponds to the maximal effectiveness.

Thereby, formula (2) provides the parametric model taking into account the heaviness of the tail in terms of $\alpha(t)$ and the dependence structure by θ.

An increasing level u induces clusters with smaller sizes. It may lead to the necessity to select a smaller cache size or to a less efficient utilization of the cache.

4 Performance Comparison of Different Caching Rules

We compare the CCR, LRU and TTL caching rules by simulation. Following [8] we use a mixture of the Moving Maxima (MM) and the Poisson renewal processes to model a common IRT process regarding all objects of the catalog of different types.

The MM process $\{\tau_{i,t}\}$ as IRT model of the ith object type satisfies

$$\tau_{i,t} = \max_{j=0,\ldots,m_i} \{\alpha_j Z_{t-j}\}, \quad t \in \mathbb{Z},$$

with nonnegative constants $\{\alpha_j\}$ such that $\sum_{j=0}^{m_i} \alpha_j = 1$ and iid standard Fréchet distributed r.v.s $\{Z_t\}$ with distribution function $F(x) = \mathbb{P}\{Z_i \leq u\} = e^{-1/u}$. The distribution of $\tau_{i,t}$ is also Fréchet. The MM process is a m_i-dependent Markov chain where m_i determines the popularity duration. The MM process models IRTs of short-term news that are of public interest for a limited time. The Poisson process with intensity λ_i models objects like scientific and culture articles which may attract interest within a long time independently. Each object of equal size $s = 1$ from the catalog has an own $(m_i, \{\alpha_j\})$ or λ_i value as unique IRT model parameter.

The MM processes generate the correlation and the cluster structure of such common IRT process that has been generated here by 90% MM and 10% Poisson renewal processes. The corresponding popularity process that is the popularity $p_i(t)$ of each requested object o_{j_i} calculated by (3) is given in Fig. 1. In (3) $J_{i,t}$ is calculated in a cross-window with $N_t = 300$ requests. The number of objects in the catalog was taken as $N = 100$.

We compare the CCR, the LRU and the TTL policies for such simulated IRT processes. For each object o_{j_i} we propose TTL timers $\{t_i\}$ depending on its popularity index $p_i(t)$ and the mean IRT $\mathbb{E}(Y_i)$ of the overall IRT process, i.e.

$$t_i = h\, \mathbb{E}(Y_i)\, p_i(t), \quad 0 < h < \infty. \tag{4}$$

h is a scalability parameter. The TTL timers are larger for highly popular objects. In (4) t_i is proportional to the popularity of the ith object in $[0, t]$.

In Fig. 4 we estimate the extremal index θ of the popularity process by the intervals estimator proposed in [17]. This allows us to estimate the effectiveness (2) and the cache size as the reciprocal $C = 1/\theta$ equal to the mean cluster size as proposed in [7]. Taking $\widehat{\theta} = 0.22$ it is easy to calculate the approximate mean cache size $\widehat{C} = 5$.

In Fig. 5 we show the hit probabilities for the TTL, LRU and CCR policies depending on the cache size C for $s = 1$. The hit probability is estimated as

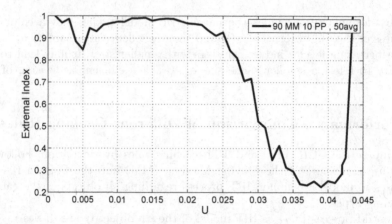

Fig. 4. The intervals estimate $\widehat{\theta}$ of the extremal index averaged over 50 samples against the threshold u: the estimate $\widehat{\theta} = 0.22$ corresponds to the stability interval of the plot by threshold u.

the ratio of the number of requests hitting the cache and the total number of requests. For small cache sizes the best hit probability is provided by the CCR scheme with a threshold u corresponding to the stability interval of the plot $(u, \widehat{\theta})$ and both the TTL and CCR work similar if h and u are relatively small. Small u generates large clusters. Then the CCR stores more objects in the same manner as TTL irrespectively of their popularity processes. If h and u are small, then the inter-cluster time for large clusters is of similar small scale as the TTL timers. For large caches and long timers TTL is better than CCR. This means a long-term placement of many objects in the large cache which is not effective. For large caches the CCR hit probability reaches a stability level

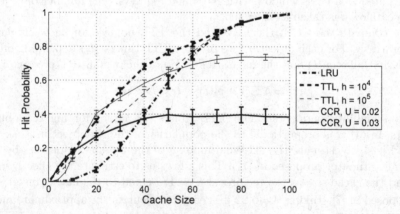

Fig. 5. The hit probabilities of the CCR, LRU and the TTL policies averaged over 50 samples against the cache size C, where horizontal lines indicate standard deviations.

that is lower than the corresponding TTL value due to the limited cluster size and the impossibility to store a larger number of objects than the cluster size. A minimal C corresponding to the stability level of the hit probability may be taken as a sufficient cache size.

5   Conclusion

The paper addresses the caching of popular multimedia and Web contents in Internet. We have extended the investigation of the Cluster Caching Rule (CCR) recently proposed in [7,8]. Assuming correlated inter-request time processes and fixed object sizes, we have studied here the caching of popular contents when the popularity of the stored objects may change over time. The LRU, CCR and TTL caching rules have been compared by a simulation study.

The following results have been obtained:

1. cache effectiveness has been introduced as new quality metrics;
2. regarding a TTL based policy, TTL timers based on popularity indices have been proposed;
3. the CCR policy has a better hit probability than TTL regarding relatively small cache sizes and thresholds u corresponding to the stability interval of the extremal index plot (u, θ);
4. the LRU policy is worse than both the CCR and TTL rule when the cache size is moderate and it works similar to TTL for large caches.

Regarding a fog computing environment based on interconnected powerful SBC boards (cf. [19,20]), optimized caching strategies for popular objects that implement the sketched approaches on a small memory are currently a very important research issue (cf. [21]). Consequently, the adoption of a dynamic version of the proposed CCR policy is a topic of our future research.

Acknowledgments. The first author acknowledges the financial support by DAAD scholarship 91619901.

References

1. Che, H., Tung, Y., Wang, Z.: Hierarchical web caching systems: modeling, design and experimental results. IEEE JSAC **20**(7), 1305–1314 (2002)
2. Lee, D., Choi, J., Kim, J.-H., Noh, S.H., Min, S.L., Cho, Y., Kim, C.S.: LRFU: a spectrum of policies that subsumes the least recently used and least frequently used policies. IEEE Trans. Comput. **50**(12), 1352–1362 (2001)
3. Berger, D.S., Gland, P., Singla, S., Ciucu, F.: Exact analysis of TTL cache networks: the case of caching policies driven by stopping times. In: 2014 ACM International Conference on Measurement and Modeling of Computer Systems, SIG-METRICS 2014, pp. 595–596 (2014)
4. Fofack, N.C., Nain, P., Neglia, G., Towsley, D.: Analysis of TTL-based cache networks. In: 6th International Conference on Performance Evaluation Methodologies and Tools (VALUETOOLS), pp. 1–10 (2012)

5. Friecker, C., Robert, P., Roberts, J.: A versatile and accurate approximation for LRU cache performance. In: Proceedings of ITC 2012, pp. 1–8 (2012)
6. Markovich, N.M.: Modeling clusters of extreme values. Extremes **17**(1), 97–125 (2014)
7. Markovich, N.: A cluster caching rule in next generation networks. In: Vishnevsky, V., Kozyrev, D. (eds.) DCCN 2015. CCIS, vol. 601, pp. 305–313. Springer, Heidelberg (2016). doi:10.1007/978-3-319-30843-2_32
8. Markovich, N.M., Krieger, U.R.: A caching policy driven by clusters of high popularity. In: 7th IEEE International Workshop on TRaffic Analysis and Characterization (TRAC 2016), 5–9 September, Paphos, Cyprus (2016)
9. Rizzo, L., Vicisano, L.: Replacement policies for a proxy cache. IEEE/ACM Trans. Netw. **8**(2), 158–170 (2000)
10. Breslau, L., Cao, P., Fan, L., Phillips, G., Shenker, S.: Web caching and Zipf-like distributions: evidence and implications. In: IEEE Proceedings of Eighteenth Annual Joint Conference of the IEEE Computer and Communications Societies (INFOCOM 1999), vol. 1, pp. 126–134 (1999)
11. Jelenković, P.R., Radovanović, A.: The persistent-access-caching algorithms. Random Struct. Algorithms **33**, 219–251 (2008)
12. Jelenković, P.R.: Asymptotic approximation of the move-to-front search cost distribution and least-recently-used caching fault probabilities. Ann. Appl. Probab. **9**, 430–464 (1999)
13. Jelenković, P.R., Radovanović, A.: Least-recently-used caching with dependent requests. Theor. Comput. Sci. **326**(1–3), 293–327 (2004)
14. Jelenković, P.R., Radovanović, A.: Asymptotic optimality of the static frequency caching in the presence of correlated requests. Oper. Res. Lett. **37**(5), 307–311 (2009)
15. Osogami, T.: A fluid limit for a cache algorithm with general request processes. Adv. Appl. Probab. **42**, 816–833 (2010)
16. Dehghan, M., Massoulie, L., Towsley, D., Menasche, D., Tay, Y.C.: A utility optimization approach to network cache design, pp. 1–11 (2016). arXiv: 1601.06838v1
17. Ferro, C.A.T., Segers, J.: Inference for clusters of extreme values. J. Roy. Statist. Soc. Ser. B **65**, 545–556 (2003)
18. Leadbetter, M.R., Lingren, G., Rootzén, H.: Extremes and Related Properties of Random Sequence and Processes. Springer, Heidelberg (1983)
19. Großmann, M., Eiermann, A., Renner, M.: Hypriot cluster lab: an ARM-powered cloud solution utilizing docker. In: 23rd International Conference on Telecommunications (ICT 2016), 16–18 May, Thessaloniki, Greece (2016)
20. Großmann, M., Eiermann, A.: Security of distributed container based service clustering with hypriot cluster lab. In: Proceedings of ITC 28, September 12–16, Würzburg, Germany (2016)
21. Pahl, C., Lee, B.: Containers and clusters for edge cloud architectures - a technology review. In: 3rd International Conference on Future Internet of Things and Cloud (FiCloud), 24–26 August 2015, pp. 379–386 (2015)

Transient Change Detection in Mixed Count and Continuous Random Data and the Cyber-Physical Systems Security

Igor Nikiforov[✉]

Université de Technologie de Troyes, UTT/ICD/LM2S, UMR 6281 CNRS,
12, rue Marie Curie, CS 42060, 10004 CEDEX Troyes, France
nikiforo@utt.fr
http://www.utt.fr

Abstract. The problem of sequential transient change detection is considered in the paper. The original contribution of this paper is twofold: first, a mixed count/continuous statistical model with abrupt changes is considered in the paper; second, a new sequential test for such a mixed count/continuous statistical model is designed and studied. The theoretical findings are applied to the problem of cyber-physical attack detection.

Keywords: Mixed count and continuous random data · Sequential change detection · Minimax criterion · Cyber-physical attacks · Cyber-physical systems

1 Introduction and Motivation

The problem of cyber-physical systems security is of great importance nowadays. A typical distributed cyber-physical system (networked control systems, SCADA, etc.) is composed of several physical and cyber layers. These layers are connected by different computer networks (WAN, LAN, VPN, etc.) The recent studies have established that the cyber-physical systems are vulnerable to cyber-physical attacks, when both, physical and cyber, components are sabotaged by attackers (see for details [1–3]).

A typical feature of cyber-physical systems is the presence of mixed count and continuous parallel data flows. The count data represent the number of events $N(t)$ occurring during a fixed time interval $(0, t]$ (for example, the number of requests per second). The data with continuous state space describe the physical parameters $\{X_t\}_{t \geq 1}$ like temperature, pressure, position/speed, etc., usually in discrete time $t = 1, 2, \ldots$. The theory and tests for sequential detection are well-developed for the observations with continuous state space and for the observations with discrete states (also for some types of point processes). To the best of our knowledge, the theory of sequential detection is practically not developed for the case of mixed count/continuous statistical models with abrupt changes.

© Springer International Publishing AG 2016
V.M. Vishnevskiy et al. (Eds.): DCCN 2016, CCIS 678, pp. 57–63, 2016.
DOI: 10.1007/978-3-319-51917-3_6

2   Sequential Detection of Transient Changes

Let ν be the number of the first post-change observation. It is assumed that the changepoint ν is unknown and not necessarily random. Let P_k and $\mathbb{E}_k$ denote the probability measure of $\{X_t\}_{t \geq 1}$ and its expectation when $\nu = k$ and let P_∞ and $\mathbb{E}_\infty$ denote the same when $\nu = \infty$, i.e., there is no change. This means that $X_t \sim \mathsf{P}_0$ for every $t < \nu$ and $X_t \sim \mathsf{P}_1$ for every $t \geq \nu$ under the measure P_ν and $X_t \sim \mathsf{P}_0$ for every $t \geq 1$ under the measure P_∞. A sequential change detection test consists in calculating the stopping time T at which the change-point ν is detected. In the classical abrupt change detection, the post-change period is assumed to be infinitely long. The conventional (Shiryaev-type, Lorden-type and Pollak-type) criteria of optimality involve the minimization of the average detection delay for a given value of false alarms (see details in [4]). For instance, the minimax Lorden-type criterion of optimality based on the minimization of the worst-worst-case average detection delay is given by [5]:

$$\inf_{T \in \mathbb{C}_\gamma} \left\{ \mathsf{ESADD}(T) = \sup_{\nu \geq 1} \operatorname{esssup} \mathbb{E}_\nu [(T - \nu + 1)^+ | \mathcal{F}_\nu] \right\} \tag{1}$$

over the class

$$\mathbb{C}_\gamma = \{T : \mathbb{E}_\infty(T) \geq \gamma\}$$

of stopping times T.

Unfortunately, such criteria of optimality (like (1)) are not adequate for the detection of transient changes of duration L (i.e., the changes of short duration) because the detection of changes after their disappearance or with the detection delay greater than a prescribed value L is considered as missed. Moreover, for the safety-critical applications, it is no matter if the true duration of the post-change period is greater than L. The changes should be detected with the delay which satisfies the following condition $T - \nu + 1 \leq L$ due to safety requirements. The penalty function related to the detection delay is quite nonlinear.

The conventional (Shiryaev-type, Lorden-type or Pollak-type) criterion warranties that some large detection delays can be compensated with some short detection delays and, hence, the (worst-worst-case) average detection delay will be optimal. In safety-critical applications, such philosophy does not work: a detection delay greater than L cannot be compensated with a detection delay shorter than L.

Motivated by safety-critical applications, we use through this paper the criterion of optimality introduced in [6,7], which involves the minimization of the worst-case conditional probability of missed detection (under the assumption that no change occurs during the "preheating" period (i.e. it is assumed that $\nu \geq L$))

$$\inf_{T \in \mathbb{C}_\alpha} \left\{ \overline{\mathbb{P}}_{\mathrm{md}}(T; L) = \sup_{\nu \geq L} \mathbb{P}_\nu (T - \nu + 1 > L | T \geq \nu) \right\} \tag{2}$$

over the class

$$\mathbb{C}_\alpha = \left\{ T : \overline{\mathbb{P}}_{\mathrm{fa}}(T; m) = \sup_{\ell \geq L} \mathbb{P}_0 (\ell \leq T < \ell + m - 1) \leq \alpha \right\},$$

where $\overline{\mathbb{P}}_{md}$ denotes the worst-case probability of missed detection and $\overline{\mathbb{P}}_{fa}$ stands for the worst-case probability of false alarm within any time window of length m.

3 Variable Threshold Window Limited CUSUM Test

The motivation and rationalities of the Window Limited (WL) CUSUM test as a solution to the transient change detection problem can be found in [6,7]. Let us first consider that the pre-change density is f_0 and the post-change density is f_θ. Because the conventional CUSUM test can be interpreted as a set of parallel open-ended sequential probability ratio tests (SPRTs), which are activated at each time n with the upper threshold h and the lower threshold $-\infty$ [4]:

$$T_k = \begin{cases} \min\{n \geq k : S_k^n \geq h\} \\ \infty \text{ if no such } n \text{ exists} \end{cases}, \quad S_k^n = \sum_{t=k}^{n} \log \frac{f_\theta(X_t)}{f_0(X_t)},$$

where $k = 1, 2, \ldots$, the stopping time T_{CS} is defined as

$$T_{\mathrm{CS}} = \inf\{T_k \mid k = 1, 2, \ldots\}. \tag{3}$$

The rationality of the WL CUSUM test is due to the fact that any detection with a delay greater than L is considered as missed. Hence, the WL CUSUM test uses at each moment only L last observations. To get a more general stopping time, we consider now the following definition of the truncated SPRT with the upper variable threshold $h_1, \ldots, h_L$ and the lower threshold $-\infty$ (see [7])

$$T_k = \begin{cases} \min\{k \leq n \leq k + L - 1 : S_k^n \geq h_{n-k+1}\} \\ \infty \text{ if no such } n \text{ exists} \end{cases}, \tag{4}$$

$$S_k^n = \sum_{t=k}^{n} \log \frac{f_{\theta_{t-k+1}}(X_t)}{f_0(X_t)}, \quad k = 1, 2, \ldots \tag{5}$$

Putting together the above mentioned equations, we get the stopping time of the Variable Threshold Window Limited CUmulative SUM (VTWL CUSUM) test [7]:

$$T_{\mathrm{VTWL}} = \inf\left\{n \geq L : \max_{1 \leq k \leq L}\left[S_{n-k+1}^n - h_k\right] \geq 0\right\}, \tag{6}$$

$$S_{n-k+1}^n = \sum_{t=n-k+1}^{n} \log \frac{f_{\theta_{k-n+t}}(X_t)}{f_0(X_t)}. \tag{7}$$

Let us consider the time instant n. Considering the parallel truncated SPRTs

$$T_1, T_2, \ldots, T_n,$$

we get a set of stopping times. All these tests, consequently activated at each time m, $1 \leq m \leq n$, accumulate the statistics in the "direct" time. Say, the test activated at time m calculates the Log-Likelihood Ratios (LLR)

$$S_m^m, S_m^{m+1}, S_m^{m+2}, \ldots.$$

From a practical point of view, it is more convenient to consider the on-line detection algorithm in the "inverse" time, i.e., to re-write Eqs. (3) and (4) for a sliding window $[n - L + 1; n]$. In the other words, the procedure of observation is stopped and a transient change is declared at the first time instant n when

$$S_{n-k+1}^n \geq h_k$$

for some k such that $1 \leq k \leq L$. These LLRs S_{n-k+1}^n are calculated by using the replicas of the profile:

$$(\theta_1, \ldots, \theta_{L-1}, \theta_L),$$
$$(\theta_1, \ldots, \theta_{L-1}),$$
$$\ldots$$
$$(\theta_1)$$

in the sliding window $[n - L + 1; n]$.

4   Mixed Continuous-Discrete Data Flows: Problem Statement

Let us formalize the transient change detection problem considered in this paper as follows. We sequentially observe n_c parallel independent sequences $\{X_{i,t}\}_{t \geq 1}$ of random variables (also independent) with absolutely continuous distributions F_i, $i = 1, \ldots, n_c$. We also sequentially observe n_d parallel sequences $\{N_{i,t}\}_{t \geq 1}$ of independent random variables with discrete distributions P_i, $i = 1, \ldots, n_d$. Therefore, the generative model of the continuous distributions with transient changes is given by:

$$X_{i,t} \sim \begin{cases} F_{i,0} & \text{if } 1 \leq t < \nu \\ F_{i,\theta_{i,t-\nu+1}} & \text{if } \nu \leq t \leq \nu + L - 1 \end{cases}, \tag{8}$$

where $F_{i,\theta}$ is the parameterized cumulative distribution function during the transient change period L and $(\theta_{i,1}, \ldots, \theta_{i,L})$ is the set of known parameters defining the dynamic profile of the transient change. Without loss of generality, it is assumed that the pre-change parameter is $\theta_{i,0} = 0$. Hence, the pre-change cumulative distribution function is denoted by $F_{i,0}$.

Analogously, the generative model of the discrete distributions with transient changes is given by:

$$N_{i,t} \sim \begin{cases} P_{i,0} & \text{if } 1 \leq t < \nu \\ P_{i,\lambda_{i,t-\nu+1}} & \text{if } \nu \leq t \leq \nu + L - 1 \end{cases}, \tag{9}$$

where $P_{i,\lambda}$ is the parameterized probability mass function of the discrete random variable $N_{i,t}$.

5   Mixed Continuous-Discrete Data Flows: FMA Test

The originality of this paper with respect to previous publications is the co-existence of parallel continuous and discrete data flows. In this case, the LLR for the mixed count/continuous statistical model is given by

$$S_{n-k+1}^n = \sum_{t=n-k+1}^{n} \sum_{i=1}^{n_c} \log \frac{f_{i,\theta_{i,k-n+t}}(X_{i,t})}{f_{i,0}(X_{i,t})} + \sum_{t=n-k+1}^{n} \sum_{i=1}^{n_d} \log \frac{P_{i,\lambda_{i,k-n+t}}(N_{i,t})}{P_{i,0}(N_{i,t})}. \tag{10}$$

The optimization of the VTWL CUSUM test in the case of arbitrary distribution of the independent observations $\{X_t\}_{t\geq 1}$ is considered in [8]. It follows from [8] that the optimal tuning of the VTWL CUSUM test leads to the Finite Moving Average (FMA) test. Applying this test to the LLR given by (10), we get:

$$T_{\text{FMA}} = \inf\left\{ n \geq L : S_{n-L+1}^n \geq h_{\text{FMA}} \right\}. \tag{11}$$

As it follows from [8], to calculate the worst-case conditional probability of missed detection $\overline{\mathbb{P}}_{\text{md}}(T_{\text{FMA}}; L)$ and the worst-case probability of false alarm $\overline{\mathbb{P}}_{\text{fa}}(T_{\text{FMA}}; m)$ within any time window of length m, it is necessary to know two Cumulative Distribution Functions (CDF). The first CDF of the LLR S_{n-L+1}^n corresponds to the pre-change period:

$$x \mapsto F_{S,\infty}(x) = \mathbb{P}_0\left(S_{n-L+1}^n < x\right), \quad n \geq L \tag{12}$$

under the measure P_∞. The second CDF of the LLR S_{n-L+1}^n corresponds to the measure P_ν. The worst-case conditional probability of missed detection $\overline{\mathbb{P}}_{\text{md}}(T_{\text{FMA}}; L)$ is upper bounded by $\mathbb{P}_\nu\left(S_\nu^{L+\nu-1} < h_{\text{FMA}}\right)$, hence, the second CDF is defined as follows:

$$x \mapsto F_{S,\nu}(x) = \mathbb{P}_\nu\left(S_\nu^{L+\nu-1} < x\right), \quad \nu \geq L \tag{13}$$

Hence, to calculate the above-mentioned probabilities, it is necessary to study the LLR as a function of continuous state space data, represented by $\{X_t\}_{t\geq 1}$ and discrete (countable) data, represented by the number of events $\{N_t\}_{t\geq 1}$ per sampling period and to define the CDF of the LLRs. As it follows from [8]

$$\overline{\mathbb{P}}_{\text{md}}(T_{\text{FMA}}; m_\alpha) \leq F_{S,\nu}(h_{\text{FMA}}) \text{ and } \overline{\mathbb{P}}_{\text{fa}}(T_{\text{FMA}}; L) \leq 1 - [F_{S,\infty}(h_{\text{FMA}})]^{m_\alpha}. \tag{14}$$

Let us define the following random values ξ (continuous) and ζ (discrete):

$$\xi = \sum_{t=n-k+1}^{n} \sum_{i=1}^{n_c} \log \frac{f_{i,\theta_{i,k-n+t}}(X_{i,t})}{f_{i,0}(X_{i,t})} \tag{15}$$

$$\zeta = \sum_{t=n-k+1}^{n} \sum_{i=1}^{n_d} \log \frac{P_{i,\lambda_{i,k-n+t}}(N_{i,t})}{P_{i,0}(N_{i,t})}. \tag{16}$$

It can be considered (without loss of generality) that

$$p_j = \mathbb{P}(\zeta = x_j), \quad j \in J \text{ avec } x_i \neq x_j \text{ pour } i \neq j.$$

Hence, let us define the random value $\vartheta = \xi + \zeta$, its CDF is given by the following formula

$$F_\vartheta(z) = \mathbb{P}(\vartheta \leq z) = \mathbb{P}(\xi + \zeta \leq z) = \sum_{j \in J} F_\xi(z - x_j) \mathbb{P}(\zeta = x_j) = \sum_{j \in J} F_\xi(z - x_j) p_j. \quad (17)$$

Therefore, we calculate the CDFs $x \mapsto F_{S,\infty}(x)$ and $x \mapsto F_{S,\nu}(x)$ by using equation (17). These CDFs permit us to estimate the worst-case conditional probability of missed detection $\overline{\mathbb{P}}_{\text{md}}(T_{\text{FMA}}; L)$ and the worst-case probability of false alarm $\overline{\mathbb{P}}_{\text{fa}}(T_{\text{FMA}}; m)$.

Example 1. To illustrate the above-mentioned algorithm, let us consider that $n_c = 1$ and $n_d = 1$. Let us assume that

$$X_t \sim \begin{cases} \mathcal{N}(0,1) \text{ if } 1 \leq t < \nu \\ \mathcal{N}(\theta,1) \text{ if } \nu \leq t \leq \nu + L - 1 \end{cases}, \quad (18)$$

where $\mathcal{N}(\theta, 1)$ is the normal distribution with mean $\theta > 0$ and variance $\sigma^2 = 1$ and

$$N_t \sim \begin{cases} \Pi(\lambda_0) \text{ if } 1 \leq t < \nu \\ \Pi(\lambda_1) \text{ if } \nu \leq t \leq \nu + L - 1 \end{cases}, \quad (19)$$

where $\Pi(\lambda)$ is the Poisson distribution with mean λ, $\lambda_0 < \lambda_1$. The LLRs are represented as follows:

$$\xi = \sum_{t=n-L+1}^{n} \log \frac{f_\theta(X_t)}{f_0(X_t)} = \frac{\theta}{\sigma^2} \sum_{t=n-L+1}^{n} \left(X_t - \frac{\theta}{2}\right) \quad (20)$$

and

$$\zeta = \sum_{t=n-L+1}^{n} \log \frac{p_{\lambda_1}(N_t)}{p_{\lambda_0}(N_t)} = (\lambda_1 - \lambda_0) \sum_{t=n-L+1}^{n} (N_t - (\lambda_0 - \lambda_1)) \quad (21)$$

The distribution of the random variable ξ is normal. If the independent random variables $N_{n-L+1}, \ldots, N_n$ have Poisson distributions with means λ_t, $t = n - L + 1, \ldots, n$, then the sum $\sum_{t=n-L+1}^{n} N_t$ has the Poisson distribution with mean $\sum_{t=n-L+1}^{n} \lambda_t$. As it follows from (17), the CDF $z \mapsto F_\vartheta(z)$ of the random variable ϑ represents a weighted sum of the normal CDFs, where the weight coefficients are defined by the resulting Poisson distribution.

References

1. Do, V.L., Fillatre, L., Nikiforov, I.: Sequential monitoring of SCADA systems against cyber/physical attacks. In: 9th IFAC Symposium on Fault Detection, Supervision and Safety for Technical Processes (SAFEPROCESS 2015), Paris, France, vol. **48**, pp. 746–753. Elsevier, September 2015
2. Do, V.L., Fillatre, L., Nikiforov, I.: Sequential detection of transient changes in stochastic-dynamical systems. J. de la Société Française de Statistique (JSFdS) **156**(4), 60–97 (2015)

3. Do, V.L., Fillatre, L., Nikiforov, I., Willett, P.: Security of SCADA systems against cyber-physical attacks. IEEE Aerosp. Electron. Syst. Mag. (2016, in print)
4. Tartakovsky, A., Nikiforov, I., Basseville, M.: Sequential Analysis: Hypothesis Testing and Changepoint Detection. CRC Press, Taylor & Francis Group, Boca Raton (2014)
5. Lorden, G.: Procedures for reacting to a change in distribution. Ann. Math. Stat. **42**(6), 1897–1908 (1971)
6. Guépié, B.K., Fillatre, L., Nikiforov, I.: Sequential detection of transient changes. Seq. Anal. **31**(4), 528–547 (2012)
7. Guépié, B.K., Fillatre, L., Nikiforov, I.: Detecting an abrupt change of finite duration. In: Conference Record of the Forty Sixth Asilomar Conference on Signals, Systems and Computers (ASILOMAR), pp. 1930–1934. IEEE (2012)
8. Guépié, B.K., Fillatre, L., Nikiforov, I.: Detecting a suddenly arriving dynamic profile of finite duration. IEEE Trans. Inf. Theor. (2015, submitted)

Performance Modeling of Finite-Source Cognitive Radio Networks Using Simulation

Janos Sztrik[1]([⊠]), Tamás Bérczes[1], Hamza Nemouchi[1], and Agassi Melikov[2]

[1] Faculty of Informatics, University of Debrecen, Debrecen, Hungary
{sztrik.janos,berczes.tamas}@inf.unideb.hu, nemouchih@gmail.com
[2] Azerbaijan National Academy of Sciences, Baku, Azerbaijan
agassi.melikov@gmail.com

Abstract. This paper deals with performance modeling of radio frequency licensing. Licensed users (Primary Users - PUs) and normal users (Secondary Users - SUs) are considered. The main idea, is that the SUs are able to access to the available non-licensed radio frequencies.

A finite-source retrial queueing model with two non-independent frequency bands (considered as service units) is proposed for the performance evaluation of the system. A service unit with a priority queue and another service unit with an orbit are assigned to the PUs ans SUs, respectively. The users are classified into two classes: the PUs have got a licensed frequency, while the SUs have got a frequency band, too but it suffers from the overloading. We assume that during the service of the non-overloaded band the PUs have preemptive priority over SUs. The involved inter-event times are supposed to be independent, hypo-exponentially, hyper-exponentially, lognormal distributed random variables, respectively, depending on the different cases during simulation.

The novelty of this work is that we create a new model to analyze the effect of distribution of inter-event time on the mean and variance of the response time of the PUs and SUs.

As the validation of the simulation program a model with exponentially distributed inter-event times is considered in which case a continuous time Markov chain is introduced and by the help of MOSEL (MOdeling Specification and Evaluation Language) tool the main performance measures of the system are derived. In several combinations of the distribution of the involved random variables we compare the effect of their distribution on the first and second moments of the response times illustrating in different figures.

Keywords: Finite source queuing systems · Simulation · Cognitive radio networks · Performance evaluation

1 Introduction

Cognitive radio has emerged as a promising technology to realize dynamic spectrum access and increase the efficiency of a largely under utilized spectrum. In a

© Springer International Publishing AG 2016
V.M. Vishnevskiy et al. (Eds.): DCCN 2016, CCIS 678, pp. 64–73, 2016.
DOI: 10.1007/978-3-319-51917-3_7

cognitive radio network (CRN), a cognitive or secondary users (SUs) are allowed to use the spectrum by primary users (PUs) as long as the PUs do not use it. This operation is called opportunistic spectrum access, see for example [1,2]. To avoid interference to PUs, SUs must intelligently release the unlicensed spectrum if a licensed user appears as it was treated in [3,4].

In this paper we introduce a finite-source queueing model with two (non independent) frequency channels. According to the CRN modeling the users are divided into two types: the Primary Users (PUs) have got a licensed frequency, which does not suffer from overloading feature. The Secondary Users (SUs) have got a frequency band too, but suffers from overloading. A newly arriving SU request can use the band of PUs (which is not licensed for SUs) if the band of SUs is engaged, in the cognitive way: the non-licensed frequency must be released by the SU when a PU request appears. In our environment the band of the PUs is modeled by a queue where the requests has preemptive priority over the SUs requests. The band of the SUs is described by a retrial queue: if the band is free when the request arrives then it is transmitted. Otherwise, the request goes to the orbit if both bands are busy. We assume that the radio transmission is not reliable, it will fail with a probability p for both channels. If a failure happens then the request retransmission process starts immediately, see for example [3,4].

Hence, it should be noted that the novelty of this work is that we create a new model to analyze the effect of distribution of inter-event time on the mean and variance of the response time of the PUs and SUs. In several combinations of the distribution of the involved random variables and using simulation we compare the effect of their distribution on the first and second moments of the response times illustrating in different figures.

2   System Model

Figure 1 illustrates a finite source queueing system which is used to model the considered cognitive radio network. The queueing system contains two interconnected, not independent sub-systems. The first part is for the requests of the PUs. The number of sources is denoted by N_1. In order to analyze the effect of the distribution, these sources generate high priority requests with hypo-exponentially, hyper-exponentially and lognormally distributed inter-request times with the same rate λ_1 or with the same mean $1/\lambda_1$. The generated requests are sent to a single server unit (Primary Channel Service - PCS) with preemptive priority queue. The service times are supposed to be also hypo-exponentially, hyper-exponentially and lognormally distributed with the same rate μ_1 or with the same mean $1/\mu_1$.

The second part is for the requests of the SUs. There are N_2 sources, the inter-request times and service times of the single server unit (Secondary Channel Service - SCS) are assumed to be hypo-exponentially, hyper-exponentially and lognormally distributed random variables with rate λ_2 and μ_2, respectively.

A generated high priority packet goes to the primary service unit. If the unit is idle, the service of the packet begins immediately. If the server is busy with

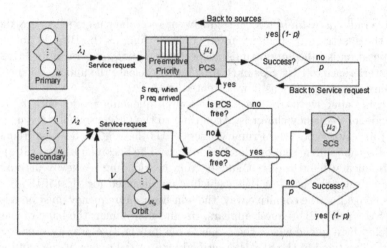

Fig. 1. A priority and a retrial queue with components

a high priority request, the packet joins the preemptive priority queue. When the unit is engaged with a request from SUs, the service is interrupted and the interrupted low priority task is sent back to the SCS. Depending on the state of secondary channel the interrupted job is directed to either the server or the orbit. The transmission through the radio channel may produce errors, which can be discovered after the service. In the model this case has a probability p, and the failed packet is sent back to the appropriate service unit. When the submission, is successful (probability $1-p$), the requests goes back to the source.

In case of requests from SUs. If the SCS is idle, the service starts, if the SCS is busy, the packet looks for the PCS. In case of an idle PCS, the service of the low priority packet begins at the high priority channel (PCS). If the PCS is busy the packet goes to the orbit. From the orbit it retries to be served after an exponentially distributed time with parameter ν. The same transmission failure with the same probability can occur as in the PCS segment.

To create a stochastic process describing the behavior of the system, the following notations are introduced

- $k_1(t)$ is the number of high priority sources at time t,
- $k_2(t)$ is the number of low priority sources at time t,
- $q(t)$ denotes the number of high priority requests in the priority queue at time t,
- $o(t)$ is the number of requests in the orbit at time t,
- $y(t) = 0$ if there is no job in the PCS unit, $y(t) = 1$ if the PCS unit is busy with a job coming from the high priority class, $y(t) = 2$ when the PCS unit is servicing a job coming from the secondary class at time t,
- $c(t) = 0$ when the SCS unit is idle and $c(t) = 1$, when the SCS is busy at time t.

It is easy to see that

$$k_1(n) = \begin{cases} N_1 - q(t), & y(t) = 0, 2 \\ N_1 - q(t) - 1 & y(t) = 1 \end{cases}$$

$$k_2(n) = \begin{cases} N_2 - o(t) - c(t), & y(t) = 0, 1 \\ N_2 - o(t) - c(t) - 1 & y(t) = 2 \end{cases}$$

In the case of exponentially distributed inter-event time a continuous-time Markov chain can be constructed and the main steady-state performance measures can be obtained, as it was carried out in [4]. The numerical result obtained in this paper were the test result for the validation of the simulation outputs.

However, in this paper we deal with more general situation allowing non-exponentially distributed times. For the sake of easier understanding the input parameters are collected in Table 1.

Table 1. List of simulation parameters

Parameter	Maximum	Value at t
Active primary sources	N_1	$k_1(t)$
Active secondary sources	N_2	$k_2(t)$
Primary generation rate		λ_1
Secondary generation rate		λ_2
Requests in priority queue	$N_1 - 1$	$q(t)$
Requests in orbit	$N_2 - 1$	$o(t)$
Primary service rate		μ_1
Secondary service rate		μ_2
Retrial rate		ν
Error probability		p

3   Simulation Results

In order to estimate the mean and variance of the response times of the requests, the batch means method is used which is the most popular confidence interval techniques for the output analysis of a steady-state simulation, see for example [5–7].

There are many possible combinations of the cases, but due to the page limitation we considered only the following sample results showing the effect of the distributions on the mean and variance of the corresponding response times.

For the easier understanding the numerical values of parameters are collected in Table 2.

Table 2. Numerical values of model parameters

No.	N_1	N_2	λ_1	λ_2	μ_1	μ_2	ν	p
Figs. 2 and 3	10	50	x-axis	0.03	1	1	20	0.1
Figs. 4 and 5	10	50	x-axis	0.03	1	1	20	0.1
Figs. 6 and 7	10	50	0.02	x-axis	1	1	20	0.1
Fig. 8	10	50	0.02	0.03	1	1	x-axis	0.1
Fig. 9	10	50	0.02	x-axis	1	1	20	0.1

Figures 2 and 3 show that the distribution of the inter-arrival time of the primary packets with the same mean has no effect on the mean and variance of response time of the secondary users, they depend only on their mean supposing that the inter-request time of the SUs and the service time of both servers units are exponentially distributed. It is the consequence of [8] in which it was proved that the steady-state distribution is insensitive to the distribution of the source times, depending only on their means.

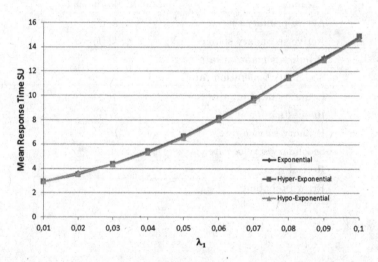

Fig. 2. The effect of inter-request time distribution of the PUs on the mean response time of SUs vs λ_1

The other operation mode is where the service time at the primary server is hyper-exponentially, hypo-exponentially and lognormally distri-buted with the same mean supposing that the inter-arrival time of PUs and SUs, and the service time of the secondary server are exponentially distributed.

Figures 4 and 5 show that the value of the mean response time and variance is greater when the service time is hypo-exponentially distributed, also the mean response time of the secondary users. In the case the service time is lognormally distributed is approximately the same when it is hypo-exponentially distributed.

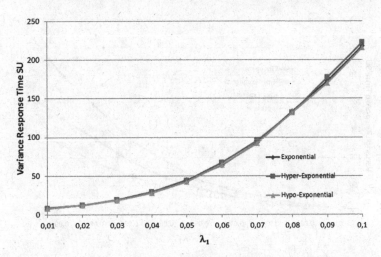

Fig. 3. The effect of inter-request time distribution of the PUs on the variance of response time of SUs vs λ_1

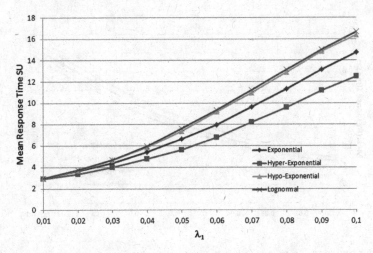

Fig. 4. The effect of service time distribution of the PUs on the mean response time of SUs vs λ_1

In Figs. 6 and 7, the inter-request time for the PUs and SUs is exponentially distributed. In this cases, figures show the effect of the SU's inter-request arrival time on the mean and variance response time of the SUs knowing that the service time of SCS is exponentially, hypo-exponentially, hyper-exponentially and lognormally distributed with the same mean. The value of the squared coefficient of variation for the hypo-exponentially distribution is always less than one and for the hyper-exponentially is always greater than one, therefore the mean and variance of response time of SUs when the service time is hyper-exponentially

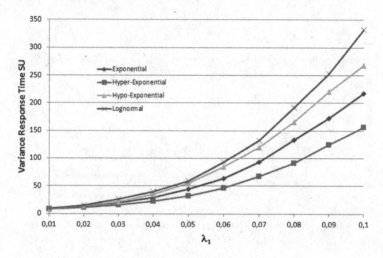

Fig. 5. The effect of service time distribution of the PUs on the variance of response time of SUs vs λ_1

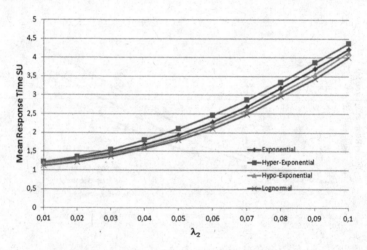

Fig. 6. The effect of service time distribution of the SUs on the mean response time of SUs vs λ_2

is greater than the mean response time of SUs when the service time is hypo-exponentially distributed.

On Fig. 8 the service time distribution of SCS is exponentially, hypo-exponentially, hyper-exponentially and lognormally distributed with the same mean. The service time of PCS and the inter-arrival time of both PUs, SUs are exponentially distributed. Figure shows the effect of the time spent in orbit on the mean response time of the SUs, it was modeled by a variable retrial rate. The result confirms the expectation that is increasing retrial rate involves shorter response times.

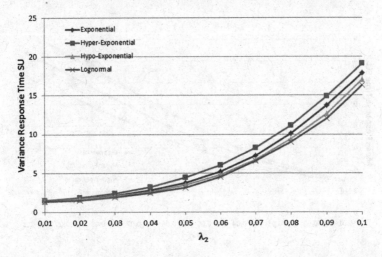

Fig. 7. The effect of service time distribution of the SUs on the variance of response time of SUs vs λ_2

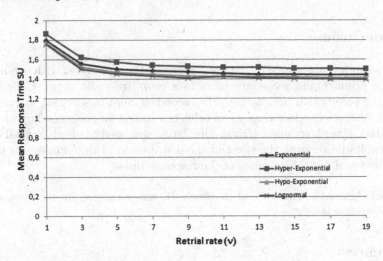

Fig. 8. The effect of service time distribution of the SUs on the mean response time of SUs vs the retrial rate ν

On the last figure, we assume that the service time of the PCS is exponentially, hypo-exponentially, hyper-exponentially and lognormally distributed with the same mean. The service time of the SCS and the inter-request time of PUs and SUs are exponentially distributed. The figure shows the effect of the inter-request time of the SUs on the mean response time of SUs. Here again we get what we expected that is increasing arrival intensity involves longer response times.

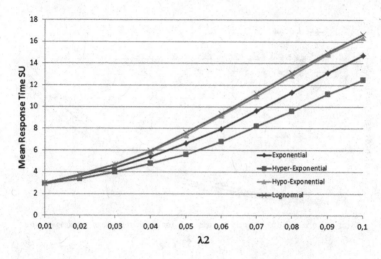

Fig. 9. The effect of service time distribution of the PUs on the mean response time of SUs vs λ_2

4   Conclusions

In this paper a finite-source retrial queueing model was proposed with two bands servicing primary and secondary users in a cognitive radio network. Primary users have preemptive priority over the secondary ones in servicing at primary channel. At the secondary channel an orbit is installed for the secondary packets finding the server busy upon arrival. Simulation was used to obtain several sample examples illustrating the effect of the distribution of the inter-events times on the first and second moments of the response times.

Acknowledgments. The work of Nemouchi H. was supported by the Stipendium Hungaricum Scholarship.

References

1. Wong, E.W., Foch, C.H., Adachi, F.: Analysis of cognitive radio spectrum decision for cognitive radio networks. IEEE J. Sel. Areas Commun. **29**, 757–769 (2011)
2. Devroye, N., Vu, M., Tarock, V.: Congnitive radio networks. IEEE Sign. Proces. Mag. **25**, 12–23 (2008)
3. Gao, S., Wang, J.: Performance analysis of a cognitive radio network based on preemptive priority and guard channels. Int. J. Comput. Math. GCOM-2013-0621-A (2014)
4. Sztrik, J., Bérczes, T., Almási, B., Kuki, A., Wang, J.: Performance modeling of finite-source cognitive radio networks. Acta Cybern. **22**, 617–631 (2016)
5. Law, A.M., Kelton, W.D.: Simulation Modeling and Analysis. McGraw-Hill College, New York (1991)

6. White, A.J., Schmidt, J.W., Bennett, G.K.: Analysis of Queueing Systems. Academic Press, INC, New York (1975)
7. Carlstein, E.W., Goldsman, D.: The use of subseries values for estimating the variance of a general statistic from a stationary sequence. Ann. Stat. **14**, 1171–1179 (1986)
8. Sztrik, J.: On the finite-source G/M/r queues. Eur. J. Oper. Res. **20**, 261–268 (1985)

Performance Measures and Optimization of Queueing System with Reserve Server

Valentina Klimenok[1], Alexander Dudin[1], Vladimir Vishnevskiy[2(✉)],
Vladimir Shumchenya[3], and Achyutha Krishnamoorthy[4]

[1] Department of Applied Mathematics and Computer Science,
Belarusian State University, 220030 Minsk, Belarus
{klimenok,dudin}@bsu.by
[2] Institute of Control Sciences of Russian Academy of Sciences, Moscow, Russia
vishn@inbox.ru
[3] National Academy of Sciences of Belarus, Minsk, Belarus
fridenn@tut.by
[4] Cochin University of Science and Technology, Cochin, India
achyuthacusat@gmail.com

Abstract. In this paper, we consider a single-server queueing system with infinite buffer and reserve server which can be used for modeling energy saving schemes in some real information transmission and processing systems. An arriving customer is serviced by the main server until the end of the service time or the expiration of the limited time defined by the timer which is set up at the beginning of the service. If the service of a customer has not yet completed while the timer has expired then the reserve server joins to the service of the customer. This allows to avoid too large delays in a system with reasonable energy saving.

Keywords: Single-server queueing system · Reserve server · Stationary distribution · Performance measures · Optimization

1 Introduction

Problem of the optimal selection of the minimal number of the servers required to guarantee the desired quality of customer's service is in the focus of queueing theory from the early beginning. However, when this problem is solved for some real world system, the new problem arises. Due to the stochastic nature of the arrival process, sometimes a certain part of servers is idle and the problem of the optimal use this idle time is also very important. If we consider the server as a machine in cloud computing system, it is known, see, e.g., reference in [1], that the idle server consumes up to 65% of energy consumed by the working server. Therefore, it is necessary to switch off the idle servers. In this way we arrive to idea of the reserved servers which are switched on only if there is a lot of customers in the system. Otherwise, they are switched off. Papers [1–3] are devoted to the problem of minimizing the energy consumption in the data center

V.M. Vishnevskiy et al. (Eds.): DCCN 2016, CCIS 678, pp. 74–88, 2016.
DOI: 10.1007/978-3-319-51917-3_8

while maintaining an acceptable level of customer service based on temporarily keeping a certain part of servers in reserve. A similar model is considered in [4]. As an early work (in Russian) in this subject we can mention the book [5] where the system with reserved server was analysed. The paper [6] deals with a finite capacity queueing system with one main server who is supported by a backup server. The queueing model assumes Markovian arrivals and phase type services. In all these models, it is assumed that and a threshold-type server backup policy with two pre-determined lower and upper thresholds is applied. In recent paper [7], a very general model with reserved servers and hysteresis strategy of control is analyzed. Short survey of the state of art in the related field is presented there as well.

In this paper, we consider another mechanism of server reservation. This mechanism does not use information about the current queue length. It assumes the change of the customer's service rate after expiration of a certain random time since the service beginning. The reserve server does not provide the service by itself. It just helps to the main server.

The rest of the paper is organized as follows. In Sect. 2 the model under study is described. The process of the system states is defined in Sect. 3. The steady-state analysis of the model is performed in Sect. 4 and in this section we also display some key system performance measures. The stationary sojourn time distribution of an arbitrary customer is presented in Sect. 5. Illustrative examples of numerical optimization are discussed in Sect. 6. Some concluding remarks are given in Sect. 7.

2   Mathematical Model

We consider a single-server queue with infinite waiting room and stationary Poisson input flow with the rate λ. The service times are independent random values having the phase type (PH) distribution with an irreducible representation (β, S). This means the following. Service time is interpreted as the time until the continuous time Markov chain $m_t, t \geq 0$, with state space $\{1, \ldots, M+1\}$ reaches the single absorbing state $M + 1$. Transitions of the chain $m_t, t \geq 0$, within the state space $\{1, \ldots, M\}$ are defined by the sub-generator S while the intensities of transitions into the absorbing state are defined by the vector $\mathbf{S}_0 = -S\mathbf{e}$. At the service beginning epoch, the state of the process $m_t, t \geq 0$, is chosen within the state space $\{1, \ldots, M\}$ according to the probabilistic row vector β. It is assumed that the matrix $S + \mathbf{S}_0\beta$ is an irreducible one. The service rate is defined as $\mu = -(\beta S^{-1}\mathbf{e})^{-1}$, the mean service time is calculated as $b_1 = \mu^{-1}$. For more information about the PH type distribution, see, e.g., [8].

Besides the main server, there is a reserve server in the system. This server connects to the service of a current customers, if the service time of the customer exceeds a certain time limit, which is set up on the timer and is defined as a random variable having the PH distribution with irreducible representation (τ, T). The underlying process $\eta_t, t \geq 0$, of the service time has state space $\{1, 2, \ldots, R+1\}$, where $R+1$ is an absorbing state. The transitions rate of the underlying process to the absorbing state is defined by the vector $\boldsymbol{T}_0 = -T\mathbf{e}$, the timer rate is defined as $\kappa = -(\boldsymbol{\tau}T^{-1}\mathbf{e})^{-1}$, the mean time until the timer expiration is $\tau_1 = \kappa^{-1}$.

After the reserve server connects to the service of a customer, both servers begin the sharing service of the customer. The service continues with PH service phase in which the timer expired but with another rate. To reflect the latter fact, we assume that at the moment of the timer expiration the sub-generator S is changed to $\tilde{S} = \alpha S, \alpha > 0$. The value α can be greater than one (the system manager aims to quickly finish the service that lasts too long) or less than one (after the timer expiration, the customer in service becomes less important or resources assigned for customer service exhaust and the service is continued in stand by mode). If $\alpha = 1$, the model reduces to the standard $M/PH/1$ queue.

3 Process of the System States

Let at the moment t

- i_t be the number of customers in the system, $i_t \geq 0$,
- $r_t = 0$, if the reserve server is idle, or $r_t = 1$, if the reserve server is busy at the moment t;
- m_t be the state of the underlying process of the service, $m_t = \overline{1, M}$;
- η_t be the state of the underlying process of the timer, $\eta_t = \overline{1, R}$.

The process of the system states is described by a regular irreducible continuous time Markov chain $\xi_t, t \geq 0$, with state space

$$\Omega = \{(0); (i, 0, m, \eta), i \geq 1, m = \overline{1, M}, \eta = \overline{1, R}; (i, 1, m), i \geq 1, m = \overline{1, M}\}.$$

In the following we will suppose that states of the chain $\xi_t, t \geq 0$, are enumerated as follows. States under fixed value of the components i, r are enumerated in the lexicographic order. Denote the obtained set as $\Omega_{i,r}$. Order the sets $\Omega_{i,r}$ as follows:

$$(0), \Omega_{1,0}, \ \Omega_{1,1}, \ \Omega_{2,0}, \ \Omega_{2,1}, \ \Omega_{3,0}, \ \Omega_{3,1} \dots$$

Let $Q_{i,j}$ be the matrix of the transition rates of the chain $\xi_t, t \geq 0$, from the states corresponding to the value i of the denumerable component to the states corresponding to the value j of this component, $i, j \geq 0$.

Theorem 1. *Infinitesimal generator Q of the Markov chain $\xi_t, t \geq 0$, has the following block structure:*

$$Q = \begin{pmatrix} Q_{0,0} & Q_{0,1} & O & O & O & \cdots \\ Q_{1,0} & Q_0 & Q_1 & O & O & \cdots \\ O & Q_{-1} & Q_0 & Q_1 & O & \cdots \\ O & O & Q_{-1} & Q_0 & Q_1 & \cdots \\ \vdots & \vdots & \vdots & \vdots & \vdots & \ddots \end{pmatrix},$$

where

$$Q_{0,0} = -\lambda, \quad Q_{0,1} = (\lambda\boldsymbol{\beta} \otimes \boldsymbol{\tau} \,|\, O_{1\times M}), \quad Q_{1,0} = \begin{pmatrix} S_0 \otimes \mathbf{e}_R \\ \tilde{S}_0 \end{pmatrix},$$

$$Q_{-1} = \begin{pmatrix} S_0\boldsymbol{\beta} \otimes \mathbf{e}_R\boldsymbol{\tau} & O_{MR\times M} \\ \tilde{S}_0\boldsymbol{\beta} \otimes \boldsymbol{\tau} & O_M \end{pmatrix}, \quad Q_0 = \begin{pmatrix} -\lambda I_{MR} + S \oplus T & I_M \otimes T_0 \\ O_{M\times MR} & -\lambda I_M + \tilde{S} \end{pmatrix},$$

$$Q_1 = \lambda I_{M(R+1)}.$$

Here $\otimes$ is a symbol of Kronecker's product of matrices; $\oplus$ is a symbol of Kronecker's sum of matrices; $\mathbf{e}$ is a column vector of 1s, I is an identity matrix; O is zero matrix. If needed, the size of the vector $\mathbf{e}$ and the matrices I, O is indicated by the suffix.

Proof of the theorem is implemented by analyzing the rates of transition of the multi-dimensional Markov chain $\xi_t, t \geq 0$.

The generator Q has three-diagonal block structure and, for $i > 1$, the blocks $Q_{i,j}$ depend on the values i, j only via the difference $i - j$. It means that

Corollary 1. *The Markov chain $\xi_t, t \geq 0$, is a Quasi Birth-and-Death process, see [8].*

4   Stationary Distribution. Performance Measures

Theorem 2. *A necessary and sufficient condition for existence of the stationary distribution of the Markov chain $\xi_t, t \geq 0$, is the fulfillment of the inequality*

$$\lambda < \mathbf{x}\boldsymbol{\mu}, \tag{1}$$

where the row vector $\mathbf{x}$ is defined as the unique solution of the system of the linear algebraic equation

$$\mathbf{x}(S \oplus T)[I - \mathbf{e}(\boldsymbol{\beta} \otimes \boldsymbol{\tau})] = \mathbf{0}, \tag{2}$$

$$\mathbf{x}\mathbf{e} - \mathbf{x}(I_M \otimes T_0)\tilde{S}^{-1}\mathbf{e} = 1, \tag{3}$$

and the column vector $\boldsymbol{\mu}$ is calculated as $\boldsymbol{\mu} = -(S \oplus T)\mathbf{e}$.

Proof. Since the chain under consideration is a Quasi Birth-and-Death process, then, according to [8], a necessary and sufficient condition for existence of its stationary distribution is the fulfillment of the inequality

$$\mathbf{z}Q_{-1}\mathbf{e} > \mathbf{z}Q_1\mathbf{e} \tag{4}$$

where the vector $\mathbf{z}$ is the unique solution of the following system of linear algebraic equations:

$$\mathbf{z}(Q_{-1} + Q_0 + Q_1) = \mathbf{0}, \quad \mathbf{z}\mathbf{e} = 1. \tag{5}$$

Represent the vector $\mathbf{z}$ in the form $\mathbf{z} = (\mathbf{x}, \mathbf{y})$, where $\mathbf{x}$ and $\mathbf{y}$ are of size MR and M respectively. Then inequality (4) is written in the form

$$\mathbf{x}(S_0 \otimes \mathbf{e}_L) + \mathbf{y}\tilde{S}_0 > \lambda, \tag{6}$$

and system (5) as

$$\mathbf{x}(S_0\beta \otimes \mathbf{e}\tau + S \oplus T) + \mathbf{y}(\tilde{S}_0\beta \otimes \tau) = \mathbf{0}, \tag{7}$$

$$\mathbf{x}(I_M \otimes T_0) + \mathbf{y}\tilde{S} = \mathbf{0}, \tag{8}$$

$$\mathbf{x}\mathbf{e} + \mathbf{y}\mathbf{e} = 1. \tag{9}$$

Express the vector $\mathbf{y}$ via the vector $\mathbf{x}$ using (8) and substitute the resulting expression into inequality (6) and Eqs. (7), (9). Then, after simple algebraic transformations, we obtain inequality (1) and system (2), (3).

Corollary 2. *In case of exponentially distributed service and timer times, the necessary and sufficient condition (1)–(3) for existence of the stationary distribution of the Markov chain ξ_t, $t \geq 0$, reduces to the following inequality:*

$$\lambda < \frac{\alpha\mu}{\alpha\mu + \kappa}(\mu + \kappa).$$

In what follows we will assume that inequality (1) holds.

Let us order the stationary probabilities of the Markov chain in accordance with the arrangement procedure defined above and form row vectors $\mathbf{p}_i$, $i \geq 0$, of the probabilities corresponding to the value i of the denumerable component of the chain.

The vectors $\mathbf{p}_i, i \geq 0$, satisfy Chapman-Kolmogorov's equations (equilibrium equations)

$$(\mathbf{p}_0, \mathbf{p}_1, \mathbf{p}_2, \dots)Q = \mathbf{0}, \quad (\mathbf{p}_0, \mathbf{p}_1, \mathbf{p}_2, \dots)\mathbf{e} = 1.$$

To solve this infinite size system, we used the numerically stable algorithm developed in [9] for calculation of the stationary distribution of multi-dimensional quasi-Toeplitz Markov chain. We can use this algorithm because the Quasi Birth-and-Death process, which describes the operation of the queue under consideration, is a partial case of a quasi-Toeplitz Markov chain. In our case, the general algorithm from [9] is reduced to the following one.

Algorithm.

- Calculate the matrix G as the minimal nonnegative solution of the matrix equation

$$Q_{-1} + Q_0 G + Q_1 G^2 = O.$$

- Calculate the matrix G_0 from the equation

$$Q_{1,0} + Q_0 G_0 + Q_1 G G_0 = O,$$

from which

$$G_0 = -(Q_0 + Q_1 G)^{-1} Q_{1,0}.$$

- Calculate the matrices $\bar{Q}_{i,l}$, $l = i, i+1, i \geq 0$, using the formulas

$$\bar{Q}_{i,l} = \begin{cases} Q_{0,0} + Q_{0,1}G_0, \ i = 0, \ l = 0, \\ Q_{0,1}, \ i = 0, \ l = 1, \\ Q_0 + Q_1 G, \ l = i, \ i \geq 1, \\ Q_1, \ l = i+1, \ i \geq 1. \end{cases}$$

- Calculate the matrices Φ_i, $i \geq 0$, using the recurrent formulas

$$\Phi_0 = I, \Phi_i = \Phi_{i-1}\bar{Q}_{i-1,i}(-\bar{Q}_{i,i})^{-1}, \ i \geq 1.$$

- Calculate the vector $\mathbf{p}_0$ as the unique solution of the system of linear algebraic equations

$$\mathbf{p}_0(-\bar{Q}_{0,0}) = \mathbf{0}, \ \mathbf{p}_0 \sum_{i=0}^{\infty} \Phi_i \mathbf{e} = 1.$$

- Calculate the vectors $\mathbf{p}_i$, $i \geq 1$, as follows $\mathbf{p}_i = \mathbf{p}_0\Phi_i$, $l \geq 1$.

Having the stationary distribution of the system states, $\mathbf{p}_i$, $i \geq 0$, been calculated we can find a number of stationary performance measures of the considered system. When calculating the performance measures, we can avoid of calculation of the infinity sums using the following result.

Theorem 3. *The vector generating function* $\mathbf{P}(z) = \sum_{i=1}^{\infty} \mathbf{p}_i z^i$, $|z| \leq 1$, *satisfies the following equation:*

$$\mathbf{P}(z)(Q_1 z^2 + Q_0 z + Q_{-1}) = z(\mathbf{p}_1 Q_{-1} - z\mathbf{p}_0 Q_{0,1}). \tag{10}$$

Proof. Equation (10) is obtained by the multiplication of the ith equation of the system $(\mathbf{p}_0, \mathbf{p}_1, \mathbf{p}_2, \dots)Q = \mathbf{0}$ in (10) by z^i and summing over $i \geq 1$.

Let us denote $f^{(n)}(z)$ the nth derivative of the function $f(z)$, $n \geq 1$, and $f^{(0)}(z) = f(z)$.

Formula (10) can be used to calculate the factorial moments $\mathbf{P}^{(m)}(1)$, of the number of customers in the system. But the problem of calculating the value of the vector generating function $\mathbf{P}(z)$ and its derivatives at the point $z = 1$ from Eq. (10) is non-trivial one because the matrix $Q_1 z^2 + Q_0 z + Q_{-1}$ is singular at the point $z = 1$. To solve this problem, the following computational procedure was elaborated.

Corollary 3. *The mth, $m \geq 0$, derivatives of the vector generating function* $\mathbf{P}(z)$ *at the point $z = 1$ are recursively calculated as the solution of the system of linear algebraic equations*

$$\begin{cases} \mathbf{P}^{(m)}(1)Q^{(0)}(1) = \boldsymbol{\Gamma}^{(m)}(1) - \sum_{l=0}^{m-1} C_m^l \mathbf{P}^{(l)}(1)Q^{(m-l)}(1), \\ \mathbf{P}^{(m)}(1)Q^{(1)}(1)\mathbf{e} = \frac{1}{m+1}[\boldsymbol{\Gamma}^{(m+1)}(1) - \sum_{l=0}^{m-1} C_{m+1}^l \mathbf{P}^{(l)}(1)Q^{(m+1-l)}(1)]\mathbf{e}. \end{cases} \tag{11}$$

where

$$\boldsymbol{\Gamma}^{(m)}(1) = \begin{cases} \mathbf{p}_1 Q_{-1} - \mathbf{p}_0 Q_{0,1}, & m = 0, \\ \mathbf{p}_1 Q_{-1} - 2\mathbf{p}_0 Q_{0,1}, & m = 1, \\ -2\mathbf{p}_0 Q_{0,1}, & m = 2, \\ O, & m > 2, \end{cases}$$

$$Q^{(m)}(1) = \begin{cases} Q_{-1} + Q_0 + Q_1, & m = 0, \\ Q_0 + 2Q_1, & m = 1, \\ 2Q_1, & m = 2, \\ O, & m > 2. \end{cases}$$

The proof of the corollary is based on the technique very similar to the one outlined in paper [10] and is omitted here.

Calculating the stationary distribution and using formula (11), we can calculate a number of important performance measures of the system. Formulas for calculating some performance measures are given below.

- Probability that the main server is idle $P_{idle}^{(1)} = p_0$.
- Mean number of customers in the system $L = \mathbf{P}'(1)\mathbf{e}$.
- Variance of the number of customers in the system $V = [\mathbf{P}''(1) + \mathbf{P}'(1)]\mathbf{e} - L^2$.
- Probability that, at an arbitrary moment, the main server serves a customer without the help of the reserve server $(P^{(0)})$ and with the help of the reserve server $(P^{(1)})$.

$$P^{(0)} = \mathbf{P}(1)\begin{pmatrix} \mathbf{e}_{MR} \\ \mathbf{0}_M^T \end{pmatrix}, \quad P^{(1)} = 1 - p_0 - P^{(0)}.$$

- Probability that a customer will be served with the help of the reserve server (P_{help}), and probability that a customer will be served without the help of the reserve server$(P_{no-help}.)$

$$P_{help} = -(\boldsymbol{\beta} \otimes \boldsymbol{\tau})(S \oplus T)^{-1}(I_M \otimes T)\mathbf{e}, \quad P_{no-help} = 1 - P_{help}.$$

5   Sojourn Time Distribution

Let $\tilde{v}_t$ be the residual service time of a customer being at the main server at time t. Let also

$$\tilde{V}(0, m, \eta, x) = \lim_{t \to \infty} P\{i_t > 0, r_t = 0, m_t = m, \eta_t = \eta, \tilde{v}_t < x\}, \ m = \overline{1, M}, \ \eta = \overline{1, R};$$

$$\tilde{V}(1, m, x) = \lim_{t \to \infty} P\{i_t > 0, r_t = 1, m_t = m, \tilde{v}_t < x\}, \ m = \overline{1, M}, \ x \geq 0.$$

Introduce the notation for the Laplace-Stieltjes transforms:

$$\tilde{v}(0, m, \eta, u) = \int_0^\infty e^{-ux} d\tilde{V}(0, m, \eta, x), \ \tilde{v}(1, m, u) = \int_0^\infty e^{-ux} d\tilde{V}(1, m, x), \ Re\,u \geq 0.$$

Denote as $\tilde{\mathbf{v}}(0, u)$ and $\tilde{\mathbf{v}}(1, u)$ column vectors formed by these transforms ordered in the lexicographical order of the components (m, η) in the first case and the component $\tilde{m}$ in the second case.

Theorem 4. *The vectors of the Laplace-Stieltjes transforms of the residual service time of a customer is calculated as*

$$\tilde{\mathbf{v}}(0, u) = (uI - S \oplus T)^{-1}[\boldsymbol{S}_0 \otimes \mathbf{e}_R + (uI - \tilde{S})^{-1}\tilde{\boldsymbol{S}}_0 \otimes \boldsymbol{T}_0], \qquad (12)$$

$$\tilde{\mathbf{v}}(1, u) = (uI - \tilde{S})^{-1}\tilde{\boldsymbol{S}}_0. \qquad (13)$$

Proof. The proof is based on the probabilistic interpretation of the Laplace-Stieltjes transform. We assume that, independently on the system operation, the stationary Poisson input of so called catastrophes arrives. Let u, $u > 0$, be the rate of this flow. Then components of the vectors $\tilde{\mathbf{v}}(0, u)$ and $\tilde{\mathbf{v}}(1, u)$ can be interpreted as follows. $\tilde{v}(0, m, \eta, u)$ is a probability that, at an arbitrary time, the main server works without the help of the reserve server, the underlying processes of the service and the timer are in the states m and η respectively and catastrophes will not arrive during the residual service time of the customer under service. $\tilde{v}(1, m, u)$ is a probability that, at an arbitrary time, the main server works with the help of the reserve server, the underlying processes of the service is in the state m and catastrophes will not arrive during the residual service time of the customer under service.

Taking into account the probabilistic interpretation, we write $\tilde{\mathbf{v}}(0, u)$ as

$$\tilde{\mathbf{v}}(0, u) = \int\limits_0^\infty e^{-ut} e^{(S \oplus T)t}(\boldsymbol{S}_0 \otimes \mathbf{e}_R)dt +$$

$$\int\limits_0^\infty e^{-ut} \int\limits_0^t e^{(S \oplus T)x}(I_M \otimes \boldsymbol{T}_0)dx e^{\tilde{S}(t-x)}\tilde{\boldsymbol{S}}_0 dt. \qquad (14)$$

Calculate the integrals in the right hand side of (14). In the obvious way we obtain expression for the first integral:

$$\int\limits_0^\infty e^{-ut} e^{(S \oplus T)t}(\boldsymbol{S}_0 \otimes \mathbf{e}_R)dt = (uI - S \oplus T)^{-1}(\boldsymbol{S}_0 \otimes \mathbf{e}_R). \qquad (15)$$

After tedious algebra we obtain the expression for the double integral in (14):

$$\int\limits_0^\infty e^{-ut} \int\limits_0^t e^{(S \oplus T)x}(I_M \otimes \boldsymbol{T}_0)dx e^{\tilde{S}(t-x)}\tilde{\boldsymbol{S}}_0 dt =$$

$$= [uI - (S \oplus T)]^{-1}[(uI - \tilde{S})^{-1} \otimes I_R](\tilde{\boldsymbol{S}}_0 \otimes \boldsymbol{T}_0). \qquad (16)$$

Substituting (15), (16) into (14), we get (12).

Further, using the probabilistic interpretation of the Laplace-Stieltjes transform, we write $\tilde{\mathbf{v}}(1, u)$ as

$$\tilde{\mathbf{v}}(1, u) = \int_0^\infty e^{-ut} e^{\tilde{S}t} \tilde{\mathbf{S}}_0 dt.$$

Calculating the integral, we get formula (13).

Corollary 4. *The vectors of the means of the residual service time of a customer is calculated as*

$$\tilde{\mathbf{t}}_0 = -(S \oplus T)^{-1} [I + \tilde{S}^{-1} \otimes T] \mathbf{e},$$

$$\tilde{\mathbf{t}}_1 = -\tilde{S}^{-1} \mathbf{e}.$$

The proof follows from the formulas $\tilde{\mathbf{t}}_0 = -\tilde{\mathbf{v}}'(0, 0)$, $\tilde{\mathbf{t}}_1 = -\tilde{\mathbf{v}}'(1, 0)$.

Corollary 5. *The Laplace-Stieltjes transform of the service time of a customer is calculate by the formula*

$$v(0, u) = (\boldsymbol{\beta} \otimes \boldsymbol{\tau}) \tilde{\mathbf{v}}(0, u).$$

Corollary 6. *Mean service time is calculated by the formula*

$$\bar{t} = -(\boldsymbol{\beta} \otimes \boldsymbol{\tau})(S \oplus T)^{-1} [I + \tilde{S}^{-1} \otimes T] \mathbf{e}.$$

The proof follows from the formula $\bar{t} = -v'(0, 0)$.

Theorem 5. *The Laplace-Stieltjes transform of the sojourn time of a customer has the following form:*

$$v(u) = p_0 v(0, u) + \mathbf{P}(v(0, u)) \begin{pmatrix} \tilde{\mathbf{v}}(0, u) \\ \tilde{\mathbf{v}}(1, u) \end{pmatrix}.$$

The proof is based on the total probability formula:

$$v(u) = p_0 v(0, u) + \sum_{i=1}^\infty \mathbf{P}_i \begin{pmatrix} \tilde{\mathbf{v}}(0, u) \\ \tilde{\mathbf{v}}(1, u) \end{pmatrix} v^i(0, u),$$

from which the required formula follows immediately.

Corollary 7. *The mean sojourn time of a customer is calculated as*

$$\bar{v} = p_0 \bar{t} + \mathbf{P}(1) \begin{pmatrix} \tilde{\mathbf{t}}_0 \\ \tilde{\mathbf{t}}_1 \end{pmatrix} + L\bar{t}.$$

The proof follows from the formula $\bar{v} = -v'(0)$.

6   Examples of Numerical Optimization

In this section, we solve numerically the problem of finding the minimum of the economic criterion, which is an average charge per unit time under the steady-state operation of the system:

$$J = a\bar{v} + c_0 P^{(0)} + c_1 P^{(1)},\tag{17}$$

where a is a charge, which is paid for one customer staying in the system per unit time, $c_0(c_1)$ is a cost of using the mode without the help of the reserve server (with the help of the reserve server) per unit time.

Experiment 1. Our aim is to find numerically the optimal value of the mean time to the timer expiration, τ_1, that provides the minimum to the cost criterion (17) under different values of the coefficient α.

We consider the following input data. The input rate $\lambda = 2$. The service time by the main server has hyper-exponential distributions of order 2. It is defined by the vector $\beta = (0.05, 0.95)$ and the matrix $S = \begin{pmatrix} -0.2 & 0.0 \\ 0.0 & -3.8 \end{pmatrix}$. The service rate $\mu = 2$ and the mean service time $b_1 = 0.5$. The coefficient of variation $c_{var} = 9.52632$.

The time to the timer expiration has Erlangian distribution of order 2. It is defined by the vector $\tau = (1.0, 0.0)$ and the matrix $T = \begin{pmatrix} -2 & 2 \\ 0 & -2 \end{pmatrix}$. The rate of timer $\kappa = 1$, the mean time to the timer expiration $\tau_1 = 1$ and the coefficient of variation $c_{cor} \doteq 0.5$.

Let us vary the mean time to the timer expiration, τ_1, in the interval $[0.05, 15]$ by scaling the time. Note that the coefficients of variation do not change under such a scaling.

We fix the cost coefficients a, c_0, c_1 setting them equal to 1, 4, 90 respectively.

According to the description of the system, the matrix $\tilde{S}$, which defines the service rates in the case when the service is carried out with the help of the reserve server, is defined as $\tilde{S} = \alpha S$, $\alpha > 0$. We will consider three values of the coefficient α: $\alpha = 1.5, 2, 3.5$.

The curves in Fig. 1 depict the dependence of the cost criterion J on the mean time to the timer expiration, $\tau_1 = 1$. Table 1 contains the values of the criterion in this example. The optimal value J^* of the cost criterion for each value of the parameter α is printed in bold face.

In this example we also interested in the behavior of the performance measures $\bar{v}$, $P^{(0)}$, $P^{(1)}$, entering into expression (17) for the cost criterion. These performance measures as functions of τ_1 for different values of α are depicted in Figs. 2, 3 and 4.

As it is seen from Fig. 1 and Table 1, under $\alpha = 1.5$ the criterion J reaches a minimum at the point $\tau_1 = 2$ and this minimum is equal to 37.04. If $\alpha = 2$ and $\alpha = 3.5$, the criterion reaches a minimum at the point $\tau_1 = 3$ and the values of J at this point are equal to 24.7 and 14.6 respectively. Note that the mean

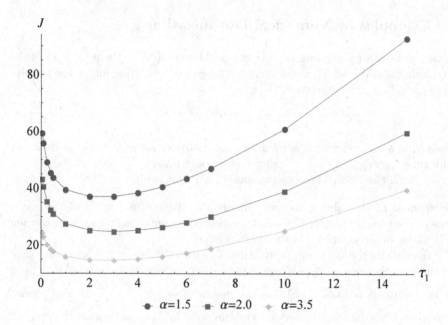

Fig. 1. Cost criterion J as a function of τ_1 for $\alpha = 1.5$, 2, 3.5

Table 1. Values of the cost criterion J for different τ_1 and α

$\alpha\backslash\tau_1$	0.05	0.1	0.25	0.4	0.5	1	2	3	4	5	6	7	10	15
1.5	59	56	49	45	44	39	**37.04**	37.07	38	41	43	47	61	93
2.0	43	40	35	32	31	27	25.1	**24.7**	25.2	26	28	30	39	60
3.5	24	23	20	19	18	16	14.7	**14.6**	15	16	17	19	25	40

service time by the main server equals $b_1 = 0.5$. It means that under the optimal strategy the reserve server rarely connects to the service due to excessive cost of its use.

Table 2 contains the relative negative profit from the use of non-optimal value of τ_1. We see that for the data under consideration the maximal negative profit is more than 170 %. At the same time, one can see that the curves at the point of minimum is rather elastic. This allow to choose as suboptimal any values of τ_1 belonging to the region $(1, 4)$.

Experiment 2. In this experiment, we solve numerically the problem of optimal choice of the mean time to the timer expiration, τ_1, that provides the minimum value to the cost criterion (17) under different values of the cost coefficient c_1. We consider the following values of this coefficient: $c_1 = 90$, 180, 270. The others cost coefficients are as follows: $a = 1, c_0 = 4$. Parameter $\alpha = 1.5$. The input rate, service times distribution and distribution of time to the timer expiration are the same as in Experiment 1.

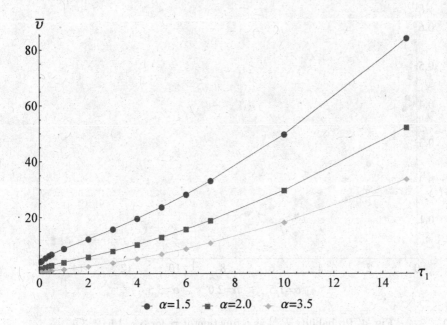

Fig. 2. Mean sojourn time $\bar{v}$ as a function of τ_1 for $\alpha = 1.5$, 2, 3.5

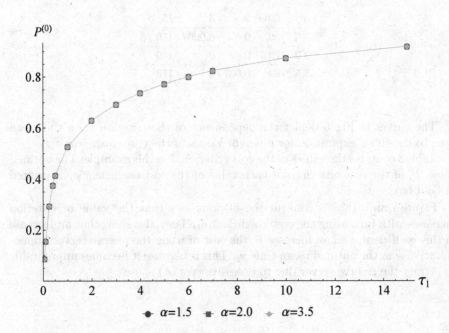

Fig. 3. Probability $P^{(0)}$ as a function of τ_1 for $\alpha = 1.5$, 2, 3.5

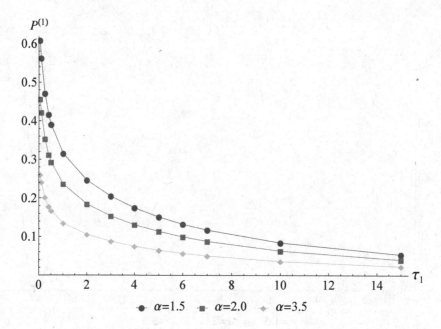

Fig. 4. Probability $P^{(1)}$ as a function of τ_1 for $\alpha = 1.5$, 2, 3.5

Table 2. Relative negative profit from the use of non-optimal value of τ_1 (in percents)

$\alpha \backslash \tau_1$	0.05	2	3	15
1.5	59	0	0.0008	150
2.0	24.7	0.02	0	142
3.5	64	0.007	0	173

The curves in Fig. 5 depict the dependence of the criterion J on the mean time to the timer expiration for different values of the cost coefficient c_1.

Table 3 contains the values of the cost criterion J in this example. The optimal value J^* of the cost criterion for each value of the cost coefficient c_1 is printed in bold face.

Figure 5 and Table 3 confirm the obvious fact that the value of criterion increases with increasing the cost coefficient c_1. They also show that an increase in this coefficient, i.e., an increase in the cost of using the reserve server implies an increase in the optimal mean time τ_1. This is because it becomes unprofitable to attract the reserve server due to excessive cost of its use.

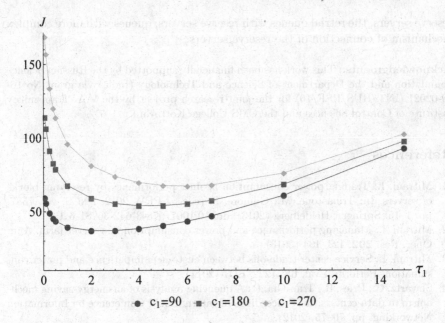

Fig. 5. Cost criterion J as a function of τ_1 for $c_1 = 90, 180, 270$

Table 3. Values of the cost criterion J for different τ_1 and c_1

$c_1 \backslash \tau_1$	0.05	0.1	0.25	0.4	0.5	1	2	3	4	5	6	7	10	15
90	59	56	49	45	44	39	**37**	37.06	38	41	43	47	61	93
180	114	106	91	83	79	68	59	55	**54**	54.1	55	58	68	98
270	169	157	134	120	114	96	81	74	70	68	**67**	68	76	102

7 Conclusion

In this paper, the single-server queuing system with a timer defining the limit service time is studied. In case of exceeding this time limit the reserve server connects to the service of a customer. This allows to increase the service rate. The process of the system operation is described in terms of Quasi Birth-and-Death process. We derive the necessary and sufficient condition for existence of the stationary regime in the system, calculate its stationary distribution and main performance measures. We derive formulas for the Laplace-Stieltjes of the sojourn time distribution and the mean sojourn time. We present the results of numerical experiments directed to the choice of the optimal value of the mean time to the timer expiration.

The results can be used for the modeling and optimization of energy savings in some real-life data transmission and processing systems. Further research in this area can be directed to the investigation of multi-server queues with several reserve servers, the retrial queues with reserve servers, queues with more complex mechanism of connection of the reserve servers.

Acknowledgments. This work has been financially supported by the Russian Science Foundation and the Department of Science and Technology (India) via grant No 16-49-02021 (INT/RUS/RSF/16) for the joint research project by the V.A. Trapeznikov Institute of Control Sciences and the CMS College Kottayam.

References

1. Mitrani, I.: Trading power consumption against performance by reserving blocks of servers. In: Tribastone, M., Gilmore, S. (eds.) EPEW 2012. LNCS, vol. 7587, pp. 1–15. Springer, Heidelberg (2013). doi:10.1007/978-3-642-36781-6_1
2. Mitrani, I.: Managing performance and power consumption in a server farm. Ann. Oper. Res. **202**, 121–134 (2013)
3. Mitrani, I.: Service center trade-offs between customers impatience and power consumation. Perform. Eval. **68**, 1222–1231 (2011)
4. Shwartz, C., Pries, R., Tran-Gia, P.: A queueing analysis of an energy-saving mechanism in data centers. In: Proceedings of International Conference on Information Networking, pp. 70–75 (2012)
5. Gortsev, A.M., Nazarov, A.A., Terpugov, A.F.: Control and Adaptation in Queueing Systems. Tomsk University Press, Tomsk (1978). (in Russian)
6. Chakravarthy, S.R., Agnihothri, S.R.: A server backup model with Markovian arrivals and phase type services. Eur. J. Oper. Res. **184**, 584–609 (2008)
7. Kim, C.S., Dudin, A., Dudin, S., Dudina, O.: Hysteresis control by the number of active servers in queueing system with priority service. Perform. Eval. **101**, 20–33 (2016)
8. Neuts, M.F.: Matrix-Geometric Solutions in Stochastic Models. The Johns Hopkins University Press, Baltimore (1981)
9. Klimenok, V.I., Dudin, A.N.: Multi-dimensional asymptotically quasi-Toeplitz Markov chains and their application in queueing theory. Queueing Syst. **54**, 245–259 (2006)
10. Dudin, A., Klimenok, V., Lee, M.H.: Recursive formulas for the moments of queue length in the $BMAP/G/1$ queue. IEEE Commun. Lett. **13**, 351–353 (2009)

Reliability of a k-out-of-n System with a Repair Facility – Essential and Inessential Services

M.K. Sathian[1], Viswanath C. Narayanan[2(⊠)], Vladimir Vishnevskiy[3],
and Achyutha Krishnamoorthy[4]

[1] Department of Mathematics, Panampilly Memorial Government College,
Chalakudy, Thrissur, India
sathianmkk@yahoo.com
[2] Department of Mathematics, Government Engineering College,
Thrissur 680 009, India
narayanan_viswanath@yahoo.com
[3] V. A. Trapeznikov Institute of Control Sciences of Russian Academy of Sciences,
65 Profsoyuznay Street, Moscow 117997, Russia
vishn@inbox.ru
[4] Department of Mathematics, Cochin University of Science and Technology,
Kochi 682022, India
achyuthacusat@gmail.com

Abstract. In this paper we study reliability analysis of a k-out-of-n system with a repair facility which provides an essential and several inessential service with given probabilities. At the epoch the system starts, all components are in operational state. Service to failed components are in the order of their arrival. When a component is selected for repair, we assume that the server may select it either for a service that turns out to be different from what is exactly needed for it, which we call the inessential service with the probability p or for desired service, called essential service with probability (1-p). Once the inessential service process starts, a random clock is assumed to start ticking which decides the event to follow: if the clock realises first (still inessential service going on) the components ongoing service is stopped and it is replaced with a new component. On the other hand if the inessential service gets completed before the realisation of the random clock, then the component moves for the essential service immediately. The life-time of a component, the essential service time and the random clock time have independent exponential distributions and the inessential service time is assumed to follow a phase type distribution. The steady state distribution of the system has been obtained explicitly and several important performance measures derived and verified numerically. The extension of the results reported to the case of more than one essential service is worth examining. This has applications in medicine, biology and several other fields of activity.

Keywords: k-out-of-n system · Essential service · Inessential service

1 Introduction

At several occasions, customers while seeking for a particular type of service are met with some other type which is not desirable for him/her. For example,

© Springer International Publishing AG 2016
V.M. Vishnevskiy et al. (Eds.): DCCN 2016, CCIS 678, pp. 89–97, 2016.
DOI: 10.1007/978-3-319-51917-3_9

consider a person admitted to a hospital. If for some unwanted reason, he/she get wrongly diagnosed and may receive a service which is unwanted for him/her. Another example is a mechanic wrongly diagnoses a car problem leading to offer an unwanted service. There are several other examples which, we are sure, the readers can think from the day-to-day life. In many such occasions, it is probable that the unwanted service which is being offered after a wrong diagnosis finishes without eternal damage and the customer receives the required service. These real world phenomena have motivated us to develop a queuing model, which we describe as follows. We consider a k-out-of-n system, where the components are subjected to failure. When selected for a repair, with probability p, the failed component may receive an unwanted service (which we call the inessential service). If the inessential service finishes before a random duration, we assume that the component has overcome the wrong diagnosis and hence proceeds for the essential service and become as good as a new one thereafter. If the random duration is over while the component still receiving the unwanted service, we assume that its ongoing service is stopped and it is replaced with a new component. Our model differs from the vacation queuing models studied in Madan et al. [1] or Ayyappan et al. [2], where at each service start, a customer can choose one of the two kinds of services. The difference is that in our model, a customer who got selected for inessential service may leave the system either without completing any service or receiving both types of services. Saravanarajan and Chandrasekaran [3] studies a vacation queuing model with system breakdowns, where the customers can choose one of the two type of services offered. This model allows a customer to remain in the system for another service, joining the tail of the queue, after a service completion. However, a customer can't receive two back to back services of different kind and it can't leave the system in between an ongoing service. Hence our model differs from this model also.

This paper has been arranged as follows: In Sect. 1, we give the technical description of the model. In Sect. 3, the steady state distribution has been found. Several important system performance measures have been derived in Sect. 4 and Sect. 5 presents the results from a numerical study of the performance measures.

2   Technical Description of the Model

We consider a k-out-of-n system with a single server repair facility. At the epoch the system starts, all components are in operational state. The life-time of a component follows an exponential distribution with parameter λ/i, when i components are operational. Service to failed components is in the order of their arrival. When a component is selected for repair, it may get selected for an inessential service with probability p and with probability $(1 - p)$, it may be taken for desired service, called the essential service. Once the inessential service process starts, the failed component either completes the service there and moves for the essential service or is replaced by a new component. A random clock is assumed to start ticking the moment the inessential service starts, which decides the event to follow: if the clock realises first (still the inessential service is going on) the failed component's ongoing service is stopped and it is replaced with a new component. On the other hand if the inessential service gets completed before the realisation of the random clock, then the component moves for the essential service immediately. After a successful repair (the essential service) the component is assumed to be as good as a new component.

The essential service time of a failed component is exponentially distributed with parameter μ and the service time of failed components in inessential service has a phase type distribution with representation (α, S) of order m. We assume that $S^0 = -Se$.

The random clock time is assumed to be exponentially distributed with parameter δ.

2.1 The Markov Chain

Let $N(t) = $ at time t number of failed components in the system.

$$J(t) = \begin{cases} 0, \text{if the failed component getting essential service,} \\ i, \text{if a failed component getting } i^{\text{th}} \text{ phase of inessential service,} \end{cases}$$

where $i = 1, 2, \ldots, m$.

Then $\{X(t), t \geq 0\}$ where $X(t) = (N(t), J(t))$ is a continuous time Markov chain with state space $\{(0,0)\} \cup \{1, 2, \ldots, n-k+1\} \times \{0, 1, 2, \ldots, m\}$.

The generator matrix of the Markov chain $\{X(t), t \geq 0\}$ is

$$Q = \begin{bmatrix} A_{00} & B_0 & & & & \\ B_1 & A_1 & A_0 & & & \\ & A_2 & A_1 & A_0 & & \\ & & \cdot & \cdot & \cdot & \\ & & & \cdot & \cdot & \cdot \\ & & & & A_2 & A_1 & A_0 \\ & & & & & A_2 & \widetilde{A_1} \end{bmatrix}$$

$$A_{00} = [-\lambda]; B_0 = \left[(1-p)\lambda \ \ p\lambda\alpha \right]; B_1 = \begin{bmatrix} \mu \\ \delta e \end{bmatrix}$$

$$A_1 = \begin{bmatrix} -(\mu + \lambda) & 0 \\ S^0 & S - (\delta + \lambda)I_m \end{bmatrix}; A_0 = [\lambda I_{m+1}]; A_2 = \begin{bmatrix} (1-p)\mu & p\mu\alpha \\ (1-p)\delta e & p\delta e\alpha \end{bmatrix}$$

$$\widetilde{A_1} = \begin{bmatrix} -\mu & 0 \\ S^0 & S - \delta I_m \end{bmatrix}$$

where $\alpha = (\alpha_1, \alpha_2 \ldots, \alpha_m)$ with $\alpha_1 + \alpha_2 + \ldots + \alpha_m = 1$.

We also define $\beta = ((1-p) \ \ p\alpha)$.

3 Steady State Distribution

Since this system is finite, it is stable. Let

$$\pi = (\pi(0), \pi(1), \ldots, \pi(n-k+1))$$

with
$$\pi(i) = (\pi(i,0), \pi(i,1), \pi(i,2), \ldots \pi(i,m)), \, 1 \le i \le n - k + 1$$
be the steady state probability vector of the system $\{X(t), t \ge 0\}$. Then it satisfies the equations $\boldsymbol{\pi}Q = 0$ and $\boldsymbol{\pi}\boldsymbol{e} = 1$.

The equation $\boldsymbol{\pi}Q = 0$ gives rise to

$$\pi(0)A_{00} + \pi(1)B_1 = 0 \tag{1}$$
$$\pi(0)B_0 + \pi(1)A_1 + \pi(2)A_2 = 0 \tag{2}$$
$$\pi(i-1)A_0 + \pi(i)A_1 + \pi(i+1)A_2 = 0, 2 \le i \le n - k \tag{3}$$
$$\pi(n-k)A_0 + \pi(n-k+1)\widetilde{A}_1 = 0. \tag{4}$$

Since $A_{00} = [-\lambda]$ and $B_1 = A_2 \boldsymbol{e}$, from (1) it follows that

$$\lambda\pi(0) = \pi(1)A_2 \boldsymbol{e}. \tag{5}$$

Since $B_0 = \lambda\beta$, equation (2) becomes

$$\pi(0)\lambda\beta + \pi(1)A_1 + \pi(2)A_2 = 0. \tag{6}$$

Using (5) we can write this equation as

$$\pi(1)B_1\beta + \pi(1)A_1 + \pi(2)A_2 = 0. \tag{7}$$

We notice that $B_1\beta = A_2$ and hence equation (7) becomes

$$\pi(1)(A_1 + A_2) + \pi(2)A_2 = 0. \tag{8}$$

Post multiplying equation (8) with $\boldsymbol{e}$, we get

$$\pi(1)(A_1 + A_2)\boldsymbol{e} + \pi(2)A_2\boldsymbol{e} = 0 \tag{9}$$

but $(A_1 + A_2)\boldsymbol{e} = -A_0\boldsymbol{e} = -\lambda\boldsymbol{e}$. Hence (9) becomes

$$\pi(1)\lambda\boldsymbol{e} = \pi(2)A_2\boldsymbol{e}. \tag{10}$$

We notice that $A_2 = A_2\boldsymbol{e}\beta$, which transforms equation (8) in to

$$\pi(1)(A_1 + A_2) + \pi(2)A_2\boldsymbol{e}\beta = 0. \tag{11}$$

Substituting for $\pi(2)A_2\boldsymbol{e}$ from (10) in (11), we get

$$\pi(1)(A_1 + A_2) + \pi(1)\lambda\boldsymbol{e}\beta = 0.$$

That is
$$\pi(1)(A_1 + A_2 + \lambda\boldsymbol{e}\beta) = 0. \tag{12}$$

Equation (12) shows that $\pi(1)$ is a constant multiple of the steady state vector φ of the generator matrix $A_1 + A_2 + \lambda\boldsymbol{e}\beta$. That is

$$\pi(1) = \eta\varphi \tag{13}$$

where η is a constant.

Equation (3) for $i = 2$ gives

$$\pi(1)A_0 + \pi(2)A_1 + \pi(3)A_2 = 0. \tag{14}$$

Since $A_2 = A_2\,e\beta$, equation (14) becomes

$$\pi(1)A_0 + \pi(2)A_1 + \pi(3)A_2\,e\beta = 0. \tag{15}$$

Post multiplying with e, we get

$$\pi(1)\lambda e + \pi(2)A_1 e + \pi(3)A_2 e = 0. \tag{16}$$

Using (10) the above equation can be written as

$$\pi(2)A_2\,e + \pi(2)A_1\,e + \pi(3)A_2\,e = 0$$
$$\text{i.e.,}\quad \pi(2)(A_1 + A_2)\,e = -\pi(3)A_2\,e$$
$$\text{i.e.,}\quad \pi(2)\lambda e = \pi(3)A_2\,e. \tag{17}$$

In the light of equation (17), equation (15) becomes,

$$\pi(1)A_0 + \pi(2)A_1 + \pi(2)\lambda\,e\beta = 0$$
$$\text{i.e.,}\quad \pi(1)A_0 + \pi(2)(A_1 + \lambda\,e\beta) = 0$$

which implies that

$$\pi(2) = -\pi(1)A_0(A_1 + \lambda\,e\beta)^{-1}.$$

That is

$$\pi(2) = -\eta\,\varphi A_0(A_1 + \lambda\,e\beta)^{-1}. \tag{18}$$

Post-multiplying equation (3) with e and proceeding in the same lines as we derived equation (17), we can derive that

$$\pi(i+1)A_2\,e = \pi(i)\lambda e, \text{ for } 3 \le i \le n - k. \tag{19}$$

Equation (19) then transforms equation (3) as

$$\pi(i-1)A_0 + \pi(i)A_1 + \pi(i)\lambda\,e\beta = 0, 3 \le i \le n - k,$$

which implies that

$$\pi(i) = -\pi(i-1)A_0(A_1 + \lambda\,e\beta)^{-1}, 2 \le i \le n - k \tag{20}$$

which in turn gives

$$\pi(i) = (-1)^{i-1}\eta\,\varphi(A_0(A_1 + \lambda\,e\beta)^{-1})^{i-1}, 2 \le i \le n - k. \tag{21}$$

We notice that $\widetilde{A}_1\,e = -A_2\,e$; post-multiplying equation (4) with e, we get

$$\pi(n-k)\lambda e = \pi(n-k+1)A_2\,e. \tag{22}$$

From equation (4), we can also write

$$\pi(n - k + 1) = -\pi(n - k)A_0(\tilde{A}_1)^{-1}. \tag{23}$$

Using (21) for $i = n - k$, (23) becomes

$$\pi(n - k + 1) = (-1)^{n-k}\eta\boldsymbol{\varphi}(A_0(A_1 + \lambda\boldsymbol{e}\beta)^{-1})^{n-k+1}A_0(\tilde{A}_1)^{-1}. \tag{24}$$

Hence, we have the following theorem for the steady state distribution:

Theorem 1. *The steady state distribution* $\boldsymbol{\pi} = (\pi(0), \pi(1), \ldots, \pi(n - k + 1))$ *of the Markov chain* $\{X(t), t \geq 0\}$ *is given by*

$$\pi(0) = \frac{1}{\lambda}\eta\boldsymbol{\varphi}B_1,$$
$$\pi(1) = \eta\boldsymbol{\varphi},$$
$$\pi(i) = (-1)^{i-1}\eta\boldsymbol{\varphi}(A_0(A_1 + \lambda\boldsymbol{e}\beta)^{-1})^{i-1}, 2 \leq i \leq n - k$$
$$\pi(n - k + 1) = (-1)^{n-k}\eta\boldsymbol{\varphi}(A_0(A_1 + \lambda\boldsymbol{e}\beta)^{-1})^{n-k-1}A_0(\tilde{A}_1)^{-1},$$

where $\boldsymbol{\varphi}$ *is the steady state vector of the generator matrix* $A_1 + A_2 + \lambda\boldsymbol{e}\beta$ *and* η *is a constant, which can be found from the normalizing condition* $\boldsymbol{\pi}\boldsymbol{e} = 1$.

4   System Performance Measures

1. Fraction of time the system is down,

$$P_{\text{down}} = \sum_{j=0}^{m} \pi(n - k + 1, j).$$

2. System reliability,

$$P_{\text{rel}} = 1 - P_{\text{down}} = 1 - \sum_{j=0}^{m} \pi(n - k + 1, j).$$

3. Average number of failed components in the system,

$$N_{\text{fail}} = \sum_{i=1}^{n-k+1} i \left(\sum_{j=0}^{m} \pi(i, j) \right).$$

4. Expected rate at which failed components are taken for essential service:

$$E_{\text{es}} = (1 - p)\lambda\pi(0) + \sum_{i=2}^{n-k+1} (1 - p)\mu\pi(i, 0) + \sum_{i=2}^{n-k+1} (1 - p)\delta \left(\sum_{j=1}^{m} \pi(i, j) \right).$$

5. Expected rate at which failed components are taken for inessential service

$$E_{\text{in es}} = p\lambda\pi(0) + \sum_{i=2}^{n-k+1} p\mu\pi(i,0) + \sum_{i=2}^{n-k+1} p\delta\left(\sum_{j=1}^{m} \pi(i,j)\right).$$

6. Expected rate at which new components were bought:

$$E_{C.R} = \sum_{i=1}^{n-k+1} \delta\left(\sum_{j=1}^{m} \pi(i,j)\right).$$

7. Expected rate at which failed components that start with inessential service subsequently moving to essential service before clock realisation:

$$E_{\text{IN E}} = \sum_{i=1}^{n-k+1}\sum_{j=2}^{m+1} \pi(i,j)S^0(j-1,1).$$

8. Fraction of time server is idle:

$$P_{\text{idle}} = \pi(0).$$

9. Fraction of time server is busy:

$$P_{\text{busy}} = 1 - \pi(0).$$

5   Numerical Study of the System Performance Measures

Notice that if a component is selected for inessential service, it is either replaced by a new component (if the random clock realises before completion of the inessential service) or is got repaired (if the inessential service completes before the random clock realises). Hence a component getting selected for inessential service according to probability p affects the system reliability only through an increase in the repair time by a random amount of time (minimum of inessential service time and random clock time). Table 1 shows that very high reliability is maintained in the system, which decreases slightly as the probability p that a failed component receives an undesired service initially, increases. The decrease in the average rate at which components directly receive essential service with an increase in p, as seen in Table 2, was expected. According to the modelling assumption, if the random clock expires during an inessential service, the component receiving the inessential service is replaced with a new component. Hence, as the probability p increases, more components will get selected for inessential service, which leads to an increase in the replacement rate as seen in Table 3.

Since the inessential service is not helping the system in any way whatsoever, one would expect the optimal value for the probability p as to be zero. However in a situation where the possibility for inessential service can't be avoided, one would like to know its harm through some number. For this purpose, we have constructed a cost function as follows:

Table 1. Variation in system reliability

p	$n = 45$	$n = 50$	$n = 55$	$n = 60$
0.001	0.999985933	0.999985933	0.999997139	0.999999404
0.007	0.999985814	0.999985814	0.999997139	0.999999404
0.03	0.999985576	0.999985576	0.99999702	0.999999404
0.07	0.999985039	0.999985039	0.999996901	0.999999344
0.09	0.999984741	0.999984741	0.999996841	0.999999344
0.3	0.999981642	0.999981642	0.999996066	0.999999166
0.7	0.999973893	0.999973893	0.99999404	0.999998629
0.9	0.999968886	0.999968886	0.999992669	0.999998271
0.99	0.999966323	0.999966323	0.999991954	0.999998093

Table 2. Average rate at which components are taken for essential service

p	$n = 45$	$n = 50$	$n = 55$	$n = 60$
0.001	3.99649739	3.99649739	3.99654222	3.99655128
0.007	3.97557425	3.97579312	3.97583771	3.97584677
0.03	3.89581704	3.89603281	3.89607692	3.89608598
0.07	3.75561023	3.7558198	3.75586271	3.75587177
0.09	3.68479681	3.6850028	3.68504548	3.68505406
0.3	2.91331673	2.91348505	2.91352081	2.91352844
0.7	1.30956876	1.30964661	1.30966437	1.30966842
0.9	0.44612866	0.446155071	0.44616127	0.446162701
0.99	0.045032669	0.04503531	0.045035943	0.045036085

Table 3. Average rate at which components were bought

p	$n = 45$	$n = 50$	$n = 55$	$n = 60$
0.001	0.001538355	0.00153844	0.001538457	0.001538461
0.007	0.010768481	0.010769077	0.010769199	0.010769224
0.03	0.046150584	0.04615318	0.046153713	0.046153817
0.07	0.107684486	0.107690714	0.107691996	0.107692257
0.09	0.138451293	0.138459414	0.138461098	0.138461441
0.3	0.461498737	0.46153	0.461536676	0.461538808
0.7	1.07679927	1.076895	1.07691669	1.0769217
0.9	1.38443172	1.38457251	1.38460553	1.38461328
0.99	1.52286148	1.52302587	1.52306497	1.52307427

Let C_1 be the cost per unit time incurred if the system is down, C_2, be the repair cost per unit time for essential service per failed component, C_3 is the cost incurred towards the time loss due to wrong diagnosis with failed components and consequent realisation of random clock before inessential service completion. C_4 is the extra cost incurred on failed components that start with inessential service subsequently moves to essential service before clock realisation, C_5 be the repair cost per unit time for inessential service.

Expected cost per unit time

$$= C_1 \cdot P_{\text{down}} + C_2 \cdot E_{\text{es}} + C_3 \cdot E_{\text{C.R}} + C_4 \cdot E_{\text{IN E}} + C_5 \cdot E_{\text{ines}}.$$

Table 4 presents the variation in cost function as the probability p increases for different component failure rates. In all the cases studied, the optimum value of p was found zero as was expected. The table also shows that as the component failure rate increases, the cost function also increases.

General parameters for Tables 1–3 are as follows: $\lambda = 4$, $\mu = 3.2$, $\delta = 5$,
$$S = \begin{bmatrix} -18 & 4 & 6 \\ 5 & -18 & 5 \\ 7 & 4 & -19 \end{bmatrix}, \alpha = (0.4, 0.3, 0.3).$$

Table 4. Variation in cost $C_1 = 9500$, $C_2 = 2600$, $C_3 = 4000$, $C_4 = 1600$ $C_5 = 3000$

P	$\lambda = 4$	$\lambda = 4.5$	$\lambda = 5$	$\lambda = 6$
0.001	10411.8662	11709.9268	12982.9131	15048.7344
0.007	10483.1328	11782.5352	13053.7373	15112.8564
0.03	10756.3047	12060.7119	13324.9258	15358.3809
0.07	11231.2598	12543.8252	13795.2471	15784.0264
0.09	11468.6729	12789.0537	14029.7676	15996.2021
0.3	13958.7285	15304.4629	16466.709	18199.4043
0.7	18682.9082	20028.3828	20973.9453	22276.6172
0.9	21032.8359	22349.3281	23158.7715	24260.2168
0.99	22087.1016	23384.0664	24126.6035	25141.5078

Acknowledgements. Second and fourth authors acknowledge Department of Science & Technology, Government of India under grant INT/RUS/RMES/P-3/2014 dated 15-4-2015; The third author's research is financially supported by the Russian Science Foundation (research project No.16-49-02021).

References

1. Madan, K.C., Al-Rawi, S.R., Al-Nasser, A.D.: On Mx/(G1G2)/1/G(BS)/Vs vacation queue with two types of general heterogeneous service. J. Appl. Math. Decis. Sci. **3**, 123–135 (2005)
2. Ayyappan, G., Sathiya, K., Subramanian, A.M.G.: M[X]/G/1 Queue with two types of service subject to random breakdowns, multiple vacation and restricted admissibility. Appl. Math. Sci. **7**(53), 2599–2611 (2013)
3. Saravanarajan, M.C., Chandrasekaran, V.M.: Analysis of M/G/1 feedback queue with two types of services, bernoulli vacations and random breakdowns. Int. J. Math. Oper. Res. **6**(5), 567–588 (2014)

Tractable Distance Distribution Approximations for Hardcore Processes

Pavel Abaev[1(✉)], Yulia Gaidamaka[1], Konstantin Samouylov[1], and Sergey Shorgin[2]

[1] RUDN University, 6 Miklukho-Maklaya st, Moscow 117198, Russia
{pabaev,ygaidamaka,ksam}@sci.pfu.edu.ru
[2] Federal Research Center 'Computer Science and Control',
Russian Academy of Sciences, 44-2 Vavilova st, Moscow 119333, Russia
sshorgin@ipiran.ru

Abstract. The Poisson point process (PPP) is widely used in performance analysis of wireless communications technologies as a basic model for random deployment of communicating entities. The reason behind widespread use of PPP is analytical tractability in terms of closed-form distributions of distances to the n-th neighbour needed for performance analysis. At the same time, the process allows for infinitesimally close distances between communicating stations not only contradicting the reality but presenting fundamental difficulties in analysis when used with power-law propagation models. As an alternative suggested in the literature ad free of abovementioned deficiencies are the hardcore processes where a certain separation distance between points is always presumed. Unfortunately, no closed-form expressions for distance distributions is available for these processes. We study distance distributions of Matern hardcore process and propose analytical approximations based on acyclic phase type distributions. The nature of approximation as a mixture of exponentials allows for their use in analytical performance analysis. Results for a range of process intensities are reported.

Keywords: Hardcore process · Approximation · Poisson process · Distance distribution · SINR

1 Introduction

The stochastic geometry has recently attracted attention from the wireless research community as a tool used to study performance metrics of modern and forthcoming wireless technologies [1,5,6]. The first step in modeling of wireless systems is the choice of the spatial model for location of nodes. Owning to unpredictable in advance locations of subscribers on the plane, random spatial point processes are often used for this purpose. Depending on specific purposes examples include Poisson point process, cluster Poisson point process, hardcore processes [4].

© Springer International Publishing AG 2016
V.M. Vishnevskiy et al. (Eds.): DCCN 2016, CCIS 678, pp. 98–109, 2016.
DOI: 10.1007/978-3-319-51917-3_10

In [10] indoor scenarios with grid aligned rooms (Fig. 1) that are typical for shopping malls and office buildings are investigated. These scenarios include communications between access points and devices scattered over the rooms, as well as direct device-to-device (D2D) communications. While providing a comprehensive analysis of SINR for Uplink scenario (Fig. 1a) based on both analytical and simulation modelling the authors confined the study of Downlink and D2D scenarios (Fig. 1b, c) to simulation only. In Uplink scenario location of an access point is always fixed in the center of a room making the model quite simple as only the location of devices is random. So, for the pair access point-device it is sufficient to operate with only two random values coordinates of the device on a plane.

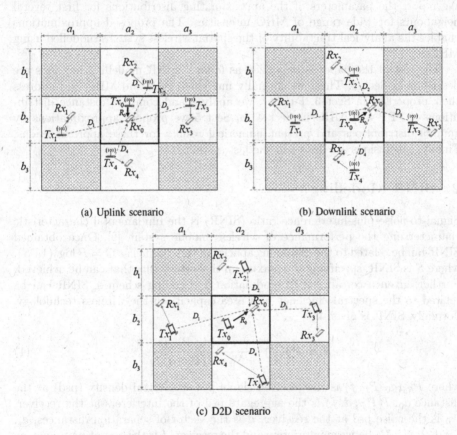

(a) Uplink scenario   (b) Downlink scenario

(c) D2D scenario

Fig. 1. Indoor communication scenarios

The reason for limitations of analytical methods is the complexity of formula derivation process, since there dependences of the distance between communicating devices and interfering with adjacent clusters appear. Therefore, consideration of several random variables describing the coordinates of the location of devices in the cluster (room) is required, which is not a trivial task. In this case assuming that clusters can be represented by circles of radius $\frac{a}{2}$ a hardcore process could be applied. As there are no closed-form expressions for distance distributions in hardcore processes, they would not offer the opportunity of analytical modelling. However, this constraint can be overcome with the help of distance approximations that allow construction of analytical models based on empirical data.

In this paper we study distance distributions in MHC showing that approximation by the thinned Poisson point process lead to significant errors in terms of both distance distributions and interference, especially, for first several neighbours providing the critical contribution to the overall interference. We then propose to approximate these distance distributions by the gamma distributions. We report the parameters of the approximating distributions for first several neighbours for wide range of MHC intensities. The proposed approximations enables the analytical tractability of the spatial wireless systems modelled using MHCs.

The rest of the paper is organized as follows. SINR modelling methods are described in Sect. 2. Then, we formally introduce PPP and MHC and address their properties in Sect. 3. Further, we analyze and compare distance distributions of PPP and MHC in Sect. 4. In Sect. 5 we propose approximations by gamma distributions and present numerical results for these approximations. Finally, conclusions are drawn in the last section.

2   SINR Modelling

Signal-to-noise-plus-interference ratio (SINR) is the fundamental characteristic characterizing the performance of wireless mobile system [9]. Once obtained SINR can be related to the Shannon capacity of the channel as $C = B\log_2(1+S)$, where S is SINR, specifying the maximum theoretical rate that can be achieved. Further, given a certain set of modulation and coding schemes, SINR can be related to the spectral efficiency and area capacity of the wireless technology. Formally, SINR is given by

$$S(\boldsymbol{d}, \boldsymbol{P}_T, f) = \frac{P_{R_0}(d_0, P_T, f)}{N_0 + I(\boldsymbol{P}_T, \boldsymbol{d}, f)}, \tag{1}$$

where $P_{R_0}(d_0, P_T, f)$ is the received signal power spectral density (psd) at the distance d_0, $I(\boldsymbol{P}_T, \boldsymbol{d}, f)$ is the aggregate psd of the interferers at the receiver, N_0 is the noise psd at the receiver, $\boldsymbol{d}$ is the vector of separation distances, d_i, $i = 1, 2, \ldots, M$, between interferers and the receiver, f is the operating frequency and M is the number of interfering nodes.

In order to model interference and SINR in spatial wireless systems PPP if universally used. The reason is that distance distributions in (1) has closed-form of generalized gamma distribution [7]. However, its usage is associated with one fundamental problem. First, it allows the distance to be of infinitesimally small length, i.e. zero. This is unrealistic as users are always separated by some distance. Conventional propagation models used on top of these distances, e.g.

$Ad^{-\gamma}$ predict the received signal strength (and, thus, interference) to approach infinity [8]. This also implies that the mean interference does not exist as the integral $E[I] = \int_0^\infty f(x)1/x^{-\gamma}dx$, where $f(x)$ is the distance to the first neighbour, γ is the path loss exponent, diverges which is unrealistic. This feature also drives the interference which is the sum of signals from N neighbours to follow heavy-tailed distribution (α-stable with $\alpha < 2$), not the Normal one ($\alpha = 2$).

In Fig. 2 distributions of SINR in the network model are shown. The distributions were obtained for three point processes: MHC, PPP and so-called shifted PPP, where distances are adjusted to coincide with MHC in mean value of SINR.

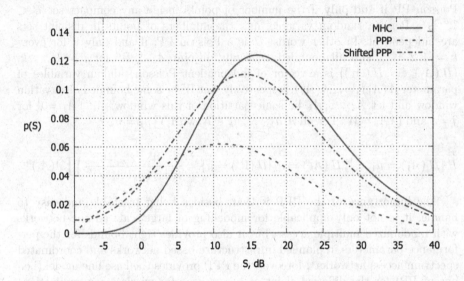

Fig. 2. Distributions of SINR in the network model.

More realistic model would be to use hardcore processes where there exist a minimum separation between neighbours. Example of such processes is Poisson hardcore process also known as Matern hardcore process (MHC) that can be obtained by special thinning procedure of the original Poisson process, see e.g. [2]. For these processes, however, there is no closed form expression for the distance distribution to the n-th neighbour preventing their usage in analytical assessment of spatial wireless networks. Very often, a thinned version of the Poisson process for approximations of the distance to the n-th neighbour.

3   Poisson and Hardcore Spatial Processes

The PPP is used to model or abstract a network composed of a possibly infinite number of nodes randomly and independently coexisting in a finite or infinite service area.

The Matern HCPP (MHC) conditions on having a minimum distance r between any two points of the process, and is obtained by applying dependent thinning to a PPP. That is, starting from a PPP, the HCPP is obtained by assigning a random mark uniformly distributed in $[0, 1]$ to each point in the PPP, then deleting all points that coexist within a distance less than the hard core parameter r from another point with a lower mark. Hence, only the points that have the lowest mark within their r-neighborhood distance are retained. As a result, no two points with a separation less that r will coexist in the constructed HCPP.

Poisson point process (PPP) is a PP $\Pi = \{x_i; \; i = 1, 2, 3, \ldots\} \subset \mathbb{R}^d$ is a Poisson PP if and only if the number of points inside any compact set $B \subset \mathbb{R}^d$ is a Poisson random variable, and the numbers of points in disjoint sets are independent. In other words: Π is a Poisson PP, if and only if for every $k = 1, 2, 3, \ldots$ and all bounded, mutually disjoint $A_i \subset \mathbb{R}^d$ for $i = 1, \ldots, k$, $(\Pi(A_1), \ldots, \Pi(A_k))$ is a vector of independent Poisson random variables of parameter $\Lambda(A_1), \ldots, \Lambda(A_k)$, respectively. Let W be some bounded observation window and let $A_1, \ldots, A_k$ be some partition of this window: $A_i \cap A_j = \emptyset$ for $j \neq i$ and $\bigcup_i A_i = W$. For all $n, n_1, \ldots, n_k \in N$ with $\sum_i n_i = n$,

$$P\{\Pi(A_1) = n_1, \ldots, \Pi(A_k) = n_k | \Pi(W) = n\} = \frac{n!}{n_1! \ldots n_k!} \frac{1}{\Lambda(W)^n} \prod_i \Lambda(A_i)^{n_i}.$$

The importance of the PPP is that, besides being tractable and easy to handle, it is not only applicable for modelling of large-scale ad-hoc networks with randomized multiple access, but it also provides tight bounds for the performance parameters in planned infrastructure-based networks and coordinated spectrum access networks. Moreover, the PPP provides the base line model (i.e., parent PP) for the different point processes used for wireless communications systems [6].

A Hardcore PP is a repulsive point process where no two points of the process coexist with a separating distance less than a predefined hard core parameter r. A PP $\Pi = \{x_i; \; i = 1, 2, 3, \ldots\} \subset \mathbb{R}^d$ is an HCPP if and only if $\Pi = \{x_i; i = 1, 2, 3, \ldots\} \subset \mathbb{R}^d$, $\forall x_i, x_j \in \Pi$, $j \neq i$, where $r \geq 0$ is a predefined hard core parameter.

In [7] an expression for distribution of distances from an arbitrary point to the n-nearest neighbor is provided in a form of gamma distributions (2):

$$f(r, n)\, dr = \frac{2(\pi\lambda)^n}{(n-1)!} r^{2n-1} e^{-\lambda\pi r^2}, r > 0, n = 1, 2, \ldots \tag{2}$$

A Hardcore PP can be derived from a parent Poisson PP by using the thinning procedure that is described in the algorithm shown below. As the result of thinning a new process of a reduced intensity is obtained. This process is called Matern Hardcore PP (MHC). As the thinning is dependent (points are removed with regards to the distance between them), MHC does not bare the property of independence what results in infeasibility of analytical expressions derivation.

Algorithm 1. Thinning algorithm

Generate $k \sim Pois(\Lambda)$
for $i = 1 : k$ **do**
    generate $p_i := (x_i \ R_{[0;a]}, y_i \ R_{[0;b]}, m_i \ R_{[0;1]})$
end for
for $i = 1 : k$ **do**
    for $j = i + 1 : k$ **do**
        if $\|p_i - p_j\| < r$ **then**
            if $m_i < m_j$ **then**
                $p_j := [\,]$
            else
                $p_i := [\,]$
            end if
        end if
    end for
end for

First, in steps 1–4 we construct a parent PPP of intensity $\Lambda = \lambda \cdot S$, where S is the area of the plane (Fig. 3a). Each point of the process is given a random mark m_i. Then for each pair of points we evaluate the distance between them. If the distance is less then predefined hardcore parameter r, the point with the smaller mark m is retained while the other is removed. Thus, as all the possible pairs are handled, we get a MHC (Fig. 3b).

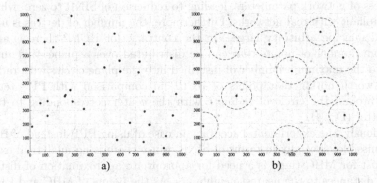

Fig. 3. Parent PPP (a) and derived MHC (b).

Although the problem of infinitesimally small distances between communicating stations have been known for years, the Poisson distributions of mobile nodes in the plane have been assumed in a number of fundamental studies of mobile wireless network performance. There are a number of indirect methods proposed to avoid it. For example, the authors usually assume (see e.g. [2,8]) that there is a certain circular area of rather small radius r around the receiver where no point of PPP may occur. However, they also revealed that the choice of

r is arbitrary and may affect the resulting SINR performance of the system. For realistic inter-node distances PPP with thinned intensity provides rather poor approximation of the MHP [2].

4   Analysis of Distance Distributions

In this section, we provide a comparative analysis of distance distributions in Poisson and Hardcore PPs. According to [3], the contribution of the devices that are located at the distance of more than $2r$ to the value of interference for both processes is practically identical. For this reason, we focus our study on distances from a target transmitter device to the nearest interfering ones. Besides, without taking into account transmission power of interfering devices the closer an interfering device to the receiver the more it impacts the signal transmission. The distributions of distances to the 3 nearest neighbours for both PPP and HCPP of the same intensity are shown in Fig. 3.

In this paper we consider a plane of area S and a set of point scattered over the plane that may be interpreted as a network service area and its communicating devices consequently. For the sake of simplicity we assume that the environment corresponds to an urban area, where path-loss exponent approximately equals to 4. We also consider that the network is based on Carrier Sensing Multiple Access (CSMA) technology meaning that the points represent active sources of target and interfering signals at an arbitrary instance of time.

The fundamental distinction between the two processes is that the distance between the target and interfering sources in PPP may be infinitesimally small regardless of network parameters, leading to reduction of SINR to zero, which is not compliant with real networks. Furthermore, the number of devices is unlimited and may be arbitrary large. On the contrary, for HCPP there is always maximum number of points that can be distributed over a plane: this number equals to the maximum number of inscribed into the plane circles with radiuses r. It is worth noting that points of HCPP in comparison with PPP tend to tighten around the circle of radius r with the center corresponding to target transmitter (Fig. 4).

Obviously, the experimental accuracy in case of using PPP instead of HCPP is inadmissible, specially in terms of SINR. Since there is no analytical expressions exist for MHC there is a need for appropriate approximation of distributions of distances to the nearest neighbours. As the points of MHC and PPP at the distance of more than $2r$ from target transmitter generate the identical interference, approximation of only points at the distance of less than $2r$ is required. Thus, we need to identify the number of the nearest neighbours N that should be approximated. Common sense guides us to suppose that N should not be more than 9 as it is impossible to locate more non-intersecting circles of radius r over a circle of radius $(2r + r)$ where $2r$ stands for the approximation distance and r for a circle the center of which may be located on the edge of the $2r$-circle. So 9 approximated nearest neighbours is always sufficient to obtain a close approximation of MHC. To reduce computational complexity we propose to use (3), where φ is significance level

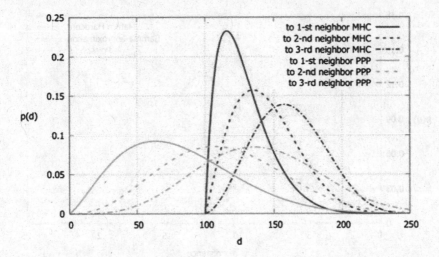

Fig. 4. Distributions of distances to the three nearest neighbours in PPP and HCPP.

$$N = \inf \mathbb{N} \left\{ \forall \eta, \eta \in \mathbb{N} \Rightarrow \int_{2r}^{\infty} f(\rho, \eta) \, d\rho < \varphi \right\}, \tag{3}$$

where $f(\rho, \eta) = \frac{2(\pi \lambda_{MHC})^{\eta}}{(\eta - 1)!} \rho^{2\eta - 1} e^{-\lambda_{MHC} \pi \rho^2}$.

The less the value of φ the more accurate and resource-intensive calculation. In this paper we consider a square plane 10^3 meters on side. The solution of the inequation (3) with $\varphi = 0.05$ for range of intensities can be provided as follows: $N = 6$ for $\varLambda = 10^{-4} : 5.5 \times 10^{-4}$ and $N = 7$ for $\varLambda = 5.5 \times 10^{-4} : 10^{-3}$. For smaller intensities approximation is meaningless since a MHC consisting of a few points may be sufficiently well approximated by PPP of the same intensity.

5   An Approximation Method and Numerical Results

In this section we propose an approximation for distribution of distances for MHC process. We impose the following two requirements on the prospective approximation: (i) the model shall be able to approximate the empirical densities with any given accuracy and (ii) the approximation should be suitable for analytical analysis of wireless communications systems.

In this paper we propose an approximation method that allows to construct a Matern-like point process. The method includes aprroximation of N nearest neighbours using EM-type algorithm to find the parameters of gamma distributions. In more details, for a given intensity of an approximated MHC, we calculate the intensity of parent PPP that is given by (4).

$$\lambda_{PPP} = -\frac{\ln \left(1 - \lambda_{MHC} \pi r^2 \right)}{\pi r^2}. \tag{4}$$

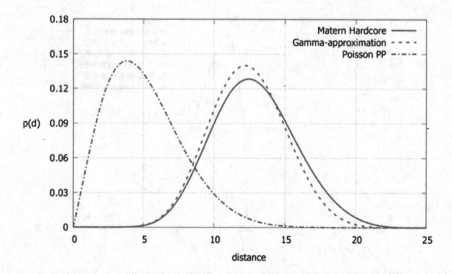

Fig. 5. Distributions of distance to the first nearest neighbour.

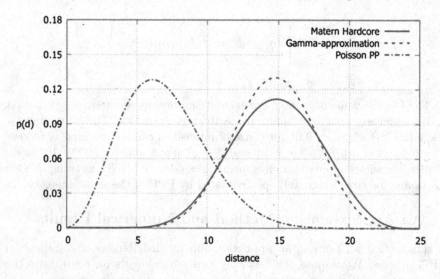

Fig. 6. Distributions of distance to the second nearest neighbour.

Then we calculate empirical distributions of distances to N nearest neighbours and subtract r. As we obtain the shifted distributions, we use EM-type algorithm to approximate them with a set of gamma distributions. As a result, we have N pairs of scale k and shape θ parameters of approximating gamma distributions. Approximation of distance distributions for three nearest neighbours is presented in Figs. 5, 6 and 7.

The final approximating point process may be constructed by generating $N_{AP} \sim Pois\,(\Lambda)$ points on the plane, where N distances from a target trans-

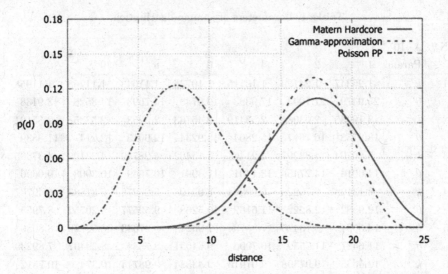

Fig. 7. Distributions of distance to the third nearest neighbour.

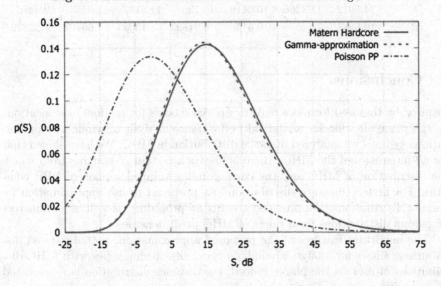

Fig. 8. SINR distributions in source HCPP and approximating PPs.

mitter to the nearest neighbours have approximated gamma distributions with addition of the shift of r. The last $(N_{AP} - N)$ points are modeled using the analytical expressions for n-neighbours in PPP. Approximation in terms of SINR is shown in Fig. 8.

Parameters of approximating gamma distributions with parameters k and θ for the square plane with area of 10^6 meters are presented in Table 1, where $N\#$ is the neighbour number.

Table 1. Parameters of gamma distribution

N#	$\lambda \cdot 10^{-4}$							
	Param	1	2	4	6	8	9	10
1	k	1.26347	1.24129	1.13116	1.10736	1.11643	1.11258	1.11958
	θ	25.9479	21.1063	17.9393	16.0743	14.7693	14.3528	13.9138
2	k	3.19935	2.93065	2.53419	2.45763	2.5253	2.5335	2.55191
	θ	18.9758	16.7617	15.2804	13.9234	12.6013	12.214	11.8539
3	k	5.80764	4.83297	4.2146	4.15462	4.28739	4.33109	4.37885
	θ	14.6016	14.2745	13.1401	11.8991	10.7669	10.3908	10.0696
4	k	8.13106	6.75228	6.09364	6.04108	6.20278	6.31883	6.49767
	θ	12.9582	12.8222	11.6162	10.5268	9.59577	9.20752	8.7955
5	k	10.6094	8.73425	7.85794	7.9624	8.27033	8.52779	8.8086
	θ	11.6487	11.7559	10.7996	9.64551	8.7384	8.29497	7.89239
6	k	12.5335	9.97398	9.07916	9.43584	9.98714	10.374	10.7512
	θ	11.1871	11.7331	10.7011	9.36291	8.34988	7.87798	7.47487
7	k	14.5912	11.3266	10.2119	10.7367	11.5137	11.9753	12.4476
	θ	10.6288	11.4675	10.6156	9.21823	8.13033	7.66462	7.25372

6   Conclusions

Inspired by the need for more realistic spatial models for random user locations on the plane allowing for analytical performance analysis of wireless communications systems we analyzed distance distribution in MHC. We have shown that for all intensities of the MHCs there is always a certain N starting from which the distribution of MHC and the corresponding thinned version of PPP coincides. For first N distance distributions we proposed to use approximation by gamma distributions. We provided the fitting procedure as well as parameters of gamma distributions for a range of MHC point densities.

The important features of the proposed approximations is that gamma distributions allows for analytical analysis of wireless technologies with MHC distributions of user on the plane. Indeed, the distance distribution is represented by a specific mixture of exponentials having rational Laplace transform.

Acknowledgements. The reported study was supported by the Russian Science Foundation, research project No. 16-11-10227.

References

1. Andrews, J., Ganti, K., Haenggi, M., Jindal, N., Weber, S.: A primer on spatial modeling and analysis in wireless networks. IEEE Commun. Mag. **48**(11), 156–163 (2010)

2. Baccelli, F., Blaszczyszyn, B.: Stochastic Geometry and Wireless Networks, Part I: Theory. Now Publishers, Delft (2009)
3. Busson, A., Chelius, G.: Point process for interference modeling in CSMA/CA ad-hoc networks. Research report RR-6624, Inria, pp. 1–11, September 2008
4. Cox, D., Isham, V.: Point Processes. Monographs on Statistics and Applied Probability. Chapman and Hall/CRC, Boca Raton (1980)
5. ElSawy, H., Hossain, E., Haenggi, M.: Stochastic geometry for modeling, analysis, and design of multi-tier and cognitive cellular wireless networks: a survey. IEEE Comm. Surv. Tutorials **15**(3), 996–1019 (2013)
6. Haenggi, M., Andrews, J., Bacelli, F., Dousse, O., Franceschetti, M.: Stochastic geometry and random graphs for the analysis and design of wireless networks. IEEE JSAC **27**(11), 1029–1046 (2009)
7. Moltchanov, D.: Distance distributions in random networks. Ad Hoc Netw. **10**(6), 1146–1166 (2012). Elsevier
8. Petrov, V., Moltchanov, D., Koucheryavy, Y.: Interference and SINR in dense terahertz networks. In: Proceedings of VTC-Fall, pp. 1437–1442 (2015)
9. Rappaport, T.: Wireless Communications: Principles and Practice. Prentice Hall, Upper Saddle River (2002)
10. Samuylov, A., Moltchanov, D., Gaidamaka, Y., Begishev, V., Kovalchukov, R., Abaev, P., Shorgin, S.: Sir analysis in square-shaped indoor premises. In: ECMS, pp. 1–6, July 2016

The Total Capacity of Customers in the Infinite-Server Queue with MMPP Arrivals

Ekaterina Lisovskaya[1], Svetlana Moiseeva[1(✉)], and Michele Pagano[2]

[1] Tomsk State University, Tomsk, Russia
{ekaterina_lisovs,smoiseeva}@mail.ru
[2] University of Pisa, Pisa, Italy
m.pagano@iet.unipi.it

Abstract. In the paper, the infinite-server queueing system with a random capacity of customers is considered. In this system, the total capacity of customers is analysed by means of the asymptotic analysis method with high-rate Markov Modulated Poisson Process arrivals. It is obtained that the stationary probability distribution of the total customer capacity can be approximated by the Gaussian distribution. Parameters of the approximation is also derived in the paper.

Keywords: Infinite-server queueing system · Customer with random capacity · Markovian Modulated Poisson Process

1 Introduction

In the design of messages processing and transmission systems, determining the memory capacity required for information storage is a relevant open issue [1,2]. The total capacity is a random quantity and in queueing theory it is given by the sum of the lengths of all messages, which are waiting in the buffer or currently processed by servers.

Since in real systems customers are heterogeneous (for instance in computer networks packet size may vary from a few tens of bytes to 1500 bytes in case of Ethernet links), this paper focuses on the analysis of queueing system (QS) with random customer capacity. The main classes of models used for such models and their applications in real information systems are given [1,2].

There are several works on the study of such systems with Poisson arrival process and service time independent of the customer capacity. For example, in [3] for systems with limited total capacity the generalization of the Erlang problem is considered in stationary conditions. Moreover, in [4,5], the stationary distribution of the customers number and the probability of losses are obtained for systems with limited memory. Finally, in [9], the authors consider systems with service time depended on the customers capacity or with waiting time restrictions, an assumption very relevant for real-time applications.

It is important to point out that most known results are obtained for queueing systems and networks with Poisson arrivals. Unfortunately, it has been proved

© Springer International Publishing AG 2016
V.M. Vishnevskiy et al. (Eds.): DCCN 2016, CCIS 678, pp. 110–120, 2016.
DOI: 10.1007/978-3-319-51917-3_11

that the Poisson model is suitable only for few cases of modern telecommunication streams [6] and, in general, the correlation among arrivals must be taken
into account. Therefore, many researches use more complex models of arrivals,
such as Markovian Arrival Processes (MAPs) [7] or semi-Markov processes [8].

The main contribution of this paper consists in extending previous works on
random capacity customers to the case of correlated arrivals. In more detail, the
problem statement is formally defined in Sect. 2 and in Sect. 3 the corresponding
Kolmogorov differential equations are derived. Then Sect. 4 presents the results
of the asymptotic analysis, focusing on first and second order approximations.
Finally, the main findings are summarized in Sect. 5.

2 Problem Statement

In this paper, the MMPP/GI/∞ QS with random capacity customers is studied. The arrival process is a Markov Modulated Poisson Process (MMPP), a
widely-used special case of MAP [7]. The system has an unlimited number of
servers and service times on each server are i.i.d. with distribution function
$B(x)$. All customers have a random capacity $\nu > 0$ with the probability distribution $G(y) = P\{\nu < y\}$ and the customers capacity are independent. Moreover,
we assume that service time and customers capacity are mutually independent.
After the service, customers leave the system and carry out the capacity.

Let us denote the number of customers in the system and the total customers
capacity at time t by $i(t)$ and $V(t)$, respectively. We consider two-dimensional
stochastic process $\{i(t), V(t)\}$, which is not Markovian. Therefore, we propose
the dynamic screening method for its investigation [10].

For the screened process construction, we fix some point in time T. We assume
that the customer arrived in the system at time $t < T$ creates a point in the
screened process with probability

$$S(t) = 1 - B(T - t)$$

or does not create it with probability $1 - S(t)$. We name the points occurred in
the screened process before t as customers in the screened process at time t.

Let us denote the customers number in the screened process at the moment
t by $n(t)$. Then, if at the initial moment $t_0 < T$ the system is empty, we have
the following equality at the moment T:

$$P\{i(T) = m\} = P\{n(T) = m\}.$$

Note that this method exactly determines the characteristics of the process
$V(t)$ since the screened process contains only customers which do not finish the
service at the moment T.

3 The System of Kolmogorov Differential Equations

Let us consider the three-dimensional Markovian process $\{k(t), n(t), V(t)\}$,
where $k(t)$ identifies the state of the modulating Markov chain of the MMPP

input process at time t $(1 \leq k(t) \leq K)$, which is defined through the infinitesimal generator matrix $\mathbf{Q}$ and rate matrix $\mathbf{\Lambda}$:

$$\mathbf{Q} = \begin{bmatrix} q_{11} & q_{12} & \cdots & q_{1K} \\ q_{21} & q_{22} & \cdots & q_{2K} \\ \cdots & \cdots & \cdots & \cdots \\ q_{K1} & q_{K2} & \cdots & q_{KK} \end{bmatrix}, \quad \mathbf{\Lambda} = \begin{bmatrix} \lambda_1 & 0 & \cdots & 0 \\ 0 & \lambda_2 & \cdots & 0 \\ \cdots & \cdots & \cdots & \cdots \\ 0 & 0 & \cdots & \lambda_K \end{bmatrix}.$$

Denoting the probability distribution of this process by $P(k, n, z, t) = P\{k(t) = k, n(t) = n, V(t) < z\}$, we can write the corresponding system of Kolmogorov differential equations taking into account the formula of total probability:

$$\begin{aligned} P(k, n, z, t + \Delta t) &= P(k, n, z, t)(1 - \lambda_k)(1 - q_{kk}) \\ &+ P(k, n, z, t)\lambda_k \Delta t (1 - S(t)) \\ &+ \lambda_k \Delta t S(t) \int_0^z P(k, n - 1, z - y, t) \, dG(y) \\ &+ \sum_{\nu \neq k} q_{\nu k} \Delta t P(\nu, n, z, t) + o(\Delta t), \\ &\text{for } k = 1 \ldots K, \ n = 0, 1, 2, \ldots, z > 0. \end{aligned} \tag{1}$$

From (1), we obtain the system of Kolmogorov differential equations

$$\begin{aligned} \frac{\partial P(k, n, z, t)}{\partial t} &= \lambda_k S(t) \left[\int_0^z P(k, n - 1, z - y, t) \, dG(y) \right] + \\ &\sum_{\nu} q_{\nu k} P(\nu, n, z, t) \\ &\text{for } k = 1 \ldots K, \ n = 0, 1, 2, \ldots, z > 0. \end{aligned}$$

We introduce a partial characteristic function of the form:

$$\begin{aligned} H(k, u_1, u_2, t) &= M\{\exp(ju_1 n(t) + ju_2 V(t))\} = \\ &\sum_{n=0}^{\infty} e^{ju_1 n} \int_0^{\infty} e^{ju_2 z} P(k, n, z, t) \, dz \\ &\text{for } k = 1 \ldots K, \ n = 0, 1, 2, \ldots, z > 0. \end{aligned}$$

Considering that

$$\begin{aligned} \sum_{n=0}^{\infty} e^{ju_1 n} \int_0^{\infty} e^{ju_2 z} \int_0^z P(k, n - 1, z - y, t) \, dG(y) \, dz = \\ e^{ju_1} H(k, u_1, u_2, t) G^*(u_2), \end{aligned}$$

where

$$G^*(u_2) = \int_0^{\infty} e^{ju_2 y} \, dG(y), \tag{2}$$

we can write the following system of equations:

$$\frac{\partial H(k,u_1,u_2,t)}{\partial t} = \lambda_k S(t) H(k,u_1,u_2,t) \left[e^{ju_1} G^*(u_2) - 1\right]$$
$$\text{for } k = 1 \ldots K.$$

We write this system in the form of a matrix equation

$$\frac{\partial \mathbf{H}(u_1,u_2,t)}{\partial t} = \mathbf{H}(u_1,u_2,t) \left[\mathbf{\Lambda} S(t)\left(e^{ju_1} G^*(u_2) - 1\right) + \mathbf{Q}\right] \tag{3}$$

with the initial condition

$$\mathbf{H}(u_1,u_2,t_0) = \mathbf{r}, \tag{4}$$

where

$$\mathbf{H}(u_1,u_2,t) = [H(1,u_1,u_2,t), H(2,u_1,u_2,t), \ldots, H(K,u_1,u_2,t)],$$

and

$$\mathbf{r} = [r(1), r(2), \ldots, r(K)],$$

is the row vector of the stationary distribution of the modulating Markov chain:

$$\begin{cases} \mathbf{rQ} = 0, \\ \mathbf{re} = 1, \end{cases} \tag{5}$$

$\mathbf{e}$ being a column vector with all entries equal to 1.

4   The Asymptotic Analysis Method

The exact solution of the Eq. (3) is, in general, not available, but it is possible to get asymptotic results in case of heavy loads. To this aim we will use the asymptotic analysis method under the condition of an infinitely growing arrival rate. Let us substitute $\mathbf{\Lambda} = N\mathbf{\Lambda}^1$ and $\mathbf{Q} = N\mathbf{Q}^1$ into the Eq. (3), where N is some parameter which will be used for the asymptotic analysis ($N \to \infty$ in theoretical studies).

Then, the Eq. (3) takes the form

$$\frac{1}{N}\frac{\partial \mathbf{H}(u_1,u_2,t)}{\partial t} = \mathbf{H}(u_1,u_2,t)\left[\mathbf{\Lambda}^1 S(t)\left(e^{ju_1} G^*(u_2) - 1\right) + \mathbf{Q}^1\right] \tag{6}$$

with the initial condition

$$\mathbf{H}(u_1,u_2,t_0) = \mathbf{r}. \tag{7}$$

4.1   The First-Order Asymptotic Analysis

The main result is summarized by the following lemma.

Lemma. *The first-order asymptotic characteristic function of the probability distribution of the process* $\{k(t), n(t), V(t)\}$ *has the form*

$$\mathbf{H}(u_1, u_2, t) = \mathbf{r} \exp\left\{ N\lambda \left[ju_1 + ju_2 a_1 \right] \int_{t_0}^{t} S(\tau) \, d\tau \right\},$$

where the row vector $\mathbf{r}$ *is defined by the system of linear equations* (5), λ *denotes the average rate*

$$\lambda = \mathbf{r}\Lambda\mathbf{e}$$

and a_1 *is the mean of the random variable defining the customer capacity*

$$a_1 = \int_0^\infty y \, dG(y).$$

Proof. Let us perform the substitutions

$$\varepsilon = \frac{1}{N}, u_1 = \varepsilon w_1, u_2 = \varepsilon w_2, \mathbf{H}(u_1, u_2, t) = \mathbf{F_1}(w_1, w_2, t, \varepsilon) \tag{8}$$

in the expressions (5) and (6).

Then the problem (5)–(6) takes the form

$$\varepsilon \frac{\partial \mathbf{F_1}(w_1, w_2, t, \varepsilon)}{\partial t} = \mathbf{F_1}(w_1, w_2, t, \varepsilon) \left[\mathbf{\Lambda^1} S(t) \left(e^{j\varepsilon w_1} G^*(\varepsilon w_2) - 1 \right) + \mathbf{Q^1} \right] \tag{9}$$

with the initial condition

$$\mathbf{F_1}(w_1, w_2, t, \varepsilon) = \mathbf{r}. \tag{10}$$

Let us find the asymptotic solution (where $\varepsilon \to 0$) of the problem (8)–(9), i.e. the $\mathbf{F_1}(w_1, w_2, t) = \lim_{\varepsilon \to 0} \mathbf{F_1}(w_1, w_2, t, \varepsilon)$.

Step 1. Letting $\varepsilon \to 0$ in (9), we obtain

$$\mathbf{F_1}(w_1, w_2, t) \mathbf{Q^1} = 0.$$

Comparing this equation with the first one in (5), we can conclude that $\mathbf{F_1}(w_1, w_2, t)$ can be expressed as

$$\mathbf{F_1}(w_1, w_2, t) = \mathbf{r}\Phi_1(w_1, w_2, t), \tag{11}$$

where $\Phi_1(w_1, w_2, t)$ is some scalar function which satisfies the condition

$$\Phi_1(w_1, w_2, t_0) = 1.$$

Step 2. Let us multiply (9) by vector $\mathbf{e}$, substitute (11), divide the results by ε and perform the asymptotic transition $\varepsilon \to 0$. Then, taking into account that $\mathbf{Q}^1 \mathbf{e} = 0$ and $\mathbf{r}\mathbf{e} = 1$, we obtain the following differential equation for the function $\Phi_1(w_1, w_2, t)$

$$\frac{\partial \Phi_1(w_1, w_2, t)}{\partial t} = \Phi_1(w_1, w_2, t)[\lambda S(t)(jw_1 + jw_2 a_1)]. \tag{12}$$

The solution of (12) with the initial condition gives

$$\Phi_1(w_1, w_2, t) = \exp\left\{\lambda(jw_1 + jw_2 a_1) \int_{t_0}^{t} S(\tau)\, d\tau\right\}.$$

Substituting this expression into (11), we obtain

$$\mathbf{F_1}(w_1, w_2, t) = \mathbf{r}\exp\left\{\lambda(jw_1 + jw_2 a_1) \int_{t_0}^{t} S(\tau)\, d\tau\right\}.$$

Using substitutions (8), we can write the asymptotic (as $\varepsilon \to 0$) equality:

$$\mathbf{H}(u_1, u_2, t) = \mathbf{F_1}(w_1, w_2, t, \varepsilon) \approx \mathbf{F_1}(w_1, w_2, t) = \mathbf{r}\Phi_1(w_1, w_2, t)$$

$$= \mathbf{r}\exp\left\{\lambda\left[j\frac{u_1}{\varepsilon} + j\frac{u_2}{\varepsilon}a_1\right] \int_{t_0}^{t} S(\tau)\, d\tau\right\}$$

$$= \mathbf{r}\exp\left\{N\lambda[ju_1 + ju_2 a_1] \int_{t_0}^{t} S(\tau)\, d\tau\right\}.$$

The proof is complete.

Corollary. *When $t = T$ we obtain the characteristic function of the process $\{i(t), V(t)\}$ in the steady state regime*

$$H(u_1, u_2, t) = \exp\{N\lambda b_1[ju_1 + ju_2 a_1]\},$$

where

$$b_1 = \int_{-\infty}^{T} S(\tau)\, d\tau = \int_{-\infty}^{T} (1 - B(T - \tau))\, d\tau = \int_{0}^{\infty} (1 - B(\tau))\, d\tau$$

denotes the mean service time.

4.2   The Second-Order Asymptotic Analysis

The main result is summarized by the following theorem.

Theorem. *The second-order asymptotic characteristic function of the probability distribution of the process* $\{k(t), n(t), V(t)\}$ *has the form*

$$
\mathbf{H}\left(u_1, u_2, t\right) = \mathbf{r} \exp \left\{ N\lambda \left(ju_1 + ju_2a_1\right) \int\limits_{t_0}^{t} S\left(\tau\right) d\tau + \right.
$$

$$
\frac{(ju_1)^2}{2} \left(N\lambda \int\limits_{t_0}^{t} S\left(\tau\right) d\tau + N\kappa \int\limits_{t_0}^{t} S^2\left(\tau\right) d\tau \right) +
$$

$$
\frac{(ju_2)^2}{2} \left(N\lambda a_2 \int\limits_{t_0}^{t} S\left(\tau\right) d\tau + N\kappa a_1^2 \int\limits_{t_0}^{t} S^2\left(\tau\right) d\tau \right) +
$$

$$
\left. j^2 u_1 u_2 \left(N\lambda a_1 \int\limits_{t_0}^{t} S\left(\tau\right) d\tau + N\kappa a_1 \int\limits_{t_0}^{t} S^2\left(\tau\right) d\tau \right) \right\},
$$

where

$$
\kappa = 2\mathbf{g}\left(\mathbf{\Lambda}^1 - \lambda\mathbf{I}\right)\mathbf{e},
$$

and the row vector $\mathbf{g}$ *satisfies the linear matrix system*

$$
\mathbf{g}\mathbf{Q}^1 = \mathbf{r}\left(\lambda\mathbf{I} - \mathbf{\Lambda}^1\right),
$$

$$
\mathbf{g}\mathbf{e} = 1.
$$

Proof. Denote by $\mathbf{H_2}\left(u_1, u_2, t\right)$ a multi-dimensional function that satisfies the equation

$$
\mathbf{H}\left(u_1, u_2, t\right) = \mathbf{H_2}\left(u_1, u_2, t\right) \exp \left\{ N\lambda \left(ju_1 + ju_2a_1\right) \int\limits_{t_0}^{t} S\left(\tau\right) d\tau \right\}. \tag{13}
$$

Substituting this expression into (6) and (7), we obtain the following problem:

$$
\frac{1}{N} \frac{\partial \mathbf{H_2}\left(u_1, u_2, t\right)}{\partial t} + \lambda \left(ju_1 + ju_2a_1\right) S\left(t\right) \mathbf{H_2}\left(u_1, u_2, t\right) \tag{14}
$$

$$
= \mathbf{H_2}\left(u_1, u_2, t\right) \left[\mathbf{\Lambda}^1 S\left(t\right)\left(e^{ju_1} G^*\left(u_2\right) - 1\right) + \mathbf{Q}^1\right],
$$

with the initial condition

$$
\mathbf{H_2}\left(u_1, u_2, t_0\right) = \mathbf{r}. \tag{15}
$$

Let us perform the following substitutions

$$
\varepsilon^2 = \frac{1}{N}, u_1 = \varepsilon w_1, u_2 = \varepsilon w_2, \mathbf{H_2}\left(u_1, u_2, t\right) = \mathbf{F_2}\left(w_1, w_2, t, \varepsilon\right). \tag{16}
$$

Using these notations the problem (14)–(15) can be rewritten in the form

$$
\varepsilon^2 \frac{\partial \mathbf{F_2}\left(w_1, w_2, t, \varepsilon\right)}{\partial t} + \mathbf{F_2}\left(w_1, w_2, t, \varepsilon\right) \lambda \left(j\varepsilon w_1 + j\varepsilon w_2 a_1\right) S\left(t\right) \tag{17}
$$

$$
= \mathbf{F_2}\left(w_1, w_2, t, \varepsilon\right) \left[\mathbf{\Lambda}^1 S\left(t\right)\left(e^{j\varepsilon w_1} G^*\left(\varepsilon w_2\right) - 1\right) + \mathbf{Q}^1\right],
$$

with the initial condition

$$\mathbf{F_2}(w_1, w_2, t_0, \varepsilon) = \mathbf{r}. \tag{18}$$

Let us find the asymptotic solution (as $\varepsilon \to 0$) of this problem, i.e. the function $\mathbf{F_2}(w_1, w_2, t) = \lim_{\varepsilon \to 0} \mathbf{F_2}(w_1, w_2, t, \varepsilon)$.

Step 1. Letting $\varepsilon \to 0$ in (17)–(18), we obtain the following system of equations:

$$\begin{cases} \mathbf{F_2}(w_1, w_2, t)\,\mathbf{Q^1} = \mathbf{0}, \\ \mathbf{F_2}(w_1, w_2, t_0) = \mathbf{r}, \end{cases}$$

Then, using (4), we can write

$$\mathbf{F_2}(w_1, w_2, t) = \mathbf{r}\Phi_2(w_1, w_2, t), \tag{19}$$

where $\Phi_2(w_1, w_2, t)$ is some scalar function which satisfies the condition

$$\Phi_2(w_1, w_2, t_0) = 1.$$

Step 2. Using (19), the function $\mathbf{F_2}(w_1, w_2, t)$ can be represented in the expansion form

$$\begin{aligned} \mathbf{F_2}(w_1, w_2, t, \varepsilon) = \\ \Phi_2(w_1, w_2, t)\,[\mathbf{r} + \mathbf{g}\,(j\varepsilon w_1 + j\varepsilon w_2 a_1)\,S(t)] + \mathbf{O}(\varepsilon^2), \end{aligned} \tag{20}$$

where $\mathbf{g}$ is some row vector which satisfying the condition $\mathbf{ge} = 1$ and $\mathbf{O}(\varepsilon^2)$ is row vector whose elements are infinitesimals of the same order as ε^2.

Let us use the substitution (20) and the Taylor-Maclaurin expansions

$$e^{j\varepsilon w_1} = 1 + j\varepsilon w_1 + O(\varepsilon^2), \, e^{j\varepsilon w_2} = 1 + j\varepsilon w_2 + O(\varepsilon^2)$$

in (17). Considering the (2), we perform in the obtained equality of the limiting transition $\varepsilon \to 0$, we obtain matrix equation for the vector $\mathbf{g}$

$$\mathbf{gQ^1} = \mathbf{r}\left(\lambda\mathbf{I} - \mathbf{\Lambda^1}\right),$$

where $\mathbf{I}$ is diagonal unit matrix.

Step 3. We multiply the (17) by the $\mathbf{e}$, using (20) and the second-order expansions

$$e^{j\varepsilon w_1} = 1 + j\varepsilon w_1 + \frac{(j\varepsilon w_1)^2}{2} + O(\varepsilon^2),$$

$$e^{j\varepsilon w_2} = 1 + j\varepsilon w_2 + \frac{(j\varepsilon w_2)^2}{2} + O(\varepsilon^2).$$

As a result of simple transformations with the notation

$$\kappa = 2\mathbf{g}\left(\mathbf{\Lambda^1} - \lambda\mathbf{I}\right)\mathbf{e},$$

we obtain the following differential equation for the function $\Phi_2(w_1, w_2, t)$

$$\frac{\partial \Phi_2(w_1, w_2, t)}{\partial t} = \Phi_2(w_1, w_2, t) \left\{ \frac{(jw_1)^2}{2} \left(\lambda S(t) + \kappa S^2(t) \right) + \right.$$

$$\left. \frac{(jw_2)^2}{2} \left(\lambda a_2 S(t) + \kappa a_1^2 S^2(t) \right) + j^2 w_1 w_2 \left(\lambda a_1 S(t) + \kappa a_1 S^2(t) \right) \right\}, \qquad (21)$$

where $a_2 = \int\limits_0^\infty y^2 dG(y)$ is the second moment of the random customer capacity ν.

The solution of the latter equation with the available initial condition $\Phi_2(w_1, w_2, t_0) = 1$ gives the expression $\Phi_2(w_1, w_2, t)$

$$\Phi_2(w_1, w_2, t) = \exp \left\{ \frac{(jw_1)^2}{2} \left(\lambda \int\limits_{t_0}^t S(\tau) d\tau + \kappa \int\limits_{t_0}^t S^2(\tau) d\tau \right) + \right.$$

$$\frac{(jw_2)^2}{2} \left(\lambda a_2 \int\limits_{t_0}^t S(\tau) d\tau + \kappa a_1^2 \int\limits_{t_0}^t S^2(\tau) d\tau \right) +$$

$$\left. j^2 w_1 w_2 \left(\lambda a_1 \int\limits_{t_0}^t S(\tau) d\tau + \kappa a_1 \int\limits_{t_0}^t S^2(\tau) d\tau \right) \right\},$$

and substituting in (19) we obtain

$$\mathbf{F_2}(w_1, w_2, t) = \mathbf{r} \exp \left\{ \frac{(jw_1)^2}{2} \left(\lambda \int\limits_{t_0}^t S(\tau) d\tau + \kappa \int\limits_{t_0}^t S^2(\tau) d\tau \right) + \right.$$

$$\frac{(jw_2)^2}{2} \left(\lambda a_2 \int\limits_{t_0}^t S(\tau) d\tau + \kappa a_1^2 \int\limits_{t_0}^t S^2(\tau) d\tau \right) + \qquad (22)$$

$$\left. j^2 w_1 w_2 \left(\lambda a_1 \int\limits_{t_0}^t S(\tau) d\tau + \kappa a_1 \int\limits_{t_0}^t S^2(\tau) d\tau \right) \right\}.$$

Performing in (22) the substitutions inverse to (16) and (13), we obtain the following expression for the asymptotic characteristic function of the number of customers of screened process and total capacity of customers at the moment t:

$$\mathbf{H}(u_1, u_2, t) = \mathbf{r} \exp \left\{ N\lambda (ju_1 + ju_2 a_1) \int\limits_{t_0}^t S(\tau) d\tau + \right.$$

$$\frac{(ju_1)^2}{2} \left(N\lambda \int_{t_0}^{t} S(\tau)\, d\tau + N\kappa \int_{t_0}^{t} S^2(\tau)\, d\tau \right) +$$

$$\frac{(ju_2)^2}{2} \left(N\lambda a_2 \int_{t_0}^{t} S(\tau)\, d\tau + N\kappa a_1^2 \int_{t_0}^{t} S^2(\tau)\, d\tau \right) +$$

$$j^2 u_1 u_2 \left(N\lambda a_1 \int_{t_0}^{t} S(\tau)\, d\tau + N\kappa a_1 \int_{t_0}^{t} S^2(\tau)\, d\tau \right) \right\},$$

where

$$\kappa = 2\mathbf{g} \left(\mathbf{\Lambda}^1 - \lambda \mathbf{I} \right) \mathbf{e},$$

and the row vector $\mathbf{g}$ satisfies the linear matrix system

$$\mathbf{g}\mathbf{Q}^1 = \mathbf{r} \left(\lambda \mathbf{I} - \mathbf{\Lambda}^1 \right),$$
$$\mathbf{g}\mathbf{e} = 1.$$

The proof is complete.

Corollary 1. *When* $t = T$ *we obtain the characteristic function of the process* $i(t), V(t)$ *in the steady state regime*

$$H(u_1, u_2, t) = \exp\left\{ N\lambda (ju_1 + ju_2 a_1) b_1 + \frac{(ju_1)^2}{2} (N\lambda b_1 + N\kappa b_2) + \right.$$

$$\left. \frac{(ju_2)^2}{2} (N\lambda a_2 b_1 + N a_1^2 \kappa b_2) + j^2 u_1 u_2 (N\lambda a_1 b_1 + N\kappa a_1 b_2) \right\}, \qquad (23)$$

where

$$b_2 = \int_{-\infty}^{T} S^2(\tau)\, d\tau.$$

From the form of function (23) it is clear that the two-dimensional process $i(t), V(t)$ is asymptotically Gaussian with the vector of mathematical expectations

$$\mathbf{a} = [N\lambda b_1, N\lambda a_1 b_1]$$

and the covariance matrix

$$\mathbf{A} = \begin{bmatrix} \sigma_1^2 & r\sigma_1\sigma_2 \\ r\sigma_1\sigma_2 & \sigma_2^2 \end{bmatrix} = \begin{bmatrix} N\lambda b_1 + N\kappa b_2 & N\lambda a_1 b_1 + N\kappa a_1 b_2 \\ N\lambda a_1 b_1 + N\kappa a_1 b_2 & N\lambda a_2 b_1 + N\kappa a_1^2 b_2 \end{bmatrix}.$$

Corollary 2. *The asymptotic characteristic function of the total customer capacity in the steady-state regime is given by a Gaussian characteristic function*

$$H(u, t) = \exp\left\{ juN\lambda a_1 b_1 + \frac{(ju)^2}{2} (N\lambda a_2 b_1 + N a_1^2 \kappa b_2) \right\},$$

with parameters $a = N\lambda a_1 b_1$ *and* $\sigma^2 = N\lambda a_2 b_1 + N a_1^2 \kappa b_2$.

5   Conclusions

In the paper, a queueing system with random customers capacity and service time independent of its capacity is considered in case of correlated arrivals, described by an MMPP process. For such system, the total customers capacity is derived by using the asymptotic analysis method in case of heavy loads. It is obtained that the stationary probability distribution of total capacity can be approximated by a Gaussian distribution and the parameters of the approximation are derived in the paper.

Acknowledgments. The work is performed under the state order of the Ministry of Education and Science of the Russian Federation (No. 1.511.2014/K).

References

1. Tikhonenko, O.M.: Destricted capacity queueing systems. Determination of their characteristics. Autom. Remote Control **58**(6), 969–972 (1997)
2. Tikhonenko, O.M.: Computer Systems Probability Analysis. Akademicka Oficyna Wydawnicza EXIT, Warsaw (2006). (in Polish)
3. Romm, E.L., Skitovich, V.V.: On certain generalization of problem of Erlang. Avtomat. i Telemekh. **6**, 164–168 (1971)
4. Tikhonenko, O.M.: Systems for servicing requests of random length with restrictions. Avtomat. i Telemekh. **10**, 126–134 (1991)
5. Tikhonenko, O.M.: Queuing systems with processor sharing and limited resources. Autom. Remote Control **5**, 803–815 (2010)
6. Heyman, D.P., Lucantoni, D.: Modelling multiple IP traffic streams with rate limits. IEEE/ACM Trans. Networking **11**, 948–958 (2003)
7. Chakravarthy, S.R.: Markovian arrival processes. In: Wiley Encyclopedia of Operations Research and Management Science (2010)
8. Moiseev, A., Nazarov, A.: Asymptotic analysis of the infinite-server queueing system with high-rate semi-Markov arrivals. In: IEEE International Congress on Ultra Modern Telecommunications and Control Systems (ICUMT 2014), pp. 507–513. IEEE Press, New York (2014)
9. Naumov, V., Samuoylov, K., Sopin, E., Yarkina, N., Andreev, S., Samuylov, A.: LTE performance analysis using queuing systems with finite resources and random requirements. In: Proceedings of 8th International Congress on Ultra Modern Telecommunications and Control Systems and Workshops, pp. 100–103 (2015)
10. Moiseev, A., Nazarov, A.: Queueing network MAP-$(GI-\infty)^K$ with high-rate arrivals. Eur. J. Oper. Res. **254**, 161–168 (2016)

On the Queue Length in the Discrete Cyclic-Waiting System of $Geo/G/1$ Type

Laszlo Lakatos[(✉)]

Eotvos Lorand University, Budapest, Hungary
lakatos1948@freemail.hu

Abstract. We consider a discrete time queueing system with geometrically distributed interarrival and general service times, with FCFS service discipline. The service of a customer is started at the moment of arrival (in case of free system) or at moments differing from it by the multiples of a given cycle time T (in case of occupied server or waiting queue). Earlier we investigated such system from the viewpoint of waiting time, actually we deal with the number of present customers. The functioning is described by means of an embedded Markov chain considering the system at moments just before starting the services of customers. We find the transition probabilities, the generating function of ergodic distribution and the stability condition. The model may be used to describe the transmission of optical signals.

Keywords: Queue length · Discrete cyclic-waiting system · $Geo/G/1$

1 Introduction

This paper continues the investigation of a single-server queueing system where an entering customer might be accepted for service at the moment of arrival or at moments differing from it by the multiples of a given cycle time T. As described in [4] such problem was motivated by the transmission of optical signals: optical signals enter a node and they should be transmitted according to the FCFS rule. The information cannot be stored, if it cannot be served at once is sent to a delay line and returns to the node after having passed it. So the signal can be transmitted at the moment of its arrival or at moments that differ from it by the multiples of time required to pass the delay line. The original problem had been raised in connection with the landing of airplanes, later it appeared to be an exact model for the transmission of optical signals where because of the lack of optical RAM the fiber delay lines are used.

First this system was considered from the viewpoint of number of present customers in the case of Poisson arrivals and exponentially distributed service time distribution [3]. By using Koba's results [1,2] in [4] we investigated the distribution of waiting time for the continuous time model. [5] solved this problem for the discrete time case if the service time had geometrical distribution. Finally, [6] considered the waiting time problem in the case of general discrete service time distribution.

© Springer International Publishing AG 2016
V.M. Vishnevskiy et al. (Eds.): DCCN 2016, CCIS 678, pp. 121–131, 2016.
DOI: 10.1007/978-3-319-51917-3_12

We have also showed that these models give possibility for the numerical optimization of cycle length.

In this paper we investigate the distribution of queue length for the discrete system with geometrical interarrival and general service time distribution.

2   The Theorem

We investigate a service system where the service may start at the moment of arrival (if the system is free) or at moments differing from it by the multiples of a given cycle time T (in the case of busy server or waiting queue). The service is realized according to the FCFS discipline. So, the service process is not continuous: during the "busy period" there are idle intervals required to reach the starting position, during them there is no real service.

Let the service of the nth customer begin at t_n, and let us consider the number of customers at moment just before the service begins. Then the number of customers is determined by the recursive formula

$$N_{t_{n+1}-0} = \begin{cases} \Delta_n - 1, & \text{if } N_{t_n-0} = 0, \\ \\ N_{t_n-0} - 1 + \Delta_n, & \text{if } N_{t_n-0} > 0, \end{cases}$$

where Δ_n is the number of customers arriving at the system for $[t_n, t_{n+1})$. In [3] we showed that these values form a Markov chain.

Theorem. *Let us consider a discrete queueing system in which the interarrival time has geometrical distribution with parameter r, the service time has general distribution with probabilities q_i $(i = 1, 2, \ldots)$. The service of a customer may start upon arrival or (in case of busy server or waiting queue) at moments differing from it by the multiples of a given cycle time T (equal to n time units) according to the FCFS discipline. Let us define an embedded Markov chain whose states correspond to the number of customers in the system at moments $t_k - 0$, where t_k is the moment of beginning of service of the k-th one. The matrix of transition probabilities has the form*

$$\begin{bmatrix} a_0 & a_1 & a_2 & a_3 & \ldots \\ a_0 & a_1 & a_2 & a_3 & \ldots \\ 0 & b_0 & b_1 & b_2 & \ldots \\ 0 & 0 & b_0 & b_1 & \ldots \\ \vdots & \vdots & \vdots & \vdots & \ddots \end{bmatrix} \tag{1}$$

its elements are determined by the generating functions

$$A(z) = \sum_{i=0}^{\infty} a_i z^i = Q_1 + z\frac{r}{1-r}Q_1 + z\sum_{k=1}^{\infty}(1 - r + rz)^{kn} \tag{2}$$

$$\times \left\{ \sum_{i=(k-1)n+2}^{kn+1} q_i + \sum_{i=kn+2}^{\infty} q_i(1-r)^{i-kn-1} - \sum_{i=(k-1)n+2}^{\infty} q_i(1-r)^{i-(k-1)n-1} \right\},$$

$$Q_k = \sum_{i=k}^{\infty} q_i (1-r)^i;$$

$$B(z) = \sum_{i=0}^{\infty} b_i z^i = \sum_{k=0}^{\infty} \sum_{j=1}^{n} q_{kn+j} (1-r+rz)^{kn+j}$$

$$\times \left\{ \frac{r}{1-(1-r)^n} \frac{1-(1-r)^{j-1}(1-r+rz)^{j-1}}{1-(1-r)(1-r+rz)} (1-r+rz)^{n-j+1} \right.$$

$$\left. + \frac{r(1-r)^{j-1}}{1-(1-r)^n} \frac{1-(1-r)^{n-j+1}(1-r+rz)^{n-j+1}}{1-(1-r)(1-r+rz)} \right\}. \tag{3}$$

The generating function of ergodic distribution $P(z) = \sum_{i=0}^{\infty} p_i z^i$ *has the form*

$$P(z) = \frac{p_0[zA(z) - B(z)] + p_1 z[A(z) - B(z)]}{z - B(z)}, \tag{4}$$

where

$$p_1 = \frac{1 - a_0}{a_0} p_0,$$

$$p_0 = \frac{a_0[1 - B'(1)]}{a_0 + A'(1) - B'(1)}. \tag{5}$$

The ergodicity condition is

$$\sum_{i=1}^{\infty} q_i \left\lceil \frac{i}{n} \right\rceil < \frac{1}{1 - (1-r)^n} \sum_{i=1}^{\infty} q_i (1-r)^{i-1 \ (\mathrm{mod}\ n)}. \tag{6}$$

3   The Proof of Theorem

The matrix of transition probabilities is given in the matrix (1), the generating functions of transition probabilities are found in the following section. Denote the ergodic probabilities by p_i ($i = 0, 1, \ldots$) and introduce the generating function $P(z) = \sum_{i=0}^{\infty} p_i z^i$. According to the theory of Markov chains we have

$$p_j = p_0 a_j + p_1 a_j + \sum_{i=2}^{j+1} p_i b_{j-i+1} \qquad (j \geq 1), \tag{7}$$

$$p_0 = p_0 a_0 + p_1 a_0. \tag{8}$$

From (7) and (8) one can obtain the expression

$$P(z) = \frac{p_0 [zA(z) - B(z)] + p_1 z [A(z) - B(z)]}{z - B(z)}.$$

This expression includes two unknown probabilities p_0 and p_1 from the desired distribution, by (8) p_1 can be expressed by p_0,

$$p_1 = \frac{1 - a_0}{a_0} p_0,$$

and p_0 can be found from the condition $P(1) = 1$, i.e.

$$p_0 = \frac{a_0[1 - B'(1)]}{a_0 + A'(1) - B'(1)}.$$

By using the corresponding (9) and (10) values we obtain

$$a_0 + A'(1) - B'(1)$$

$$= \frac{1}{1-(1-r)^n} \left\{ -\sum_{i=1}^{\infty} q_i(1-r)^{i-1 \,(\mathrm{mod}\ n)}[1 - (1-r)^{\lceil \frac{i}{n} \rceil n}] \right.$$

$$\left. + \sum_{i=1}^{\infty} q_i(1-r)^{i-1 \,(\mathrm{mod}\ n)} \right\}$$

$$= \frac{1}{1-(1-r)^n} \sum_{i=1}^{\infty} q_i(1-r)^{i-1 \,(\mathrm{mod}\ n)}(1-r)^{\lceil \frac{i}{n} \rceil n} > 0,$$

consequently the numerator must be positive, too; so the condition

$$1 - B'(1) > 0$$

must be fulfilled. This leads to the ergodicity condition $B'(1) < 1$, i.e.

$$\sum_{k=0}^{\infty}\sum_{j=1}^{n} q_{kn+j}[1 + (k+1)nr] - \frac{nr}{1 - (1-r)^n}\sum_{k=0}^{\infty}\sum_{j=1}^{n} q_{kn+j}(1-r)^{j-1} < 1,$$

which can be written in the form (6).

4   The Generating Functions of Transition Probabilities

Concerning the transition probabilities we have to distinguish two cases: at the moment when the service of a customer begins the next one is present or not. First we find the generating function $A(z)$ corresponding to the case when the next customer is not there yet, then we find the generating function $B(z)$ for the case when the next customer is present, too.

4.1   The Generating Function $A(z)$

This possibility appears at the states 0 and 1. Assume that the service time of first customer is equal to u, the second customer appears v time after stating its service. The probability of event $\{u - v = \ell\}$ is

$$P\{u - v = \ell\} = \sum_{k=\ell+1}^{\infty} q_k(1-r)^{k-\ell-1}r \qquad (\ell = 1, 2, \ldots).$$

We are interested in the number of customers appearing during intervals whose lengths are the multiples of n, i.e. $\ell \in [(i-1)n+1, in]$, the generating functions are represented by the tables, if ℓ changes from 1 till n (we will not write the factors $r(1-r+rz)^n$)

$$q_2 \; q_3(1-r) \; q_4(1-r)^2 \; \ldots \; q_n(1-r)^{n-2} \; q_{n+1}(1-r)^{n-1} \; q_{n+2}(1-r)^n \quad \ldots$$
$$q_3 \qquad q_4(1-r) \; \ldots \; q_n(1-r)^{n-3} \; q_{n+1}(1-r)^{n-2} \; q_{n+2}(1-r)^{n-1} \ldots$$
$$q_4 \qquad\qquad \ldots \; q_n(1-r)^{n-4} \; q_{n+1}(1-r)^{n-3} \; q_{n+2}(1-r)^{n-2} \ldots$$

$$\vdots \qquad\qquad \vdots \qquad\qquad \vdots$$

$$q_n(1-r) \qquad q_{n+1}(1-r)^2 \qquad q_{n+2}(1-r)^3 \quad \ldots$$
$$q_n \qquad\quad q_{n+1}(1-r) \qquad q_{n+2}(1-r)^2 \quad \ldots$$
$$q_{n+1} \qquad\quad q_{n+2}(1-r) \quad \ldots$$

for the following columns it is continued as

$$q_{n+2}(1-r)^n \quad q_{n+3}(1-r)^{n+1} \; \ldots \; q_{2n}(1-r)^{2n-2} \; q_{2n+1}(1-r)^{2n-1} \ldots$$
$$q_{n+2}(1-r)^{n-1} \; q_{n+3}(1-r)^n \quad \ldots \; q_{2n}(1-r)^{2n-3} \; q_{2n+1}(1-r)^{2n-2} \ldots$$
$$q_{n+2}(1-r)^{n-2} \; q_{n+3}(1-r)^{n-1} \; \ldots \; q_{2n}(1-r)^{2n-4} \; q_{2n+1}(1-r)^{2n-3} \ldots$$

$$\vdots \qquad\qquad \vdots \qquad\qquad \vdots \qquad\qquad \vdots$$

$$q_{n+2}(1-r)^3 \quad q_{n+3}(1-r)^4 \quad \ldots \; q_{2n}(1-r)^{n+1} \quad q_{2n+1}(1-r)^{n+2} \quad \ldots$$
$$q_{n+2}(1-r)^2 \quad q_{n+3}(1-r)^3 \quad \ldots \; q_{2n}(1-r)^n \qquad q_{2n+1}(1-r)^{n+1} \quad \ldots$$
$$q_{n+2}(1-r) \qquad q_{n+3}(1-r)^2 \quad \ldots \; q_{2n}(1-r)^{n-1} \quad q_{2n+1}(1-r)^n \qquad \ldots$$

and

$$q_{2n+2}(1-r)^{2n} \quad q_{2n+3}(1-r)^{2n+1} \; \ldots \; q_{3n}(1-r)^{3n-2} \; q_{3n+1}(1-r)^{3n-1} \ldots$$
$$q_{2n+2}(1-r)^{2n-1} \; q_{2n+3}(1-r)^{2n} \quad \ldots \; q_{3n}(1-r)^{3n-3} \; q_{3n+1}(1-r)^{3n-2} \ldots$$
$$q_{2n+2}(1-r)^{2n-2} \; q_{2n+3}(1-r)^{2n-1} \; \ldots \; q_{3n}(1-r)^{3n-4} \; q_{3n+1}(1-r)^{3n-3} \ldots$$

$$\vdots \qquad\qquad \vdots \qquad\qquad \vdots \qquad\qquad \vdots$$

$$q_{2n+2}(1-r)^{n+3} \quad q_{2n+3}(1-r)^{n+4} \; \ldots \; q_{3n}(1-r)^{2n+1} \; q_{3n+1}(1-r)^{2n+2} \ldots$$
$$q_{2n+2}(1-r)^{n+2} \quad q_{2n+3}(1-r)^{n+3} \; \ldots \; q_{3n}(1-r)^{2n} \quad q_{3n+1}(1-r)^{2n+1} \ldots$$
$$q_{2n+2}(1-r)^{n+1} \quad q_{2n+3}(1-r)^{n+2} \; \ldots \; q_{3n}(1-r)^{2n-1} \; q_{3n+1}(1-r)^{2n} \quad \ldots$$

etc., if ℓ changes from $n+1$ till $2n$ (we omit the factor $r(1-r+rz)^{2n}$)

$$q_{n+2} \; q_{n+3}(1-r) \; q_{n+4}(1-r)^2 \; \ldots \; q_{2n}(1-r)^{n-2} \; q_{2n+1}(1-r)^{n-1} \ldots$$
$$q_{n+3} \qquad q_{n+4}(1-r) \; \ldots \; q_{2n}(1-r)^{n-3} \; q_{2n+1}(1-r)^{n-2} \ldots$$
$$q_{n+4} \qquad\qquad \ldots \; q_{2n}(1-r)^{n-4} \; q_{2n+1}(1-r)^{n-3} \ldots$$

$$\vdots \qquad\qquad \vdots$$

$$q_{2n}(1-r) \qquad q_{2n+1}(1-r)^2 \quad \ldots$$
$$q_{2n} \qquad\quad q_{2n+1}(1-r) \quad \ldots$$
$$q_{2n+1} \qquad \ldots$$

for the following columns it is continued as

$$
\begin{array}{llll}
q_{2n+2}(1-r)^n & q_{2n+3}(1-r)^{n+1} & \ldots q_{3n}(1-r)^{2n-2} & q_{3n+1}(1-r)^{2n-1} \ldots \\
q_{2n+2}(1-r)^{n-1} & q_{2n+3}(1-r)^n & \ldots q_{3n}(1-r)^{2n-3} & q_{3n+1}(1-r)^{2n-2} \ldots \\
q_{2n+2}(1-r)^{n-2} & q_{2n+3}(1-r)^{n-1} & \ldots q_{3n}(1-r)^{2n-4} & q_{3n+1}(1-r)^{2n-3} \ldots \\
\vdots & \vdots & \qquad\vdots & \qquad\vdots \\
q_{2n+2}(1-r)^3 & q_{2n+3}(1-r)^4 & \ldots q_{3n}(1-r)^{n+1} & q_{3n+1}(1-r)^{n+2} \ldots \\
q_{2n+2}(1-r)^2 & q_{2n+3}(1-r)^3 & \ldots q_{3n}(1-r)^n & q_{3n+1}(1-r)^{n+1} \ldots \\
q_{2n+2}(1-r) & q_{2n+3}(1-r)^2 & \ldots q_{3n}(1-r)^{n-1} & q_{3n+1}(1-r)^n \quad \ldots
\end{array}
$$

etc., if ℓ changes from $2n+1$ till $3n$ (we omit the factor $r(1-r+rz)^{3n}$)

$$
\begin{array}{llll}
q_{2n+2} \; q_{2n+3}(1-r) \; q_{2n+4}(1-r)^2 & \ldots q_{3n}(1-r)^{n-2} & q_{3n+1}(1-r)^{n-1} \ldots \\
q_{2n+3} & q_{2n+4}(1-r) & \ldots q_{3n}(1-r)^{n-3} & q_{3n+1}(1-r)^{n-2} \ldots \\
& q_{2n+4} & \ldots q_{3n}(1-r)^{n-4} & q_{3n+1}(1-r)^{n-3} \ldots \\
& & \qquad\vdots & \qquad\vdots \\
& & q_{3n}(1-r) & q_{3n+1}(1-r)^2 \quad \ldots \\
& & q_{3n} & q_{3n+1}(1-r) \quad \ldots \\
& & & q_{3n+1} \quad \ldots
\end{array}
$$

etc.

Summing up the elements of columns for $r(1-r+rz)^n$, $r(1-r+rz)^{2n}$, $r(1-r+rz)^{3n}$, ... we get the coefficients

$$
\sum_{i=2}^{n+1} q_i \frac{1-(1-r)^{i-1}}{1-(1-r)} + \sum_{i=n+2}^{\infty} q_i(1-r)^{i-n-1}\frac{1-(1-r)^n}{1-(1-r)},
$$

$$
\sum_{i=n+2}^{2n+1} q_i \frac{1-(1-r)^{i-n-1}}{1-(1-r)} + \sum_{i=2n+2}^{\infty} q_i(1-r)^{i-2n-1}\frac{1-(1-r)^n}{1-(1-r)},
$$

$$
\sum_{i=2n+2}^{3n+1} q_i \frac{1-(1-r)^{i-2n-1}}{1-(1-r)} + \sum_{i=3n+2}^{\infty} q_i(1-r)^{i-3n-1}\frac{1-(1-r)^n}{1-(1-r)}, \ldots
$$

and, in general, for $r(1-r+rz)^{kn}$

$$
\sum_{i=(k-1)+2}^{kn+1} q_i \frac{1-(1-r)^{i-(k-1)n-1}}{1-(1-r)} + \sum_{i=kn+2}^{\infty} q_i(1-r)^{i-kn-1}\frac{1-(1-r)^n}{1-(1-r)},
$$

which canceling r is

$$
\sum_{i=(k-1)n+2}^{kn+1} q_i - \sum_{i=(k-1)n+2}^{kn+1} q_i(1-r)^{i-(k-1)n-1} + \sum_{kn+2}^{\infty} q_i(1-r)^{i-kn-1}
$$

$$
- \sum_{i=kn+2}^{\infty} q_i(1-r)^{i-(k-1)n-1}
$$

$$
= \sum_{i=(k-1)n+2}^{kn+1} q_i + \sum_{i=kn+2}^{\infty} q_i(1-r)^{i-kn-1} - \sum_{i=(k-1)n+2}^{\infty} q_i(1-r)^{i-(k-1)n-1}.
$$

Taking into account that the probability of event during the service of a customer a new one does not arrive

$$\sum_{i=1}^{\infty} q_i(1-r)^i = Q_1 = a_0,$$

and the probability of zero waiting time is

$$\sum_{i=1}^{\infty} q_i(1-r)^{i-1}r = \frac{r}{1-r}Q_1,$$

for the generating function $A(z)$ we obtain the expression

$$A(z) = Q_1 + z\frac{r}{1-r}Q_1 + z\sum_{k=1}^{\infty}(1-r+rz)^{kn}$$

$$\times \left\{ \sum_{i=(k-1)n+2}^{kn+1} q_i + \sum_{i=kn+2}^{\infty} q_i(1-r)^{i-kn-1} - \sum_{i=(k-1)n+2}^{\infty} q_i(1-r)^{i-(k-1)n-1} \right\}.$$

Its derivative is

$$A'(z) = \frac{rQ_1}{1-r} + \sum_{k=1}^{\infty}(1-r+rz)^{kn}$$

$$\times \left\{ \sum_{i=(k-1)n+2}^{kn+1} q_i + \sum_{i=kn+2}^{\infty} q_i(1-r)^{i-kn-1} - \sum_{i=(k-1)n+2}^{\infty} q_i(1-r)^{i-(k-1)n-1} \right\}$$

$$+znr\sum_{k=1}^{\infty} k(1-r+rz)^{kn-1}$$

$$\times \left\{ \sum_{i=(k-1)n+2}^{kn+1} q_i + \sum_{i=kn+2}^{\infty} q_i(1-r)^{i-kn-1} - \sum_{i=(k-1)n+2}^{\infty} q_i(1-r)^{i-(k-1)n-1} \right\}$$

and at $z=1$ gives

$$A'(1) = \frac{rQ_1}{1-r} + \sum_{k=1}^{\infty}\sum_{i=(k-1)n+2}^{kn+1} q_i - \frac{Q_2}{1-r}$$

$$+nr\sum_{k=1}^{\infty} k \sum_{i=(k-1)n+2}^{kn+1} q_i - nr\sum_{k=1}^{\infty}\sum_{i=(k-1)n+2}^{\infty} q_i(1-r)^{i-(k-1)n-1}. \tag{9}$$

4.2   The Generating Function $B(z)$

At the beginning of service of first customer the second customer is present, too. Let $x = u - \left[\dfrac{u-1}{n}\right] n$ ($[x]$ denotes the integer part of x), and let y be the mod T interarrival time ($1 \le y \le n$). The time elapsed between the starting moments of two successive customers is

$$\left[\frac{u-1}{n}\right] n + y \quad \text{if} \quad x \le y \quad \text{and} \quad \left(\left[\frac{u-1}{n}\right] + 1\right) n + y \quad \text{if} \quad x > y.$$

One can easily see that y has truncated geometrical distribution with probabilities

$$P\{y = \ell\} = \frac{(1-r)^{\ell-1}r}{1-(1-r)^n} \qquad (\ell = 1, 2, \ldots, n),$$

the generating function of entering customer for a time slice is $1 - r + rz$. The generating functions of entering customers depending on the service time and the mod T interarrival time are given in the tables (the rows correspond to the mod T interarrival and the columns to the service times):

$$
\begin{array}{llll}
q_1(1-r+rz) & q_2(1-r+rz)^{n+1} & \ldots\ q_{n-1}(1-r+rz)^{n+1} & q_n(1-r+rz)^{n+1} \\
q_1(1-r+rz)^2 & q_2(1-r+rz)^2 & \ldots\ q_{n-1}(1-r+rz)^{n+2} & q_n(1-r+rz)^{n+2} \\
q_1(1-r+rz)^3 & q_2(1-r+rz)^3 & \ldots\ q_{n-1}(1-r+rz)^{n+3} & q_n(1-r+rz)^{n+3} \\
\vdots & \vdots & \vdots & \vdots \\
q_1(1-r+rz)^{n-1} & q_2(1-r+rz)^{n-1} & \ldots\ q_{n-1}(1-r+rz)^{n-1} & q_n(1-r+rz)^{2n-1} \\
q_1(1-r+rz)^n & q_2(1-r+rz)^n & \ldots\ q_{n-1}(1-r+rz)^n & q_n(1-r+rz)^n
\end{array}
$$

the following n columns

$$
\begin{array}{lll}
q_{n+1}(1-r+rz)^{n+1} & q_{n+2}(1-r+rz)^{2n+1} & \ldots\ q_{2n}(1-r+rz)^{2n+1} \\
q_{n+1}(1-r+rz)^{n+2} & q_{n+2}(1-r+rz)^{n+2} & \ldots\ q_{2n}(1-r+rz)^{2n+2} \\
q_{n+1}(1-r+rz)^{n+3} & q_{n+2}(1-r+rz)^{n+3} & \ldots\ q_{2n}(1-r+rz)^{2n+3} \\
\vdots & \vdots & \vdots \\
q_{n+1}(1-r+rz)^{2n-1} & q_{n+2}(1-r+rz)^{2n-1} & \ldots\ q_{2n}(1-r+rz)^{3n-1} \\
q_{n+1}(1-r+rz)^{2n} & q_{n+2}(1-r+rz)^{2n} & \ldots\ q_{2n}(1-r+rz)^{2n}
\end{array}
$$

and

$$
\begin{array}{lll}
q_{2n+1}(1-r+rz)^{2n+1} & q_{2n+2}(1-r+rz)^{3n+1} & \ldots\ q_{3n}(1-r+rz)^{3n+1} \\
q_{2n+1}(1-r+rz)^{2n+2} & q_{2n+2}(1-r+rz)^{2n+2} & \ldots\ q_{3n}(1-r+rz)^{3n+2} \\
q_{2n+1}(1-r+rz)^{2n+3} & q_{2n+2}(1-r+rz)^{2n+3} & \ldots\ q_{3n}(1-r+rz)^{3n+3} \\
\vdots & \vdots & \vdots \\
q_{2n+1}(1-r+rz)^{3n-1} & q_{2n+2}(1-r+rz)^{3n-1} & \ldots\ q_{3n}(1-r+rz)^{4n-1} \\
q_{2n+1}(1-r+rz)^{3n} & q_{2n+2}(1-r+rz)^{3n} & \ldots\ q_{3n}(1-r+rz)^{3n}
\end{array}
$$

etc. Summing up the elements in the columns, then considering these sums shifted by n (i.e. the sums of columns corresponding to the service times $q_j, q_{n+j}, q_{2n+j}, \ldots (1 \le j \le n)$) for a concrete deviation j the generating function equals

$$
\frac{r}{1-(1-r)^n} \frac{1-(1-r)^{j-1}(1-r+rz)^{j-1}}{1-(1-r)(1-r+rz)}(1-r+rz)^{n-j+1}
$$
$$
\times \sum_{k=0}^{\infty} q_{kn+j}(1-r+rz)^{kn+j}
$$
$$
+\frac{r(1-r)^{j-1}}{1-(1-r)^n} \frac{1-(1-r)^{n-j+1}(1-r+rz)^{n-j+1}}{1-(1-r)(1-r+rz)} \sum_{k=0}^{\infty} q_{kn+j}(1-r+rz)^{kn+j},
$$

and $B(z)$ will be

$$B(z) = \sum_{k=0}^{\infty} \sum_{j=1}^{n} q_{kn+j}(1 - r + rz)^{kn+j}$$
$$\times \left\{ \frac{r}{1-(1-r)^n} \frac{1-(1-r)^{j-1}(1-r+rz)^{j-1}}{1-(1-r)(1-r+rz)}(1 - r + rz)^{n-j+1} \right.$$
$$\left. + \frac{r(1-r)^{j-1}}{1-(1-r)^n} \frac{1-(1-r)^{n-j+1}(1-r+rz)^{n-j+1}}{1-(1-r)(1-r+rz)} \right\}.$$

Its derivative at $z = 1$ is

$$B'(1) = \frac{r}{1-(1-r)^n} \sum_{j=1}^{n} \sum_{k=0}^{\infty} q_{kn+j} \cdot$$
$$\left\{ [1 - (1-r)^n]\frac{[(k+1)n+1]r^2+r(1-r)}{r^2} - \frac{nr^2(1-r)^{j-1}}{r^2} \right\}.$$

After some arithmetics we obtain

$$B'(1) = \sum_{j=1}^{n} \sum_{k=0}^{\infty} q_{kn+j} \left\{ r\frac{(k+1)nr^2 + r^2 + r - r^2}{r^2} - \frac{nr}{1-(1-r)^n}(1-r)^{j-1} \right\}$$
$$= \sum_{k=0}^{\infty} \sum_{j=1}^{n} q_{kn+j}[(k+1)nr + 1] - \frac{nr}{1-(1-r)^n} \sum_{k=0}^{\infty} \sum_{j=1}^{n} q_{kn+j}(1-r)^{j-1}.$$

$$(10)$$

Remark 1. In [6] and the present paper we characterized the same discrete cyclic-waiting system, so between their characteristics there exists certain connection. One can check the coincidence of stability condition (in the two cases they are written in different forms), between the zero probabilities there is valid the relation

$$p_0^{(w)} = \left(p_0^{(q)} + p_1^{(q)} \right) \sum_{i=1}^{\infty} q_i(1-r)^{i-1}. \tag{11}$$

Here the upper index w corresponds to the waiting time, the upper index q to the queue length.

We clarify this expression. Consider a moment just before starting the service of a customer and let the system be free or let there be present one customer. The probability of this event is $p_0^{(q)} + p_1^{(q)}$. One starts the service of the actual customer and it takes i time units. The waiting time for the next customer will be zero if during the first $i - 1$ time slices no customer enters and on the last time slice either no customer enters (the server becomes free) or a new customer appears. So, it is not important that during this time slice a further customer arrives or not since either the server becomes free or the service of new one can be started on the following time slice, in such sense it will be taken for service without waiting.

Remark 2. Our formulas for the generating functions of transition probabilities, p_0, $P(z)$ and the stability condition in the case of geometrical service time distribution (i.e. it is i time slices with probability $(1 - q)q^{i-1}$) give the following formulas.

$P(z)$ has the form (4), in the concrete case it is

$$P(z) = p_0 \frac{zA(z) - B(z) + \frac{rz}{(1-r)(1-q)}[A(z) - B(z)]}{z - B(z)}.$$

The generating functions of transition probabilities are

$$A(z) = \sum_{i=0}^{\infty} a_i z^i =$$

$$\frac{(1-r)(1-q)}{1-q(1-r)} + z\frac{r(1-q)}{1-q(1-r)} + z\frac{rq(1-r+rz)^n(1-q^n)}{[1-q(1-r)][1-q^n(1-r+rz)^n]},$$

$$B(z) = \sum_{k=1}^{\infty} b_i z^i$$

$$= \frac{1-(1-r)^n(1-r+rz)^n}{1-(1-r)(1-r+rz)} \frac{r(1-r+rz)}{1-(1-r)^n}$$

$$+\frac{1-q^n(1-r)^n(1-r+rz)^n}{1-q(1-r)(1-r+rz)} \frac{rq(1-r+rz)[(1-r+rz)^n-1]}{[1-(1-r)^n][1-q^n(1-r+rz)^n]}.$$

By using these expressions for $A'(1)$ and $B'(1)$ one gets

$$A'(1) = \frac{r}{1 - q(1-r)} + \frac{nr^2q}{(1-q^n)[1-q(1-r)]}$$

and

$$B'(1) = 1 - \frac{nr(1-r)^n}{1-(1-r)^n} + \frac{nr^2q[1-q^n(1-r)^n]}{(1-q^n)[1-q(1-r)][1-(1-r)^n]}.$$

For the probability of free state we had the expression

$$p_0 = \frac{a_0[1 - B'(1)]}{a_0 + A'(1) - B'(1)}.$$

Substituting the corresponding values we have

$$a_0 + A'(1) - B'(1) = \frac{nr(1-r)^n(1-q)}{[1-(1-r)^n][1-q(1-r)]} > 0,$$

so $1 - B'(1) > 0$ must be fulfilled, it leads to the inequality

$$\frac{rq[1-(1-r)^n]}{(1-q)(1-q^n)(1-r)^n} < 1.$$

The probability of free state is

$$p_0 = 1 - r - \frac{rq(1-r)[1-q^n(1-r)^n]}{(1-r)^n(1-q^n)[1-q(1-r)]},$$

the condition $p_0 > 0$ gives the same ergodicity condition.

Finally, we show the validity of (11) in the case of geometrical service time distribution. In [4] we obtained the formula

$$p_0^{(w)} = \left[1 - \frac{rq[1-(1-r)^n]}{(1-q)(1-q^n)(1-r)^n}\right]\frac{1-q}{1-q(1-r)}.$$

From one side we have

$$\sum_{i=1}^{\infty} q_i(1-r)^{i-1} = \sum_{i=1}^{\infty}(1-q)q^{i-1}(1-r)^{i-1} = \frac{1-q}{1-q(1-r)}.$$

From another side

$$p_0^{(q)} + p_1^{(q)} = p_0^{(q)} + \frac{1-a_0}{a_0}p_0^{(q)} = \frac{p_0^{(q)}}{a_0},$$

$$a_0 = \frac{(1-r)(1-q)}{1-q(1-r)}.$$

Consequently, it is enough to show that $p_0^{(q)}/a_0$ coincides with the expression in the brackets for $p_0^{(w)}$. One has

$$\begin{aligned}
\frac{p_0^{(q)}}{a_0} &= \left[1 - r - \frac{rq(1-r)[1-q^n(1-r)^n]}{(1-r)^n(1-q^n)[1-q(1-r)]}\right]\frac{1-q(1-r)}{(1-r)(1-q)} \\
&= \left[1 - \frac{rq[1-q^n(1-r)^n]}{(1-r)^n(1-q^n)[1-q(1-r)]}\right]\frac{1-q(1-r)}{1-q} \\
&= 1 + \frac{rq}{1-q} - \frac{rq[1-q^n(1-r)^n]}{(1-q)(1-q^n)(1-r)^n} = 1 - \frac{rq[1-(1-r)^n]}{(1-q)(1-q^n)(1-r)^n},
\end{aligned}$$

which proves our assertion.

References

1. Koba, E.V.: On a GI/G/1 queueing system with repetition of requests for service and FCFS service discipline. Dopovidi NAN Ukrainy **6**, 101–103 (2000). (in Russian)
2. Koba, E.V., Pustova, S.V.: Lakatos queuing systems, their generalization and application. Cybern. Syst. Anal. **48**, 387–396 (2012)
3. Lakatos, L., Szeidl, L., Telek, M.: Introduction to Queueing Systems with Telecommunication Applications. Springer, Berlin (2013)
4. Lakatos, L., Efroshinin, D.: Some aspects of waiting time in cyclic-waiting systems. In: Dudin, A., Klimenok, V., Tsarenkov, G., Dudin, S. (eds.) BWWQT 2013. CCIS, vol. 356, pp. 115–121. Springer, Heidelberg (2013). doi:10.1007/978-3-642-35980-4_13
5. Lakatos, L., Efrosinin, D.: A discrete time probability model for the waiting time of optical signals. Commun. Comput. Inf. Sci. **279**, 114–123 (2014)
6. Lakatos, L.: On the waiting time in the discrete cyclic–waiting system of $Geo/G/1$ type. In: Vishnevsky, V., Kozyrev, D. (eds.) DCCN 2015. CCIS, vol. 601, pp. 86–93. Springer, Heidelberg (2016). doi:10.1007/978-3-319-30843-2_9

Optimal Control of $M(t)/M/K$ Queues with Homogeneous and Heterogeneous Servers

Dmitry Efrosinin[1,2(✉)] and Michael Feichtenschlager[3]

[1] Institute of Control Sciences, Profsoyuznaya str., 65, 117997 Moscow, Russia
dmitry.efrosinin@mail.ru
http://www.ipu.ru
[2] RUDN University, Miklukho-Maklaya str., 6, 117198 Moscow, Russia
http://www.rudn.ru
[3] Johannes Kepler University Linz, Altenbergerstrasse 69, 4040 Linz, Austria
michael-feichtenschlager@gmx.at
http://www.jku.at

Abstract. The paper deals with a multi-server controllable queueing system $M(t)/M/K$ with time-dependent and, in particular, with periodic arrival rates. The models with homogeneous and heterogeneous servers are of interest. In latter case the fastest free server allocation mechanism is assumed and the preemption is allowed. The control problem consists in evaluation of the optimal number of servers during some specified stages and is solved by finite horizon dynamic programming approach. To calculate the transient solutions we use a forth-order Runge-Kutta method for the system with a truncated queue length. The results are compared with corresponding queues operating in a stationary regime. It is shown that the optimal control policies are also time dependent and periodic as arrival rates and heterogeneous systems are superior in performance comparing to the homogeneous ones.

Keywords: Time-dependent arrival rate · Controllable queueing system · Dynamic programming approach · Forth-order Runge-Kutta method

1 Introduction

Many queueing systems are subject to time-dependent changes in system parameters. This feature is very important to cover the problems with seasonality and periodicity of stochastic processes. Particularly it happens with an arrival rate which is used for modelling of arrivals of calls and inquires at call centres, arrival of packets at routers of the telecommunication systems, of time changing air traffic at airports, different arrival rates of trucks to the warehouses, goods

D. Efrosinin—This work was funded by the Russian Foundation for Basic Research, Project No. 16-37-60072 mol_a_dk. The reported study was funded within the Agreement No. 02.a03.21.0008 dated 24.11.2016 between the Ministry of Education and Science of the Russian Federation and RUDN University.

V.M. Vishnevskiy et al. (Eds.): DCCN 2016, CCIS 678, pp. 132–144, 2016.
DOI: 10.1007/978-3-319-51917-3_13

depot or seaports and so on. A very good literature overview on this subject can be found in [4]. This paper surveys and classifies the results on performance evaluation approaches for time-dependent queueing systems and their applications and identifies the links between different approaches. The performance analysis of multi-server queueing system subject to breakdowns was studied in [1]. There are several approaches to analyse such systems. The dynamic behaviour of Markovian queueing systems is described by a system of Kolmogorov differential equations (KDEs). Analytical solution of such systems exists only for special cases. Numerical approaches are based on a Runge-Kutta method. The systems with an infinite buffer are normally approximated by using a finite buffer system. The numerical solution of KDEs is used for example for the performance evaluation of a $M(t)/M/1/N$ system in [3].

In many cases the time-dependent system parameters must be combined with some controllable problems. The paper [5] deals with optimal allocation of such resources as beds in emergency departments of a hospital. For the mathematical modelling the queueing system with losses of the type is used. It was shown that the periodic variation of arrival rates makes a hysteretic policy time-dependent and periodic with the same period. To find the optimal decisions dynamic programming is used. The same approach was used in [2] for the multi-server queueing system with a controllable number of homogeneous servers.

The contribution of the present work is an evaluation of the optimal number of servers in multi-server queueing system with homogeneous and heterogeneous servers and time-dependent arrival rate. The discretization of a continuous-time Markov process is performed to apply the Runge-Kutta method and the iterative dynamic programming algorithm over a finite horizon. The decisions are chosen at specified moments of time which divide the observation time interval into so called stages. The number of available servers is assumed to be a constant within each stage. The paper provides comparison analysis of stationary and transient solutions as well as homogeneous and heterogeneous systems.

The rest of the paper is organized as follows. In Sect. 2 we describe a mathematical model and formulate a optimization problem for transient and stationary case. Section 3 deals with a description of a time-dependent arrival rate. In Sect. 4 a forth-order Runge-Kutta method is adopted for the model under study. The recursive dynamic programming algorithm is shown in Sect. 5. Some illustrative numerical examples are discussed in Sect. 6. Conclusions are given in Sect. 7.

2   Mathematical Model

Consider the controllable multi-server queueing system $M(t)/M/K$ with K servers. This system features Poisson arrival stream with time dependent arrival rate $\lambda(t)$. The servers are assumed to be heterogeneous with servers intensities $\mu_j, j = 1, 2, \ldots, K$. In special case when all intensities are equal, $\mu_j = \mu, j = 1, 2, \ldots, K$, we get the homogeneous system. The control consists in specification of the number of servers $K(t)$ at any decision epoch which will be specified later.

Denote by $N(t)$ the number of customers in the system at time t. The dynamics of the system is described by means of the controllable continuous-time inhomogeneous Markov chain $\{N(t)\}_{t \geq 0}$ with a set of states $E = \{n; n \in \mathbb{N}_0\}$ and set of control actions $A = \{1, 2, \ldots, K\}$. Define additional cost structure with the following components: c_1 – the waiting cost per unit of time for each customer in the queue, $c_{2,j}$ – the idle state cost per unit of time when the server j is idle. In homogeneous case it is assumed that $c_{2,j} = c_2$, $j = 1, 2, \ldots, K$. The servers are enumerated in such a way that

$$\mu_1 \geq \mu_2 \geq \cdots \geq \mu_K, \quad c_{2,1} \geq c_{2,2} \geq \cdots \geq c_{2,K}. \tag{1}$$

In accordance with the given cost structure, the mean total cost criterion is denoted by

$$J^f(n) = \mathbb{E}^f[\int_0^T c(N(t), K(t), \lambda(t), t)dt | N(0) = n]. \tag{2}$$

Here f is a Markov control policy which depends on the current state and time only, i.e. $K(t) = f(t, n(t))$, the expectation $\mathbb{E}^f$ is taken with respect to the probability distribution $\mathbb{P}^f$ over the state-action sequence under control policy f. The immediate cost function $c(N(t), K(t), \lambda(t), t)$ is defined as

$$c(N(t), K(t), \lambda(t), t) = c_1 \sum_{k=K(t)+1}^{\infty} (k - K(t)) 1_{\{N(t)=k\}} \tag{3}$$

$$+ \sum_{k=0}^{K(t)} \sum_{j=k+1}^{K(t)} c_{2,j} 1_{\{N(t)=k\}}.$$

The substitution of (3) into (2) yields the relation for the mean total cost in form

$$J^f(n) = \int_0^T \eta(n, K(t), \lambda(t), t)dt \tag{4}$$

$$= \int_0^T \left[c_1 \bar{Q}(n, K(t), \lambda(t), t) + \sum_{k=0}^{K(t)} \sum_{j=k+1}^{K(t)} c_{2,j} \pi_k(n, K(t), \lambda(t), t) \right] dt.$$

The first term by c_1 at the right hand side of (4) stands for the mean number of customers in the queue at time t with $K(t)$ servers and initial state $N(0) = n$, the second term stands for the mean idle state costs. We wish to minimize the functional $J^f(n)$ over all control policies and find optimal policy f^* that achieves the minimal cost $J^*(n)$, i.e.

$$J^*(n) := J^{f^*}(n) = \min_f J^f(n). \tag{5}$$

The solution of proposed optimization problem can be performed numerically. To realize some iterative algorithm the continuous time model must be converted to a discrete one. We divide a time interval $[0, T]$ into I equally spaced periods. The mean total cost functional in this case can be rewritten a follows,

$$J^f(n) = \sum_{i=1}^{I} \eta(n, K(i), \lambda(i), i) \tag{6}$$

$$= \sum_{i=1}^{I} \left[c_1 \bar{Q}(n, K(i), \lambda(i), i) + \sum_{k=0}^{K(i)} \sum_{j=k+1}^{K(i)} c_{2,j} \pi_k(n, K(i), \lambda(i), i) \right],$$

where $K(i)$ is a number of servers at period i, $\bar{Q}(n, K(i), \lambda(i), i)$ is a mean number of customers in the queue at period i, $\pi_k(n, K(i), i)$ – probability of k customers in the system with $K(i)$ servers at period i with initial state n.

The transient solution of the problem will be compared with a stationary one. In this case the long-run average cost per unit of time

$$\eta(K, \lambda(t)) = c_1 \bar{Q}(K, \lambda(t)) + \sum_{k=0}^{K} \sum_{j=k+1}^{K} c_{2,j} \pi_k(K, \lambda(t)) \tag{7}$$

must be minimized over $K(t)$ for any fixed value $\lambda(t)$. The substitution of the stationary state probabilities of the infinite buffer system into (7) yields the relation

$$\eta(K(t), \lambda(t)) = \left[c_1 \prod_{j=1}^{K(t)} \frac{\lambda(t)}{\sum_{k=1}^{j} \mu_k} \frac{\lambda(t) \sum_{k=1}^{K(t)} \mu_k}{(\lambda(t) - \sum_{k=1}^{K(t)} \mu_k)^2} \right. \tag{8}$$

$$\left. + \sum_{k=0}^{K(t)} \sum_{j=k+1}^{K(t)} c_{2,j} \frac{\lambda(t)^k}{\prod_{l=1}^{k} \sum_{j=1}^{l} \mu_j} \right] \pi_0(K(t), \lambda(t)),$$

$$\pi_0(K(t), \lambda(t)) = \tag{9}$$

$$\left[\sum_{k=0}^{K(t)-1} \frac{\lambda(t)^k}{\prod_{l=1}^{k} \sum_{j=1}^{l} \mu_j} + \prod_{j=1}^{K(t)} \frac{\lambda(t)}{\sum_{k=1}^{j} \mu_k} \frac{\sum_{k=1}^{K(t)} \mu_k}{\sum_{k=1}^{K(t)} \mu_k - \lambda(t)} \right]^{-1}.$$

3   Arrival Rate

We have chosen a similar arrival rate $\lambda(t)$ as in the paper from [2]. The authors have studied there incoming and service of airplanes of an airport modelled via a $M(t)/M/K/N$ queuing system. In the queueing system under study the condition

$$\lambda(t) < \sum_{j=1}^{K} \mu_j \tag{10}$$

is a necessary one, since there is no cost relation for customers who get rejected. That means that the maximum number of server K can handle the average arrival rate of users. The data for $\lambda(t)$ is given in Table 1.

The arrival rate will be divided into three equidistant stages. It means that each stage lasts eight hours, which is a normal working shift cycle. At the beginning of a stage, the number of customers n in the system is known. This value

Table 1. Values for the arrival rate $\lambda(t)$

Time in hour	Input intensity
1–5	$8.75 + 4.25\cos(t/1.6)$
6	4.5
7	5.0
8	5.5
9	6.5
10–17	7.0
18–21	10.0
22–24	13.0

will be called initial state of the current stage. A natural question, one can ask is, how many servers (in this context workers) should be hired at current and following stage(s) so that the expected costs are minimized. Obviously the number of necessary operators are depending on the initial state n.

4   Fourth-Order Explicit Runge-Kutta Method

Since there is no way to solve the system of Kolmogorov forward equations

$$\pi'(t) = \pi(t)A(t) \tag{11}$$

analytically a numerical algorithm is needed to get an approximate solution. For this task we have used the standard fourth order explicit Runge-Kutta procedure which is a widely used one-step method. It considers differential equations of the form

$$y'(t) = f(t, y(t)) \quad \forall t \in (0, T) \tag{12}$$

with given initial condition

$$y(t_0) = y_0. \tag{13}$$

Algorithm 1. *The explicit fourth-order Runge-Kutta method.*
Step 1. Computation of five parameters $\kappa_1, \kappa_2, \kappa_3, \kappa_4$ and κ:

$$\kappa_1 = f(t_n, y_n)\Delta t$$
$$\kappa_2 = f(t_n + \frac{\Delta t}{2}, y_n + \frac{\kappa_1}{2})\Delta t$$
$$\kappa_3 = f(t_n + \frac{\Delta t}{2}, y_n + \frac{\kappa_2}{2})\Delta t$$
$$\kappa_4 = f(t_n + \Delta t, y_n + \kappa_3)\Delta t$$
$$\kappa = \frac{1}{6}(\kappa_1 + 2\kappa_2 + 2\kappa_3 + \kappa_4)$$

Step 2. Evaluation of y_{n+1} by the recursive relation,

$$y_{n+1} = y_n + \kappa.$$

For solving the Kolmogorov forward equations (11) interpret $f(t)\pi(t)$ as $f(\pi(t), t)$, choose a appropriate step size Δt and apply this Runge-Kutta method directly on the function $f(\pi(t), t)$. Obviously the error between the calculated and the real solution gets less if one selects a smaller step size. On the other hand a greater step size means less computing time. We have chosen $\Delta t = 0.005$. This value for the step size seems to have a reasonable balance between computing time and computation error.

To solve the system (11) we use a truncation of the buffer capacity by assuming that N is the maximum allowable number of customers in the system.

5   Optimisation Problem

Let $T = 24\,\mathrm{h}$ be an observation cycle and a finite horizon for the dynamic programming. The decision epochs occur each $8\,\mathrm{h}$, hence we get $S = 3$ stages with $\frac{I}{S}$ periods i within each stage s. A strategy at a decision epoch $d = (S-s)I/S+1$ which depends on a current stage s is donated by $f(d, n)$, where n stands as before for the initial state. A strategy is equal for any period i from the interval

$$d \leq i \leq d + \frac{I}{S} - 1.$$

Denote by $V_n(s)$ the optimal cost function for s stages left which we refer to as value function:

$$V_n(s) = \min_f \mathbb{E}^f \left[\sum_{i=1}^{I} c(N(i), K(i), \lambda(i), i) \mid N(s) = n \right], \ s = 1, 2, \ldots, S, \ n \in E.$$

$$(14)$$

The minimum must be taken over tail policies

$$(f(1, K(1)), f(I/S+1, K(I/S+1)), \ldots, f((S-1)I/S+1, K((S-1)I/S+1)).$$

Obviously, $V_n(S) = J^*(n)$.

Algorithm 2. *The finite horizon dynamic programming:*
Step 1. Backward recursion: $V_n(0) = 0, n \in E$, *and*

$$V_n(s) = \min_{1 \leq k \leq K} \left\{ r(n, k, s) + \sum_{m=0}^{N} p_{nm}(k, s) V_m(s - 1) \right\}, \qquad (15)$$

where $r(n, k, s)$ is the total average cost per stage,

$$r(n, k, s) = \sum_{i=d}^{d+\frac{I}{S}-1} \eta(n, k, \lambda(i), i), \ s = 1, 2, \ldots, S, \qquad (16)$$

and the transition probabilities between the stages are defined as

$$p_{nm}(k,s) = \mathbb{P}\Big[N(d+I/S) = m \Big| N(d) = n, f(d,n) = k\Big], \tag{17}$$

$$s = 2, \ldots, S. \tag{18}$$

Step 2. Any Markov policy f^* that satisfies

$$f^*(d,n) = \underset{1 \le k \le K}{\arg\min}\Big\{r(n,k,s) + \sum_{m=0}^{N} p_{nm}(k,s)V_m(s-1)\Big\} \tag{19}$$

is an optimal control policy.

The values $p_{nm}(k,s)$ in (17) have to be interpreted in the following way. They stand for the probability to be in state m at the beginning of the next stage under the condition that the initial state of the previous stage was n.

Algorithm 3. *The following basic steps are involved into the computation procedure:*

Step 1. Compute the state probabilities $\pi(n,k,\lambda(i),i)$ for each n, k and i via the Runge-Kutta fourth order method.

Step 2. Compute the cost function $\eta(n,k,\lambda(i),i)$ for each n, k and i.

Step 3. Compute $r(n,k,s)$ for each n, k and s by accumulating $\eta(n,k,\lambda(i),i)$ over all periods i within the corresponding stage.

Step 4. Compute the transition probabilities $p_{nm}(k,s)$ for each n, m, k and s via Runge-Kutta fourth order method.

Step 5. Evaluation of the optimal strategy for any s and n by means of Algorithm 2.

The number of servers $k^* = f(s,n)$ defined by (19) for which the expression in the right hand side of (15) is minimal is called the best or optimal strategy at stage s given the initial state is n.

Remark 1. Notice that in this queueing model rejecting of customers is not a valid option. If, for example, the initial state n at the current stage is the capacity of the buffer plus 2, the best strategy cannot be one server. However if the one choose the waiting room capacity high enough, restrict n up to this value and condition (10) is clearly fulfilled there will be no dropping of users.

6   Numerical Realisation and Results

The main goal in this paper is to compare the operating costs for the $M(t)/M/K$ queue between homogeneous and heterogeneous servers when optimal policies are used. Further the difference between transient and stationary solution will be contrasted for both philosophies. For this computations the following assumptions are used:

1. The maximum number of servers is $K = 6$.
2. The buffer capacity is $N - K = 10$.
3. The waiting cost $c_1 = 10$.
4. The service rate of the server in homogeneous case is $\mu = 4$.
5. The service intensities $\mu_j, j = 1, 2, \ldots, K$, of heterogeneous servers are listed in Table 2.
6. The idle state cost in homogeneous case is $c_2 = 2.1$.
7. The idle state costs $c_{2,j}, j = 1, 2, \ldots, K$, for heterogeneous servers are listed in the Table 2.

To get comparability with the homogeneous case we have chosen the service rates and operator idle costs for heterogeneous servers so that they satisfy the following conditions,

$$\sum_{j=1}^{K} \mu_j = K\mu, \quad \sum_{j=1}^{K} \frac{c_{2,j}}{\mu_j} = \frac{c_2}{\mu} \tag{20}$$

together with the ordering (1). That means that the server 1 is the fastest one but has the highest mean standing costs. Followed by server 2 and so forth. This is a quite reasonable assumption because a faster operator needs more resources. If this server is idle more costs are generated as for a slower one. In heterogeneous case the fastest free server policy is used for the allocation mechanism. If more then one operator is free and a new customer enters the queue the fastest free server will be entrusted with this task. This must be considered in the calculation of $\eta(n, k, \lambda(i), i)$. To calculate the best stationary policy the long-run average cost $\eta(k, \lambda(i))$ must be minimized for $k, 1 \leq k \leq K$. For homogeneous servers the service intensities μ_j as well as idle state costs $c_{2,j}$ in homogeneous case must be set to be equal as was discussed before.

Table 2. Values of μ_j and $c_{2,j}$ for heterogeneous servers

Server	Idle state cost	Service rate
1	$c_{2,1} = 0.4371$	$\mu_1 = 480/49$
2	$c_{2,2} = 0.9\, c_{2,1}$	$\mu_2 = \mu_1/2$
3	$c_{2,3} = 0.85\, c_{2,2}$	$\mu_3 = \mu_1/3$
4	$c_{2,4} = 0.8\, c_{2,3}$	$\mu_4 = \mu_1/4$
5	$c_{2,5} = 0.75\, c_{2,4}$	$\mu_5 = \mu_1/5$
6	$c_{2,6} = 0.7\, c_{2,5}$	$\mu_6 = \mu_1/6$

In Figs. 1(a) and (b) we illustrate the mean number of customers in the buffer $\bar{Q}(n, K(t), \lambda(t), t)$ in homogeneous and heterogeneous cases for different number of servers $K(t)$. The queue length for $K(t) = 4, 5, 6$ servers are not shown in these figures, since the values are very small (especially when heterogeneous servers are used). In Fig. 1(a) and (b) the initial state $N(t) = n$ at time $t = 0$ is set to be zero. That means the queueing system is at start empty. The mean number of waiting customers, which is needed in (6), is calculated via the formula

$$\bar{Q}(n, K(i), \lambda(i), i) = \sum_{k=K(i)+1}^{N} (k - K(i))\pi_k(n, K(i), \lambda(i), i), \qquad (21)$$

where $K(i)$ is fixed in period i, k is a state at period i and N is the maximum number of customers in the system. The state probabilities in this expression are the solution of the system (11) performed by the fourth-order Runge-Kutta method for each given number of servers $K(i)$ at each period i.

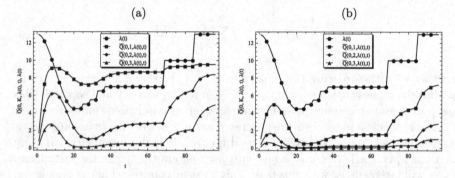

Fig. 1. $\bar{Q}(n, K, \lambda(t), t)$ for (a) homogeneous and (b) heterogeneous system

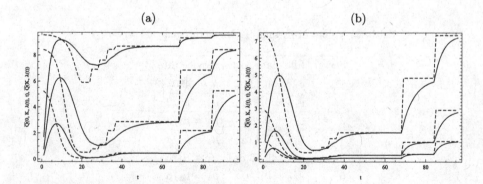

Fig. 2. Transient/stationary $\bar{Q}(n, K, \lambda(t), t)/\bar{Q}(K, \lambda(t))$ for (a) homogeneous and (b) heterogeneous system

Figures 2(a) and (b) illustrate how the mean queue length (21) differs from the stationary solution in the homogeneous and heterogeneous case which can be calculated by expression from (8). Again the mean queue length for 4, 5 and 6 servers are not shown because of the small values. The continuously plotted lines belong to the transient and the dashed lines to the stationary solution.

These pictures illustrate clearly the behaviour of the queuing system. When the transient solution is off and $\lambda(t)$ is a constant value for a certain period it converges to the stationary result as time goes by. This is not astonishing because a stationary queuing system can be interpreted as a long running transient system with a constant arrival rate.

To compare the minimum expected costs between homogeneous/heterogeneous servers in stationary/transient case simply evaluate $\eta(n, k^*, \lambda(i), i)$ and $\eta(k^*, \lambda(i))$ for each period i and corresponding optimal number of servers k^*. The calculated values of the optimal policy are listed in Table 3. To get

Table 3. Optimal policy f

Initial state at stage one	Homogen transient	Homogen stationary	Heterogen transient	Heterogen stationary
0	4	5	6	6
1	4	5	6	6
2	5	5	6	6
3	5	5	6	6
4	5	5	6	6
5	5	5	6	6
6	5	5	6	6
7	5	5	6	6
8	5	5	6	6
9	5	5	6	6
10	6	5	6	6
Stage two				
0	4	4	4	4
1	4	4	4	4
2	4	4	4	4
3	4	4	4	4
4	4	4	4	4
5	4	4	4	4
6	4	4	5	4
7	4	4	5	4
8	4	4	6	4
9	4	4	6	4
10	5	4	6	4
Stage three				
0	5	5	6	6
1	5	5	6	6
2	5	5	6	6
3	5	5	6	6
4	5	5	6	6
5	5	5	6	6
6	5	5	6	6
7	5	5	6	6
8	5	5	6	6
9	6	5	6	6
10	6	5	6	6

more server variety in the transient solutions one can increase the waiting room capacity and adjust c_1 and c_2.

Figures 3(a), (b) and 4(a), (b) show the minimum expected costs $\eta(n, K(i), \lambda(i), i)$ in homogeneous and heterogeneous case. The first one deals with a

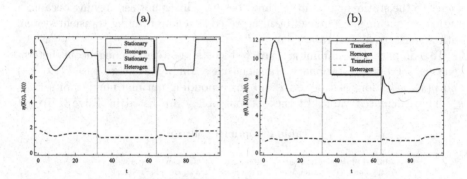

Fig. 3. (a) $\eta(K(t), \lambda(t))$ in stationary case and (b) $\eta(n, K(t), \lambda(t), t)$ in transient case for $n = 0$

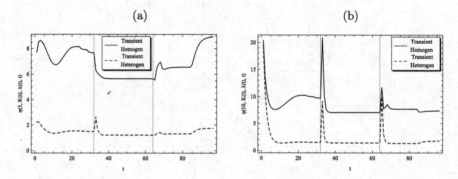

Fig. 4. $\eta(n, K(t), \lambda(t), t)$ in transient case for (a) $n = 5$ and (b) $n = 10$

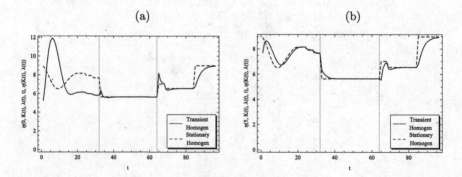

Fig. 5. $\eta(n, K(t), \lambda(t), t)$ with homogeneous servers for (a) $n = 0$ and (b) $n = 5$

stationary case. Here the initial state n is without any significance. The second, third, fourth one is dedicated to the transient solution of the optimizing problem with initial state n at each stage equal to 0, 5 and 10. In the following pictures the stages should be interpreted severally.

As one would expect, the queuing system with heterogeneous servers is superior in terms of running costs. A similar gap to the homogeneous costs as in Figs. 3(b), 4(a) and (b) can be seen for different initial states n at the end of every stage.

The Figs. 5(a), (b) and 6(a), (b) deal with homogeneous and number Figs. 7(a), (b) and 8(a) and (b) with heterogeneous operators. They picture the minimum expected costs $\eta(n, K(i), \lambda(i), i)$ and $\eta(K(i), \lambda(i))$ which were induced in the stationary and transient case respectively for chosen initial states.

(a)   (b)

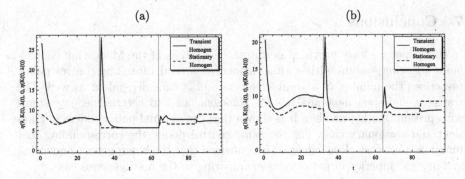

Fig. 6. $\eta(n, K(t), \lambda(t), t)$ with homogeneous servers for (a) $n = 9$ and (b) $n = 10$

(a)   (b)

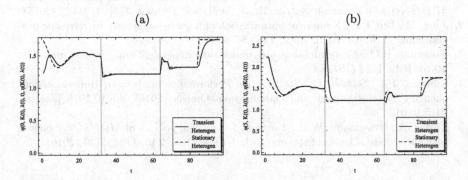

Fig. 7. $\eta(n, K(t), \lambda(t), t)$ with heterogeneous servers for (a) $n = 0$ and (b) $n = 5$

At a close look at Figs. 5(a) up to 8(b) one can see that the costs induced by the transient queuing system converges to the expenses of the stationary model if and only if the computed optimal policies for this stage are the same.

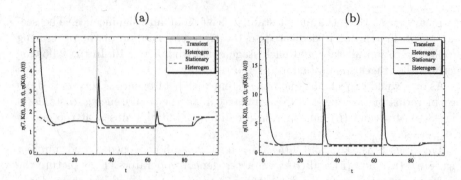

Fig. 8. $\eta(n, K(t), \lambda(t), t)$ with heterogeneous servers for (a) $n = 7$ and (b) $n = 10$

7   Conclusions

In this paper we have provided performance analysis of the Markovian controllable queueing system with a time-dependent arrival rate. The control policy prescribes the number of allowable servers and is time-dependent as well. The transient and stationary analysis for homogeneous and heterogeneous systems with preemption is provided. It is shown that the optimal policy differs in transient and stationary case. For the same control policy the corresponding cost functions take very close values. It is confirmed that the heterogeneous queueing systems are superior in performance comparing to the homogeneous case.

References

1. Ghimire, S., Ghimire, R.P., Thapa, G.B.: Performance evaluation of unreliable $M(t)/M(t)/n/n$ queueing system. Br. J. Appl. Sci. Technol. **7**(4), 412–422 (2015)
2. Jung, M., Lee, E.S.: Numerical optimization of a queueing system by dynamic programming. J. Math. Anal. Appl. **141**, 84–93 (1989)
3. Koopman, B.O.: Air-terminal queues under time-dependent conditions. Oper. Res. **20**(6), 1089–1114 (1972)
4. Schwarz, J.A., Selinka, G., Stolletz, R.: Performance analysis of time-dependent queueing systems: survey and classification. Omega (2016). doi:10.1016/j.omega.2015.10.013
5. Tirdad, A., Grassmann, W.K., Tavakoli, J.: Optimal policies of $M(t)/M/c/c$ queues with two different levels of servers. Eur. J. Oper. Res. **249**, 1124–1130 (2016)

Algorithmic and Software Tools for Optimal Design of New Generation Computer Networks

Yuriy Zaychenko[✉] and Helen Zaychenko

Institute for Applied System Analysis, NTUU "KPI", Kiev, Ukraine
zaychenkoyuri@ukr.net, syncmaster@bigmir.net

Abstract. Algorithmic and software tools for new generation networks (NGN) design are elaborated and presented in this paper. The tools are based on original methods and algorithms suggested by authors and include algorithms for solution of numerous tasks: channels capacities assignment, flows distribution, survivability analysis and structural synthesis. The elaborated models and algorithms take into account the specificity of NGN technology. The results of experimental investigations and practical implementation of the suggested tools are presented and discussed.

Keywords: New generation networks · Capacities assignment · Flows distribution · Topological optimization · Survivability analysis

1 Introduction

Burst increase in the volume of the traffic in the global computer networks due to millions of mobile users and the increasing demands on the quality of information transfer caused the appearance and implementation of new generation of computer networks (NGN), in particular communication network with MPLS technology [1]. Distinctive features of these networks are: (1) the high data transfer rate; (2) the presence of various classes of flows (users); (3) introduction of quality of service (QoS), namely, the average delay in the delivery of packets and a packets loss ratio. The emergence of a new generation of networks put on the agenda for the development of appropriate models and methods of performance analysis and optimization of networks in terms of quality. These methods are used in the optimal design of NGN networks. The aim of this work is to develop and investigate models and algorithmic and software tools for analysis and optimization of next-generation networks.

2 Analytic Models for Estomation QoS NGN Networks

To solve the problems of the analysis and optimization of NGN network using Quality of Service (QoS) indicators must first be developed analytical models for assessing the quality indices for different classes of service, depending on the

© Springer International Publishing AG 2016
V.M. Vishnevskiy et al. (Eds.): DCCN 2016, CCIS 678, pp. 145–161, 2016.
DOI: 10.1007/978-3-319-51917-3_14

intensity of the input flows, channel capacities, flow distribution (FD) over communication channels. These models should take into account all the specificities of NGN networks the different classes of service (flows) and their priority service.

The main QopS indicators are: packets transfer delay (PTD) of different classes and packets loss ratio (PLR) [2]. Consider analytical models of these QoS.

Let it be one channel in which N flows are transmitted with priorities P_i. Denote flow with priority i in channel (r, s) as f_{rs}^i, channel capacity - μ_{rs}. For convenience set priorities in such order: $p_0 > p_1 > \ldots > p_N$.

For obtaining analytical models for average delay of packets of k-th priority introduce the following assumptions:

1. Input flows in node of all classes are Poisson with intensity $h_{ij}^{(k)}$.
2. Service time in channels (r, s) is distributed by exponential law with parameter intensity μ_{rs} (Mbit/s), where μ_{rs} is capacity of the channel (r, s).
3. Service time of a packet in different channels are statistically independent random variables.

Under such assumptions using queue theory write down the expression for delay in the channel (r, s) for flows of different priorities (classes) [1,2]: p_0:

$$t_{rs}^0 = \frac{f_{rs}^0}{(\mu_{rs} - f_{rs}^0) \cdot \mu_{rs}}, \tag{1}$$

$$t_{rs}^j = \frac{\sum\limits_{k=0}^{j} f_{rs}^k}{(\mu_{rs} - \sum\limits_{k=0}^{j-1} f_{rs}^k) \cdot (\mu_{rs} - \sum\limits_{k=0}^{j} f_{rs}^k)}. \tag{2}$$

Let be given demand matrix for flow transmission of the l-th priority $H_l = ||h_{ij}^l||$. Using these expressions in [1] the final expression was obtained for average delay of k-th priority flow in a network:

$$T_{mean,k} = \frac{1}{H_\Sigma^{(k)}} \sum\limits_{(r,s) \in E} \frac{f_{rs}^{(k)} \sum\limits_{i=1}^{k} f_{rs}^{(i)}}{(\mu_{rs} - \sum\limits_{i=1}^{k-1} f_{rs}^{(i)}) \cdot (\mu_{rs} - \sum\limits_{i=1}^{k} f_{rs}^{(i)})}, \tag{3}$$

where $H_\Sigma^{(k)} = \sum\limits_{j=1}^{n} \sum\limits_{i=1}^{n} h_{ij}^{(k)}$ is total intensity of input flow of the k-th priority (class); $f_{rs}^{(i)}$ is the flow of i-th priority in the (r, s).

Now obtain the expression for packets loss probability of different classes.

The probability of packets loss of the k-th class (priority) is equal to the probability of the state when all the virtual channels allotted for k-th class of flow in the channel (r, s) are occupied [2]:

$$P_{losses(r,s)}^{(k)} = P_0 \cdot \left(\frac{f_{rs}^{(k)}}{\mu}\right)^{n_k} \cdot \frac{1}{n_k!} \cdot \left(\frac{f_{rs}^{(k)}}{n_k \mu}\right)^{N_k}, \tag{4}$$

where μ is capacity of the base channel, e.g., $\mu_1 = 2.048$ Mbit/s; n_k is a number of channels in the link (r, s) allotted for transmission of the k-th flow; N_k is a buffer size in LSR assigned for queue of k-th class of packets; P_0 is a normalizing multiplier.

Then the loss ratio of the k-th class of packets will be equal to:

$$PLR_k = 1 - \prod_{(r,s) \in E} \left(1 - P^{(k)}_{losses(r,s)}\right). \tag{5}$$

These models are used for the solution of problems of performance analysis and optimization of NGN networks considered in this paper: (1) flows distribution (FD) problem; (2) channels capacities choice and flows distribution (CCFD); (3) NGN networks survivability analysis and optimization; (4) NGN structure optimization under constraints on QoS indicators.

Consider the general statement algorithms for these solution.

3 Flows Distribution Problem

3.1 Problem Statement

Let be given NGN network as a graph $G(X, E)$, where $X = \{x_j\}$ is a set of nodes (routers), $E = \{(r, s)\}$ is a set of channels, capacities of channels are given $\{\mu_{rs}\}$ and matrices of demands for transmission of all the classes of flows $H(k) = ||h_{ij}(k)||, i, j = \overline{1, n}$, where $h_{ij}(k)$ is an intensity of information flow of k-th class of service (CoS) to be transmitted from the node x_i to node x_j (Mbit/s). It's demanded to find such transmission routes and flows distribution for all the classes $F(k) = [f_{rs}(k)]$, under which the constraints on average packets delay

$$T_{mean,k} \leq T_{k,preassigned}, k = \overline{1, K}, \tag{6}$$

and constraint on the packets loss ratio

$$PLR_k \leq PLR_{k,preassigned}, k = \overline{1, K}, \tag{7}$$

will be fulfilled.

3.2 The Algorithm of FD Problem Solution

The algorithm consists of K stages (by number of classes of service) at each of which the distribution of k-th class of flow is determined $F(k)$ using constraints 6 and 7.

1 Stage
0 step. Initialize $F_1(0) = 0; H_1(0) = 0$.
This stage consists of $2C_n^2 = n(n-1)$ iterations at each of which flow distribution from next demand h_{ij} is searched. $i, j = \overline{1, n}, i \neq j$.
<u>1 iteration</u>

1. Find the initial conditional metrics: $l_{rs}(1) = \lambda \frac{\partial T_{mean,1}}{\partial f_{rs}^{(1)}} + (1-\lambda) \frac{\partial PLR_1}{\partial f_{rs}^{(1)}}$, where $\lambda \in [0;1]$.

   As an initial value of λ its possible to take $\lambda = 0.5$.

2. Determine the shortest paths in the chosen metrics between all the nodes - $\pi_{ij}^{min}(1)$.

3. Choose the first demand in matrix $H_1 = ||h_{ij}^1||$, for example $h_{i_1 j_1}$. Find the shortest path $\pi_{i_1 j_1}^{min}$ and distribute flow from the demand $h_{i_1 j_1}$ and find the initial flows distribution:

$$f_{rs}^{(1)}(1) = \begin{cases} f_{rs}^{(1)}(0) + h_{i_1 j_1} = h_{i_1 j_1}, & \text{if } (r,s) \in \pi_{i_1 j_1}^{min}, \\ f_{rs}^{(1)}(0) = 0, & \text{otherwise.} \end{cases} \tag{8}$$

End of the first iteration. Go to the second iteration.

r-th iteration

Let $(r-1)$ iteration was completed, flows distribution of the first $(r-1)$ demands $H^{(1)}$ FD $f_{rs}^{(1)}(r-1)$ were found.

1. Determine the conditional metrics:

$$l_{rs}^{(1)}(r) = \lambda \frac{\partial T_{mean,1}}{\partial f_{rs}^{(1)}} + (1-\lambda) \frac{\partial PLR_1}{\partial f_{rs}^{(1)}} | f_{rs} = f_{rs}^{(1)}(r-1). \tag{9}$$

2. Choose the next demand $h_{i_r j_r}$ from matrix $H(1)$ and find the shortest path $\pi_{i_r j_r}^{min}$ in the metrics $l_{rs}^{(1)}(r)$.

   Distribute the flow from demand $h_{i_r j_r}$ over the path $\pi_{i_r j_r}^{min}$ and find the new flow $F_1(r)$:

$$f_{rs}^{(1)}(r) = \begin{cases} f_{rs}^{(1)}(r-1) + h_{i_r j_r}^{\alpha}, & \text{if } (r,s) \in \pi_{i_r j_r}^{min}, \\ f_{rs}^{(1)}(r-1), & \text{otherwise.} \end{cases} \tag{10}$$

where $h_{i_r j_r}^{\alpha} = \min\{h_{i_r j_r}; Q_{res}(\pi_{i_r j_r}^{min})\}$ is a portion of $h_{i_r j_r}$ which is transmitted by the path $\pi_{i_r j_r}^{min}$.

End of r-th iteration.

The rest of iterations of the first stage are fulfilled similarly up to the full exhausting of demands in matrix $H(1)$. Denote the obtained flow $F_1 = [f_{rs}^{(1)}]$.

Check up the fulfillment of the following constraints:

$$T_{mean}(F_1) \leq T_{1,preassigned}, \tag{11}$$

$$PLR(F_1) \leq PLR_{1,preassigned}. \tag{12}$$

If the constraints (11) and (12) are fulfilled then end of the first stage and go to the next stage 2. Otherwise, perform the additional optimization of the flow F_1. If after this step at least one of the constraints (11), (12) won't be fulfilled, then this problem is unsolvable at the given channels capacities.

k Stage

Let $k-1$ stages were performed and flows distribution from the first $(k-1)$ demands $F_1, F_2, \ldots, F_{k-1} = [f_{rs}^{(k-1)}]$ were found. Find the distribution of the k-th class of flows. The stage consists of $n(n-1)$ iterations like the first stage.

1. Take the first demand $h_{i_1j_1}$ out of matrix $H_k = ||h_{ij}^k||$. Find the shortest path in the metrics (9) $\pi_{i_1j_1}^{min}(k)$.

2. Determine the capacity reserve of the path $\pi_{i_1j_1}^{(k)\,min}$:

$$Q_{res}(\pi_{i_1j_1}^{min}) = \min_{(r,s)\in\pi_{i_1j_1}^{min}(k)} \left\{ \mu_{rs} - \sum_{i=1}^{k-1} f_{rs}^{(i)} \right\} - \varepsilon.$$

3. Distribute the flow of demand $h_{i_1j_1}^{(k)}$ with value $h_{i_1j_1}^{\alpha}$, where

$h_{i_1j_1}^{\alpha} = \min\{h_{i_1j_1}^{(k)}; Q_{res}(\pi_{i_1j_1}^{min})\}$ and calculate new flow value:

$$f_{rs}^{(k)}(1) = \begin{cases} f_{rs}^{(k)}(0) + h_{i_1j_1}^k, & \text{if } (r,s) \in \pi_{i_1j_1}^{min}(k), \\ f_{rs}^{(k)}(0), & \text{otherwise.} \end{cases}$$

End of the first iteration.

The following iterations are performed similarly for the rest of demands in matrix $H(k) = ||h_{ij}(k)||, i,j = \overline{1,n}$.

In result obtain flows distribution of the k-th class $F(k) = [f_{rs}(k)]$.

Check the fulfillment of the constraints:

$$T_{mean}(F_k) \le T_{k,preassigned}, \tag{13}$$

$$PLR(F_k) \le PLRk, preassigned. \tag{14}$$

If both constraints are fulfilled then STOP, the end of algorithm.

In case if $T_{mean}(F_k) > T_{k,preassigned}$ then the corresponding FD problem is unsolvable under given channels capacities and requirements on given values of QoS $T_{k,preassigned}$.

4   Problem of Optimal Capacities Choice and Flows Distribution

For ensuring of transmission of all the input flows with given values QoS under arbitrary demand matrices it's necessary to solve combined problem of traffic engineering in which the optimal channels capacities and flows distributions of all classes should be found simultaneously.

4.1   Problem Statement

MPLS network structure is given as an oriented graph $G = (X, E)$, where $X = \{x_j\}, j = \overline{1,n}$ is a set of nodes (routers), $E = \{(r,s)\}$ is a set of channels, set of channels capacities $D = \{d_1, d_2, \ldots, d_k\}$ and their costs of unit length $C = \{c_1, c_2, \ldots, c_k\}$.

Let it also be given demand matrices of input flows of corresponding classes $H = ||h_{ij}^{(k)}||, i,j = \overline{1,n}, k = \overline{1,K}$ and constraints on average packets delay for the k-th flow $T_{mean,k}, k \in K_1 \subset K$, and the constraint on packets loss ratio for different classes of flows.

It's required to choose such channels capacities $\{\mu_{rs}^{(0)}\}$ and to find the flows distributions of all the classes $F(k) = [f_{rs}(k)]$ for which total cost of NGN network would be minimal and the established constraints on given values of QoS be fulfilled completely:

Find

$$\min C_\Sigma = \sum_{(r,s)\in E} C_{rs}(\mu_{rs}), \tag{15}$$

under constraints

$$T_{mean,k}(F(k), \mu_{rs}) \leq T_{preassigned,k}, k = \overline{1, K}, \tag{16}$$

$$PLR(F_k) \leq PLR_{k,preassigned}. \tag{17}$$

Describe the algorithm of solution the problem of capacities choice and flows distribution (CCFD) for NGN network. It consists of preliminary stage and finite number of iterations [2].

At the preliminary stage find initial channels capacities $\mu_{rs}(0)$ and flows distributions of all classes $F(k)$. Then go to the first iteration.

$l + 1$ iteration

Let l iterations be already performed and current capacities $\{\mu_{rs}(l)\}$, flows distributions $F_k(l) = [f_{rs}^{(k)}(l)]$ and total network cost $C_\Sigma(l)$ were found.

The goal of iteration $(l+1)$ is the optimization of channels capacities and flows distribution by criterion of total cost minimization C_Σ and check of optimality condition.

1. For given values of capacities $\mu_{rs}(l)$ solve the problem FD and find new flows distributions of all classes $F_{(k)}(l + 1) = [f_{rs}^{(k)}(l + 1)], k = \overline{1, K}$.
2. For new flows $F_{(k)}(l + 1)$ solve the problem of optimal capacities choice (CC) and find new channels capacities $\mu_{rs}(l+1)$ and total network cost $C_\Sigma(l+1) = \sum_{(r,s)\in E} \mu_{r0} C_{rs}(l + 1)$
3. Compare if $|C_\Sigma(l) - C_\Sigma(l + 1)| < \varepsilon$, where ε is given accuracy, then end of algorithm. Found capacities $\{\mu_{rs}(l + 1)\}$ and flows distribution of all classes $F_k(l + 1)$ are optimal, otherwise $l := l + 1$ and go to the next iteration.

5   Survivability Analysis

We'll consider the system survivability as its ability to preserve its functioning and to ensure the fulfillment of its main functions (perhaps in the shortened amount) under given QoS. As the main MPLS network function is the transmission of different classes of packets flows so we'll estimate the survivability level as maximal flow value to be transmitted in a network under its channels and nodes failures under given values of QoS factors [3].

5.1   Survivability Analysis Problem Statement

Let it be given MPLS network which is defined by oriented graph $G = \{X, E\}$. Assume that in network K classes of service (CoS) are transmitted due to so-called demand matrices $H(k) = ||h_{ij}(k)||, i = \overline{1, N}, j = \overline{1, N}$ (Mbits per sec). For each flow class CoS the corresponding QoS are introduced. It's necessary to determine the survivability factors for a given network.

In papers [2,3] the following complex factor was suggested for survivability analysis of MPLS networks

$$P\{H_\Sigma^\Phi(1) \geq r\%H_\Sigma^0(1)\}, P\{H_\Sigma^\Phi(2) \geq r\%H_\Sigma^0(2)\}, \ldots, P\{H_\Sigma^\Phi(k) \geq r\%H_\Sigma^0(k)\}, \tag{18}$$

where $H_\Sigma^0(k)$ is k-th class flow value in the faultless state; $H_\Sigma^\Phi(k)$ is real flow value of class k in case of failures; $r = (50 \div 100)$; $k = \overline{1, K}$; $P\{H_\Sigma^0(k) \geq r\%H_\Sigma^0(k)\}$ is the probability that flow value of k-th class transmitted in a network would be not less than fraction r of nominal flow value in the faultless state $H_\Sigma^0(k)$.

5.2   The Algorithm of MPLS Networks Survivability Analysis

Let MPLS network $G = (X, E)$ be considered consisting of n elements: channels and nodes exposed to influence of environment due to which they may fail. It's assumed that reliability characteristics of network elements such as readiness coefficients of channels $k_{\Gamma(r,s)}$ and nodes k_{Γ_i} are known. Consider the following network failure states:

1. Failure of one channel: Z_1;
2. Failure of one node: Z_2;
3. Failure of two channels: Z_3;
4. Simultaneous failure of one channel and one node: Z_4;
5. Failure of three channels: Z_5.

Assuming failures of network elements to be statistically independent events we may determine the probability of each state $P(Z_i)$. For example, if Z_i is the failure state of the channel (r_i, s_i) then

$$P(Z_i) = \left(1 - K_{\Gamma_{r_i, s_i}}\right) \prod_{(r,s) \neq (r_i, s_i)} K_{\Gamma_{r,s}} \prod_{i=1}^{n} K_{\Gamma_i}, \tag{19}$$

where $K_{\Gamma_{r,s}}$ is a probability of faultless state of the channel (r, s), $(r, s) \neq (r_i, s_i)$; $1 - K_{\Gamma_{r_i, s_i}}$ is a probability of the channel (r_i, s_i) failure.

In [3] MPLS networks survivability estimation algorithm was suggested, which consists of the following steps.

1. Compute the total flow value in the faultless state for all classes of service (CoS) $H_\Sigma^{(0)}(1), H_\Sigma^{(0)}(2), \ldots, H_\Sigma^{(0)}(K)$.
2. Simulate network different failure states: Z_1, Z_2, Z_3, Z_4, Z_5. For each failure state calculate the corresponding probability according to (19).

3. Find the maximal flow value for all CoS in the state $Z_j : H_{\Sigma}^{\Phi}(k, z_j), k = \overline{1, K}$. For it we use specially developed algorithm of finding maximal flow [1,2].

4. Calculate the complex survivability index for each CoS:
   for CoS k

$$P\{H_{\Sigma}^{\Phi}(k) \geq r\%H_{\Sigma}^{0}(k)\} = \sum_{Z_i} P(Z_i), \tag{20}$$

where summing in (20) is performed over all states Z_i such that $H_{\Sigma}^{\Phi}(k) \geq r\%H_{\Sigma}^{(0)}(k)$; $H_{\Sigma}^{(0)}(k)$ is nominal flow value of the class k in network faultless state; $H_{\Sigma}^{\Phi}(k)$ is real flow value of the class k in case of failures; $r = (50 \div 100)$; $k = \overline{1, K}$.

The found dependencies $P\{H_{\Sigma}^{\Phi}(1) \geq r\%H_{\Sigma}^{0}(1)\}, P\{H_{\Sigma}^{\Phi}(2) \geq r\%H_{\Sigma}^{0}(2)\}, \ldots,$ $P\{H_{\Sigma}^{\Phi}(k) \geq r\%H_{\Sigma}^{0}(k)\}$, are further presented as curves in coordinates $P\{H_{\Sigma}^{\Phi}(k)\} - r\%H_{\Sigma}^{0}$ and by its change we may estimate the survivability of the corresponding network.

6 The Network Survivability Optimization Problem Statement

In the process of network design after analysis of its survivability the problem arises to ensure the desired survivability level. Naturally, this problem may be solved by the way of reserving its channels and nodes and the structural optimization demanding the additional expenses. Consider the corresponding problem statement of network structural optimization by survivability indices.

Let it be MPLS network which as earlier is defined by oriented graph $G = \{X, E\}$, where $X = \{x_j\}$ is a set of network nodes, $E = \{(r, s)\}$ is a set of channels; μ_{rs} are channels capacities.

The reliability characteristics of channels and nodes are given, namely readiness coefficients for channels $K_{\Gamma_{rs}}$ and nodes K_{Γ_i} and corresponding failure probabilities $P_{failure, rs} = 1 - K_{\Gamma_{rs}}$.

For each class k quality of service (QoS) is given as a of mean delay time value $T_{mean, k}$. Let judging from functional purpose the following values of survivability indicators are established for k flow class: $P_{0,preassigned}^{(k)}, P_{2,preassigned}^{(k)}, \ldots, P_{5,preassigned}^{(k)}$.

It is demanded to determine such network structure for which the following requirements on survivability level will be ensured:

$$P\{H_{\Sigma}^{\Phi}(k) \geq r\%H_{\Sigma}^{(0)}(k)\} \geq P_{r,preassigned}^{(k)}, r = (50 \div 100), k = \overline{1, K}, \tag{21}$$

and the additional costs would be minimal: $C_{\Sigma} = \sum C_{rs}^{res}(\mu_{rs}) \rightarrow \min$.

The achievement of the desired survivability level we'll ensure by introducing corresponding reservation of the most responsible channels and nodes. For reservation efficiency estimation we propose to introduce the following index for channels

$$\alpha_{r_i s_i} = -\frac{\Delta P(Z_i)}{C_{r_i s_i}}, \tag{22}$$

where Z_i is a state of failure of the channel (r_i, s_i); $\Delta P(Z_i)$ is a probability change of the state Z_i in case of its reservation, $C_{r_i s_i}$ is a cost of this reservation.

The value $\Delta P(Z_i)$ is estimated by the following formula:

$$P_{res}(Z_i) - P(Z_i) =$$

$$= -(1 - P_{failure, r_i s_i}) \cdot P_{failure, r_i s_i} \prod_{(r,s) \neq (r_i, s_i)} K_{\Gamma_{r,s}} \cdot \prod_{i=1}^{n} K_{\Gamma_i} =$$

$$= -(1 - P_{failure, r_i s_i}) \cdot P(Z_i), \quad (23)$$

where $P_{failure, r_i s_i}$ is failure probability of the channel (r_i, s_i).

The similar expressions are used for estimation of nodes reservation.

The index $\alpha_{r_i s_i}$ is used for selection the proper elements nodes and channels to be reserved in the first turn. In [3] the following method of MPLS network optimization by survivability level is suggested.

6.1   Method of MPLS Network Optimization by Survivability

The algorithm consists of finite number of iterations. On each iteration the next element (node or channel) is reserved.

k iteration

1. For each channel and node the index $\alpha_{r_i s_i}$ is calculated by formula (22).
2. Select channel (r^*, s^*) such that $\alpha_{r^* s^*} = \max_{(r^*, s^*)} \alpha_{r_i s_i}$.
3. Reserve channel (r^*, s^*) and recalculate survivability indices for all the flow classes after reservation using the following formula:

$$P^H\{H_\Sigma^\Phi(k) \geq r\% H_\Sigma^{(0)}\} = P\{H_\Sigma^\Phi(k) \geq r\% H_\Sigma^{(0)}\} + |\Delta P(Z_i^*)|, \quad (24)$$

where $\Delta P(Z_i^*)$ is a probability change of the state Z_i after the channel (r^*, s^*) reservation.

4. Check the fulfillment of condition (25):

$$P^H\{H_\Sigma^\Phi(k) \geq r\% H_\Sigma^{(0)}\} \geq P_{r, preassigned}^{(k)}, r = (50 \div 100), k = \overline{1, K}. \quad (25)$$

If the conditions (25) are fulfilled for each r and for all the classes K, then stop otherwise go to $(k+1)$ iteration.

The described iterations are repeated until the condition (6) would be true. The algorithm converges by finite number of iterations not exceeding $m + n$, where m is a number of channels n is a number of nodes.

7   NG Network Structure Optimization

The final stage of NGN performance analysis and optimization is network structure synthesis (optimization). Its statement is follows.

There are given a set of network nodes $X = \{x_j\}, j = \overline{1,n}$- MPLS routers (so-called LRS Label Switching Routers), their locations over regional area, channels capacities $D = \{d_1, d_2, \ldots, d_k\}$ and their costs per unit length $C = \{c_1, c_2, \ldots, c_k\}$, classes of service (CoS) are determined, matrices of input demands for the k-th CoS are known $H(k) = \|h_{ij}(k)\|; i, j = \overline{1,n}; k = \overline{1,K}$, where $h_{ij}(k)$ is the intensity of flow of the k-th CoS which is to transfer from node i to node j in sec (Mbits/s).

Additionally the constraints are introduced on the Quality of Service (QoS) for each class k as constraint on mean PTD (packets transfer delay) $T_{inp,k}, k = \overline{1,K}$, (16) and packets loss ratio (17).

It's demanded to find network structure as a channels set $E = \{(r, s)\}$, choose channels capacities $\{\mu_{rs}\}$ and find flows distributions of all classes so that ensure the transmission demands of all classes $H(k)$ in full volume with mean delays T_{av}, not exceeding the given values $T_{preassigned,k}$ and by this the constraint on packets loss ratio (PLR) should be fulfilled and total network cost be minimal [1,4].

Let's construct the mathematical model of this problem.

It's demanded to find such network structure $E = \{(r, s)\}$ for which

$$\min_{\{\mu_{rs}\}} C_\Sigma(M) = \sum_{(r,s) \in E} C_{rs}(\{\mu_{rs}\}), \tag{26}$$

under constraints

$$T_{av}(\{\mu_{rs}\}; \{f_{rs}\}) \leq T_{preassigned}, k = \overline{1,K}, \tag{27}$$

$$PLR_k(\{\mu_{rs}\}; \{f_{rs}\}) \leq PLR_{k,preassigned}, k = \overline{1,K}, \tag{28}$$

where $PLR_k(\{\mu_{rs}\}; \{f_{rs}\})$ is packets loss ratio for the k-th flow, $PLR_{k,preassigned}$ given constraint on its value.

This problem belongs to so-called NP-complete optimization problems. For its solution general genetic algorithm was developed using two operators: crossover and mutation [4].

Define matrix of channels $K = \|k_{ij}\|$, where

$$k_{ij} = \begin{cases} 1, \text{ if } \exists(i - j), \\ 0, \text{ otherwise, i.e. } \neg\exists(i - j), \end{cases} \tag{29}$$

for each network structure. Then initial population of different structures in given class of multi-connected structures is generated with connectivity coefficient 2. For synthesis we'll use semi-uniform crossover which is grounded for small population size.

Parents (structures $E_i(k)$ are chosen randomly with probability inverse proportional to cost $C_\Sigma(E_i(k))$, each parent is determined by matrix $K^i, i = 1, 2$. In the process of semi-uniform crossover each descendant receives exactly a half of quantity of parents genes. The crossover mask is presented as the following matrix $M = \|m_{ij}\|$:

$$m_{ij} = \begin{cases} 0, \text{ if } p \geq p_0, \\ 1, \text{ if } p < p_0, \end{cases}$$

where parameter $p_0 = 0,5$ and $p \in [0;1]$ is a random value.

This process of crossover may be written so:

$$E(k)' = \|e(k)'_{ij}\| = \begin{cases} (i-j)^1, k^1_{ij} = 1, \text{if } m_{ij} = 0; \\ (i-j)^2, k^2_{ij} = 1, \text{if } m_{ij} = 1. \end{cases}$$

In this crossover only one descendant is generated that to maximize the algorithm productivity. In case if after crossover isolated sub-graphs were generated then with direct channels they are connected to a root. Further for generated descendant-structure $E(k)'$ capacities assignment and flows distribution problems are solved described in Sect. 4 and network total cost is calculated. Then using cost value C_Σ decision is made whether to include new structure $E(k)'$ in the set of local optimal structures (new population) Π or not.

After crossover operation the mutations are performed. Note that basic GA algorithm suggested in [1] uses unconditional mutations. Mutations consist in deleting or adding new channels into network structure.

In the process of algorithm perfection in pair with unconditional mutation the following schemes of crossover probability change were applied and investigated [4]:

- *Deterministic.* Implementation of deterministic scheme of crossover probability change is based on hypothesis that at different stages of genetic search crossover may be more or less significant, therefore as a function of crossover probability change should be used non-monotonic function. That's why in this problem we used the function $\sigma(t) = |\sin(t)|$, where $0 \le t \le T$.
- *Adaptive.* Define adaptive crossover as an operator whose probability decreases if population is sufficiently homogeneous and increases in case if the population is heterogeneous. As the measure of homogeneity/ heterogeneity we chose the value

$$C_\Delta = \max C_\Sigma(E_i(k)) - C_\Sigma(E_j(k)), i \in [1, \ldots, n], j \in [1, \ldots, n], i \ne j,$$

where n is a population size (in this case $n = 3$).

It's reasonable to suggest that in case of similar individuals in population the crossover operation will be inefficient and vice versa. Therefore in adaptive crossover the law of crossover probability change takes the form:

$$\sigma(t) = \begin{cases} \sigma(t-1)\lambda, & \text{if } C_\Delta > C^*, \\ \sigma(t-1)/\lambda, & \text{if } C_\Delta < C^*, \\ \sigma(t-1), & \text{if } C_\Delta = C^*, \end{cases} \tag{30}$$

where C^* is threshold value, $\lambda = 1$, 1 is a learning coefficient.

After experiments with real problems of NG network structure optimization it was found that a *combination of dynamic adaptive crossover and unconditional mutation: proved to be the most successful* and ensured the performance increase (cost cut) by 20–22% [4]. Thus, the hypothesis about some positive properties of crossover which mutation operator doesn't have was confirmed. Note that application of crossover is grounded only if a population is heterogeneous, i.e. the individuals differ from each other sufficiently.

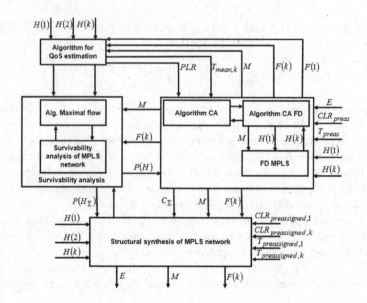

Fig. 1. Structure of software kit "NGN NetBuilder"

8   Structure and Functions of Software Kit "NGN NetBuilder"

In accordance with developed algorithms a software kit "NGN Builder" was designed. Software kit consists of the following functional modules:

– Module of estimation of QoS indicators: packets average delay $T_{mean,k}$ and PLR_k;
– Module of analysis and performance optimization of NGN including software products flows distribution (FD) and capacities assignment (CA) and flows distribution;
– Module of NGN survivability analysis and optimization;
– Module of structural synthesis of NG networks

Structure of software complex "NGN Builder" and information connections between modules are presented in Fig. 1. Apart from main functional modules software complex contains service modules, including interface module for interconnection with designer. This software module allows to put out the synthesized structure at map, to perform design procedures, to present by indication of user at the display flows distribution and transmission routes, the channels characteristics, their loads.

For software kit "NGN NetBuilder" was developed and installed a data bank consisting of several data bases (DB) including:

1. DB of input flows;
2. DB of network channels and nodes with their characteristics;
3. DB of flows distribution of different classes over channels.

All functional modules interact with each other by data bases. As the DB management system MySQL Server is used.

The software kit "NGN NetBuilder" enables to solve wide scope of practical problems of NG networks analysis and optimal design under constraints on the values of QoS indicators: packets transfer delays and packets loss ratio *PLR* for different classes of service, analyze survivability and perform structure optimization under constraint on survivability level.

9   Experimental Results

The experimental investigations of methods and algorithms of software kit were carried out. Some of them are presented below.

Algorithms FD and CA FD were explored at the global Ukrainian network whose structure is presented in the Fig. 2. Three classes of service were introduced. During experiments in the algorithm FD elements of demand matrix were increased by multiplying of corresponding elements on a coefficient k. Under the increase of input demands the delays in data transfer were also increased for all classes of service.

The dependence of average packets transfer delay for the first class (K_1) is presented in the Fig. 3, and for the second class (K_2) - in the Fig. 4. In the next experiments the algorithm CCFD was investigated. In these experiments the dependence of network cost versus transfer delay for different CoS was explored. The corresponding dependence for class 2 is presented in the Fig. 5. As it was expected the less is constraint on transfer delay the greater would be the network cost. This may be explained thereby for decrease of values T_{mean} it's necessary to increase channels capacities (Fig. 6).

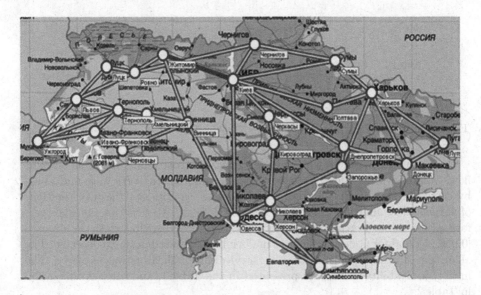

Fig. 2. Structure of a global investigated network

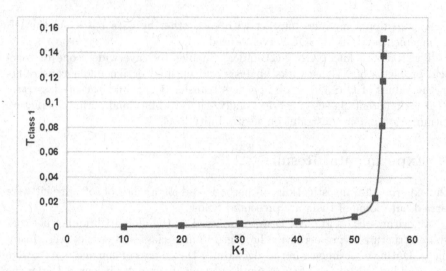

Fig. 3. Dependence PTD for first class flow (CoS_1) versus loading coefficient K_1

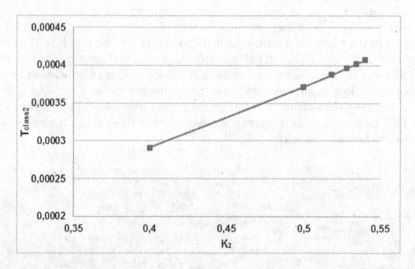

Fig. 4. Dependence PTD for second class flow (CoS_2) versus loading coefficient K_2

In the next experiment the dependence network cost versus variation of input flows volumes (intensities). For this initial demand matrices H_k were multiplied by corresponding coefficient k. In the following experiments the investigations of survivability analysis algorithms were carried out. All experiments were performed for channels readiness coefficients uniformly distributed in the interval [0,9; 0,95], and for nodes coefficients uniformly distributed in the [0,95; 0,99]. The algorithm of network optimization by survivability indicators was explored. The initial survivability values are presented in the Table 1, and after optimization in Table 2.

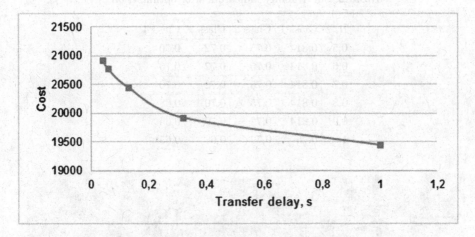

Fig. 5. Dependence network cost versus constraint on PTD for second class flow

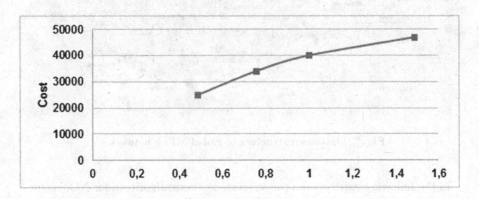

Fig. 6. Dependence network cost versus input flow intensity

Table 1. Initial survivability indicators for different CoS before optimization.

H,%	Class 1	Class 2	Class 3	Class 4
0,5	0,385	0,385	0,385	0,385
0,6	0,385	0,385	0,385	0,385
0,7	0,385	0,385	0,385	0,385
0,8	0,385	0,385	0,385	0,385
0,9	0,385	0,385	0,385	0,357
1	0,385	0,385	0,385	0,357

Table 2. Survivability indicators after optimization.

H,%	Class 1	Class 2	Class 3	Class 4
0,5	0,814	0,75	0,72	0,70
0,6	0,814	0,75	0,72	0,70
0,7	0,814	0,75	0,72	0,70
0,8	0,814	0,75	0,70	0,68
0,9	0,814	0,72	0,70	0,65
1	0,75	0,72	0,68	0,65

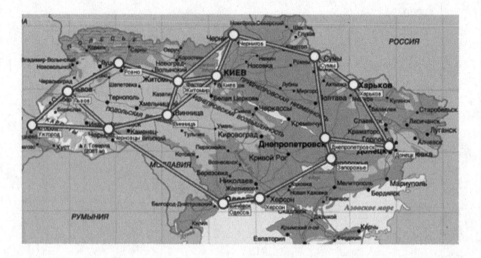

Fig. 7. Optimized structure of global MPLS network

In the following experiments the suggested genetic algorithm for NGN structural optimization was investigated. During experiments the matrices of input flows and constraints on QoS indicators were varied. Analysing the obtained structures after varying the intensities of input flows in the range 0, 2–2, 0 of nominal values it was detected the stability of network back bone structure under sufficiently wide variation of input load.

In the Fig. 7 the optimized structure obtained with application of suggested genetic algorithm which realizes a combination of dynamic adaptive crossover and unconditional mutation is presented. Note that cut in total network cost for optimized structure in comparison with initial structure equals to 14250 thousand \$−10023 thousand \$ = 4227 thousand \$, i.e. the cost after optimization is by 30% less than the cost of initial network structure. Additionally the suggested algorithm productivity increased by 22% in comparison with basic genetic algorithm [4].

10   Conclusion

The instrumental kit of algorithms and programs "NGN Builder" for analysis and optimization of NGN networks is described. The developed software kit is based on the original methods and algorithms.

The developed kit enables to solve wide range of NG networks design problems, incl. optimal flows distribution, optimal capacities choice and flows distribution, survivability analysis and network structure synthesis under constraints on the set QoS values.

The experimental investigations of suggested algorithms were carried out.

The developed kit enables to cut design time, and decrease the costs of NG networks to be designed.

References

1. Zaychenko, H., Zaychenko, Y.P.: MPLS Networks modeling, analysis and optimization. Kiev. NTUU "KPI", 240 p (2008). (in Russian)
2. Zaychenko, H.: Complex of models and algorithms for performance optimization of MPLS networks. Syst. Res. Inf. Technol. #4, pp. 58–71 (2007). (in Russian)
3. Zaychenko, Y., Zaychenko, H.: New generation computer networks survivability analysis and optimization. In: Vishnevsky, V., Kozyrev, D., Larionov, A. (eds.) DCCN 2013. CCIS, vol. 279, pp. 73–81. Springer, Heidelberg (2014). doi:10.1007/978-3-319-05209-0_6
4. Zaychenko, H.Y., Anikiev, A.S.: Efficiency of genetic algorithm application for MPLS network structure synthesis. Vestnik of ChNTU 1., Chercassy, pp. 176–182 (2008). (in Russian)

One Problem of the Risk Control

A.M. Andronov[(⊠)] and T. Jurkina

Transport and Telecommunication Institute, Lomonosov Str., 1, Riga LV-1019, Latvia
lora@mailbox.riga.lv, ju_ta@mits.lv

Abstract. A problem of supply order is considered. It is necessary to determine an order which gives the minimal probability of the upsetting of the supply. A corresponding probabilistic model is elaborated. The reduced gradient method is used for the minimization. Numerical example illustrates the efficiency of the suggested approach.

Keywords: Normal distribution · Reduced gradient method · Planned reward · Risk minimization

1 Introduction

We consider the following problem of the risk. A firm wishes to order some product. The full size of the order equals m^*. For that the firm has the money resources C. There exists n suppliers with numbers $i = 1, \ldots, n$. The i-th supplier is characterized by the following indices: a_i - maximal size of the supply, c_i - the cost of the product per unit. A product quality is different for the different suppliers and is a random variable. As result, production unit of the i-th supplier gives a random reward R_i. Additionally, R_i has normal distribution with mean r_i and standard deviation σ_i. Vector $R = (R_1, R_2, \ldots, R_n)$ has multivariate distribution with mean $r = (r_1, r_2, \ldots, r_n)$ and covariance matrix $\sigma = (\sigma_{i,j})_{n \times n}$ where $\sigma_{i,i} = \sigma_i^2$. The firm is planning getting a reward r^*, at least. It is necessary to determined supply plan $m = (m_1, m_2, \ldots, m_n)$, which satisfies the request size m^* and the disposed sum C, and gives the minimal probability that gotten reward $S = R_1 m_1 + R_2 m_2 + \ldots + R_n m_n$ will be less then r^*.

Let us give a mathematical setting of the described problem. The reward of the firm is the random variable $S = m_1 R_1 + m_2 R_2 + \ldots + m_n R_n = m^T R$. We suppose that it has normal distribution with mean and variance, calculated by formulas

$$E(S) = r_1 m_1 + r_2 m_2 + \ldots + r_n m_n = r^T m,$$

$$V(S) = m^T \sigma m = \sum_{i=1}^{n} \sum_{j=1}^{n} m_i \sigma_{i,j} m_j. \tag{1}$$

If $\Phi(z)$ is the cumulative distribution function of the standard normal distribution, then the probability doesn't receive the planned reward r^*

$$P\{S \leq r^*\} = \Phi\left(\frac{r^* - r^T m}{\sqrt{V(S)}}\right). \tag{2}$$

© Springer International Publishing AG 2016
V.M. Vishnevskiy et al. (Eds.): DCCN 2016, CCIS 678, pp. 162–167, 2016.
DOI: 10.1007/978-3-319-51917-3_15

We must minimize this probability under restrictions:

$$m_1 + m_2 + \ldots + m_n = m^*,$$

$$m_1 c_1 + m_2 c_2 + \ldots + m_n c_n \leq C, \tag{3}$$

$$0 \leq m_i \leq a_i, \quad i = 1, \ldots, n.$$

It is possible to simplify the formulas if to consider weights $w = \left(\frac{m_1}{m^*}, \frac{m_2}{m^*}, \ldots, \frac{m_n}{m^*}\right)$ instead of $m_1, m_2, \ldots, m_n$. Setting $C^* = C/m^*$, $\rho = r^*/m^*$, $\alpha_i = c_i/m^*$, and $D(w) = V(S)(m^*)^{-2} = w^T \sigma w$, have the problem:
Minimize

$$f(w) = P\left\{\frac{S}{m^*} \leq \rho\right\} = \Phi\left\{\frac{\rho - r^T w}{\sqrt{D(w)}}\right\} \tag{4}$$

under restrictions

$$w_1 + w_2 + \ldots + w_1 = 1,$$

$$w_1 c_1 + w_2 c_2 + \ldots + w_n c_n \leq C^*,$$

$$0 \leq w_i \leq \alpha_i, \quad i = 1, \ldots, n. \tag{5}$$

We will solve this problem by reduced gradient method [1].

2   Reduced Gradient Method

Using vector-matrix notations

$$A = \begin{pmatrix} 1 & 1 & \ldots & 1 \\ c_1 & c_2 & \ldots & c_n \end{pmatrix}, \quad \begin{pmatrix} 1 \\ C^* \end{pmatrix}, \quad w = (w_1 \ w_2 \ \ldots \ w_n)^T,$$

we rewrite two upper equations as

$$Aw = b. \tag{6}$$

Let us declare w_i and w_j, $i \neq j$, as basic variables, the remaining - non-basic ones, and denote $w_B = (w_i, w_j)$, $w_N = \{w_1, \ldots, w_n\} - (w_i, w_j)$ so $w = (w_B \ w_N)$. Let B be the submatrix of A corresponding to the basic variables, and N be the analogous matrix for non-basic variables. We suppose that the basic is such, that matrix B is nonsingular matrix. Then we have

$$Bw_b + Nw_N = b,$$

$$w_b = B^{-1}(b - Nw_N) = \bar{b} - \bar{N}w_N,$$

$$where \ \bar{b} = B^{-1}b, \ \bar{N} = B^{-1}N.$$

As the basic variables dependent on non-basic variables only, we have for the aim function

$$\bar{f}(w_N) = f(w_B, w_N) = f(B^{-1}(b - N\omega_N)w_N). \tag{7}$$

The gradient is calculated with respect to the chain rule [2]:

$$\frac{\partial}{\partial w_N}\bar{f}(w_N) = \frac{\partial}{\partial w_N}f(w_B, w_N)|_{w_B=\bar{b}B^{-1}w_N} + \left(\frac{\partial}{\partial w_N}w_B\right)\frac{\partial}{\partial w_B}f(w_B, w_N) =$$

$$\frac{\partial}{\partial w_N}f(w_B, w_N)|_{w_B=\bar{b}-\bar{N}w_N} + (-B^{-1}N)^T\frac{\partial}{\partial w_B}f(w_B, w_N)|_{w_B=\bar{b}-\bar{N}w_N}. \quad (8)$$

The gotten vector is called the reduced gradient $u_N(w)$:

$$u_N(w) = \frac{\partial}{\partial w_N} == \bar{f}(w_N) =$$

$$\frac{\partial}{\partial w_N}f(w_B, w_N)|_{w_B=\bar{b}-\bar{N}w_N} - \bar{N}^T\frac{\partial}{\partial w_B}f(w_B, w_N)|_{w_B=\bar{b}-\bar{N}w_N}. \quad (9)$$

Now the gradient optimization works as follows. Let current value $w = (w_B, w_N)$ be gotten. We go to direction $y = (y_B, y_N)$, where y_N is determined as follows:

$$y_i = 0 \quad if \; > 0 \quad and \quad w_i = 0, \quad or \quad u_i < 0 \quad and \quad w_i = \alpha_i,$$

$$y_i = -u, \quad otherwise,$$

and for the basic variables $y_B = -\bar{N}w_N$.

If $y_N = 0$, then the optimum is reached. Otherwise we find value $\theta > 0$, which minimizes function $g(w) = Pr(w+\theta y)$. The non-negation condition requests that

$$\theta \leq \theta_{max} = \min_{y_i < 0}\left\{-\frac{w_i}{y_i}\right\}.$$

This one-dimension procedure gives optimal value θ^* on the interval $(0, \theta_{max})$. Two cases are possible here.

(1) $\theta^* < \theta_{max}$. The procedure is continued with new point $w' = w + \theta^* y$ and the same base B.

(2) $\theta^* < \theta_{max}$. In this case one basic variable, let with number s, is cancelled. If it is non-basic variable then the base doesn't change and all is repeated. If it is a basic variable then the base is changed. Any non-basic variable with number t is introduced on the base, if new matrix B will be nonsingular.

3   Reduced Gradient of the Aim Function

We have:

$$\frac{\partial}{\partial w_i}\Phi\left(\frac{\rho - r^T w}{\sqrt{D(w)}}\right) = \frac{\partial}{\partial w_i}\int_{-\infty}^{\frac{\rho-r^T w}{\sqrt{D(w)}}}\frac{1}{\sqrt{2\pi}}exp\left(-\frac{1}{2}z^2\right)dz$$

$$= \frac{1}{\sqrt{2\pi}} exp\left(-\frac{1}{2}\left(\frac{\rho - r^T w}{\sqrt{D(w)}}\right)^2\right) \frac{\partial}{\partial w_i}\left(\frac{\rho - r^T w}{\sqrt{D(w)}}\right)$$

$$= \frac{1}{\sqrt{2\pi}} exp\left(-\frac{1}{2}\left(\frac{\rho - r^T w}{\sqrt{D(w)}}\right)^2\right).$$

$$\left(-\frac{1}{\sqrt{D(w)}} r_i - (\rho - r^t w)\frac{1}{2}D(w)^{-3/2}\frac{\partial}{\partial w_i}\left(\sum_{k=1}^{n}\sum_{j=1}^{n} w_k \sigma_{k,j} w_j\right)\right)$$

$$= \frac{1}{\sqrt{2\pi}} exp\left(-\frac{1}{2}\left(\frac{\rho - r^T w}{\sqrt{D(w)}}\right)^2\right).$$

$$\left(-\frac{1}{\sqrt{D(w)}} r_i - (\rho - r^T w)\frac{1}{2}D(w)^{-3/2}2\sum_{j=1}^{n} \sigma_{i,j} w_j\right).$$

Therefore the full gradient is of the form

$$\nabla f(w) = \frac{\partial}{\partial w} f(w) = \frac{1}{\sqrt{2\pi}} exp\left(-\frac{1}{2}\left(\frac{\rho - r^T w}{\sqrt{D(w)}}\right)^2\right).$$

$$\left(-\frac{1}{\sqrt{D(w)}} r - (\rho - r^T w)D(w)^{-3/2}\sigma w\right).$$

The sub-gradient $\frac{\partial}{\partial w_N} f(w)$ is the following:

$$\frac{\partial}{\partial w_N} f(w) = \frac{1}{\sqrt{2\pi}} exp\left(-\frac{1}{2}\left(\frac{\rho - r^T w}{\sqrt{D(w)}}\right)^2\right).$$

$$\left(-\frac{1}{\sqrt{D(w)}} r_N - (\rho - r^T w)D(w)^{-3/2}\sigma w_N\right), \qquad (10)$$

where r_N and w_N are subvectors of r and w, corresponding to non-basic variables.

The sub-gradient $\frac{\partial}{\partial w_B} f(w)$ is calculated analogously by means of change r_N and w_N by r_B and w_B. Also the reduced gradient is calculated.

4   Numerical Example

Below there is presented relative initial data, corresponding to unit production: the disposed financial sum $C^* = 15$, the number of the suppliers $n = 5$, the vector of the costs per unit of the production $c = (9\ 12\ 15\ 18\ 21)$, the vector of reward means $r = (10\ 15\ 20\ 25\ 30)$. A restrictions on maximal sizes of the supply $\{a_i\}$ is absent. Rewards from various suppliers are multivariate random vector with standard deviation $\sigma = (4\ 7\ 9\ 12\ 17)$. Minimal desired reward $r^* = 25$.

Iterative procedure begins with uniform distributed orders $w = (0.2\ 0.2\ 0.2\ 0.2\ 0.2)$. Matrix A and vector b from (6) are the following:

$$A = \begin{pmatrix} 1 & 1 & 1 & 1 & 1 \\ 9 & 12 & 15 & 18 & 21 \end{pmatrix}, \quad b = \begin{pmatrix} 1 \\ 15 \end{pmatrix}.$$

Initially the basic variables are w_1 and w_2, so

$$B = \begin{pmatrix} 1 & 1 \\ 9 & 12 \end{pmatrix}, \quad N = \begin{pmatrix} 1 & 1 & 1 \\ 15 & 18 & 21 \end{pmatrix},$$

$$B^{-1} = \begin{pmatrix} 4 & -1/3 \\ -3 & 1/3 \end{pmatrix}, \quad \bar{N} = B^{-1}N = \begin{pmatrix} -1 & -2 & -3 \\ 2 & 3 & 4 \end{pmatrix}.$$

Table 1 below contains stepwise results of the gradient optimization: the current vector of the variables w, corresponding aim's function value P, and further the numbers i and j of basic variables, the step length along gradient θ for the next step. We see that minimal probability equals 0.717 and is reached by vector $(0.5\ 0\ 0\ 0\ 0.5)$, but it is a local minimum only. The less value of the probability is reached for the degenerative case when the third supplier is used only, namely vector $(0\ 0\ 1\ 0\ 0)$ gives probability 0.711. As the sequence of basic solutions $(\epsilon\ 0\ 1 - 2\epsilon\ 0\ \epsilon)$ and $(0\ \epsilon\ 1 - 2\epsilon\ 0\ \epsilon)$ trends to $(0\ 0\ 1\ 0\ 0)$ when ϵ tends to 0, we can get probability P arbitrarily close to 0.711.

Now we suppose that restrictions on maximal sizes of the supply from the third and fifth suppliers exist: $\alpha_3 = 0.8, \alpha_5 = 0.4$. In this case vector $(0.10\ 0.8\ 0\ 0.1\ 0)$ gives optimal solution with probability $P(0.10\ 0\ 0.8\ 0\ 0.10) = 0.750$. Other solution grows worse: $P(0.4\ 0\ 0.2\ 0\ 0.4) = 0.756$, $P(0\ 0.5\ 0\ 0.5\ 0.10) = 0.764$. The situation is changes for various values of the desired reward r^*. For example, if restrictions on maximal sizes of the supply α_i absents and the firm wish cover its own expenses at least, i.e. $r^* = C^* = 15$, then the optimal solution is $(0.169\ 0.226\ 0.224\ 0.197\ 0.184)$ or $(0.182\ 0.208\ 0.216\ 0.216\ 0.178)$. This solution insures the probability of the expenses covering 0.145. It is close

Table 1. Protocol of gradient optimization

Step	1	2	3	4	5	7	8
w_1	0.200	0.247	0.436	0.464	0.492	0.500	0.500
w_2	0.200	0.136	0.048	0.009	0.009	0.000	0.000
w_3	0.200	0.201	0.055	0.056	0.000	0.000	0.000
w_4	0.200	0.203	0.001	0.002	0.002	0.000	0.000
w_5	0.200	0.213	0.459	0.468	0.496	0.500	0.500
P	0.851	0.846	0.730	0.719	0.717	0.717	
i	1	1	1	1	2	2	2
j	2	5	2	3	3	3	3
θ	0.02	0.02	0.01	0.01	0.01	0.01	0.01

to uniform distributed order $w = (0.2\ \ 0.2\ \ 0.2\ \ 0.2\ \ 0.2)$, which gives probability 0.149. The previous optimal solutions $(0.5\ \ 0.0\ \ 0.0\ \ 0.0\ \ 0.5)$ and $(0\ \ 0\ \ 1\ \ 0\ \ 0)$ give worse results 0.283 and 0.289, correspondingly.

More involved situation arises, if we take into account dependence of the rewards for various suppliers.

5 Conclusions

Considered examples show that it is impossible to foresee a solution, that insures the minimal risk. The probability theory gives us such possibilities.

References

1. Minoux, M.: Programmation mathematique. Theorie et algorithmes. Bordas, Paris (1989)
2. Turkington, D.A.: Matrix Calculus & Zero-One Matrices. Statistical and Econometric Applications. Cambridge University Press, Cambridge (2002)

Analysis of the Throughput in Selective Mode of Transport Protocol

Vladimir Kokshenev[1], Pavel Mikheev[2], Sergey Suschenko[2(✉)],
and Roman Tkachyov[2]

[1] F5 Networks Inc., Seattle, USA
doka.patrick@gmail.com
[2] National Research Tomsk State University, Lenina str., 36, 634050 Tomsk, Russia
ssp.inf.tsu@gmail.com

Abstract. The proposed model is a virtual connection managed by the transport protocol with a forward error correction mechanism for selective repeat mode in the form of Markov chain with discrete time. The analysis of the impact of protocol parameters window size and the duration of the timeout of waiting confirmation, the likelihood of distortion of the segments in the individual links of the transmission path data, the duration of the round-trip delay, the parameters of mechanism to restore the distorted segments (without retransmissions) on throughput of a transport connection. In the area of protocol parameters, the characteristics of the transmission channel and parameters of the forward error correction mechanism found in the area of superiority of the management procedures of the transport protocol with forward error correction over the classic procedure with decision feedback on the criterion of the throughput of a transport connection. The expediency of applying of the method of forward error correction for transport links with large round-trip delay.

Keywords: Transport protocol · Data path · Forward error correction · Markov chain · Throughput of a transport connection · Window size · Duration of the timeout · Round-trip delay · Loss rate

1 Introduction

The most important indicator of the quality of interaction and networking applications used software and hardware of computer networks is the throughput transport links. This operating characteristic is largely determined by the transport protocol, its parameters—the width of the window and the duration of the timeout [1,2], as well as additional mechanisms to increase performance by reducing the number of retransmissions of distorted data [3–5]. Simulation of the subscriber connection and the analysis of its potential ability to perform is in the [2–10], and other works. But the results were obtained only for single-link data path [6–8], or with significant restrictions on the protocol parameters [9,10]. Known technology of forward error correction (FEC) are used as the transport protocol typically as services for the lower level network architecture [3–5]. A comprehensive analysis of

© Springer International Publishing AG 2016
V.M. Vishnevskiy et al. (Eds.): DCCN 2016, CCIS 678, pp. 168–181, 2016.
DOI: 10.1007/978-3-319-51917-3_16

the advantages and effectiveness of the methods of advanced error correction performed only on a qualitative level, as well as numerically for a number of special cases and not allowed to identify areas of possible application methods in the area of protocol parameters and characteristics of the transport connection. Modern transport protocols contain a wide variety of congestion control mechanisms [11]. There is a wide range of studies [11–21] in the field of control parameters of the transport protocol with the aim of preventing and bypass congestion, focused on building models of diagnostic over various indicators [11] and adaptation of protocol parameters and mechanisms of error correction to changing network load and connectivity, the level of losses, activity interactive subscribers, etc. conditions of data transmission. However, the implementation of control mechanisms to bypass congestion based on available bandwidth of the transport connection at the current and predictable to changing the values of protocol parameters. Thus, the potential of the transport protocol using the methods of forward error correction has not been studied yet. There is no analytical dependence of the overall effect of protocol parameters, the characteristics of the transmission channel data and parameters of the correction method, the resulting operating characteristics of the transport connection. Not studied the effect of correlations between the duration of round-trip delay and protocol parameters on the throughput of the transmission path of the managed data transport protocol. In addition, the process of data transmission in computer networks are essentially discrete in nature [22], due to the pipeline transfer mechanism in the network of limited size segments and the use of algorithms with decision feedback at different levels of network architecture, however, most of the results [2–7, 11–20] based on models with continuous time, which leads to the narrowing of the field of their applicability.

The paper presents a mathematical model of the data transfer process from the FEC in the phase of information transfer in the form of a Markov chain with discrete time, analytically found stationary distribution of probabilities of states for the mode of selective reject, [2] obtained analytical expressions for the bandwidth on the basis of which the analysis of potential capabilities of the transport connection is performed.

2   Transport Connection Model

Let's consider the process of transferring data between subscribers of transport protocol based on the algorithm with decision feedback and operating in a selective reject mode [2]. An example of such reliable family of protocols is dominant in modern computer networks, the TCP protocol [1]. We believe that interactive subscribers have unlimited data stream for transmission, and the exchange is performed by data units of the transport protocol (segments) of equal length. We believe that the area of hops along the transmission path data have the same speed in both directions, and the length of loop segment in a separate link is t. In general, the path length from the sender to the destination of the transfer information flow, and the length of the return path that are transmitted to received acknowledgment segments can be different. We believe that the data length of the transmission path, expressed in hop number of sites in the forward direction is equal $D_n \geq 1$. Reverse path delivers the confirmation to the sender about the correct reception of the blocks of the sequence of segments has length $D_o \geq 1$. Set the probability of segment distortions in communication channels for forward $R_n(d)$, $d = \overline{1, D_n}$ and reverse $R_o(d)$, $d = \overline{1, D_o}$ directions of transmission of each part of hops. Then the reliability of segments transmission along the path from the sender to the destination and back will be $F_n = \prod_{d=1}^{Dn}(1 - R_n(d))$ and $F_o = \prod_{d=1}^{Do}(1 - R_o(d))$ respectively. We believe that the loss of segments due to the absence in the buffer memory in tract's nodes does not occur. We believe that data transmission by the sender is implemented by blocks containing B segments of which $1 \leq A \leq B$—are informative, and $B - A$—additional (redundant) of the same length. Loss (distortion) to $B - A$ arbitrary segments in the block allows you to restore all of the block segments for RAID-arrays of the fourth level [23]. Flow control mechanism is implemented by sliding window [1,2] protocol parameters with a window width $w \geq 1$, expressed in a number of blocks. We believe that the verification of the correctness of the received target segments blocks are transferred in each segment of the oncoming flow. In case of impossibility of direct recovery of the transmitted segment blocks (the distortion more $B - A$ segments in the block) entire block is retransmitted. Then, the process of information transfer in virtual connection managed by the transport protocol, can be described by Markov process with discrete time (with the cycle duration t) due to the fact that the time between the receipt of confirmation has the geometric distribution with parameter F_o. This model is a generalization of the formalizations of the data transfer process, proposed in [3–6], in case of a transport connection of arbitrary length and the mechanism of forward error correction. The area of possible states of the Markov's chain determined by the duration of the timeout of waiting confirmation S, expressed in number of cycles of duration t. The size of the time-out is associated with the length of the tract, the width of the window and a block size of the inequalities $S \geq wB + 1$, $S \geq D_n + D_o + B - 1$. It is obvious that the sum of the lengths of the forward and backward paths can be interpreted as a circular delay of a single segment $D = D_n + D_o$, expressed in the durations t (excluding losses of protocol blocks in the transmission along the path). The cyclic delay for block of segments will be $D + B - 1$. The states of the Markov chain $i = \overline{0, wB}$ corresponds to the size of the queue transmitted, but not confirmed segments in the stream, and the states $i = \overline{wB + 1, S - 1}$—the time during which the sender is not active and is waiting to receive confirmation about the correctness of receiving the transmitted sequence from w blocks of segments. From the zero state in the $(D + B - 2)$-th sender is moving with every cycle t with a probability of determine event. In the states $i \geq D + B - 2$ after the expiration of the next discrete cycle t the sender begins to receive confirmation and, depending on the results of the delivery units of the segments according to the technology of forward error correction, the sender transmits new blocks of segments (with positive confirmation) or re-distorted (not allowing to direct recovery). The completion of the cycle of the host state $D + B - 2$ corresponds to the time of bringing the first set of segments to the destination and receiving an acknowledgment. Further growth of the state occurs with a probability of distortion of the confirmation $1 - F_o$ in the reverse path. In states $i \geq D + B - 2$ in the selective reject mode the confirmation gives conversion to $(D-1)$-th state when $w \geq K + 2$, $K = \lfloor \frac{D-2}{B} \rfloor$, where $\lfloor \cdots \rfloor$—means "integer part" of the fraction, or in the state kB, $k = \overline{1, w-1}$ where $w \leq K + 1$. Due to the fact that in states $i \geq wB$ the sender stops sending blocks of segments, obtaining confirmation when $w \geq K + 2$ in the states $i = \overline{(w+1)B - 1, (w + K + 1)B - 2}$, $k = \overline{1, K}$ leads to a transition in state $D - kB - 1$, $k = \overline{1, K}$. When $w \leq K + 1$ out of state $i = \overline{D + (w - K)B - 2, D + (w - k + 1)B - 3}$, $k = \overline{1, K}$ there is a transition to state kB, $k = \overline{1, w-1}$. In states $i = \overline{(w + K + 1)B - 1, S - 2}$ with arbitrary width of the window there is a transition to zero state, since the size of the queue transmitted but unconfirmed information segments come clear. In the state $S - 1$ the timeout period of waiting for acknowledgment from the recipient runs out, of the correctness of the received blocks and segments and there is unconditional conversion to zero state.

3   The State Probabilities for Selective Rejection Mode with Forward Error Correction Mechanism

The transition probabilities π_{ij} from the initial state i in the resulting j Markov's chain that describes the process of transmission of information flow with the technology of direct correction error in selective reject mode have the form:

$$
\pi_{ij} = \begin{cases}
1, & j = i + 1, i = \overline{0, D + B - 3}; \\
1 - F_o, & j = i + 1, i = \overline{D + B - 2, S - 2}; \\
F_o, & j = D - 1, i = \overline{D + B - 2, (w + 1)B - 2}, w \geq K + 2; \\
F_o, & j = kB, i = \overline{D + (w - k)B - 2, D + (w - k + 1)B - 3}, \\
& k = \overline{1, w - 1}, w \leq K + 1; \\
F_o, & j = D - kB - 1, i = \overline{(w + k)B - 1, (w + k + 1)B - 2}, \\
& k = \overline{1, w - 1}, w \geq K + 2; \\
F_o, & j = 0, i = \overline{(w + K + 1)B - 1, S - 2}; \\
1, & j = 0, i = S - 1.
\end{cases}
$$

A variety of types of solutions for system of equilibrium equations for probabilities of states of Markov's chain is determined by the relationships between the protocol parameters w, S, the size of block parameter B and the total length of the tract D. Since the duration of the timeout must exceed the window width and be no shorter than the round trip delay of block segments ($S \geq D + B - 1$), there are five variants of the solutions for various areas of applications of protocol parameters.

For the protocol-related parameters associated with total path length of inequalities of the form

$$
w \geq K + 2, \quad S \geq D + (w + 1)B - 2, \tag{1}
$$

the system of local equations can be written as follows:

$$P_0 = P_{S-1} + F_o \sum_{i=B(w+K+1)-1}^{S-2} P_i; \tag{2}$$

$$P_i = P_{i-1}, \ i = \overline{1, D-KB-2}, \overline{D-(K-k)B, D-(K-k-1)B-2}, \ k = \overline{0, K}; \tag{3}$$

$$P_{D-kB-1} = P_{D-kB-2} + F_o \sum_{i=(w+k)B-1}^{(w+k+1)B-2} P_i, \quad k = \overline{1, K}; \tag{4}$$

$$P_{D-1} = P_{D-2} + F_o \sum_{i=D+B-2}^{(w+1)B-2} P_i; \tag{5}$$

$$P_i = P_{i-1}(1 - F_o), \quad i = \overline{D+B-1, S-1}. \tag{6}$$

Taking into account the condition of normalization the solution of this system is determined by the relations:

$$P_i = P_0, \quad i = \overline{0, D-KB-2};$$

$$P_i = P_0(1 - F_o)^{-kB}, \ i = \overline{D-(K-k+1)B-1, D-(K-k)B-2}, \ k = \overline{1, K};$$

$$P_i = P_0(1 - F_o)^{D-1-(w+k)B}, \quad i = \overline{D-1, D+B-2};$$

$$P_i = P_0(1 - F_o)^{i-(w+K+1)+1}, \quad i = \overline{D+B-1, S-1};$$

$$P_0 = F_o\left[1-(1-F_o)^B\right](1-F_o)^{B(w+K)} \Big/ \left\{ \left(1-(1-F_o)^B\right)\left[(D-KB-1)\times \right.\right.$$

$$F_o(1-F_o)^{B(w+K)} + (1-F_o)^{D-1}\left(1+F_o(B-1)-(1-F_o)^{S-D-B+2}\right)\right]$$

$$+ F_o(1-F_o)^{wB}\left(1-(1-F_o)^{KB}\right)\bigg\}. \tag{7}$$

If the window width w prevails over the total paths length of the data transmission, and the range of the duration of the timeout S interval has limitations

$$w \geq K+2, \quad (w+1)B \leq S \leq D+(w+1)B-2, \tag{8}$$

the balance Eqs. (2) and (3) for states $i = \overline{0, D-2}$ are converted to

$$P_0 = P_{S-1}; \tag{9}$$

$$P_i = P_{i-1}, \quad i = \overline{1, D-(M+1)B-2}, \overline{D-(M-k+1)B, D-(M-k)B-2},$$
$$k = \overline{0, M+1};$$

$$P_{D-(M+1)B-1} = P_{D-(M+1)B-2} + F_o \sum_{i=(w+M+1)B-1}^{S-2} P_i;$$

$$P_{D-kB-1} = P_{D-kB-2} + F_o \sum_{i=(w+k)B-1}^{(w+k+1)B-2} P_i, \quad k = \overline{1, M}.$$

Here $M = \lfloor S/B \rfloor - w - 1$—the distance, expressed as a length of the block B, between the states of Markov's chain $S - 2$ and $(w + 1)B - 1$, and corresponding non-sender activity. The probabilities of the Markov's chain are taking the following form:

$$P_i = P_0, \quad i = \overline{0, D - (M + 1)B - 2};$$

$$P_i = P_{D-kB-1}, \quad i = \overline{D - (M - k + 1)B, D - (M - k)B - 2}, \quad k = \overline{0, M};$$

$$P_{D-kB-1} = P_0(1 - F_o)^{B(w+k)-S}, \quad k = \overline{1, M + 1};$$

$$P_i = P_0(1 - F_o)^{D+B-S-1}, \quad i = \overline{D - 1, D + B - 2};$$

$$P_i = P_0(1 - F_o)^{i-S+1}, \quad i = \overline{D + B - 1, S - 1};$$

$$P_0 = F_o(1-F_o)^S \left(1-(1-F_o)^B\right) \bigg/ \bigg\{ BF_o(1-F_o)^{B(w+1)} \left(1-(1-F_o)^{B(M+1)}\right)$$

$$+ \left(1-(1-F_o)^B\right) \left[\left(D-(M+1)B\right) F_o(1-F_o)^S + (1-F_o)^{D+B-1}\right. \tag{10}$$

$$\times \left. \left(1+F_o(B-1)-(1-F_o)^{S-D-B+1}\right)\right] \bigg\}. \tag{11}$$

For window width w and the length of the timeout duration S of type

$$w \geq K + 2, \quad wB + 1 \leq S < (w + 1)B, \tag{12}$$

balance equation converted to

$$P_i = P_{i-1}, \quad i = \overline{1, D - 2}, \overline{D, D + B - 2};$$

$$P_{D-1} = P_{D-2} + F_o \sum_{i=D+B-2}^{S-2} P_i.$$

The solution takes the form:

$$P_i = P_0, \quad i = \overline{0, D - 2};$$

$$P_i = P_0(1 - F_o)^{D+B-S-1}, \quad i = \overline{D - 1, D + B - 2};$$

$$P_i = P_0(1 - F_o)^{i-S+1}, \quad i = \overline{D + B - 1, S - 1};$$

$$P_0 = \frac{F_o(1 - F_o)^S}{(DF_o - 1)(1 - F_o)^S + [1 + F_o(B - 1)](1 - F_o)^{D+B-1}}. \tag{13}$$

Under restrictions on the protocol parameters

$$1 \leq w \leq K + 1, \quad S \geq D + (w + 1)B - 2, \tag{14}$$

the balance Eqs. (3) and (4) can be rewritten as:

$$P_0 = P_{S-1} + F_o \sum_{i=D+(w+1)B-2}^{S-2} P_i;$$

$$P_i = P_{i-1}, \quad i = \overline{kB+1, (k+1)B-1}, \overline{(w-1)B+1, D+B-2}, \quad k = \overline{0, w-2};$$

$$P_{kB} = P_{kB-1} + F_o \sum_{i=D+(w-k)B-2}^{D+(w-k+1)B-3} P_i, \quad k = \overline{1, w-1}.$$

The state probabilities in this case have subset $(i = \overline{(w-1)B, D+B-2})$ of values, and invariant to the number of states:

$$P_i = P_0(1-F_o)^{-kB}, \quad i = \overline{kB, (k+1)B-1}, \quad k = \overline{0, w-2};$$

$$P_i = P_0(1-F_o)^{-(w-1)B}, \quad i = \overline{(w-1)B, D+B-2};$$

$$P_i = P_0(1-F_o)^{i-D-wB+2}, \quad i = \overline{D+B-1, S-1};$$

$$P_0 = F_o(1-F_o)^{B(w-1)}\left[1-(1-F_o)^B\right] \Big/ \left\{\left[1+(D-(w-2)B-2)F_o\right.\right.$$

$$\left.\left. -(1-F_o)^{S-D-B+2}\right]\left[1-(1-F_o)^B\right] + F_o(1-F_o)^B\left[1-(1-F_o)^{B(w-1)}\right]\right\}. \quad (15)$$

In the case of interval limitation on both the protocol parameters

$$1 \le w \le K+1, \quad D+B-1 \le S \le D+(w+1)B-2 \tag{16}$$

Equation (2) takes the form (9) and the Eqs. (3) and (4) is converted to the following:

$$P_i = P_{i-1}, \quad i = \overline{1, (w-N-1)B-1}, \overline{kB, (k+1)B-1}, \overline{wB, D+B-2}$$
$$k = \overline{w-N, w-1};$$

$$P_{kB} = P_{kB-1} + F_o \sum_{i=D+(w-k)B-2}^{D+(w-k+1)B-3} P_i, \quad k = \overline{w-N, w-1};$$

$$P_{(w-N-1)B} = P_{(w-N-1)B-1} + F_o \sum_{i=D+(N+1)B-2}^{S-2} P_i.$$

Here $N = \lfloor(S-D+1)/B\rfloor - 1$—the distance between the last conditional recurrent state of Markov's chain $S-2$ and a state $D+B-2$, in which the sender begins to receive acknowledgements. Local balance equations solution will be determined by the dependencies of two subsets $(i = \overline{0, (w-N-1)B-1}, \overline{(w-1)B, D+B-2})$ of the probabilities of state's values that do not depend on number of the state:

$$P_i = P_0, \quad i = \overline{0, (w - N - 1)B - 1};$$

$$P_i = P_0(1 - F_o)^{D+(w-k)B-S-1}, \quad i = \overline{kB, (k+1)B-1}, \ k = \overline{w-N-1, w-2};$$

$$P_i = P_0(1 - F_o)^{D+B-S-1}, \quad i = \overline{(w-1)B, D+B-2};$$

$$P_i = P_0(1 - F_o)^{i-S+1}, \quad i = \overline{D+B-1, S-1};$$

$$P_0 = F_o(1-F_o)^S\left[1-(1-F_o)^B\right] \Big/ \bigg\{ BF_o(1-F_o)^{D+2B-1}\left[1-(1-F_o)^{BN}\right]$$

$$+ \left[1-(1-F_o)^B\right]\bigg\{(1-F_o)^S\left[F_o(1+B(w-N-1))-1\right] + (1-F_o)^{D+B-1}$$

$$\times \left[1+F_o(D-B(w-2)-2)\right]\bigg\}\bigg\}. \tag{17}$$

Thus, the stationary probability distribution of Marcov circuit states with different relationships between the width of the window w, duration of a time-out S, the total length of the transmission path D and data block size B (1), (8), (12), (14), (16) determined by is the relations (7), (11), (13), (15) and (17) respectively.

4 Throughput Analysis of Selective Repeat Mode

The most important operating characteristic of the Protocol is its throughput, defined by the parameters of the data transmission channel and the mechanism of forward error correction, and overhead as well as the peculiarities of procedure of transmission control [1,2]. Normalized performance of a transport connection is determined by the average number delivered to the recipient undistorted segments (including selective repeat mode [2] and a direct mechanism to restore the distorted segments) mean time between two successive arrivals of acknowledgement [3–6]. As the time between arrivals of acknowledgments distributed according to a geometric law with parameter F_o, mean time between arrivals of acknowledgements in the duration of the cycle t will be $\bar{T} = 1/F_o$. Then for the selective procedure of rejection throughput under $S \geq D + (w+1)B - 2$ will be determined by the relationship:

$$Z(w,S) = F_o\Phi\left\{ \sum_{k=1}^{w} Ak \sum_{i=D+kB-2}^{D+(k+1)B-3} P_i + Aw \sum_{i=D+(w+1)B-2}^{S-1} P_i \right\},$$

and for $\max\{D+B-1, wB+1\} \leq S \leq D+(w+1)B-2$—by relation:

$$Z(w,S) = F_o\Phi\left\{ \sum_{k=1}^{L} Ak \sum_{i=D+kB-2}^{D+(k+1)B-3} P_i + A(L+1) \sum_{i=D+(L+1)B-2}^{S-1} P_i \right\}.$$

Here $\Phi = \sum_{i=A}^{B} C_B^i F_n^i (1 - F_n)^{B-i}$—the probability of a direct reduction unit segments with errors (the reliability of delivery of the unit segments to the destination without retransmission), and $L = \lfloor (S - D + 2)/B \rfloor - 1$—is the time between the duration of the timeout and the time of arrival of the acknowledgement. Given the variability of expression for the probabilities of Markov's chain here in various connections between protocol parameters and a round-trip delay obtained functional dependence of this indicator with an accuracy factor P_0:

$$
Z(w,S) = \begin{cases}
\dfrac{P_0 A\Phi}{(1-F_o)^{(w+K)B-D+1}} \left\{ \dfrac{1-(1-F_o)^{wB}}{1-(1-F_o)^B} - w(1-F_o)^{S-B-D+2} \right\}, \\
\quad w \geq K+2, \quad S \geq D+(w+1)B-2; \\
\dfrac{P_0 A\Phi}{(1-F_o)^{(w-1)B}} \left\{ \dfrac{1-(1-F_o)^{wB}}{1-(1-F_o)^B} - w(1-F_o)^{S-B-D+2} \right\}, \\
\quad 1 \leq w \leq K+1, \quad S \geq D+(w+1)B-2; \\
\dfrac{P_0 A\Phi}{(1-F_o)^{S-D-B+1}} \left\{ \dfrac{1-(1-F_o)^{B(L+1)}}{1-(1-F_o)^B} - (L+1)(1-F_o)^{S-B-D+2} \right\}, \\
\quad \max\{D+B-1, wB+1\} \leq S \leq D+(w+1)B-2.
\end{cases}
$$

Hence it is easy to see that when $A = B = 1$ we get the known result [10]. For absolutely reliable reverse data path ($F_o = 1$) the throughput is determined by the relation $Z(w,S) = \frac{A\Phi}{B}$ when $w \geq K+2$ and by the dependence $Z(w,S) = \frac{A\Phi}{D-1-(w-2)B}$ when $w \leq K+1$. In the case of unlimited window size ($w = \infty$) performance of a transport connection will take the form $Z(w,S) = \frac{AF_o\Phi}{[1-(1-F_o)^B](1+(B-1)F_o)}$. For timeout of minimal duration $S = D+B-1$ the performance of the transport connection is invariant to the size of the window $Z(w,S) = \frac{AF_o\Phi}{D+B-1}$. When $w \leq K+1$ and limitless duration of timeout ($S = \infty$) the throughput is converted to:

$$
Z(w,\infty) = A\Phi P_0 \left[1-(1-F_o)^{wB} \right] \Bigg/ \left\{ F_o(1-F_o)^B \left[1-(1-F_o)^{B(w-1)} \right] \right.
$$

$$
\left. + \left[1 + (D-(w-2)B-2)F_o \right] \left[1-(1-F_o)^B \right] \right\}.
$$

If $S = D+(w+1)B-2$, then

$$
Z(w,S) = \frac{P_0 A\Phi \left\{ 1-(1-F_o)^{wB} \left(1+w\left[1-(1-F_o)^B \right] \right) \right\}}{(1-F_o)^{B(w-1)} \left[1-(1-F_o)^B \right]}.
$$

From the numerical results shown in Fig. 1 it can be seen that the dependence of the throughput size when the window $1 \leq w < K+1$ has a slight increase, in the area $w = K+1$—the sharp increase and then when $w > K+2$—the saturation to the limit values. The duration of the round trip delay D the throughput has an inverse (symmetric) dependency. The performance of a transport connection depends on duration timeout and has the character of a curve with saturation, and with the growth S quickly goes for extreme performance.

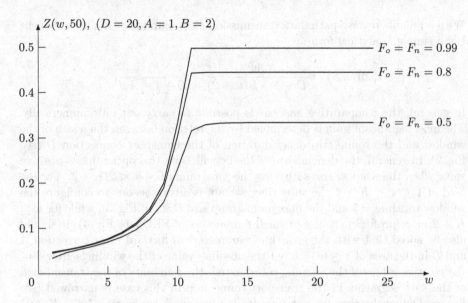

Fig. 1. Dependence of the transport protocol throughput from the window width

5 Discussion of the Advantages of Forward Error Correction Mechanism

The most important task of effective use of technology of forward error correction is the choice of parameters of block length of a sequence of segments B and the number of redundant segments $B - A$ in the block for error correction, providing the fastest possible transport connection. It is obvious that the presence of excess segments in a transmitted sequence increases the probability of delivery to the recipient information segments in the group. However, this is achieved by the growth of overheads in the form of time transfer of redundant data. In this connection there is the task of searching for ranges of values of characteristics of transport connection (D, F_o, F_n), parameters of transport protocol (w, S) and the mechanism of forward error correction (A, B), providing superior management procedures with forward error correction to the classical protocol procedure without FEC. Let's do comparative analysis of the protocol procedures with and without the use of FEC mechanism. Comparison of the control procedure is performed under conditions of equal intensities of subscriber streams offered for transfer $\lambda = Aw$. We define the benefit in speed from the use of FEC mechanism compared with the classical protocol procedure, decision feedback in the form of:

$$\Delta(w, S) = Z(D, w, S, A, B) - Z(D, Aw, S, 1, 1).$$

With a reliable reverse path data transmission $F_o = 1$ and $w \leq K+1$ the benefit has a simple analytical form:

$$\Delta(w, S) = \frac{A\Phi}{D - 1 - B(w - 2)} - \frac{F_n}{D + 1 - Aw}.$$

In general, the comparative analysis is possible to carry out only numerically. The most significant gain is determined by the relation between the width of the window and the round-trip delay duration of the transport connection D (see Fig. 2). In general, the dependence of the benefit from this option has a positive value when the window size satisfying the constraints $1 \leq w < 2(K+2)$. For the field of $1 \leq w < K+1$ the subscriber stream is advantageous to configure the window width $w = 1$ and the maximum group size $B < D$ (Fig. 3), while for $w \geq K+2$ more profitable to use of small parameters of FEC (see Fig. 4). It should also be noted that with the growth of parameters of forward error correction A and B in the area of $1 \leq w < K+1$ the absolute value of the winnings grow, but the positive values of the winning coordinate is the reliability of the transmission of the data segments in the transport connection in this case is narrowed (see Fig. 3). Obviously, that for not fully loaded transport connection ($w \leq K+1$) the use of FEC mechanism would be most beneficial. This is due to the fact that during idle periods of the sender waiting for acknowledgements sender can download the data transmission path of the mandrel redundant segments and thereby reduce the probability of re-transmission with virtually no increase in overhead.

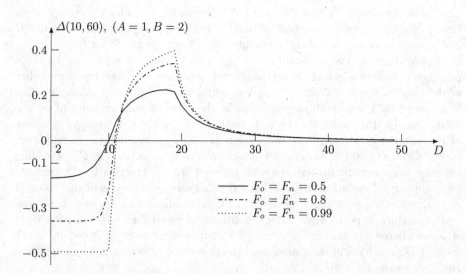

Fig. 2. Dependence of the benefit from the round trip delay duration

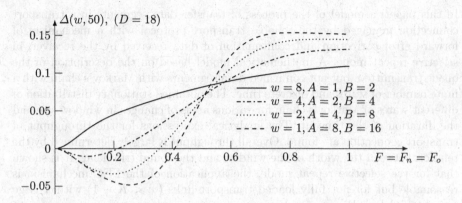

Fig. 3. Dependence of the benefit from the reliability of data transmission for $w > K+2$

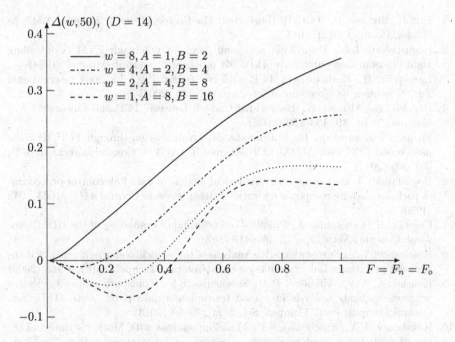

Fig. 4. Dependence of the benefit from the reliability of data transmission for $w < K+1$

6   Conclusions

In this paper, a model of the process of transfer data segments in a transport connection managed by the reliable transport protocol with a mechanism of forward error correction and confirmation of data received by the receiver, in selective repeat mode. A mathematical model based on the description of the queue transmitted, but not confirmed data segments with Markov's chain with a finite number of states and discrete time. The obtained stationary distribution of different states of Markov's chain for various areas of changes in window size and the duration of the timeout. The analytical expressions for the throughput of transport connection are found. Overall throughput is largely determined by the relation between the width of the window and the round trip delay. It is shown that for the selective repeat mode, the application of the FEC mechanism is reasonable but for not fully loaded transport links ($w < K + 1$) with a large round trip delay (D).

Acknowledgments. The work is performed under the state order of the Ministry of Education and Science of the Russian Federation (No. 1.511.2014/K)

References

1. Fall, K., Stevens, R.: TCP/IP Illustrated: The Protocols, vol. 1, 2nd edn. Addison-Wesley, Boston (2012). 1017 p.
2. Boguslavskii, L.B.: Upravlenie potokami dannykh v setyakh EVM (Controlling Data Flows in Computer Networks), 168 p. Energoatomizdat, Moscow (1984)
3. Lundqvist, H., Karlsson, G.: TCP with end-to-end FEC. In: 2004 International Zurich Seminar on Communications, pp. 152–156 (2004)
4. Barakat, C., Altman, E.: Bandwidth tradeoff between TCP and link-level FEC. Comput. Netw. **39**, 133–150 (2002)
5. Shalin, R., Kesavaraja, D.: Multimedia data transmission through TCP/IP using hash based FEC with AUTO-XOR scheme. ICTACT J. Commun. Technol. **3**(3), 604–609 (2012)
6. Boguslavskij, L.B., Gelenbe, E.: Analytical models of data link control procedures in packet-switching computer networks. Autom. Remote Control **41**(7), 1033–1042 (1980)
7. Gelenbe, E., Labetoulle, J., Pujolle, G.: Performance evaluation of the HDLC protocol. Comput. Netw. **2**(4/5), 409–415 (1978)
8. Kokshenev, V.V., Suschenko, S.P.: Analysis of the asynchronous performance management procedures link transmission data. Comput. Technol. **15**(5), 61–65 (2008)
9. Kokshenev, V.V., Mikheev, P.A., Suschenko, S.P.: Transport protocol selective acknowledgements analysis in loaded transmission data path. Vestn. TSU. Ser. Control Comput. Facil. Comput. Sci. **3**(24), 78–94 (2013)
10. Kokshenev, V.V., Suschenko, S.P.: Modeling sessions with Markov's chains: Theory of probability, random processes, mathematical statistics and applications. In: Proceedings of the International Scientific Conference Dedicated to the 80th Anniversary of Professor G.A. Medvedev. Minsk on 23–26 February 2015, RIVS 2015, Minsk, pp. 311–316 (2015)

11. Callegari, C., Giordano, S., Pagano, M., Pepe, T.: A survey of congestion control mechanisms in Linux TCP. In: Vishnevsky, V., Kozyrev, D., Larionov, A. (eds.) DCCN 2013. CCIS, vol. 279, pp. 28–42. Springer, Heidelberg (2014). doi:10.1007/978-3-319-05209-0_3

12. Arvidsson, A., Krzesinski, A.E.: A Model of a TCP link. In: In Proceedings of 5th International Teletraffic Congress Specialist Seminar (2002)

13. Altman, E., Avrachenkov, K., Barakat, C.: A stochastic model of TCP/IP with stationary random loss. ACM SIGCOMM Comput. Commun. Rev. **30**(4), 231–242 (2000)

14. Olsen, Y.: Stochastic modeling and simulation of the TCP protocol. Uppsla Dissertations in Mathematics 28, 94 p. (2003)

15. Kassa, D.F.: Analytic models of TCP performance. Ph.D. thesis, University of Stellenbosch, 199 p. (2005)

16. Bogoiavlenskaia, O.: Discrete model of TCP congestion control algorithm with round dependent loss rate. In: Balandin, S., Andreev, S., Koucheryavy, Y. (eds.) ruSMART 2015. LNCS, vol. 9247, pp. 190–197. Springer, Heidelberg (2015). doi:10.1007/978-3-319-23126-6_17

17. Giordano, S., Pagano, M., Russo, F., Secchi, R.: Modeling TCP startup performance. J. Math. Sci. **200**(4), 424–431 (2014)

18. Kravets, O.Ya.: Mathematical modeling of parameterized TCP protocol. Autom. Remote Control **74**(7), 1218–1224 (2013)

19. Wang, J., Wen, J., Han, Y., Zhang, J., Li, C., Xiong, Z.: Achieving high throughput and TCP Reno fairness in delay-based TCP over large networks. Front. Comput. Sci. **8**(3), 426–439 (2014)

20. Nikitinskiy, M.A., Chalyy, D.Ju.: Performance analysis of trickles and TCP transport protocols under high-load network conditions. Autom. Control Comput. Sci. **47**(7), 359–365 (2013)

21. Kokshenev, V., Suschenko, S.: Analytical model of the TCP Reno congestion control procedure through a discrete-time Markov chain. In: Vishnevsky, V., Kozyrev, D., Larionov, A. (eds.) DCCN 2013. CCIS, vol. 279, pp. 124–135. Springer, Heidelberg (2014). doi:10.1007/978-3-319-05209-0_11

22. Ivanovskii, V.B.: Properties of output flows in discrete queuinge systems. Autom. Remote Control Part 1 **45**(11), 1413–1419 (1984)

23. Tannenbaum, A.: Modern Operating Systems. Pearson, Piter (2002)

A Cyclic Queueing System with Priority Customers and T-Strategy of Service

Anatoly Nazarov and Svetlana Paul[✉]

Tomsk State University, Lenina pr. 36, Tomsk, Russia 634050
nazarov.tsu@gmail.com, paulsv82@mail.ru

Abstract. We review the queuing system, the input of which is supplied with the Poisson process of priority customers and N number of the Poisson processes of non-priority customers. Durations of service for both priority and non-priority customers have a distribution functions of $A(x)$ and $B_n(x)$ for applications from priority flow and for customers from n flow $(n = 1 \ldots N)$ respectively. By using methods of systems with server vacations and asymptotic analysis in conditions of a large load we have found the asymptotic probability distribution of a value of an unfinished work. It is shown that this distribution is exponential.

Keywords: Cyclic queueing system with server vacations · Priority customers · Asymptotic analysis · Exponential distribution

1 Introduction

A cyclic queueing systems with priority customers are mathematical models of telecommunication systems, which are pretty common in practice [1]. Method of research of such systems is her decomposition and research of system with server vacations. In real systems "vacations" are considered as a temporal suspension of service either for device other applications or for its breakdown or repair [2].

The mathematical model under study is presented in the Sect. 2. Kolmogorov equations for the investigated processes can be found in the Sect. 3. Asymptotic analysis of the obtained equations is performed in the Sect. 4.

2 Mathematical Model

Let's review the queuing system with one service device, the input of which is supplied with the Poisson process of priority information with the intensity of τ and a N number of the Poisson processes of non-priority information with the intensity of λ_n, where $n = 1, N$ [3]. The flows of non-priority customers will be called λ_n-flows, and the flow of priority information will be called τ-flow. Let's assume that the intensity of a τ-flow is substantially lower than the total intensity of λ_n-flows. Applications of each λ_n-flow form a queue with an unlimited number of waiting seats.

© Springer International Publishing AG 2016
V.M. Vishnevskiy et al. (Eds.): DCCN 2016, CCIS 678, pp. 182–193, 2016.
DOI: 10.1007/978-3-319-51917-3_17

Device visits queues in a cyclic order, starting from a first queue and finishing with the N, then the cycle repeats. Duration of a visit is random and has the distribution function of $T_n(x)$. During that time device receives customers of a λ_n-flow for service, duration of which has the distribution function of $B_n(x)$. If there are no customers in the queue when device addresses it or if device has already serviced all applications in the queue, device is still addressing the queue till the end of a visit, duration of which is determined by the distribution function of $T_n(x)$.

τ-flow customers form their queue with an unlimited number of waiting seats as well. If the system receives an application from a τ-flow, device stops the service of a common application and instantly starts servicing priority customers for a time, duration of which is distributed by the function of $A(x)$. Upon finishing the service of a priority customer device resumes the service of non-priority customers. If during a service of a priority customer device receives another priority application device services all priority application and only then returns to the queue of a non-priority customers and resumes servicing them.

Cyclic system research method is it's decomposition and the research of a system with server vacations. Let's review the queuing system with one service device and two queues with an unlimited number of waiting seats. The system receives a Poisson process of priority customers with the intensity of τ (τ-flow) and a Poisson process of non-priority customers with the intensity of λ (λ-flow).

The system functions in cyclic mode, the cycle of which consists of two consecutive intervals. During the first interval customers of a λ-flow are serviced at the device. If there are no customers in the queue at the start of an interval or if device has serviced all customers that were in the queue during that interval, device still remains in this mode, waiting for customers to come. At the end of an interval device goes on a "vacation" during a second interval of the said cycle. Customers of a λ-flow that were received during a vacation are accumulating in the queue and wait till device returns to servicing them [4].

Let's assume that durations of these intervals are random and are determined by distribution functions of $T_1(x)$ and $T_2(x)$ respectively. During the first interval device services customers for a random time with a distribution function of $B(x)$. If a server vacation interrupted the service of a common application then after vacation device resumes the service of this application. When the system receives priority customers of a τ-flow, then regardless of where the device previously was (in a service mode or on a vacation) it starts servicing priority customers for duration of time, that has distribution function of $A(x)$. After a device has serviced all priority customers it either resumes the service of non-priority application or returns to a vacation. In order to research the waiting time in systems with server vacations we have to find the characteristic function and the probability distribution of value $V(t)$ of an unfinished work on servicing all non-priority customers that were in a system at the time t. Let's denote

$V(t)$ the volume of work on servicing all non-priority customers that were in a system at the time t.

$Y(t)$ the volume of work on servicing all priority customers that were in a system at time t.

$k(t)$ device status: 1 device is available for λ-flow applications, 2 device is on a vacation, 3 device is servicing priority customers, having interrupted the interval of being available for a common information (1st device status), 4 device is servicing priority information, having interrupted a vacation.

$z(t)$ the remaining time of a vacation or servicing priority information.

3 Kolmogorov Equations

Let's review the Markov process $\{V(t), Y(t), k(t), z(t)\}$ and set up a direct system of Kolmogorov differential equations for a following probability distribution:

$$P_k(v, z, t) = P\{V(t) < v, k(t) = k, z(t) < z\}, k = 1, 2,$$

$$P_k(v, y, z, t) = P\{V(t) < v, Y(t) < y, k(t) = k, z(t) < z\}, k = 3, 4.$$

Let's assume that a system functions in a stationary mode, then:

$$-(\lambda + \tau) P_1(v, z) + \lambda \int_0^v B(v - x) dP_1(x, z) + \frac{\partial P_2(v, 0)}{\partial z} T_1(z)$$

$$+\frac{\partial P_3(v, 0, z)}{\partial y} + \frac{\partial P_1(v, z)}{\partial v} + \frac{\partial P_1(v, z)}{\partial z} - \frac{\partial P_1(v, 0)}{\partial z} = 0,$$

$$-(\lambda + \tau) P_2(v_1, z) + \lambda \int_0^v B(v - x) dP_2(x, z) + \frac{\partial P_1(v, 0)}{\partial z} T_2(z)$$

$$+\frac{\partial P_4(v, 0, z)}{\partial y} + \frac{\partial P_2(v, z)}{\partial z} - \frac{\partial P_2(v, 0)}{\partial z} = 0,$$

$$-(\lambda + \tau) P_3(v, y, z)$$

$$+\lambda \int_0^v B(v - x) dP_3(x, y, z) + \tau \int_0^y A(y - x) dP_3(v, x, z)$$

$$+P_1(v, z) A(y)\tau + \frac{\partial P_3(v, y, z)}{\partial y} - \frac{\partial P_3(v, 0, z)}{\partial y} = 0,$$

$$-(\lambda + \tau) P_4(v, y, z)$$

$$+\lambda \int_0^v B(y - x) dP_4(x, y, z) + \tau \int_0^y A(y - x) dP_4(v, x, z)$$

$$+P_2(v, z) A(y)\tau + \frac{\partial P_4(v, y, z)}{\partial y} - \frac{\partial P_4(v, 0, z)}{\partial y} = 0.$$

Let's introduce functions

$$H_k(u,z) = \int_0^\infty e^{-uv} dP_k(v,z), k = 1,2,$$

$$H_k(u,y,z) = \int_0^\infty e^{-uv} dP_k(v,y,z), k = 3,4,$$

for which we will rewrite a system of Kolmogorov equations in a following form:

$$
\begin{cases}
[\lambda\beta(u) - (\lambda+\tau) + u]\,H_1(u,z) - uP_1(0,z) \\
+\frac{\partial H_2(u,0)}{\partial z}T_1(z) + \frac{\partial H_3(u,0,z)}{\partial y} + \frac{\partial H_1(u,z)}{\partial z} - \frac{\partial H_1(u,0)}{\partial z} = 0, \\
[\lambda\beta(u) - (\lambda+\tau)]\,H_2(u,z) \\
+\frac{\partial H_1(u,0)}{\partial z}T_2(z) + \frac{\partial H_4(u,0,z)}{\partial y} + \frac{\partial H_2(u,z)}{\partial z} - \frac{\partial H_2(u,0)}{\partial z} = 0, \\
[\lambda\beta(u) - (\lambda+\tau)]\,H_3(u,y,\ z) + H_1(u,z)A(y)\tau \\
+\tau\int_0^y A(y-x)dH_3(u,x,z) + \frac{\partial H_3(u,y,z)}{\partial y} - \frac{\partial H_3(u,0,z)}{\partial y} = 0, \\
[\lambda\beta(u) - (\lambda+\tau)]\,H_4(u,y,\ z) + H_2(u,z)A(y)\tau \\
+\tau\int_0^y A(y-x)dH_4(u,x,z) + \frac{\partial H_4(u,y,z)}{\partial y} - \frac{\partial H_4(u,0,z)}{\partial y} = 0.
\end{cases}
\tag{1}
$$

Here

$$\beta(u) = \int_0^\infty e^{-uv} dB(v), \int_0^\infty e^{-uv} d\frac{\partial P_1(v,z)}{\partial v} = \beta(u)H_k(u,z),$$

$$\int_0^\infty e^{-uv} d\left(\int_0^v B(v-x)dP_k(x,y,z)\right) = \beta(u)H_k(u,y,z).$$

$P_1(0,z)$ is probability of situation where device stays in the service mode, and there are no customers in the system [5]. Let's make changes in the third and the fourth equations of a system (1)

$$H_3(u,y,z) = H_1(u,z)H_3(u,y), H_4(u,y,z) = H_2(u,z)H_4(u,y),$$

$$
\begin{cases}
[\lambda\beta(u) - (\lambda+\tau)]\,H_3(u,y) + A(y)\tau \\
+\tau\int_0^y H_3(u,y-x)dA(x) + \frac{\partial H_3(u,y)}{\partial y} - \frac{\partial H_3(u,0)}{\partial y} = 0, \\
[\lambda\beta(u) - (\lambda+\tau)]\,H_4(u,y) + A(y)\tau \\
+\int_0^y H_4(u,y-x)dA(x) + \frac{\partial H_4(u,y)}{\partial y} - \frac{\partial H_4(u,0)}{\partial y} = 0.
\end{cases}
\tag{2}
$$

Let's take a Laplace-Stieltjes transform of the equations of system (2), by denoting:

$$G_k(u,v) = \int_0^\infty e^{-yv} dH_k(u,y), \alpha(v) = \int_0^\infty e^{-yv} dA(y).$$

Then we have:

$$
\begin{cases}
[\lambda\beta(u) - (\lambda+\tau) + \tau\alpha(v) + v]\,G_3(u,y) + \alpha(v)\tau - \frac{\partial H_3(u,0)}{\partial y} = 0, \\
[\lambda\beta(u) - (\lambda+\tau) + \tau\alpha(v) + v]\,G_4(u,y) + \alpha(v)\tau - \frac{\partial H_4(u,0)}{\partial y} = 0.
\end{cases}
$$

Solution of the last system exists for $v = v(u)$ when

$$[\lambda\beta(u) - (\lambda + \tau) + \tau\alpha(v) + v] = 0.$$

Then

$$v(u) = (\lambda + \tau) - \lambda\beta(u) - \tau\alpha(v(u)). \qquad (3)$$

We have

$$\frac{\partial H_3(u,0)}{\partial y} = \frac{\partial H_4(u,0)}{\partial y} = \alpha(v)\tau.$$

Now let's make changes in the first and the second equations of a system (1)

$$\frac{\partial H_3(u,0,z)}{\partial y} = \alpha(v)\tau H_1(u,z), \qquad \frac{\partial H_4(u,0,z)}{\partial y} = \alpha(v)\tau H_2(u,z),$$

we'll get

$$[\lambda\beta(u) - (\lambda + \tau) + u] H_1(u,z) - uP_1(0,z)$$

$$+\frac{\partial H_2(u,0)}{\partial z}T_1(z) + \alpha(v)\tau H_1(u,z) + \frac{\partial H_1(u,z)}{\partial z} - \frac{\partial H_1(u,0)}{\partial z} = 0,$$

$$[\lambda\beta(u) - (\lambda + \tau)] H_2(u,z)$$

$$+\frac{\partial H_1(u,0)}{\partial z}T_2(z) + \alpha(v)\tau H_2(u,z) + \frac{\partial H_2(u,z)}{\partial z} - \frac{\partial H_2(u,0)}{\partial z} = 0.$$

Considering that (3), we can write down this

$$\begin{cases} [-v(u) + u] H_1(u,z) - uP_1(0,z) \\ +\frac{\partial H_2(u,0)}{\partial z}T_1(z) + \frac{\partial H_1(u,z)}{\partial z} - \frac{\partial H_1(u,0)}{\partial z} = 0, \\ -v(u)H_2(u,z) + \frac{\partial H_1(u,0)}{\partial z}T_2(z) + \frac{\partial H_2(u,z)}{\partial z} - \frac{\partial H_2(u,0)}{\partial z} = 0. \end{cases} \qquad (4)$$

The last system will be solved by using a method of asymptotic analysis in conditions of a large load [6,7].

4   Method of Asymptotic Analysis

Let's denote S the bandwidth value of a system with vacations, b first moment of a random value, which is determined by a function of servicing time of non-priority customer $B(x)$, ε is a small positive parameter, which in theoretical research is considered to be $\varepsilon \to 0$. Let's make changes in the system (4) of equations.

$$\lambda = (1 - \varepsilon)S/b, u = \varepsilon w, P_1(0,z) = \varepsilon\pi_1(z,\varepsilon).$$

$$H_k(u,z) = F_k(w,z,\varepsilon), H_k(u,y,z) = F_k(w,y,z,\varepsilon),$$

S is the systems load. It's value will be found below. We'll get

$$\begin{cases} [-v(\varepsilon w) + \varepsilon w] F_1(w,z,\varepsilon) - \varepsilon^2 w\pi_1(z,\varepsilon) \\ +\frac{\partial F_2(w,0,\varepsilon)}{\partial z}T_1(z) + \frac{\partial F_1(w,z,\varepsilon)}{\partial z} - \frac{\partial F_1(w,0,\varepsilon)}{\partial z} = 0, \\ -v(\varepsilon w)F_2(w,z,\varepsilon) \\ +\frac{\partial F_1(w,0,\varepsilon)}{\partial z}T_2(z) + \frac{\partial F_2(w,z,\varepsilon)}{\partial z} - \frac{\partial F_2(w,0,\varepsilon)}{\partial z} = 0. \end{cases} \qquad (5)$$

Theorem 1. *The limit value for $\varepsilon \to 0$ $F_k(w) = F_k(w, \infty)$ solutions $F_k(w, z, \varepsilon)$ of the system* (5) *has the following form*

$$F_k(w, z) = R_k(z)\Phi(w),$$

where $S = \frac{T_1}{T_1+T_2}(1 - \tau a)$,

$$\begin{cases} R_1 = R_1(\infty) = \frac{T_1}{T_1+T_2}(1 - \tau a) = S, \\ R_2 = R_2(\infty) = \frac{T_2}{T_1+T_2}(1 - \tau a) = 1 - S - \tau a. \end{cases}$$

Asymptotic characteristic function

$$\Phi(w) = \frac{S\gamma}{S\gamma + w}$$

is a Laplace-Stieltjes transform of an exponentially distributed random function with the parameter of γ type.

$$\gamma = \left(S\frac{b_2}{2b} + \tau \left(\frac{S}{1 - \tau a} \right)^2 \frac{a_2}{2} + \frac{\Delta}{1 - \tau a} \right)^{-1}.$$

Here

$$\Delta = -\frac{T_1 T_2}{T_1 + T_2} R_1 R_2 + \frac{R_1 R_2}{T_1 + T_2} \left\{ T_2 \frac{T_1^{(2)}}{2T_1} + T_1 \frac{T_2^{(2)}}{2T_2} \right\},$$

b and b_2 - are the starting moments of a first and second orders of time of a service of non-priority customer, a and a_2 are the starting moments of a first and second orders of time of a service of priority customer. $\int_0^\infty (1 - T_k(x))dx = T_k$ is the average time of staying in the corresponding mode. $T_k^{(2)}$ is the second initial moment of the time of devices staying in mode k. Lets find an asymptotic function of volume of work on servicing all customers

$$H(u) = \frac{(S - \lambda b)\gamma}{(S - \lambda b)\gamma + u}.$$

Proof. We perform the proof by three stages.

Stage 1. By tending ε to zero $\varepsilon \to 0$ in (5) we will obtain:

$$\begin{cases} \frac{\partial F_1(w,z)}{\partial z} - \frac{\partial F_1(w,0)}{\partial z} + \frac{\partial F_2(w,0)}{\partial z} T_1(z) = 0, \\ \frac{\partial F_2(w,z)}{\partial z} - \frac{\partial F_2(w,0)}{\partial z} + \frac{\partial F_1(w,0)}{\partial z} T_2(z) = 0. \end{cases}$$

We have

$$v(0) = (\lambda + \tau) - \lambda \beta(0) - \tau \alpha(v(0)) = 0.$$

where

$$\alpha(0) = \int_0^\infty dA(y) = 1, \beta(0) = \int_0^\infty dB(v) = 1.$$

We will seek solution for this system in this form

$$F_k(w, z) = R_k(z)\Phi(w).$$

We will obtain following system

$$\begin{cases} R_1'(z) - R_1'(0) - R_2'(0)T_1(z) = 0, \\ R_2'(z) - R_2'(0) - R_1'(0)T_2(z) = 0. \end{cases}$$

solution for which we will write down in this form:

$$\begin{cases} R_1(z) = \int\limits_0^z (R_1'(0) - R_2'(0)T_1(x))dx, \\ R_2(z) = \int\limits_0^z (R_2'(0) - R_1'(0)T_2(x))dx. \end{cases}$$

$\{R_1(z), R_2(z)\}$ is a two-dimensional distribution of devices mode and value of remaining time of devices staying in that mode. By tending z to infinity $z \to \infty$, considering that $T(\infty) = 1$, we will obtain:

$$\begin{cases} R_1(\infty) = \int\limits_0^\infty (R_1'(0) - R_2'(0)T_1(x))dx, \\ R_2(\infty) = \int\limits_0^\infty (R_2'(0) - R_1'(0)T_2(x))dx. \end{cases}$$

For improper integral to be convergent it is necessary that the following condition is met

$$R_1'(0) - R_2'(0)T_1(\infty) = 0,$$

then we will obtain

$$R_1'(0) = R_2'(0) = R'(0).$$

We have

$$R_k(z) = R'(0) \int\limits_0^z (1 - T_k(x))dx.$$

Let's denote

$$\int\limits_0^\infty (1 - T_k(x))dx = T_k$$

as the average time of staying in the corresponding mode.
    Then, from the one side

$$R_1(\infty) + R_2(\infty) + \tau\alpha = 1,$$

and from the other

$$R_1(\infty) + R_2(\infty) + \tau\alpha$$

$$= R'(0) \int_0^\infty (1 - T_1(x))dx + \int_0^\infty (1 - T_2(x))dx + \tau\alpha$$

$$= R'(0)(T_1 + T_2) + \tau\alpha.$$

We will obtain

$$R'(0) = \frac{1 - \tau\alpha}{(T_1 + T_2)}.$$

Then

$$\begin{cases} R_1(\infty) = \frac{1-\tau\alpha}{T_1+T_2} \int_0^\infty (1 - T_1(x))dx = \frac{T_1(1-\tau\alpha)}{T_1+T_2}, \\ R_2(\infty) = \int_0^\infty (R_2'(0) - R_1'(0)T_2(x))dx = \frac{T_2(1-\tau\alpha)}{T_1+T_2}. \end{cases}$$

Stage 2. We will denote the expansion for the function $v(\varepsilon w)$

$$v(\varepsilon w) = \varepsilon w \frac{S}{1 - \tau\alpha} - (\varepsilon w)^2 \frac{S\frac{b_2}{2b} + \frac{S}{w} + \tau(\frac{S}{1-\tau\alpha})^2 \frac{a_2}{2}}{1 - \tau\alpha} + O(\varepsilon^2).$$

In the Eq. (5) let's tend z to infinity $z \to \infty$

$$\begin{cases} [-v(\varepsilon w) + \varepsilon w] F_1(w, \varepsilon) - \varepsilon^2 w \pi_1(\varepsilon) + + \frac{\partial F_2(w,0,\varepsilon)}{\partial z} - \frac{\partial F_1(w,0,\varepsilon)}{\partial z} = 0, \\ -v(\varepsilon w) F_2(w, \varepsilon) + \frac{\partial F_1(w,0,\varepsilon)}{\partial z} - \frac{\partial F_2(w,0,\varepsilon)}{\partial z} = 0. \end{cases}$$

Let's sum the equations of the last system

$$(-v(\varepsilon w) + \varepsilon w)F_1(w, \varepsilon) - v(\varepsilon w)F_2(w, \varepsilon) = \varepsilon^2 w \pi_1(\varepsilon).$$

We will substitute the following expansion in the system (5)

$$F_k(w, z, \varepsilon) = \Phi(w)\{R_k(z) + j\varepsilon w f_k(z)\} + O(\varepsilon^2). \tag{6}$$

We will substitute expansions in the equality which we got from above

$$\Phi(w)\left(\varepsilon w(R_1 - S) + + (\varepsilon w)^2 \left(f_1 + S\frac{b_2}{2b} + \frac{S}{w} + \tau\left(\frac{S}{1-\tau\alpha}\right)^2 \frac{a_2}{2} - \frac{S}{1-\tau\alpha}[f_1 + f_2]\right)\right)$$

$$= \varepsilon^2 w \pi_1(\varepsilon),$$

where

$$f_k = f_k(\infty), k = 1, 2.$$

Then we will obtain

$$S = R_1,$$

$$\Phi(w)(\varepsilon w)^2 \left(f_1 + S\frac{b_2}{2b} + \frac{S}{w} + \tau\left(\frac{S}{1-\tau\alpha}\right)^2 \frac{a_2}{2} - \frac{S}{1-\tau\alpha}[f_1 + f_2]\right) = \varepsilon^2 w \pi_1(\varepsilon).$$

By dividing both parts of the equations by ε^2 and by tending ε^2 to zero $\varepsilon \to 0$ we will obtain:

$$\Phi(w) = \frac{\pi}{S + w\left(f_1 + S\frac{b_2}{2b} + \tau\left(\frac{S}{1-\tau\alpha}\right)^2 \frac{a_2}{2} - \frac{S}{1-\tau\alpha}[f_1 + f_2]\right)}.$$

Let's assume that $w = 0$ and considering that

$$\Phi(0) = 1,$$

we will obtain:

$$\pi_1 = S, \quad \pi_1 = \lim_{\varepsilon \to 0} \pi_1(\varepsilon^2).$$

Then

$$\Phi(w) = \frac{S}{S + w\left(f_1 + S\frac{b_2}{2b} + \tau\left(\frac{S}{1-\tau\alpha}\right)^2 \frac{a_2}{2} - \frac{S}{1-\tau\alpha}[f_1 + f_2]\right)}.$$

Stage 3. We will substitute the expansion (6) in the system (5)

$$\begin{cases} \varepsilon w \left(f_1'(z) - f_1'(0) + f_2'(0)T_1(z)\right) + \varepsilon w \left(1 - \frac{S}{1-\tau\alpha}\right) R_1(z) = O(\varepsilon^2), \\ \varepsilon w \left(f_2'(z) - f_2'(0) + f_1'(0)T_2(z)\right) - \varepsilon w \frac{S}{1-\tau\alpha} R_2(z) = O(\varepsilon^2). \end{cases}$$

By dividing both equations by ε and be tending to zero $\varepsilon \to 0$, we will obtain:

$$\begin{cases} f_1'(z) - f_1'(0) + f_2'(0)T_1(z) + \left(1 - \frac{S}{1-\tau\alpha}\right) R_1(z) = 0, \\ f_2'(z) - f_2'(0) + f_1'(0)T_2(z) - \frac{S}{1-\tau\alpha} R_2(z) = 0. \end{cases}$$

Let's write down

$$\begin{cases} f_1(z) = \int_0^z \left\{f_1'(0) - f_2'(0)T_1(x) - R_1(x)\frac{R_2}{1-\tau\alpha}\right\} dx, \\ f_2(z) = \int_0^z \left\{f_2'(0) - f_1'(0)T_2(x) + R_2(x)\frac{R_1}{1-\tau\alpha}\right\} dx. \end{cases}$$

Let's find $f_1(\infty)$, $f_2(\infty)$

$$\begin{cases} f_1(\infty) = \int_0^\infty \left\{f_1'(0) - f_2'(0)T_1(x) - R_1(x)\frac{R_2}{1-\tau\alpha}\right\} dx, \\ f_2(z) = \int_0^\infty \left\{f_2'(0) - f_1'(0)T_2(x) + R_2(x)\frac{R_1}{1-\tau\alpha}\right\} dx. \end{cases}$$

It is necessary that the following is true

$$\begin{cases} f_1'(0) - f_2'(0)T_1(\infty) - R_1(\infty)\frac{R_2}{1-\tau\alpha} = 0, \\ f_2'(0) - f_1'(0)T_2(\infty) + R_2(\infty)\frac{R_1}{1-\tau\alpha} = 0. \end{cases}$$

Then

$$f_1'(0) - f_2'(0) = \frac{R_1 R_2}{1 - \tau\alpha}, \tag{7}$$

where

$$R_1 = R_1(\infty), \quad R_2 = R_2(\infty).$$

Let's write down:

$$\begin{cases} f_1(\infty) = f_2'(0) \int\limits_0^\infty (1 - T_1(x))\, dx + \frac{R_2}{1-\tau\alpha} \int\limits_0^\infty (R_1 - R_1(x))\, dx, \\ f_2(\infty) = f_1'(0) \int\limits_0^\infty (1 - T_2(x))\, dx - \frac{R_1}{1-\tau\alpha} \int\limits_0^\infty (R_2 - R_2(x))\, dx. \end{cases}$$

Let's denote

$$\Delta_k = \int\limits_0^\infty (R_k - R_k(x))\, dx.$$

$$\begin{cases} f_1(\infty) = f_2'(0)T_1 + \frac{R_2}{1-\tau\alpha}\Delta_1, \\ f_2(\infty) = f_1'(0)T_2 + \frac{R_1}{1-\tau\alpha}\Delta_2. \end{cases} \tag{8}$$

Let's review the expression in the denominator separately

$$R_2 f_1(\infty) - R_1 f_2(\infty).$$

Considering (7) and (8) we will obtain and denote the following

$$R_1 f_2(\infty) - R_2 f_1(\infty)$$

$$= -R_1 f_1'(0)T_2 + \frac{R_1^2}{1 - \tau\alpha}\Delta_2 + R_2 f_2'(0)T_1 + \frac{R_2^2}{1 - \tau\alpha}\Delta_1$$

$$= -R_1 R_2 \frac{T_1 T_2}{T_1 + T_2} + \frac{R_1 R_2}{T_1 + T_2}\left(T_1 \frac{T_2^{(2)}}{2T_2} + T_2 \frac{T_1^{(2)}}{2T_1}\right) = \Delta. \tag{9}$$

Here

$$\Delta_k = \int\limits_0^\infty (R_k - R_k(x))\, dx$$

$$= R_k \int\limits_0^\infty \left\{ 1 - \frac{1}{T_k} \int\limits_0^z (1 - T_k(x))\, dx \right\} dz$$

$$= R_k \left\{ 1 - \frac{1}{T_k} \int\limits_0^z (1 - T_k(x))\, dx \right\} z \Bigg|_{z=0}^\infty$$

$$- R_k \int\limits_0^\infty z\, d\left\{ 1 - \frac{1}{T_k} \int\limits_0^z (1 - T_k(x))\, dx \right\}$$

$$= R_k \int_0^\infty z \frac{1}{T_k} \left(1 - T_k(z)\right) dz$$

$$= R_k \int_0^\infty \frac{1}{T_k} \left(1 - T_k(z)\right) d\frac{z^2}{2} = R_k \frac{1}{T_k} \left(1 - T_k(z)\right) \left.\frac{z^2}{2}\right|_{z=0}^\infty$$

$$-R_k \int_0^\infty \frac{1}{T_k} \frac{z^2}{2} d\left(1 - T_k(z)\right) = R_k \frac{1}{T_k} \int_0^\infty \frac{z^2}{2} d\left(T_k(z)\right) = R_k \frac{T_k^{(2)}}{2T_k}.$$

Where

$$\int_0^\infty \left\{ 1 - \frac{1}{T_k} \int_0^z \left(1 - T_k(x)\right) dx \right\} dz$$

is the average value of the remaining time of staying in mode k. $T_k^{(2)}$ is the second initial moment of the time of devices staying in mode k. $R_k(z)$ is written down in this form:

$$R_k(z) = (1 - \tau\alpha) \frac{T_k}{T_1 + T_2} \left\{ \frac{1}{T_k} \int_0^z \left(1 - T_k(x)\right) dx \right\}.$$

Then $\Phi(w) = \frac{S\gamma}{S\gamma + w}$ is a Laplace-Stieltjes transform of an exponentially distributed random function with the parameter of γ type.

$$\gamma = \left(S\frac{b_2}{2b} + \tau \left(\frac{S}{1 - \tau a}\right)^2 \frac{a_2}{2} + \frac{\Delta}{1 - \tau a} \right)^{-1}.$$

where Δ is determined by Eq. (9). The theorem was proved.

By making backward substitutions, we will tend z to infinity and ε to zero, $z \to \infty$, $\varepsilon \to 0$, and well get the characteristic function of a value of an unfinished work

$$H(u) = \sum_k F(w, \varepsilon) = \sum_k \Phi(w) R_k(\infty) + o(\varepsilon) = H(u) = \frac{(S - \lambda b)\gamma}{(S - \lambda b)\gamma + u}.$$

5   Conclusions

In this work we have researched mathematical model of the system with server vacations. By using method of asymptotic analysis under large load, we have found asymptotic distribution of probabilities of a value of an unfinished work $V(t)$. The found function and an exponential probability distribution let us perform research of a virtual time of waiting in cyclic systems by reviewing models with vacations once again.

Acknowledgments. The work is performed under the state order of the Ministry of Education and Science of the Russian Federation (No. 1.511.2014/K).

References

1. Pechinkin, A.V., Sokolov, I.A.: Queueing system with an unreliable device in discrete time. J. Inform. Appl. **5**(4), 6–17 (2005). (in Russian)
2. Saksonov, E.A.: Method of the calculation of probabilities of modes for one-line queueing systems with the server vacation. J. Autom. Tele-Mech. **1**, 101–106 (1995). (in Russian)
3. Nazarov, A.A., Terpugov, A.F.: Queueing Theory: Educational Material. NTL, Tomsk (2004). (in Russian)
4. Nazarov, A.A., Paul, S.V.: Research of queueing system with the server vacation that is controlled by T-strategy. In: Proceedings of International Science Conference Theory of Probabilies, Random Processes, Mathematical Statistics and Applications, pp. 202–207 (2015). (in Russian)
5. Nazarov, A., Paul, S.: A number of customers in the system with server vacations. In: Vishnevsky, V., Kozyrev, D. (eds.) DCCN 2015. CCIS, vol. 601, pp. 334–343. Springer, Heidelberg (2016). doi:10.1007/978-3-319-30843-2_35
6. Moiseeva, E.A., Nazarov, A.A.: Research of RQ-system MMP—GI—1 by using method of asymptotic analysis under large load. TSUs herald/messenger. Adm. Calc. Tech. Inform. **4**(25), 83–94 (2013). (in Russian)
7. Nazarov, A.A., Moiseev, A.N.: Analysis of an open non-Markovian GI(GI—∞)K queueing network with high-rate renewal arrival process. Probl. Inf. Transm. **49**(2). doi:10.1134/S0032946013020063

Comparative Analysis of Reliability Prediction Models for a Distributed Radio Direction Finding Telecommunication System

Dmitry Aminev[1], Alexander Zhurkov[2], Sergey Polesskiy[2], Vladimir Kulygin[2], and Dmitry Kozyrev[1,3(✉)]

[1] V.A. Trapeznikov Institute of Control Sciences of Russian Academy of Sciences, Moscow, Russia
aminev.d.a@ya.ru, kozyrevdv@gmail.com
[2] National Research University Higher School of Economics, Moscow, Russia
petrovyc@gmail.com, {spolessky,vkulygin}@hse.ru
[3] RUDN University, Moscow, Russia

Abstract. We consider the problem of reliability assurance of a local ground-based distributed radio direction finding system (RDFS), which consists of a local dispatching center (LDC) and unattended radio terminals (URT), which are up to several hundred kilometers apart from the LDC and are connected to the LDC via communication channels. The performance criteria of the RDFS are defined according to its topology and structure. Requirements on the mean time between failures (MTBF) and the availability factor are imposed. A methodic has been developed for determining the reliability parameters both in approximate analytical form and in the form of a formalized simulation model that takes into account different hierarchy levels of the system from the topology of the network and communication channels to the printed board assemblies and individual types of electronic components. Simulation and calculation of reliability measures was performed using an automated system for reliability calculation of electronic modules and reconfigurable manufacturing calculation (ASONIKA). The weak spots (least reliable elements) of the RDFS have been revealed and recommendations were given to ensure the reliability of individual elements and the RDFS as a whole. The composition of spare parts for LDC, URT equipment and communication channels is proposed.

Keywords: Reliability model · Diagnostics · Radio technical system · Communications network · System topology · Communication channel · Radio direction finding · Radar · Methodic · Link

1   Introduction

Despite the rapid development of global navigation satellite positioning systems, video surveillance and machine vision, radar and radio direction finding is still

This research has been financially supported by the Russian Science Foundation (research project №16-49-02021).

widely used at present, for the reason that in remote and underdeveloped areas radars and does not undetectable automatic direction finders (ADF) are among the main flight facilities [1].

The radio direction finding system (RDFS) is distributed over large area (Fig. 1), has a multi-level star topology and consists of equipment of the local dispatching center (LDC), communication channels and unattended radio terminals (URT), which, depending on the allocation conditions can be far from the LDC up to several hundred kilometers [2,3]. Therefore, the vital task is to provide reliable functioning of the RDFS [4] and work out recommendations on reservation of its elements.

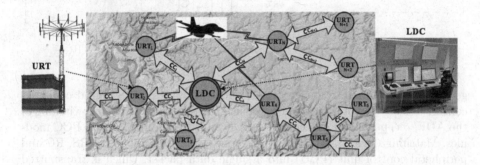

Fig. 1. Ground-based distributed radio direction finding system

As a part of solution of this problem, it is necessary to create a reliability model, for which at the initial stage it is necessary to analyze the structure and components of the RDFS, to formalize the statement of the reliability enhancement problem, to develop a method for reliability evaluation. Then it is necessary to develop a reliability block diagram, and, on the basis of an expert estimation of operational failure rates of components, to perform both analytical and numerical modeling and compare the obtained results. Based on the modeling results recommendations are to be given on the reservation of components and the composition of spare parts.

2 Structure of the Distributed RDFS

Structural diagram of URT and LDC interaction is shown in Fig. 2. Automatic Direction Finder (ADF) receives signals from the quasi-doppler (QD) antenna and generates a corresponding corner bearing quadrature voltage, from which analogue phase converter (APC) forms a bearing value as a phase shift. Phase code converter (PCC) generates a bearing angle code out of the phase shift. Remote signaling (RS) device is a receiver, which indicates breakdowns (hardware temperature, fire, smoke) to the URT and performs the following functions: receives and stores in its buffer the status signals form ADF and URT hardware,

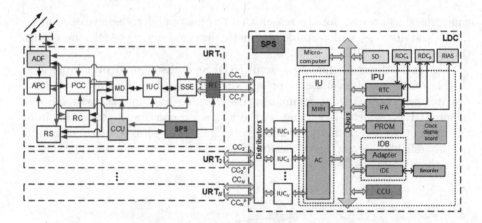

Fig. 2. Structural scheme of interaction between NSD and the TIR

generates information existence signals and transmits the RS digital code and control signals. Remote control (RC) device performs remote on/off switching of the ADF, reception and processing of ADF information in APC and PCC modules. Matching device (MD) collects code messages from the PCC, RS, RC and peripheral control unit (PCU) into a single data packet, which is transmitted to the links through the interface unit with the channel (IUC) and secondary sealing equipment (SSE). PCU monitors the performance of URT components while in operation and while performing repair or maintenance works. Secondary power system (SPS) provides power to the other elements of the scheme. Non-end URTs are equipped with a retransmitter (RT), which provides data exchange between end URTs and the LDC via CCs.

LDC hardware is designed to: collect bearing information from URT via communication channels (CC); with the help of bearing values determine the location of an object at the moment of radio contact of its on-board transmitter; provide control over URT via the CC; perform bearings membership test; display and record the air situation; perform the automated control of the technical state of URT and LDC hardware.

LDC hardware consists of a micro-computer with management firmware, service data storage device (SD), information processing unit (IPU), the reference indicator of air situation (RIAS), remote dispatching controller (RDC), interface unit with a channel (IUC) to interact with an available communication system through distributors.

IPU is designed to: collect information on bearing and signals coming from the IUC and input it to the microcomputer; prepare the information on air situation to display it on RIAS; perform URT remote control and reception of signals; record and store the dynamic information displayed on RIAS; store service data in accordance with which the input information is conversed; perform oversight. IPU includes an interface unit (IU), a remote technical controller (RTC), an image forming apparatus (IFA), an information documenting block (IDB), a configurable PROM and a central control unit (CCU). IU provides interface between a CC and a microcomputer, and control over the map data recording in PROM. IU consists of interfaces hardware (IH) and the map information input hardware (MIIH). IH allows to receive/transmit code messages between the microcomputer and the IUC. IDB provides (i) connection between a recording and storing device and a microcomputer; (ii) conversion of "Common Bus" interface to the "Q-Bus" interface. IDB consists of the information documenting equipment (IDE) and adapter boards. IDE operates in two modes "recording" and "displaying". IFA is a part of the display equipment and provides: (i) interface between RIAS and a microcomputer, (ii) RIAS backup, (iii) generation of the current time code. PROM is designed for storing programs and constants, according to which the input information is transformed. Output of programs and constants is performed upon request from the microcomputer. CCU refers to the control and diagnostic equipment and is designed to provide timely information on the technical condition of LDC. The objects of control are CCU, RTC, PROM, IDB, IU. The rest of the LDC is controlled by the microcomputer's CPU. To manage the CPU control and to ensure the monitoring and diagnostic coordination between channeling equipment and communication channels, CCU has a "semi-active" access to the "Q-bus" that allows to automatically transfer data from CCU into the microcomputer for displaying it RIAS and form the microcomputer to the CCU for displaying it on its front panel.

3   Problem Statement

The initial data for reliability assessment of a distributed RDFS is: its topology, LDC and URT structure, type of CC, information about operational failure rate of all printed board assemblies and communication channels, clear criteria of the system efficiency, working time schedule and operating conditions. Performance criteria of the distributed RDFS at the top level of its hierarchy are defined by the requirements for the coverage area, that is, the territory where the RDFS can determine the coordinates of an emitting object using radio direction-finders with covering radius R_{ADF} (Fig. 3).

RDFS coverage area is divided nominally into 4 quadrants with respect to the cardinal directions in relation to LDC. In general case each quadrant contains URT (denoted by U on the diagram, and U* stands for intermediate URT equipped with a repeater). At that the number of URTs for each quadrant may be different—m_1, m_2, m_3, m_4.

Since the RDFS is spatially distributed, it is reasonable to consider 2 operation states working state and state of failure. While in the working state, all the RDFS units operate adequately without faults. The condition for the failure state is disability of all the URTs in at least one quadrant. Mathematically, these criteria can be expressed as follows:

$$Q_0 = \overline{\{U_{1...m1}^1 \& U_{1...m2}^2 \& U_{1...m3}^3 \& U_{1...m4}^4\}} \times \Delta t_i, \tag{1}$$

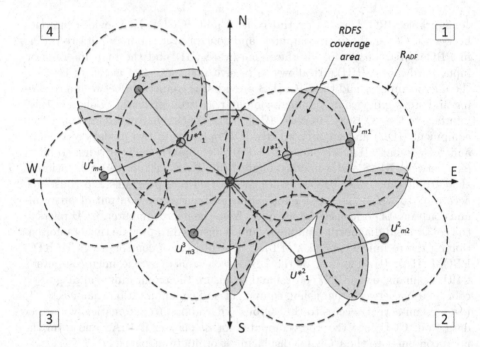

Fig. 3. Distributed RDFS coverage area

where Q_0 is a failure function, Δt_i is a time period during which the failure occurs.

Mean time before failures (MTBF) shall be at least 500 h under day-and-night service. There can be used telegraph, telephone, radio relay, tropospheric, fiber optic and cellular communications links as communication channels. Printed board assemblies of all the system components are 170 × 200 mm in size and have an average degree of integration (20–30 elements with 2–3 microchips each). Restoration of the RDFS should be ensured through a cold redundancy (replacement of components with ones from the set of spare parts), and the restoration time τ_B should not exceed 30 min.

4   Reliability Assessment Method

We propose a reliability prediction method for the RDFS, which consists of six consolidated procedures. This method allows to carry out a comprehensive assessment of reliability measures together with analysis of the obtained results, search for the most critical nodes and release of recommendations. This reliability assessment method is represented in the form of integrated definition language (IDEF0) diagram in Fig. 4.

Consider the sequence of actions performed by a researcher in accordance with the described method. It includes performing the following activities:

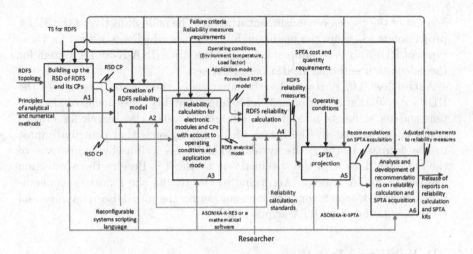

Fig. 4. IDF0 diagram of reliability assessment method

- **Activity A1.** Construction of a reliability structure diagram (RSD) of RDFS and its components. RSD is based on the analysis of requirements of technical specifications for the RDFS, the failure criterion, and the reliability measures requirements. The necessary data for building up the RSD are: RDFS topology, a list of system elements (specification).

- **Activities A2.** Building of a formalized reliability model is performed: an analytical model, formed on the basis of logical-probabilistic method; a simulation model, formed on the basis of the numerical Monte-Carlo method using a complex systems description language. The formalized model is built on the basis of RSD and the reliability measures data listed in the technical statement (TS). Rules for building the simulation model are given in [5] and for the analytical one – in [6,8]. To avoid inaccuracies, it is recommended to build the simulation model directly in the automated system for reliability calculation for functionally complex systems - ASONIKA-K-RES.

- **Activities A3.** Top-down and bottom-up calculation of reliability characteristics is performed first for electronic components, then for electronic modules and components of RDFS, with account of temperature conditions, load factors and other parameters that may affect reliability characteristics, as well as with account of the average daily cycles of application modes of RDFS elements.

- **Activities A4.** Complex RDFS reliability calculation is performed based on the formalized reliability model, which describes the behavior of the RDFS in different states and the reliability characteristics of individual components calculated while performing functions A2 and A3, respectively, including the use of MathCad and ASONIKA-K-RES computing software.

- **Activities A5.** Calculation and optimization of individual, group, or multi-level [7,9] sets of spare parts, tools and accessories (SPTA) is performed, relevant to the current configuration and operation model of the RDFS with account of the chosen replenishment strategy. The main objective of the SPTA projection is to ensure the maintainability factors which are defined for these types of RDFS. It is recommended to use the ASONIKA-K-SPTA system for the calculation and optimization of SPTA sets.

- **Activities A6.** A comprehensive analysis of the obtained results of the RDFS calculation is performed, including reliability assessment of the system and its elements on different hierarchy levels. Also search for critical nodes is carried out by the means of identifying ones that contribute most to the decrease in the entire system reliability as well as by comparison of the obtained data with the specified one by the TS. Besides the acquisition analysis of SPTA is done. According to the results the report documentation is released, which contains information on the on RDFS reliability and recommendations for SPTA acquisition.

5   Reliability Model

Using logical connections of the operating RDFS components needed for successful system functioning (see Sects. 2 and 3), the topology and structure of hardware components, as well as the basics of logical-probabilistic method, the RDFS reliability model can be represented in the form of reliability structure diagram (RSD) shown in Fig. 5.

Also Fig. 5 depicts RSD of RDFS hardware components and a state diagram of the system birth-death process.

From Fig. 5(b) we conclude that failure of any of the URT components leads to failure of the whole URT except for a failure of one of the CCs or RDC. Communication channels are cold standby connected. At that all the URT modules are in hot standby with a critical failure condition failure of $N-1$ URT modules.

Analysis of RDFS RSD (see Fig. 5(a)) shows that LDC failure leads to the failure of the whole complex, and failure of end and non-end URTs in each quadrant leads to the system failure only in case of failure of n out of m URTs.

It is assumed in the analytical model for reliability measures assessment that operating time and recovery time of RDFS RSD elements are exponentially distributed, and failures and repairs of RDFS elements are statistically independent events [12].

First we consider an arbitrary separate group of redundant elements shown in Fig. 5 (assume that LDC is excluded from consideration, and thus the subject of study are four areas of the reliability structure diagram). Birth and death scheme can be represented as a simple state diagram (see Fig. 5(d)), where state S0 corresponds to the absence of failures in the group, state S_1—the presence of a failure in the group, S_i—the presence of i failures in the group, S_n—failure of the entire group (when all the elements of the group fail—both the main and the redundant ones).

Consider a sequence of four sections, each containing two groups with redundancy. Denote: λ—URT failure rate, λ^* - URT* failure rate, μ and μ^*— corresponding recovery rates. Failure flows of redundant groups of URT and

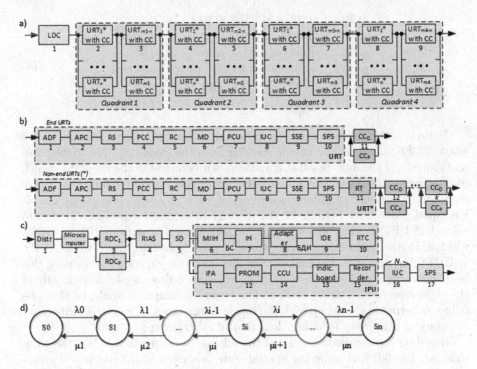

Fig. 5. RSD for topology of (a) RDFS, (b) URT complex, and (c) LDC and a state diagram of the birth-death process (d)

URT* are simple Poisson flows so we can use the principle of superposition. Indeed, now each state of the birth-death process in Fig. 5(d) (except for the first and last states) implies a failure either of an URT, or an URT* or a restoration of an URT or an URT*, or no changes. It is noteworthy that the situation, is taken into account when the system is in the i-th state and we consider the possible events that can happen in a short time Δt. We write an event S_i as follows:

$$S_i = A_i^{\text{URT}} \cup A_i^{\text{URT*}} \cup B_i^{\text{URT}} \cup B_i^{\text{URT*}} \cup D_i^{\text{URT}} \cup D_i^{\text{URT*}} \qquad (2)$$

Failures of URT or URT* correspond to outcomes A, restorations of URT and URT* correspond to outputs B, and zero changes correspond to outputs D. According to the extended axiom of addition, we obtain:

$$P(S_i) = P(A_i^{\text{URT}}) + P(A_i^{\text{URT*}}) + P(B_i^{\text{URT}}) + P(B_i^{\text{URT*}}) + P(D_i^{\text{URT}}) + P(D_i^{\text{URT*}}) \ (3)$$

Thus, the expression above remains valid, but now every event A, B and D should be presented as the union of the corresponding events in each section of the reliability structure diagram:

$$A_i^{\mathrm{URT}} = A_{i,1}^{\mathrm{URT}} \cup A_{i,2}^{\mathrm{URT}} \cup A_{i,3}^{\mathrm{URT}} \cup A_{i,4}^{\mathrm{URT}}$$
$$A_i^{\mathrm{URT}*} = A_{i,1}^{\mathrm{URT}*} \cup A_{i,2}^{\mathrm{URT}*} \cup A_{i,3}^{\mathrm{URT}*} \cup A_{i,4}^{\mathrm{URT}*}$$
$$B_i^{\mathrm{URT}} = B_{i,1}^{\mathrm{URT}} \cup B_{i,2}^{\mathrm{URT}} \cup B_{i,3}^{\mathrm{URT}} \cup B_{i,4}^{\mathrm{URT}}$$
$$B_i^{\mathrm{URT}*} = B_{i,1}^{\mathrm{URT}*} \cup B_{i,2}^{\mathrm{URT}*} \cup B_{i,3}^{\mathrm{URT}*} \cup B_{i,4}^{\mathrm{URT}*} \tag{4}$$
$$D_i^{\mathrm{URT}} = D_{i,1}^{\mathrm{URT}} \cup D_{i,2}^{\mathrm{URT}} \cup D_{i,3}^{\mathrm{URT}} \cup D_{i,4}^{\mathrm{URT}}$$
$$D_i^{\mathrm{URT}*} = D_{i,1}^{\mathrm{URT}*} \cup D_{i,2}^{\mathrm{URT}*} \cup D_{i,3}^{\mathrm{URT}*} \cup D_{i,4}^{\mathrm{URT}*}$$

Naturally, the question arises as to how to define the failure criterion for the whole RDFS. Consider the circuit starting from the events S_0. The system has no failures and the only possible transition can be made to the event S_1, which may represent the denial of URT or URT* on any section of reliability structure diagram. Occurrence of failure (recovery) of URT in any section of the circuit has equal probability because of the identity of elements (the same is true in respect of URT*), thus, we cannot say exactly in which section URT or URT* will fail. Further failures and recoveries will not only be interleaved with of URT or URT*, but also with sections of reliability structure diagram. Eventually, this leads us to the last condition of Sn, which is the occurrence of failure in any of the eight reserve containing groups. The logical assumption would be that the failure occurred in the group with the lowest redundancy rate, as the probability of failure of this group is higher than that of other groups.

Based on the detailed analysis of the scheme of birth and death of the complex (see Fig. 5d) and using the spatial state method and the method of structure diagrams, we construct the analytical models for calculating the probability of no-failure operation ($R_{\mathrm{RDFS}}(t)$), availability factor (A_{RDFS}) and MTBF ($T_{0\,\mathrm{RDFS}}$).

The model is as follows:

$$R_{\mathrm{RDFS}}(t) = \left[e^{-\lambda_{\mathrm{LDC}} \cdot t} \right] \tag{5}$$

$$\times \left[\exp\left\{ -\frac{\lambda_{\mathrm{URT}n1*} \cdot t}{\displaystyle\sum_{s=0}^{n}\sum_{i=0}^{s} \frac{1}{(i+1)! C_{n-s+i+1}^{i+1} \gamma^i}} \right\} \exp\left\{ -\frac{\lambda_{\mathrm{URT}m1} \cdot t}{\displaystyle\sum_{s=0}^{n}\sum_{i=0}^{s} \frac{1}{(i+1)! C_{m-s+i+1}^{i+1} \gamma^i}} \right\} \right]_1$$

$$\times \left[\exp\left\{ -\frac{\lambda_{\mathrm{URT}n2*} \cdot t}{\displaystyle\sum_{s=0}^{n}\sum_{i=0}^{s} \frac{1}{(i+1)! C_{n-s+i+1}^{i+1} \gamma^i}} \right\} \exp\left\{ -\frac{\lambda_{\mathrm{URT}m2} \cdot t}{\displaystyle\sum_{s=0}^{n}\sum_{i=0}^{s} \frac{1}{(i+1)! C_{m-s+i+1}^{i+1} \gamma^i}} \right\} \right]_2$$

$$\times \left[\exp\left\{ -\frac{\lambda_{\mathrm{URT}n3*} \cdot t}{\displaystyle\sum_{s=0}^{n}\sum_{i=0}^{s} \frac{1}{(i+1)! C_{n-s+i+1}^{i+1} \gamma^i}} \right\} \exp\left\{ -\frac{\lambda_{\mathrm{URT}m3} \cdot t}{\displaystyle\sum_{s=0}^{n}\sum_{i=0}^{s} \frac{1}{(i+1)! C_{m-s+i+1}^{i+1} \gamma^i}} \right\} \right]_3$$

$$\times \left[\exp \left\{ -\frac{\lambda_{\mathrm{URT}n4*} \cdot t}{\displaystyle\sum_{s=0}^{n}\sum_{i=0}^{s} \frac{1}{(i+1)! C_{n-s+i+1}^{i+1} \gamma^i}} \right\} \exp \left\{ -\frac{\lambda_{\mathrm{URT}m4} \cdot t}{\displaystyle\sum_{s=0}^{n}\sum_{i=0}^{s} \frac{1}{(i+1)! C_{m-s+i+1}^{i+1} \gamma^i}} \right\} \right]_4$$

where: λ_{LDC} – LDC failure rate, 1/h; t – operating time; $\lambda_{\mathrm{URT}nX*}$ – failure rate of a n-th non-end URT of quadrants 1, 2, 3 and 4, 1/h; $\lambda_{\mathrm{URT}mX*}$ – failure rate of an n-th end URT of quadrants 1,2, 3 and 4, 1/h; $\gamma = \frac{\lambda}{\mu}$ – duty ratio; $\mu = \frac{1}{\tau_B}$ – recovery rate, 1/h; C_{N-s+i}^{i+1}—number of combinations; $N = n+1$ – total number of elements; n (or m) – number of redundant elements.

The model of A_{RDFS} is as follows:

$$A_{\mathrm{RDFS}} = \left[\frac{T_{0\,\mathrm{LDC}}}{T_{0\,\mathrm{LDC}} + \tau_{B\,\mathrm{LDC}}} \right] \cdot \left[\left(1 - \prod_{j=1}^{N} \frac{\tau_{B\,j}/T_{0\,j}}{1 + \tau_{B\,j}/T_{0\,j}} \right) \left(1 - \prod_{j=1}^{N} \frac{\tau_{B\,j}/T_{0\,j}}{1 + \tau_{B\,j}/T_{0\,j}} \right) \right]_1$$

$$\times \left[\left(1 - \prod_{j=1}^{N} \frac{\tau_{B\,j}/T_{0\,j}}{1 + \tau_{B\,j}/T_{0\,j}} \right) \cdot \left(1 - \prod_{j=1}^{N} \frac{\tau_{B\,j}/T_{0\,j}}{1 + \tau_{B\,j}/T_{0\,j}} \right) \right]_2$$

$$\times \left[\left(1 - \prod_{j=1}^{N} \frac{\tau_{B\,j}/T_{0\,j}}{1 + \tau_{B\,j}/T_{0\,j}} \right) \cdot \left(1 - \prod_{j=1}^{N} \frac{\tau_{B\,j}/T_{0\,j}}{1 + \tau_{B\,j}/T_{0\,j}} \right) \right]_3 \qquad (6)$$

$$\times \left[\left(1 - \prod_{j=1}^{N} \frac{\tau_{B\,j}/T_{0\,j}}{1 + \tau_{B\,j}/T_{0\,j}} \right) \cdot \left(1 - \prod_{j=1}^{N} \frac{\tau_{B\,j}/T_{0\,j}}{1 + \tau_{B\,j}/T_{0\,j}} \right) \right]_4$$

where: $T_{0\,\mathrm{LDC}}$ – LDC mean time to failure, h.; $\tau_{B\,\mathrm{LDC}}$ – LDC mean time to recovery, h.; $T_{0\,j}$ – MTBF of a j-th element, h.; $\tau_{B\,j}$ – mean time to recovery of a j-th element, h.; m – number of redundant elements; N – total number of elements.

The model of $T_{0\,\mathrm{RDFS}}$ is as follows:

$$\frac{1}{T_{0\,\mathrm{RDFS}}} = \frac{1}{\left[\frac{1}{\lambda_{\mathrm{LDC}}} \right]} + \left[\frac{1}{T_{0\,\mathrm{GR\,URT}_{n*}}} + \frac{1}{T_{0\,\mathrm{GR\,URT}_{m1}}} \right]_1$$

$$+ \left[\frac{1}{T_{0\,\mathrm{GR\,URT}_{n*}}} + \frac{1}{T_{0\,\mathrm{GR\,URT}_{m1}}} \right]_2 + \left[\frac{1}{T_{0\,\mathrm{GR\,URT}_{n*}}} + \frac{1}{T_{0\,\mathrm{GR\,URT}_{m1}}} \right]_3$$

$$+ \left[\frac{1}{T_{0\,\mathrm{GR\,URT}_{n*}}} + \frac{1}{T_{0\,\mathrm{GR\,URT}_{m1}}} \right]_4 \qquad (7)$$

where: λ_{LDC} – LDC failure rate, 1/h; $T_{0\,\mathrm{GR\,URT}_{n*}}$ – MTBF of a group of non-end URTs, h.; $T_{0\,\mathrm{GR\,URT}_{m1}}$ – MTBF of a group of end URTs, h.

Calculation of $T_{0\,\mathrm{GR\,URT}_{n*}}$ or $T_{0\,\mathrm{GR\,URT}_{m1}}$ is performed according to the following mathematical model:

$$T_{0\,\mathrm{GR\,URT}_{n*}}\,(T_{0\,\mathrm{GR\,URT}_{m1}}) = \cfrac{1}{\left(\prod\limits_{j=1}^{N}(1+\lambda_j\tau_{Bj}) - \prod\limits_{j=1}^{N}(\lambda_j\tau_{Bj})\Big/\sum\limits_{j=1}^{N}\frac{1}{\tau_{Bj}}\prod\limits_{j=1}^{N}(\lambda_j\tau_{Bj})\right)} \quad (8)$$

where: λ_j – failure rate of a j-th element, 1/h; τ_{Bj} – mean time to recovery of a j-th element, h.; m – number of redundant elements; N – total number of elements.

Though it is possible to build a general formalized simulation model for RDFS RSD (see Fig. 5a) but it will be cumbersome. For that matter we give an example of a model that precisely corresponds to the topology of Fig. 2 with the number of URTs in quadrants $m_1 = 1$, $m_2 = 2$, $m_3 = 3$, m_4. RSD for such RDFS is shown in Fig. 6.

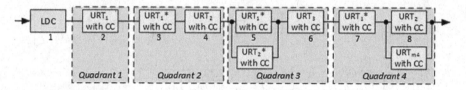

Fig. 6. RSD of RDFS with the number of URT equal to $m_1 = 1$, $m_2 = 2$, $m_3 = 3$, $m_4 = 2$

Using the description language for reconfigurable systems simulation models [5,10], we construct a formal model of the RDFS. Its simplified form can be represented as follows:

$$F(x) = \{\mathrm{LDC}\ \&\ n*(\text{non-end URT})\ \&\ m*(\text{end URT})\ \&\ \mathrm{CC}\} \quad (9)$$

where: LDC = {Distributors & IUC & IH & ... & RIAS}; URT = {QD antenna & ADF & ... & SPS & RT (for non-end)}; CC = {Telephone | Telegraph | Radio Relay | Fiber optic | Mobile}.

Figure 7a shows a simulation model of the upper-level of RDFS in the form of a block diagram describing the RDFS functioning in accordance with the RSD shown in Fig. 6. Based on the efficiency criteria, LDC and all four sets of URTs (for the four quadrants, respectively), are connected via a logical "AND", which means that RDFS fail in case of failure of any of these elements.

URT* simulation model in Fig. 7a describes the URT* functioning in accordance with RSD shown in Fig. 5b. According to the URT* failure criteria, five types of communication channels connected via logical "OR" means the implementation of the reliability criterion of at least one communication channel; the output of the logical "OR" is connected with the rest of URT* via a logical "AND", which means that the failure occurs when any of the elements or all of the communication channels fail. The initial data for the numerical model are the failure rate and the mean time to recovery, besides there are behavior patterns of components in case of failure. A fragment of a simulation model in the form of a pseudo code describing the operation of a RDFS communication channel is shown in Fig. 7b.

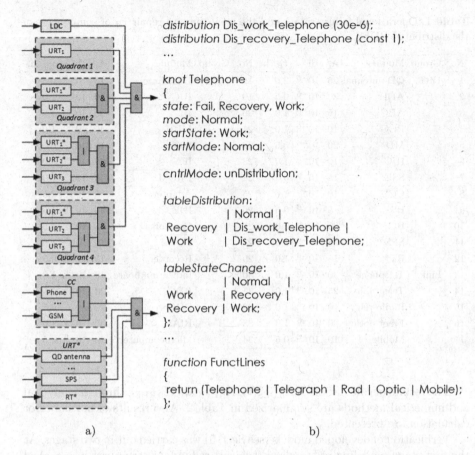

distribution Dis_work_Telephone (30e-6);
distribution Dis_recovery_Telephone (const 1);
...

knot Telephone
{
state: Fail, Recovery, Work;
mode: Normal;
startState: Work;
startMode: Normal;

cntrlMode: unDistribution;

tableDistribution:
                | Normal |
Recovery | Dis_work_Telephone |
Work | Dis_recovery_Telephone;

tableStateChange:
                | Normal |
Work | Recovery |
Recovery | Work;
};
...

function FunctLines
{
 return (Telephone | Telegraph | Rad | Optic | Mobile);
};

a)   b)

Fig. 7. Representation of RDFS simulation model in the form of a block diagram (a) and a fragment of pseudocode for a communication channel simulation (b)

6   Example of Reliability Calculation for RDFS Topology

As an example of reliability calculation we carry out a comparative analysis of reliability measures of a restorable geographically distributed system using RDFS RSD topology shown in Fig. 6, using analytical models describing the reliability structure diagram in Sect. 5 - models (1)–(4), respectively, and the numerical simulation model (see Fig. 7) [5, 6].

All the input data for the reliability calculation of RDFS part (see Fig. 5b,c), namely the failure rate and mean time to repair are given in Table 1 [11]. Assume the operating time equal to 24 h, since such systems run continuously and on a round-the-clock basis.

In numerical calculation example, the values of operational failure rate and mean time to repair for each element of the RDFS RSD were assigned with the help of analogies methods and expert evaluation.

Table 1. Operational failure rate λ and the mean time to repair τ_B of components of the distributed RDFS

No.	Group	Module	λ, 1/h	τ_B, h.	No.	Group	Module	λ, 1/h	τ_B, h.
1	URT	QD antenna	$25 \cdot 10^{-6}$	2,0	18	LDC	Distributors	$15 \cdot 10^{-6}$	0,5
2		ADF	$30 \cdot 10^{-6}$	0,5	19		IUC	$10 \cdot 10^{-6}$	1,0
3		APC	$15 \cdot 10^{-6}$	1,0	20		IH	$20 \cdot 10^{-6}$	1,0
4		PCC	$10 \cdot 10^{-6}$	1,0	21		MIIH	$40 \cdot 10^{-6}$	1,0
5		MD	$20 \cdot 10^{-6}$	0,5	22		RTC	$10 \cdot 10^{-6}$	0,5
6		IUC	$10 \cdot 10^{-6}$	1,0	23		IFA	$50 \cdot 10^{-6}$	1,0
7		SSE	$15 \cdot 10^{-6}$	0,5	24		PROM	$20 \cdot 10^{-6}$	1,0
8		CCU	$10 \cdot 10^{-6}$	1,0	25		Adapter	$5 \cdot 10^{-6}$	0,5
9		RS	$5 \cdot 10^{-6}$	0,5	26		IDE	$40 \cdot 10^{-6}$	1,0
10		RC	$20 \cdot 10^{-6}$	1,0	27		Controller	$5 \cdot 10^{-6}$	1,0
11		SPS	$15 \cdot 10^{-6}$	1,0	28		CCU	$10 \cdot 10^{-6}$	1,0
12		RT*	$15 \cdot 10^{-6}$	1,0	29		Recorder	$15 \cdot 10^{-6}$	0,5
13	Link	Telephone	$30 \cdot 10^{-6}$	1,0	30		Indication board	$10 \cdot 10^{-6}$	1,0
14		Telegraph	$25 \cdot 10^{-6}$	1,0	31		RDC1	$20 \cdot 10^{-6}$	0,5
15		Radio relay	$10 \cdot 10^{-6}$	1,0	32		RDCr	$20 \cdot 10^{-6}$	0,5
16		Fiber optic	$20 \cdot 10^{-6}$	1,0	33		RIAS	$30 \cdot 10^{-6}$	1,0
17		Mobile	$10 \cdot 10^{-6}$	0,5	34		Microcomputer	$20 \cdot 10^{-6}$	0,5

The results of calculation of $R_{\mathrm{RDFS}}(t)$, A_{RDFS} and $T_{0\,\mathrm{RDFS}}$ through analytical and numerical methods are summarized in Table 2. Also results of mutual error calculation are presented.

Verification of developed models (see Sect. 5) was carried out in two stages. At the first stage modeling of standard (typical) redundant structures was carried out, for which analytical formulas are known without assumptions. According to the simulation results the measure values have been obtained with an error less than 0.5–1% relative to the analytical models, which is due to an error of a finite number of experiments. At the second stage, the calculation of reliability measures for RDFS RSD topology shown in Fig. 6 was performed. The calculation results are shown in Table 2.

Table 2. Comparison of reliability measures of a distributed RDFS

Calculation method	Reliability measures			Mutual calculation error,%		
	$R_{\mathrm{RDFS}}(t)$	A_{RDFS}	$T_{0,\mathrm{RDFS}}$	$\Delta(R_{\mathrm{RDFS}}(t))$	$\Delta(A_{\mathrm{RDFS}})$	$\Delta(T_{0,\mathrm{RDFS}})$
Analytical method (structure graph method)	0,9740966	0,9991321	915	2,509337	0,0232334	31,2568
Numerical method (method of Monte Carlo simulation)	0,99854	0,9989	629			

To build an accurate analytical model of RDFS reliability assessment (see Fig. 6) it is necessary to analyze 2^{18} possible states of the system, taking into account the sequence of failures, therefore we obtained only a lower-bound estimate of $R_{RDFS}(t)$. The developed models allow to fully describe the algorithms and fault criteria of RDFS. To verify the models experts from RDFS developing enterprises were involved, which gave the expert confirmation of model adequacy.

There were obtained expected results for three reliability measures that can be considered closer to the truth, compared with analytical models, where admittedly the assumptions where made leading to undercount in one case and overstating in other case. For RDFS in estimation of $R_{RDFS}(t)$ the 2.5% difference was obtained, while the error for $T_{0\,RDFS}$ is 31% which is a substantial difference.

While checking the stability and correctness of analytical and numerical models is supported by figures (see Fig. 8a,b).

As can be seen from Fig. 8a the increase in the number of end URTs over 4 stops to give a substantial contribution to $T_{0\,RDFS}$, but the increase in the number of redundant end URTs in quadrant 2 of RDFS leads to a substantial increase of T_0—almost 3 times.

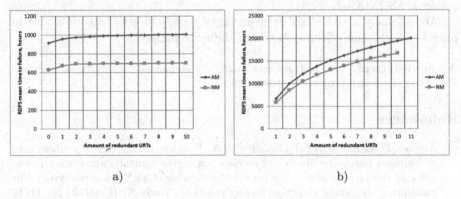

Fig. 8. Plot of mean time between failures of the RDFS (a) and its second quadrant (b) against the number of redundant end URTs in quadrant 2

As for the analysis of quantitative values of parameters of reliability of RDFS, it shows the need to improve indicators of recovery through creation of an effective system of repair, namely making up a set of components by own sets of SPTA. As it can be seen from the experience of designing similar systems, the best option is a two-tier system of SPTA, while single set of SPTA is provided for URT and URT*, SPTA group kit is located near LDC. The main parameters that influence the composition of SPTA are the replenishment strategy (for similar systems the most economically efficient strategy is a continuous strategy of replenishment) and the failure rate, obtained in our case through the results of expert estimation. On the basis of this the following components will preliminary become parts of SPTA - 2, 5, 10, 13, 14, 16, 20, 21, 23, 24, 26, 31–34 (see Table 1).

7   Conclusion

Performance criteria for a distributed radio direction finding telecommunication system (RDFS) were defined according to its topology and structure. There was developed a unified IDEF0-diagram of RDFS reliability assessment method. Comparative analysis of analytical and numerical models for RDFS reliability prediction was carried out.

A methodic has been developed for determining the reliability parameters both in approximate analytical form and in the form of a formalized simulation model that takes into account different hierarchy levels of the system from the topology of the network and communication channels to the printed board assemblies and individual types of electronic components. The analysis of both models showed that the difference in mean time to failure of the whole RDFS is significant and is not less than 31%. It is primarily due to strong assumption in analytical calculations for redundant groups with restoration contrary to a more precise description of the algorithm of system functioning and failures in the framework of numerical model. A formalized model was built for its implementation in the automated system for reliability simulation of functionally complex systems (ASONIKA-K-RES).

An optimally matched set of spare parts, tools and attachments is able to provide a high value of the system availability factor.

Acknowledgment. This research was financially supported by the Russian Science Foundation (research project No. 16-49-02021).

References

1. Aminev, D.A., Zhurkov, A.P., Kozyrev, A.A., Uvaysov, S.U.: Algoritmy raboty programmnogo obespechenija mikroprocessornyh sistem kontrolja apparatury pelengatornoj pozicii [The algorithms used in the software of microprocessor systems for monitoring equipment direction finding position]. Trudy NIIR, vol. 3, pp. 11–17 (2014). (in Russian)
2. Aminev, D.A., Zhurkov, A.P., Kozyrev, A.A.: Algoritm kontrolja apparatury mestnogo dispetcherskogo punkta nazemnoj lokal'noj radiopelengacionnoj sistemy nabljudenija [The control algorithm for local control point equipment of RDF system]. Trudy NIIR, vol. 4, pp. 77–78 (2015). (in Russian)
3. Zhurkov, A.P., Aminev, D.A., Guseva, P.A., Miroshnichenko, S.S., Petrosjan P.A.: Analysis of the possibilities of self-diagnosis approaches to distributed electronic surveillance system. Systems of Control, Communication, Security, vol. 4, pp. 114–122 (2015). (in Russian). http://sccs.intelgr.com/archive/2015-04/06-Zhurkov.pdf. Accessed 20 Nov 2015
4. Vishnevsky, V.M., Kozyrev, D.V., Semenova, O.V.: Redundant queuing system with unreliable servers. In: International Congress on Ultra Modern Telecommunications and Control Systems and Workshops, pp. 283–286. IEEE Xplore (2015)
5. Tikhmenev, A.N., Zhadnov, V.V.: Imitacionnoe modelirovanie v zadachah ocenki nadezhnosti otkazoustoichivyh sredstv [Simulation problems in assessing the reliability of fault-tolerant electronic means]. Nadezhnost, vol. 1, no. 44, pp. 32–43 (2013). (in Russian)

6. Gertsbakh, I., Shpungin, Y., Vaisman, R.: Ternary Networks: Reliability and Monte Carlo. Springer Briefs in Electrical and Computer Engineering. Springer, Heidelberg (2014)
7. Golovin, I.N., Chuvarygin, B.V., Shura-Bura, A.E.: Raschet i optimizacia komplektov zapasnih elementov radioelektronnyh sistem [Calculation and optimization of sets of spare components of radioelectronic systems]. Moscow: Radio i svyaz [Radio and Communications], 176 p. (1984). (in Russian)
8. Lisnianski, A., Frenkel, I. (eds.): Recent Advances in System Reliability: Signatures, Multi-state Systems and Statistical Inference. Springer Series in Reliability Engineering. Springer, Heidelberg (2012)
9. Antonov, A.V., Plyaskin, A.V., Tataev, K.N.: K voprosu rascheta nadezhosti rezervirovannyh struktur s uchetom starenia elementov [To the problem of reliability calculation of redundant structures with account to aging of elements]. Nadezhnost. vol. 1, no. 44, pp. 55–61 (2013). (in Russian)
10. Tikhmenev A.N.: Problemi rascheta pokazatelei dostatochnosti i optimizacii zapasov v sistemah ZIP [Problems of sufficiency factor calculation and optimization of reserves in SPTA systems], vol. 3, no. 38, pp. 53–60 (2011)
11. Belyaev, Y.K., Bogatyrev, V.A., Bolotin, V.V., et al.: Nadezhnost tekhnicheskikh system [Reliability of technical systems]. In: Ushakov, I.A. (ed.) Moscow: Radio i svyaz [radio and Communications], 608 p. (1985). (in Russian)
12. IEC 61078 (2006–01) Analysis techniques for dependability – Reliability block diagram and Boolean methods

Low-Priority Queue and Server's Steady-State Existence in a Tandem Under Prolongable Cyclic Service

Victor Kocheganov[(⊠)] and Andrei Zorine

Institute of Information Technology, Mathematics and Mechanics,
N. I. Lobachevsky State University of Nizhny Novgorod,
Gagarina av. 23, 603950 Nizhny Novgorod, Russia
kocheganov@gmail.com

Abstract. A mathematical model of a tandem of queuing systems is considered. Each system has a high-priority input flow and a low-priority input flow which are conflicting. In the first system, the customers are serviced in the class of cyclic algorithms. The serviced high-priority customers are transferred from the first system to the second one with random delays and become the high-priority input flow of the second system. In the second system, customers are serviced in the class of cyclic algorithms with prolongations. Low-priority customers are serviced when their number exceeds a threshold. A mathematical model is constructed in form of a multidimensional denumerable discrete-time Markov chain. Conditions of low-priority queue stationary distribution existence were found.

Keywords: Tandem of controlling queuing systems · Cyclic algorithm with prolongations · Conflicting flows · Multidimensional denumerable discrete-time Markov chain

1 Introduction

An enormous amount of work has been done on the problem of conflicting traffic flows control at crossroad by the moment. In the queuing theory literature one can find following algorithms investigated: fixed duration cyclic algorithm, cyclic algorithm with a loop, cyclic algorithm with changing regimes, etc [1–6]. However, in a real-life situations cars pass several consecutive crossroads on their way rather then only one. In other words, an output flow of cars from the first intersection forms an input flow of cars of the next intersection. Hence, the second input flow no longer has an *a priori* known simple probabilistic structure (for example, a non-ordinary Poison flow), and knowledge about the service algorithm should be taken into account to deduce formation conditions of the first output flow.

One can find several works about tandems of intersections. In [7] a computer-aided simulation of adjacent intersections was carried out. In [8] a mathematical model of two intersections in tandem governed by cyclic algorithms was investigated and stability conditions were found. In this paper we assume that the first

© Springer International Publishing AG 2016
V.M. Vishnevskiy et al. (Eds.): DCCN 2016, CCIS 678, pp. 210–221, 2016.
DOI: 10.1007/978-3-319-51917-3_19

intersection is governed by a cyclic algorithm while the second intersection is governed by a cyclic algorithm with prolongations. The low-priority queue on the second intersection and necessary conditions of its stationary state existence take central place of this paper. This work continues studying in paper [10].

2   The Problem Settings

Consider a queuing system with a scheme shown in Fig. 1. There are four input flows of customers Π_1, Π_2, Π_3, and Π_4 entering the single server queueing system. Customers in the input flow Π_j, $j \in \{1, 2, 3, 4\}$ join a queue O_j with an unlimited capacity. For $j \in \{1, 2, 3\}$ the discipline of the queue O_j is FIFO (First In First Out). Discipline of the queue O_4 will be described later. The input flows Π_1 and Π_3 are generated by an external environment, which has only one state. Each of these flows is a nonordinary Poisson flow. Denote by λ_1 and λ_3 the intensities of bulk arrivals for the flows Π_1 and Π_3 respectively. The probability generating function of number of customers in a bulk in the flow Π_j is

$$f_j(z) = \sum_{\nu=1}^{\infty} p_\nu^{(j)} z^\nu, \quad j \in \{1, 3\}. \tag{1}$$

We assume that $f_j(z)$ converges for any $z \in \mathbb{C}$ such that $|z| < (1 + \varepsilon)$, $\varepsilon > 0$. Here $p_\nu^{(j)}$ is the probability of a bulk size in flow Π_j being exactly $\nu = 1, 2, \ldots$. Having been serviced the customers from O_1 come back to the system as the Π_4 customers. The Π_4 customers in turn after service enter the system as the Π_2 ones. The flows Π_2 and Π_3 are conflicting in the sense that their customers can't

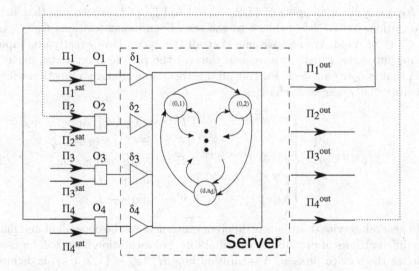

Fig. 1. Scheme of the queuing system as a cybernetic control system

be serviced simultaneously. This implies that the problem can't be reduced to a problem with fewer input flows by merging the flows together.

In order to describe the server behavior we fix positive integers d, n_0, n_1, ..., n_d and we introduce a finite set $\Gamma = \{\Gamma^{(k,r)}: k = 0, 1, \ldots, d; r = 1, 2, \ldots n_k\}$ of states server can reside in. At the state $\Gamma^{(k,r)}$ the server stays during a constant time $T^{(k,r)}$. Define disjoint subsets Γ^{I}, Γ^{II}, Γ^{III}, and Γ^{IV} of Γ as follows. In the state $\gamma \in \Gamma^{\mathrm{I}}$ only customers from the queues O_1, O_2 and O_4 are serviced. In the state $\gamma \in \Gamma^{\mathrm{II}}$ only customers from the queues O_2 and O_4 are serviced. In the state $\gamma \in \Gamma^{\mathrm{III}}$ only customers from queues O_1, O_3, and O_4 are serviced. In the state $\gamma \in \Gamma^{\mathrm{IV}}$ only customers from queues O_3 and O_4 are serviced. We assume that $\Gamma = \Gamma^{\mathrm{I}} \cup \Gamma^{\mathrm{II}} \cup \Gamma^{\mathrm{III}} \cup \Gamma^{\mathrm{IV}}$. Set also $^1\Gamma = \Gamma^{\mathrm{I}} \cup \Gamma^{\mathrm{III}}$, $^2\Gamma = \Gamma^{\mathrm{I}} \cup \Gamma^{\mathrm{II}}$, $^3\Gamma = \Gamma^{\mathrm{III}} \cup \Gamma^{\mathrm{IV}}$.

The server changes its state according to the following rules. We call a set $C_k = \{\Gamma^{(k,r)}: r = 1, 2, \ldots n_k\}$ the k-th cycle, $k = 1, 2, \ldots, d$. For $k = 0$ the state $\Gamma^{(0,r)}$ with $r = 1, 2, \ldots, n_0$ is called a prolongation state. Put $r \oplus_k 1 = r + 1$ for $r < n_k$, and $r \oplus_k 1 = 1$ for $r = n_k$ ($k = 0, 1, \ldots, d$). In the cycle C_k we select a subset C_k^{O} of input states, a subset C_k^{I} of output states, and a subset $C_k^{\mathrm{N}} = C_k \setminus (C_k^{\mathrm{O}} \cup C_k^{\mathrm{I}})$ of neutral states. After the state $\Gamma^{(k,r)} \in C_k \setminus C_k^{\mathrm{O}}$ the server switches to the state $\Gamma^{(k,r \oplus_k 1)}$ within the same cycle C_k. After the state $\Gamma^{(k,r)}$ in C_k^{O} the server switches to the state $\Gamma^{(k,r \oplus_k 1)}$ if number of customers in the queue O_3 at switching instant is greater than a predetermined threshold L. Otherwise, if the number of customers in the queue O_3 is less than or equal to L then the new state is the prolongation one $\Gamma^{(0,r_1)}$ where $r_1 = h_1(\Gamma^{(k,r)})$ and $h_1(\cdot)$ is a given mapping of $\bigcup_{k=1}^{d} C_k^{\mathrm{O}}$ into $\{1, 2, \ldots, n_0\}$. After the state $\Gamma^{(0,r)}$ if the number of customers in O_3 is not above L the state of the same type $\Gamma^{(0,r_2)}$ is chosen where $r_2 = h_2(r)$ and $h_2(\cdot)$ is a given mapping of the set $\{1, 2, \ldots, n_0\}$ into itself; in the other case the new state is $\Gamma^{(k,r_3)} \in C_k^{\mathrm{I}}$ where $\Gamma^{(k,r_3)} = h_3(r)$ and $h_3(\cdot)$ is a given mapping of $\{1, 2, \ldots, n_0\}$ to $\bigcup_{k=1}^{d} C_k^{\mathrm{I}}$. We assume that each prolongation state $\Gamma^{(0,r)}$ belongs to the set $^2\Gamma$ and that relations $C_k^{\mathrm{O}} \subset {}^2\Gamma$ and $C_k^{\mathrm{I}} \subset {}^3\Gamma$ hold. We also assume that all the cycles have exactly one input and output state. Finally, we assume that all the prolongation states make a cycle, that is $h_2(r) = r \oplus_0 1$. Putting all together, we introduce a function which formalizes the server state changes:

$$h(\Gamma^{(k,r)}, y) = \begin{cases} \Gamma^{(k,r \oplus_k 1)} & \text{if } \Gamma^{(k,r)} \in C_k \setminus C_k^{\mathrm{O}} \text{ or} \\ & \quad (\Gamma^{(k,r)} \in C_k^{\mathrm{O}}) \wedge (y > L); \\ \Gamma^{(0,h_1(\Gamma^{(k,r)}))} & \text{if } \Gamma^{(k,r)} \in C_k^{\mathrm{O}} \text{ and } y \leqslant L; \\ \Gamma^{(0,r \oplus_0 1)} & \text{if } k = 0 \text{ and } y \leqslant L; \\ h_3(r) & \text{if } k = 0 \text{ and } y > L. \end{cases} \qquad (2)$$

In general, service durations of different customers can be dependent and may have different laws of probability distributions. So, saturation flows will be used to define the service process. A saturation flow Π_j^{sat}, $j \in \{1, 2, 3, 4\}$, is defined as a virtual output flow under the maximum usage of the server and unlimited number of customer in the queue O_j. The saturation flow Π_j^{sat}, $j \in \{1, 2, 3\}$

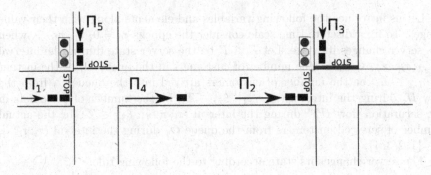

Fig. 2. A tandem of crossroads, the physical interpretation of the queuing system under study

contains a non-random number $\ell(k, r, j) \geqslant 0$ of customers in the server state $\Gamma^{(k,r)}$. In particular, $\ell(k, r, j) \geqslant 1$ for $\Gamma^{(k,r)} \in {}^j\Gamma$ and $\ell(k, r, j) = 0$ for $\Gamma^{(k,r)} \notin {}^j\Gamma$. Let $\mathbb{Z}_+$ be the set of non-negative integer numbers. If the queue O_4 contains $x \in \mathbb{Z}_+$ customers the saturation flow Π_4^{sat} also contains the x customers. Finally, in the state $\Gamma^{(k,r)}$ every customer from queue O_4 with probability $p_{k,r}$ and independently of others ends servicing and joins Π_2 to go to O_2. With the complementary probability $1 - p_{k,r}$ the customer stays in O_4 until the next time slot. In the next time slot it repeats its attempt to join Π_2 with a proper probability.

A real-life example of just described queuing system is a tandem of two consecutive crossroads (Fig. 2). The input flows are flows of vehicles. The flows Π_1 and Π_5 at the first crossroad are conflicting; Π_2 and Π_3 at the second crossroad are also conflicting. Every vehicle from the flow Π_1 after passing first road intersection joint the flow Π_4 and enters the queue O_4. After some random time interval the vehicle arrives to the next road intersection. Such a pair of crossroads is an instance of a more general queuing model described above.

3 Mathematical Model

The queuing system under investigation can be regarded as a cybernetic control system that helps to rigorously construct a formal stochastic model [8]. The scheme of the control system is shown in Fig. 1. There are following blocks present in the scheme: (1) the external environment with one state; (2) input poles of the first type—the input flows Π_1, Π_2, Π_3, and Π_4; (3) input poles of the second type—the saturation flows Π_1^{sat}, Π_2^{sat}, Π_3^{sat}, and Π_4^{sat}; (4) an external memory—the queues O_1, O_2, O_3, and O_4; (5) an information processing device for the external memory—the queue discipline units δ_1, δ_2, δ_3, and δ_4; (6) an internal memory—the server (OY); (7) an information processing device for internal memory—the graph of server state transitions; (8) output poles— the output flows Π_1^{out}, Π_2^{out}, Π_3^{out}, and Π_4^{out}. The coordinate of a block is its number on the scheme.

Let us introduce the following variables and elements along with their value ranges. To fix a discrete time scale consider the epochs $\tau_0 = 0$, τ_1, τ_2, ... when the server changes its state. Let $\Gamma_i \in \Gamma$ be the server state during the interval $(\tau_{i-1}; \tau_i]$, $\varkappa_{j,i} \in \mathbb{Z}_+$ be the number of customers in the queue O_j at the instant τ_i, $\eta_{j,i} \in \mathbb{Z}_+$ be the number of customers arrived into the queue O_j from the flow Π_j during the interval $(\tau_i; \tau_{i+1}]$, $\xi_{j,i} \in \mathbb{Z}_+$ be the number of customers in the saturation flow Π_j^{sat} during the interval $(\tau_i; \tau_{i+1}]$, $\bar{\xi}_{j,i} \in \mathbb{Z}_+$ be the actual number of serviced customers from the queue O_j during the interval $(\tau_i; \tau_{i+1}]$, $j \in \{1, 2, 3, 4\}$.

The server changes its state according to the following rule:

$$\Gamma_{i+1} = h(\Gamma_i, \varkappa_{3,i}) \tag{3}$$

where the mapping $h(\cdot, \cdot)$ is defined by Formula (2).

Let $\varphi_1(\cdot, \cdot)$ and $\varphi_3(\cdot, \cdot)$ be defined by series expansions

$$\sum_{\nu=0}^{\infty} z^{\nu} \varphi_j(\nu, t) = \exp\{\lambda_j t(f_j(z) - 1)\}$$

with functions $f_j(z)$ defined by (1), $j \in \{1, 3\}$. The function $\varphi_j(\nu, t)$ equals the probability of $\nu = 0, 1, \ldots$ arrivals in the flow Π_j during time $t \geq 0$. If $\nu < 0$ the value of $\varphi_j(\nu, t)$ is set to zero.

Mathematical model in more details can be found in work [10]. From now on we focus on low-priority customers in the queue O_3.

4   The Low-Priority Queue

Here we will consider the stochastic sequence

$$\{(\Gamma_i(\omega), \varkappa_{3,i}(\omega)); i = 0, 1, \ldots\}, \tag{4}$$

which includes the number of low-priority customers $\varkappa_{3,i}(\omega)$ in the queue O_3. In this section we will report several results concerning this stochastic sequence.

Let $\Gamma^{(k,r)} \in \Gamma$ and $x_3 \in \mathbb{Z}_+$. Denote by $H_{-1}(\Gamma^{(k,r)}, x_3)$ the set of all server states γ such that $h(\gamma, x_3) = \Gamma^{(k,r)}$ and put $r \ominus_k 1 = r - 1$ for $n_k \geq r > 1$, and $r \ominus_k 1 = n_k$ for $r = 1$ ($k = 0, 1, \ldots, d$). Then formula (2) makes it possible to define the mapping $H_{-1}(\Gamma^{(k,r)}, x_3)$ explicitly:

$$H_{-1}(\Gamma^{(k,r)}, x_3) = \begin{cases} \{\Gamma^{(k_1,r_1)}, \Gamma^{(0,r \ominus_0 1)}\} & \text{if } (k = 0) \wedge (x_3 \leq L), \\ \{\Gamma^{(k,r \ominus_k 1)}, \Gamma^{(0,r_2)}\} & \text{if } (\Gamma^{(k,r)} \in C_k^{\text{I}}) \wedge (x_3 > L), \\ \{\Gamma^{(k,r \ominus_k 1)}\} & \text{if } (\Gamma^{(k,r)} \in C_k^{\text{O}}) \vee (\Gamma^{(k,r)} \in C_k^{\text{N}}); \\ \varnothing & \text{if } (k = 0) \wedge (x_3 > L) \\ & \quad \text{or } (\Gamma^{(k,r)} \in C_k^{\text{I}}) \wedge (x_3 \leq L) \end{cases} \tag{5}$$

where $h_1(\Gamma^{(k_1,r_1)}) = r$ and $h_3(r_2) = \Gamma^{(k,r)}$.

Let's define for $\gamma \in \Gamma$ and $x_3 \in Z_+$ values

$$Q_{3,i}(\gamma, x) = \mathbf{P}(\{\omega\colon \Gamma_i(\omega) = \gamma, \varkappa_{3,i}(\omega) = x_3\}).$$

Theorem 1 concerns generating functions and corrects ones in paper [10]. Suppose k and r are such that $\Gamma^{(k,r)} \in \Gamma$. Let's define partial probability generating functions

$$\mathfrak{M}^{(3,i)}(k, r, v) = \sum_{w=0}^{\infty} Q_{3,i}(\Gamma^{(k,r)}, w)v^w,$$

$$q_{k,r}(v) = v^{-\ell(k,r,3)} \sum_{w=0}^{\infty} \varphi_3(w, T^{(k,r)})v^w.$$

and auxilary functions

$$\tilde{\alpha}_i(k, r, v) = \sum_{x_3=0}^{\ell(k,r,3)} \sum_{\gamma \in H_{-1}(\tilde{\gamma}, x_3)} Q_{3,i}(\gamma, x_3) \sum_{a=0}^{\ell(k,r,3)-x_3} \varphi_3(a, T^{(k,r)})$$

$$- \sum_{x_3=0}^{\ell(k,r,3)} \sum_{\gamma \in H_{-1}(\tilde{\gamma}, x_3)} Q_{3,i}(\gamma, x_3)v^{x_3-\ell(k,r,3)} \sum_{w=0}^{\ell(k,r,3)-x_3} \varphi_3(w, T^{(k,r)})v^w,$$

$$\alpha_i(0, r, v) = \tilde{\alpha}_i(0, r, v) + q_{0,r}(v) \times \sum_{x_3=0}^{L} \Big[Q_{3,i}(\Gamma^{(k_1,r_1)}, x_3)$$

$$+ Q_{3,i}(\Gamma^{(0,r\ominus_0 1)}, x_3) \Big] v^{x_3}, \quad \Gamma^{(0,r)} \in \Gamma,$$

$$\alpha_i(k, r, v) = \tilde{\alpha}_i(k, r, v) - q_{k,r}(v) \sum_{x_3=0}^{L} \Big[Q_{3,i}(\Gamma^{(k,r\ominus_k 1)}, x_3)$$

$$+ Q_{3,i}(\Gamma^{(0,r_2)}, x_3) \Big] v^{x_3} + q_{k,r}(v) \sum_{x_3 \geqslant 0} Q_{3,i}(\Gamma^{(0,r)}, x_3)v^{x_3}, \quad \Gamma^{(k,r)} \in C_k^{\mathrm{I}},$$

$$\alpha_i(k, r, v) = \tilde{\alpha}_i(k, r, v), \quad \Gamma^{(k,r)} \in C_k^{\mathrm{O}} \cup C_k^{\mathrm{N}}.$$

Theorem 1. *Following recurrent w.r.t. $i \geqslant 0$ relations take place for the partial probability generating functions:*

1. $\Gamma^{(0,r)} \in \Gamma$, $r = \overline{1, n_0}$

$$\mathfrak{M}^{(3,i+1)}(0, r, v) = \alpha_i(0, r, v);$$

2. $\Gamma^{(k,r)} \in \Gamma$, $k = \overline{1,d}$, $r = \overline{1,n_k}$

$$\mathfrak{M}^{(3,i+1)}(k,r,v) = q_{k,r}(v) \times \mathfrak{M}^{(3,i)}(k, r \ominus_k 1, v) + \alpha_i(k,r,v);$$

The last result (Theorem 2) is new and concerns low-priority queue and server's steady-state existence.

Theorem 2. *For Markov chain (4) to have stationary distribution $Q_3(\gamma, x)$, $(\gamma, x) \in \Gamma \times \mathbb{Z}_+$ it is necessary that following inequality takes place*

$$\min_{k=\overline{1,d}} \frac{\lambda_3 f_3'(1) \sum_{r=1}^{n_k} T^{(k,r)}}{\sum_{r=1}^{n_k} \ell(k,r,3)} < 1.$$

Proof. Let's assume that stationary distribution $Q_3(\gamma, x)$, $(\gamma, x) \in \Gamma \times \mathbb{Z}_+$, exists. Then choosing this distribution as the initial one imposes existence of following limits:

$$\lim_{i \to \infty} Q_{3,i}(\gamma, w) = Q_3(\gamma, w),$$

which are equal to stationary probabilities of corresponding states.

After defining generating functions

$$\mathfrak{M}^{(3)}(k,r,v) = \sum_{w=0}^{\infty} Q_3(\gamma, w) v^w,$$

for the stationary distribution similar relations can be derived as in Theorem 1:

1. $\Gamma^{(0,r)} \in \Gamma$, $r = \overline{1,n_0}$
$$\mathfrak{M}^{(3)}(0,r,v) = \alpha(0,r,v); \tag{6}$$

2. $\Gamma^{(k,r)} \in \Gamma$, $k = \overline{1,d}$, $r = \overline{1,n_k}$

$$\mathfrak{M}^{(3)}(k,r,v) = q_{k,r}(v) \times \mathfrak{M}^{(3)}(k, r \ominus_k 1, v) + \alpha(k,r,v); \tag{7}$$

where

$$\tilde{\alpha}(k,r,v) = \sum_{x_3=0}^{\ell(k,r,3)} \sum_{\gamma \in \mathbb{H}_{-1}(\tilde{\gamma}, x_3)} Q_3(\gamma, x_3) \sum_{a=0}^{\ell(k,r,3)-x_3} \varphi_3(a, T^{(k,r)})$$

$$- \sum_{x_3=0}^{\ell(k,r,3)} \sum_{\gamma \in \mathbb{H}_{-1}(\tilde{\gamma}, x_3)} Q_3(\gamma, x_3) v^{x_3 - \ell(k,r,3)} \sum_{w=0}^{\ell(k,r,3)-x_3} \varphi_3(w, T^{(k,r)}) v^w,$$

$$\alpha(0,r,v) = \tilde{\alpha}(0,r,v) + q_{0,r}(v) \times \sum_{x_3=0}^{L} \left[Q_3(\Gamma^{(k_1,r_1)}, x_3) \right.$$

$$\left. + Q_3(\Gamma^{(0,r \ominus_0 1)}, x_3) \right] v^{x_3}, \quad \Gamma^{(0,r)} \in \Gamma,$$

$$\alpha(k, r, v) = \tilde{\alpha}(k, r, v) - q_{k,r}(v) \sum_{x_3=0}^{L} \Big[Q_3(\Gamma^{(k,r\ominus_k 1)}, x_3)$$
$$+ Q_3(\Gamma^{(0,r_2)}, x_3) \Big] v^{x_3} + q_{k,r}(v) \sum_{x_3 \geqslant 0} Q_3(\Gamma^{(0,r)}, x_3) v^{x_3}, \quad \Gamma^{(k,r)} \in C_k^{\mathrm{I}},$$

$$\alpha(k, r, v) = \tilde{\alpha}(k, r, v), \quad \Gamma^{(k,r)} \in C_k^{\mathrm{O}} \cup C_k^{\mathrm{N}}$$

Taylor expansion of $q_{k,r}(v)$ gives

$$q_{k,r}(v) = v^{-\ell(k,r,3)} \exp\left(\lambda_3 T^{(k,r)}(f_3(v) - 1)\right)$$
$$= 1 + (\lambda_3 T^{(k,r)} f_3'(1) - \ell(k,r,3))(v - 1) + O((v - 1)^2)).$$

Summing all the relations (6) and (7) one can find

$$\sum_{k=0}^{d} \sum_{r=1}^{n_k} \mathfrak{M}^{(3)}(k, r, v)$$

$$= \sum_{r=1}^{n_0} \alpha(0, r, v) + \sum_{k=1}^{d} \sum_{r=1}^{n_k} \left[q_{k,r}(v) \mathfrak{M}^{(3)}(k, r \ominus_k 1, v) + \alpha(k, r, v) \right]$$

$$= \sum_{k=1}^{d} \sum_{r=1}^{n_k} q_{k,r}(v) \mathfrak{M}^{(3)}(k, r \ominus_k 1, v) + \sum_{k=1}^{d} \sum_{r=1}^{n_k} \alpha(k, r, v) + \sum_{r=1}^{n_0} \alpha(0, r, v). \quad (8)$$

Similarly lets expand $\sum_{k=1}^{d} \sum_{r=1}^{n_k} \alpha(k, r, v)$ and $\sum_{r=1}^{n_0} \alpha(0, r, v)$. First of all

$$\tilde{\alpha}(k, r, v)$$

$$= \sum_{x_3=0}^{\ell(k,r,3)} \sum_{\gamma \in \mathbb{H}_{-1}(\tilde{\gamma}, x_3)} Q_3(\gamma, x_3) \sum_{w=0}^{\ell(k,r,3)-x_3} \varphi_3(w, T^{(k,r)})(1 - v^{w-(\ell(k,r,3)-x_3)})$$

$$= -(v - 1) \sum_{x_3=0}^{\ell(k,r,3)} \sum_{\gamma \in \mathbb{H}_{-1}(\tilde{\gamma}, x_3)} Q_3(\gamma, x_3)$$

$$\times \sum_{w=0}^{\ell(k,r,3)-x_3} \varphi_3(w, T^{(k,r)})(w - (\ell(k,r,3) - x_3)) + O((v - 1)^2).$$

In particular, $\ell(k, r, 3)$ equals to zero for $k = 0$, that implies $\tilde{\alpha}(0, r, v) = O((v - 1)^2)$.

And now we are ready to expand further:

$$\sum_{r=1}^{n_0} \alpha(0,r,v)$$

$$= \sum_{r=1}^{n_0} q_{0,r}(v) \times \sum_{x_3=0}^{L} \left[Q_3(\Gamma^{(k_1,r_1)}, x_3) + Q_3(\Gamma^{(0,r\ominus_0 1)}, x_3) \right] v^{x_3} + O((v-1)^2)$$

$$= \sum_{r=1}^{n_0} (1 + (\lambda_3 T^{(0,r)} f_3'(1) - \ell(0,r,3))(v-1))$$

$$\times \sum_{x_3=0}^{L} \left[Q_3(\Gamma^{(k_1,r_1)}, x_3) + Q_3(\Gamma^{(0,r\ominus_0 1)}, x_3) \right] v^{x_3} + O((v-1)^2),$$

$$\sum_{k,r:\ \Gamma^{(k,r)} \in C_k^{\mathrm{I}}} \alpha(k,r,v) = \sum_{k,r:\ \Gamma^{(k,r)} \in C_k^{\mathrm{I}}} q_{k,r}(v) \left[\mathfrak{M}^{(3)}(0,r_2,v) \right.$$

$$\left. - \sum_{x_3=0}^{L} (Q_3(\Gamma^{(k,r\ominus_k 1)}, x_3) + Q_3(\Gamma^{(0,r_2)}, x_3)) v^{x_3} \right]$$

$$+ \sum_{k,r:\ \Gamma^{(k,r)} \in C_k^{\mathrm{I}}} \tilde{\alpha}(k,r,v) = \sum_{k,r:\ \Gamma^{(k,r)} \in C_k^{\mathrm{I}}} (1 + (\lambda_3 T^{(k,r)} f_3'(1) - \ell(k,r,3))(v-1))$$

$$\times \left[\mathfrak{M}^{(3)}(0,r_2,v) - \sum_{x_3=0}^{L} (Q_3(\Gamma^{(k,r\ominus_k 1)}, x_3) + Q_3(\Gamma^{(0,r_2)}, x_3)) v^{x_3} \right]$$

$$- (v-1) \sum_{k,r:\ \Gamma^{(k,r)} \in C_k^{\mathrm{I}}} \sum_{x_3=0}^{\ell(k,r,3)} \sum_{\gamma \in \mathbb{H}_{-1}(\tilde{\gamma}, x_3)} Q_3(\gamma, x_3)$$

$$\times \sum_{w=0}^{\ell(k,r,3)-x_3} \varphi_3(w, T^{(k,r)})(w - (\ell(k,r,3) - x_3)) + O((v-1)^2),$$

$$\sum_{k,r:\ \Gamma^{(k,r)} \in C_k^{\mathrm{O}} \cup C_k^{\mathrm{N}}} \alpha(k,r,v) = \sum_{k,r:\ \Gamma^{(k,r)} \in C_k^{\mathrm{O}} \cup C_k^{\mathrm{N}}} \tilde{\alpha}(k,r,v)$$

$$= -(v-1) \sum_{k,r:\ \Gamma^{(k,r)} \in C_k^{\mathrm{O}} \cup C_k^{\mathrm{N}}} \sum_{x_3=0}^{\ell(k,r,3)} \sum_{\gamma \in \mathbb{H}_{-1}(\tilde{\gamma}, x_3)} Q_3(\gamma, x_3)$$

$$\times \sum_{w=0}^{\ell(k,r,3)-x_3} \varphi_3(w, T^{(k,r)})(w - (\ell(k,r,3) - x_3)) + O((v-1)^2).$$

It's important to mention, that any input system state corresponds to one and only one prolongation system state. That is why substitution of calculated expressions into (8) gives:

$$0 = O((v-1)^2) + (v-1) \sum_{k=1}^{d} \sum_{r=1}^{n_k} (\lambda_3 T^{(k,r)} f_3'(1) - \ell(k,r,3)) \mathfrak{M}^{(3)}(k, r \ominus_k 1, v)$$

$$+ (v-1) \sum_{r=1}^{n_0} (\lambda_3 T^{(0,r)} f_3'(1) - \ell(0,r,3)) \times \sum_{x_3=0}^{L} \Big[Q_3(\Gamma^{(k_1,r_1)}, x_3)$$

$$+ Q_3(\Gamma^{(0,r\ominus_0 1)}, x_3) \Big] v^{x_3} + (v-1) \sum_{k,r:\ \Gamma^{(k,r)} \in C_k^I} (\lambda_3 T^{(k,r)} f_3'(1) - \ell(k,r,3))$$

$$\times \Big[\mathfrak{M}^{(3)}(0, r_2, v) - \sum_{x_3=0}^{L} (Q_3(\Gamma^{(k,r\ominus_k 1)}, x_3) + Q_3(\Gamma^{(0,r_2)}, x_3)) v^{x_3} \Big]$$

$$- (v-1) \sum_{k,r:\ \Gamma^{(k,r)} \in C_k^O \cup C_k^N} \sum_{x_3=0}^{\ell(k,r,3)} \sum_{\gamma \in \mathbb{H}_{-1}(\tilde{\gamma}, x_3)} \Big(Q_3(\gamma, x_3)$$

$$\times \sum_{w=0}^{\ell(k,r,3)-x_3} \varphi_3(w, T^{(k,r)})(w - (\ell(k,r,3) - x_3)) \Big), \qquad (9)$$

dividing by $(v-1)$ and then sending v to 1 from the left one continues

$$0 = \sum_{k,r:\ \Gamma^{(k,r)} \in C_k^I} (\lambda_3 T^{(k,r)} f_3'(1) - \ell(k,r,3)) \times \Big[\mathfrak{M}^{(3)}(k, r \ominus_k 1, 1)$$

$$- \sum_{x_3=0}^{L} Q_3(\Gamma^{(k,r\ominus_k 1)}, x_3) + \mathfrak{M}^{(3)}(0, r_2, 1) - \sum_{x_3=0}^{L} Q_3(\Gamma^{(0,r_2)}, x_3) \Big]$$

$$+ \sum_{k,r:\ \Gamma^{(k,r)} \in C_k^O \cup C_k^N} (\lambda_3 T^{(k,r)} f_3'(1) - \ell(k,r,3)) \mathfrak{M}^{(3)}(k, r \ominus_k 1, 1)$$

$$+ \sum_{r=1}^{n_0} \lambda_3 T^{(0,r)} f_3'(1) \times \sum_{x_3=0}^{L} \Big[Q_3(\Gamma^{(k_1,r_1)}, x_3) + Q_3(\Gamma^{(0,r\ominus_0 1)}, x_3) \Big]$$

$$+ \sum_{k=1}^{d} \sum_{r=1}^{n_k} \sum_{x_3=0}^{\ell(k,r,3)} \sum_{\gamma \in \mathbb{H}_{-1}(\tilde{\gamma}, x_3)} Q_3(\gamma, x_3) \sum_{w=0}^{\ell(k,r,3)-x_3} \varphi_3(w, T^{(k,r)})(\ell(k,r,3) - x_3 - w).$$

$$(10)$$

Fixing $v = 1$ in corresponding generating functions relations one can get

$$\mathfrak{M}^{(3)}(0, r, 1) = \sum_{x_3=0}^{L} \Big(Q_3(\Gamma^{(k_1,r_1)}, x_3) + Q_3(\Gamma^{(0,r\ominus_0 1)}, x_3) \Big),$$

$$\mathfrak{M}^{(3)}(k, r, 1) = \mathfrak{M}^{(3)}(k, r \ominus_k 1, 1) + \mathfrak{M}^{(3)}(0, r_2, 1)$$

$$- \sum_{x_3=0}^{L} \Big(Q_3(\Gamma^{(k,r\ominus_k 1)}, x_3) + Q_3(\Gamma^{(0,r_2)}, x_3) \Big), \quad \Gamma^{(k,r)} \in C_k^I,$$

$$\mathfrak{M}^{(3)}(k, r, 1) = \mathfrak{M}^{(3)}(k, r \ominus_k 1, 1), \quad \Gamma^{(k,r)} \in C_k^O \bigcup C_k^N.$$

The last equation gives $\mathfrak{M}^{(3)}(k, n_k, 1) = \mathfrak{M}^{(3)}(k, n_k \ominus_k 1, 1) = \ldots = \mathfrak{M}^{(3)}(k, 1, 1) = M_k$. Using these facts, let's simplify expression (10)

$$0 = \sum_{k,r:\ \Gamma^{(k,r)} \in C_k^I} (\lambda_3 T^{(k,r)} f_3'(1) - \ell(k, r, 3)) \times M_k$$

$$+ \sum_{k,r:\ \Gamma^{(k,r)} \in C_k^O \cup C_k^N} (\lambda_3 T^{(k,r)} f_3'(1) - \ell(k, r, 3)) M_k$$

$$+ \sum_{r=1}^{n_0} \lambda_3 T^{(0,r)} f_3'(1) \times \sum_{x_3=0}^{L} \left[Q_3(\Gamma^{(k_1, r_1)}, x_3) + Q_3(\Gamma^{(0, r \ominus_0 1)}, x_3) \right]$$

$$+ \sum_{\substack{k,r:\ \Gamma^{(k,r)} \in \\ \in C_k^O \cup C_k^N}} \sum_{x_3=0}^{\ell(k,r,3)} \sum_{\gamma \in \mathbb{H}_{-1}(\tilde{\gamma}, x_3)} Q_3(\gamma, x_3) \sum_{w=0}^{\ell(k,r,3)-x_3} \varphi_3(w, T^{(k,r)})(\ell(k, r, 3) - x_3 - w).$$

After grouping terms we get:

$$0 = \sum_{k=1}^{d} \left(M_k \times \sum_{r=1}^{n_k} (\lambda_3 T^{(k,r)} f_3'(1) - \ell(k, r, 3)) \right)$$

$$+ \sum_{r=1}^{n_0} \lambda_3 T^{(0,r)} f_3'(1) \times \sum_{x_3=0}^{L} \left[Q_3(\Gamma^{(k_1, r_1)}, x_3) + Q_3(\Gamma^{(0, r \ominus_0 1)}, x_3) \right]$$

$$+ \sum_{\substack{k,r:\ \Gamma^{(k,r)} \in \\ \in C_k^O \cup C_k^N}} \sum_{x_3=0}^{\ell(k,r,3)} \sum_{\gamma \in \mathbb{H}_{-1}(\tilde{\gamma}, x_3)} Q_3(\gamma, x_3) \sum_{w=0}^{\ell(k,r,3)-x_3} \varphi_3(w, T^{(k,r)})(\ell(k, r, 3) - x_3 - w).$$

Assumption that expression $\sum_{r=1}^{n_k} \ell(k, r, 3) / \lambda_3 f_3'(1) \sum_{r=1}^{n_k} T^{(k,r)}$ for any $k = \overline{1; d}$ is less than or equal to 1, leads to impossible conclusion $Q_3(\Gamma^{(0,1)}, 0) = 0$. Theorem is proved.

Acknowledgments. This work was fulfilled as a part of State Budget Research and Development program No. 01201456585 "Mathematical modeling and analysis of stochastic evolutionary systems and decision processes" of N.I. Lobachevsky State University of Nizhni Novgorod and was supported by State Program "Promoting the competitiveness among world's leading research and educational centers".

References

1. Neimark , Y.I., Fedotkin M.A., Preobrazhenskaja A.M.: Operation of an automate with feedback controlling traffic at an intersection. Izvestija of USSR Academy of Sciences. Tech. Cybern. (5), 129–141 (1968)
2. Fedotkin, M.A.: On a class of stable algorithms for control of conflicting flows or arriving airplanes. Probl. Control Inf. Theory **6**(1), 13–22 (1977)

3. Fedotkin, M.A.: Construction of a model and investigation of nonlinear algorithms for control of intense conflict flows in a system with variable structure of servicing demands. Lith. Math. J. **17**(1), 129–137 (1977)
4. Litvak, N.V., Fedotkin, M.A.: A probabilistic model for the adaptive control of conflict flows. Autom. Remote Control **61**(5), 777–784 (2000)
5. Proidakova, E.V., Fedotkin, M.A.: Control of output flows in the system with cyclic servicing and readjustments. Autom. Remote Control **69**(6), 993–1002 (2008)
6. Afanasyeva, L.G., Bulinskaya, E.V.: Mathematical models of transport systems based on queueing theory. Tr. Mosc. Inst. Phys. Technol. **4**, 6–21 (2010)
7. Yamada, K., Lam T.N.: Simulation analysis of two adjacent traffic signals. In: Proceedings of the 17th Winter Simulation Conference, pp. 454–464. ACM, New York (1985)
8. Zorin, A.V.: Stability of a tandem of queueing systems with Bernoulli noninstantaneous transfer of customers. Theory Probab. Math. Stat. **84**, 173–188 (2012)
9. Kocheganov, V.M., Zorine, A.V.: Probabilistic model of tandem of queuing systems under cyclic control with prolongations. In: Proceedings of International Conference "Probability Theory, Stochastic Processes, Mathematical Statistics and Applications" (Minsk, 23–26 February 2015), pp. 94–99 (2015)
10. Kocheganov, V., Zorine, A.V.: Low-priority queue fluctuations in tandem of queuing systems under cyclic control with prolongations. In: Vishnevsky, V., Kozyrev, D. (eds.) DCCN 2015. CCIS, vol. 601, pp. 268–279. Springer, Heidelberg (2016). doi:10.1007/978-3-319-30843-2_28

On Regenerative Envelopes for Cluster Model Simulation

Evsey Morozov[1,2]($\boxtimes$), Irina Peshkova[1], and Alexander Rumyantsev[1,2]

[1] Petrozavodsk State University, Lenin Street 33, Petrozavodsk 185910, Russia
emorozov@karelia.ru, iaminova@petrsu.ru, ar0@krc.karelia.ru
[2] Institute of Applied Mathematical Research of the Karelian Reseach Centre
of R.A.S., Pushkinskaya Street 11, Petrozavodsk 185910, Russia

Abstract. We continue to develop a novel approach for confidence estimation of the stationary measures in the model describing high performance multiserver queueing systems, such as high performance clusters (HPC). We call this model *cluster model*. This model is described by a stochastic process, and in the framework of the approach, we construct two envelopes, minorant and majorant regenerative processes for the queue size process in the original system. These envelopes have classical regenerations while the original process may not be regenerative or its regenerations happen too rare to be useful for statistical estimation. It allows to construct confidence intervals for the steady-state queue size of the cluster model. We use simulation to illustrate the applicability of the approach and give recommendations how to select the predefined parameters of the envelopes to increase the efficiency of estimation. As simulation shows, the constructed envelopes allow to estimate the mean stationary queue size in the original system with a given accuracy in an acceptable time.

Keywords: Regenerative envelopes · High performance cluster · Queue size estimation

1 Introduction

Regenerative simulation approach is well known as one of the powerful tools for high performance queueing systems simulation with a complicated correlation structure [1–4]. It allows to apply well-developed classical statistical methods based on the special form of the Central Limit Theorem (CLT) for confidence estimation of QoS parameters in the case when data are dependent. At the same time the efficiency of the approach depends strongly on the accuracy of estimation which, in turn, depends on the frequency of the regenerations.

In this work we continue to develop regenerative envelopes method proposed in [5] with focus on regenerative estimation of classical multiserver system (not necessary regenerative) by means of construction of majorant and minorant regenerative system (called *envelopes*). In this work instead we focus on simulation and estimation of a cluster model. (This possibility has been mentioned in previous work [5].) In general, this new approach can be effectively applied for

© Springer International Publishing AG 2016
V.M. Vishnevskiy et al. (Eds.): DCCN 2016, CCIS 678, pp. 222–230, 2016.
DOI: 10.1007/978-3-319-51917-3_20

estimation of QoS parameters, such as queue size in the case, when the number of servers is large and classical regenerations do not exist, or are too rare to be useful in practice to provide the desired accuracy in an acceptable simulation time.

The main idea of the method is to construct the two new systems: *majorant* and *minorant* regenerative systems (called regenerative envelopes). In fact we obtain a regeneration point when a basic (Markov) process hits a predefined compact set. At these instances we transform the remaining service times in an appropriate way to obtain classical regeneration in each envelop. The queue size processes in the new majorant and minorant systems regenerate in classical sense and have i.i.d. regenerative cycles with i.i.d. lengths, however, at the regeneration instant, need not to be in an empty state as in conventional construction. Then we apply a monotonicity property of the queue size process (which is based on the concept of *coupling*) to construct confidence estimate of the steady-state queue size in the original system. It is worth mentioning, that there is a proximity of the queue size processes in the majorant and original systems. Moreover in some experiments these processes merge in a neighborhood of regeneration points. Potentially it allows to estimate the performance measures in (generally non-regenerative) cluster model with a very high precision. It is important to note that, unlike classical multiserver system $M/M/m$, the performance measures in the new cluster model are not explicitly available, and we rely on the simulation to select the most frequent regenerations applied for estimation.

In summary, the present work is a development of previous paper [5] with focus on the application of the new method to the cluster model. The novelty of this work is the description of regeneration points for cluster model. The main contribution of this work is a detailed analysis of the high performance cluster simulation based on regenerative envelopes.

This paper is organised as follows. In Sect. 2 we describe the cluster model in detail and present a modified Kiefer–Wolfowitz-type recursion which recursively defines m-dimensional workload process in this model. Moreover, we present recently proved stability criteria of this model. In Sect. 3 we give monotonicity properties of the basic processes, queue-size and number of customers, in the cluster model. Then we describe construction of the envelopes with focus on regenerations of the majorant system. In Sect. 4.1 we present the main notions of the regenerative simulation method including confidence estimation based on the corresponding variant of Central Limit Theorem. Then, in Sect. 4.2, we demonstrate results of simulation of the cluster model for some governing distributions, and also give estimates of the stationary queue size. Other performance measures of the cluster model are discussed as well. Obtained results show some advantages of the method in estimation of non-regenerative queueing system, in particular, the cluster model.

2 Description of the Model and Stability Criterion

Consider the cluster model Σ with FCFS service discipline which has m servers working in parallel. (In what follows we will suppress serial index to denote a generic element of an i.i.d sequence.) Denote by $\tau_i = t_{i+1} - t_i$ the interarrival times (with arrival rate $\lambda = 1/\mathsf{E}\tau$), where t_i is the arrival instant of the ith customer, $i \geqslant 1$. Customer i requires N_i servers simultaneously for service time S_i (with service rate $\mu = 1/\mathsf{E}S$). Note that $\{N_i\}$ is an i.i.d. sequence with distribution of a generic sequence member $\mathbf{p} = \{p_k := \mathsf{P}(N = k), \, k = 1, \ldots, m\}$.

The following Kiefer–Wolfowitz-type recursion describes the dynamics of the workload of the cluster model

$$
W_{i+1} = R\big(\overbrace{W_{i,N_i} + S_i - \tau_i, \ldots, W_{i,N_i} + S_i - \tau_i}^{N_i \text{ components}},
$$
$$
W_{i,N_i+1} - \tau_i, \ldots, W_{i,m} - \tau_i\big)^+, \tag{1}
$$

where operator R puts the components of the vector in an increasing order and $W_{i,j}$ is the remaining work allocated to the jth most busy server at the arrival epoch t_i of customer i. In other words, customer i occupies the N_i least busy servers. Note that if $N = 1$ w.p. 1, then (1) defines the workload vector for a classical GI/G/m multiserver system [6], that is

$$
W_{i+1} = R(W_{i,1} + S_i - \tau_i, W_{i,2} - \tau_i, \ldots, W_{i,m} - \tau_i)^+.
$$

Denote $\nu_i\,(Q_i)$ the number of customers (queue size) at instant t_i^- in the cluster model. All basic processes in this system are regenerative with regeneration points defined as arrival epochs to an empty system. When the mean regeneration period is finite, $\mathsf{E}T < \infty$, the regenerative process is called positive recurrent, and it is the most important step to establish stability of the system [1]. The following stability (positive recurrence) criterion has been proved in [7]

$$
\frac{\lambda}{\mu}C < 1, \tag{2}
$$

where

$$
C := \sum_{i=1}^{m} \frac{1}{i} \sum_{j=i}^{m} (\mathbf{p}^{*i})_j \sum_{t=m-j+1}^{m} p_t, \tag{3}
$$

and $(\mathbf{p}^{*i})_j$ is the j-th component of i-convolution of $\mathbf{p}$ with itself, that is $(\mathbf{p}^{*i})_j = \mathsf{P}(N_1 + \cdots + N_i = j)$.

For a 2-server model, the criterion (2) becomes [8]

$$
\frac{\lambda}{\mu}\left(1 - \frac{p_1^2}{2}\right) < 1. \tag{4}
$$

3   Monotonicity and Construction of Regenerative Envelopes

In the regenerative envelopes method, we construct two queueing systems, *majorant (or upper)* system $\overline{\Sigma}$ and *minorant (or lower)* system $\underline{\Sigma}$. We endow the corresponding variables in the system $\overline{\Sigma}$ ($\underline{\Sigma}$) with overline (underline). The majorant and minorant systems have the same input sequence $\{\tau_i\}$ as in Σ, and appropriately enlarged, respectively, shortened service times $\{\overline{S}_i\}$ and $\{\underline{S}_i\}$, $i \geqslant 1$. These transformations occur at the epochs when each system hits the corresponding fixed compact set, implying regeneration. In turn, it allows to perform regenerative confidence estimation of the steady-state performance measures of the discrete-time process $\{X_n, n \geqslant 0\}$ which describes the dynamics of the original stochastic system [2–4], even if the original system is not regenerative. (More on regenerative estimation see in [5].)

The procedure is based on the following monotonicity result which holds true for the cluster model as well as for a classical $GI/G/m$ model [1]. Assume the initial workloads ordered as $\underline{W}_1 \leqslant W_1 \leqslant \overline{W}_1$, and the service times are ordered as $\underline{S}_i \leqslant S_i \leqslant \overline{S}_i$, $i \geqslant 1$. (In fact we apply coupling method which allows to replace stochastic relations by relations w.p.1.) Then

$$\underline{\nu}_i \leqslant \nu_i \leqslant \overline{\nu}_i, \ \underline{Q}_i \leqslant Q_i \leqslant \overline{Q}_i, \ i \geqslant 1. \tag{5}$$

To define the aforementioned compact set, we suggest the following methods:

(1) Fix integer Q_0 (the number of customers in the queue), N_0 (the number of servers required by the customer at the head of the queue) and constants $0 \leqslant a \leqslant b \leqslant \infty$, and define

$$\overline{\beta}_{n+1} = \inf\Big\{k > \overline{\beta}_n : \overline{Q}_k = Q_0 > 0, \ N_{k-Q_0} = N_0,$$

$$\overline{S}_i(k) \in [a, b], \ i \in \overline{\mathcal{M}}_k, \ \mathcal{M}_{\overline{\beta}_n} \bigcap \mathcal{M}_k = \varnothing\Big\}, \ n \geqslant 0, \tag{7}$$

where $\overline{\mathcal{M}}_n = \{i : t_i \leqslant t_n < \overline{z}_i\}$ is the set of the customers served in the system $\overline{\Sigma}$ at instant t_n, $\overline{S}_i(n)$ is the remaining service time of customer i at instant t_n and $\overline{z}_i := t_i + \overline{W}_{i,N_i} + \overline{S}_i$ is the departure instant of ith customer.

(2) Fix integer m_0 (the number of busy servers, if the queue is empty) and define

$$\overline{\beta}_{n+1} = \inf\Big\{k > \overline{\beta}_n : \overline{Q}_k = 0, \ \sum_{i \in \overline{\mathcal{M}}_k} N_i = m_0 \leqslant m,$$

$$\overline{S}_i(k) \in [a, b], \ i \in \overline{\mathcal{M}}_k, \ \mathcal{M}_{\overline{\beta}_n} \bigcap \mathcal{M}_k = \varnothing\Big\}, \ n \geqslant 0. \tag{8}$$

At each such instant $\overline{\beta}_n$ we replace the remaining times $\overline{S}_i(k), i \in \overline{\mathcal{M}}_k$, by the upper bound b. The condition $\mathcal{M}_{\overline{\beta}_n} \bigcap \mathcal{M}_k = \varnothing$ means that all customers *being served* at instant $\overline{\beta}_n$ have left the system before instant $\overline{\beta}_{n+1}$. It is easy to see that $\overline{\mathbf{X}}_n := \{\overline{Q}_n, \overline{S}_i(n), i \in \overline{\mathcal{M}}_n\}, \ n \geqslant 1$, is a Markov process, and that distribution of $\overline{\mathbf{X}}_{\overline{\beta}_n}$ is independent of n and pre-history $\{\overline{\mathbf{X}}_k, \ k < \overline{\beta}_n\}, \ n \geqslant 1$. In other words, the process $\{\overline{\mathbf{X}}_n, \ n \geqslant 1\}$ regenerates at the instants $\{\overline{\beta}_k\}$ and has the iid *regeneration cycles* $\{\overline{\mathbf{X}}_k, \overline{\beta}_n \leqslant k < \overline{\beta}_{n+1}\}$ with iid cycle lengths $\overline{\beta}_{n+1} - \overline{\beta}_n, \ n \geqslant 1$.

We define the moments $\underline{\beta}_n, n \geqslant 0$ in a similar way (with possibly other values a, b, Q_0, N_0, m_0). When the queue-size process in the minorant system hits, at some arrival instant t_k, the corresponding compact set, we replace the current remaining service times $\underline{S}_i(k), i \in \mathcal{M}_k$, by a. (More details can be found in [5].)

It is easy to show that the constructed moments $\{\overline{\beta}_n\}, \{\underline{\beta}_n\}$ are regenerative instants of the corresponding basic processes, and moreover, these replacements keep the monotonicity of the performance measures of interest. We stress that, at the regeneration instant $\overline{\beta}_n$, all the remaining service times of the customers being served become equal to b, and customers in $\mathcal{M}_{\overline{\beta}_n}$ can be treated a single *large* customer occupying m_0 servers. Note that for a classical $GI/G/m$ system the cases (1) and (2) coincide, since

$$N = 1 \text{ w.p. } 1, \quad \sum_{i \in \mathcal{M}_k} N_i = \nu_k \text{ and } Q_k = \max(0, \nu_k - m).$$

In general, selection of the constants a, b, Q_0, N_0, m_0 must be done in such a way to obtain a trade-off between the frequency of regenerations and the difference between the upper and lower systems. In this regard we recall the well-known result for classical $M/M/m$ system: the value $\lceil \lambda/\mu \rceil$ of the stationary number of customers in the system ν is the most probable, see, for instance, [1]. However, in the cluster model, stationary distributions of the number of customers in the system ν, the stationary queue size Q and the number of the busy servers are not available in an explicit form. Nevertheless, as our simulation confirms, it is surprising that the state $\nu = \lceil \lambda/\mu \rceil$ is still the most probable in the cluster model as well. This fact can be used to speed-up regenerative estimation of the performance measures of general (non-Markovian) cluster model.

4   Simulation and Estimation

4.1   Regenerative Simulation

Define the regenerative instants $\{\beta_k, k \geqslant 1\}$ for a discrete-time regenerative process $\{X_n, n \geqslant 0\}$. The regenerative cycles $\mathcal{C}_k := \{X_j : \beta_k \leqslant j < \beta_{k+1}\}$, $k \geqslant 0$, are iid with the iid cycle lengths $\alpha_k := \beta_{k+1} - \beta_k$ ($\beta_0 := 0$). We discuss the confidence estimation of steady-state characteristic $r = \mathsf{E}X$. Consider the iid sequence

$$Y_k = \sum_{i=\beta_k}^{\beta_{k+1}-1} X_i, \qquad V_k = Y_k - r\alpha_k, \qquad k \geqslant 0,$$

Applying CLT we obtain the following asymptotic $100(1 - 2\gamma)\%$ confidence interval for the unknown r:

$$\left[\hat{r}_k - \frac{z_{1-\gamma} S(k)}{\hat{\alpha}_k \sqrt{k}}, \ \hat{r}_k + \frac{z_{1-\gamma} S(k)}{\hat{\alpha}_k \sqrt{k}} \right], \tag{9}$$

where $\hat{r}_k = \hat{Y}_k/\hat{\alpha}_k$, $\hat{Y}_k(\hat{\alpha}_k)$ is the sample mean of $\mathsf{E}Y(\mathsf{E}\alpha)$, the estimate $S^2(k) \to \sigma^2 = \mathsf{E}V_1^2$, $k \to \infty$, w.p.1, the quantile $z_{1-\gamma}$ is defined as $\mathsf{P}(\mathbb{N}(0,1) \leqslant z_{1-\gamma}) = 1 - \gamma$, and $\mathbb{N}(0,1)$ is standard normal variable. It is important to stress that, unlike classic estimation, in this case we deal with the iid groups $\{Y_k\}$ related to regeneration cycles, but not original data $\{X_i\}$, and simulation with a given accuracy requires in general much more simulation time.

4.2   Regenerative Estimation of Cluster Model

To study the performance measures of the cluster model, we perform the following numerical experiment. We take $m = 100$ and generate a sequence of $n = 10^5$ Poisson arrivals (with rate $\lambda = 1$). We use the Zipf's law, the discrete analogue of the heavy-tailed Pareto law, to model distribution $\{p_i\}$:

$$p_i = p_0/i, i = 1, \ldots, m, \text{where } p_0 = (p_1 + \cdots + p_m)^{-1}.$$

Then we select $\mu = 0.9\lambda C$, where C is defined in (3). It gives $\lceil \lambda/\mu \rceil = 4$. We calculate the following performance measures: the number of customers in the system, the number of customers in the queue, the number of customers being served, the number of busy servers. The histograms of the results are shown on Fig. 1. It is seen that the most frequent number of customers in the system equals 4, and it corresponds to our conjecture. However, in the most of the cases (≈ 0.15) the queue size $Q = 0$, and the (empirical) probability $\mathsf{P}(Q = i)$ decreases for $i > 0$. This effect of (geometrical) decreasing of probabilities $P(\nu = k)$ for k large has been established as a property of the matrix-geometric solution for the cluster model, see [7,9]. Now to increase the frequency of regenerations, we select $Q_0 = 0$ both for the upper and for the lower models, and set $m_0 = 100$, as the most frequent value. Next, we select (a,b) as $(0, 2.4)$ for the upper system, and $(2.2, \infty)$ for the lower system. Thus, the regeneration instants defined by (8) become in our setting

$$\overline{\beta}_{n+1} = \inf\left\{k > \overline{\beta}_n : \overline{Q}_k = 0, \sum_{i \in \overline{\mathcal{M}}_k} N_i = 100,\right.$$

$$\left.\overline{S}_i(k) \in (0, 2.4), i \in \overline{\mathcal{M}}_k, \mathcal{M}_{\overline{\beta}_n} \bigcap \overline{\mathcal{M}}_k = \varnothing\right\}, n \geqslant 0.$$

$$\underline{\beta}_{n+1} = \inf\left\{k > \underline{\beta}_n : \underline{Q}_k = 0, \sum_{i \in \underline{\mathcal{M}}_k} N_i = 100,\right.$$

$$\left.\underline{S}_i(k) \in (2.2, \infty), i \in \underline{\mathcal{M}}_k, \mathcal{M}_{\underline{\beta}_n} \bigcap \underline{\mathcal{M}}_k = \varnothing\right\}, n \geqslant 0.$$

We apply regenerative confidence estimations for the regenerative envelopes (with $\gamma = 0.05$), and for the original system as well, because in this case it has classical regenerations when $\nu_i = 0$. The results of the experiments are shown on Fig. 2. It is seen that the regenerative envelop method slightly outperforms (in terms of acquired accuracy) the confidence interval constructed for original system using classical regenerations. This important effect shows the advantage

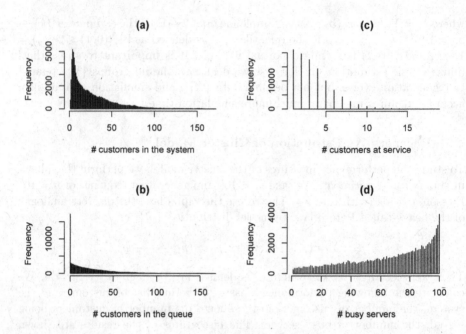

Fig. 1. Histograms of the performance measures for $m = 100$ servers, Poisson arrivals with rate $\lambda = 1$, exponential service time with rate $\mu = 0.3$ and $p_i = p_0/i, i = 1, \ldots, s$: (a) customers in the system, (b) customers in the queue, (c) customers at service, (d) busy servers.

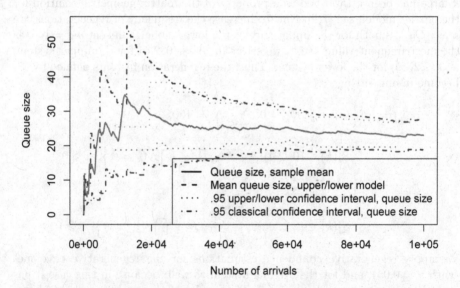

Fig. 2. Confidence intervals for the systems with $m = 100$ servers, Poisson arrivals with rate $\lambda = 1$, exponential service time with rate $\mu = 0.3$ and $p_i = p_0/i, i = 1, \ldots, m$.

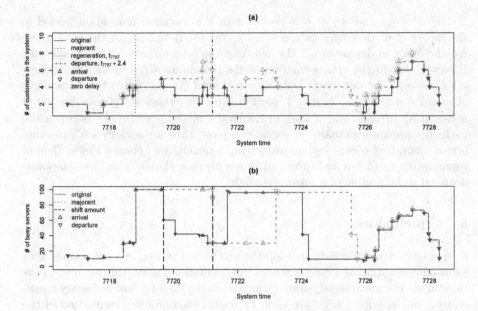

Fig. 3. Coupling of the majorant and original systems with $m = 100$ servers, Poisson arrivals with rate $\lambda = 1$, exponential service time with rate $\mu = 0.3$ and $p_i = p_0/i, i = 1, \ldots, m$ in a neighborhood of the regeneration epoch $t_{7787} + 0$ (a): number of customers in the system vs. system time; (b): number of busy servers vs. system time.

of the method of regenerative envelops. However, the nature of such phenomenon requires a separate investigation.

Now we study in detail the paths of the queue size in the majorant system (possessing regenerations) and in the original system, in a neighborhood of a regeneration point, see Fig. 3. We consider the first regeneration point of this trajectory, that is, just after arrival of customer 7787. It turns out that the this instant is the first regeneration point of the majorant system, that is $\overline{\beta}_1 = 7787$. At instant $t_{\overline{\beta}_1}$, the number of busy servers equals 100, and the residual service times of customers $\mathcal{M}_{\overline{\beta}_1} = \{7783, 7784, 7785, 7787\}$ belong to interval $[0, 2.4]$. At that, the residual service times $\{\overline{S}_i(\overline{\beta}_1), i \in \mathcal{M}_{\overline{\beta}_1}\}$ in the system $\overline{\Sigma}$ take maximal value 2.4. Then the customers from the set $\mathcal{M}_{\overline{\beta}_1}$ depart system simultaneously at instant $t_{\overline{\beta}_1} + 2.4$ (see Fig. 3(a)). By this transformation, the path of majorant system dominates the corresponding path of the original system because of the mentioned monotonicity property. However, as our experiment shows, long interarrival time period between $t_{7793} \approx 7722.79$ and $t_{7794} \approx 7726.03$, allows to unload the system in such a way that the majority of servers become free (see Fig. 3(b)). As a result, a sequence of non-waiting arrivals appears (see Fig. 3(a), after arrival of customer 7794). In turn, it leads to *merging* (or coupling) of the paths of the system $\overline{\Sigma}$ and original system. We stress that both systems stay not empty between the regeneration epoch $t_{\overline{\beta}_1}$ and the instant of the coupling.

We mention an analogy of this period with the so-called renovation period in the theory of *renovating events*, see [1,10]. We also note that, by the monotonicity, the order of departures is the same for the original and majorant systems. However, due to the transformation of the remaining service times at regeneration instant, the departure instants are shifted (to the right) by the amount of time $2.4 - \min_{i \in \mathcal{M}_{\overline{\beta}_1}} \{\overline{S}_i(\overline{\beta}_1)\}$, see Fig. 3(b). Nevertheless, by combination of non-waiting arrivals and a long interarrival time, this difference between the paths disappears eventually. It shows one more time an analogy with the theory of renovating events and an interesting interrelation between the method of regenerative envelopes and theory of Harris Markov chains. This interrelation is assumed to be a topic of a future research.

5   Conclusion

In this work we continue to develop the method of regenerative envelopes with focus on estimation of a cluster model. As simulation shows, this method allows to estimate the mean steady-state queue size in the original (not necessary regenerative) model with a high precision by means of confidence estimation of the upper and lower (envelopes) regenerative systems.

Acknowledgments. This work is supported by Russian Foundation for Basic research, projects No. 15–07–02341, 15–07–02354, 15–07–02360, 15–29–07974, 16–07–00622 and by the Program of strategic development of Petrozavodsk State University.

References

1. Asmussen, S.: Applied Probability and Queues. Wiley, New York (1987)
2. Glynn, P.: Some topics in regenerative steady-state simulation. Acta Appl. Math. **34**, 225–236 (1994)
3. Glynn, P., Iglehart, D.: Conditions for the applicability of the regenerative method. Manag. Sci. **39**, 1108–1111 (1993)
4. Sigman, K., Wolff, R.W.: A review of regenerative processes. SIAM Rev. **35**, 269–288 (1993)
5. Morozov, E., Nekrasova, R., Peshkova, I., Rumyantsev, A.: A regeneration-based estimation of high performance multiserver systems. In: Gaj, P., Kwiecień, A., Stera, P. (eds.) CN 2016. CCIS, vol. 608, pp. 271–282. Springer, Heidelberg (2016). doi:10.1007/978-3-319-39207-3_24
6. Kiefer, J., Wolfowitz, J.: On the theory of queues with many servers. Trans. Am. Math. Soc. **78**, 1–18 (1955)
7. Rumyantsev, A., Morozov, E.: Stability criterion of a multiserver model with simultaneous service. Ann. Oper. Res. 1–11 (2015). doi:10.1007/s10479-015-1917-2
8. Chakravarthy, S., Karatza, H.: Two-server parallel system with pure space sharing and Markovian arrivals. Comput. Oper. Res. **40**(1), 510–519 (2013)
9. He, Q.-M.: Fundamentals of Matrix-Analytic Methods. Springer, New York (2014)
10. Borovkov, A.: Asymptotic Methods in Queueing Theory. Wiley, New York (1984)

Two Asymptotic Conditions in Queue with MMPP Arrivals and Feedback

Agassi Melikov[1], Lubov Zadiranova[2], and Alexander Moiseev[2(✉)]

[1] Institute of Control Systems ANAS, Baku, Azerbaijan
agassi@science.az
[2] Tomsk State University, Tomsk, Russia
zhidkovala@mail.ru, moiseev.tsu@gmail.com

Abstract. We consider infinite-server queue with Markov modulated Poisson arrivals and feedback. Asymptotic analysis of the aggregate arrival process is made under conditions of frequent changing of the underlying chain states of the arrival process and increasing service time. It is proved that the aggregate arrival process is Poisson asymptotically. Parameter of the Poisson approximation is obtained. Applicability area of the asymptotic results is derived by means of numerical experiments.

Keywords: Infinite-server queue · Asymptotic analysis · Feedback · Aggregate arrival process

1 Introduction

Queuing models have been investigated by many authors due to their different applications in production, communications, banking, computer systems and other areas. Due to present rapid development of these systems there is a need to expand the modification of queuing systems, as well as design and develop new methods of their investigation. It is known that the queuing system with feedback [1] can be used to describe socio-economic processes [2,3], as well as afterservice processes in information and telecommunication systems [4–9].

Models of queuing systems with feedback are of two types: models with instantaneous feedback and models with delayed feedback. In the available literature, both types of models are investigated separately. Recently in [10–12], Markov models with both types of feedbacks were investigated and both exact and approximate methods to calculate their performance metrics are developed. Detailed review of the queueing models with both types of feedback might be found in [12]. Infinite-server queue with feedback and Poisson arrivals were considered in the articles [13,14]. In this paper, we study the aggregate arrival process in infinite-server queue with Markov modulated Poisson arrivals and feedback.

V.M. Vishnevskiy et al. (Eds.): DCCN 2016, CCIS 678, pp. 231–240, 2016.
DOI: 10.1007/978-3-319-51917-3_21

2   Mathematical Model

Consider an infinite-server queue with arrivals as Markov modulated Poisson process (MMPP) which underlying chain $k(t)$ has a finite number of states: $1, 2, \ldots, K$. The MMPP is determined by given matrix of infinitesimals $\mathbf{Q} = \|q_{ij}\|$, $i, j = 1, 2, \ldots, K$, and by matrix of conditional rates $\mathbf{\Lambda} = \mathrm{diag}\{\lambda_k\}$, $k = 1, 2, \ldots, K$ [15]. Let $\mathbf{R}$ be a vector of stationary probability distribution of states of Markov chain $k(t)$ which is determined by the system of equations

$$\begin{cases} \mathbf{Re} = 1, \\ \mathbf{RQ} = 0. \end{cases} \tag{1}$$

When a customer arrives at the system, it occupies any free server. Service time is distributed according to the exponential law with a parameter μ. When service is finished, the customer leaves the system with a probability $(1 - r)$ or goes into the system again for additional service with a probability r. So, the arrivals in the system is an aggregate arrival process which contains "pure" arrivals from MMPP and feedback arrivals. The problem of the study is an analysis of the aggregate arrival process.

Let us denote the following: $i(t)$ is a number of busy servers at the time instant t, $m(t)$ is a number of customers in aggregate arrival process that have come into the system during time interval $[0, t)$, $k(t)$ is a state of the underlying Markov chain of the MMPP at the time instant t. Three-dimensional process $\{k(t), i(t), m(t)\}$ is Markov. For its probability distribution $P(k, i, m, t) = P\{k(t) = k, i(t) = i, m(t) = m\}$, we can write down Kolmogorov differential equation system in the following form

$$\frac{\partial P(k, i, m, t)}{\partial t} = -\lambda_k P(k, i, m, t) - i\mu P(k, i, m, t) + \lambda_k P(k, i - 1, m, t) +$$

$$\mu(i + 1)(1 - r)P(k, i + 1, m, t) + \mu i r P(k, i, m - 1, t) + \sum_{\nu=1}^{K} P(\nu, i, m, t)q_{\nu k} \tag{2}$$

for $k = 1, 2, \ldots, K$, $i, m = 0, 1, 2, \ldots$

We introduce partial characteristic functions [15]:

$$H(k, u, s, t) = \sum_{i=0}^{\infty} \sum_{m=0}^{\infty} e^{jui} e^{jsm} P(k, i, m, t),$$

where $j = \sqrt{-1}$. Then we can write the system (2) in the form of differential matrix equation

$$\frac{\partial \mathbf{H}(u, s, t)}{\partial t} + j\mu \left[re^{js} - 1 + (1 - r)e^{-ju} \right] \frac{\partial \mathbf{H}(u, s, t)}{\partial u} =$$

$$\mathbf{H}(u, s, t) \left[\left(e^{j(u+s)} - 1 \right) \mathbf{\Lambda} + \mathbf{Q} \right], \tag{3}$$

where $\mathbf{H}(u, s, t) = [H(1, u, s, t), \ldots, H(K, u, s, t)]$.

3   Asymptotic Analysis

Consider the system MMPP/M/∞ described above. Let us fix a matrix of infinitesimals as $\mathbf{Q}^{(1)}$. Let S be a positive value. Denote $\mathbf{Q} = S \cdot \mathbf{Q}^{(1)}$. It is clear that the stationary probability distributions of state of the underlying Markov chain $k(t)$ coincide for the cases both with infinitesimal matrices $\mathbf{Q}^{(1)}$ and $\mathbf{Q}$. In other words, this distribution does not depend on parameter S, but the intensity of the transitions of Markov chain $k(t)$ from one state to another increases while value of the parameter S increases. This corresponds us to the condition of frequent changing of the underlying chain states of the arrival process.

Let us obtain the steady-state asymptotic characteristic function of the number of busy servers under conditions of frequent changing of the underlying chain states of the arrival process and increasing service time. To do this, we set $s = 0$ in the Eq. (3) and make a transition to the steady state ($t \to \infty$), then we derive the following equation

$$j(1 - r)\left(e^{-ju} - 1\right)\mathbf{H}(u)' = \mathbf{H}(u)\left[\left(e^{j(u)} - 1\right)\mathbf{\Lambda} + S\mathbf{Q}^{(1)}\right]. \qquad (4)$$

Theorem 1. *The steady-state asymptotic characteristic function of the number of busy servers in infinite-server queue with Markov modulated Poisson arrivals, exponential service times and feedback has the form*

$$h(u) = \mathrm{E}\left\{e^{jui(t)}\right\} = \exp\left\{\frac{ju\kappa}{\mu(1 - r)}\right\}$$

under the conditions of frequent changing of the underlying chain states of the arrival process and increasing service time. Here $\kappa = \mathbf{R}\mathbf{\Lambda}\mathbf{e}$.

Proof. We denote

$$\mu = \varepsilon, u = \varepsilon y, \frac{1}{S} = \varepsilon, \mathbf{H}(u) = \mathbf{F}(y, \varepsilon).$$

Let us rewrite the Eq. (10), taking into account these notations:

$$j\varepsilon(1 - r)(e^{-j\varepsilon y} - 1)\frac{\partial \mathbf{F}(y, \varepsilon)}{\varepsilon \cdot \partial y} = \mathbf{F}(y, \varepsilon)\left[\left(e^{j(\varepsilon y)} - 1\right)\mathbf{\Lambda} + S\mathbf{Q}^{(1)}\right]. \qquad (5)$$

Dividing right and left side of the Eq. (10) by S and making a transition $S \to \infty$, we obtain the system of equations

$$0 = \mathbf{F}(y)\mathbf{Q}^{(1)}.$$

Its solution has the form

$$\mathbf{F}(y) = \mathbf{R}\Phi(y), \qquad (6)$$

where $\Phi(y)$ is a some scalar function.

To determine the form of this function, we multiply both sides of the Eq. (5) by the vector $\mathbf{e}$:

$$j\varepsilon(1-r)(e^{-j\varepsilon y}-1)\frac{\partial \mathbf{F}(y,\varepsilon)}{\varepsilon \cdot \partial y}\mathbf{e} = \mathbf{F}(y,\varepsilon)\left[\left(e^{j(\varepsilon y)}-1\right)\mathbf{\Lambda}\right]\mathbf{e}.$$

Performing an asymptotic transition $\varepsilon \to 0$ and taking into account the form (11), we derive the following differential equation

$$(1-r)\Phi'(y) = j\Phi(y)\mathbf{R\Lambda e}.$$

Solving this equation under the initial condition $\Phi(0) = 1$, we obtain

$$\Phi(y) = \exp\left\{\frac{jy\kappa}{(1-r)}\right\}. \tag{7}$$

Therefore, the solution of Eq. (5) has the form

$$\mathbf{F}(y) = \mathbf{R}\exp\left\{\frac{jy\kappa}{(1-r)}\right\}. \tag{8}$$

We derive the following steady-state asymptotic characteristic function of the number of busy servers under conditions of frequent changing of the underlying chain states of the arrival process and increasing service time:

$$h(u) = \mathrm{E}\left\{e^{jui(t)}\right\} = \mathbf{H}(u,0,\varepsilon)\mathbf{e} \approx \mathbf{F}(u,0,\varepsilon)\mathbf{e} = \exp\left\{\frac{ju\kappa}{\mu(1-r)}\right\}.$$

The theorem is proved.

Let us investigate the number of customers in the aggregate arrival process under conditions of frequent changing of the underlying chain states of the arrival process and increasing service time.

Theorem 2. *Then the asymptotic characteristic function of the number of customers in the aggregate arrival process $h(s,t) = \mathrm{E}\left\{e^{jsm(t)}\right\}$ has the form*

$$h(s,t) = \exp\left\{\frac{\kappa t\left(e^{js}-1\right)}{1-r}\right\} \tag{9}$$

under the conditions of frequent changing of the underlying chain states of the arrival process and increasing service time.

Proof. We denote

$$\mu = \varepsilon, u = \varepsilon y, \frac{1}{S} = \varepsilon, \mathbf{H}(u,s,t) = \mathbf{F}(y,s,t,\varepsilon).$$

Let us rewrite the Eq. (3), taking into account these notations:

$$\frac{\partial \mathbf{F}(y,s,t,\varepsilon)}{\partial t} = j\varepsilon\left[1 - re^{js} - (1-r)e^{-j\varepsilon y}\right]\frac{\partial \mathbf{F}(y,s,t,\varepsilon)}{\varepsilon \cdot \partial y} +$$
$$\mathbf{F}(y,s,t,\varepsilon)\left[\left(e^{j(\varepsilon y+s)}-1\right)\mathbf{\Lambda} + S\mathbf{Q}^{(1)}\right]. \tag{10}$$

Dividing right and left sides of the Eq. (10) by S and making a transition $S \to \infty$, we obtain the system of equations

$$0 = \mathbf{F}(u, s, t)\mathbf{Q}^{(1)}.$$

Its solution has the form

$$\mathbf{F}(u, s, t) = \mathbf{R}\Phi(u, s, t), \tag{11}$$

where $\Phi(u, s, t)$ is a some scalar function.

To determine the form of this function, we multiply both sides of the Eq. (10) by the vector $\mathbf{e}$:

$$\frac{\partial \mathbf{F}(y, s, t, \varepsilon)}{\partial t}\mathbf{e} = j\left[1 - re^{js} - (1 - r)e^{-j\varepsilon y}\right]\frac{\partial \mathbf{F}(y, s, \dot{t}, \varepsilon)}{\cdot \partial y}\mathbf{e} +$$
$$\mathbf{F}(y, s, t, \varepsilon)\left[\left(e^{j(\varepsilon y + s)} - 1\right)\mathbf{\Lambda} + S\mathbf{Q}^{(1)}\right]\mathbf{e}.$$

Performing an asymptotic transition $\varepsilon \to 0$ and taking into account the form (11), we derive the following differential equation

$$\frac{\partial \Phi(y, s, t)}{\partial t} = jr\left(1 - e^{js}\right)\frac{\partial \Phi(y, s, t)}{\partial y} + \Phi(y, s, t)\kappa\left(e^{js} - 1\right). \tag{12}$$

To solve this equation, we use the method of characteristics [16]. As the first step, we write the following corresponding system of characteristics equations:

$$\frac{dt}{1} = \frac{dy}{jr\left(e^{js} - 1\right)} = \frac{d\Phi(y, s, t)}{\kappa\left(e^{js} - 1\right)\Phi(y, s, t)}.$$

As the next step, we find two first integrals of the system. One of them can be found from the equation

$$\frac{dt}{1} = \frac{dy}{jr\left(e^{js} - 1\right)}.$$

Solving this equation, we obtain

$$y = jr\left(e^{js} - 1\right)t - C_1.$$

The second integral can found from the equation

$$\frac{dy}{jr\left(e^{js} - 1\right)} = \frac{d\Phi(y, s, t)}{\kappa\left(e^{js} - 1\right)\Phi(y, s, t)}.$$

Its solution gives the expression

$$C_2 = \frac{\kappa y}{jr} - \ln |\Phi(y, s, t)| \, .$$

Finally, taking into account the initial condition $\Phi(y, s, 0) = \exp\left\{\frac{j\kappa y}{1-r}\right\}$, we obtain the solution of the Eq. (12)

$$\Phi(y, s, t) = \exp\left\{\frac{j\kappa}{1-r}\left(-jr\left(e^{js} - 1\right)t + y\right)\right\}$$

$$= \exp\left\{\frac{\kappa t\left(e^{js} - 1\right)}{1-r} + \frac{j\kappa y}{1-r}\right\}.$$

So, we obtain that the solution of the Eq. (10) has the form

$$\mathbf{F}(y, s, t) = \mathbf{R}\exp\left\{\frac{\kappa t\left(e^{js} - 1\right)}{1-r} + \frac{j\kappa y}{1-r}\right\}. \tag{13}$$

Assuming here $y = 0$, we derive the following asymptotic approximation of the characteristic function of the number of customers in the aggregate arrival process under conditions of frequent changing of the underlying chain states of the arrival process and increasing service time:

$$h(s, t) = \mathrm{E}\left\{e^{jsm(t)}\right\} = \mathbf{H}(0, s, t)\mathbf{e} \approx \mathbf{F}(0, s, t)\mathbf{e} = \exp\left\{\frac{\kappa t\left(e^{js} - 1\right)}{1-r}\right\}.$$

The theorem is proved.

This theorem shows that the aggregate arrival process in the system has the Poisson distribution with parameter $\frac{\kappa t}{1-r}$.

4   Numerical Analysis

We have made numerical experiments using simulations in a goal to obtain the applicability area of the Poisson approximation (9) for considered models. We use the Kolmogorov distance [17] for estimation of precision:

$$d = \max_{n=0,1,\ldots}\left|\sum_{i=0}^{n}(P_i - \tilde{P}_i)\right|,$$

where $P_i, i = 0, 1, \ldots$ are relative frequencies of the distribution constructed on base of simulation results, $\tilde{P}_i, i = 0, 1, \ldots$ are the probabilities given by the Poisson approximation (9).

Let us consider one of numerical examples. Let parameters of the MMPP arrivals be the following:

$$\Lambda = \begin{pmatrix} 0.2 & 0 & 0 \\ 0 & 1 & 0 \\ 0 & 0 & 5 \end{pmatrix}, \quad Q = S \cdot \begin{pmatrix} -2 & 1 & 1 \\ 7 & -8 & 1 \\ 3 & 3 & -6 \end{pmatrix}.$$

So, the fundamental rate of this arrival process is equal to $\kappa = 1$. Let the parameter r be equal 0.5.

We assume that the approximation result is acceptable if the Kolmogorov distance is not more than 0.03. In the Table 1, the Kolmogorov distances for considered model with different values of parameters S and μ are presented. Acceptable results are marked as boldface.

On the Fig. 1, you can see the comparison of the probability distribution constructed on base of simulation results with the Poisson approximation (9) for various values of parameters μ and S. Increasing the frequency of changing underlying chain states (figure a) or increasing the service time (figure b) separately don't give us a sufficient accuracy but using both of these conditions (figure c) makes the accuracy acceptable. Moreover, further changing of both parameters give us the result that is near the exact one (figure d).

So, we can make a conclusion that if we take into account the condition of growing the frequency of changing of arrival process states in addition to the condition of growing the service time, it allows to extend the applicability area (upper bound of parameter μ and lower bound of parameter S). If we use only one of these conditions, it may be not give a possibility to obtain enough precision of the Poisson approximation (see the first row and the first column of the Table 1).

Table 1. Kolmogorov distances between and Poisson approximation distribution based on simulation for various values of exponential service parameter μ and parameter S of state changing frequency of MMPP arrivals

$\mu \setminus S$	1	2.5	5	10	25	50	100
1.0	0.1316	0.0921	0.0789	0.0763	0.0724	0.0673	0.0529
0.5	0.0867	0.0562	0.0455	0.0374	0.0337	**0.0297**	**0.0273**
0.25	0.0621	0.0397	**0.0296**	**0.0246**	**0.0216**	**0.0183**	**0.0154**
0.10	0.0481	**0.0285**	**0.0190**	**0.0114**	**0.0108**	**0.0068**	**0.0067**
0.010	0.0379	**0.0174**	**0.0109**	**0.0081**	**0.0054**	**0.0039**	**0.0029**

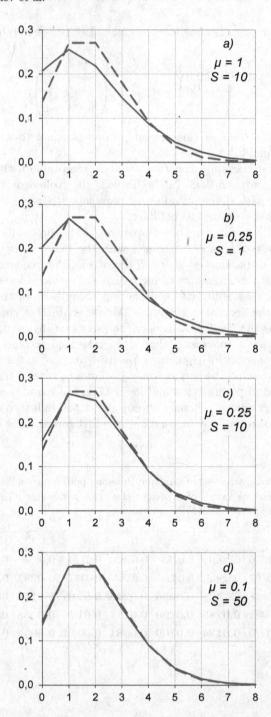

Fig. 1. Comparison of distribution based on simulation results (solid line) with the Poisson approximation (dashed line) for various values of μ and S

5   Conclusions

In the paper, an asymptotic characteristic function of the number of customers in the aggregate arrival process is obtained for infinite-server queue with MMPP arrivals and a feedback under conditions of frequent changing of the underlying chain states of the arrival process and increasing service time. It is shown that the probability distribution of the number of customers in the aggregate arrival process under these conditions is Poisson asymptotically. Obtained numerical results show that the range of applicability of the asymptotic method is increased when we use both asymptotic conditions instead of single one [18].

Results of the paper are similar to the earlier obtained results for queueing systems with Poisson and renewal arrival processes and feedback [13,14]. Some more complicate asymptotic analysis for queueing networks with MAP arrivals (but only about number of busy servers) can be found in [19].

Acknowledgments. The work is performed under the state order of the Ministry of Education and Science of the Russian Federation (No. 1.511.2014/K).

References

1. Takács, L.: A single-server queue with feedback. Bell Syst. Tech. J. **42**(5), 505–519 (1963)
2. Morozova, A.S., Moiseeva, S.P., Nazarov, A.A.: Investigation of economic-mathematical model of the effect of price discounts for regular customers profit commercial organization. Tomsk State Univ. J. **293**, 49–52 (2006). (in Russian)
3. Zakhorolnaya, I.A., Moiseeva, S.P.: Mathematical model of the process of changes in income from the sale of complementary products. In: Proceedings of the 10th International Conference on Financial and Actuarial Mathematics and Event-convergence of Technology (FAMET-2011), pp. 157–170 (2011). (in Russian)
4. Wan-Chun, C.: A computer processing queueing system with feedback. Inf. Control **16**, 473–486 (1970)
5. Foley, R.D., Disney, R.L.: Queues with delayed feedback. Adv. Appl. Probab. **15**(1), 162–182 (1983)
6. Pekoz, E.A., Joglekar, N.J.: Poisson traffic flow in a general feedback. Appl. Probab. **39**(3), 630–636 (2002)
7. Zaryadov, I.S., Korolkov, A.V., Milovanov, T.A., Sherbanskaya, A.A.: Mathematical model of calculating and analysis of characteristics of systems with generalized and repeating service. T-Comm. Telecommun. Transp. **8**(6), 16–20 (2014). (in Russian)
8. Zaryadov, I.S., Sherbanskaya, A.A.: Time characteristics of queuing system with renovation and reservice. Bull. PFUR Ser. Math. Inf. Sci. Phys. **2**, 61–65 (2014)
9. D'Avignon, G.R., Disney, R.L.: Queues with instantaneous feedback. Manag. Sci. **24**(2), 168–180 (1977)
10. Melikov, A.Z., Ponomarenko, L.A., Kuliyeva, K.H.N.: Calculation of the characteristics of multichannel queuing system with pure losses and feedback. J. Autom. Inf. Sci. **47**(6), 19–29 (2015)
11. Melikov, A.Z., Ponomarenko, L.A., Kuliyeva, K.N.: Numerical analysis of the queuing system with feedback. Cybern. Syst. Anal. **51**(4), 566–573 (2015)

12. Melikov, A., Ponomarenko, L., Rustamov, A.: Methods for analysis of queueing models with instantaneous and delayed feedbacks. In: Dudin, A., Nazarov, A., Yakupov, R. (eds.) ITMM 2015. CCIS, vol. 564, pp. 185–199. Springer, Heidelberg (2015). doi:10.1007/978-3-319-25861-4_16

13. Morozova, A.S., Moiseeva, S.P., Nazarov, A.A.: Research of QS with repeating requests and unlimited number of servers by means of a method of limit decomposing. Comput. Technol. 5(13), 88–92 (2005)

14. Moiseeva, S.P., Morozova, A.S., Nazarov, A.A.: The probability distribution of two-dimensional flow applications in the infinite-server queuing with feedback. Tomsk State Univ. J. 16(Suppl.), 125–128 (2006). (in Russian)

15. Nazarov, A.A., Moiseeva, S.P.: Method of Asymptotic Analysis in Queuing Theory. NTL, Tomsk (2006). (in Russian)

16. Delgado, M.: The lagrange-charpit method. SIAM Rev. 39(2), 298–304 (1997)

17. Kolmogorov, A.: Sulla determinazione empirica di una legge di distribuzione. Giornale dell' Intituto Italiano degli Attuari 14, 83–91 (1933)

18. Zadiranova, L.A., Moiseeva, S.P.: Investigation of the aggregate arrival process of QS with feedback by the method of asymptotic analysis. Proc. Tomsk State Univ. 297, 99–105 (2015). (in Russian)

19. Moiseev, A., Nazarov, A.: Queueing network MAP-$(GI-\infty)^K$ with high-rate arrivals. Euro. J. Oper. Res. 254, 161–168 (2016)

Applications of Augmented Reality Traffic and Quality Requirements Study and Modeling

A. Koucheryavy, M. Makolkina$^{(\boxtimes)}$, and A. Paramonov

The Bonch-Bruevich Saint-Petersburg State University of Telecommunications,
Saint-Petersburg, Russia
akouch@mail.ru, makolkina@list.ru, alex-in-spb@yandex.ru
https://www.sut.ru/

Abstract. The further development of communication networks appears today on the basis of the concept of the Internet of Things. At the same time gaining popularity technology "augmented reality" that allows you to manage different objects and processes in networks. Sharing the "augmented reality" and the concept of the Internet of Things technology, requires the development of the new service model and traffic pattern and establishment of a new approach of the Quality of Experience estimation.

Keywords: Augmented reality · Internet of Things · Quality of Experience · Service model · Traffic pattern · D2D technology

1 Introduction

Today in the market of telecommunications there was a new type of service - it is services augmented reality (AR, Augmented Reality) [1,2]. Primal problem of technology augmented reality not to create the new world, and to improve existing for the account strengthenings of feelings of the user such as hearing, vision, sense of smell, knowledge, etc. The augmented and virtual realities in essence different concepts. The virtual reality replaces the actual world synthetic, artificially created environment, being in which, the person does not see what occurs around. At that time as AR supplements a real, but does not replace it. It allows the user to see the actual world with the virtual objects combined or imposed atop actual world [3,4]. Each person, not very well who is he, the doctor or the climber, in search of the nearest subway station in the big city will feel all advantages of new technology which allows to impose computer graphics on environmental space under review the person, and by that gives the chance to glance in buildings and to see through walls. So, walking down the street augmented reality wearing glasses which look as the routine couple of points, is available to you graphics with various information in dependence from your turn of the head and orientation of a look, audio, video, data on objects coincide with what you see at present. So, the main idea of augmented reality consists in imposing of graphics, audio, other sensory data and feelings over the existing world around in real time scale [5]. Thus, it is possible to mark out three main properties AR technologies:

© Springer International Publishing AG 2016
V.M. Vishnevskiy et al. (Eds.): DCCN 2016, CCIS 678, pp. 241–252, 2016.
DOI: 10.1007/978-3-319-51917-3_22

(1) A combination of real and virtual objects in uniform space;
(2) Interactivity in real time;
(3) 3D objects.

In the modern applications of augmented reality it is possible to allocate on to smaller measure six classes: medicine; assembly, maintenance and repair the difficult technique; addition of information of private and common character to existing to objects; control of robots, unmanned aerial vehicles, etc.; games and entertainments; military.

2 Applications of Augmented Reality

The range of application of technology of augmented reality in medicine is extensive and bears a number of benefits for doctors in the composite cases demanding fast reaction and acceptance decisions. As it is known a lot of time leaves on diagnostics, poll and survey the patient, here the doctor possesses as if "x-ray vision", and looking on particular parts of a body of the patient sees results of a computer tomography or magnetic and resonance tomography, etc. Applications can carry also training character, prompting that else the patient needs to ask or appoint, in dependences on its complaints and diagnosis.

Other class of applications of augmented reality is an assembly, technical service, repair of the difficult technique. So in case of breakage in the cold winter on to the deserted road any woman can make primary car repairs. It puts on augmented reality glasses, via them carries out diagnostics, i.e. it looks on some detail of the car and it information is displayed. as has to look in a normal duty this detail as to check it also in a case to repair malfunctions. Similar applications are very popular also at offices, for example when replacing paper or a cartridge in the printer. When targeting on it of points, the user sees the short instruction from what party what to press and what it will lead to. The following class unites applications which add information to objects of various character, sometimes, this class is called by "the summary and visualization" [6,8]. For example, in library, passing on ranks, to you information is displayed about the books standing on shelves, thereby, there is no need last or will bend that to get each of them. Other application helps teachers and teachers to trace progress of students. So looking at the student, through points of added realities, the teacher sees whether this student is allowed to examination, whether it is frequent it attended classes what GPA of progress, whether that the student came to examination, what ticket it extended as he spends at University, personal much time achievements, victories at competitions of projects, etc. Similar applications are popular in construction, architecture and design of rooms. Users can in actual time to add or clean the virtual object the existing space, to estimate dimensions, planning, spaciousness, color schemes, etc. To estimate the projected skyscraper will how successfully fit into the built-up area, what look will open from windows. Control of robots and unmanned aerial vehicles. Management of technique on distance is the complex challenge, there are problems with a transfer delay the operating information, reaction rate of the robot, distortion of information at to transfer [9,10]. Therefore there was an idea to operate not directly, and to control the virtual version of the robot. The user plans and defines operations of the robot with the help of a manipulation the local virtual version of the robot in actual scale time. Results are displayed immediately on actual object. After that, as the plan is checked and is chosen suitable, the user gives command to actual to the robot to implement the specified plan. These applications allow to avoid delays when performing sequence of actions and not to interrupt process, also to predict results of the taken actions on environmental objects, that most to prevent undesirable consequences.

Games and entertainments one of the most inventive classes of applications augmented reality [7]. So games on searching of things or tracings of participants games in the actual world can be supplied with three-dimensional graphics and the virtual hints, I make game unpredictable and fascinating. Also, having, for example the image, poster or photo of the player in soccer, the application distinguishes who it and provides you everything the available information on this player, the best goals personal achievements for which clubs are visualized by means of three-dimensional graphics, this athlete and many other things played. Of course theatrical sceneries, color decisions, lighting receptions can change and be imposed on alive acting, expanding and deepening space. Applications of augmented reality for military stand a separate class industries. Helmets of augmented reality are for a long time developed, which are actively used in simulators of flights by military planes various models in a wide criteria range and tasks. In investigation devices are used with support of AR which can impose drawings or a view from the satellite or information from the unmanned aerial vehicle immediately in the field of vision soldier. Apparently from the above, there is a lot of applications of augmented reality, they are very different and set contradictory tasks for telecom operators [11–13].

The greatest interest and popularity at users bought applications from a class "summary and visualization" and "assembly, maintenance and repair of equipment", therefore for model building of a traffic, model of interaction of separate elements, definitions of the main indexes of functioning of data of systems, were chosen applications of these groups.

3   Service Model

According to the majority of definitions, providing service augmented reality consists in introduction in the field of perception person of padding information. Generally, appointment to facilitate the solution of various tasks, the bound to perception, the analysis and management. In Fig. 1 the example of AR of the star chart is given (StarChart iPhone). Difference from just card or an interactive map consists in that, that its display is bound to position of the device in space (its targeting in a particular point of a palate), i.e. it is added with results information processing of the user.

In more general case process of providing service DR it is possible to consider as interactive interaction of the user with the applied functions realizing the

Fig. 1. An example of augmented reality when looking at the starry sky (StarChart iPhone).

analysis of a condition of its environment and providing it padding information, Fig. 2.

As the purpose of AR is a granting to the user padding context-sensitive information, that quality of its granting it is necessary to consider from the point of view of extent of achievement of this purpose that it is possible to characterize:

- degree of compliance to the provided padding information to needs of the user (compliance to purpose, volume, specification, etc.);
- degree of a susceptibility of the provided data (video this, graphics, sound, tables, the text and other elements of the interface with user, quality of their representation);
- timeliness of providing padding information.

Realization of service AR can be various, depending on the used technical means. In a set of these tools surely the subscriber device which can represent enters the smartphone, the tablet personal computer, multimedia glasses, a helmet, etc., allowing to organize the user interface, data acquisition about an environment and having rather high efficiency of computing devices. In particular, all functionality of AR can be realized as a part of the client application of such device. However, in many applications, for example, it is inexpedient or it is impossible to store all padding information and also to carry out all data

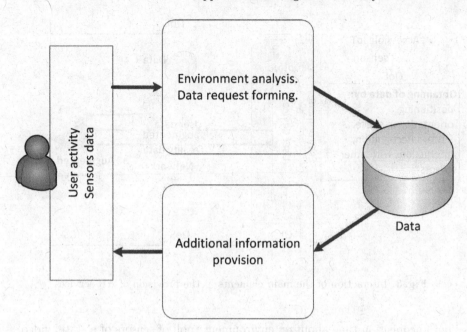

Fig. 2. User interaction with the application function

processing user resources of the mobile device. Therefore, following the AR element is infocommunication making (communication network), which provides delivery of padding information to the user, databases and perhaps servers the carrying-out part of functions on processing information of the user, Fig. 3. At existence in the user's environment the sensors of the Internet of Things (IoT) capable to provide the useful information, the D2D technology providing can be used direct connection of the subscriber device with them.

The basic elements providing quality of perception services by the user are the device of the user of AR, the service server AR, databases and communication network. All elements interact through a network communications (in the presence of IoT devices also D2D). In this system the main problem of ensuring quality of service is distribution functionality and data between the client application of the device user, server of service and databases. This distribution it is reflected in a run time of functions on data processing, time deliveries of data through a communication network and on the traffic made in a network. Degree of compliance of padding information to requirements the user and degree of its susceptibility are defined by the organization services, existence of the required information and the organization of user interface. Service AR is interactive therefore timeliness providing information is one of the most important factors, defining its quality. Timeliness is characterized by time between an event the bound to changes of a condition of an environment or the user and the event characterizing availability to the user padding information. This time (delay) is defined by a row components, such as:

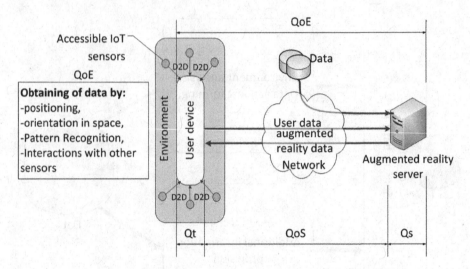

Fig. 3. Interaction of the main elements in the provision of AR services

- data acquisition time about an environment (poll of sensors of a state, video, etc.) and their processings;
- time of delivery of data for the service server (if it is necessary);
- data processing time service server;
- time of delivery of data to the user;
- data presentation time.

At the organization of service with participation of the server data are requested and transferred to the user at change of a condition of an environment (change of an environment in the field of vision/perception of the user). Depending on functionality of service it can occur at change of position in space of the user (terminal) or some objects that is equivalent to change of a set of objects in the field of perception which require providing padding data. Identification of change can be made, for example, on the basis of the analysis of data on coordinates of the device and its orientation in space, discernments of objects by the analysis of video of data, etc.

4   Traffic Pattern

For the description of the traffic made by service it is necessary to connect volume given by the user and to the user of data at changes of his environment.

- service space model;
- model of an environment of the user;
- behavior model;

We will understand informational model of physical three-dimensional space in which there can be a user of service as space of service. The informational model includes the description of some objects which are in this space $X = \{\bar{x}_1, \bar{x}_2, \bar{x}_n\}$, where n total number of objects. The model of an environment of the user is a subspace service spaces, i.e. part of space restricted by opportunities perceptions (model of these opportunities). Environment, as a rule, it is attached to position of the user in space of service and includes in a set of objects $X^{(U)} = \{\bar{x}_1^{(U)}, \bar{x}_2^{(U)}, \bar{x}_k^{(U)}\}$, where k number of the objects which are to the area perceptions of the user. The behavior model describes changes of position of the user and it environments in service space. Changes in the user's environment can occur as owing to movements of the user, and movements of objects in service space. The change caused emergence in an environment of the user of new object $\bar{x}_i$, leads to inquiry data on this object.

The algorithm of realization of service has to provide realization following functions:

- identification of an event of change of an environment and calculation change parameters;
- information request about change of an environment;
- data acquisition and their display.

The possible chart of data exchange is given in Fig. 4.

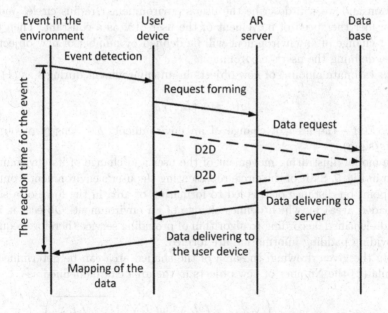

Fig. 4. Data exchange diagram when providing AR services

As an environment also as well as space of service is restricted (area of perception) physical three-dimensional space, changes in it can be described as a stream of events of the bound to emergence in it objects. Objects can enter through its borders owing to movement of these borders or objects. In case of the former, movement of borders is bound to behavior of the user, and in the second to behavior of objects. In that and other case quality of functioning of system will depend on its ability in due time to serve events of this stream. Thus, the problem of ensuring quality of service can be considered as the choice of parameters of system (efficiency, a channel capacity, distribution of its functionality) from the characteristic of a stream of events and load of the system made by this stream.

Properties of a stream of events substantially define properties of a data flow between elements of system. For example, in system of positioning on a district map such stream is defined by events of change of coordinates of the user and is defined by characteristics of his driving, in system it will decide by their orientation in space on use of points of DR and it is possible events the bound to data transmission about objects under review or their characteristics.

It is apparent that characteristics of a stream will depend on distribution and characteristics of objects in service space, and also characteristics of driving of the user. Let's make an assumption that objects in space of service are distributed in a random way (form the Poisson field) and are not mobile, only the user is mobile. Then, change of position of the user is equivalent to change of its environment. Taking into account properties of space of service and an environment, this change can be described by the volume or the area. Let's consider 2D option and we will describe the user's environment r radius circle, and we will consider the speed of movement of the user of v as a constant, Then that during t change of an environment will be defined by number of new objects in the area defining the user's environment.

Let's estimate amount of new objects in an environment during t as (1)

$$n(t) = \tilde{S}\left(L(t)\right)\rho \qquad (1)$$

where $\tilde{S}L(t)$ - the area of change of an environment; ρ - density of objects (objects/sq.m).

The model illustrating movement of the user and change of its environment is given in Fig. 5. Shift of the circle representing the user's environment from the initial point on distance of L is led to formation of area in the form of a sickle (the shaded area) by which defines change of an environment. Objects it is in the field identified, according to algorithm of providing service therefore requests for providing padding information are formed.

From the given drawing, the area of the shaded area can be determined by a formula (2) the Number of new objects in the area can be defined as

$$\tilde{S}(L) = \pi r^2 - 2\left(r^2 arccos\left(\frac{L}{2r}\right) - \left(\frac{L}{2}\right)\sqrt{r^2 - \left(\frac{L}{2}\right)^2}\right) \qquad (2)$$

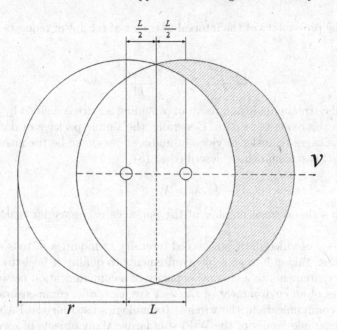

Fig. 5. Movement of the user and change of its environment

The number of the new objects in the field can be defined as (3)

$$n(t) = \tilde{S}(L)\rho \tag{3}$$

where ρ - density of objects (objects/sq.m). Considering this process in dynamics, i.e. when driving the user the stream of events (requests of data) takes place. Intensity of a stream of events (requests for data) can be defined as number of objects in a small increment of the area of the considered figure

$$\lambda_r = \frac{d\tilde{S}(L)}{dL}\rho v \tag{4}$$

Where ρ - density of objects (objects/sq.m); v - speed of movement (m/s). Derivative of expression (2) in a point $L = 0$

$$\frac{d\tilde{S}(L)}{dL} = [L = 0] = 2r \tag{5}$$

Then taking into account (4) and (5) we got (6)

$$\lambda_r = 2r\rho v \tag{6}$$

Considering properties of the Poisson field accepted for model the amount of objects in some restricted area is casual, distributed under the law of Poisson and depends only on the area (or volume) the considered area. Therefore, for the accepted model the stream of inquiries will represent the elementary stream for which the probability of the interval of time t of receipt of requests k will be defined as (7)

$$p_k = \frac{(\lambda_r t)^k}{k!} e^{-\lambda_r t} = \frac{(2r\rho vt)^k}{k!} e^{-2r\rho vt} \tag{7}$$

The traffic stream made as a result of providing service is defined by a stream of replies to the requests of data. Generally, the simple package of data, and a stream of packages (transfer of video or audio of data) can be the answer both. Intensity of this stream can be described as (8)

$$\lambda_r = \lambda_s \eta \tag{8}$$

Where η - the average number of the packages necessary for realization of inquiry.

By transfer of video data can exceed intensity of inquiries in tens and hundreds of times. Taking into account requirements to quality it leads to essential growth of requirements to a channel capacity of a communication network. The physical sizes of an environment of the user are, as a rule, commensurable with a radius of communication, the wireless technologies used for the PAN organization, for example family of the WiFi standards. Many objects of services AR (elements of city infrastructure, vehicles, household appliances) can be equipped with clusters of access and necessary data which it can be provided to users. Therefore, use of the D2D technologies can be the possible decision providing essential decrease in a traffic on a communication network. In that case the traffic of data can be delivered immediately from object to the terminal of the user of Fig. 4 (dashed line). Intensity of this stream will be defined as (9)

$$\lambda_r = \lambda_s \eta_{D2D} \tag{9}$$

Where $\eta_{D2D} = (1 - \gamma)\eta$, γ - a share of the objects of an environment supporting the D2D technology.

Certainly, application of D2D technologies is possible only when objects of service are the physical objects mentioned above which I can be equipped with the corresponding communication centers. Rather wide range of services nevertheless demands interaction with remote databases and problem solving of ensuring quality.

5   Conclusions

1. Rendering of services of AR demands differentiated approach to definition of their indexes of quality depending on purpose of service.
2. Quality of providing service DR is characterized by degree of compliance to purpose, degree of a susceptibility of the provided data and timeliness of providing information.
3. Properties of the traffic made by service are defined by behavior of the user, an environment of the user and way of data presentation.

4. Intensity of a traffic of services AR when using multimedia data it is comparable to a traffic of services of transfer of video that taking into account interactivity of service and requirements to quality leads to increase in requirements to a channel capacity of a communication network.
5. One of ways of reduction of requirements to a channel capacity of a network can be application of the D2D technologies which use leads to short circuit of a traffic, between the terminal of the user and objects of an environment, passing networks of telecom operators.

Acknowledgments. The reported study was supported by grants from the RFBR, research project No. 16-37-00209 mol a Development of the principles of integration the Real Sense technology and Internet of Things.

References

1. Bergenti, F., Gotta, D.: Augmented reality for field maintenance of large telecommunication networks. In: Conference and Exhibition of the European Association of Virtual and Augmented Reality (2014)
2. Sorensen, L., Skouby, K.E.: Use scenarios 2020 - a worldwide wireless future. Visions and research directions for the Wireless World/Outlook. In: Wireless World Research Forum-N4, July 2009
3. Ganapati, P.: How it Works: Augmented Reality. Wired, 25 August 2009. http://www.wired.com/gadgetlab/2009/08/total-immersion/
4. Rocha, R.: Tech's new reality. Montreal Gazette. 23 October 2009. http://www.montrealgazette.com/technology/Tech+reality/2138659/story.html
5. Wortham, J.: UrbanSpoon Makes It Easier to 'Scope' Out Restaurants. New York Times. 10 October 2009. http://bits.blogs.nytimes.com/2009/10/14/urbanspoon-makes-it-easier-to-scope-out-restaurants/
6. Ensha, A.: Another Augmented-Reality App for the iPhone. New York Times. 15 October 2009. http://gadgetwise.blogs.nytimes.com/2009/10/15/augmented-reality-apps-continute-to-roll-out/
7. Inbar O.: Top 10 reality demos that will revolutionize video games. Games Alfresco. 3 March 2008. http://gamesalfresco.com/2008/03/03/top-10-augmented-reality-demos-that-will-revolutionize-video-games/
8. Parr, B.: Easter Egg: Yelp Is the iPhone's First Augmented Reality App. Mashable. 27 August 2009. http://mashable.com/2009/08/27/yelp-augmented-reality
9. Koucheryavy, A., Vladyko, A., Kirichek, R.: State of the art and research challenges for public flying ubiquitous sensor networks. In: Balandin, S., Andreev, S., Koucheryavy, Y. (eds.) ruSMART 2015. LNCS, vol. 9247, pp. 299–308. Springer, Heidelberg (2015). doi:10.1007/978-3-319-23126-6_27
10. Dao, N., Koucheryavy, A., Paramonov, A.: Analysis of Routes in the Network Based on a Swarm of UAVs. In: Kim, K., Joukov, N. (eds.) Information Science and Applications (ICISA) 2016. LNEE, vol. 376, pp. 1261–1271. Springer, Heidelberg (2016). doi:10.1007/978-981-10-0557-2_119
11. Paramonov, A., Koucheryavy, A.: M2M traffic models and flow types in case of mass event detection. In: Balandin, S., Andreev, S., Koucheryavy, Y. (eds.) NEW2AN 2014. LNCS, vol. 8638, pp. 294–300. Springer, Heidelberg (2014). doi:10.1007/978-3-319-10353-2_25

12. Shafig, M.Z., et al.: A first look at cellular machine-to-machine traffic: large scale measurement and characterization. In: 12th ACM SIGMETRICS Performance International Conference. June 11–15, London, England, UK (2012)
13. Kirichek, R., Vladyko, A., Zakharov, M., Koucheryavy, A.: Model networks for internet of things and SDN. In: Kirichek, R., Vladyko, A., Zakharov, M., Koucheryavy, A. (eds.) 18th International Conference on Advanced Communication Technology (ICACT), pp. 76–79 (2016)

Rate of Convergence to Stationary Distribution for Unreliable Jackson-Type Queueing Network with Dynamic Routing

Elmira Yu. Kalimulina[(⊠)]

V. A. Trapeznikov Institute of Control Sciences, Russian Academy of Sciences,
Profsoyuznaya Street 65, Moscow 11799, Russia
elmira.yu.k@gmail.com

Abstract. In this paper we consider a Jackson type queueing network with unreliable nodes. The network consists of $m < \infty$ nodes, each node is a queueing system of M/G/1 type. The input flow is assumed to be the Poisson process with parameter $\Lambda(t)$. The routing matrix $\{r_{ij}\}$ is given, $i, j = 0, 1, ..., m$, $\sum_{i=1}^{m} r_{0i} \leq 1$. The new request is sent to the node i with the probability r_{0i}, where it is processed with the intensity rate $\mu_i(t, n_i(t))$. The intensity of service depends on both time t and the number of requests at the node $n_i(t)$. Nodes in a network may break down and repair with some intensity rates, depending on the number of already broken nodes. Failures and repairs may occur isolated or in groups simultaneously. In this paper we assumed if the node j is unavailable, the request from node i is send to the first available node with minimal distance to j, i.e. the dynamic routing protocol is considered in the case of failure of some nodes. We formulate some results on the bounds of convergence rate for such case.

Keywords: Dynamic routing · Queueing system M/G/1 · Unreliable network · Jackson network · Convergence rate

1 Introduction

Queueing systems and networks are the most suitable mathematical tools for modelling and performance evaluation of complex systems such as modern computer systems, telecommunication networks, transport, energy and others [1–5]. The reliability is another important factor for quality assessment of these systems [6]. So models with unreliable elements are a subject of great interest last years. A large number of research papers study queueing systems with unreliable servers. [7–12]. The less ones consider queueing networks. In this paper we analyse the performance characteristics of an open queueing network, whose nodes are subject to failure and repair. This assumption is often missed in theoretical papers, but it's essential for applications, for example, for telecommunication, sensor, ad-hoc, mesh and other kind of networks.

E.Y. Kalimulina—The work is partially supported by RFBR, research projects No. 14-07-31245 and No. 15-08-08677.

© Springer International Publishing AG 2016
V.M. Vishnevskiy et al. (Eds.): DCCN 2016, CCIS 678, pp. 253–265, 2016.
DOI: 10.1007/978-3-319-51917-3_23

This work is motivated by a practical task of modelling of modern telecommunication networks. We consider the mathematical model of the queueing network as a set of connected nodes that can break down and repair. We propose a modification of the classical model of an open queueing network (see e.g. [13, Chap. 2]), based on the principle of dynamic routing.

A strong mathematical definition of the term "dynamic routing" doesn't exist. It originally appeared in telecommunication industry. Dynamic routing is the technology that enables active network nodes (called routers) to perform many vital functions: detection, maintaining and modification of routes with considering of a network's topology, as well as some functions of routes calculation and their estimations. In difference from static routing technology routes are calculated dynamically using any one of a number of dynamic routing protocols. Dynamic routing is crucial technology for reliable packet transmission in case of failures. Protocols are used by routers to detect nodes availability and find routes available for packets forwarding over network. Figure 1 shows a simple example of four-node internetworking transmission. The transmission between Local network 1 and Local network 4 uses the link Router D - Router C. Assume the transmission facility between Gateway Routers C and D has failed. This renders the link between C and D unusable and the data transmission between Local network 1 and 2 is impossible (for static routing). The dynamic routing protocols use a route redistribution scheme and dynamically define new paths for transmission between Router D and Router C via B and A (dotted line). Thus network availability can be significantly enhanced through dynamic routing, an alternate route of more length is a redundant communications link between nodes D and C.

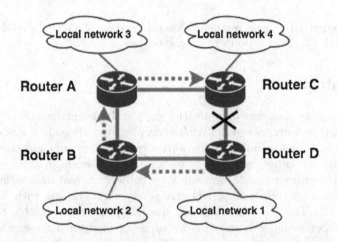

Fig. 1. A four-node internetwork with some route redundancy.

There are some math research papers where queueing networks with dynamic routing were considered. Queueing networks with constant routing matrix were considered in papers [14,15], each node there was modelled as a multichannel system, principle of dynamic routing was a random selection of a channel at the node. Kelly in his papers (see. [16]) considered a network as a set of parallel queues with several types of requests incoming into it, and the dynamic routing principle is the selection of particular queue depending on the type of request. We will adhere to interpret the dynamic routing as it's defined in telecoms and we will understand it as reconstruction of the route depending on the availability/unavailabiity of a specific node in the network, sending the requests (message) via alternate paths. In terms of queueing networks models it means the change the values of a routing matrix $\{r_{ij}\}$. A request will be rerouted to the available node in the case of failure of node j, i.e., this concept is as close as possible to this definition in telecommunication networks [18].

It is noted that other approaches can be applied to the problem of estimation of networks reliability. For example, the Erdos-Renyi random graphs model can be used to analyse the network connectivity (Erdos-Renyi graph) [19].

But as our task is the performance evaluation and analysis of traffic flows in networks taking into account reliability, we use queueing theory models and a model of Jackson network. There are some researches on unreliable queueing networks. Several algorithms for modifying the routing matrix $\{r_{ij}\}$ in the case of failures of nodes were described in [17], the common idea which they are based on is the principle of blocking of requests and repeated service after nodes recovery. The result related to the rate of convergence to the stationary distribution for unreliable network is given in [20,21]. In this paper we give some results for unreliable networks similarly as it was done in [21], but we propose another approach to the modification of the route matrix $\{r_{ij}\}$ and consider in a more general model for network nodes. There are several main classical problems in the analysis of queueing systems(networks) models that have to been considered: stability, ergodicity and probability of overflow. These problems are strongly connected with the transient behaviour of networks and the speed of convergence to stationary regime, so here we will concentrate on this problem for unreliable network systems.

2 Process Definition

2.1 The Classic Jackson Network

The classical model of Jackson network consists of m nodes ($m < \infty$), $M :=$ $\{1, \ldots, m\}$. Each node is a $M/M/1$ queueing system with the "first-come-first-served" discipline of service. Incoming requests are indistinguishable. The external flow into the network is a Poisson process with the parameter $\Lambda(t)$, depending on time. Incoming request is sending for the service to the node j c with probability r_{0j}, $\sum_{j=1}^{m} r_{0j} = r \leq 1$. A request is served at the node j with intensity $\mu_j(n_j)$, where n_j - the numbers of requests at the node j. $X_j(t)$ is denoted as the number of requests at the node j at time t, so the system state at time t is characterised with a vector $X(t) = (X_1(t), X_2(t), ..., x_m(t))$. The unique stationary distribution for the process $X(t)$ exists and is defined as

$$C_i = 1 + \sum_{n=1}^{\infty} \frac{\lambda_i^n}{\prod_{y=1}^{n} \mu_i(y)}, i = 1, 2, ..., m, \tag{1}$$

if the system of equations for a network traffic

$$\lambda_i = \Lambda * r_{0i} + \sum_{j=1}^{m} \lambda_j r_{ji}, i = 1, 2, ..., m \tag{2}$$

has the unique solution [17].

Degradable Networks and Some Routing Mechanisms. If we consider the servers at the nodes to be unreliable, the network's performance is degraded because a subset of the operating nodes is not available. Different routing mechanisms were considered by Daduna [17]. He provided the most extensive research of this subject. A short survey from [17] about the methods of routing is provided here.

The first method of routing is a blocking after service (BAS). It assumes that after completing the service at node at time t a customer chooses the next destination node according to his routing instruction. If at time t the destination node is not able to accept further customers, the customer stays and blocks current server until the situation at the destination node has changed and the customer can enter the node. The current node is blocked during this waiting period, i.e., it cannot start serving another customer, who might be waiting in the waiting room. If several nodes are simultaneously blocked by the same node j, then the first blocked - first unblocked- rule usually determines the order in which the nodes will be unblocked when a departure occurs from the destination node j.

The second method is blocking before present service. A customer at node i selects the following destination node j according to the routing rule before he starts receiving service at node i at time t. If node j is full at time t, node i immediately becomes blocked. When a departure occurs from node j, node i becomes unblocked and service begins. However, as soon as the destination node j becomes full again during the customer's service at i, the service is interrupted and node i becomes blocked again. Depending upon whether the customer at the blocked node i is allowed to occupy the position in front of the server or not, one can distinguish the two cases BBS-server occupied and BBS-server not occupied. Of course this is only meaningful when node i has finite waiting capacity.

The third one is the repetitive service. A customer after being served at node i chooses the next destination node j according to the routing instruction. If node j is full, the customer stays at node i to obtain another service. When this additional service expires the customer either selects his destination node anew according to his routing instruction (repetitive service - random destination) or the previously chosen node j remains his fixed destination. In this case, the customer's service at node i has to be repeated until at the end of a service, node j is able to receive another customer (repetitive service - fixed destination).

The last mechanism is the skipping of unavailable set of nodes. Customers are not allowed to enter to the node from the certain set, skipped it and immediately performs the jump to the next one according to the routing matrix.

We provided the another approach called the dynamic routing.

2.2 Unreliable Jackson Network with Dynamic Routing

It is assumed now that nodes at the network are unreliable and may break down or repair. Failures can be both individual and in a group (as in models in [17, 20, 21]). We will refer to $M_0 = \{0, 1, 2, ..., m\}$ as the set of nodes, where "0" is the "external node" (entry and exit from the network) and to $D \subset M$ as the subset of failed nodes, $I \subset M \setminus D$ the subset of working nodes, nodes from I may break down with the intensity $\alpha^D_{D \cup I}(n_i(t))$. Nodes from $H \subset D$ may recover with the intensity $\beta^D_{D \setminus H}(n_i(t))$. It is assumed the routing matrix (s_{ij}) is given. Additionally the adjacency matrix for our network (s_{ij}) is considered:

$$s_{ij} = \begin{cases} 1, & \text{if } r_{ij} \neq 0, \\ 0, & \text{if } r_{ij} = 0. \end{cases}$$

Now we can consider all possible paths of the network graph. To find them we need to calculate the following matrix: $(s_{ij})^2$, $(s_{ij})^3$, ..., $(s_{ij})^m$, $m < \infty$, $(s_{ij})^1 = (s_{ij})$. The matrix $(s_{ij})^m$ has the following property: the element in row i and column j is the number of paths from node i in the unit j of length m (including $(m-1)$ transitional nodes) [23].

We take the following routing scheme for network nodes from the subset D (we call it as "dynamic routing without blocking"). Only transitions to $M_0 \setminus D$ are possible for nodes from D:

$$r^D_{ij} = \begin{cases} 0, & \text{if } j \in D, i \neq j, \\ r_{ij} + r_{ik}/s^p_{ik}, & \text{if } j \notin D, k \in D \\ \exists\, i \to j \to i' \to j' \to ... \to i'' \to k : \underbrace{s^1_{ij} * s^1_{ji'} * s^1_{i'j'} * ... * s^1_{i''k}}_{p+1} \neq 0, \\ \quad \text{where } p = \min\{2, 3, ..., m : s^p_{ik_{k \in D}} \neq 0\}, \\ r_{ii} + \sum\limits_{\substack{k \in D \\ s^p_{ik} = 0\ \forall\ 1 < p \leq m}} r_{ik}, & \text{if } i \in M_0 \setminus D, i = j, \end{cases}$$

where s^p_{ik} - element of a matrix $(s_{ij})^p$.

The routing matrix is changed according to the same way for the input flow:

$$\Lambda r^D_{0j} = \begin{cases} \Lambda r_{0j}, & \text{if } j \in M \setminus D, \\ \Lambda(r_{0j} + r_{0k}/s^p_{0k} * \underbrace{(s^1_{0j} * s^1_{ji'} * s^1_{i'j'} * ... * s^1_{i''k})}_{p+1}), & \text{if } j \notin D, k \in D \\ 0, & \text{otherwise.} \end{cases}$$

Futher we will refer to The modified routing matrix as $R^D = (r_{ij}^D)$. The intensities of failures and recoveries depend on the state of nodes and does not depend on network load and are defined as follows [17]:

$$\alpha(D, I) = \frac{\psi(D \cup I)}{\psi(D)},$$

$$\beta(D, H) = \frac{\phi(D)}{\phi(D \setminus H)},$$

where ψ, ϕ are positive functions with domain on all subsets of set of nodes and taking only finite values for finite sets ($\psi(\emptyset) := 1$, $\phi(\emptyset) := 1$).

A more general model than in [20] is considered for network nodes. It is assumed that each network node is a queueing system type $M/G/1$. The system's dynamic will be described by a continuous in time random process $X(t)$ taking values from the following enlarged state space $\mathbb{E}$:

$$\tilde{\mathbf{n}} = ((n_1, z_1), (n_3, z_2), ..., (n_m, z_m), D) \in \{\mathbb{Z}_+ \times \{R_+ \cup 0\}\}^m \times |D| = \mathbb{E},$$

where n_i is the number of requests at the node i, z_i - may be considered as the time elapsed from the beginning of service for the current request i or as remaining service time of the customer in service, $|D|$ - the cardinality of set D. Intensity rates $\mu_i(n_i, z_i)$ depend on both the number of requests at nodes $n_i(t)$ and time $z_i(t)$. If $z_i(t)$ - the time elapsed from the beginning of service for the current request at time t, the conditional probability of the absence of any events occurring in a fixed interval of time $[t, t + \Delta t)$ (= {no new request} $\cup$ {no current service finished at all nodes} under the condition that the current value of the process $X(t)$):

$$\exp\left(-\int_t^{t+\Delta t} \left(\Lambda(s) + \sum_{i=1}^m \mu_i(n_i(t), z_i(t) + s)\right) ds\right),$$

which, if Δt is small enough, equals to [24, Chap. 2–4]

$$1 - \int_t^{t+\Delta t} \left(\Lambda(s) + \sum_{i=1}^m \mu_i(n_i(t), z_i(t) + s)\right) ds + O(\Delta t)^2,$$

if $\Delta t \to 0$ terms with $O((\Delta t)^2)$ are negligible in comparison with the terms without Δt or with Δt in a power one. The probability of any jump (finishing of service in one of the nodes or new request arriving in a network) is defined similarly:

$$\mu_j(n_j(t), z_j(t))\Delta t \left(1 - \int_t^{t+\Delta t} \left(\Lambda(s) + \sum_{i \neq j}^m \mu_i(n_i(t), z_i(t) + s)\right) ds + O(\Delta t)^2\right), \quad (3)$$

$$\Lambda(t)\Delta t \left(1 - \int_t^{t+\Delta t} \left(\sum_{i=1}^m \mu_i(n_i(t), z_i(t) + s)\right) ds + O(\Delta t)^2\right). \quad (4)$$

The following transitions in a network are possible:

$$T_{ij}\tilde{\mathbf{n}} := (D, n_1, \cdots, n_i - 1, \cdots, n_j + 1 \cdots, n_m),$$
$$T_{0j}\tilde{\mathbf{n}} := (D, n_1, \cdots, n_j + 1, \cdots, n_m),$$
$$T_{i0}\tilde{\mathbf{n}} := (D, n_1, \cdots, n_i - 1, \cdots, n_m),$$
$$T_H\tilde{\mathbf{n}} := (D \setminus H, n_1, \cdots, n_m),$$
$$T^I\tilde{\mathbf{n}} := (D \cup I, n_1, \cdots, n_m).$$

Definition 1. *The markov process* $\mathbf{X} = (X(t), t \geq 0)$ *is called unreliable queueing network if it's defined by the following infinitesimal generator:*

$$\begin{aligned}
\tilde{\mathbf{Q}}f(\tilde{\mathbf{n}}) = &\sum_{j=1}^{m}[f(T_{0j}\tilde{\mathbf{n}}) - f(\tilde{\mathbf{n}})]\Lambda(t)r_{0j}^D \\
&+ \sum_{i=1}^{m}\sum_{j=1}^{m}[f(T_{ij}\tilde{\mathbf{n}}) - f(\tilde{\mathbf{n}})]\mu_i(n_i(t), z_i(t))r_{ij}^D \\
&+ \sum_{I \subset M}[f(T^I\tilde{\mathbf{n}}) - f(\tilde{\mathbf{n}})]\alpha(D, I) \\
&+ \sum_{H \subset M}[f(T^I\tilde{\mathbf{n}}) - f(\tilde{\mathbf{n}})]\beta(D, H) \\
&+ \sum_{j=1}^{m}[f(T_{j0}\tilde{\mathbf{n}}) - f(\tilde{\mathbf{n}})]\mu_j(n_i(t), z_i(t))r_{j0}^D.
\end{aligned} \qquad (5)$$

3   Main Results

Like the classical and the Jackson network with blocking cases [17] the existence of a stationary distribution for an unreliable network with dynamic routing may be proved.

Theorem 1. *It is assumed the following conditions for unreliable network from the Definition 1*

$$(1) \inf_{n_j,t} \mu_j(n_j, z_j) > 0 \quad \forall\, j,$$

(2) time of service and time between new arrivals are independent random variables,
(3) routing matrix R^D *is reversible,*
then the stationary distribution for unreliable networks is defined by formulae

$$\pi(\tilde{\mathbf{n}}) = \pi(D, n_1, n_2, \cdots, n_m) = \frac{1}{C}\frac{\psi(D)}{\phi(D)}\prod_{i=1}^{m}\frac{1}{C_i}\frac{\lambda_i^{n_i}}{\prod_{k=1}^{n_i}\mu_i(k)}$$

where

$$C_i = \sum_{n=0}^{\infty}\frac{\lambda_i^{n_i}}{\prod_{y=1}^{n}\mu_i(y)}, \quad \lambda_i = \sum_{j=0}^{m}\Lambda * r_{ji}.$$

The main result for the convergence rate is formulated in terms of the spectral gap for unreliable queueing network. The preliminary notations and results on the spectral gap: there is a Markov process $\mathbf{X} = (X_t, t \geq 0)$ with the matrix of transition intensities $Q = [q(\mathbf{e}, \mathbf{e}')]_{e,e' \in \mathbb{E}}$, with stationary distribution π and an infinitesimal generator given by

$$\mathbf{Q}f(\mathbf{e}) = \sum_{e' \in \mathbb{E}} (f(\mathbf{e}') - f(\mathbf{e}))q(\mathbf{e}, \mathbf{e}').$$

The usual scalar product on $L_2(\mathbb{E}, \pi)$ is defined as

$$\langle f, g \rangle_{pi} = \sum_{e \in \mathbb{E}} f(\mathbf{e})g(\mathbf{e})\pi(\mathbf{e}). \tag{6}$$

The spectral gap for $\mathbf{X}$ is

$$Gap(\mathbf{Q}) = \inf\{-\langle f, \mathbf{Q}f \rangle_\pi : \|f\|_2 = 1, \langle f, 1 \rangle_\pi = 0\}.$$

The main result for a network is formulated in the following theorems:

Theorem 2. *If $\mathbf{X}$ is a markov process with infinitesimal generator $\mathbf{Q}$, it is assumed that $\mathbf{Q}$ is bounded, the minimal intensity of service is strictly positive $\inf_{n_j,t} \mu_j(n_j, z_j) > 0$ and the routing matrix (r_{ij}^D) is reversible, then $Gap(\mathbf{Q}) > 0$, if the following condition is true: for any $i = 1, \cdots, m$, for the birth and death process, corresponding to the node i with parameters λ_i and $\mu_i(n_i, z_i)$ the spectral gap is strictly positive $Gap_i(\mathbf{Q}_i) > 0$.*

Theorem 3. *If $\mathbf{X}$ is a markov process with a bounded infinitesimal generator $\mathbf{Q}$, positive minimal intensity of service $\inf_{n_j,t} \mu_j(n_j, z_j) > 0$ and reversible routing matrix (r_{ij}^D), then $Gap(\mathbf{Q}) > 0$ iff for any $i = 1, \cdots, m$, the distribution $\pi = (\pi_i), i \geq 0$ has light tails, i.e. the following condition $\inf_k \frac{\pi_i(k)}{\sum_{j>k} \pi_i(j)} > 0$.*

Theorem 4. *(Corollary from [22]). If $\mathbf{X}$ is unreliable queueing network with dynamic routing from Definition 1 with infinitesimal generator $\mathbf{Q}$ and transition probabilities matrix P_t. It is assumed that routing matrix (r_{ij}^D) is reversible and $(r_{ij}^D)^k > 0$ dor $k \geq 1$. If the distribution π_i has light tails for any $i = 1, \cdots, m$, then the following conditions are equivalent.*

- *for any $f \in L_2(\mathbb{E}, \pi)$*

$$\|P_t f - \pi(f)\|_2 \leq e^{-Gap(\mathbf{Q})t}\|f - \pi(f)\|_2, t > 0,$$

- *for any $\mathbf{e} \in \mathbb{E}$ the constant $C(\mathbf{e}) > 0$ exists such, that*

$$\|\delta_{\mathbf{e}} - \pi(f)\|_{TV} \leq C(\mathbf{e})e^{-Gap(\mathbf{Q})t}, t > 0.$$

4 The Numerical Example

We consider two numerical examples of network state probabilities calculation:

Example 1: The network consists of three nodes, each node is a system with two servers (see Fig. 2).

Example 2: The network consists of two nodes, each node is a system with three servers (see Fig. 3).

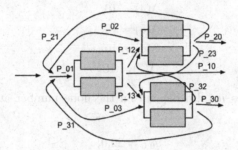

Fig. 2. Network with two-servers nodes

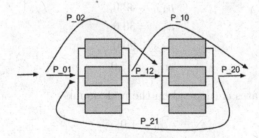

Fig. 3. Network with three-servers nodes

We use the following initial data for calculations in Example 1:
the number of nodes: $m = 3$,
the transition probabilities matrix:

$$P_{ij} = \begin{pmatrix} 0.03 & 0.57 & 0.35 & 0.05 \\ 0.1 & 0.002 & 0.398 & 0.5 \\ 0.35 & 0.25 & 0.15 & 0.25 \\ 0.2 & 0.25 & 0.3 & 0.25 \end{pmatrix}$$

The matrix P_{ij} has a size of $\times(m+1)x(m+1)$. The first row of matrix (P_{0j}, $j = 0, 1, 2, 3$) defines the probabilities the application received by the network, will be sent for service to node j, the first column gives the probability that the application leaves the system.

The intensities of input flow λ (depend on the number of customers in the system):

$$\lambda(0) = 10,$$
$$\lambda(1) = 10,$$
$$\lambda(2) = 10,$$
$$\lambda(3) = 10,$$
$$\lambda(4) = 10,$$
$$\lambda(5) = 10,$$
$$\lambda(6) = 10,$$
$$\lambda(7) = 10.$$

The service rates μ (also depend on the node number and the number of customers):

$$\mu_{0,0} = 20.0,$$
$$\mu_{0,1} = 10.0,$$
$$\mu_{0,2} = 5.0,$$
$$\mu_{0,3} = 7.0,$$
$$\mu_{1,0} = 32.0,$$
$$\mu_{1,1} = 30.0,$$
$$\mu_{1,2} = 20.0,$$
$$\mu_{1,3} = 17.0.$$

Server failure rates α (depend on the node number):

$$\alpha_0 = 1.0,$$
$$\alpha_1 = 2.0,$$
$$\alpha_2 = 3.0.$$

Recovery rates β (depend on the node number):

$$\beta_0 = 2.0,$$
$$\beta_1 = 4.0,$$
$$\beta_2 = 6.0.$$

The calculation results for Example 1. The stationary probabilities for Example 1:

$$P_0(0,0,0) = 0,340448527,$$
$$P_0(0,0,1) = 0,010639016,$$
$$P_0(0,0,2) = 0,003546339,$$
$$P_0(0,1,0) = 0,048635504,$$
$$P_0(0,1,1) = 0,015198595,$$
$$P_0(0,1,2) = 0,005066198,$$
$$P_0(0,2,0) = 0,162118346,$$
$$P_0(0,2,1) = 0,050661983,$$
$$P_0(0,2,2) = 0,016887328,$$
$$P_0(1,0,0) = 0,017022426,$$
$$P_0(1,0,1) = 0,005319508,$$
$$P_0(1,0,2) = 0,001773169,$$
$$P_0(1,1,0) = 0,024317752,$$
$$P_0(1,1,1) = 0,007599297,$$
$$P_0(1,1,2) = 0,002533099,$$
$$P_0(1,2,0) = 0,081059173,$$
$$P_0(1,2,1) = 0,025330992,$$
$$P_0(1,2,2) = 0,008443664,$$
$$P_0(2,0,0) = 0,017022426,$$
$$P_0(2,0,1) = 0,005319508,$$
$$P_0(2,0,2) = 0,001773169,$$
$$P_0(2,1,0) = 0,024317752,$$
$$P_0(2,1,1) = 0,007599297,$$
$$P_0(2,1,2) = 0,002533099,$$
$$P_0(2,2,0) = 0,081059173,$$
$$P_0(2,2,1) = 0,025330992,$$
$$P_0(2,2,2) = 0,008443664.$$

Based on these results can be obtained by other characteristics of the network, such as: - The probability of denial of service (the probability that all sites are occupied) = 0.0084; - Availability factor of the system (the system is completely free) = 0.34.

References

1. Lakatos, L., Szeidl, L., Telek, M.: Introduction to Queueing Systems with Telecommunication Applications, Springer Science & Business Media, Mathematics, 388 pages. Springer, Heidelberg (2012)

2. Ghosal, A., Gujaria, S.C., Ghosal, R.: Network Queueing Systems: With Industrial Applications. South Asian Publishers, 1 January 2004. Computer network protocols, 138 pages (2004)

3. Daigle, J.: Queueing Theory with Applications to Packet Telecommunication. Technology & Engineering, p. 316. Springer Science & Business Media, Berlin (2006)

4. Alexander, T.: Analysis of fork/join and related queueing systems. ACM Comput. Surv. **47**(2), 71 (2014). doi:10.1145/2628913. Article 17

5. Lakshmi, C., Sivakumar, A.I.: Application of queueing theory in health care: a literature review. Oper. Res. Health Care **2**(1–2), 25–39 (2013)

6. Sterbenz, J.P.G., Çetinkaya, E.K., Hameed, M.A., et al.: Evaluation of network resilience, survivability, and disruption tolerance: analysis, topology generation, simulation, and experimentation. Telecommun. Syst. **52**, 705 (2013). doi:10.1007/s11235-011-9573-6

7. Hou, I.-H., Kumar, P.R.: Queueing systems with hard delay constraints: a framework for real-time communication over unreliable wireless channels. Queueing Syst. **71**(1), 1510–1577 (2012). doi:10.1007/s11134-012-9293-y

8. Madhu Jain, G.C., Sharma, R.S.: Unreliable server M/G/1 queue with multi-optional services and multi-optional vacations. Int. J. Math. Oper. Res. **5**(2), 145–169 (2013). doi:10.1504/IJMOR.2013.052458

9. Ba-Rukab, O.M., Tadj, L., Ke, J.-C.: Binomial schedule for an M/G/1 queueing system with an unreliable server. Int. J. Modell. Oper. Manage. **3**(3–4), 206–218 (2013). doi:10.1504/IJMOM.2013.058326

10. Tadj, L., Choudhury, G., Rekab, K.: A two-phase quorum queueing system with Bernoulli vacation schedule, setup, and N-policy for an unreliable server with delaying repair. Int. J. Serv. Oper. Manage. **12**(2), 139–164 (2012). doi:10.1504/IJSOM.2012.047103

11. Bama, S., Afthab Begum, M.I., Fijy Jose, P.: Unreliable Mx/G/1 queueing system with two types of repair. Int. J. Innov. Res. Dev. **4**(10) (2015)

12. Klimenok, V., Vishnevsky, V.: Unreliable queueing system with cold redundancy. In: Gaj, P., Kwiecień, A., Stera, P. (eds.) CN 2015. CCIS, vol. 522, pp. 336–346. Springer, Heidelberg (2015). doi:10.1007/978-3-319-19419-6_32

13. Chen, H., Yao, D.D.: Fundamentals of Queueing Networks. In: Performance, Asymptotics, and Optimization Springer, Book on Stochastic Modelling and Applied Probability, vol. 46 (2001). doi:10.1007/978-1-4757-5301-1

14. Vvedenskaya, N.D.: Configuration of overloaded servers with dynamic routing. Probl. Inf. Transm. **47**, 289 (2011). doi:10.1134/S0032946011030070

15. Sukhov, Y., Vvedenskaya, N.D.: Fast Jackson networks with dynamic routing. Probl. Inf. Transm. **38**, 136 (2002). doi:10.1023/A:1020010710507

16. Kelly, F.P., Laws, C.N.: Dynamic routing in open queueing networks: Brownian models, cut constraints and resource pooling. Queueing Syst. **13**, 47–86 (1993)

17. Sauer, C., Daduna, H.: Availability formulas and performance measures for separable degradable networks. Econ. Qual. Control **18**(2), 165–194 (2003)

18. Marco, C.: Dynamic Routing in Broadband Networks. Springer, Heidelberg (2012)

19. Yavuz, F., Zhao, J., Yagan, O., Gligor, V.: Toward k-connectivity of the random graph induced by a pairwise key predistribution scheme with unreliable links. IEEE Trans. Inf. Theor. **61**(11), 6251–6271 (2015). doi:10.1109/TIT.2015.2471295

20. Lorek, P., Szekli, R.: Computable bounds on the spectral gap for unreliable Jackson networks. Adv. Appl. Probab. **47**, 402–424 (2015)

21. Lorek, P.: The exact asymptotic for the stationary distribution of some unreliable systems, [math.PR] arXiv:1102.4707 (2011)

22. Chen, M.F.: Eigenvalues, Inequalities, and Ergodic Theory. Springer, Heidelberg (2005)
23. Cormen, T.H., Leiserson, C.E., Rivest, R.L., Stein, C.: Introduction to Algorithms, 3rd edn. The MIT Press, Cambridge (2009)
24. Saaty, T.L.: Elements of Queueing Theory with Applications. Dover Publications, New York (1983)

On the Method of Group Polling upon the Independent Activity of Sensors in Unsynchronized Wireless Monitoring Networks

Alexander Shtokhov[1], Ivan Tsitovich[1,2(✉)], and Stoyan Poryazov[3]

[1] National Research University Higher School of Economics, Moscow, Russia
[2] Institute for Information Transmission Problems (Kharkevich Institute) RAS,
Moscow, Russia
cito@iitp.ru
[3] Institute of Mathematics and Informatics BAS, Sofia, Bulgaria

Abstract. It is considered the model of large monitoring networks with working independently sensors for an alarm signalization. Outlined in the previous papers the method of group polling for alarming sensors identification used the time synchronization. The last condition is very strong for vide distributed monitoring networks. Recently proposed method of group polling for the alarming sensors identification in unsynchronized wireless monitoring network is investigated. Based on numerical simulations, it is found that the group polling method may be effective for unsynchronized networks with thousands or more sensors and the decoding algorithm may be realized on-time using parallel executions. Recommended number of the code signal repetitions is proposed.

Keywords: Wireless Sensor Network · Sensor for an alarm signalization · Group polling · Unsynchronized time

1 Introduction

Wireless sensor networks (WSN) based on LTE communication protocols, satellite systems, and wireless computer networks WiFi and WiMAX substantially stimulates the development of the monitoring and telemetric systems. Such systems are widely employed in various branches due to the progress in microelectronics, which led to the creation of cheap sensors that transmit the parameters of objects. Modern WSN systems are used for monitoring of emergencies, operation of technical objects, voltage sensors in power lines, ecological problems, urban transport, systems of the distant control of sensors of water, gas, and electric power consumption, payment terminals, etc. Note the advent of corporate networks of distributed monitoring, for example, of gas and oil pipelines. The stationary systems for monitoring and telemetry are supplemented with dynamic ones. Various types of sensors are mounted on ambulance cars and taxies, are incorporated in the positioning systems, and transmit data on the position and state of object ([1,2]).

© Springer International Publishing AG 2016
V.M. Vishnevskiy et al. (Eds.): DCCN 2016, CCIS 678, pp. 266–278, 2016.
DOI: 10.1007/978-3-319-51917-3_24

In the paper [4], it was proposed the method of group polling for detecting of alarming sensors in the monitoring network where properties of this method were investigated under the assumption of independent activity of the alarming sensors. It is supposed that the monitoring network is very large and contains thousands of sensors but all sensors synchronize in time their alarm signals. The last demand is difficult for its practical realization. But proposed in [4] method ensures the fulfilment of a short time of an alarming sensors detection, i.e. if t is a number of sensors in the network then the detection time is $O(\log t)$. In the paper [3], it is proposed a generalization of this method onto a case of unsynchronized by time alarming signal sending. The goal of our paper consists in investigating the properties of this method by numerical modeling for optimal parameters of the algorithms finding.

The formulation of the problem is presented in Sect. 2. It is described the algorithm of WSN output signal modeling when alarming sensors begin their signals in random time moments and take into account digitization in time of the output signal.

In the next section we present the calculated results with variations in such model parameters as the number of sensors in network t, a number of the sensor code symbol repeats k for the methods of the alarming sensors identification. As in [4], we examine the characteristics P_1—the detection of redundant alarming sensors is not an error, P_2—the probability of incorrect identification of activity of at least one sensor, and $\overline{s}$—the mean number of identified sensors.

In contrast with [4] we may use codes with varied length. This problems is illustrated in the Sect. 5. Finally, we present conclusions and recommendations on the group polling of sensors.

2   Setting of the Problem

We will follow, in general, the notations of [3]. The sensor network contains t sensors. We develop a polling strategy aimed at the fastest identification of s sensors that are ready for data transmission and named as alarming sensors. We assume that $s \ll t$ (a relatively small number of the active sensors in the network, really no more then 5). Such a scenario is typical of the sensor network that is located at a relatively large area where the probability of local emergency is relatively low. In contrast with [4], now a time is continuous. This means the following. The i-th alarming sensor begins to send the signal in a random time u_i, $i = 1, \ldots, s$, and is sending it during a time U. After the time $u_i + U$ it repeats the signal over and over while the network supervisor does not identify its and switches its working state. After this moment the sensor does not alarm. Therefore, the number of simultaneously alarming sensors is no less then the number of sensors with u_i such that $|u_i - u_j| \leq O(U)$ and may be small.

The alarming signal is dropped onto N short time intervals with the length Δ (therefore, $U = N\Delta$), in every of this intervals the sensor may send or not send the special signal. This interprets as sending 1 or 0, respectively. Therefore, every sensor has its unique code $\mathbf{a} = (a^1, \ldots, a^N)$ and for the i-th sensor its code is denoted by $\mathbf{a}_i$. The first question is how to construct the vectors $\mathbf{a}_i, i = 1, \ldots, t$. According with [5], it is proposed the procedure for the Boolean matrix $\mathbf{A} = (\mathbf{a}_i, i = 1, \ldots, t)$ generation with near to optimal properties, where a_i^j are independent random numbers 0 or 1 with a proper probability p^0 for 1 in the matrix.

Using a minimum number of steps N, we must identify the active sensors in such a way that the mean probability of the false identification of one sensor does not exceed a predetermined level (the averaging is performed with respect to a priori uniform distribution of alarming sensors on the set $T = \{1, \ldots, t\}$ denoted by $\mathbf{P}$). As in [4], we use two criterions for an admissibility of alarming sensors identification denoted by values P_1 and P_2.

The state of t sensors to be alarming or not is described using variables $x_1, \ldots, x_t$ that can be 0 or 1 (a passive sensor, which is not alarming, and an active sensor, which is alarming, respectively). The variables with numbers $i_1, \ldots, i_s$ are unities, and the remaining variables are zeros. Let S be the ordered set of alarming sensors, i.e. $S = \{i_1, \ldots, i_s\}$. Let $\hat{S}$ be the set of identified sensors than $P_1 = 1 - \mathbf{P}(\hat{S} \supseteq S)$ and $P_1 = 1 - \mathbf{P}(\hat{S} = S)$.

In the group polling, we simultaneously receive signals from several sensors. The j-th column of $\mathbf{A}$ gives us the group of sensors involving in the j-th polling (if $a_i^j = 1(0)$ then the i-th sensor is (is not) involved in the polling). When the group contains at least one alarming sensor, we receive the signal that is interpreted as 1. If the group does not contain alarming sensors, we do not receive signals and the result is 0. Thus, response of the sensors of the j-th group is represented as

$$f_j = (a_1^j \wedge x_1) \vee \cdots \vee (a_t^j \wedge x_t), \tag{1}$$

where $\wedge$ is the Boolean product and $\vee$ is the Boolean sum.

We assume that data transmission errors are possible in the network. This means that the value f_j is known with a certain error. In each polling session, the result is distorted regardless of the remaining polling sessions in accordance with the stochastic transition matrix

$$\mathbf{W} = \begin{pmatrix} 1 - \beta_0, & \beta_0 \\ \beta_1, & 1 - \beta_1 \end{pmatrix} \tag{2}$$

where β_0 is the probability of false zero (i.e., detection of 1 instead of 0 as the output signal) and β_1 is the probability of false unity (i.e., detection of 0 instead of 1). Therefore, the result of the j-th polling is g_j, which is 0 or 1 in accordance with matrix $\mathbf{W}$ regardless of the results in the remaining sessions provided that the values of f_j are fixed.

In our case the problem is more difficult. Firstly, when we analyze results $g_1, \ldots, g_N$ we do not know a real sequence $\mathbf{a}_i$ for the i-th sensor since the vector may have a cyclic shift on the researched time interval. Therefore, for the i-th sensor we have the set of its possible codes: the code $\mathbf{a}_i$ and all its cyclic shifts. Therefore, the matrix $\mathbf{A}$ is well for alarming sensors detection if the Hamming distance is sufficiently large not only between its rows but between all different cyclic shifts its rows also. In the terms of the paper, it means that the number of "sensors" is tN. Secondly, from continuity of the time, the time u_i may not coincide with beginning of an interval of the time quantization and the result f_j may be generated uncertainly. By this reason, we propose to repeat every signal in the sensor code k times. Therefore, every code has the length Nk and an optimal value of k needs be found.

We need to obtain an algorithm of identification of the set S of the alarming sensors based on the observations $g_1, \ldots, g_{Nk}$ and take into account that, in contrast with [4], random values $g_1, \ldots, g_{Nk}$ are dependent variables now. The theory of [4] does not give an specified recommendation for this case and we use numerical methods to find a method for alarming sensors identification in our case.

Since we consider continuity of a time, a received signal is described as a function of the time $f(u)$. Function values may be 0 or 1 and indicate receiving data from alarming sensors or absence of a signal at the time u. The code $\mathbf{a}_i$ for the i-th alarming sensor we have is a function $a_i(u)$ of the time now and it is calculated by the following way

$$a_i(u) = \begin{cases} 0, & \text{if } u < u_i, \\ a_i^j, & \text{if } u_i + (j-1)k\Delta \leq u < u_i + jk\Delta, j = 1, \ldots, N, \\ a_i^j, & \text{if } u_i + (N+j-1)k\Delta \leq u < u_i + (N+j)k\Delta, \\ & \quad j = 1, \ldots, N, \\ & \text{and so on.} \end{cases} \quad (3)$$

In this formula, u_i is the time to start alarm by the i-th sensor and has the uniform distribution on the interval $[0, \ Nk\Delta]$. After the time $u_i + Nk\Delta$ the sensor repeats his signal as on the interval $[u_i, \ u_i + Nk\Delta]$ while the sensor is not identified and stopped. We bound the time by $2Nk\Delta$ for providing at list one full time interval for all alarming sensors to send their signals.

Thus, we get response of the sensors at an arbitrary time is represented as

$$f(u) = (a_1(u) \wedge x_1) \vee \cdots \vee (a_t(u) \wedge x_t). \quad (4)$$

The result output continuous signal is dropped onto short time intervals with the length Δ, as a result, the continuous function $f(u)$ drops onto the group of observations $(f_1, f_2 \ldots)$, that can be 0, 1 or nil.

The value nil we introduce by the following reason. We observe 0 value of the function $f(u)$ for a time interval of digitation in the situation of absence of signals during the time from the interval, that means $f(u) \equiv 0$ on the interval. We observe 1 value of the function $f(u)$ at the interval if $f(u) \equiv 1$ on the interval. When the function $f(u)$ at the interval changes its value (from 0 to 1, or 1 to 0), i.e. $\exists u_1, u_2 : f(u_1) \neq f(u_2)$, we can not interpret the received signal correctly. This situation is a conflict and we should mark such signals as nil. Thus, for the j-th Δ-interval

$$f_j = \begin{cases} 1, & \text{if } f(u) \equiv 1, \\ 0, & \text{if } f(u) \equiv 0, \\ nil, & \text{if } \exists u_1, u_2 : \ f(u_1) \neq f(u_2). \end{cases} \quad (5)$$

This conflicts is determined by the functions $a_i(u)$ for alarming sensors also. By the same way construct discrete values $\hat{a}_i(j)$ that can be also 0, 1 or nil (i is a number of the sensor and j is a number of the corresponding time interval). Then, the result of an output signal for the j-th interval

$$f_j = (\hat{a}_1(j) \wedge x_1) \vee \cdots \vee (\hat{a}_t(j) \wedge x_t),$$

providing that $1 \vee nil \equiv 1$ and $0 \vee nil \equiv nil$.

Finally, results 0 or 1 of $g_j, j = 1, \ldots, 2Nk$, we transform in accordance with the matrix $\mathbf{W}$ and we get the vector of observation $\hat{\mathbf{g}} = (\hat{g}_1, \ldots, \hat{g}_{2Nk})$. If $g_j = nil$ then $\hat{g}_j$ equals 0 or 1 with probability 0.5 and the observations $g_j = nil$ can not help for an alarming sensor detection. Therefore, we lose a part of information in contrast with the case from [4] and need to have longer sensors codes for an alarming sensor identification with the same quality. By this reason, we investigate a portion of loss information depend on k.

3   Algorithm of Alarming Sensors Detection

Our algorithm of alarming sensors detection are based on the factor analysis with the aid of the maximum likelihood method. As described in the previous section, we analyze the output signal $\hat{\mathbf{g}}$. We use a data window of the width Nk and the window shifts from position $l = 0$, i.e. we use the data $(\hat{g}_1, \ldots, \hat{g}_{Nk})$, to $l = Nk$. In the position l we use the data $(\hat{g}_{1+l}, \ldots, \hat{g}_{Nk+l})$. For every sensor i and any the window position l we calculate $L(i, l, m)$ where m is the code shift.

We introduce vectors $\tilde{a}_i(j)$ for indicating positions in the code where the output result may be incorrect. It is calculated by the following: if $a_i(j) = a_i(j + 1)$ then $\tilde{a}_i((j - 1)k + 1) = \tilde{a}_i((j - 1)k + 2) = \cdots = \tilde{a}_i(jk) = \tilde{a}_i(jk + 1) = a_i(j)$ else $\tilde{a}_i((j - 1)k + 1) = \tilde{a}_i((j - 1)k + 2) = \cdots = \tilde{a}_i(jk - 1) = a_i(j)$ and $\tilde{a}_i(jk) = \tilde{a}_i(jk + 1) = nil$, $j = 1, \ldots, N - 1$. For $j = N$ if $a_i(N) = a_i(1)$ then $\tilde{a}_i((N - 1)k + 1) = \tilde{a}_i((N - 1)k + 2) = \cdots = \tilde{a}_i(Nk) = \tilde{a}_i(1) = a_i(N)$ else $\tilde{a}_i((N - 1)k + 1) = \tilde{a}_i((N - 1)k + 2) = \cdots = \tilde{a}_i(Nk - 1) = a_i(N)$ and $\tilde{a}_i(Nk) = \tilde{a}_i(1) = nil$.

Based on following data:

- $x_{00}(i, l, m)$ is the number of observations in the result output signal for which $\tilde{a}_i((j + m) \bmod Nk) = 0$ and the output result $\hat{g}_{(j+l) \bmod Nk} = 0, j = 1, \ldots, Nk$;
- $x_{11}(i, l, m)$ is the number of observations in the result output signal for which $\tilde{a}_i((j + m) \bmod Nk) = 1$ and the output result $\hat{g}_{(j+l) \bmod Nk} = 1, j = 1, \ldots, Nk$;
- $x_{01}(i, l, m)$ is the number of observations in the result output signal for which $\tilde{a}_i((j + m) \bmod Nk) = 0$ or nil and the output result $g_{(j+l) \bmod Nk}$ is 0 or nil, $j = 1, \ldots, Nk$;
- $x_{10}(i, l, m)$ is the number of observations in the result output signal for which $\tilde{a}_i((j + m) \bmod Nk) = 1$ and the output result $g_{(j+l) \bmod Nk}$ is 1 or nil, $j = 1, \ldots, Nk$.

In accordance with [4], the contribution of numbers $x_{00}(i, l, m)$, $x_{10}(i, l, m)$, $x_{01}(i, l, m)$, $x_{11}(i, l, m)$ to $L(i, l, m)$ are with their weights

$$a_{10} = \log \frac{\beta_1}{1 - \beta_0 - p^*(1 - \beta_0 - \beta_1)},$$

$$a_{11} = \log \frac{1 - \beta_1}{\beta_0 - p^*(1 - \beta_0 - \beta_1)},$$

$$a_{01} = \log \frac{\hat{p}(1 - \beta_0 - \beta_1) + \beta_0}{p^*(1 - \beta_0 - \beta_1) + \beta_1},$$

$$a_{00} = \log \frac{1 - \beta_0 - \hat{p}(1 - \beta_0 - \beta_1)}{1 - \beta_0 - p^*(1 - \beta_0 - \beta_1) + \beta_1},$$

where $\hat{p} = 1 - (1 - p_0)^s$, $p^* = 1 - (1 - p_0)^s$.

Next, we calculate

$$L(i, l, m) = a_{00}x_{00}(i, l, m) + a_{01}x_{01}(i, l, m) + a_{10}x_{10}(i, l, m) + a_{11}x_{11}(i, l, m).$$

In contrast with [4], the codes $\tilde{a}_i$ are not independent random variables and $L(i, l, m)$ is an approximation of the logarithm of the likelihood ratio only now.

Based on the error of the first kind, we calculate threshold L_0. We conclude that the i-th sensor is alarming if

$$\max_{l=0,\ldots,Nk} \max_{m=1,\ldots,Nk} L(i, l, m) > L_0, \tag{6}$$

otherwise the i-th sensor does not be alarming.

4   Numerical Results

In this section, we outline results the numerical simulations of alarming sensors detection. We investigate the model with $s = 2$ (the simplest case when signals from alarming sensors are tangled). The parameters of noise $\beta_0 = \beta_1 = 0.01$. To evaluate the result we used a parallel execution of the numerical simulation on a cluster of 16 servers with 24-core on each server. As it is found in [3], the computational complexity is so big that for $t = 1000$ calculating time is several hours.

In this connection we used a fft-based algorithm for the pattern matching to reduce the calculating time. We find the minimal number of mismatches in the result $\hat{g}_i$ and the sensor code $\tilde{a}_i$ using a polynomials multiplication ([3]). The main idea of the algorithm is to use the convolution of polynomial coefficients to calculate the most suitable cyclic shift of $\tilde{a}_i$. This approach reduces the asymptotic complexity to $O(Nk \log(Nk))$ instead of $O((Nk)^2)$.

As parameters of the algorithm quality we use estimations of P_1 and P_2 and the mean value of detected sensors $\bar{s}$; $\bar{s} - s$ is the mean value of "false alarms".

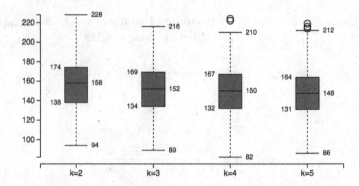

Fig. 1. Box plot of nil signals distribution for $t = 1000$, $N = 152$, $s = 2$ based on 1000 simulations

It is found that the mean number of output signals with uncertainty in their generation is similar to N (see Fig. 1). Therefore, for $k = 1$ results of our methods of alarming sensors detection will be very poor (P_1 and P_2 near to 1).

A number of nil signals for the i-th sensor depends on the code $\mathbf{a}_i$ and may vary in a long interval (see Fig. 1). Every change in the code $\mathbf{a}_i$ gives two nil signals in the corresponding extended code $\widetilde{\mathbf{a}}_i$. Therefore, it is interesting to investigate codes with variable length instead of the Nk-length codes with the same properties of alarming sensors detection. As it is found in [3], the computational complexity of a sensor detection strongly depends on its code length and more accurate extending of the code may reduce a computational time with the similar values of the alarming sensors detection quality. Because we identify sensors separately and code lengths are known, the previous algorithm may be used where L_0 is non constant but a linear function of the code length now.

For $k = 2$ we have unsatisfactory results also because the number of additionally identified sensors is big (see Table 1) (we expect $\overline{s}$ near of 2 but get essentially larger number). This means that a number of non nil components in the codes $\widetilde{\mathbf{a}}$ is small and a satisfactory detection of alarming sensors is impossible.

Table 1. Results for $k = 2$.

t	N	M	P_1	P_2	$\overline{s}$
500	144	1008	0.012	0.676	13.25
1000	152	1008	0.014	0.659	19.01
5000	172	1008	0.0001	0.726	72.79

For $k \geq 3$ we outline results of numerical simulations below.

For different values of t and k the algorithms were simulated M times and the results outline in the Tables 2, 3 and 4. In the tables, M is a number of

Table 2. Results for $t = 500$, $N = 144$.

k	M	P_1	P_2	$\bar{s}$
3	1000	0.004	0.007	1.999
4	2000	0.0025	0.004	1.999
5	3000	0.0023	0.004	1.999

Table 3. Results for $t = 1000$, $N = 152$.

k	M	P_1	P_2	$\bar{s}$
3	1000	0.001	0.010	2.009
4	2000	0.0015	0.006	2.003
5	3000	0.0017	0.0047	2.001

Table 4. Results for $t = 5000$, $N = 172$.

k	M	P_1	P_2	$\bar{s}$
3	1000	0.000	0.009	2.009
4	2000	0.000	0.0045	2.004
5	3000	0.000	0.0030	2.003

independent simulations our method of alarming sensors identification, P_1, P_2, $\bar{s}$ are estimated values of the method quality.

It is followed from Tables 2, 3 and 4 that a probability of alarming sensor lost is less than 1% for $t = 500$, is near 0.001 for $t = 1000$, and is practically 0 for $t = 5000$ when the code length is 432, 456, and 516 consequently. The probability decreases slowly when the number of code repetitions grows. The values $\bar{s} - s$ is practically 0 for all cases. By this reason we may recommended $k = 3$ as the number of code repetitions.

In contrast with [6], summands in the statistic $L(i, l, m)$ are not independent random variables. By this reason, it is difficult to corroborate the value L_0 and properties of the algorithm based on the theory of large deviations. For avoiding this disadvantage we propose the second method of the logarithm likelihood ratio estimating; we exchange summing onto maximum.

The total interval of observations $(\hat{g}_{1+l}, \ldots, \hat{g}_{Nk+l})$ is dropped onto N segments with length k. For every segment we calculate

- $y_{00}(i, l, m)$ is the number of segments where it is existed at least one observation with $\tilde{a}_i((j + m) \bmod Nk) = 0$ and $\hat{g}_{(j+l) \bmod Nk} = 0$, j is from the segment;
- $y_{11}(i, l, m)$ is the number of segments where it is existed at least one observation with $\tilde{a}_i((j + m) \bmod Nk) = 1$ and $\hat{g}_{(j+l) \bmod Nk} = 1$, j is from the segment;

- $y_{01}(i, l, m)$ is the number of segments where all $\tilde{a}_i((j + m) \bmod Nk)$ are 0 or *nil* with conditions on $\hat{\mathbf{g}}$: $\hat{g}_{(j+l) \bmod Nk}$ are 0 or *nil* if $\tilde{a}_i((j+m) \bmod Nk) = 0$ and may be any if $\tilde{a}_i((j + m) \bmod Nk) = nil$;
- $y_{10}(i, l, m)$ is the number of segments where all $\tilde{a}_i((j + m) \bmod Nk)$ are 1 or *nil* with conditions on $\hat{\mathbf{g}}$: $\hat{g}_{(j+l) \bmod Nk}$ are 1 or *nil* if $\tilde{a}_i((j+m) \bmod Nk) = 1$ and may be any if $\tilde{a}_i((j + m) \bmod Nk) = nil$.

Next,

$$L_2(i, l, m) = a_{00}y_{00}(i, l, m) + a_{01}y_{01}(i, l, m) + a_{10}y_{10}(i, l, m) + a_{11}y_{11}(i, l, m).$$

The values $y_{00}(i, l, m)$, $y_{01}(i, l, m)$, $y_{10}(i, l, m)$, $y_{11}(i, l, m)$ are independent random variables now.

We take decision as in (6) based on $L_2(i, l, m)$ but L_0 is calculated anew.

The Tables 5, 6 and 7 correspond to the method of alarming sensors identification based on the statistic L_2.

We get results similar to analogous ones for the first method of alarming sensors detection if $k = 3$. If we use $k = 4$ or 5 then the results for P_2 and $\bar{s}$ are worse.

In general, this results correspond to analogous results in [4]; with increasing of t the algorithms give more accurate results. The value N need be calculated as in [4] but for tNk as a number of sensors. Minimal value of k with satisfactory results of alarming sensors identification is 3. Therefore, the code length for

Table 5. Results for $t = 500$, $N = 144$.

k	M	P_1	P_2	$\bar{s}$
3	1008	0.002	0.006	2.003
4	2016	0.0025	0.1736	3.31
5	3024	0.0030	0.1859	3.26

Table 6. Results for $t = 1000$, $N = 152$.

k	M	P_1	P_2	$\bar{s}$
3	1008	0.001	0.006	2.004
4	2016	0.0015	0.197	4.92
5	3024	0.0017	0.209	4.39

Table 7. Results for $t = 5000$, $N = 172$.

k	M	P_1	P_2	$\bar{s}$
3	1008	0.000	0.004	2.004
4	2016	0.0001	0.211	6.85
5	3024	0.00006	0.214	6.11

$t = 500$ is 432, that is near to poll every sensor separately and a group polling is not effective in this case. In contrast, for $t = 5000$ the code length is 516 that is more than 9 times less than for polling every sensor separately.

5   Sensors Coding with Varied Length

In this section we analyze advantages of codes with a varied length as a method to reduce code lengths and as a consequence to reduce computational complexity of alarming sensors determination. In synchronized in time WSN we can not use codes with varied length because we destroy the main advantage of such networks. In contrast, in unsynchronized in time WSN we can use codes with varied length without losses because we check an activity for every sensor separately from each other.

The method of coding consists in the following. For the i-th sensor the full extended code we obtain by the formula (3) from the code $\mathbf{a}_i$ and obtain the discrete analogous $\hat{a}_i$. Now we add to the code $\mathbf{a}_i$ additional symbols 0 and 1 between symbols 0 and 1 in the code $\mathbf{a}_i$ and symbols 1 and 0 between symbols 1 and 0 respectively. All previous symbols of $\mathbf{a}_i$ are repeated $k - 2$ times. The new code $\tilde{\mathbf{a}}_i^r$ is analogous of $\tilde{\mathbf{a}}_i$ with k repetitions of code symbols and have the same property that a number of correctly interpreted symbols is at least $k - 2$.

We illustrate the method on Figs. 2 and 3 for 3 sensors with codes $\mathbf{a}_1 = (1, 0, 1, 0)$, $\mathbf{a}_2 = (1, 1, 1, 0)$, and $\mathbf{a}_3 = (1, 1, 0, 0)$ and $k = 3$.

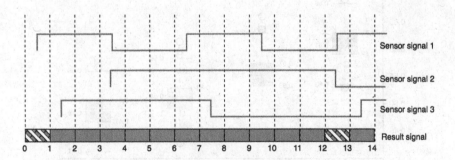

Fig. 2. Full extended codes for $k = 3$

On the Fig. 2, it is shown the full extended codes $\tilde{\mathbf{a}}_1 = (1, 1, 1, 0, 0, 0, 1, 1, 1, 0, 0, 0)$ (started in the first interval), $\tilde{\mathbf{a}}_2 = (1, 1, 1, 1, 1, 1, 1, 1, 1, 0, 0, 0)$ (started in the fourth interval), and $\tilde{\mathbf{a}}_3 = (1, 1, 1, 1, 1, 1, 0, 0, 0, 0, 0, 0)$ (started in the second interval). On the Fig. 3, it is shown the codes with varied length where $\tilde{\mathbf{a}}_1^r = (1, 1, 1, 0, 0, 0, 1, 1, 1, 0, 0, 0)$ (has the same number of symbols as in the full code), $\tilde{\mathbf{a}}_2^r = (1, 1, 1, 1, 1, 0, 0, 0)$ (has 8 symbols instead of 12 symbols in the full code), $\tilde{\mathbf{a}}_1^r = (1, 1, 1, 1, 0, 0, 0, 0)$ (has 8 symbols instead of 12 symbols in the full code). Therefore, effectiveness of coding with varied length depends on the code

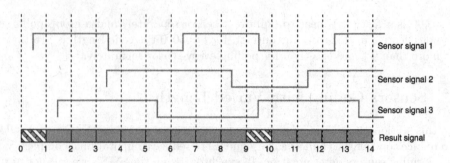

Fig. 3. Codes with varied length for $k = 3$

a. In a bottom of the figures, we illustrate the corresponding output vector $\hat{\mathbf{g}}$ with *nil* or non *nil* values.

Further we outline the results of numerical simulation of effectiveness of coding with varied length. As a parameter of effectiveness we use the value $\varkappa = \frac{\bar{L}}{kN}$ where $\bar{L}$ is the mean value of code lengthes. On the Figs. 4 and 5 we illustrate $\varkappa$ and variation of values $\frac{L}{kN}$ where L is a code length. Parameters are the following: $t = 5000$, $N = 172$, s_0 is a number of alarming sensors.

Computational complexity of alarming sensor detection reduce by 2 or more times slowly depending on s_0.

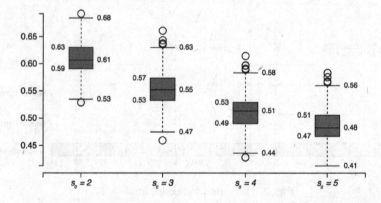

Fig. 4. Box plot of code lengthes distribution for $k = 3$ based on 1000 simulations

We analyze the nil signals distribution when coding with varied length is used. Under the same parameters as used for the Fig. 1 we simulate output signals $\hat{\mathbf{g}}$. It is followed from Fig. 6 that a quality of noninformative observations (nil-signals) is less than for the case with fix length codes. Therefore, coding with varied length may give more accurate results of observations of WSN.

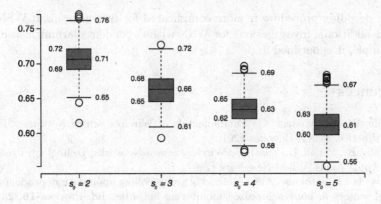

Fig. 5. Box plot of code lengthes distribution for $k = 4$ based on 1000 simulations

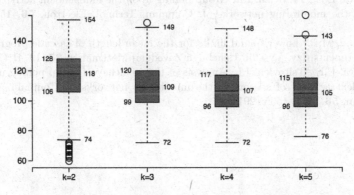

Fig. 6. Box plot of nil signals distribution of coding with varied length for $t = 1000$, $N = 152$, $s = 2$ based on 1000 simulations

6   Conclusions

The main result of the paper consists in that for unsynchronized in time WSN may be constructed a group polling with $O(\log t)$ time for alarming sensors detection. The constant before $\log t$ is more than 3 times that the analogous constant for the synchronized WSN. Therefore, the group polling method may be effective for WSN with thousands or more sensors.

Recommended number of the code signal repetitions k is 3 or 4. If codes with varied length are used then recommended number of the code signal repetitions is 1, 2, or 3 as an analogous of $k = 3$ or 2, 3, or 4 as an analogous of $k = 4$.

Computational complexity of alarming sensor detection is so big in contrast with [4] that the algorithm optimization is necessary. An optimization may be done by different ways as it was described in [3] and includes sensors coding with varied length.

The decoding procedure is more complicated for unsynchronized WSN and requires additional investigations for WSN with dependent alarming sensors as, for example, it is outlined in [6].

References

1. Dargie, W., Poellabauer, C.: Fundamentals of Wireless Sensor Networks: Theory and Practice. Wiley, Hoboken (2010)
2. Sohraby, K., Minoli, D., Znati, T.: Wireless Sensor Networks: Technology, Protocols, and Applications. Wiley, New York (2007)
3. Tsitovich, I.I., Shtokhov, A.N.: Method of group polling upon the independent activity of sensors in nonsynchronized monitoring networks. Inf. Process. **16**, 237–245 (2016)
4. Malikova, E.E., Tsitovich, I.I.: Group polling upon the independent activity of sensors in the monitoring networks. J. Commun. Technol. Electron. **56**, 1556–1563 (2011)
5. Malyutov, M.B.: Lower bound limits for the mean length of cascade programming of experiments. Izv. Vysshih Uchebnyh Zavedenij Matematika **27**, 19–41 (1983)
6. Malikova, E.E., Tsitovich, I.I.: Analysis of the efficiency of group polling upon the dependent activity of sensors in the monitoring networks. J. Commun. Technol. Electron. **56**, 1552–1555 (2011)

A Noising Method for the Identification of the Stochastic Structure of Information Flows

Andrey Gorshenin[1(✉)] and Victor Korolev[1,2]

[1] Institute of Informatics Problems, Federal Research Center
"Computer Science and Control" of the Russian Academy of Sciences,
Moscow, Russia
agorshenin@frccsc.ru
[2] Faculty of Computational Mathematics and Cybernetics,
Lomonosov Moscow State University, Moscow, Russia
victoryukorolev@yandex.ru

Abstract. The paper demonstrates a way for application of a methodology for the stochastic analysis of random processes based on the method of moving separation of finite normal mixtures to analyze the non-negative time series. We suggest to noise the initial data by adding i.i.d. normal random variables with known parameters. Then the one-dimensional distributions of observed processes are approximated by finite location-scale mixtures of normal distributions. The finite normal mixtures are convenient approximations to general location-scale normal mixtures or normal variance-mean mixtures which are limit laws for the distributions of sums of a random number of independent random variables or non-homogeneous and non-stationary random walks and hence, are reasonable asymptotic approximations to the statistical regularities in observed real processes. This approach allows to analyze the regularities in the variation of the parameters and capturing the low-term variability in the case of complex internal structure of data. An implementation of the methodology is shown by the examples of the intensity for the simulated information system.

Keywords: Noisy data · Moving separation of mixtures · Finite normal mixtures · Information systems

1 Introduction

One of the most important indicator in the characterization of the fine structure of the processes in various modern information systems is the intensity of events, traffic, etc. (see, for example, papers [1,2]). The values of the intensity are positive, so the results of the classical statistical techniques based on the assumption normality of the distribution of observations (such as, say, the classical ANOVA) may be inadequate. The paper presents a method of the statistical analysis of the stochastic structure of random processes based on the noising of initial time series so that the procedure of moving separation of finite normal mixtures [3] can be correctly applied to positive time series.

V.M. Vishnevskiy et al. (Eds.): DCCN 2016, CCIS 678, pp. 279–289, 2016.
DOI: 10.1007/978-3-319-51917-3_25

The key idea of the methodology is based on the noising of the initial data by adding independent and identically-distributed (i.i.d.) normal random variables with known parameters to refine the output. The one-dimensional distribution of the noisy sample is approximated by finite location-scale mixtures of normal distributions. It is well-known that finite normal mixtures are convenient approximations to general location-scale normal mixtures or normal variance-mean mixtures which are limit laws for the distributions of sums of a random number of independent random variables or non-homogeneous and non-stationary random walks and hence, are reasonable asymptotic approximations to the statistical regularities in observed real processes [4].

A similar approach with the noise benefit is known in the statistical signal processing as a stochastic resonance [5–8]. The base numerical method for finding values of the unknown parameters of the model in the suggested methodology is the EM algorithm [9]. It is known [10] that the noising increases the average convergence speed of the EM algorithm.

The paper proposes a way to improve the quality of the structural analysis of the unknown processes in the real information systems. The approach allows to analyze the regularities in the variation of the parameters and capturing the low-term variability in the case of complex internal structure of data. An implementation of the methodology is shown by the examples of the intensity for the simulated information system.

2   Approximation of the Initial Data

The typical form of the analyzed data is demonstrated on the Fig. 1.

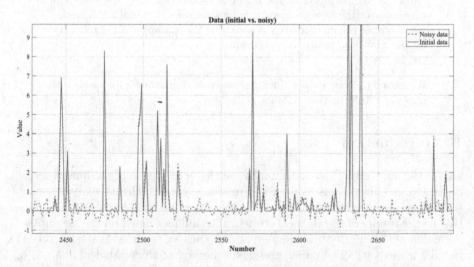

Fig. 1. The comparison of the noisy and initial data. (Color figure online)

The values of intensities (initial data) are nonnegative (see the solid red line on the Fig. 1), so the approximations by the finite normal mixtures with the support ℝ can be incorrect. For solving this problem, we suggest noising of the initial sample by adding i.i.d. normal random variables with known parameters. The blue dotted line on the Fig. 1 corresponds to the new "noisy" sample. The changing of the distribution of the initial sample under noising is shown on the Fig. 2. The left graph (Fig. 2a) corresponds to the initial data whereas the right one (Fig. 2b) demonstrates the sample distribution of the noisy data.

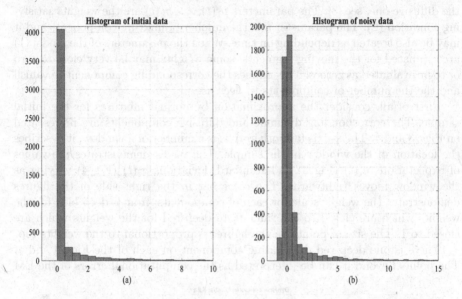

Fig. 2. The histograms of the initial sample (a) and the noisy data (b).

There is a unique peak for the initial data histogram (Fig. 2a), all observations are nonnegative. For the noisy histogram (Fig. 2b) the structure of data changes, there are negative elements too.

To analyze the changes of the stochastic process forming the initial data, the problem of statistical estimation of unknown parameters of the model should be solved for a moving sample segment (which is called a window) of a fixed length forming the sub-samples to be further analysed. Estimating parameters for the windows, one can derive the time series of these parameters. The resulting time series of the parameters will allow for the analysis of temporal changes in the behavior of the so-called diffusive and the dynamical components in the process.

Assume that the cumulative density function for a given window centered at the time moment t can be represented as

$$F_t(x) = \sum_{i=1}^{k} \frac{p_i(t)}{\sigma_i(t)\sqrt{2\pi}} \int_{-\infty}^{x} \exp\left\{ -\frac{(t - a_i(t))^2}{2\sigma_i^2(t)} \right\} dt, \tag{1}$$

where

$$\sum_{i=1}^{k} p_i(t) = 1, \quad p_i(t) \geqslant 0 \tag{2}$$

for all $x \in \mathbb{R}$, $a_i(t) \in \mathbb{R}$, $\sigma_i(t) > 0$, $i = 1, \ldots, k$.

The model (1) is called a finite location-scale normal mixture. The parameters $a_1(t), \ldots, a_k(t)$ are associated with the dynamic component of the internal variability of the process, and the parameters $\sigma_1(t), \ldots, \sigma_k(t)$ are associated with the diffusive one, see [3]. The parameters $p_1(t), \ldots, p_k(t)$ are the weights satisfying condition (2). The parameter k is the number of mixture components and it may be also treated as depending on time. When the parameters of the model (1) are estimated for the moving segments, some weights may be very close to zero or to be evaluated as zeroes. This implies the corresponding component to vanish and the the number of components to decrease.

First of all, consider the approximation by normal mixtures for the initial sample. The corresponding dynamic and diffusive components are represented on Figs. 3 and 4. The x-axis for both graphs is a number of a window, it describes the location of the window in the sample. The y-axis demonstrates the values of expectations $a_i(t)$ (Fig. 3) and standard deviations $\sigma_i(t)$ (Fig. 4) varying as the window moves rightwards. The color bar in the right side of the figures demonstrates the weight scale for each of components: from a deep blue for the weights which are closed to the value 0, to deep red for the weights which are closed to 1. The size of points on the figures is proportional to the weights too.

There is one deep red dominating component on each of the Figs. 3 and 4. The points beyond it can be interpreted as the computational errors of the EM

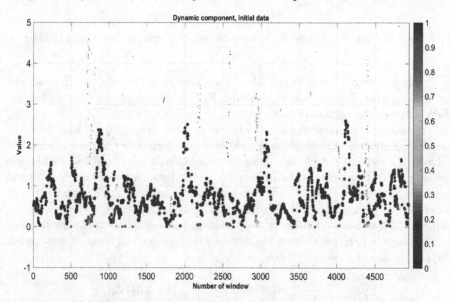

Fig. 3. The dynamic component of the initial data. (Color figure online)

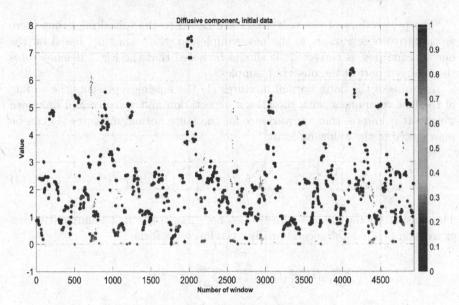

Fig. 4. The diffusive component of the initial data. (Color figure online)

algorithm. The statistical structure of the process cannot be analyzed in this trivial situation. Moreover, it is complicated to represent the set of points on the Fig. 4 as the curve which corresponds to the diffusive component.

So, the arising problems for the analysis of the initial sample are demonstrated. In the next section we describe a procedure of the noising and compare the results for the initial and modified data.

3 Noising of Data

Suppose that the cumulative density function of each observation X_j can be represented in form (1), that is

$$X_j \sim \sum_{i=1}^{k} p_i \, \Phi \left(\frac{x - a_i}{\sigma_i} \right)$$

The noising implies the following replacement of the original observations:

$$X_j \to X_j + \varepsilon_j,$$

for all $j = 1, \ldots, N$, where N means a sample size, and $\varepsilon_i \sim \mathcal{N}(0, \sigma^2)$ (a normal distribution with expected value 0 and standard deviation σ). As the example we use noising with the value σ equals 1 % of the sample standard deviation. The choice of the value for the σ is a difficult problem due to the necessity to keep initial stochastic structure of the data. The noisy data is shown on the Fig. 1 by the blue dashed line.

The peak values of the noisy data are closed to the initial ones, but there are negative observations in the new sample. So, the technique, based on the normal mixtures, is correct. It is should be noted that the Fig. 1 demonstrates the enlarged part of the observed samples.

In terms of the finite normal mixtures (1) the noising represents the adding of the new component with the known expectation and variance and unknown weight. It is known, that the variance for the finite normal mixtures (1) can be represented in the following form:

$$\sum_{i=1}^{k} p_i(t) \left[a_i(t) - \sum_{i=1}^{k} p_i(t) a_i(t) \right]^2 + \sum_{i=1}^{k} p_i(t) \sigma_i^2(t). \tag{3}$$

The noising by the $\varepsilon_i \sim \mathcal{N}(0, \sigma^2)$ does not change the first summand in the expression (3), but the second one has the following form

$$\sum_{i=1}^{k} p_i(t)(\sigma_i^2(t) + \sigma^2) = \sum_{i=1}^{k} p_i(t) \sigma_i^2(t) + \sigma^2.$$

We use the independence of the corresponding random variables, so

$$\mathbb{D}(X_j + \varepsilon_j) = \sigma_j^2 + \sigma^2.$$

So, it can be simply removed from the resulting approximating mixture (it should be removed from the diffusive component while the dynamic one does not change). It is visually demonstrated on Figs. 5 and 6.

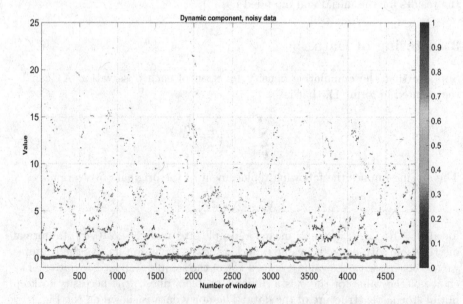

Fig. 5. The dynamic component of the noisy data. (Color figure online)

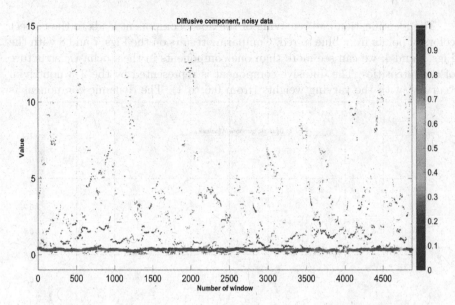

Fig. 6. The diffusive component of the noisy data. (Color figure online)

The "solid" red line on the Figs. 5 and 6 corresponds to the noisy component. Due to the computational errors, the expectation close to the 0 on the Fig. 5 (but it is not equals exactly). The red line on the Fig. 6 is the estimation of the noisy standard deviation σ.

The blue lines represents the fine stochastic structure of the initial process. These results can be used for the analysis. Note that the settings of the computational procedure for the initial and noisy sample are the same: width of the window equals 120 observations, the number of components in the mixture is 3, the computational accuracy ε equals 10^{-8}:

$$\max \left| \theta^{(m)}(t) - \theta^{(m-1)}(t) \right| < \varepsilon,$$

where $\theta^{(m)}(t)$ is a vector of estimated parameters at the m-th iteration step:

$$\theta^{(m)}(t) = (a_i^{(m)}(t), \sigma_i^{(m)}(t), p_i^{(m)}(t)), \quad i = 1, \dots, k.$$

The Figs. 7 and 8 demonstrates the results for the initial sample after removing of the noisy component. The mean and standard deviation σ are known, so the closest values (in a sense of some metrics, for example, L^1-norm) can be interpreted as the estimators of the noisy component at each position of the windows. The new weights $\widetilde{p}_i(t)$ of the another components should be defined as follows:

$$\widetilde{p}_i(t) = \frac{p_i(t)}{1 - p(t)},$$

where $p(t)$ is the weight of the noisy component at time moment t. So, the sum of the $\widetilde{p}_i(t)$, $i = 1, \dots, k-1$ equals 1 for all t (see (2)).

The identification and removing of the noisy component leads to the correct colors of points from blue to red. Comparing results on the Figs. 7 and 8 with the Figs. 3 and 4, we can see more than one components in the stochastic structure of the intensities. The diffusive component is represented by the 2–3 nontrivial "curves" with the varying weights (from 0.6 to 1). The dynamic component is

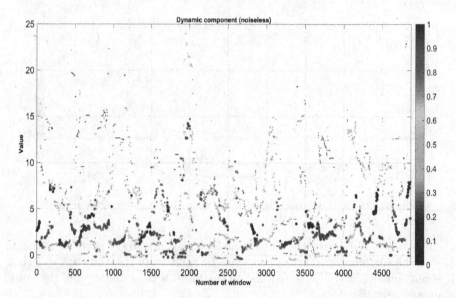

Fig. 7. The noiseless dynamic component of the initial data. (Color figure online)

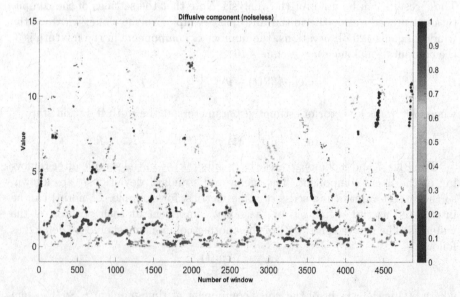

Fig. 8. The noiseless diffusive component of the initial data. (Color figure online)

represented by the two main lines: one of them corresponds to the expectations no more than 5 with the weights 0.6–0.9, the second component is determined by the expectations 5–20 with the weights 0.1–0.4.

4   Conclusions

The paper empirically demonstrates the efficiency of the suggested noising methodology to refine the output of the method of moving separation of finite normal mixtures. The key problem of the further researches is a formulation of the convenient conditions for the parameter σ of the noisy component. The results of the paper [10], based on the special sets and theorems for the conditional expectations, are difficult for using in practice, primarily in a sense of the automatization of the analysis. One of the possible way is based on the application of some information criteria (for example, the Akaike information criterion [11] and the Kullback-Leibler divergence [12]).

So, the appropriate optimization problem should be solved. The extremum of the log-likelihood function of the finite normal mixture (1) can be found by the following way:

$$F = \log \prod_{j=1}^{n} \left(\sum_{i=1}^{k} \frac{p_i}{\sigma_i \sqrt{2\pi}} e^{-\frac{(x_j - a_i)^2}{2\sigma_i^2}} + \frac{p_{k+1}}{\sigma \sqrt{2\pi}} e^{-\frac{x_j^2}{2\sigma^2}} \right) =$$

$$= \sum_{j=1}^{n} \log \left(\sum_{i=1}^{k} \frac{p_i}{\sigma_i \sqrt{2\pi}} e^{-\frac{(x_j - a_i)^2}{2\sigma_i^2}} + \frac{p_{k+1}}{\sigma \sqrt{2\pi}} e^{-\frac{x_j^2}{2\sigma^2}} \right).$$

The derivative with the respect to the unknown parameter σ has the following form

$$F'_\sigma \sim \sum_{j=1}^{n} \left(\sum_{i=1}^{k} \frac{p_i}{\sigma_i \sqrt{2\pi}} e^{-\frac{(x_j - a_i)^2}{2\sigma_i^2}} + \frac{p_{k+1}}{\sigma \sqrt{2\pi}} e^{-\frac{x_j^2}{2\sigma^2}} \right)^{-1} \times$$

$$\times \frac{p_{k+1}}{\sigma^2 \sqrt{2\pi}} e^{-\frac{x_j^2}{2\sigma^2}} \left[\frac{x_j^2}{\sigma^2} - 1 \right].$$

So, the main problem can be formulated as that of finding of an appropriate penalty function.

Moreover, our approach can be useful for an analysis of the increments of the initial data too. Since new process is an increment of the basic process, then $a_i(t)$ is the expected value of the increment, i.e., the "trend" component. So, $a_i(t)$ is the expected value of the random variable whose distribution is just the i-th component of mixture (1). By construction, this random variable is the

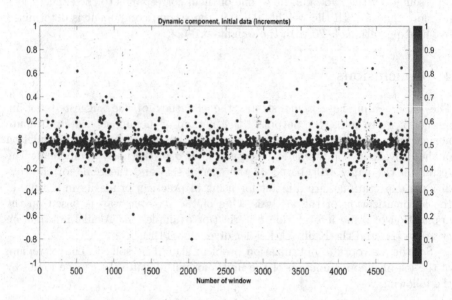

Fig. 9. The dynamic component of the initial data increments. (Color figure online)

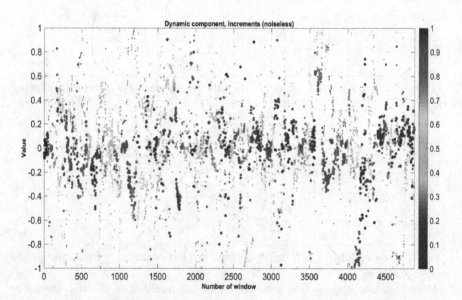

Fig. 10. The noiseless dynamic component of the initial data increments. (Color figure online)

increment of the initial process at the unit time interval, that is, $a_i(t)$ is the mean velocity of the variation of the i-th component. Thus, the set of pairs $(a_1(t), p_1(t)), \ldots (a_k(t), p_k(t))$ determines the distribution of the velocities over local trends at time t.

At last, we briefly demonstrate the results for the dynamic component of the increments. The adding, identification and removing of the noisy component leads to the substantial form of the multivariate volatility [3]. Comparing results on the Fig. 9 with the Fig. 10, we can see more than one components in the stochastic structure of the initial data increments after the "noising – denoising" procedure. The dynamic component is mainly represented by the non-trivial (non-zero) curve (see Fig. 10) with the weight more than 0.5–0.6. The same results can be represented for the diffusive component too.

The further investigations should be focused on both the finding of an appropriate penalty function and the methods of noising.

Acknowledgments. The research is partially supported by the Russian Foundation for Basic Research (projects 15-37-20851 and 16-07-00736).

References

1. Korolev, V.Yu., Chertok, A.V., Korchagin, A.Yu., Gorshenin, A.K.: Probability and statistical modeling of information flows in complex financial systems based on high-frequency data. Inf. Appl. **7**(1), 12–21 (2013)
2. Gorshenin, A.K., Korolev, V.Yu., Zeifman, A.I., Shorgin, S.Y., Chertok, A.V., Evstafyev, A.I., Korchagin, A.Yu.: Modelling stock order flows with non-homogeneous intensities from high-frequency data. In: AIP Conference Proceedings, vol. 1558, pp. 2394–2397 (2013)
3. Korolev, V.Yu.: Probabilistic and Statistical Methods of Decomposition of Volatility of Chaotic Processes. Moscow University Publishing House, Moscow (2011)
4. Korolev, V.Yu.: Generalized hyperbolic laws as limit distributions for random sums. Theor. Prob. Appl. **58**(1), 63–75 (2014)
5. Brey, J.J., Prados, A.: Stochastic resonance in a one-dimension ising model. Phys. Lett. A. **216**, 240–246 (1996)
6. Bulsara, A.R., Gammaitoni, L.: Tuning in to noise. Phys. Today **49**(3), 39–45 (1996)
7. Gammaitoni, L., Hänggi, P., Jung, P., Marchesoni, F.: Stochastic resonance. Rev. Mod. Phys. **70**, 223–287 (1998)
8. Kosko, B., Mitaim, S.: Stochastic resonance in noisy threshold neurons. Neural Netw. **16**(5), 755–761 (2003)
9. Dempster, A., Laird, N., Rubin, D.: Maximum likelihood from incomplete data via the EM algorithm. J. Roy. Stat. Soc. Ser. B **39**(1), 1–38 (1977)
10. Osoba, O., Mitaim, S., Kosko, B.: The noisy Expectation-Maximization algorithm. Fluctuation Noise Lett. **12**(3), 1350012 (2013)
11. Akaike, H.: A new look at the statistical model identification. IEEE Trans. Autom. Control **19**(6), 716–723 (1974)
12. Kullback, S., Leibler, R.A.: On information and sufficiency. Ann. Math. Stat. **22**(1), 79–86 (1951)

Efficiency of Redundant Multipath Transmission of Requests Through the Network to Destination Servers

V.A. Bogatyrev[✉] and S.A. Parshutina

ITMO University, Kronverksky Pr. 49, 197101 St. Petersburg, Russia
vladimir.bogatyrev@gmail.com, svetlana.parshutina@gmail.com
http://en.ifmo.ru

Abstract. It is not uncommon that delay-sensitive requests cannot be processed repeatedly in case of delivery failures, especially in real-time systems, which results in a strong need to enhance reliability of sending of requests. This can be achieved through concurrent transmission of copies of a request over multiple routes in a given network to a number of similar destination nodes. However, the increase in the initial flow of requests leads to the rise of the network load and the average residence time and potentially to the excess of the ultimate residence time. In the research, the usefulness of redundant distribution of requests through the network was estimated, with the maximized probability of successful delivery and the minimized average residence time. It was found possible to determine the optimal redundancy order for a given set of parameters, namely the intensity of the flow of incoming requests and the bit error rate.

Keywords: Reliability · Request · Redundancy · Multipath routing · Queueing systems

Constantly growing complexity of distributed computing systems, which is ahead of the pace of the increasing reliability of storage, processing, and transmission devices, gives rise to the need for developing methods to ensure high reliability, availability, and security of distributed resources for data processing and storage [1–8]. High availability of distributed resources is particularly important for real-time systems, which are sensitive to delays. It is commonly provided by introducing redundant elements into the structure of a given network and by designing the network in the way that lets arrange redundant data transmission and processing. In delay-sensitive systems, sending requests twice or more times to the destination, where they are somehow processed, is often impossible or pointless. This is due to rapid changes in the state of such systems, occurring under complex internal or external influence. As a result, methods to introduce redundancy into such systems turn out to be topical and prospective for in-depth exploration.

The proposed way of redundant data transmission should be employed when it is not possible to transmit data (a request, a data packet) through the network

© Springer International Publishing AG 2016
V.M. Vishnevskiy et al. (Eds.): DCCN 2016, CCIS 678, pp. 290–301, 2016.
DOI: 10.1007/978-3-319-51917-3_26

and/or process it at the destination node (a server) more than once. This approach relies on the mechanism of multipath data transmission, particularly on multipath routing, which implies at least two appropriate routes (paths) between the sender and the receiver, or receivers – in case of multicast transmission. In this paper, the receiver is a server belonging to a specified group of servers (or a given cluster). The servers have similar principal characteristics and serve incoming requests in the similar manner.

Multipath routing is widely used for multicast transmission, when data packets undergo fragmentation and the resulting fragments are sent over different paths to be brought together again at the destination. This is intended to hasten the process of transmission of large data packets and to secure them, especially when it comes to the secret data. Packet fragmentation based on multipath routing is mostly applied to load balancing.

Alternatively, using several paths for data transmission proves beneficial for enhancing reliability and fault-tolerance of the system. The traditional approach thereto suggests that one route is regarded as the main, the "best" path from some point of view, i.e. a given criterion, or metrics: bandwidth, average residence time, router and link load, and so on. Other routes are considered as alternate ones, used in case the main path fails.

Another approach is presented in [9], where requests – more precisely, numerous copies of each request – are sent over multiple paths concurrently. There is no need to choose the "best" path, because all of them are regarded appropriate, based on some metrics, with no ranking. On the one hand, redundant distribution of requests through the network causes the increased network load and the higher average residence time as well as possible excess of the ultimate residence time. On the other hand, sending copies of a request over multiple paths results in the enhanced chance that at least one copy will be delivered successfully. The goal of current research is to estimate the efficiency of redundant distribution of requests through the network when their repeated transmission is impossible, and therefore, to discover if there is a way to resolve this technical contradiction.

1   Problem Statement

Current study focuses on the process of redundant distribution of requests through a given network with the rigid requirements for the maximum period of time during which the requests are to be processed. In the known systems without redundancy, a request is sent from the source of requests to the single receiver, which might be a switching node or a server. In case of redundant systems, two or more copies of a request are transmitted from the sender to the multiple receivers (in this paper – servers) over a number of routes, as shown in Fig. 1.

1.1   System Design

It is assumed that there are n appropriate paths to deliver (several copies of) a request to their destination point within the system under consideration – one of

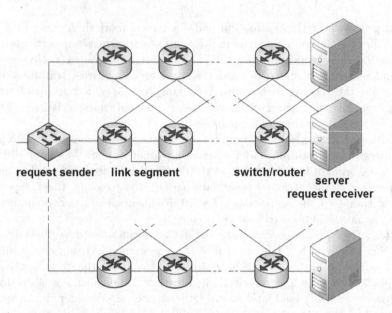

Fig. 1. A possible network configuration for redundant data transmission. There is a number of routes (paths) between the sender of requests and their receivers (in this paper – destination servers). In general case, these paths may overlap, having in common several intermediate network devices, like switches or routers.

the servers which comprise a specified cluster. In this research, all those paths are considered disjoint for two reasons. In the first place, addressing routes with no common nodes facilitates modeling and further calculations, particularly in the primary stage of such study. Secondly, it is done of practical interest, because the failure of a common node leads to the failure of two or several routes at once. It is a usual thing in the real world that current routing protocols accept a limited number of common nodes, e.g. no more than one per two paths, like Multipath Optimized Link State Routing Protocol (MOLSR).

Each path i leading from the source of requests to one of the destination servers includes a set of d_i switches or routers K_{ij} and d_i+1 link segments j between them; all units of network equipment have specified characteristics of reliability and performance. Each node K_{ij} can be represented as a single-channel non-preemptive M/M/1 queueing system with the infinite queue [10,11], where the letter M refers to the Markovian arrival process (the first position of M) and the Markovian service process (the second one).

Any errors and faults of switching nodes and link segments as well as of the destination servers lead to the situation when requests cannot be delivered and served. Suppose that the bit error rate B_{ij}, which is a measure of transmission errors for the link segment j of the path i, is known. It should be noted that some requests will never be processed – not only those which have been lost on the way to a destination server, because of faults of network equipment, but also those which have arrived at one of them but contained bit errors. Besides, it is possible that a request reaches one of the destination servers, with no bit errors, but finds it unavailable due to a fault or temporary shutdown. What is more, the server might be overloaded with the requests which have arrived earlier than the one in question and been waiting for service in the queue of the server. Thus, the probability that the server is ready to execute incoming requests, or the probability of server availability P_0, introduced in [9], ought to be taken into account. Also, it should be noted that by the time the server is ready to process the request, the latter will probably be irrelevant, as it has resided in the system too long (more than the ultimate residence time).

Copies of requests are distributed through the network over k $(k = 2, ..., n)$ routes, where k is the order (rate) of redundancy, or the redundancy order. It is said that a request is delivered successfully if at least one copy of it arrives at the destination, contains no bit errors, and is placed into the server queue or executed immediately, within the specified time limit. Here, we admit that no copies of the request are discarded after one of them has been delivered successfully. This leads to the upper (pessimistic) estimation of the average residence time, or the *time in the system*, needed for error-free transmission of a request from the source to the destination.

1.2   Probability of Error-Free Transmission

To start with, let us estimate distribution of requests through the network on the basis of server availability P_0 and the bit error rate B_{ij}, which is in fact the probability of bit errors which might occur while transmitting a request over the link segment j of the path i.

The probability of successful transmission of a data packet over the path i can be calculated as following:

$$R_i = \prod_{j=1}^{d_i}(1 - B_{ij})^N, \tag{1}$$

where d_i is the number of link segments of the path i and N is the average data packet length in bits. Provided that all routes under consideration are identical, the bit error rate can be regarded a constant and (1) can be replaced with

$$R = (1 - B)^{Nd}. \tag{2}$$

When requests are transmitted over k redundant paths concurrently, the probability of error-free transmission of at least one copy of a request over the path is

$$P_c = 1 - (1 - P_0 R)^k. \tag{3}$$

Here, P_0 is the probability that the server is available, or ready to serve incoming requests, meaning that it is powered up and in working order and there is enough space in its queue for the requests coming from the network. In this paper, we suggest that all servers of the interest, i.e. belonging to the same group (cluster) and serving requests from the same source, have similar server availability P_0.

1.3 Average Residence Time

Let us estimate the average residence time for data packets transmitted through the network to the destination servers, depending on the intensity of the request flow Λ. It is a period of time during which data packets reside in the system, starting from the moment when they are initially sent to the network by the source node until the moment when they are placed in the queue of one of the destination servers. The delay T_i in the path i is defined as the sum of delays in all segments of that path:

$$T_i = \sum_{j=1}^{d_i} T_{ij}, \tag{4}$$

where T_{ij} is the delay in the segment j of the path i, or the average time needed to transmit a data packet through the node j of the path i (with respect to the speed of transmission in the output port):

$$T_{ij} = \sum_{j=1}^{d_i} \frac{v_{ij}}{1 - \frac{v_{ij}k\Lambda}{n}}. \tag{5}$$

Here, v_{ij} is the average time of transmission in the segment j of the path i and computed as follows:

$$v_{ij} = \frac{N}{L_{ij}}, \tag{6}$$

where N is the average length of data packets and L_{ij} is the speed of transmission.

When a data packet – i.e. its copies – is sent over k identical routes simultaneously, the average residence time in the system T is defined as equal to the delay in the path over which the packet has been transmitted with no bit errors and accepted for further processing by the destination server. Assessing the average residence time for that path in the described above way gives an upper (conservative) estimate. This is because the estimation does not take into account the probability that in at least one path used to transmit copies of data packets, the delivery time of the copy under consideration can be less than the resulting value of T, due to the stochastic nature of the processes of servicing.

1.4 Criteria of Efficient Delivery

The efficiency of redundant distribution of requests through the network depends on the average residence time T, which is to be minimized, and the probability of error-free transmission P_c, which is to be maximized. The combination of these two single criteria produces a new multiplicative criterion $Mult_1$:

$$Mult_1 = \frac{P_c}{T}. \tag{7}$$

For the situation when the ultimate residence time t_0 is known, another multiplicative criterion can be defined, namely

$$Mult_2 = P_c(t_0 - T),\tag{8}$$

which describes the average stock time between the ultimate and the average residence time for the packets that have been delivered successfully.

Last but not least, let us estimate the efficiency of redundant transmission of requests through the network based on the following additive criterion, with the normalized residence time:

$$Add = P_c + \frac{t_0 - T}{t_0 - T_{min}},\tag{9}$$

where t_0 is the ultimate residence time, while T_{min} is the minimal residence time.

It can be seen that the major indicator of efficiency of distributed systems, including those with multipath routing, which is embodied in the given scalar additive and multiplicative criteria, aggregates the average time that requests reside in the system for and the rate at which they arrive at the destination node successfully, i.e. containing no bit errors.

In general, the efficiency of data transmission in distributed systems and networks is to be described with a set of indicators, which are – apart from the indicators mentioned above, namely the average residence time and the probability of error-free transmission, – the probability of packet loss, the rate of availability, and the rate of operational availability.

In real-time systems, when assessing the efficiency of servicing of requests, it is also important to take into account the probability that the average residence time does not exceed the ultimate residence time t_0.

In systems with link aggregation support, when copies of each packet are sent through multiple channels, comprising a group of redundant routes, the probability that the residence time of at least one copy does not exceed the ultimate residence time t_0 is calculated as following [12]:

$$d = 1 - \frac{k}{n}\Lambda v \exp\left(-t_0\left(v^{-1} - \frac{k}{n}\Lambda\right)\right),\tag{10}$$

where Λ is the intensity of the flow of incoming requests, v is the average time needed to transmit a packet of N-bit length at a bit rate of L. Equation (10) takes into consideration the increase in the initial intensity of the flow of requests for each of n paths ($k/n\Lambda$), with the redundancy order of k (k copies are spawned for each request).

The probability of timely delivery D of at least one of k copies, being sent via different channels [12–15]:

$$D = 1 - (1 - d)^k.\tag{11}$$

Here, timeliness implies that the time of a packet transmission is less than or equal to t_0.

The indicators discussed in [12] define the probability of timely redundant delivery of packets through one of k channels, under the assumption that those channels are ideal in terms of reliability and inerrancy of transmission.

To evaluate the efficiency of data transmission in real-time systems comprehensively, in view of unreliability of redundant communication links, an aggregate probabilistic indicator is proposed. The indicator is to take into account

- the probability that on entering the system, an incoming request finds it in working order;
- the probability that k channels which are selected for transferring requests in the redundant manner, are completely faultless;
- the probability that t_0 is not exceeded in at least one of k channels used for redundant data transmission, when sending a copy of each request;
- the probability that at least one copy of each packet, sent through one of k channels under consideration, contains no bit errors on reaching the destination node.

When aggregating N channels, which is proposed in [12], the informal criterion K has a physical meaning because it corresponds to the probability that at least one of k copies of a packet, transferred through k out of n links, will be delivered in a timely way.

$$K = \sum_{n=1}^{M} P_n \left(1 - (1 - pRd)^k\right), \qquad (12)$$

where P_n is the probability that a request enters the system at the moment when n out of M aggregated channels are in working order. For each of them, p is the probability of faultlessness, d is the probability of the residence time being less than the ultimate residence time t_0, R is the probability that a packet of N-bit length is transferred successfully, with no bit errors.

$$R = (1 - B)^N, \qquad (13)$$

Here, B is the bit error rate.

The discussed system of aggregate quality indicators, including the probability of timely and error-free redundant delivery of packets, introduced in [12], is focused on their transmission through redundant channels. The use of the considered criteria in assessing the efficiency of decisions on arranging multipath routing and their optimization requires development of models that take into account the peculiarities of multipath routing, in particular simulation models.

2   Calculations and Discussion

The efficiency of redundant distribution of requests (data packets) through the network to the destination servers is evaluated depending on the redundancy order and according to the multiplicative criteria $Mult_1$ and $Mult_2$ and the additive criterion Add.

Suppose that there are ten disjoint routes between the sender of requests and the receivers (servers), $n = 10$. Each route contains $d = 5$ intermediate nodes (switches or routers) and $d + 1$ link segments between those nodes.

For simplicity, we assume that the average time of transmission v_{ij} is the same for all segments and is equal to v. Let us define values of the following parameters of the model, to perform calculations:

– the average length of data packets $N = 2048\,(bit)$,
– the speed of transmission $L_{ij} = 100\,(Mbit/s)$,
– and, as a result, the average time of transmission $v = 2.048 \times 10^{-5}$ (s).

Considering the *ideal* conditions, when there are no queues in the switching nodes of any route, we can define the minimal time that a request resides in the system:

$$T_{\min} = dv. \tag{14}$$

Thus, $T_{\min} = 1.024 \times 10^{-4}$. With respect to a series of experimental estimations, the ultimate residence time can be initialized as $t_0 = 7.6 \times 10^{-4}$.

Suppose that the probability of server availability is reasonably high, taking into account the fact that the considered servers might fail or be overloaded with incoming requests from time to time: $P_0 = 0.9$.

The intensity of the flow of requests Λ varies over time. Let us set the following values for Λ (1/s):

$$\Lambda_1 = 9 \times 10^3, \ \Lambda_2 = 1 \times 10^4, \ \Lambda_3 = 2 \times 10^4, \ \Lambda_4 = 3 \times 10^4.$$

Further estimations demonstrate that for the given values of other parameters, the plausible values of the bit error rate B are in the range 10^{-7} to 10^{-5}. The results of calculations, shown in Fig. 2, are derived from the bit error rate $B = 10^{-7}$.

Figure 2 illustrates the efficiency of redundant distribution of requests over k routes, which is computed on the basis of the multiplicative criterion $Mult_1(a)$, the multiplicative criterion $Mult_2(b)$, and the additive criterion $Add(c)$. The displayed curves 1–4 in each of the three graphs correspond to the intensities of the request flow Λ_1–Λ_4.

It can be seen that sending data packets through the network in the redundant way increases the efficiency of transmission but only up to a certain point, which is the maximum value of each criterion in question. This appears to be an evidence that there is a need to solve the problem of optimization, in order to determine the redundancy order that provides the maximum efficiency across all proposed criteria. It is particularly important for systems with the limited residence time.

It is important to note that the higher the intensity of the request flow Λ is, the lesser the values of all criteria under consideration are – $Mult_1$, $Mult_2$, and Add. With the increase of Λ, the maximum values of those criteria correspond to the lesser values of the order of redundancy k, as shown in Fig. 2 and in Table 1.

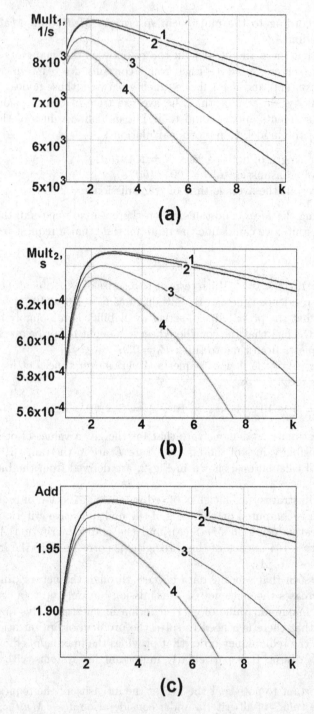

Fig. 2. The efficiency of redundant data transmission estimated on the basis of the multiplicative criteria $Mult_1(a)$ and $Mult_2(b)$, the additive criterion $Add(c)$.

Table 1. The redundancy order depending on the intensities of the flow of requests $\Lambda_1 = 9 \times 10^3$, $\Lambda_2 = 1 \times 10^4$, $\Lambda_3 = 2 \times 10^4$, $\Lambda_4 = 3 \times 10^4$ (1/s) and the bit error rate $B = 10^{-7}$ (calculations are performed using Mathcad 15 environment)

	$Mult_1$	$Mult_2$	Add
Λ_1	2.09	2.86	2.86
Λ_2	2.05	2.80	2.80
Λ_3	1.73	2.45	2.46
Λ_4	1.54	2.23	2.24

In case the redundancy order k is a whole number or its fractional part is insignificant, the value of k indicates how many paths (and hence copies of a request) are needed to transmit the request through the network in the optimal manner, i.e. as soon as possible and in the most reliable way (see the multiplicative criterion $Mult_1$ for Λ_1 and Λ_2 in Table 1). Otherwise, one should use some sort of a mixed strategy when the decision on the redundancy order is made on the probabilistic basis. Thus, it is advisable to use either two or three routes when transmitting requests through the given network — in case of the multiplicative criterion $Mult_2$ and the additive criterion Add (Table 1). For instance, if the fractional part of k is approximately equal to 0.5 ($Mult_1$ – Λ_4, $Mult_2$ – Λ_3, Add – Λ_3), it is possible to decide for two or three paths with the probability of 0.5.

In practice, however, it makes sense to employ generalized assessments, derived from all criteria. For example, it is advisable to send requests over two or three routes for the intensities of the flow of requests Λ_3 and Λ_4, even though the multiplicative criterion $Mult_1$, taken separately, denotes that a single path should be chosen for this purpose.

3   Conclusion

The results of the conducted study support the initial hypothesis that introducing redundancy in the process of data transmission in delay-sensitive systems, under conditions of possible transmission errors and server unavailability for processing incoming requests, is likely to enhance those systems' efficiency.

The efficiency is understood in terms of the increased probability of errorless delivery of requests (data packets) and the decreased average residence time. Delivery failures happen because of faults of network equipment and server unavailability due to faults and temporary shutdown and due to the fact that the server can be overloaded with the already received requests.

The scope of efficiency of redundant distribution of requests through the network to the destination servers was determined.

On the basis of the proposed additive and multiplicative criteria, it was found that there existed the optimal redundancy order, or the optimal number of routes needed to transmit (copies of) requests through the network in the redundant manner.

It was discovered that the redundancy order depended on the intensity of the flow of incoming requests as well as on the probability of transmission errors, or the bit error rate.

The results of the research may appear useful for those who need to decide for the redundancy order when sending data packets through the network, with the purpose to improve reliability of data transmission and accelerate distribution of requests to the destination servers.

Acknowledgments. The study is a part of the research project "Methods for designing the key systems of information infrastructure" (State Registration Number is 615869) of ITMO University, St. Petersburg, Russia.

References

1. Bogatyrev, V.A.: Protocols for dynamic distribution of requests through a bus with variable logic ring for reception authority transfer. Autom. Control Comput. Sci. **33**(1), 57–63 (1999)
2. Bogatyrev, V.A.: On interconnection control in redundancy of local network buses with limited availability. Eng. Simul. **16**(4), 463–469 (1999)
3. Bogatyrev, V.A.: An interval signal method of dynamic interrupt handling with load balancing. Autom. Control Comput. Sci. **34**(6), 51–57 (2000)
4. Bogatyrev, V.A.: Fault tolerance of clusters configurations with direct connection of storage devices. Autom. Control Comput. Sci. **45**(6), 330–337 (2011)
5. Bogatyrev, V.A.: Exchange of duplicated computing complexes in fault tolerant systems. Autom. Control Comput. Sci. **45**(5), 268–276 (2011). doi:10.3103/S014641161105004X. http://link.springer.com/article/10.3103/S01464116110 5004X
6. Bogatyrev, V.A., Bogatyrev, S.V., Golubev, I.Y.: Optimization and the process of task distribution between computer system clusters. Autom. Control Comput. Sci. **46**(3), 103–111 (2012). doi:10.3103/S0146411612030029. http://link.springer.com/article/10.3103/S0146411612030029
7. Bogatyrev, V.A., Bogatyrev, A.V.: Functional reliability of a real-time redundant computational process in cluster architecture systems. Autom. Control Comput. Sci. **49**(1), 46–56 (2015)
8. Arustamov, S.A., Bogatyrev, V.A., Polyakov, V.I.: Back up data transmission in real-time duplicated computer systems. In: Abraham, A., Kovalev, S., Tarassov, V., Snášel, V. (eds.) IITI 2016. AISC, vol. 451, pp. 103–109. Springer, Heidelberg (2016). doi:10.1007/978-3-319-33816-3_11
9. Bogatyrev, V.A., Parshutina, S.A.: Redundant distribution of requests through the network by transferring them over multiple paths. In: Vishnevsky, V., Kozyrev, D. (eds.) DCCN 2015. CCIS, vol. 601, pp. 199–207. Springer, Heidelberg (2016). doi:10.1007/978-3-319-30843-2_21
10. Aliev, T.: The synthesis of service discipline in systems with limits. In: Vishnevsky, V., Kozyrev, D. (eds.) DCCN 2015. CCIS, vol. 601, pp. 151–156. Springer, Heidelberg (2016). doi:10.1007/978-3-319-30843-2_16
11. Aliev, T.I., Rebezova, M.I., Russ, A.A.: Statistical methods for monitoring travel agencies. Autom. Control Comput. Sci. **49**(6), 321–327 (2015)

12. Bogatyrev, V.A., Bogatyrev, S.V.: Redundant data transmission using aggregated channels in real-time network. J. Instrum. Eng. **59**(9), 735–740 (2016). doi:10.17586/0021-3454-2016-59-9-735-740

13. Bogatyrev, V.A., Bogatyrev, A.V.: Optimization of redundant routing requests in a clustered real-time systems. Inf. Technol. **21**(7), 495–502 (2015)

14. Bogatyrev, V.A., Bogatyrev, A.V.: The model of redundant service requests real-time in a computer cluster. Inf. Technol. **22**(5), 348–355 (2016)

15. Bogatyrev, V.A.: Increasing the fault tolerance of a multi-trunk channel by means of inter-trunk packet forwarding. Autom. Control Comput. Sci. **33**(2), 70–76 (1999)

The Fault-Tolerant Structure of Multilevel Secure Access to the Resources of the Public Network

Vladimir Kolomoitcev$^{(\boxtimes)}$ and V.A. Bogatyrev$^{(\boxtimes)}$

Department of Computation Technologies, ITMO University,
49 Kronverksky Pr., 197101 St. Petersburg, Russia
dekskornis@gmail.com, vladimir.bogatyrev@gmail.com
http://www.ifmo.ru

Abstract. The paper presents the evaluation of the effectiveness of the structural organization of the system of multi-level secure access to external network resources. We conducted a comparative analysis and optimization of the pattern of access 'Direct connection', with its various forms of implementation during the organization of a secure connection of end-node internal network to the resources located in the external network. The study was conducted on the basis that each security element is included in the pattern of the secure access is able to detect and eliminate the threats of the other elements of the system of protection. Pattern of access 'Direct connection' in a general form has four variants of construction, differing from each other by mutual arrangement of the key elements: firewall with packet-filtering, firewall with adaptive detailed packet inspection and the router. It was a mathematical model to calculate the reliability of the ways of construction of the pattern of access. It is shown that the most reliable way of construction of pattern of access is one that includes a single group of routers for the entire system. Ways are not very different from each other reliability value that include two groups of routers on the overall system.

Keywords: Firewalls · Corporate networks · Information security · Fault tolerance · Access pattern · Reliability · Networking

1 Introduction

Modern computer networks, both corporate and public, have sophisticated structures. In such networks, there are some very serious challenges of information security. They may be at risk of unauthorized access, denial-of-service (DoS) nodes, the loss of transmitted information, as well as threats of violations of privacy that could lead to significant economic and other losses [1].

Threats can be both external - as a result of remote network attacks, and internal - by various stowing software or hardware. To eliminate the challenges of information security some actions can be taken and means of information security located on various levels of the network can be used.

© Springer International Publishing AG 2016
V.M. Vishnevskiy et al. (Eds.): DCCN 2016, CCIS 678, pp. 302–313, 2016.
DOI: 10.1007/978-3-319-51917-3_27

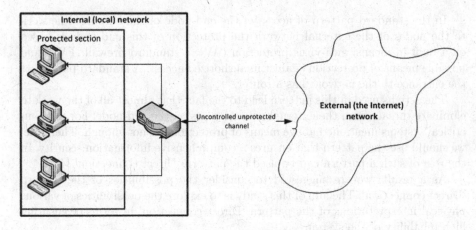

Fig. 1. The standard pattern of access node to the external network

The principles of organization of a secure connection of the corporate network to public network are among the most important elements for ensuring information security. They have a significant impact on the safety and reliability of the network. However, it is worth remembering that the most effective security techniques usually imply some significant costs.

Currently, information security means that are often used are: firewall, protection against unauthorized access, anti-virus, encryption solutions of data stored on disk and transmitted to the information transmission channel, any organizational and technical actions to protect information and other means.

In this study, we investigate possibilities of the pattern for the organization secure access to external network resources, taking into account the requirements set out in the guidance documents (legislative and legal documents) on information security. The study is aimed at a choice of rational options for creating protection system, with ensuring high levels of reliability [2,3].

Mechanisms of distribution requests in clusters are known. They are beyond the scope of this paper, so we will ignore the costs of dispatching requests.

2 Object and Objectives of the Study

The pattern which is regarded in this paper is focused on improving the level of protection devices of the network. The key challenge of the pattern is to organize secure access to poorly protected and/or uncontrolled portions of the network. This pattern allows reducing the threat of DDoS-attacks, unauthorized access to a network node, listening to the information channel and penetration of malicious software [4].

The pattern under consideration is based on a standard network access pattern to the resources of the external network: the node 'Internal (local) network' - Routers - 'External network (the Internet)' (Fig. 1). This approach minimizes the degree of possible reorganization of existing corporate network [5–7].

In the standard pattern of access of the end-node of the corporate network to the nodes of the external network the protection of this end-node is based on a built-in means: anti-virus protection (AV), a standard firewall (FW), and possible means of protection against unauthorized access. A standard pattern at the entrance to the network has a router.

The actions used in this pattern lead to the fact that almost all of the work to eliminate threats from the external network rests on the end-node. For mission-critical systems mentioned above means of protection are not enough. Therefore, we should use the pattern that ensures a comprehensive information security. In the role of such a pattern can be used the pattern 'Direct connection' [4].

As a result, we are suggested to consider the possibilities of the pattern 'Direct connection'. The aim of this study is to explore the possibilities of various physical interpretations of the pattern 'Direct connection' in terms of ensuring high reliability of the system.

3   The Basic Version Pattern of Access 'Direct Connection'

Using the pattern 'Direct connection' involves minimal changes in the architecture of the corporate network, as well as minimal additional financial costs to implement it. The structure of the pattern 'Direct connection' is presented in Fig. 2.

In this pattern, at the entrance to the internal network (just after the router) is set firewall with packet-filtering (FW-1). Firewall with packet-filtering is needed to filter the packets incoming to corporate network based on the addresses of sender and receiver of packets, ports numbers and static rules created by the administrator. This type of firewalls allows easily separate the resources of the network and block the access from outside to the important areas of the corporate network. Among the positive advantages of firewalls with filtering-packets is their smaller influence on performance of network than of other firewalls. However this type of firewalls provides the worst degree of the network security as it

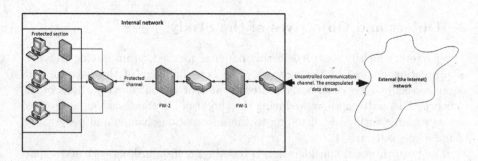

Fig. 2. The pattern 'Direct connection'

is not able to analyse payload and parameters of incoming packets. FW-1 eliminates spam, reduces the total load of the channel within the network, as well as reduces the risk of DDoS-attacks.

Firewall with adaptive detailed packet inspection (FW-2) is placed next to the FW-1 for a deeper analysis of the contents of packages. Firewall with adaptive detailed packet inspection, in addition to advantages of firewall with filtering-packets, is able to recognize sessions between applications, block the packets that break the rules of TCP/IP, counteract scan of resources or break out/slow down the connection and prevent injection of data. In some case, it can also check incoming data transmitted to corporate network for malicious information. The disadvantages of using this type of firewall are high costs, difficulty of installing the rules of work, supporting right parameters and (depending on 'analysis depth' of packets) decrease of network performances.

Often available as part of pattern routers can carry functional FW-1 [8]. However, the computing system architecture is not known in advance (which will be implemented by the pattern of access). The router and the firewall will take account of what how the different elements of the system, where the router (as an element) is intended primarily for communication with parts of the system together, and the FW-1 - filter traffic in a computer system. Otherwise, routers (for communicating between a firewall in this pattern) may be combined into a single cluster and replaced with a cluster of firewall with packet-filtering.

Once the data have passed the FW-2, they are (potentially 'clean-data') must be received to the desired end-node. On end-node there are local antivirus (AV) (with personal firewall) installed [8], as well as some systems of protection against unauthorized access (UAA), and some secure data storage in order to reduce the negative effects of potential insider attacks. In this pattern of access, the channel data is to be protected, thereby reducing or even prevents the possibility of influence an intruder on data flowing in the channel.

To improve the overall network protection from DDoS-attacks, data loss or destruction and other threats, mission critical nodes should be reserved, and for the data stored on them, backups are created.

4   Ways of Construction the Network Infrastructure Pattern of Access 'Direct Connection'

For qualitative and uninterrupted operation of the network, you must do a backup of system components. Network architecture of the pattern 'Direct Connection' has three main components: firewall with packet-filtering, firewall with adaptive detailed packet inspection and routers that connect all the elements of pattern together. There are four possible ways for the construction of this pattern (depending on the network architecture). They are:

- Way 1 - three groups of routers for connection all devices of system and nodes from eternal network;
- Way 2 - two groups of routers. The first group connects both groups of firewalls and target end-nodes. The second group is needed for connection FW-1 with resources from eternal network;

– Way 3 - two groups of routers. The first group connects both groups of firewalls and the nodes from eternal network. The other group connects FW-2 and end-nodes;
– Way 4 - one group of routers to connect eternal network, firewalls and end-nodes together.

Possible ways of constructing a network infrastructure pattern 'Direct Connection' presented in Fig. 3.

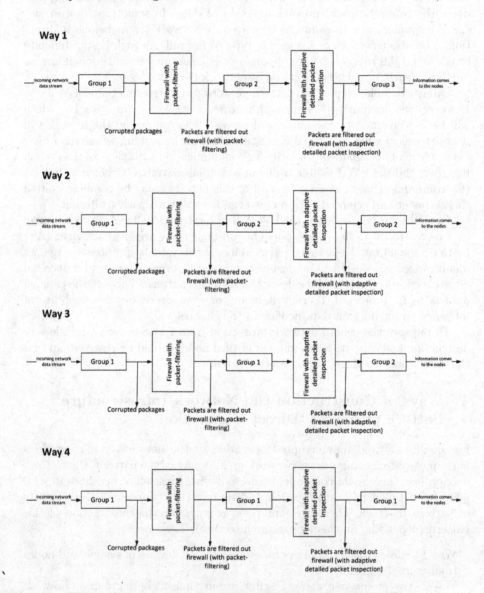

Fig. 3. Variations network architecture pattern 'Direct connection'

The ways of connection protection means into fault tolerant clusters in the pattern of access 'Direct connection' (shown in other works) suggest that each of means can eliminate only certain field of threats [8]. This study shows systems that use the pattern of access 'Direct connection' with common field of threats elimination for different protection means (each mean of the system can eliminate a part of threats that can be eliminated by another mean of the used pattern).

5   Estimation of Reliability of the System

Estimation of possible ways pattern 'Direct connection' requires a search for multiplicity of redundant nodes in each group. Required reach the highest possible level of reliability of the system, given the limitations imposed on the cost of implementation [9–12].

As today the challenge of memory costs is not critical any more (the same as its overflow), every mean (router, firewall and others) has its own memory of different volume inside. As the result, each node of the network can be examined as a 'queue network' of $M/M/1$ type with unlimited queue.

In assessing the reliability assume that failures of nodes are independent, and the flow of failures is exponentially distributed (as is customary in the calculation of reliability). The impact of malicious actions to reduce the reliability and security of the system is not considered in this study.

Reliability of the system consisting of several nodes is defined as the product of the reliability of each of the elements (groups) of the system [13]. Reliability of the proposed patterns is equal:

$$P_1 = P_{01} \cdot P_{m1} \cdot P_{02} \cdot P_{m2} \cdot P_{03};$$
$$P_{2-3} = P_{01} \cdot P_{m1} \cdot P_{02} \cdot P_{m2}; \qquad (1)$$
$$P_4 = P_{01} \cdot P_{m1} \cdot P_{m2}.$$

Where $P_{m1} = (1 - (1 - r_1)^{n_1})$, $P_{m2} = (1 - (1 - r_2)^{n_2})$. Assuming that the routers in each group are the same: $P_{0i} = (1 - (1 - r_0)^{n_{0i}})$. Here $r_j = e^{-\lambda_j t}$ and $\lambda_0, \lambda_1, \lambda_2$ - failure rate of routers, FW-1 and FW-2; n_{0i} - the number of routers in the i-th group; n_1 - the number of FW-1; n_2 - the number of FW-2.

Costs for the implementation of the ways of construction of the scheme are shows in (2) and defined as:

$$C_{1-4} = c_0 \cdot \sum_i n_{0i} + c_1 \cdot n_1 + c_2 \cdot n_2. \qquad (2)$$

Here c_0, c_1, c_2 - the cost of router, FW-1, FW-2.

Estimation of protection systems includes finding the distribution of each type of node that provides maximum reliability of the entire system considering the limitation of the cost of implementation: $C_1 \leq C$, $C_2 \leq C,...,C_4 \leq C$; and compliance steady state conditions of service [14–18].

After passing through the router the input flow is filtered and, thus, the density of the flow on the FW-1 is lower than the router. The same happens with the input flow received at the FW-2. After passing through the FW-1 a certain proportion of the input flow is filtered and to the FW-2 is received smaller input flow. As a result of limitations imposed on the capacity of each element of the system will be equal to:

For Way 1:

$$\begin{cases} L_1 < 1; \\ L_2 < 1; \\ d_2 \cdot R/n_{02} < 1; \\ L_4 < 1; \\ d_4 \cdot R/n_{03} < 1; \\ d_5 \cdot R/n_{03} < 1. \end{cases}$$

For Way 2:

$$\begin{cases} L_1 < 1; \\ L_2 < 1; \\ d_2 \cdot R/n_{02} < 1; \\ L_4 < 1; \\ d_4 \cdot R/n_{02} < 1; \\ d_5 \cdot R/n_{02} < 1. \end{cases}$$

For Way 3:

$$\begin{cases} L_1 < 1; \\ L_2 < 1; \\ d_2 \cdot R/n_{01} < 1; \\ L_4 < 1; \\ d_4 \cdot R/n_{02} < 1; \\ d_5 \cdot R/n_{02} < 1. \end{cases}$$

For Way 4:

$$\begin{cases} L_1 < 1; \\ L_2 < 1; \\ d_2 \cdot R/n_{01} < 1; \\ L_4 < 1; \\ d_4 \cdot R/n_{01} < 1; \\ d_5 \cdot R/n_{01} < 1. \end{cases}$$

Here $L_1 = R/n_{01}$; $L_2 = d_1 \cdot \lambda \cdot V_1/n_1$; $L_4 = d_3 \cdot \lambda \cdot V_2/n_2$; $R = \lambda \cdot V_0$ where d_i - the proportion of the filtered input flow of previously placed node; V_0, V_1, V_2 - average service time of request in routers, FW-1 and FW-2; λ - the arrival rate of requests; n_{0i} - the number of routers in the i-th group; n_i - the number of FW-1; n_2 - the number of FW-2.

The proportion of the filtered input flow of previously placed node can be obtained by the Eqs. (3)–(7).

After first router:
$$d_1 = (1 - A_0 \cdot p_0). \tag{3}$$

After first router and FW-1:
$$d_2 = 1 - (p_1 \cdot (p_1 \cdot (A_1 - l_{10}) + p_0 \cdot (A_0 - l_{10}) + l_{10} \cdot (1 - \bar{p}_0 \cdot p_1)). \tag{4}$$

After first router, FW-1 and second router:
$$d_3 = 1 - (p_1 \cdot (p_1 \cdot (A_1 - l_{10}) + (1 - \bar{p}_0^2) \cdot (A_0 - l_{10}) + l_{10} \cdot (1 - \bar{p}_0^2 \cdot p_1)). \tag{5}$$

After first router, FW-1, second router and FW-2:
$$d_4 = 1 - (p_1 \cdot M_{e1} + (1 - \bar{p}_0^2) \cdot (R_{emp}) + (p_2 \cdot M_{e2}) + (l_{10} - l_{00}) \cdot$$
$$\cdot (1 - \bar{p}_0 \cdot \bar{p}_1 \cdot \bar{p}_2)) + (l_{20} - l_{00}) \cdot (1 - \bar{p}_0^2 \cdot \bar{p}_2) + (l_{21} - l_{00}) \cdot \tag{6}$$
$$\cdot (1 - \bar{p}_1 \cdot \bar{p}_2) + l_{00} \cdot (1 - \bar{p}_0^2 \cdot \bar{p}_1 \cdot \bar{p}_2)).$$

After first router, FW-1, second router, FW-2 and third router:
$$d_5 = 1 - (p_1 \cdot M_{e1} + (1 - \bar{p}_0^3) \cdot (R_{emp}) + (p_2 \cdot M_{e2}) + (l_{10} - l_{00}) \cdot$$
$$\cdot (1 - \bar{p}_0 \cdot \bar{p}_1 \cdot \bar{p}_2)) + (l_{20} - l_{00}) \cdot (1 - \bar{p}_0^3 \cdot \bar{p}_2) + (l_{21} - l_{00}) \cdot \tag{7}$$
$$\cdot (1 - \bar{p}_1 \cdot \bar{p}_2) + l_{00} \cdot (1 - \bar{p}_0^3 \cdot \bar{p}_1 \cdot \bar{p}_2)).$$

where $R_{emp} = (A_0 - l_{20} - l_{10} + l_{00})$; $M_{e1} = (A_1 - l_{21} - l_{10} + l_{00})$; $M_{e2} = (A_2 - l_{21} - l_{20} + l_{00})$. At the same time A_0, A_1, A_2 - respectively, the proportion of threats (errors) in the input stream [20], the router detected with a probability p_0; FW-1 with a probability p_1; FW-2 with a probability p_2.

Results of reliability calculation, depending on the constraints imposed on the system throughput determined by a known the arrival rate - λ are shown in Fig. 4, when:

- $r_0 = 0.85$, $r_1 = 0.9$;
- $V_0 = 0.025$ s, $V_1 = 0.04$ s, $V_2 = 0.075$ s;
- $c_0 = 10$ cu, $c_1 = 20$ cu, $c_2 = 35$ cu, $C = 500$ cu;
- $p_0 = 0.85$, $p_1 = p_2 = 0.899$;
- $A_0 = 0.07$, $A_1 = 0.15$, $A_2 = 0.26$;
- $l_{00} = 0.04$, $l_{10} = 0.04$, $l_{20} = 0.06$, $l_{21} = 0.12$.

As shown in Fig. 4, for small values of the arrival rate, the reliability of each of the ways of the pattern 'Direct Connection' is approximately equal. However, with increasing the arrival rate is detected, the fourth way of pattern is more reliable than other ways, and the first - the least reliable. Also from Fig. 4 shows that if you want to choose one of two ways contain two groups of routers, it is a bit more reliable to take the 'Way 2' than 'Way 3'.

As a result, we can conclude that if you want to use the most reliable way of the pattern, it is best to choose the 'Way 4'. In the case where raises the question of maximizing the system reliability and to select one of two options - the 'Way 2' or 'Way 3' then there is no much difference which of them to use.

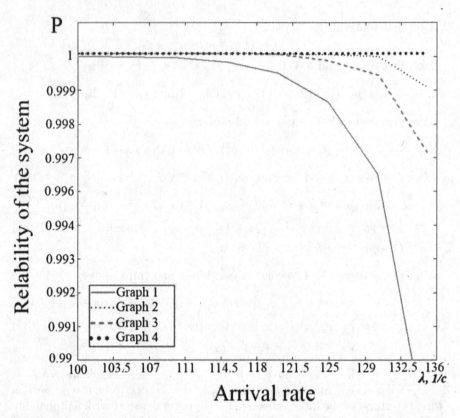

Fig. 4. Reliability pattern of access at a fixed time of operation of the computer system based on the arrival rate of requests. Sequence numbers of curves in the graphs have a direct accordance with sequence numbers of their defining formulas, namely the graphs 1-4 - ways of constructing pattern 'Direct connection'

6   Conclusion

The paper analyses possibilities of the pattern of access 'Direct Connection' that allow you to organize a secure connection between the end-node internal network and resources located in the external network. The study identified the advantages and disadvantages of the pattern 'Direct connection', depending on the way of its construction.

It has been shown that one of the ways of construction the pattern of access 'Direct connection' (using a single group of routers to the entire pattern of access), has a higher level of reliability than the others. All things being equal, we have the opportunity with a lower cost to organize the connection of external and internal networks together with the required degree of reliability of the system, when we use the way of constructing pattern of access [21,22]. On the other hand, the 'Way 4' allows an attacker to ignore the used hardware security system, if the attack on the computer system has been well-prepared. In other words, since we use in this way of pattern of access only one group of routers connected to both FW-1, FW-2, and with the end-nodes of system, the attacker can send a specially-configured packages that will ignore the FW-1 and FW-2, but immediately go to the end-nodes of the system (which means only a part of the node itself fighting with the threat), which makes its (hardware firewalls) use questionable, as a part of this pattern of access.

We also show that the second and the third variants of pattern are almost identical to each other in terms of reliability. In addition, it is worth noting that the use of two groups of routers allows partially solve the previously mentioned challenges with ignoring the attacker means of information security existing in the system. In this case, in one way of constructing pattern of access (Way 2), we are able to split the access to our network into two areas - "pre-analysis" (using the firewall with packet-filtering) and "detailed analysis" (which includes the firewall with adaptive detailed packet inspection and / or information security means available on the end-node). Thereby, it enables at least partially analysis the input data flow for threats by hardware.

For another way of constructing of the pattern of access (Way 3), access to our network will also be divided into two zones - "hardware firewalls" and "end-nodes". In such a case, an attacker would first have access to the used set of firewalls, and then, at least after passing the strongest of them (because they are located in order to increase their features) - to the end-nodes of the system.

The way of pattern that includes three groups of routers (Way 1) also allows to split access network area (but with mandatory hit input flow to each of the security mean used in the pattern). However, it significantly reduces the reliability of the computer system.

Thus, when implementing the pattern of access 'Direct connection' in question of reliability, it is best to use a way of its construction using a common pool of routers for the entire system. At the same time, the use of two groups of routers increases the degree of information security of the corporate network access to an external network resource by reducing the reliability of the computer system.

In this study we published works on finding the fastest (minimum average residence time of the request in the system) pattern configuration [4,8,19].

Acknowledgments. The work is partially supported by Government of St. Petersburg grant.

References

1. Aliev, T.I., Rebezova, M.I., Russ, A.A.: Statistical methods for monitoring travel agencies. Autom. Control Comput. Sci. **49**(6), 321–327 (2015)
2. Bogatyrev, V.A., Bogatyrev, S.V., Golubev, I.Y.: Optimization and the process of task distribution between computer system clusters. Autom. Control Comput. Sci. **46**(3), 103–111 (2012)

3. Arustamov, S.A., Bogatyrev, V.A., Polyakov, V.I.: Back Up Data Transmission in Real-Time Duplicated Computer Systems. In: Abraham, A., Kovalev, S., Tarassov, V., Snášel, V. (eds.) IITI 2016. AISC, vol. 451, pp. 103–109. Springer, Heidelberg (2016). doi:10.1007/978-3-319-33816-3_11

4. Kolomoitcev, V.S.: A comparative analysis of approaches to organizing of secure connection of the corporate network nodes to the public network. Cybern. Program. (2), 46–58 (2015). http://en.e-notabene.ru/kp/article_14349.html

5. Whitmore, J.J.: A method for designing secure solutions. IBM Syst. J. **40**(3), 747–768 (2001)

6. Peisert, S., Talbot, E., Bishop, M.: Turtles all the way down: a clean-slate, ground-up, first-principles approach to secure systems. In: Proceedings of 2012 New Security Paradigms Workshop (NSPW 2012), Bertinoro, Italy, pp. 15–26 (2012)

7. Ellison, R.J., Fisher, D.A., Linger, R.C., Lipson, H.F., Longstaff, T.A., Mead, N.R.: Survivability: protecting your critical systems. IEEE Internet Comput. **3**(6), 55–63 (1999)

8. Kolomoitcev, V.S.: Choice of option for implementation of the multilevel secure access to the external network. Sci. Tech. J. Inf. Technol. Mech. Opt. **16**(1), 115–121 (2016)

9. Bogatyrev, V.A., Bogatyrev, A.V.: Functional reliability of a real-time redundant computational process in cluster architecture systems. Autom. Control Comput. Sci. **49**(1), 46–56 (2015)

10. Bogatyrev, V.A.: Exchange of duplicated computing complexes in fault tolerant systems. Autom. Control Comput. Sci. **45**(5), 268–276 (2011)

11. Bogatyrev, V.A.: Fault tolerance of clusters configurations with direct connection of storage devices. Autom. Control Comput. Sci. **45**(6), 330–337 (2011)

12. Bogatyrev, V.A., Bogatyrev, A.V.: The reliability of the cluster real-time systems with fragmentation and redundant service requests. Inf. Technol. **22**(6), 409–416 (2016)

13. Bogatyrev, V.A., Slastikhin, I.A.: Efficiency of redundant query execution in multi-channel service system. Sci. Tech. J. Inf. Technol. Mech. Opt. **16**(2), 311–317 (2016)

14. Bogatyrev, V.A., Parshutina, S.A.: Redundant distribution of requests through the network by transferring them over multiple paths. In: Vishnevsky, V., Kozyrev, D. (eds.) DCCN 2015. CCIS, vol. 601, pp. 199–207. Springer, Heidelberg (2016). doi:10.1007/978-3-319-30843-2_21

15. Bogatyrev, V.A.: An interval signal method of dynamic interrupt handling with load balancing. Autom. Control Comput. Sci. **34**(6), 51–57 (2000)

16. Bogatyrev, V.A.: Protocols for dynamic distribution of requests through a bus with variablelogic ring for reception authority transfer. Autom. Control Comput. Sci. **33**(1), 57–63 (1999)

17. Bogatyrev, V.A.: On interconnection control in redundancy of local network buses with limited availability. Eng. Simul. **16**(4), 463–469 (1999)

18. Aliev, T.: The synthesis of service discipline in systems with limits. In: Vishnevsky, V., Kozyrev, D. (eds.) DCCN 2015. CCIS, vol. 601, pp. 151–156. Springer, Heidelberg (2016). doi:10.1007/978-3-319-30843-2_16

19. Kolomoitcev, V.S., Bogatyrev, V.A.: Selecting multilevel structure secure access to resources external network. In: Conference of Distributed Computer and Communication Networks: Control, Computation, Communications (DCCN-2015), pp. 525–532 (2015)

20. Kolomoitcev, V.S., Bodrov, K.U., Krasilnikov, A.V.: Calculating the probability of detection and removal of threats to information security in data channels. In: 2016 XIX IEEE International Conference on Soft Computing and Measurements (SCM), St. Petersburg, Russia, pp. 25–27 (2016)
21. Ellison, R.J., Fisher, D.A., Linger, R.C., Lipson, H.F., Longstaff, T.A., Mead, N.R.: Survivable network systems: an emerging discipline. http://www.cert.org/research/97tr013.pdf
22. Kenneth, I., Stephanie, F.: A history and survey of network firewalls. University of New Mexico, p. 42 (2002)

Formation of the Instantaneous Information Security Audit Concept

I.I. Livshitz[2(✉)], D.V. Yurkin[1], and A.A. Minyaev[2]

[1] Department of Secure Communication Systems,
Federal State Budget-Financed Educational Institution of Higher Education,
The Bonch-Bruevich Saint - Petersburg State University of Telecommunications,
Prospekt Bolshevikov 22, St. Petersburg 193232, Russia
dvyurkin@ya.ru
[2] Research Department of Information Security Issues,
Federal State Budgetary Institution of Science of St. Petersburg
Institute for Informatics and Automation of the Russian Academy of Sciences,
14 Line 39, St. Petersburg 199178, Russia
livshitz.il@yandex.ru, minyaev.a@gmail.com

Abstract. This publication covers the problem of formation the concept of the instantaneous information security (IT-Security) audits, including protection against zero-day threats. Various recent materials are presented to the actual problem of counter zero-day threats notes that "any process-driven people, is unreliable. In this situation it is proposed to use not only a technical methods to counter zero-day threats, but to offer a combined method based on the concept of instantaneous IT-Security audits. Methodological basis of this concept for instantaneous audits defined both ISO 27001 and ISO 19011 standards, which extended with the set of IT-security metrics for quantify the object protection level. In the example for one variable was demonstrated an increase in the rate of growth of the ISMS level variables with known IT-Security audits process.

Keywords: Audit · Information security · Integrated management system · Information security management system · Risk management · Function · Standard

1 Introduction

Recently, the application of the Integrated Management Systems (IMS) attracts more top management attention. Nowadays there is an important problem of running the audits in IMS and particularly, realization of complex checks of different ISO standards in full scale with the essential reducing of available resources. In a greater degree this problem is illustrative of supporting IT-Security audit program, as far as negative consequences can lead to essential damage. The realization of IT-Security management systems gets more application in practice. Moving to analysis based on risks provides the increasing of

© Springer International Publishing AG 2016
V.M. Vishnevskiy et al. (Eds.): DCCN 2016, CCIS 678, pp. 314–324, 2016.
DOI: 10.1007/978-3-319-51917-3_28

interest to rational exploitation of modern risk-oriented ISO standards. Studying the problem with realization of IMS audits makes the essential interest also the search of ways of IMS audit program optimization that are based on principles of continuous adaptation in the process of incoming data during one micro cycle of IT-Security audit. It is supposed that new method of audit program optimization will let us to provide more rational acceptance of the IT-Security control solution.

2   Problem Description

The technical aspect is rather fulfilled now – a complex of means (measures) aimed at providing IT-Security (in the notation of ISO - asset), it is accepted to unite in the uniform IT-security management system (ISMS) created within all organization, subordinated to the top management and periodically estimated on certain metrics. The modern science offers various approaches for the solution of this problem; the direction of application risk-focused approach based on system of the international standards [1–3] is represented to the most perspective. To provide stable development of organizations in the context of risks of different origin, it is appear to be reasonable to apply risk-oriented standard and implement the IMS. From the point of view for an IMS audits in supposed method we should notice the necessity of solution of next important practical tasks [4,5]:

1. The task of resources allocation for audit program;
2. The task of account of factors that influence on the depth of audit-leak program, incidents, the appearance of criminal actions, revealed earlier mismatches and in this way the volume definition of audit program;
3. The task of collection of verifiable information;
4. The task to provide the auditors with special knowledge and skills either to invite engineers.

It is necessary to admit that we should be aware of recommendations PAS-99 in IMS [6–9], that allows to take into account the specific requirements of carrying out combined audits, the account of risks, flexible controlling of IMS audit program volume with the account of last results and the importance of processes [4,5,10,11,13].

3   The Base Model for the Integrated Management System Audits

The IMS model containing all the basic nature for performing audits (criteria, evidence, object surveillance audit). It allows to create assess the IT-security level [13]. The audit process provides an important component of the overall (integrated) assessing the IMS effectiveness. It allows to perform the decomposition of "general" objectives in the IMS specific objectives, such as assessing the IT-security level. The basic model for ISM audits is shown on Fig. 1.

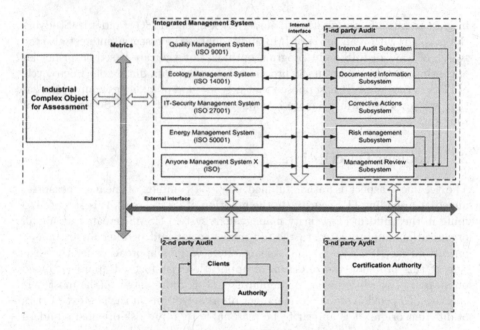

Fig. 1. Basic model of IMS audits

General explanations of the basic model of IMS audits as follows:

- Audit ISM involves the use of a single set of metrics on the functioning of different interfaces for internal audits and external audits;
- Internal audits required to take into account the external audits results; converse is also true;
- The impact on the assessment of the object is realized through the management review subsystem (in accordance with the PDCA cycle).

4   Problems of a Risk Management for Complex Industrial Facilities

Management of risk is the processes, which is carried out in the organization for the purpose of identification, identification, management and control of the events potentially capable to influence achievement of the complex industrial facilities (CIF). In the offered approach all main are considered essence: formation of internal and external aspects, a context, system of a risk management and the main types of documentary information – scales of an assessment, criteria of acceptance of risks, the register of risks, the plan of processing of risks and so forth [11–13]. The example of realization of process of management of risk for the phase "Plan" in the cycle "Plan-Do-Check-Act" (PDCA) is shown on Fig. 2. Problems of a risk management for CIF it is convenient to arrange as realization of the cycle PDCA (or Deming cycle):

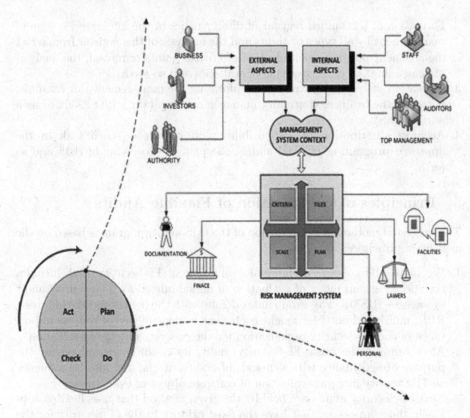

Fig. 2. Process of management of risks for the phase "Plan" in PDCA cycle

1. Plan – formation of regulatory base, development of regulations, passports of risk, definition of scales of an assessment of risks, criteria of acceptance of risks, formation summary the card of risks for the organization.
2. Do – development of actions complex for decrease in probability (alleviation of the consequences) at emergence of risks.
3. Check – control of completeness, timeliness and efficiency of realization of actions complex of a risk management for the organization.
4. Act – the analysis of productivity of actions complex of a risk management at the level of the decision-maker and formation administrative the decision for management system for the organization optimization.

The following important question: "closed circuit" of a cycle PDCA taking into account offered risk-focused approach. It is represented rational for CIF to recommend:

1. Planning and carrying out internal (including technical) audits – taking into account risk-focused standards (for example ISO 27001 or the new ISO 9001 version).

2. Formation of the unified register of discrepancies in the integrated system of management by all types of audits and the analysis of this register from a risk management position (identification of critical points of refusal, the analysis of "cascading" of risks, studying of statistics and so forth).
3. Formation of system of expeditious informing the management (for example, based on the business continuity management standard – ISO 22301 or as a part of IMS).
4. Accurate distribution of responsibility and powers on each task in the approved program of internal audits, the plan of processing of risks and so forth.

5   Principles of Organization of Flexible Audits

The suggested method of optimization of the IMS audit program is based on the next basic principles:

1. We input the concept of integral evaluation of IT-Security that includes the specific group index of evaluation of all submitted for IT-Security audit processes – RISMS. This group index defines with the help of specific indexes – RPR, multiplied on their weight coefficient in dependence of process importance in the IT-Security organization for the concrete object of evaluation.
2. After running the basic IT-Security audit, its condition is valued for the purpose of accordance with demands of audit criteria, and also its influence on IT-Security integral evaluation of concrete object of evaluation.
3. Next IT-Security audits are held by the given method that uses flexible approach: those processes, that have the most priority in the IT-Security for the concrete object of evaluation, and where the essential mismatches of last audit were revealed, are exposed of more detailed check.
4. Frequency and detail, which must be differentiated for different checked processes, comports with IT-Security too. For example, definite groups of processes, that have priority meaning in integral evaluation (for example, it depends on the model of actual threats of IT-Security), are exposed more detailed and often with audits. The processes, that have the lowest priority in the integral evaluation for the concrete object of evaluation, are checked seldom and less detailed.
5. The depth of check and frequency of audits, each time for k-audit in micro cycle PDCA, defines in dependence of oncoming function of integral evaluation for the concrete object of evaluation to some stated objective index – R_{target} for complex evaluation of concrete object of evaluation security.

In addition we should note the importance of implementation of new standard, ISO 55000 [6–8] – as many assets are not ruled in a proper manner. Accordingly, the appliance of demands of one implemented standard (for example, modern ISO 27001) substantially relieves the solution of standard problems of security, that are solved simultaneously, therefore they must be checked simultaneously within the context of combined audits of all MS in organization (for example, ISO 9001, ISO 50001, ISO 27001) [1–8].

6   Mathematical Statement of the Problem

For the evaluation of a degree of providing IT-Security system conformance on the IMS audits to presented requirements of IT-Security we use private and group IT-Security indexes. For the purposes of realizing IMS audits in the aspect of providing IT-Security we suggest to use the index of effectiveness of ISMS – R_{ISMS}, which we can calculate in each cycle of k-audit using the additive formula with the account of α-weight coefficients and index of effectiveness of each concrete process of IT-Security – R_{PR}:

$$R_{ISMS} = \sum_{i=1}^{n} \alpha_i \bullet R_{\mathrm{Pr}\,i} \tag{1}$$

in this case:

$$\sum_{i=1}^{n} \alpha_i = 1$$

In its turn, indexes of effectiveness of each concrete i-process of IT-Security – R_{PR} are calculated by additive formula with the account of β-weight coefficients and indexes of IT-Security metrics for each concrete i-process of IT-Security – K_{KPI}:

$$R_{\mathrm{Pr}\,i} = \sum_{j=1}^{m} \beta_j \bullet K_{PKIj} \tag{2}$$

in this case:

$$\sum_{j=1}^{m} \beta_j = 1$$

The coefficients of relevancy of private indexes of IT-Security, that are used by calculation of IT-Security group indexes, must be equal to 1 that provides ritualization of all indexes in additive formula above (1) and (2). Accordingly, the final index of effectiveness of ISMS – R_{ISMS} must maximize reaching 1:

$$R_{ISMS} = \sum_{i=1}^{n} \alpha_i \bullet R_{\mathrm{Pr}\,i} \rightarrow 1 \tag{3}$$

In the process of IMS audits, the constant measuring of current nonconformance for k-audit R_{ISMS} is measured as discrepancy with the objective (maximal) index:

$$\Delta R = 1 - R_{ISMS} = \sum_{i=1}^{n} [\alpha_i \bullet (1 - R_{\mathrm{Pr}\,i})] \tag{4}$$

Regarding the results of all audits, that are carried out in a strict accordance with IMS audit program, we fill in the following matrix with the account of IT-Security processes – PR, IT-Security audits – k-audits and IT-Security metrics – KPI.

7 Basic Optimization Cycle of IMS Audit Program

In terms of known audit standards (in particular [4,5]), we offer a method of multistage optimization of IMS audit processes for the CIF, which let us to provide the system of coordination, distribution of recourses and system of effective reduction of results of IMS audits till the person who takes decision [13].

This method consists of scientifically grounded and object-oriented immediate functioning of IT-Security subsystem within IMS and it differs from existing methods with cyclic continuous evaluation of effectiveness on the basis of optimal system of IT-Security numeral indexes (metrics). The offered method consists of two connected cycles of optimization of IMS audits program that differs with the existence of:

1. Basic optimization cycle, which characterizes the effective carrying out of IMS audits in terms of evaluation of efficiency for each PRi- IT-Security process, each KPIj – IT-security metric, and also it defines cycles of resources optimization in audits program: of depth ("Scope"), size of auditor's sample, number of involved auditors (engineers) and etc.
2. Fast block of evaluation of efficiency of correction measures and corrective actions in current k-audit, that touches the changes each of next process of IT-Security and next k+1 audit program. It is also provided fast transfer to evaluation of efficiency indexes of IMS – R_{ISMS} in k-audit and k+1 audit for the constant and effective optimization of all IMS audit program.

Let's consider the basic optimization cycle of IMS audit program that was built with the account of audit's formal ISO standards requirements and ISAGO standards supported with new components (see Fig. 3):

Preconditions for the start of basic optimization cycle of audit program are given below:

1. T_0 – basis period of IT-Security audits;
2. S_0 – basic (planned) cost of IT-Security audits;
3. V_0 – basic volume of IT-Security audits (number of units);
4. F_0 – basic list of functional questions of IT-Security audits;
5. O_0 – basic list of attended IT-Security audit objects.

8 Mathematical Provision for Complex Industrial Facilities Audits

With regard to the task definition is introduced unilaterally limit (more precisely, the limit of a function on the left). The number of A ∈ R is called the left limit of function $f(x)$ at the point "a", if for any positive number ε will be found corresponding to his positive number δ, such that for all x in the interval | f (x) – A | < ε (a −δ, a) the inequality, or:

$$lim_{x \to a-0} f(x) = A \Leftrightarrow \forall \varepsilon > 0 \, \exists \delta = \delta(\varepsilon) > 0 \, \forall x \in (a-\delta, a) : | f(x) - A | < \varepsilon$$

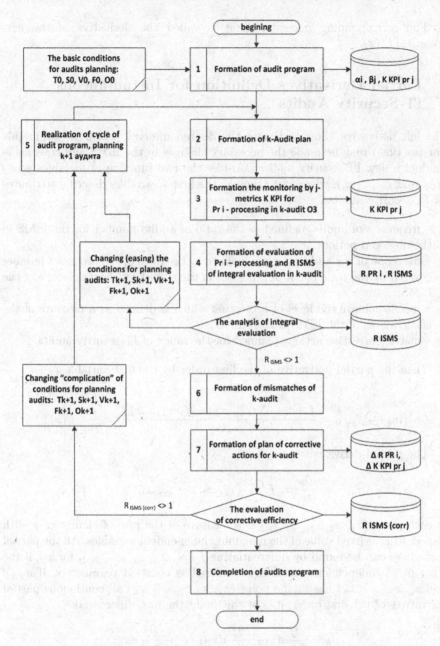

Fig. 3. Basic optimization cycle of IMS audit program

The derivative of the function $f(x)$:

$$\lim_{\Delta x \to 0} = \frac{f(x+\Delta x) - f(x)}{\Delta x} = \lim \frac{d}{dx} f(x) = f'(x)$$

The corresponding one-sided limit is called the derivative of the left-designate $f'_-(x)$.

9 Partial Derivatives Definition for Instantaneous IT-Security Audits

The left derivative allows estimate the desired interval, which is permissible (for the time) may be made the necessary changes in the ISMS and reasonable conduct a new IT-security audit. Consider the real function of variables $y = f$ $(x_1, x_2, x_3, \ldots, x_n)$, where, for example, the first 4 variables describe attributes for IT-security audits:

x_1 – frequency of audits, defined as the ratio of audits number for the ISMS in the observed period;

x_2 – the scope of the audit program, defined as the ratio of the processes number covered by the total count of processes in the stated certification scope of the ISMS;

x_3 – metric achieve the level of protection which is defined as a measure of the effectiveness of the ISMS Rbase/Rmax;

x_4 – metric corrective actions planned for the range of IT-security audits.

Then the partial derivative of the first order by the first variable x_1 is:

$$lim_{\Delta x_1 \to 0} = \frac{f(x_1 + \Delta x_1 \ldots x_k) - f(x_1 \ldots x_k)}{\Delta x_1} = \frac{\partial}{\partial x_1} f(x).$$

The partial derivative

$$\frac{\partial y}{\partial x_1} = f'x_1 (x_1, x_2, x_3, \ldots, x_n)$$

at each point $(x_1, x_2, x_3, \ldots, x_n)$ is a measure of the rate of change of y with respect to x_1 as fixed value of the remaining independent variables. All the partial derivatives can be found by differentiating $f(x_1, x_2, x_3, \ldots, x_n)$, for x_k, if the other $n - 1$ independent variables considered as constant parameters. If $y = f$ $(x_1, x_2, x_3, \ldots, x_n)$ has at the point $(x_1, x_2, x_3, \ldots, x_n)$ all continuous partial derivatives of the first order, it is at this point the first differential:

$$dy = \frac{\partial f}{\partial x_1} dx_1 + \frac{\partial f}{\partial x_2} dx_2 + \cdots + \frac{\partial f}{\partial x_n} dx_n.$$

For one changed variable x_1 (for example, the frequency of IT-Security audits) will evaluate the practical value of the partial derivative (at constant other variables), we estimate the growth rate of the ISMS security level:

$$\frac{\partial}{\partial x_1} = f'x_1 (x_1, x_2, x_3, \ldots, x_n) = \frac{\Delta R_k}{\Delta t_k}.$$

The solution of the problem - can be shown as a reduction of the period (increase frequency) IT-security audits in complex industrial object when using the left function variables. In the example of one variable x_1 demonstrated an increase in the growth rate of the protection level for known $\frac{\Delta R_k}{\Delta t\,k}$ variables ISMS audit process (see Fig. 4):

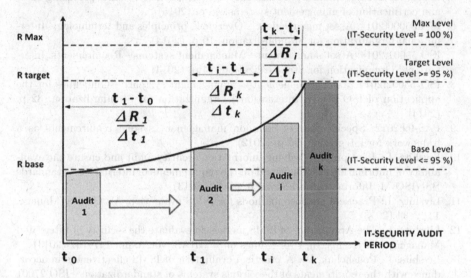

Fig. 4. Basic optimization cycle of IMS audit program

10   Conclusions

Given method of ISMS audit program optimization is based on the modern risk-oriented standards and provide the constant optimization of carrying out the IT-Security audits on the basis of joined flexible adaptive algorithms.

The proposed model for security audit based on the risk approach should be followed the importance of the "closing" principle of PDCA cycle in the creation and evaluation of management systems.

An approach that allows to use in assessing the management systems of complex industrial object reduce the period of conception of IT-security audits. This increases the speed of conducting audits process, management responses and increase the growth of IT-security protection level.

References

1. ISO/IEC 27001:2013. Information technology. Security techniques. Information security management systems. Requirements, International Organization for Standardization, 23 p. (2013)

2. ISO/IEC 27000:2014. Information technology. Security techniques. Information security management systems. Overview and vocabulary, International Organization for Standardization, 31 p. (2014)
3. ISO/IEC 27004:2009. Information technology. Security techniques. Information security management systems. Measurement, International Organization for Standardization, 55 p. (2009)
4. ISO 19011:2011. Guidelines for auditing management systems, 44 p. (2011)
5. ISO 17021:2015. Conformity assessment – Requirements for bodies providing audit and certification of management systems, 48 p. (2015)
6. ISO 55000:2014. Asset management – Overview, principles and terminology. International Organization for Standardization, 19 p. (2014)
7. ISO 55001:2014. Asset management – Management systems – Requirements. International Organization for Standardization, 14 p. (2014)
8. ISO 55002:2014. Asset management – Management systems – Guidelines for the application of ISO 55001. International Organization for Standardization, 32 p. (2014)
9. PAS-99:2012. Specification of common management system requirements as a framework for integration, 36 p. (2012)
10. Livshitz, I.: Joint problem solving information security audit and ensure the availability of information systems based on the requirements of international standards BSI/ISO M. Informatisatia i Svyaz **6**, 67–62 (2013)
11. Livshitz, I.: Practical purpose methods for ISMS evaluation. M. Quality Manage. **1**, 22–34 (2013)
12. Livshitz, I.: The Application of ISMS models to evaluate the security of Integrated Management Systems. In: Proceedings of SPIIRAS, vol. 8, pp. 147–162 (2013)
13. Livshits, I., Polishchuk, V.: A practical evaluation of ISMS effectiveness in accordance with the requirements of the various systems of standardization – ISO 27001 and STO Gazprom. In: Proceedings of SPIIRAS, vol. 3, pp. 33–44 (2015)

Computer Simulation of Average Channel Access Delay in Cognitive Radio Network

A.Yu. Grebeshkov[1]([⊠]), A.V. Zuev[1], and D.S. Kiporov[2]

[1] Chair of Automatic Telecommunications,
Povolzhskiy State University of Telecommunications and Informatics,
Leo Tolstoy str. 23, Samara 443010, Russia
grebeshkov-ay@psuti.ru
[2] Bachelor of Department of Applied Mathematics, Samara University,
Moskovskoye sh. 34, Samara 443086, Russia

Abstract. Cognitive radio (CR) is a new wireless communication concept of the future networks, that can help to use all available radio resources at a local area with a great effectiveness. Cognitive radio is based on the dynamic spectrum access (DSA) where available spectrum segments are used in an intelligent manner with help of advanced spectrum analysis and probing for unoccupied radio frequencies. An implementation of the cognitive radio networks raises an issue of the medium access control (MAC) protocol researching, in particular MAC protocol impacts on the access delay to radio channels. In this paper uncoordinated access method is studied where the event of spectrum and channel accessing is random and determined by probabilistic value from 0.1 to 0.99 named as channel availability. The subject of research was impact of channel availability on the access delay with simulation on the base ns2 program simulator with CRNC patch.

Keywords: Cognitive radio · Radio terminal device · Software-defined radio · Dynamic spectrum access · Media access control protocol · Simulator ns2

1 Introduction

Cognitive radio (CR) is a new wireless communication concept of the future networks, that can help to use all available radio resources at a local area [1–3]. Cognitive radio has a clear ability to be concerning as self-configurable platform including set of different software and hardware.

Cognitive radio is based on the dynamic spectrum access (DSA). This technology explores an opportunistic spectrum access, where available spectrum segments are used in an intelligent manner with help of advanced spectrum analysis and probing for unoccupied radio frequencies. CR and DSA forms a new paradigm for radio spectrum and radio channel access and a great challenge for the traditional radio spectrum using. Regular radio access technologies (2G, 3G, 4G/LTE, WiFi) was designed on the base of centralized principle of spectrum allocation.

© Springer International Publishing AG 2016
V.M. Vishnevskiy et al. (Eds.): DCCN 2016, CCIS 678, pp. 325–336, 2016.
DOI: 10.1007/978-3-319-51917-3_29

Now this scheme has a drawback in term of flexibility and adaptability which are the important point of advantage of cognitive radio. There are two types of users sharing a common spectrum under DSA rules:

- Primary (licensed) users who have high-priority in spectrum access and utilization within the predefined frequency bands.
- Secondary users who must access the spectrum with DSA technologies for a limited time.

The subject of the research is secondary users' character like as average channel access time delay. There is a really situation on practice when the part of frequency band previously licensed for primary user, is not being utilized by this kind of user for a short period of time. In traditional wireless systems there are no technologies that can help to use this unexpected "white space" for data transmission or receiving. Cognitive radio technologies with software defined radio (SDR) are more adaptable. But with DSA there is a time period when SDR tried to access this temporary not-in-use channel for signal transmission and packet exchange. The packet transmission delay between source and destination terminal is restricted by quality of service requirements. The average channel access time delay is the part of summarized end to end transmission delay. In this report the DSA's access delay is under investigation with computer simulation.

2   CR and SDR Standardization

Now cognitive radio technologies, special software and hardware for SDR, some aspects of CR implementations in civil and military fields are the point of innovations. Intellectual and self-learning program based control with receiver scheme reconfiguration on the base of field programmable gate array (FPGA) technology and wideband antennas supports DSA access features and bring new paradigm of the spectrum access into reality.

One of the main problem is prevention of co-channel interference in wireless communications under DSA mode. Next problem is cooperation between secondary users and prevention of channel access conflicts between terminals with DSA mode. For this reason an intensively standardization activity take place in the cognitive radio technologies and SDR fields.

Efforts in CR and SDR process of standardization are implemented by the Institute of Electrical and Electronics Engineers (IEEE), International Telecommunication Union (ITU), European Telecommunications Standards Institute (ETSI) and European Computer Manufacturers Association (ECMA) [4].

In 2004 the first CR and SDR standard IEEE 802.22 was initiated. IEEE 802.22 contained information about cognitive radio principles, technologies and DSA base description. This standard has been designed for data transmission devices in the wireless regional area network (WRAN). In WRAN radio terminal uses "white spaces" in the TV frequency spectrum.

In 2004 was initiated SDR standardization project called IEEE P1900. In 2006 IEEE 802.22 and IEEE P1900 became a part of IEEE Standards Coordinating Committee 41 (IEEE SCC41). Finally IEEE SCC41 was renamed as IEEE

Dynamic Spectrum Access Networks Standards Committee (DySPAN-SC). In the focus of DySPAN-SC is development of CR standards.

In 2012 ECMA-392 standards was finally published with information concerning physical layer and media access control layer for CR networks. A part of ECMA-392 was devoted to the personal devices features and SDR terminals function, including their operation modes into digital TV frequency band.

The efforts of the ETSI Reconfigurable Radio Systems Technical Committee are concentrated on the software-defined radio (SDR) architecture, function and use case standards.

The corpus of ETSI CR and SDR standards takes in account the regional conditions and requirements of the European regulators in telecommunications and TV white spaces (TVWS) with requirements for the digital TV signal characteristics in European Union.

In 2012 the cognitive radio systems was defined by ITU Radiocommunication Sector (ITU-R) as a radio system employing advanced technology that provides the telecommunication system the clear facilities to obtain knowledge of its radio environment with help of sensing and probing technologies and equipment, to adjust operational parameters and characteristics including physical (PHY), media access control (MAC) and network protocols attributes.

On the base of discussed standards there are three base parts in the cognitive cycle of DSA decision making [5]:

1. Analysis of the radio environment and search of the free frequencies, which is performed in the receiver by SDR terminal.
2. Dynamic spectrum access management, transmit power control, which are performed in the transmitter by SDR terminal.
3. Global feedback, enabling the transmitter to act in context of information about the radio environment feedback to it by the receiver.

MAC protocol uses data collected at the all stages. On the base of MAC layer, cognitive radio system can learn from the obtained results and probing.

The knowledge used by the CR/SDR include parameters of operational radio environment, information of location and available wireless networks at this area, existing policies of spectrum access, users' needs, preferable networks and terminal features.

Sometimes for DSA realization a special devices called "coordinators" or "arbiters" are used. These devices can collect and provide useful information like as available frequency bands, radio access technologies (RAT) in association with base stations locations, access points and user terminals, restriction to the transmission power values. The coordination of base stations' and radio terminals' positions and positions of another telecommunication systems can be obtained with global position system (GPS) or wireless systems coordinating and positioning features.

3   Research Issue in CR Data Link Layers

The subject of research will be case with uncoordinated access as a more common point in CR system for ad-hoc network. The idea is that the estimation of the average access delay [6] for uncoordinated manner of access will be an upper estimation of access delay when secondary users have to wait for a time to get access to the spectrum and radio channel resource. The subject of further research with computer modelling will be MAC protocol unit [7,8].

The issue of unoccupied acceptable channels selection is discovered. This problem solution in term of open system interconnection (OSI) model is at the data-link level where is a control access protocol that grants access to the transmission medium. As it said above, it is a medium access control protocol with specificity in the context of cognitive network and DSA.

MAC protocol operation bases on the data from the physical layer. This data used to solve the problem of recognition of temporary unoccupied radio frequencies. The next step is how to get access to the unoccupied channel. However, information from data link layer help to find the optimal direction/transmission path, indicating a list of available channels for the network layer. In return, the network layer can transfer to the link layer an information about which channel has an appropriable quality of service (QoS) for the data transferring session initiation.

The MAC protocol for uncoordinated spectrum access supports following main functions:

- The control and prevention of interference.
- Prevention of conflicts of access to the channel.
- Realization of the selection process and finding unoccupied radio channel.

The computer model of uncoordinated access method to the cognitive network based on the Cognitive Radio Cognitive Network (CRCN) patch for network simulator ns2 on the base of Linux Ubuntu 10.04 operating system [9,10] where for CR network some special features of MAC protocol are added.

4   CRNC Features for Simulation

The CRCN patch has the input data as the amount of radio terminal devices with SDR features and the overall number of radio channels. The main modelling scenario provides a description of queues and channels for each SDR scenario with help of the test scripting language (TSL) program library. Finally, the network simulator ns2 with CRCN patch has the following functionality for simulation cognitive MAC protocol unit:

1. The description of multi-channel data transmission medium.
2. The interface description for radio channel selecting.
3. A possibility to change and choice transmission power value.
4. Interference information.
5. Information about position of radio terminal devices.

The CRCN patch has settings to describe the collisions on the MAC layer. In fact, the choice of unoccupied channel depends on the MAC layer cognitive radio. Next, in the program code to the "sendDown" procedure the function "WirelessPhy" be added. This function includes the description of the frame transmission process when the frame transmits to the physical layer.

To avoid access conflicts or to reduce interference between adjacent nodes the special channel index is used in CRNC. This channel index obtained from the MAC layer or from DSA algorithm.

A particular channel can be assigned by means of simulator to specific interface radio terminal device with SDR features. The assignation may be carried out by MAC level or can be transmitted from the network level. At the network layer the routing protocol named as "ad hoc on-demand distance vector" (AODV) is used in the model.

AODV protocol [11,12] does not depend on physical layer on wireless network, but the broadcasting mode should be suppose that corresponding nodes can detect each others' broadcasts with "hello". "Hello" message can help AODV to operate independently from another underlying protocol. AODV uses dynamically establishing route tables and provides only loop-free routes for finite number of nodes in the ad-hoc network. In routing table, there are addresses of active corresponding nodes for each destination. In common case AODV can discover and find routes for data interchange quickly and correctly.

As it showed in [13] routing process in CR ad-hoc network is realized with the main goal to find a short route between source and destination nodes. An advantage of AODV prorocol is the small bandwidth for maintain routing table and AODV protocol massaging. An disadvantage is the uncertain delay of path search, because route has to be determined before transmission and packet sending. For 14 nodes in CR ad-hoc network AODV routing protocol is given the best performance regarding the packet transmission over the different channel than the Destination Sequence Distance-Vector (DSDV) protocol.

In [14] is presented results of multipath routing protocol modelling based on AODV, introduced next hop routing layer thought to the MAC layer, proposed the establishment of a plurality of next hop multi next hop (MNH) a Mesh MAC (MMAC) protocol like improved AODV. It is interesting to note that the average end-to-end delay in cognitive radio mesh network with multi next hop MAC protocol (MNH MAC), with AODV + MMAC protocol is from 0.6 to 0.7 s for 10 to 40 flows accordingly, where every flow has 100 packet per second, packet length 512 bits, number of nodes is 49 into the territory with dimension 250 m to 250 m, simulation time is 40 s. The average channel access delay was not modelled.

In [15] is presented results of compare two reactive routing protocols: AODV and Dynamic Source Routing (DSR) by using three performance metrics packet delivery ratio, average end to end delay and routing load for mobile ad hoc network. Parameters of model was 5 to 30 sources of traffic, packet rate was 5 packets per 1 second with packet size 512 bytes and topology size 500 m to 500 m. The maximum value of average end to end delay was 50 ms with 10 sources. Conclusion was when the number of sources has been increasing AODV is better for average end to end delay. The average channel access delay was not modelled too.

In conclusion [16] is said that AODV is more suitable for cognitive wireless networks compared to DSR, because DSR route discovery may lead to unpredictable packet length, which is not suitable for intermittent connectivity environment of CR network.

Due to all these results and conclusions, we are considering the AODV protocol and model of "hello" packets transition and the routing of these messages at the same time on several radio interfaces of terminal with aim to establish communication session.

Since management of multi-channel structure is performed by simulator for MAC layer, the test includes two stages. In the first stage, each node will send a packet to the upper OSI layer and provides with information about the unoccupied channel(s). In the second stage, the node will use selected channel to transmit and receive data.

Evaluating the probability of channel availability was described in [17]. In [18] a special monitoring network was proposed for classification of the channel availability.

Simulation CRNC parameters for the study of dependence between channel availability and average channel access delay with fixed packet size shown at the Table 1.

Table 1. Simulation CRNC parameters

Description	Value
Simulation tool	ns2 (CRNC patch)
Network area	100 m × 100 m
Number of nodes	20
Number of channels	1, 3, 5
Packet size	512 bytes
Channel availability (availability)	0.1 to 0.99
Simulation time	60 s

Simulation parameters for study of dependence between channel availability and average channel access delay with fixed packet size 512 bytes and variable number of nodes and channels shown at the Table 2.

Three options of multichannel structure are performed for simulation. At the first option there is 1 channel with 5, 10 and 20 nodes; at the second option there are 3 channels with 5, 10 and 20 nodes; in the end there are a 5 channels with 5, 10 and 20 nodes.

Table 2. Simulation CRNC parameters with variable number of nodes

Description	Value
Simulation tool	ns2 (CRNC patch)
Network area	100 m × 100 m
Number of nodes	5, 10, 20
Number of channels	1, 3, 5
Channel availability (availability)	0.1 to 0.99
Simulation time	60 s

5 Analysis of Computer Simulation Results

The simulation results do not provide accurate values because modeled by a random processes. In order to estimate probability p of event, where event is the case of channel occupation, the probing simulation was done with result in 14 success attempts of channel occupy during 50 tests, since probing $p = 0, 28$. The accuracy evaluation (closeness in estimation) in all simulation experiments was set to $\epsilon < 0.01$. With the 95 % confidence interval for the 30 points used for plot composition, the number of tests in the one statistical experiment was determined as 9939, round to 10000. During simulation was realized one experiment with 10000 tests for CRNC parameter at the Tables 1 and 2.

The nodes placed static in random order. These nodes selected randomly as senders or receivers data. The queue service time of each network described by the exponential distribution. The node selection is randomly. In addition, necessary to note that availability defined as the probability that a channel is available for the secondary user as result of sensing and probing process.

The process of the network model includes the creation of topology and the interaction sites. It is necessary to form the grid coordinates and the size of the model. Next code shows the initial stages of the interaction of components:

```
set tcp_(0) [$ns_ create-connection  TCP $node_(0) TCPSink $node_(1) 0]
$tcp_(0) set window_32
$tcp_(0) set packetSize_512
set ftp_(0) [$tcp_(0) attach-source FTP]
$ns_ at 2.5568388786897245 "$ftp_(0) start"
```

In this code tcp connection is created. The size of the transmitted packets is exposed to 512 bytes. As example, the node 0 sends a welcome message to node 1.

The more detailed description of simulation model is:

```
set val(chan)      WirelessChannel   ; #Channel Type
set val(prop)      TwoRayGround      ;#Radio propagation model
set val(netif)     WirelessPhy       ; #Network interface type
set val(ant)       OmniAntenna       ; #Antenna model
```

```
set val(rp)        AODV         ;#Routing Protocol
set val(ifq)       Queue/DropTail/PriQueue    ;# interface queue type
set val(ifqlen)    500      ;# max packet in ifq
set val(mac)       Mac/Macng    ;# MAC type
set val(ll)        LL           ;# link layer type
set val(nn)        5/10/20        ;# number of nodes
set val(channum)   1/3/5      ;# number of channels per radio
set val(cp)        ./random.tcl   ;# topology traffic file
set val(stop)      60           ;# simulation time
```

This code (with some modification) is used as in the model with parameters in the Table 1 as in the model with parameters in the Table 2.

The network layer transmits routing information about the available channel to the lower level. MAC-level is available in the multichannel structure.

In Fig. 1 simulation is carried out with a different number of channels (1, 3, 5) and with CRNC parameters from Table 1. Nodes get access to the channel(s) at the same time, this procedure leads to access delay and collisions in the simulated network.

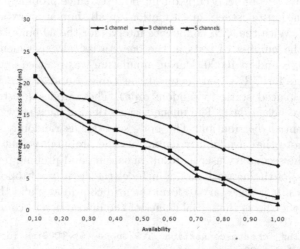

Fig. 1. Average channel access delay for 20 nodes, 512 bytes packet size

In Fig. 1, when availability value is observed from 0.1 to 0.3 approximately, there is a high-delay access scheme, but when the availability is increasing the number of collisions is reduced.

Next experiments with computer model will focus on networking scheme with fixed packet size, like 512 byte as a typical size and with CRNC parameters from Table 2. Simulation is carried out with a different number of channels (1, 3, 5) with 5, 10 and 20 nodes, which get access to the channel(s) at the same time, this procedure leads to access delay and collisions in the simulated network like in previous experiment. The code of program model is similar to program code fragment above.

In Fig. 2 for one-channel scheme, when availability value is observed from 0.1 to 0.3...0.4 approximately, a high-delay access scheme like in Fig. 1 take place. There is an area of the more faster delay decreasing for an availability between 0.4 and 0.6. Plots in Fig. 2 could not be describe as monotone decreasing function due to constant area between availability value from 0.2 to 0.4. Next we will discuss the situation in Fig. 3.

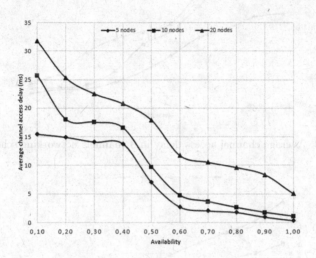

Fig. 2. Average channel access delay in one-channel networking scheme

In Fig. 3 for 3 channel networking scheme, when the availability value is observed from 0.1 to 0.3 approximately, a high-delay access scheme occurs. There is an area of the more faster delay decreasing for availability between 0.4 and 0.7 like in Fig. 2, but the gradient in Fig. 3 for the area from 0.4 to 0.6 (scheme with 10 nodes and 20 nodes) no more than the gradient in Fig. 2.

Plots in Fig. 3 could be describe as monotone decreasing function approximately, but, definitely, it is not analytical proof.

In the end, let's discuss results in Fig. 4. For 5 channels networking scheme, when availability value is observed from 0.1 to 0.9 there is no area with a high-delay access scheme like in Figs. 2 and 3. The reason of this is sufficient number of channels, which can be a kind of 'damper' for all types of demand channel access requests. Clearly, plots in Fig. 4 could be characterize as monotone strictly decreasing functions.

It is need to remark there are some unusual rises and falls in the plot in Figs. 1, 2, 3 and 4 because the process of channel selecting is a random and the channel availability is random value too. The results of simulation there is a threshold value availability for static ad-hoc network in the context of average channel access delay.

If the availability value will be at the range between 0.4...0.6 approximately than 1 and 3 channels networking scheme (and partially, 5-channels scheme) with

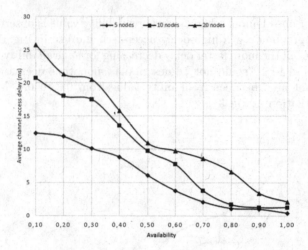

Fig. 3. Average channel access delay in 3 channels networking scheme

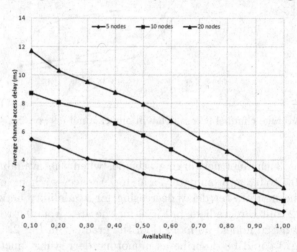

Fig. 4. Average channel access delay in 5 channels networking scheme

5,10 and 20 nodes, fixed packet size have the area of fast decreasing delay. This area could be find at the scheme with 20 nodes, 1 and 5 channels and packet size equals to 512 bytes.

If the availability value will be bigger than threshold value (approximately 0.6...0.7) than networking scheme with 3 and 5 transmission channels for 5, 10 and 20 nodes have not dramatically decreasing of delay with a large gradient as an effect of cognitive network features. In continue, it seems that using 5-channel networking scheme from an engineering point of view has not a great advantage over 3-channel networking scheme. For example, if availability value is 0.5 than for a 3-channel networking scheme with 20 nodes there is 11 ms channel access delay (see Fig. 3), and for a 5-channel networking scheme with 20 nodes there is 8 ms channel access delay (see Fig. 4). It means that the number of channels could be decrease without big losses of quality.

In the future DSA with cognitive radio principles and software configurable radio-terminals create a wide range opportunities for reconfigurable radio networks in military, social works, medicine, radio access technology new generation including cognitive wireless sensor networks.

6   Conclusions

DSA method with cognitive radio principles create a wide range opportunities for research and applications in military, social works, medicine and new generation of radio access technology. The main aim of DSA technology is to improve the spectrum efficiency using for radio terminal devices with SDR features.

The actual issue is the research of future parameters estimation of the cognitive radio networks, like as average channel access delay for secondary users. The results of computer simulation for uncoordinated access method shows that access delay value depends of availability of radio channels. There is a possibility to decrease number of accessible channels without dramatically average channel access delay changing at the ad-hoc network with not a great number of nodes but only if channel availability will be more than 0.3...0.4. In many cases a reasonable value of channel availability is 0.6. For the further studying, the problem of access delay for coordinated access method and network with high-rate nodes will be important.

References

1. Rao, K.R., Bojkovic, B.M.: Wireless Multimedia Communication Systems. Design Analysis and Implementation. CRS Press, Florida (2014)
2. Oiu, R.C., Hu, Z., Li, H., Wicks, M.C.: Cognitive Radio Communications and Networking. Principle and Practice. Wiley, United Kingdom (2012)
3. Zhang, Y., Zheng, J.: Cognitive Radio Networks: Architectures. Protocols and Standards. CRC Press, Florida (2014)
4. Filin, S., Harada, H., Murakami, H., Ishizu, K.: International standardization of cognitive radio systems. IEEE Commun. Mag. 49(3), 82–89 (2011)
5. Haykin, S.: Cognitive dynamic systems: radar, control, and radio. Proc. IEEE 100(7), 2095–2103 (2012)
6. Report ITU-R SM.2256: Spectrum Occupancy Measurement and Evaluation. SM Series. Spectrum Management. ITU Electronic Publication, Geneva (2012)
7. De Dominico, A., Calvanese, S.E.: A survey on MAC strategies for cognitive radio networks. IEEE Commun. Surv. Tutor. 14(1), 21–44 (2012)
8. Shahid, K.U., Maqsood, T.: CRN survey and a simple sequential MAC protocols for CRN learning. In: 2th International Conference on Advances in Cognitive Radio, pp. 22–27. International Academy, Research and Industry Association, Wilmington (2012)

9. Lee, P., Wey, G.: NS2 model for cognitive radio networks routing. In: Proceedings of the 1st International Symposium on Symposium on Computer Network and Multimedia Technology, CNMT 2009, pp. 1–4. IEEE Inc., New Jersey (2009)

10. Bhrugubanda, M.: A survey on simulators for cognitive radio network. IJCSIT **5**(3), 4760–4761 (2014)

11. Perkins C.E., Royer E.M.: Ad-hoc on-demand distance vector routing. In: Proceedings of the Second IEEE Workshop on Mobile Computer Systems and Applications, WMCSA 1999, pp. 90–100. IEEE Computer Society Washington, DC (1999)

12. Perkins, C.E., Belding-Royer, E.M., Das, S.R.: Ad hoc on-demand distance vector (AODV) routing. IETF Draft, draft-ietf-manet-aodv-13.txt (2003)

13. Ugale, M.R., Deshmukh, R.P., Thakare, A.N.: Analysis of routing protocol for cognitive adhoc networks. IJECCE **3**(2), 314–319 (2012)

14. Gao, H., Zeng, W.: Analysis on cognitive wireless network MAC protocol access mode based on NS2. JATIT **47**(3), 1092–1099 (2012)

15. Moses, G.J., Kumar, D.S., Varma, P.S., Supriya, N.: Simulation based performance comparison of reactive routing protocols in mobile adhoc network using NS-2. In: Kacprzyk, J. (ed.) Proceedings of the First International Conference on Advances in Computing. AISC, vol. 174, pp. 423–428. Springer, New Delhi (2012)

16. Salim, S., Moh, S.: On-demand routing protocols for cognitive ad hoc networks. EURASIP J. Wirel. Commun. Network. SpringerOpen J. **102**, 1–10 (2013)

17. Kaniezhil, R., Chandrasekar, C.: Evaluating the probability of channel availability for spectrum sharing using cognitive radio. IJERA **2**(4), 2186–2197 (2012)

18. Canberk, B., Oktug, S.: A channel availability classification for cognitive radio networks using a monitoring network. In: Proceedings of 17th IEEE ISCC2012 Symposium on Computers and Communications, pp. 690–695. IEEE Computer Society, Washington, DC (2012)

Efficiency of Redundant Service
with Destruction of Expired and Irrelevant
Request Copies in Real-Time Clusters

V.A. Bogatyrev$^{(\boxtimes)}$, S.A. Parshutina, N.A. Poptcova, and A.V. Bogatyrev

ITMO University, Kronverksky Pr. 49, 197101 St. Petersburg, Russia
vladimir.bogatyrev@gmail.com, svetlana.parshutina@gmail.com
http://en.ifmo.ru

Abstract. Possible ways of increasing the probability of timely and faultless execution of delay-sensitive requests in real-time clustered computing systems, when multiple copies of requests are created and served in different cluster nodes, are investigated. The proposed models for queueing and functional reliability prove the existence of scope of efficiency for service disciplines with redundant execution of copies of requests, when the probability of their prompt and error-free servicing can be increased significantly, despite the rise of load in the nodes. It was examined how the ways to arrange redundancy and the redundancy order affected timely and reliable servicing of requests, with possible faults and errors in the nodes. It was shown that destruction of expired copies, whose waiting time in the queue exceeded a given ultimate time, and copies which became irrelevant, after one of them had been processed, produced an essential enhancement of efficiency of the system.

Keywords: Cluster · Real-time · Reliability · Queueing systems · Request copies

1 Introduction

Reliability, security, and efficiency of the processes of handling and sending data in information and communication systems and networks are highly dependent on timely and faultless transmission of requests over redundant communication channels and execution of those requests by servers within given clusters [1–8]. It is especially important in real-time systems, dealing with delay-sensitive data.

In distributed computing systems with multipath routing and redundant communication channels, the probability of successful delivery and servicing of a request, within a specified time limit, can be increased in case multiple copies of the request are sent over different routes concurrently [9]. Under this approach and in accordance with the requirement that at least one copy of the request needs to be delivered and served, in order to regard delivery of this request successful, it is possible to arrange timely and error-free delivery of data packets without using acknowledgement-based protocols and repeated transmission.

© Springer International Publishing AG 2016
V.M. Vishnevskiy et al. (Eds.): DCCN 2016, CCIS 678, pp. 337–348, 2016.
DOI: 10.1007/978-3-319-51917-3_30

However, such redundant transmission of numerous copies of requests causes the rise of the total intensity of the flow of requests and possibly the growth of the residence (waiting) time of those copies in the system. At the same time, the stochastic nature of sending copies of data packets over different communication channels independently may result in the increase of the probability of timely delivery of at least one copy. Under certain conditions, this can lead to the reduction of the average residence (waiting) time of requests in the system. The efficiency of redundant servicing of requests in real-time systems is determined by a set of efficiency metrics, including the average residence time, the probability of data packet loss, the probability of timely execution of requests, and the probability of faultless execution of requests [3].

Thus, the choice of design solutions, used to build redundant information and communication systems, should rely on modeling for multi-criteria evaluation of the efficiency of applying redundancy in such systems, with the view of making the optimal choice of a network structure and the way of data transmission and servicing.

The efficiency of information and communication systems that can be represented by queueing models is largely determined by service disciplines [10]. There is a model for multichannel queueing systems with the common queue, described in [11], which demonstrates the efficiency of service disciplines with redundant execution of requests. If a request arrives at the very moment when all service devices (namely, servers, data channels) are not busy, it is cloned and its copies are served by several devices; otherwise, the request is put into the waiting queue.

Also, there is an analytical model for the cluster composed of a group of n single-channel queueing systems with the local queues, in which copies of each request are sent to k out of n service devices, depending on the state of the local queues [4]. The model proves that independent servicing of k copies in different nodes leads to the increase of not only the probability of errorless execution of requests but also – under certain conditions – the probability of timely execution of those requests, i.e. of at least one copy.

It should be noted that when multiple copies of requests in the queues of different nodes are processed in the redundant manner, by the time errorless computational results are received in some node (or in two or more nodes contemporaneously), the rest of the copies will continue waiting in the queues, in spite of the fact they will have become irrelevant. Therefore, on the basis of the developed model, it was shown that discarding overdue (expired) requests, namely those that have been waiting in the queues longer than the maximal allowable time, was likely to cause the significant reduction in the non-productive load of the cluster's nodes and, as a result, to enlarge the probability of timely and error-free execution of requests.

An analytical model to assess the (ultimate) residence time of requests is to be developed based on the existing solutions [1–4, 9] and to underlie a simulation model needed to estimate the efficiency of redundant transmission and execution of requests.

2   Object and Purposes of Research

The object of the current research is a group of n servers which comprise the cluster created in order to arrange redundant servicing of (copies of) requests, arriving from the multilevel network (Fig. 1). Redundant execution of copies of requests by the servers under consideration and redundant transmission of those copies through the network to the servers serve as a basis for providing timely and faultless servicing of at least one copy of each request.

Let us consider a distributed computing system in which copies of requests can be transmitted in the redundant manner, i.e. over k out of n possible routes (paths) concurrently to k servers belonging to the cluster. A request is regarded to be served successfully if at least one copy of it has been transmitted with no distortion to one of the servers in the cluster and the server has accepted and executed that copy of request within a given period of time. To enhance the performance of the system, the possibility to destroy irrelevant copies of requests, waiting in the queue, was introduced.

The aim of the study is to develop a simulation model and tools to support the design process of highly reliable distributed computing systems. The model is to provide the basis for selecting the optimal design solutions, used to create networks with redundant transmission and execution of requests, and analyzing the efficiency of those solutions. The research focuses on the assessment of efficiency and usefulness of redundant distribution of requests through the network and their redundant servicing in the cluster.

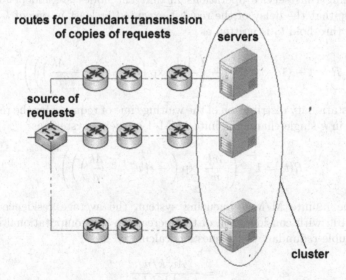

Fig. 1. The structure of a distributed computing system with redundant transmission and servicing of copies of requests

3   Analytical Model Development

Consider the reliability model for the redundant computational process, like in [4], given absolutely reliable nodes and ideal control when, by the time of executing the request, without extra delays, information about the correctness (validity) of the computations is generated.

Let us analyze the redundancy of the computational process in k nodes when the condition of timeliness for computations in the system is fulfilled at least in one of k nodes, i.e. the delay of the request in the queue is less than some limiting value t_0 at least in one of k nodes.

Taking into account the increase in the computational load for the redundant computations in k nodes, the probability that the waiting time of requests will not exceed the limit t_0 in some (particular) node is evaluated as

$$r = 1 - \frac{\Lambda v k}{n} \exp\left(-t_0\left(v^{-1} - \frac{\Lambda k}{n}\right)\right), \tag{1}$$

where the computational load of the node is $\rho = \Lambda v k/n$.

Taking into account the computational load generated by redundant computations [10], the stationary distribution of the waiting time of requests in the M/M/1 queueing system is calculated as

$$r(t) = 1 - \frac{\Lambda v k}{n} \exp\left(-t\left(v^{-1} - \frac{\Lambda k}{n}\right)\right). \tag{2}$$

Assuming that service operations in different nodes are independent, the probability that the delay of the request executed at least in one of k nodes is below the threshold t_0 is written as

$$R = 1 - (1-r)^k = 1 - \left(\frac{\Lambda v k}{n} \exp\left(-t_0\left(v^{-1} - \frac{\Lambda k}{n}\right)\right)\right)^k, \tag{3}$$

while the stationary distribution of the waiting time of requests for the redundant execution in k single-channel infinite M/M/1 queueing systems is

$$R(t) = 1 - \left(\frac{\Lambda v k}{n} \exp\left(-t\left(v^{-1} - \frac{\Lambda k}{n}\right)\right)\right)^k. \tag{4}$$

For the infinite M/M/1 queueing system, the average residence time of requests [10], with consideration of the increase in the computational load due to the k-tuple redundancy of requests is calculated as

$$w = \frac{\Lambda v^2 k/n}{1 - (\Lambda v k/n)}. \tag{5}$$

Equation (5) provides the upper estimate of the average waiting time, since it reflects the increase in the computational load due to the redundancy; however, (5) takes no account of the possibility of decreasing the average residence time owing to the fact that, in one of the nodes, the redundant request can wait less time than in the other nodes.

In systems with multipath routing and the possibility to process requests in the redundant manner, transferring packets through the network redundantly, along a number of routes, to the group of servers in a destination cluster, leads to the increase of load both in the network and in the cluster. Destroying irrelevant requests, namely their irrelevant copies, both in the network and in the cluster is a possible way to potentially reduce the negative impact of the increased load in the system, which happens due to redundant transmission and servicing of requests in the cluster [12]. A copy of the request is referred to as irrelevant if its residence time in the network exceeds the ultimate residence time, i.e. the copy is overdue, or the given number of copies of that request has already been delivered successfully. Depending on the purpose of the system, there can be different requirements, such as to deliver at least one packet (one copy of a request) to the particular server or a server that belongs to a certain group of servers in the given cluster. In case of redundant servicing of copies of requests by a group of servers in the cluster, the specified number of those copies is expected to be transmitted either to every server of the group or to some of them [13].

An analytical model for systems with cluster architecture and support for redundant execution of requests is proposed in [12]. This research focuses on the service discipline which implies that the results of calculations are registered at some point in time t, with time counting starting immediately after a copy of the request is put into the queue of a server in the cluster. The results of servicing of the request are brought by using the timer which counts down to the given moment, with respect to the ultimate time of waiting and processing requests. The existence of such discipline can be explained by the fact that the results of execution of requests in real-time operating systems are required to be available by a specific point in time. After the timer has counted down to the maximum allowable time t, the redundant copies of requests, still waiting in the queues, will become irrelevant and be therefore destroyed. Destroying irrelevant instances of requests is an efficient way to minimize the undue processing of overdue redundant copies and inefficient servicing of other requests. This makes it possible to reduce the load in the nodes and consequently to minimize the delay time in the queues.

Applying models for redundant execution in systems with support for destroying redundant copies of requests, waiting for servicing in the queues longer than t, as shown in [12], yields a lower bound of the probability that the results are delivered in time.

A model for the service discipline which assumes that redundant copies of requests in the queues are destroyed at the end of time t is introduced in [12]. Destroying overdue copies of requests causes the decrease of the intensity of copies that are waiting to be processed (the load in the nodes). If the redundancy order, or the number of spawned copies of requests, is equal to k, the intensity of copies waiting for execution increases not by a factor of k, but by a factor of $I \leqslant k$ ($I \geqslant 1$) [12].

The multiplier I is defined as the expected value of the number of nodes that process incoming requests, if the residence time of the latter is less or equal to the maximum allowable value of t [12]. In those nodes, copies of requests waiting in the queues are not destroyed.

The coefficient I is calculated as

$$I = kr. \tag{6}$$

Here, r is the probability that the waiting time in the queue of a given computing node is not more than the ultimate allowable value of t:

$$r = 1 - \rho \exp\left(-t\left(v^{-1} - \frac{\Lambda I}{n}\right)\right), \tag{7}$$

where ρ, which is the load in the node in view of destroying overdue copies in the queues, is

$$\rho = \frac{\Lambda v I}{n}. \tag{8}$$

By replacing r in (6) with (7), we derive a new equation

$$I = k\left(1 - \rho \exp\left(-t\left(v^{-1} - \frac{\Lambda I}{n}\right)\right)\right), \tag{9}$$

which can be solved using the *root* function provided by Mathcad 15:

$$s := root\left[-I + k\left[1 - \frac{\Lambda v I}{n} \exp\left[-t\left(v^{-1} - \frac{\Lambda I}{n}\right)\right]\right], I\right]. \tag{10}$$

After computing the value of the coefficient I, taking into account the k-tuple redundant servicing of copies of requests in the cluster, we can calculate the probability of not exceeding the ultimate delay t in at least one of k nodes, which is equal to (4).

For a real-time distributed computing system, the considered dependencies can serve as the basis for optimizing the redundancy order, in case of redundant multipath transmission of packets through the network and their redundant processing in the cluster [14–16].

The discussed analytical model let us consider the impact of destroying overdue requests on the probability of timely redundant execution in the single-tier cluster, with interarrival and service times being exponentially distributed. Besides, the model takes into account the possibility of destroying only overdue requests.

In the distributed computing system under consideration, overdue requests can be destroyed at different stages of transmitting their redundant copies through the network and servicing them in the cluster. This makes it difficult to design analytical models. Therefore, we can set the task of developing a simulation model, which takes into consideration the processes of redundant transferring and servicing both in the network and the cluster.

4 Simulation Model Development

Let us develop a network model to illustrate forwarding requests to the group of servers belonging to a given cluster, where those requests are to be served. Delivery of requests might fail due to their loss and bit errors, emerging in the communication links with a certain probability. It is required that requests can be sent over one or multiple routes at a time. For simplicity, we assume that

- the routes are disjoint and predefined;
- there is no loss of requests due to queue overflow;
- no acknowledgement of delivery of requests is generated;
- the servers have identical performance characteristics;
- each server can handle one request at a time;
- some constant speed of transmission of requests is maintained along the whole path, i.e. within every link segment.

We need to compute the residence time for each request in order to calculate the average residence time and thus to draw the line between those requests which are processed in a timely manner and those requests which expire.

It is expected that parameters of the proposed model, particularly the redundancy order of transmission of requests, can be easily adjusted and the model contains three units (Fig. 2):

- the source of requests with the mechanism which distributes them over multiple predefined routes;
- the routes which can be configured to distort or "lose" requests;
- the servers which handle incoming requests or reject them if they are overdue or irrelevant.

The source of requests consists of

- the generator of requests (*source*),
- the buffer for requests produced by the generator (*queue*),
- the element which delays forwarding requests over the routes (*delay*), and
- the balancer of requests, which distributes requests among the routes (*selectOutput*).

Each route (*route1*, *route2*, ..., *routeN*) is a set of the elements which are linked between each other and can simulate

- loss of requests (*loss*),
- distortion of requests due to bit errors (*damage*), and
- delays (*delay*).

The servers contain the elements, which

- buffer requests (*queue*),
- sort out expired requests (*timeouter*),
- sort out irrelevant (already handled) requests (*req_done*),

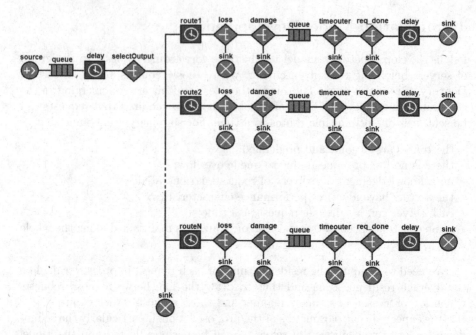

Fig. 2. The simulation model illustrating redundant transmission of copies of requests over k routes from the source to at least one of k servers in the cluster

- produce delays (*delay*), and
- destroy requests (*sink*).

The following parameters are to be defined for the source:

- the intensity of the flow of requests;
- the redundancy order of transmission, or the number of copies of the request to be sent concurrently;
- the algorithm for selecting the route for each particular incoming copy of the request;
- the algorithm for assigning numbers to the requests and their copies.

The following parameters are to be defined for each route:

- the number of hops (or transitions on the way from the source to the server);
- the speed of transmission;
- the probability of loss of requests;
- the probability of faultless delivery of requests, which is calculated as $(1-b)^{Nh}$, where b is the bit error rate for the link segment, N is the number of bits in the packet, and h is the number of link segments.

Additionally, we should introduce an equation to compute the delay depending on the size of the packet and the speed of transmission.

The following parameters are to be defined for the server:

- the queue capacity;
- the algorithm for sorting out expired requests;
- the algorithm for sorting out irrelevant requests;
- the service time of requests.

Any copy of the request should contain the information about

- the number to distinguish between different copies of the request;
- the number of bits in the packet;
- the time when the source generated the copy;
- the time when the request (one of its copies) was served;
- the route which has been selected to transmit the copy.

5 Timeliness of Results with Redundant Transmission and Redundant Execution of Copies of Requests

On the basis of the proposed model, we explored the probability of timely and faultless distribution of requests through the network and their servicing by one of the servers belonging to the certain cluster as depending on the redundancy order. The simulation process involved applying different values of such parameters as the probability of packet loss in communication channels and the bit error rate, with various values of the ultimate residence time allowed.

We considered that the size of packets varied from 1024 to 4096 bits and the speed of transmission in the channels was 10 Mbit/s. The intensity of the flow of requests generated by the source was 1000 requests per second, while the intensity of their servicing by the server varied from 0.01 to 5 seconds. Both intensities followed the exponential distribution. The simulation model was developed in AnyLogic 7 simulation environment.

Figure 3 illustrates how the probability of prompt and error-free transmission of requests through the network and their execution in the cluster depends on the redundancy order, when their copies are transmitted through the network over routes with different bit error rate. The increase in redundancy results in the higher probability of the successful delivery and execution of at least one copy of the request, but until the redundancy order is less than some particular value.

According to Fig. 3, redundant transmission and execution of requests is efficient if the redundancy order does not exceed a certain threshold value. For example, the optimal redundancy order for the bit error rate $b=0.00001$ is two, whilst it is three with $b=0.0001$.

The conducted study proves the efficiency of redundant transmission of (copies of) requests through the network over multiple routes and redundant execution of those (copies of) requests by the servers in the cluster.

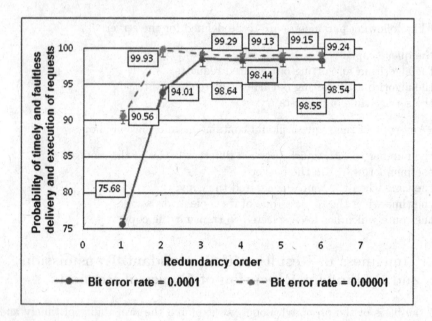

Fig. 3. The probability of timely and faultless delivery and execution of requests as depending on the redundancy order in case the bit error rate is 0.0001 (*solid line*) and 0.00001 (*dashed line*)

6 Conclusions

The proposed simulation model makes it possible to evaluate the efficiency of redundant distribution of requests over multiple routes in the network and redundant (and independent) servicing of those requests by the group of servers belonging to a certain cluster.

The efficiency of redundant transmission and execution of requests is calculated as depending on

- the probability of packet loss in the network and the bit error rate,
- the ultimate residence time of requests in the system – computed based on the analytical model considered in the current paper, and
- the algorithm for destroying copies of requests in the queue, both expired (because the waiting time of the request has exceeded the ultimate residence time) and irrelevant (due to the fact that another copy of the request has already been processed by one of the servers).

To sum up, the results of the current research prove the efficiency of redundant transmission of requests (data packets) and their redundant servicing by the servers in the cluster. Consequently, the probability of faultless delivery and execution of at least one copy of the request within a given period of time can be increased.

It was found that there exists the scope of efficiency for redundant transmission and servicing of requests. This scope depends on the bit error rate, the ultimate residence time, and the algorithm for destroying expired and irrelevant copies of requests.

The proposed models can be applied in CAD systems and underlie the choice and the optimization process of design solutions, to create fault-tolerant redundant distributed real-time computing systems, dealing with delay-sensitive requests.

Acknowledgments. The study was fulfilled as a part of the research project "Methods for designing the key systems of information infrastructure" (State Registration Number is 615869) of ITMO University, St. Petersburg, Russia.

References

1. Aliev, T.: The synthesis of service discipline in systems with limits. In: Vishnevsky, V., Kozyrev, D. (eds.) DCCN 2015. CCIS, vol. 601, pp. 151–156. Springer, Heidelberg (2016). doi:10.1007/978-3-319-30843-2_16
2. Aliev, T.I., Rebezova, M.I., Russ, A.A.: Statistical methods for monitoring travel agencies. Autom. Control Comput. Sci. **49**(6), 321–327 (2015)
3. Bogatyrev, V.A.: Exchange of duplicated computing complexes in fault tolerant systems. Autom. Control Comput. Sci. **45**(5), 268–276 (2011). doi:10.3103/S014641161105004X. http://link.springer.com/article/10.3103/S014641161105004X
4. Bogatyrev, V.A., Bogatyrev, A.V.: Functional reliability of a real-time redundant computational process in cluster architecture systems. Autom. Control Comput. Sci. **49**(1), 46–56 (2015)
5. Bogatyrev, V.A., Bogatyrev, S.V., Golubev, I.Y.: Optimization and the process of task distribution between computer system clusters. Autom. Control Comput. Sci. **46**(3), 103–111 (2012). doi:10.3103/S0146411612030029. http://link.springer.com/article/10.3103/S0146411612030029
6. Bogatyrev, V.A.: An interval signal method of dynamic interrupt handling with load balancing. Autom. Control Comput. Sci. **34**(6), 51–57 (2000)
7. Bogatyrev, V.A.: Protocols for dynamic distribution of requests through a bus with variable logic ring for reception authority transfer. Autom. Control Comput. Sci. **33**(1), 57–63 (1999)
8. Bogatyrev, V.A.: Fault tolerance of clusters configurations with direct connection of storage devices. Autom. Control Comput. Sci. **45**(6), 330–337 (2011)
9. Bogatyrev, V.A., Parshutina, S.A.: Redundant distribution of requests through the network by transferring them over multiple paths. In: Vishnevsky, V., Kozyrev, D. (eds.) DCCN 2015. CCIS, vol. 601, pp. 199–207. Springer, Heidelberg (2016). doi:10.1007/978-3-319-30843-2_21
10. Vishnevsky, V.M.: Teoreticheskie osnovy proektirovaniya komputernykh setey. Theoretical Fundamentals for Design of Computer Networks. Tekhnosfera, Moscow (2003). (in Russian)
11. Dudin, A.N., Sun', B.: A multiserver MAP/PH/N system with controlled broadcasting by unreliable servers. Autom. Control Comput. Sci. **43**(5), 247–256 (2009)
12. Bogatyrev, V.A., Bogatyrev, A.V.: The model of redundant service requests real-time in a computer cluster. Inf. Technol. **22**(5), 348–355 (2016)

13. Arustamov, S.A., Bogatyrev, V.A., Polyakov, V.I.: Back up data transmission in real-time duplicated computer systems. In: Abraham, A., Kovalev, S., Tarassov, V., Snášel, V. (eds.) IITI 2016. AISC, vol. 451, pp. 103–109. Springer, Heidelberg (2016). doi:10.1007/978-3-319-33816-3_11
14. Bogatyrev, V.A., Bogatyrev, S.V.: Redundant data transmission using aggregated channels in real-time network. J. Instr. Eng. 59(9), 735–740 (2016). doi:10.17586/0021-3454-2016-59-9-735-740
15. Bogatyrev, V.A., Bogatyrev, A.V.: Optimization of redundant routing requests in a clustered real-time systems. Inf. Technol. 21(7), 495–502 (2015)
16. Bogatyrev, V.A.: Increasing the fault tolerance of a multi-trunk channel by means of inter-trunk packet forwarding. Autom. Control Comput. Sci. 33(2), 70–76 (1999)

Stationary Waiting Time Distribution in $G|M|n|r$ with Random Renovation Policy

Ivan Zaryadov[1,2], Rostislav Razumchik[1,2]([⊠]), and Tatiana Milovanova[2]

[1] Institute of Informatics Problems of the Federal Research Center
"Computer Science and Control" Russian Academy of Sciences,
44-2 Vavilova Street, Moscow 119333, Russia
izariadov@gmail.com, rrazumchik@ipiran.ru
[2] RUDN University, 6 Miklukho-Maklaya Street, Moscow 117198, Russia
milovanova_ta@pfur.ru

Abstract. Recent recommendation RFC 7567 by IETF indicates that the problem of active queue management remains vital for modern communications networks and the development of new active queue management is required. Queueing system with renovation when customers upon service completion pushes-out other customers residing in the queue with a given probability distribution may have potential application as an alternative active queue management. In this paper one presents the analytic method for the computation of the customer's stationary waiting time distribution in $G|M|n|r$ queue with random renovation under FCFS (and non-preemptive LCFS) scheduling in the terms of Laplace-Stieltjes transform. The method is illustrated by one particular case: stationary waiting time distribution of the customer, which either received service or was pushed-out under FCFS scheduling.

Keywords: Queueing system · FCFS · LCFS · Renovation · Finite capacity

1 Introduction

This paper is devoted to the analysis of stationary waiting distribution characteristics in one special type of queueing systems – queueing systems with random renovation – under FCFS scheduling. The renovation policy implies that a customer upon service completion pushes-out ("kills") l, $l \geq 0$, other customers residing in the queue with the given probability $q(l)$. Random renovation means, that if l customers are to be removed from the queue containing $n \geq l$ customers, then the positions in the queue, from which the customers are to be removed, are chosen in a purely random way. Such a random push-out mechanism looks similar to the well-known random order of service (ROS) policy according to which the next customer to enter server is selected in a purely random way (i.e. with probability $1/N$ if N customers are present in the queue at the moment of service completion). Being an alternative policy to the classic

© Springer International Publishing AG 2016
V.M. Vishnevskiy et al. (Eds.): DCCN 2016, CCIS 678, pp. 349–360, 2016.
DOI: 10.1007/978-3-319-51917-3_31

ones (like FCFS/LCFS) and other more sophisticated size-oblivious and size-based policies (like PS, SRPT, FB), both continuous and discrete time queueing systems with ROS policy have received less attention from the research community. Yet a plenty of results is available for them. One can refer to the recent paper [1] (and references therein) for a respective short but informative review. Among the results not cited in [1] one should also mention papers [3–5], where the authors apply the theory of branching processes for analysis of the stationary waiting time characteristics in various single/multiple server systems with ROS policy (like $M|G|1|n$ with retrials, $MAP|G|1$, $SM_2|MSP|n|r$ and several others).

Having potential application as an alternative active queue management mechanism, queueing system with renovation have been a subject of extensive research (see, for example, [7–12]). In most of the considered cases customers are pushed-out either from the head of the queue or from the back. Being motivated by the recent paper [2], where the authors show that ROS policy may outperform FCFS policy, one addresses here the case when the customers are pushed-out from the queue in a random fashion.

The main performance characteristic under consideration in this paper is the stationary waiting time distribution. Below one presents the method for the computation of the stationary waiting time distribution (in terms of transforms) in $G|M|n|r$ queue under FCFS (and also suitable for non-preemptive LCFS) scheduling and with random renovation policy. The method leads to the recursive algorithm for the computation of the moments of the waiting time, but does not allow one to recover the waiting time distribution in any simple way.

2   System Description

Consider a $G|M|n|r$ queueing system. Times between arrivals are i.i.d. with the distribution function $A(x)$ and finite mean. Service times are exponentially distributed with rate μ. Any arriving customer which finds system full, is lost. Customers are served from the queue in FCFS order.

It is assumed that the random renovation policy is implemented in the system. This policy implies that a customer upon service completion pushes out l customers from the queue with probability $q(l)$, if there are more than l customers in the queue and with the probability $Q(l) = \sum_{k=l}^{r} q(k)$ empties the queue i.e. pushes out all customers. The pushed out customers leave the system and further have no effect on it. The probabilities $q(l)$, $0 \leq l \leq r - 1$, and $Q(r) = q(r)$ according to [9] are called renovation probabilities. As usual the normalization condition $Q(0) = \sum_{l=0}^{r} q(l) = 1$ must hold. Clearly $q(0)$ is the probability that the customer upon service completion leaves the system without pushing out any customers from the queue. The choice of positions in the queue from which the customers have to be pushed out is done in a purely random fashion i.e. all outcomes are equally likely. After renovation is performed, one customer (if any is left) is chosen for service from the head of the queue.

In what follows one will make use of several results, which were obtained in [10]. Specifically one considers that the following quantities are known: the stationary distribution p_i^-, $i = \overline{0, n + r - 1}$, of the number of customers in the system at arrival instants, stationary loss probability π, stationary probability p_{serv} that an arbitrary customer admitted to the system will be served. Notice that these quantities do not depend on the service order and the order in which the customers are pushed out from the queue.

In the next section one obtains several preliminary results needed for the calculation of the stationary waiting time distribution.

3 Auxiliary Quantities

Assume that at some time instant τ the queue is not empty and choose any customer in the queue. Let us call it a tagged customer. It can happen that, once the tagged customer is chosen, a number of other customers may be in front and behind it in the queue.

Let

- τ_i denote the instant of the i-th service completion;
- $N_s(t)$ denote the total number of customers in the system at instant t;
- $N_a(t)$ denote the number of arrivals in the interval $(\tau, \tau + t)$;
- $N_c(t)$ denote the number of service completions in the interval $(\tau, \tau + t)$;
- $N_f(t)$ denote the number of customers *in front* of the tagged customer at instant t including those in service;
- $N_b(t)$ denote the number of customers *behind* the tagged customer at instant t.

Due to the fact that the system is analysed in the stationary regime, without the loss of generality, one can put $\tau = 0$.

Introduce the following probabilities:

$$\pi_m(i, j; i', j') = \mathbf{P}\{N_f(\tau_m) = i', N_b(\tau_m) = j' \mid$$
$$N_f(0) = i, N_b(0) = j, N_a(\tau_m) = 0, N_s(0) = i + n\},$$
$$m = \overline{1, r - 1}, i = \overline{m, r - 1}, j = \overline{0, r - i - 1}, i' = \overline{0, i - m}, \ j' = \overline{0, j}. \quad (1)$$

Notice that due to the random renovation policy adopted in the system, each customer on service completion may remove several (or even all) other customers from the queue. The stationary probabilities $\pi_m(i, j; i', j')$ are needed to keep track of the number of customers in the queue after each service completion. The relations for $\pi_m(i, j; i', j')$ can be written out using the first step analysis. It can be verified that, due to the memoryless property of the exponential distribution, following recursive relations hold:

$$\pi_m(i, j; i', j') = \sum_{i''=i'+1}^{i-m+1} \sum_{j''=j'}^{j} \pi_{m-1}(i, j; i'', j'') \pi_1(i'', j''; i', j'),$$
$$m = \overline{2, r - 1}, \ i = \overline{m, r - 1}, \ j = \overline{0, r - i - 1}, \ i' = \overline{0, i - m}, \ j' = \overline{0, j}. \quad (2)$$

where

$$\pi_1(i,j;i',j') = q(i-i'-1+j-j')\frac{C_i^{i'+1}C_j^{j'}}{C_{i+j+1}^{i'+j'+2}},$$

$$i = \overline{1,r-1}, \quad j = \overline{0,r-i-1}, \quad i' = \overline{0,i-1}, \quad j' = \overline{0,j}. \quad (3)$$

Here and henceforth C_n^k denotes the number of k-combinations from a set of n elements.

Another important observations is that after each service completion, the tagged customer may be either "killed", or enter service or remain in the queue. Denote by χ_k the status of the tagged customer immediately after k-th service completion. It is a random variable with three possible values: $\chi_k = 0$ meaning that the tagged customer is in the queue immediately after k-th service completion; $\chi_k = 1$ means that the tagged customer is selected for service immediately after k-th service completion; $\chi_k = -1$ means that the tagged customer was "killed" at the k-th service completion. Clearly if $\chi_k \neq 0$ for some k, all other χ_n, $n > k$, are undefined.

Because one is interested in the waiting time distribution of the customer, which was admitted to the system and eventually received service, one needs to keep track of the service completion at which the customer is selected for service. In order to do this, let us introduce the following probabilities:

$$\pi_m^*(i,j) = \mathbf{P}\left\{\sum_{k=1}^m \chi_k = 1 \middle| N_f(0) = i, N_b(0) = j, N_a(\tau_m) = 0, N_s(0) = i+n\right\},$$

$$m = \overline{1,r-1}, i = \overline{m-1,r-1}, j = \overline{0,r-i-1}, \quad (4)$$

which is the stationary probability that the tagged customer was selected for service immediately after the m-th service completion, given that initially there were i customer in front and j customers behind the tagged customer, all servers were busy and until the m-th service completion there were no new arrivals. The relations for $\pi_m^*(i,j)$ can be written out using the first step analysis. It holds

$$\pi_m^*(i,j) = \sum_{i'=0}^{i-m+1} \sum_{j'=0}^{j} \pi_{m-1}(i,j;i',j')\pi_1^*(i',j'),$$

$$m = \overline{2,r-1}, \quad i = \overline{m-1,r-1}, \quad j = \overline{0,r-i-1}, \quad (5)$$

where

$$\pi_1^*(i',j') = \sum_{l=i'}^{i'+j'} q(l)\frac{C_{j'}^{l-i'}C_{i'}^{i'}}{C_{i'+j'}^l}, \quad i' = \overline{0,r-1}, \quad j' = \overline{0,r-1-i'}. \quad (6)$$

As it was mentioned above the quantities $\pi_m(i,j;i',j')$ and $\pi_m^*(i',j')$ are needed to keep track of the position of the tagged customer and the number of other customers in the queue after each service completion. Now one can proceed to the calculation of the time T the tagged customer spends in the queue under different initial conditions. Introduce the following two quantities:

$$u_{m,\,(k,l)}^{\quad(i,j)}(x)=\mathbf{P}\{N_f(x) = i - k, N_b(x) = j - l, N_c(x) = m|$$

$$N_f(0) = i, N_b(0) = j, N_a(x) = 0\},$$

$$m = \overline{1, r-1}, \; i = \overline{m - 1, r-1}, j = \overline{0, r-i-1}, \; k = \overline{0, i}, \; l = \overline{0, j}, \;\; (7)$$

$$U_{i,j}^*(x) = \mathbf{P}\left\{T < x, \sum_{k=1}^{i+n} \chi_k = 1 \middle| N_f(0) = i, N_b(0) = j, N_a(x) = 0\right\},$$

$$i = \overline{0, r-1}, \quad j = \overline{0, r-i-1}. \quad (8)$$

As each customer upon its service completion may "kill" other customers in the queue independently of its past service time, one can use the law of total probability and (2)–(6) to write out the exact expressions for the two quantities defined by (7) and (8):

$$u_{m,\,(k,l)}^{\quad(i,j)}(x) = \pi_m(i,j;i - m - k, j - l)\frac{(\mu n x)^m}{m!}e^{-\mu n x},$$

$$m = \overline{1, r-1}, \; k = \overline{0, i}, \; l = \overline{0, j}, \; i = \overline{m - 1, r-1}, \; j = \overline{0, r-1-i}, \quad (9)$$

$$u_{0,\,(k,l)}^{\quad(i,j)}(x) = 0, \; k = \overline{1, i}, \; l = \overline{1, j}, \quad\quad\quad\quad\quad\quad (10)$$

$$u_{0,\,(0,0)}^{\quad(i,j)}(x) = u_0(x) = e^{-\mu n x}, \quad\quad\quad\quad\quad\quad (11)$$

$$u_{i,j}^*(x) = \frac{dU_{k,i}^*(x)}{dx} = \sum_{m=1}^{i+1} \pi_m^*(i,j)\frac{(\mu n)^m x^{m-1}}{m!}e^{-\mu n x},$$

$$i = \overline{0, r-1}, \; j = \overline{0, r-i-1}. \quad (12)$$

As one will obtain the waiting time in terms of LST, it is convenient to have the expression for the functions $u_{k,i}^*(x)$ and $u_{m,\,(k,l)}^{\quad(i,j)}(x)$ in terms of the following transforms:

$$\tilde{u}_{m,\,(k,l)}^{\quad(i,j)}(s)=\int_0^\infty e^{-sx}u_{m,\,(k,l)}^{\quad(i,j)}(x)\,dA(x), \; \tilde{u}_{i,j}^*(s)=\int_0^\infty e^{-sx}u_{i,j}^*(x)(1-A(x))dx. \;\; (13)$$

Denote by $\alpha(s)$ and $\tilde{\alpha}(s)$ the LST and the Laplace transformation (LT) of the inter-arrival distribution $A(x)$ and by $\alpha^{(m)}(s)$ and $\tilde{\alpha}^{(m)}(s)$ – m-th derivatives of

$\alpha(s)$ and $\tilde{\alpha}(s)$ with respect to s. Substituting (9)–(12) into (13) one obtains the closed form expressions for $\tilde{u}_{k,i}^*(s)$ and $\tilde{u}_{m,(k,l)}^{(i,j)}(s)$:

$$\tilde{u}_{m,(k,l)}^{(i,j)}(s) = (-1)^m \pi_m(i,j;i-m-k,j-l)\frac{(\mu n)^m}{m!}\alpha^{(m)}(\mu n+s),$$

$$m = \overline{1,r-1}, \; k = \overline{0,i}, \; l = \overline{0,j}, \; i = \overline{m-1,r-1}, \; j = \overline{0,r-1-i}, \quad (14)$$

$$\tilde{u}_0(s) = \alpha(\mu n+s), \quad (15)$$

$$\tilde{u}_{k,i}^*(s) = \sum_{m=1}^{i+1} \pi_m^*(i,k-i-1)\left(\left(\frac{\mu n}{\mu n+s}\right)^m + (-1)^m\frac{(\mu n)^m}{(m-1)!}\tilde{\alpha}^{(m-1)}(\mu n+s)\right),$$

$$k = \overline{0,r}, \; i = \overline{0,r-1}. \quad (16)$$

In the next section it is shown how, using the introduced quantities, one can construct the recursive procedure for the calculation of the stationary waiting time of the customer, which was accepted to the system and eventually selected for service.

4 Stationary Waiting Time Distribution of the Served Customer

Denote by $W_{i,j}(x)$ the conditional stationary probability that tagged customer, residing in the queue, will be selected for service in the interval $(0,x)$, given that initially there were i customers in front (excluding those in service) and j customers behind it and all servers were busy i.e.

$$W_{i,j}(x) = \mathbf{P}\left\{T < x, \sum_{k=1}^{i+n}\chi_k = 1 \bigg| N_f(0) = i, N_b(0) = j, N_s(0) = i+n\right\},$$

$$i = \overline{0,r-1}, \; j = \overline{0,r-1-i}.$$

Using this definition, the conditional stationary waiting time distribution of the arriving customer, which finds $(i+n)$, $0 \le i \le r-1$, customers in the system and which will be eventually selected for service is equal to $W_{i,0}(x)$. In what follows one shows that $W_{i,0}(x)$ can be calculated recursively in terms of LST $\tilde{w}_{i,j}(s) = \int_0^\infty e^{-sx}dW_{i,j}(x)$.

Denote by $w_{i,j}(x) = dW_{i,j}(x)/dx$ the probability density of the $W_{i,j}(x)$. Using the first step analysis and relations (13) and (16), one can write out the following system of equations:

$$w_{0,j}(x) = u_{0,j}^*(x)[1 - A(x)] + \int_0^x u_0(y)dA(y)w_{0,j+1}(x-y), \quad j = \overline{0,r-2}, \quad (17)$$

$$w_{0,r-1}(x) = u^*_{0,r-1}(x)[1 - A(x)] + \int_0^x u_0(y)dA(y)w_{0,r-1}(x - y), \qquad (18)$$

$$w_{i,j}(x) = u^*_{i,j}(x)[1 - A(x)] +$$

$$+ \sum_{m=1}^{i} \sum_{k=0}^{i-m} \sum_{l=0}^{j} \int_0^x u_{m,(k,l)}^{(i,j)}(y)dA(y)w_{i-m-k,j-l+1}(x - y) +$$

$$+ \int_0^x u_0(x)dA(y)w_{i,j+1}(x - y), \quad i = \overline{1, r-2}, \quad j = \overline{0, r-i-2}, \quad (19)$$

$$w_{i,r-i-1}(x) = u^*_{i,r-i-1}(x)[1 - A(x)] +$$

$$+ \sum_{m=1}^{i} \sum_{k=0}^{i-m} \sum_{l=0}^{r-i-1} \int_0^x u_{m,(k,l)}^{(i,j)}(y)dA(y)w_{i-m-k,r-i-l}(x - y) +$$

$$+ \int_0^x u_0(y)dA(y)w_{i,r-i-1}(x - y), \quad i = \overline{0, r-2}, \qquad (20)$$

$$w_{r-1,0}(x) = u^*_{r-1,0}(x)[1 - A(x)] +$$

$$+ \sum_{m=1}^{r-1} \sum_{k=0}^{r-1-m} \int_0^x u_{m,(k,0)}^{(i,j)}(y)dA(y)w_{r-1-m,1}(x - y) +$$

$$+ \int_0^x u_0(y)dA(y)w_{r-1,0}(x - y). \qquad (21)$$

Multiplication of each equation by e^{-sx} and subsequent integration over all x, leads to the system of linear algebraic equations for the LST $\tilde{w}_{i,j}(s)$.

The system of equations for the LST $\tilde{w}_{0,j}(s)$, $j = \overline{0, r-1}$, follows from (17) and (18):

$$\tilde{w}_{0,j}(s) = \tilde{u}^*_{0,j}(s) + \tilde{u}_0(s)\tilde{w}_{0,j+1}(s), \quad j = \overline{0, r-2},$$

$$\tilde{w}_{0,r-1}(s) = \tilde{u}^*_{0,r-1}(s) + \tilde{u}_0(s)\tilde{w}_{0,r-1}(s),$$

Its solution can be written in the form:

$$\tilde{w}_{0,j}(s) = \sum_{m=j+1}^{r-1} \tilde{u}_{m,0}(s)[\tilde{u}_0(s)]^{m-j-1} + \tilde{u}_{r,0}(s)\frac{[\tilde{u}_0(s)]^{r-j-1}}{1 - \tilde{u}_0(s)}, \quad j = \overline{0, r-1}.$$

The expressions for the LST $\omega_{i,j}(s)$, $i = \overline{1, r-2}$, $j = \overline{0, r-i-1}$, are found from the system of equations, which follows from (19) and (20):

$$\tilde{w}_{i,j}(s) = \tilde{u}_{i,j}^*(s) + \sum_{m=1}^{i} \sum_{k=0}^{i-m} \sum_{l=0}^{j} \tilde{u}_{m,(k,l)}^{(i,j)}(s)\tilde{w}_{i-m-k,j-l+1}(s) + \tilde{u}_0(s)\tilde{w}_{i,j+1}(s),$$

$$i = \overline{1, r-2}, \quad j = \overline{0, r-i-2},$$

$$\tilde{w}_{i,r-i-1}(s) = \tilde{u}_{i,r-i-1}^*(s) + \sum_{m=1}^{i} \sum_{k=0}^{i-m} \sum_{l=0}^{r-i-1} \tilde{u}_{m,(k,l)}^{(i,j)}(s)\tilde{w}_{i-m-k,r-i-l}(s) +$$

$$+ \tilde{u}_0(s)\tilde{w}_{i,r-i-1}(s). \quad i = \overline{1, r-2}.$$

Its solution has the form

$$\tilde{w}_{i,j}(s) = \sum_{t=j}^{r-1-i} \hat{u}_{i,t}(s)[\tilde{u}_0(s)]^{t-j} + \hat{u}_{i,j}(s)\frac{[\tilde{u}_0(s)]^{r-j-i-1}}{1 - \tilde{u}_0(s)},$$

$$i = \overline{1, r-2}, \quad j = \overline{0, r-1-i},$$

where

$$\hat{u}_{i,t}(s) = \tilde{u}_{i,t}^*(s) + \sum_{m=1}^{i} \sum_{k=0}^{i-m} \sum_{l=0}^{t} \tilde{u}_{m,(k,l)}^{(i,t)}(s)\tilde{w}_{i-m-k,t-l+1}(s), \ t = \overline{j, r-1-i}.$$

The computation of $\tilde{w}_{i,j}(s)$ must be performed first by fixing i (starting from $i = 1$) and then iterating over j from $r - 1 - i$ down to 0.

The last unknown LST of $\tilde{w}_{r-1,0}(s)$ is computed from the relation

$$\tilde{w}_{r-1,0}(s) = \frac{1}{1 - \tilde{u}_0(s)} \left(\tilde{u}_{r-1,0}^*(s) + \sum_{m=1}^{r-1} \sum_{k=0}^{r-1-m} \tilde{u}_{m,(k,0)}^{(i,j)}(s)\tilde{w}_{r-1-m,1}(s) \right),$$

which follows from (21).

The unconditional stationary waiting time distribution $W(x) = \mathbf{P}\{T < x\}$ of the arriving customer which was admitted to the system and was eventually selected for service, in terms of its LST $\tilde{w}(s) = \int_0^\infty e^{-sx} dW(x)$, is equal to

$$\tilde{w}(s) = \frac{1}{(1 - \pi)p_{\text{serv}}} \sum_{i=0}^{r-1} \tilde{w}_{i,0}(s)p_{n+i}^-,$$

where p_{serv} denotes the stationary probability that an arbitrary customer admitted to the system is selected for service and p_i^- is the stationary probability that a customer on arrival finds i customers in the system. Again notice that the expressions for both p_{serv} and p_i^- have been found in [10].

5   Stationary Waiting Time Distribution of the Pushed-Out Customer

The customer admitted to the system may not be served but may be pushed out from the queue, while waiting for service. The derivation of the unconditional waiting time in terms of LST of such a customer can be performed in a completely similar manner as in Sects. 3 and 4 with minor modifications. Instead of probabilities $\pi_m^*(i,j)$ that the customer will be selected for service immediately after the m-th service completion, one must use the probabilities $\tilde{\pi}_m(i,j)$ that the customer will be pushed out from the queue immediately after the m-th service completion i.e.

$$\tilde{\pi}_m(i,j) = \mathbf{P}\left\{\sum_{k=1}^{m}\chi_k = -1 \middle| N_f(0)=i, N_b(0)=j, N_a(\tau_m)=0, N_s(0)=i+n\right\},$$

$$m = \overline{1,r-1}, i = \overline{m-1,r-1}, j = \overline{0,r-i-1}.$$

Using the first step analysis one can obtain the following recursive procedure for the computation of $\tilde{\pi}_m(i,j)$:

$$\tilde{\pi}_m(i,j) = \sum_{i'=0}^{i-m+1}\sum_{j'=0}^{j}\pi_{m-1}(i,j;i',j')\tilde{\pi}_1(i',j'),$$

$$m = \overline{2,r-1}, \quad i = \overline{m-1,r-1}, \quad j = \overline{0,r-i-1},$$

where

$$\tilde{\pi}_1(i',j') = \sum_{i''=1}^{i'+j'}q(i'')\frac{i''}{i'+j'+1} + Q(i'+j'+1),$$

$$i' = \overline{0,r-1}, \quad j' = \overline{0,r-1-i'}.$$

Next, instead of the functions $U_{i,j}^*(x)$ defined by (8) one introduces functions

$$U_{i,j}^-(x) = \mathbf{P}\left\{T < x, \sum_{k=1}^{i+n}\chi_k = 1 \middle| N_f(0)=i, N_b(0)=i-j, N_a(x)=0\right\},$$

$$i = \overline{0,r}, \quad j = \overline{0,r-1}. \quad (22)$$

Again applying the law of total probability, one gets the exact expression for (22):

$$u_{i,j}^-(x) = \frac{dU_{k,i}^-(x)}{dx} = \sum_{m=1}^{i+1}\tilde{\pi}_m(i,i-j-1)\frac{(\mu n)^m x^{m-1}}{m!}e^{-\mu nx},$$

$$i = \overline{0,r}, \quad j = \overline{0,r-1}, \quad (23)$$

which in terms of transformation

$$\tilde{u}_{i,j}^-(s) = \int_0^{\infty}u_{i,j}^-(x)(1-A(x))dx$$

has the form

$$\tilde{u}_{i,j}^-(s) = \sum_{m=1}^{i+1} \tilde{\pi}_m(j, i-j-1)\left(\left(\frac{\mu n}{\mu n+s}\right)^m + (-1)^m \frac{(\mu n)^m}{(m-1)!}\tilde{a}^{(m-1)}(\mu n+s)\right),$$

$$i = \overline{0,r}, \ j = \overline{0,r-1}. \quad (24)$$

Denote by $W_{i,j}^-(x)$ the conditional stationary probability that tagged customer, residing in the queue, will be killed in the interval $(0,x)$ while waiting for service, given that initially there were i customers in front (excluding those in service) and j customers behind it and all servers were busy i.e.

$$W_{i,j}^-(x) = \mathbf{P}\left\{T < x, \sum_{k=1}^{i+n}\chi_k = -1\middle| N_f(0) = i, N_b(0) = j, N_s(0) = i+n\right\},$$

$$i = \overline{0,r-1}, \ j = \overline{0,r-1-i}.$$

The system of equations for the computation of $W_{i,j}^-(x)$ and the system of linear algebraic equations for $W_{i,j}^-(x)$ in terms of LST $\tilde{w}_{i,j}^-(s)$ can be written out by analogy with the previous section. Thus they are omitted here and only the final solution is given below. The LST $\tilde{w}_{0,j}^-(s)$, $j = \overline{0,r-1}$, are determined by the following expression:

$$\tilde{w}_{0,j}^-(s) = \sum_{m=j+1}^{r-1}\tilde{u}_{m,0}^-(s)[\tilde{u}_0(s)]^{m-j-1} + \tilde{u}_{r,0}^-(s)\frac{[\tilde{u}_0(s)]^{r-j-1}}{1-\tilde{u}_0(s)}, \ j = \overline{0,r-1}.$$

The expressions for the LST $\tilde{w}_{i,j}^-(s)$, $i = \overline{1,r-2}$, $j = \overline{0,r-i-1}$, have the form:

$$\tilde{w}_{i,j}^-(s) = \sum_{t=j+i+1}^{r-1}\hat{u}_{m,i}(s)[\tilde{u}_0(s)]^{m-j-i-1} + \hat{u}_{r,i}(s)\frac{[\tilde{u}_0(s)]^{r-j-i-1}}{1-\tilde{u}_0(s)},$$

$$i = \overline{1,r-2}, \ j = \overline{0,r-1-i},$$

where

$$\hat{u}_{t,i}(s) = \tilde{u}_{t,i}^-(s) + \sum_{m=1}^{i}\sum_{k=0}^{i-m}\sum_{l=0}^{t-i-1}\tilde{u}_{m,\,(i,t-i-1)}^-(s)\tilde{w}_{i-m-k,t-i-l}^-(s), \ t = \overline{i+j+1,r}.$$

The computation must be performed for the fixed i (starting from $i = 1$) by iterating over j from $r-1-i$ down to 0. The LST $\tilde{w}_{r-1,0}^-(s)$ is computed from the relation

$$\tilde{w}_{r-1,0}^-(s) = \frac{1}{1-\tilde{u}_0(s)}\left(\tilde{u}_{r-1,0}^-(s) + \sum_{m=1}^{r-1}\sum_{k=0}^{r-1-m}\tilde{u}_{m,\,(r-1,0)}^-(s)\tilde{w}_{r-1-m-k,1}^-(s)\right).$$

Finally the unconditional stationary waiting time distribution of the arriving customer which was admitted to the system and but was pushed out while waiting for service in terms LST $\tilde{w}^-(s)$ is equal to

$$\tilde{w}^-(s) = \frac{1}{(1-\pi)p_{\text{loss}}} \sum_{i=0}^{r-1} \tilde{w}_{i,0}^-(s) p_{n+i}^-,$$

where p_{loss} denotes the stationary probability that an arbitrary customer admitted to the system is pushed out of the system (before receiving service) and p_i^- is the stationary probability that a customer on arrival finds i customers in the system.

6   Conclusion

All the presented results were obtained assuming the FCFS order. The method does not change, if one changes the service order to non-preemptive LCFS. One of the interesting directions of further research is to check whether the method proposed in this paper as well as the method proposed for the computation of stationary distribution can be applied in the markov-modulated case. This as shown, for example, in [14] is not always possible and thus may require additional insight into the model.

Acknowledgments. This work was supported by the Russian Foundation for Basic Research (grants 15-07-03007 and 15-07-03406).

References

1. Wouter, R., Laevens, K., Walraevens, J., Bruneel, H.: Random-order-of-service for heterogeneous customers: waiting time analysis. Ann. Oper. Res. **226**(1), 527–550 (2015)
2. Wouter, R., Laevens, K., Walraevens, J., Bruneel, H.: When random-order-of-service outperforms first-come first-served. Oper. Res. Lett. **43**(5), 504–506 (2015)
3. Pechinkin, A.V.: A queueing system with a markov input flow and the discipline of random selection of customers in a queue. Autom. Remote Control **61**(9), 1495–1500 (2000)
4. Pechinkin, A.V., Trishechkin, S.I.: An $SM_2/MSP/n/r$ system with random-service discipline and a common buffer. Autom. Remote Control **64**(11), 1742–1754 (2003)
5. Bocharov, P.P., Pechinkin, A.V.: Application of branching processes to investigate the $M/G/1$ queueing system with retrials. In: International Conference on Distributed Computer Communication Networks. Theory and Applications, pp. 20–26. Tel-Aviv (1999)
6. Bocharov, P.P., D'Apice, C., Manzo, R., Pechinkin, A.V.: Analysis of the multi-server markov queuing system with unlimited buffer and negative customers. Autom. Remote Control **68**(1), 85–94 (2007)
7. Kreinin, A.: Queueing systems with renovation. J. Appl. Math. Stoch. Anal. **10**(4), 431–443 (1997)

8. Bocharov, P.P., Zaryadov, I.S.: Queueing systems with renovation. Stationary probability distribution. Bull. Peoples Friendship Univ. Russ. Ser. Math. Inf. Sci. Phys. **1–2**, 14–23 (2007)

9. Zaryadov, I.S., Pechinkin, A.V.: Stationary time characteristics of the GI/M/n/ system with some variants of the generalized renovation discipline. Autom. Remote Control **70**(12), 2085–2097 (2009)

10. Zaryadov, I.S.: Stationary service characteristics of queueing system G/M/n/r with general renovation. Bull. Peoples Friendship Univ. Russ. Ser. Math. Inf. Sci. Phys. **2**, 3–10 (2008)

11. Zaryadov, I.S.: Queueing systems with general renovation. In: International Congress on Ultra Modern Telecommunications, pp. 1–4. IEEE Press, St. Petersburg (2009)

12. Zaryadov, I.S.: The GI/M/n/ queuing system with generalized renovation. Autom. Remote Control **71**(4), 663–671 (2010)

13. Bocharov, P.P., D'Apice, C., Pechinkin, A.V., Salerno, S.: Queueing Theory. VSP, Utrecht, Boston (2004)

14. Razumchik, R., Telek, M.: Delay analysis of a queue with re-sequencing buffer and markov environment. Queueing Syst. **82**(1–2), 7–28 (2016)

Analysis of the Packet Path Lengths in the Swarms for Flying Ubiquitous Sensor Networks

Anastasia Vybornova[✉], Alexander Paramonov, and Andrey Koucheryavy

The Bonch-Bruevich Saint-Petersburg State University of Telecommunications,
Prospekt Bolshevikov 22, St. Petersburg 193232, Russia
a.vybornova@gmail.com
http://sut.ru

Abstract. The article gives an approach for estimating the packet path length between the nodes for Flying Ubiquitous Sensor Networks. Authors show that such networks may be represented as three-dimensional swarms of nodes. The mathematical models for the ball and cubic swarms are given, as well as the simulation approach for the other shapes. Also two types of network architecture are considered – direct transfer and multi-hop.

Simulation shows that the shape of the swarm significantly impact on the average packet path length both for direct and multi-hop data transfer cases: more centralized shapes (e.g. ball) give better results (smaller path length) than less centralized (e.g. cube), and this difference becomes more significant in the multi-hop mode. Number of hops for multi-hop mode also shows the same dependency.

Keywords: Flying sensor network · Swarms of nodes · Path length · Floyd-Warshall algorithm

1 Introduction

Over the recent years a few new trends have appeared in the telecommunication area. Internet of things, Flying sensor networks, Tactile Internet, and other modern conceptions have quite different requirements, applications, and technology approaches, but also have one thing in common. All these new ideas involve millions of new telecommunication devices.

For example, Internet of Things intends to connect almost everything to the global telecommunication network. In this case not millions, but even billions new devices may be connected to the network, and the spatial density of the telecommunication units may become quite high. Tactile internet conception complements the Internet of Things with the new requirements to the Quality of Service for the new networks, especially to the latency [1].

Flying Ubiquitous Sensor Networks idea also involves numerous new flying devices with a new Quality of Service requirements to be connected to the global

© Springer International Publishing AG 2016
V.M. Vishnevskiy et al. (Eds.): DCCN 2016, CCIS 678, pp. 361–368, 2016.
DOI: 10.1007/978-3-319-51917-3_32

networks [2]. Therefore, one of the most important trends in the telecommunication fields now is latency issues research in the large groups of network nodes.

The type of such nodes (mobile phones, sensors, unmanned aerial vehicles (UAVs), or other devices) is not such important. As the large amount of nodes is considered in modern networks, we may describe the groups of such devices as a spatial geometric forms (e.g. ball, cube, cylinder and others) filled in with the nodes. For example, an office building with the numerous mobile phones, laptops, sensors may be represented as a rectangular cuboid filled in with the eventually distributed nodes, as well as a swarm of unmanned aerial vehicles may be described as a ball-shaped swarm, road sensors – as a long cylinder and so on.

Since time delay depends on the distance between nodes (among other things, such as radio channel parameters and communication protocols), in this article we investigate the distances between nodes inside the different spatial geometrical shapes. A few articles in this area may be found. For instance in [3] authors suggest a new network models for the Flying Ad-hoc Networks. In [4] the swarm of UAVs is considered as a queueing system, which allows to obtain more significant information about Quality of Service in such swarms.

In this research two types of network architecture are considered. The first is a direct transfer architecture, which is typical for the traditional infrastructure networks. In this case packets are transferred directly from the source to the destination with the radio channel. The second approach is a multi-hop network architecture, in which packets on a way from the source to the destination may go through one or more intermediated nodes. Multi-hop communication is typical for the wireless sensor networks and Device-to-Device (D2D) communication, which can be widely used in the 5G networks. For this method besides the distance between nodes, we also estimate the average number of hops.

The rest of the article is structured as follows. Section 1 gives the theoretical background of the problem, Sect. 2 describes the different models for the swarm of sensor nodes, and Sect. 3 represent the result of the modelling.

2   Theoretical Background

The problem of finding distances between random nodes inside the different geometric shapes is a well-known line picking problem. According to [5], line picking probabilities (i.e. probability distributions for the lengths of the lines between two random points) and averages line lengths are known for a number of geometrical shapes. This allows us to prove the simulation models for the different groups of nodes. For example, the probability distribution of the distance between two random points within the ball is given by:

$$P(l) = 3\frac{l^2}{R^3} - \frac{9}{4}\frac{l^3}{R^4} + \frac{3}{16}\frac{l^5}{R^6} \tag{1}$$

where R is the radius of the ball [6]. The average distance is:

$$\bar{l} = \frac{36}{35}R \qquad\qquad (2)$$

The probability function of the distance between two random points inside the unit cube has quite complex mathematical expression and is given in [7]. The average distance is given by the Robbins constant:

$$\Delta(3) = 0.66170... \qquad\qquad (3)$$

For the other spatial geometrical shapes, the probability function of the random line length is unknown, but in some cases estimation of a mean value is given. For instance, the mean distance between two points within the tetrahedron of unit volume is approximately [8]:

$$\bar{l} = 0.7308 \pm 0.0002 \qquad\qquad (4)$$

3   Model Description

The following model was used to obtain the network simulation results. We used $N = 100$ network nodes that were randomly distributed inside the different spatial shapes: ball, cube, cylinder. All shapes had the same volume equal to $8\,000\,000\,\mathrm{m}^3$ (see Figs. 1, 2 and 3).

The distances between all pairs of nodes for different shapes were found. In case of direct transfer, packets should be transferred exactly on these distances. However, in the networks with the multi-hop infrastructure such as wireless sensor networks radio coverage area is usually limited due to the energy saving purpose. In this case we can transform the distance matrix so that all the distances that is

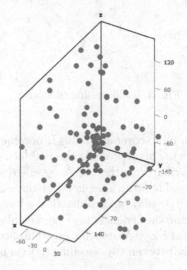

Fig. 1. Ball-shaped swarm

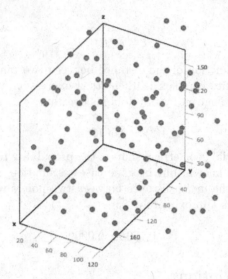

Fig. 2. Cubic swarm

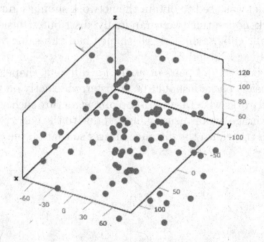

Fig. 3. Cylinder-shaped swarm

larger than the maximum radio coverage radius R become infinite (or quite large comparing with the other). This represents the inability of nodes to connect to the other nodes that is farther then the node's coverage area. The data transfer radius for our model is equal to 50 m. Therefore, we obtained the distance matrix for all the nodes. This matrix may be considered as a weighted graph, where the sensor nodes are vertices and the distances are edge weights. To find shortest paths using this graph we applied Floyd-Warshall algorithm, because it allowed to find the shortest path not only between two specific vertices, but between all pairs of vertices. As a result, we obtained the matrices of distances between nodes for the two types of network architecture: direct transfer and multi-hop.

The other aim of the simulation was number of hops estimation (for multi-hop mode). Since the exact calculation of the number of hops with Floyd-Warshall algorithm requires recursive function and therefore significant computation resources, the average number of hops was estimated for all the geometric shapes using the following formula:

$$\overline{h} = \frac{\text{Average multi-hop distance}}{\text{Average direct distance for distances less than 50 m}} \quad (5)$$

The models for cube and ball were proved by comparing the theoretical distances distribution and simulated distances distribution (direct transfer case). The model for a cylinder could not be proved, as the theoretical form of the distances between two random points within this geometrical shape is unknown.

4 Results

The simulation models show the following results. For all the geometrical shapes multi-hop distances are larger than direct distances, as it might be expected. Difference between two these distances is the smallest for the ball-shaped swarm and the largest for a cube-shaped (see Figs. 4, 5 and 6).

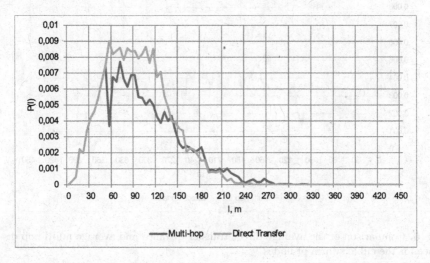

Fig. 4. Comparison of the average direct transfer distance and average multi-hop distance for the swarm of nodes inside the ball

Moreover the compare shows that for the cubic swarm multi-hop transfer distances is significantly larger than for the cylinder and ball swarms. Also the average number of hops for the cubic-shaped swarm is larger than for the other shapes (see Table 1) (Figs. 7 and 8).

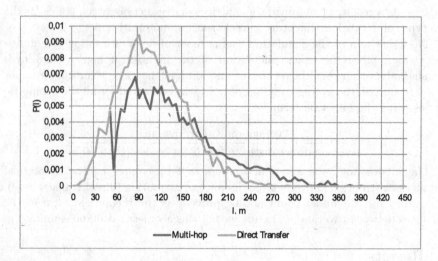

Fig. 5. Comparison of the average direct transfer distance and average multi-hop distance for the swarm of nodes within the cylinder

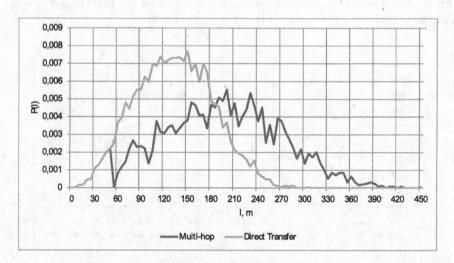

Fig. 6. Comparison of the average direct transfer distance and average multi-hop distance for the cubic swarm of nodes

Table 1. Average number of hops

Shape	Ball	Cylinder	Cube
Average number of hops	2,3	3,1	4,8

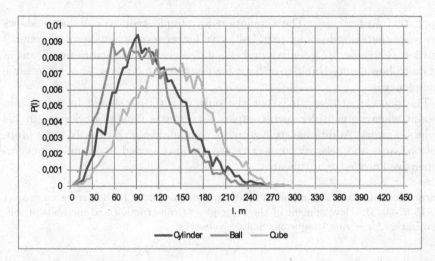

Fig. 7. Comparison of the average direct transfer distances for the different geometric figures

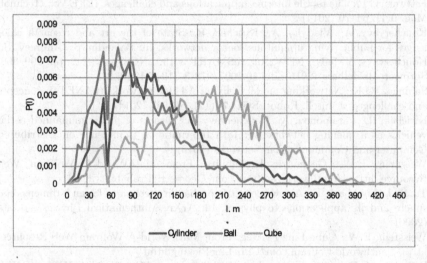

Fig. 8. Comparison of the average multi-hop distances for the different geometric figures

5   Conclusions

This article gives an approach for estimating the packet path length between the nodes in three-dimensional swarms. Unfortunately, the strict mathematical models are absent for the majority of shapes, therefore the simulation of Flying Ubiquitous Sensor Network nodes' swarms is the best solution for this purpose.

According to the obtained results we may conclude that shape of the swarm significantly impact on the average packet path length both for direct and multi-hop data transfer cases. The simulation for simple geometrical shapes shows that more centralized shapes (e.g. ball) give better results (smaller path length) than less centralized (e.g. cube), and this difference becomes more significant in the multi-hop mode. Number of hops for multi-hop mode also shows the same dependency.

The packet path length is not the only factor that should be taken into account when we consider such complex aspect as a Quality of Service in the modern network. Additional studies on the communication protocols, radio propagation issues and network connectiveness are also needed to obtain more comprehensive knowledge in this area.

Acknowledgments. The reported study was supported by RFBR, research project No. 15 07-09431a Development of the principles of construction and methods of self-organization for Flying Ubiquitous Sensor Networks.

References

1. Fettweis, G.P.: The tactile internet: applications and challenges. IEEE Veh. Technol. Mag. **9**(1), 64–70 (2014)
2. Koucheryavy, A., Vladyko, A., Kirichek, R.: State of the art and research challenges for public flying ubiquitous sensor networks. In: Balandin, S., Andreev, S., Koucheryavy, Y. (eds.) NEW2AN/ruSMART 2015. LNCS, vol. 9247, pp. 299–308. Springer, Heidelberg (2015). doi:10.1007/978-3-319-23126-6_27
3. Sahingoz, O.K.: Networking model in flying Ad Hoc networks (FANETs): concepts and challenges. J. Intell. Robot. Syst. **74**(12), 513527 (2014)
4. Kirichek, R., Paramonov, A., Koucheryavy, A.: Swarm of public unmanned aerial vehicles as a queuing network. In: 18th International Conference on Distributed Computer and Communication Networks (2015)
5. Weisstein, E.W.: Geometric Probability. From MathWorld-A Wolfram Web Resource. http://mathworld.wolfram.com/GeometricProbability.html
6. Tu, S.-J., Fischbach, E.: A new geometric probability technique for an n-dimensional sphere and its applications to physics. In: ArXiv Mathematical Physics e-prints (2000)
7. Weisstein, E.W.: Cube Line Picking. From MathWorld-A Wolfram Web Resource. http://mathworld.wolfram.com/CubeLinePicking.html
8. Weisstein, E.W.: Tetrahedron Line Picking. From MathWorld-A Wolfram Web Resource. http://mathworld.wolfram.com/TetrahedronLinePicking.html

Properties of Fluid Limit for Closed Queueing Network with Two Multi-servers

Svetlana Anulova[✉]

IPU RAN, Profsoyuznaya, 65, 117997 Moscow, Russian Federation
anulovas@ipu.rssi.ru
http://www.ipu.ru/

Abstract. A closed network consists of two multi-servers with n customers. Service requirements of customers at a server have a common cdf. State parameters of the network: for each multi-server empirical measure of the age of customers being serviced and for the queue the number of customers in it, all multiplied by n^{-1}.

Our objective: asymptotics of dynamics as $n \to \infty$. The asymptotics of dynamics of a single multi-server with an arrival process as the number of servers $n \to \infty$ is currently studied by famous scientists K. Ramanan, W. Whitt et al. Presently there are no universal results for general distributions of service requirements—the results are either for continuous or for discrete time ones; the same for the arrival process. We develop our previous asymptotics results for a network in discrete time: find equilibrium and prove convergence as $t \to \infty$.

Motivation for studying such models: they represent call/contact centers.

Keywords: Multi-server queues · GI/G/n queue · Fluid limits · Mean-field limits · Strong law of large numbers · Measure-valued processes · Fluid limit equilibrium and convergence · Switching networks · Call/contact centers

1 Introduction

1.1 Review of Investigated Contact Centers Models

In the last ten years an extensive research in mathematical models for telephone call centers has been carried out, cf. Refs. [2–6, 8–17] of [1]. The object has been expanded to more general customer contact centers (with contact also made by other means, such as fax and e-mail). In order to describe the object efficiently the state of the model must include: (1) for every customer in the queue the time that he has spent in it and (2) for every customer in the multi-server the time that he has spent after entering the service area, that is being received by one of the available servers.

S. Anulova—This work was supported by RFBR grant No. 14-01-00319 "Asymptotic analysis of queueing systems and nets".

© Springer International Publishing AG 2016
V.M. Vishnevskiy et al. (Eds.): DCCN 2016, CCIS 678, pp. 369–380, 2016.
DOI: 10.1007/978-3-319-51917-3_33

The focus of research was on a multi-server with a large number of servers, because it is typical of contact centers. One of important relating questions is the dynamics of the queue of a multi-server with a large number of servers. For such queues were found fluid limits as the number of servers tends to infinity. Notice that such a limit is a deterministic function of time with values in a certain measure space, or in a space containing such a component.

An important particular question is the convergence of the fluid limit to a stable state as time tends to infinity. For a discrete time model W. Whitt has found equilibrium states (a multitude) of the fluid model and proved the time convergence in a special case—for a primitive arrival process and for initial condition with empty multi-server and queue, [2, Sect. 7] .

1.2   A New Model for Contact Centers and Its Fluid Limit with Equilibrium Behavior

We have suggested in [1] a more suitable model for contact centers. The number of customers is fixed. Customers may be situated in two states: normal and failure. There is a multi-server which repairs customers in the failure state. The repair time/the time duration of a normal state is a random variable, independent and identically distributed for all customers. Now "the arrival process" in the multi-server does not correspond to that of the previous $G/GI/s+GI$ model. For a large number of customers and a suitable number of servers we have calculated approximately the dynamics of the normalized state of the system—its fluid limit. Now we explore the convergence of the fluid limit as time tends to infinity and find its steady-state (or equilibrium).

We confine ourselves to a discrete time model. W. Whitt has written a very interesting seminal article about fluid limits of multi-server dynamics [2], with Sects. 6 and 7 in a simple discrete case. About 150 authors have cited it and made generalizations to the continuous time. But their results do not enclose Whitt's discrete time ones. Walsh Zuñiga [3], with his results most close to including discrete time, permits discrete time only for service but his arrival process is strictly continuous.

In Whitt's article [2] the ideas of the convergences proofs are true and very lucid, and the proofs are clearly presented. We have transferred his proof technique to our new network model in [1]. Now we exploit his equilibrium technique.

2   Closed Multi-server Network with n Customers and Its Fluid Limit Equilibrium

2.1   Network Description

Consider a closed network consisting of n customers. They may be situated in two states: normal and failure. A multi-server repairs customers in the failure state. The repair time (resp., the time duration of a normal state) is a random variable, independent and identically distributed for all customers. For a large number of customers and a suitable number of servers we shall calculate the number of current failures, so much as an approximation.

Now we give a rigorous description of this model.

Consider a closed network consisting of n customers and two multi-servers. Multi-server 1 (further denoted MS1) consists of n servers (for the customers in the normal state), the time they service a customer has distribution G^1. Multi-server 2 (further denoted MS2) consists of $s_n n$ servers with a number $s_n \in (0, 1)$ (for the customers in the failure state), the time they service a customer has distribution G^2. The distributions G^1, G^2 are discrete: they are concentrated on $\{1, 2, \ldots\}$. Service times are independent for both servers and all customers. We will investigate the behavior of the net as $n \to \infty$, namely, we shall establish a stochastic-process fluid limit. It will be done only in a special case: discrete time $t = 0, 1, 2, \ldots$

We begin with a simple example of functioning of this network.

Example 1. Let at time $t = 0$ all n customers be in a normal state. Each customer switches over to the failure state according to the distribution function G^1 and tries to enter multi-server 2. The early failure customers can do it, but with time growing multi-server 2 may become fully occupied. Then the failure customers create a queue, waiting for the first available server in multi-server 2. Recall that the server becomes afresh available with time distribution G^2.

In this example and everywhere further we demand:

Assumption 1. *Customer enters service immediately upon arrival to a multi-server if there is a server available. If the servers in MS2 are all busy, the arriving customer waits in queue. Customers from queue are served in order of their arrival (FCFS) by the first available MS2 server.*

In MS1 no queue may arise—if all n its servers are occupied then all customers are in MS1, therefore no new customer can arrive.

Denote the number of customers at a moment $t = 0, 1, \ldots$ in MS1 (resp., MS2) by $B_n^1(t)$ (resp., $B_n^2(t)$) and the number of customers in the queue $Q_n(t) = n - B_n^1(t) - B_n^2(t)$. These quantities must be defined more exactly. Namely,

$$B_n^i(t) = \sum_{k=0}^{\infty} b_n^i(t, k), \ i = 1, 2, \text{ and } Q_n(t) = \sum_{0}^{\infty} q_n(t, k)$$

with $b_n^i(t, k)$ being the number of customers in the multi-server i at the moment t who have spent there time k, $i = 1, 2$, and $q_n(t, k)$ being the number of customers in the queue at the moment t who have been there precisely for time k. $b_n^i(t, k)$ may also be interpreted as the number of busy servers at time t in the multi-server i that are serving customers that have been in service precisely for time k, $i = 1, 2$.

At the same time moment $t \in \{1, 2, \ldots\}$ multiple events can take place, so we have to specify their order.

We must create a fictitious queue for the MS1—in fact this multi-server is so large (n servers), that any customer of the whole quantity n trying to enter the MS1 at once finds a free server in it.

At the time moment t the parameters b^1, b^2, q are taken from the previous time $t - 1$ and processed to the current situation.

For both multi-servers:

- first, customers in service are served;
- second, the served customers move to another multi-server queue, to the end of it;
- third, waiting customers in queue move into service of the multi-server according to Assumption 1.

Customers enter service in MS2 whenever a server is available, so that the system is work-conserving; i.e. we assume that $Q_n(t) = 0$ whenever $B_n^2(t) < s_n n$, and that $B_n^2(t) = s_n n$ whenever $Q_n(t) > 0$, $t = 0, 1, 2, \ldots$. This condition can be summarized by the equation

$$(s_n - B_n^2(t)/n)Q_n(t) = 0 \text{ for all } t \text{ and } n.$$

2.2   Fluid Limit Dynamics

Notations

Denote for $i = 1, 2$:

- $G^{i;c}(k) := 1 - G^i(k)$ and $g^i(k) := G^i(k) - G^i(k-1), k = 1, 2, \ldots$
- E^i the expectation of the time the server in MSi services a customer[1]:

$$E^i := \sum_{k=1}^{\infty} k g^i(k) = 1 + \sum_{k=1}^{\infty} G^i(k) .$$

- $\sigma_n^i(t)$ the number of service completions in MSi at time moment $t = 1, 2, \ldots$.

Symbol $\Rightarrow$ means convergence of the network state characteristics to a constant in probability as the index n denoting the number of customers tends to infinity.

Fluid Limit Dynamics

Under certain conditions, specifically $\lim_{n \to \infty} s^{(n)} = s \in (0, 1)$, the fluid limit exists and its dynamics is described below (the proof is given in [1, Theorem 1]).

As $n \to \infty$,

$$\frac{b_n^i(t, k)}{n} \Rightarrow b^i(t, k), i = 1, 2, \tag{1}$$

---

[1] for the last equality see "Expected value" in Wikipedia.

$$\frac{q_n(t,k)}{n} \Rightarrow q(t,k), \tag{2}$$

$$\frac{\sigma_n^i(t)}{n} \Rightarrow \sigma^i(t), \ i = 1, 2, \tag{3}$$

for each $t \geq 0$ and $k \geq 0$, where $(b^1, b^2, q, \sigma^1, \sigma^2)$ is a vector of deterministic functions (all with finite values).

Further, for each $t = 0, 1, \ldots$

$$\frac{B_n^i(t)}{n} \equiv \frac{\sum_{k=0}^{\infty} b_n^i(t,k)}{n} \Rightarrow B^i(t) \equiv \sum_{k=0}^{\infty} b^i(t,k), \ i = 1, 2, \tag{4}$$

$$\frac{Q_n(t)}{n} \equiv \frac{\sum_{k=0}^{\infty} q_n(t,k)}{n} \Rightarrow Q(t) \equiv \sum_{k=0}^{\infty} q(t,k), \tag{5}$$

with
$$B^1(t), B^2(t), Q(t) \geq 0, \ B^1(t) + B^2(t) + Q(t) = 1, \tag{6}$$

$$B^2(t) \leq s, \text{ and } (s - B^2(t))Q(t) = 0. \tag{7}$$

The evolution of the vector $(b^1, b^2, q, \sigma^1, \sigma^2)(t)$, $t = 0, 1, 2 \ldots$, proceeds with steps of t in the following way. As we go from time $t - 1$ to t, there are two cases, depending on whether $B^2(t-1) = s$ or $B^2(t-1) < s$.

Case 1. $B^2(t-1) = s$. In this first case, after moment $t - 1$ asymptotically all servers are busy and in general there may be a positive queue. In this case,

$$\sigma^i(t) = \sum_{k=1}^{\infty} b^i(t-1, k-1) \frac{g^i(k)}{G^{i;c}(k-1)}, \tag{8}$$

$$b^i(t,k) = b^i(t-1, k-1) \frac{G^{i;c}(k)}{G^{i;c}(k-1)}, \ k = 1, 2, \ldots, \ i = 1, 2, \tag{9}$$

$$b^1(t,0) = \sigma^2(t), \tag{10}$$

$$b^2(t,0) = \min\{\sigma^2(t), Q(t-1) + \sigma^1(t)\}, \tag{11}$$

and finally q is determined with the help of an intermediate queue q',

$$q'(t,0) = \sigma^1(t), \ q'(t,k) = q(t-1, k-1), \ k = 1, 2, \ldots : \tag{12}$$

if $\sigma^2(t) = 0$ then $q(t,k) = q'(t,k)$, $k = 0, 1, \ldots$, \qquad (13)

$$\text{if } \sigma^2(t) \geq \sum_{k=0}^{\infty} q'(t,k) \text{ then } q(t,k) = 0, \ k = 0, 1, \ldots, \tag{14}$$

$$\text{if } 0 < \sigma^2(t) < \sum_{k=0}^{\infty} q'(t,k) \text{ then with} \tag{15}$$

$$c(t) = \min\{i \in \{0, 1, \ldots\} : \sum_{k=i}^{\infty} q'(t,k) \leq \sigma^2(t)\}, \tag{16}$$

$$q(t,k) = \begin{cases} 0 \text{ for } k \geq c(t), \\ \displaystyle\sum_{i=c(t)-1}^{\infty} q'(t,i) - \sigma^2(t) \text{ for } k = c(t) - 1, \\ q'(t,k) \text{ for } k < c(t) - 1. \end{cases}$$

Case 2. $B^2(t-1) < s$. In this second case, after the time moment $t-1$ asymptotically all servers are not busy so that there is no queue. As in the first case, Eqs. (8), (9), and (10) hold. Instead of (11),

$$b^2(t,0) = \min\{s - B^2(t-1) + \sigma^2(t), \sigma^1(t)\}. \tag{17}$$

Then,

$$q(t,k) = 0 \text{ for all } k > 0 \text{ and } q(t,0) = \sigma^1(t) - b^2(t,0). \tag{18}$$

In the rest of the article the queue is described less detailed than in this Subsect. 2.2—no customer age is taken into account.

2.3   Fluid Limit Equilibrium

Consider the discrete time fluid limit for the closed network model dynamics established in our article [1].

Definition 1. *A point in the state space of deterministic fluid processes is called "equilibrium" if fluid processes after reaching this point remain in it. Deterministic fluid processes are formally described/characterized by sets (b^1, b^2, q) consisting of non-negative functions of $b^1(t,k)$, $b^2(t,k)$, $q(t), t = 0, 1, \ldots, k = 0, 1, \ldots,$ satisfying*

$$\sum_{k=0}^{\infty} b^2(t,k) \leq s, \quad \sum_{k=0}^{\infty}(b^1 + b^2)(t,k) + q(t) = 1, t = 0, 1, \ldots,$$

and equilibrium points are described/characterized by sets (b^{*1}, b^{*2}, q^*) consisting of non-negative functions $b^{*1}(k), b^{*2}(k), k = 0, 1, \ldots,$ and a non-negative number q^* satisfying

$$\sum_{k=0}^{\infty} b^{*2}(k) \leq s, \quad \sum_{k=0}^{\infty}(b^{*1} + b^{*2})(k) + q^* = 1. \tag{19}$$

If the initial condition of a fluid process is an equilibrium, then this fluid process is constant in time:

$$b^1(0,k) = b^{*1}(k), b^2(0,k) = b^{*2}(k), k = 0, 1, \ldots, q(0) = q^*$$

implies for $t = 1, 2, \ldots$

$$b^1(t,k) = b^{*1}(k), b^2(t,k) = b^{*2}(k), k = 0, 1, \ldots, q(t) = q^*.$$

For equilibrium (b^{*1}, b^{*2}, q^*) denote $B^{*i} = \sum_{k=0}^{\infty} b^{*i}(k), i = 1, 2.$

Theorem 1. *For the deterministic fluid processes there exists a single equilibrium point. The characteristics b^{*1}, b^{*2}, q^* of this equilibrium point have the form*

$$b^{*i}(k) = b^{*i}(0)G^{i;c}(k), \ k = 1, 2, \ldots, \ i = 1, 2, \ and \ q^* = 1 - (B^{*1} + B^{*2}) \quad (20)$$

with the values of $b^{*1}(0)$ and $b^{*2}(0)$ being equal and determined by the ratio of E^1 to E^2:

1. $B^{*i} = b^{*i}(0)E^i, \ i = 1, 2.$

2. *If* $\dfrac{E^2}{E^1 + E^2} > s$ *then:*

$$B^{*2} = s, \ b^{*2}(0) = \frac{s}{E^2}, \ B^{*1} = s\frac{E^1}{E^2}, \ q^* = 1 - \frac{s}{E^2}(E^1 + E^2) > 0.$$

3. *If* $\dfrac{E^2}{E^1 + E^2} < s$ *then:*

$$B^{*2} < s, \ b^{*2}(0) = \frac{1}{E^1 + E^2}, \ B^{*i} = \frac{E^i}{E^1 + E^2}, \ i = 1, 2, \ and \ q^* = 0.$$

4. *If* $\dfrac{E^2}{E^1 + E^2} = s$ *then:*

$$B^{*2} = s, \ b^{*2}(0) = \frac{1}{E^1 + E^2}, \ B^{*i} = \frac{E^i}{E^1 + E^2}, \ i = 1, 2, \ and \ q^* = 0.$$

Proof. Equation (20) follows from [2, formula (7.7)]. The amount arriving at the time step to MS2 is $b^{*2}(0)$ and it equals the amount of $(b^{*2}(k), \ k = 1, 2, \ldots)$ serviced in the time step—this is demanded by the equilibrium. But the latter amount arrives to MS1, therefore it equals $b^{*1}(0)$. Thus $b^{*1}(0) = b^{*2}(0)$.

1. $\sum_{k=0}^{\infty} G^{i;c}(k)$ is equal to the expectation corresponding to the distribution $G^i, \ i = 1, 2$. Thus

$$B^{*i} = \sum_{k=0}^{\infty} b^{*i}(k) = b^{*i}(0) + \sum_{k=1}^{\infty} b^{*i}(0)G^{i;c}(k) = \sum_{k=0}^{\infty} b^{*i}(0)G^{i;c}(k) =$$
$$b^{*i}(0)\sum_{k=0}^{\infty} G^{i;c}(k) = b^{*i}(0)E^i, \ i = 1, 2. \quad (21)$$

2. If $q^* = 0$, then $B^{*1} + B^{*2} = 1$, that is

$$b^{*2}(0)(E^1 + E^2) = 1, \quad b^{*2}(0) = \frac{1}{E^1 + E^2}.$$

Then $B^{*2} = \dfrac{E^2}{E^1 + E^2} > s$, what is impossible. This impossibility demands q^* to be positive and consequently $B^{*2} = s$. Since $B^{*2} = b^{*2}(0)E^2$

$$b^{*2}(0) = \frac{B^{*2}}{E^2} = \frac{s}{E^2} \text{ and } q^* = 1 - (B^{*1} + B^{*2}) = 1 - \frac{s}{E^2}(E^1 + E^2).$$

q^* is really positive as $\dfrac{s}{E^2}(E^1 + E^2) = s/\dfrac{E^2}{E^1 + E^2} < 1.$

3. If $q^* > 0$, then $B^{*1} + B^{*2} < 1$, that is

$$b^{*2}(0)(E^1 + E^2) < 1, \quad b^{*2}(0) < \frac{1}{E^1 + E^2}.$$

Then $B^{*2} < \dfrac{E^2}{E^1 + E^2} < s$, what is incompatible with $q^* > 0$. This incompatibility demands q^* to be 0 and consequently $B^{*1} + B^{*2} = 1$. As $B^{*1} + B^{*2} = b^{*2}(0)(E^1 + E^2)$,

$$b^{*2}(0) = \frac{B^{*1} + B^{*2}}{E^1 + E^2} = \frac{1}{E^1 + E^2} \text{ and } B^{*2} = b^{*2}(0)E^2 = \frac{E^2}{E^1 + E^2} < s.$$

4. As in previous item, $B^{*2} < \dfrac{E^2}{E^1 + E^2}$, thus $B^{*2} < s$, what is incompatible with $q^* > 0$. This incompatibility demands q^* to be 0 and consequently $B^{*1} + B^{*2} = 1$. As $B^{*1} + B^{*2} = b^{*2}(0)(E^1 + E^2)$,

$$b^{*2}(0) = \frac{B^{*1} + B^{*2}}{E^1 + E^2} = \frac{1}{E^1 + E^2} \text{ and } B^{*2} = b^{*2}(0)E^2 = \frac{E^2}{E^1 + E^2} = s$$

with

$$b^{*2}(0) = \frac{s}{E^2}.$$

2.4   Fluid Limit Convergence to Equilibrium as $t \to \infty$

W. Whitt tried to investigate convergence of the fluid limit trajectory to equilibrium in [2, Sect. 7]. No strong result for universal convergence has been presented, only in particular case—starting from an empty multi-server and an empty queue, see [2, Theorem 7.3]. We shall transfer this simple theorem to our closed network model.

With Whitt's assumption for the initial condition (no customers either in the multiserver or in the queue) and main expectations assumption (the expectation of the service time is equal to 1 and the arrival rate λ is constant and at most 1) the empty multi-server adds with time steps customers with growing age and the fluid process converges monotonically to the unique equilibrium:

for $t = 1, 2, \ldots$ $b(t, k) = \lambda G^c(k)$, $0 \le k \le t$, and $b(t, k) = 0$ for all $k > t$.

In our model this case takes the following form. If MS1 is empty and MS2 is filled with equilibrium parameters, then MS2 remains in this state, the queue decreases and MS1 adds with time steps customers of the next age with equilibrium parameters, like the single multi-server by Whitt, and the state of MS1 converges monotonically to the unique equilibrium state: for $t = 1, 2, \ldots$

$$b^1(t, k) = \begin{cases} b^{*2}(0)G^{1;c}(k), & 0 \le k < t, \\ b^1(t, k) = 0, & k \ge t. \end{cases}$$

This convergence is evident, just like in the corresponding theorem of W. Whitt.

Theorem 2. *Suppose the fluid limit satisfies at time $t = 0$ the following conditions: $B^1(0) = 0$ and $b^2(0, \cdot) = b^{*2}$ (b^{*2} is the component of equilibrium point (b^{*1}, b^{*2}, q^*), see Theorem 1). Then the fluid limit converges to the equilibrium point as $t \to \infty$. Namely:*

– *the state of MS2 remains equilibrium:*

$$b^2(t, \cdot) = b^{*2}, t = 0, 1, 2, \ldots; \tag{22}$$

– *the state of MS1 grows occupying its equilibrium state—with each time step adds the next age equilibrium parameter:*

$$b^1(0, \cdot) \equiv 0 \text{ and for } t = 1, 2, \ldots \, b^1(t, \cdot) = b^{*1}(\cdot) I_t \quad \text{with } I_t = I_{\{0,1,\ldots,t-1\}}; \tag{23}$$

– *the queue decreases—with each time step loses the amount of the previous age MS1 equilibrium parameter:*

$$q(0) = 1 - B^{*2}, \, q(t) = q(t - 1) - b^{*1}(t - 1) = 1 - B^{*2} - \sum b^{*1}(\cdot) I_t =$$
$$1 - B^{*2} - \sum_{l=0}^{t-1} b^{*1}(l), t = 1, 2, \ldots$$

Proof. According to the equilibrium of b^{*2}, as shown in the proof of Theorem 1, at the first time step ($t = 0 \to t = 1$) the multiserver 2 services $b^{*2}(0)$ customers. They proceed into the multiserver 1. And exactly so many customers proceed from the queue to the multiserver 2. Really, the queue at the time 0 is large enough:

$$q(0) = 1 - (B^1(0) + B^2(0)) = 1 - B^{*2} = B^{*1} + q^* \geq b^{*1}(0) \, (= b^{*2}(0)).$$

We finish the proof by induction. Suppose at time t which is not less than 1 the statement of the theorem in Eq. (23) holds. By virtue of equilibrium of $b^2(t, \cdot) = b^{*2}$ at the time step $t \to t + 1$ the multiserver 2 services again $b^{*2}(0)$ customers and proceeds to the state $\{0, b^{*2}(1), , b^{*2}(2), \ldots\}$. The state of MS1 at the time t $b^1(t, \cdot)$ equals to $\{b^{*1}(0), b^{*1}(1), \ldots, b^{*1}(t - 1), 0, 0, \ldots\}$. Having serviced its customers at time step $t \to t + 1$ MS1 proceeds to the statement $\{0, b^{*1}(1), \ldots, b^{*1}(t-1), b^{*1}(t), 0, 0, \ldots\}$. Now the serviced customers of MS2 proceed to MS1: the state of MS1 at time $t + 1$ $b^1(t + 1, \cdot)$ equals to

$$\{b^{*1}(0), b^{*1}(1), \ldots, b^{*1}(t), 0, 0, \ldots\}.$$

And how large is the queue? Does it have $b^{*2}(0)$ customers to fill MS2 at time $t + 1$ again up to equilibrium? The queue equals to

$$1 - \left(\sum_{l=0}^{t} b^{*1}(l) + \sum_{l=1}^{\infty} b^{*2}(l) \right) = 1 - \left(\sum_{l=0}^{t} b^{*1}(l) + \sum_{l=0}^{\infty} b^{*2}(l) - b^{*2}(0) \right) \geq$$
$$1 - (B^{*1} + B^{*2}) + b^{*2}(0) = q^* + b^{*2}(0) \geq b^{*2}(0) \text{ as } q^* \geq 0.$$

The inverse initial condition—MS1 is filled with an equilibrium expectation and MS2 is empty—does not allow such a simple proof for its convergence to the equilibrium.

3   Generalization to Changing Environment

Return to the beginning of Subsect. 2.1. Now the environment of the system is not permanent, it changes with discrete time. And the time duration of a customer's normal state (namely, its distribution) changes also—it corresponds to the current environment. The multi-server which repairs customers in the failure state is irrespective of the environment—the distribution of the customer repair time does not change. We make this restriction on the repairing multi-server for the simplicity of our research. We shall describe a generalization of the fluid limit and its equilibrium in changing environment.

There is a finite number N of changing environments, each of them is denoted by its $i \in \{1, \ldots, N\}$. In the environment $i \in \{1, \ldots, N\}$ the time duration of the normal state is marked by the corresponding i. So, instead of the previous distribution G^1 of the time MS1 services a customer, we have distributions $G^{1|i}$, $i \in \{1, \ldots, N\}$.

The change of environments is specified by a discrete time-homogeneous markov chain $u := u(t)$, $t = 0, 1, \ldots$, with state space $\{1, \ldots, N\}$ and transition probabilities

$$\mathbb{P}(u(1) = j | u(0) = i) = p_{ij}, \ i, j \in \{1, \ldots, N\}.$$

It is obvious that the fluid limit now turns to a piecewise-deterministic Markov process (see [4, Chap. 2]). Really, denote by τ_i, $i = 0, 1, 2, \ldots$, the moments when u receives its new value:

- $\tau_0 = 0$;
- τ_i is the moment when the value which u had at the moment τ_{i-1} becomes changed: $\tau_i = \min\{t > \tau_{i-1} : u(t) \neq u(\tau_{i-1})\}$, $i = 1, 2, \ldots$

Then on each interval $[\tau_i, \tau_{i+1} - 1]$ and subspace $u(\tau_i) = j$ Eqs. (1)–(3) hold with $G^{1|j}$—the distributions of customers ages converge to a fluid limit, $i = 0, 1, \ldots, j \in \{1, \ldots, N\}$.

We have generalized the distribution process describing the state of the network (it is defined in Subsects. 2.1 and 2.2) to a switching (see [5]) distribution process. We explained how this switching distribution process converges as the number of customers $n \to \infty$. Namely, it converges in a complicated way: not to the fluid limit, but to a quasi-fluid limit. Now we intend to find an equilibrium for this quasi-fluid limit.

The evolution of the process with n customers goes on in such a way. At the moment t and with $u(t) = i$ the process makes the time step to $t+1$ as described in [1], $i \in \{1, \ldots, N\}$, and then $u(t)$ makes the time step to $u(t + 1)$.

The corresponding formula is:

$$\mathbb{P}(u(t+1) = i)b^{1|i}(t+1, k+1) = \mathbb{P}(u(t) = i)p_{ii}\, b^{1|i}(t, k)\frac{G^{1|i;c}(k+1)}{G^{1|i;c}(k)}$$
$$+ \sum_{j \neq i} \mathbb{P}(u(t) = j)p_{ji}\, b^{1|j}(t, k)\frac{g^{1|j;c}(k+1)}{G^{1|j;c}(k)}. \tag{24}$$

Now we shall describe the equilibrium points of the quasi-fluid limit process.

First of all, the markov chain u must have a stable distribution, denote it by P_i, $i \in \{1, \ldots, N\}$.

And quasi-fluid limit distributions $b^{*1|i}(k)$, $i \in \{1, \ldots, N\}$, $k = 0, 1, \ldots$ must satisfy

$$
P_i b^{*1|i}(k+1) = P_i p_{ii} \, b^{*1|i}(k) \frac{G^{1|i;c}(k+1)}{G^{1|i;c}(k)}
$$
$$
+ \sum_{j \neq i} P_j p_{ji} \, b^{*1|j}(t,k) \frac{g^{1|j}(k+1)}{G^{1|j;c}(k)} . \tag{25}
$$

And quasi-fluid limit distributions $b^{*2|i}(k)$, $i \in \{1, \ldots, N\}$, $k = 0, 1, \ldots$, do not depend on i as $G^{2;i} \equiv G^2$, $i \in \{1, \ldots, N\}$. Thus b^{*2} must satisfy its original Eq. (20)

$$
b^{*2}(k) = b^{*2}(0) G^{2;c}(k), \ k = 1, 2, \ldots . \tag{26}
$$

Finally, similar to Subsect. 2.3 $b^{*1|i}(0) \equiv b^{*2}(0)$, $i \in \{1, \ldots, N\}$. To calculate $b^{*2}, b^{*1|i}$, $i \in \{1, \ldots, N\}$, q^*, we must insert $b^{*2}(0)$ into the equations system (25) and repeat the estimation of the proportion s of MS2 like in Theorem 1. Of course for the present model it is more difficult to calculate $b^{*1|i}$ and corresponding $B^{*1|i}$, $i \in \{1, \ldots, N\}$, than we made it without changing environment. But think just of bounded distributions $G^2, G^{1;i}$, $i \in \{1, \ldots, N\}$—surely in this case you can calculate the equilibrium point.

4 Conclusion

The models of call/contact centers have an important approximate description for their dynamics—fluid limit. For one multi-server and for a closed network with two multi-servers models the time convergence of their fluid limits is investigated only in a simple case. We plan to investigate the convergence for our model in a fully general case.

This and developed problems were stated and studied by famous scientists, cf. Refs. [2–6, 8–17] of [1]. We respect deeply these authors and their results, but we do not understand, why this trend is important for practical applications. To our mind really important is the calculation of the corresponding time-dependent expectation of the measure state of the multi-server(s) systems. The dispersion of the corresponding random state function is limited as $n \to \infty$ by a diffusion approximation. And time convergence of this deterministic function follows from ergodicity of the queueing process. A large class of systems possesses the ergodicity property (this depends on distributions of service requirements).

We shall investigate the described problem after consulting specialists working in applied directions.

References

1. Anulova, S.: Approximate description of dynamics of a closed queueing network including multi-servers. In: Vishnevsky, V., Kozyrev, D. (eds.) DCCN 2015. CCIS, vol. 601, pp. 177–187. Springer, Heidelberg (2016). doi:10.1007/978-3-319-30843-2_19
2. Whitt, W.: Fluid models for multiserver queues with abandonments. Oper. Res. **54**(1), 37–54 (2006). http://pubsonline.informs.org/doi/abs/10.1287/opre.1050.0227
3. Walsh Zuñiga, A.: Fluid limits of many-server queues with abandonments, general service and continuous patience time distributions. Stochastic Processes Appl. **124**(3), 1436–1468 (2014)
4. Davis, M.: Markov Models and Optimization. Monographs on Statistics and Applied Probability, vol. 49. Chapman & Hall, London (1993)
5. Yin, G., Zhu, C.: Hybrid Switching Diffusions. Properties and Applications. Springer, New York (2010)

On Strong Bounds of Rate of Convergence
for Regenerative Processes

Galina Zverkina[✉]

Moscow State University of Railway Engineering, Moscow, Russia
zverkina@gmail.com

Abstract. We give strong bounds for the rate of convergence of the regenerative process distribution to the stationary distribution in the total variation metric. For this aim we propose a new modification of the coupling method which we call *stationary coupling method*. Use of this stationary coupling method improves the classic results about the convergence rate of the distribution of the regenerative process in the case of a heavy tail. Also this method can be applied for obtaining the bounds of the rate of the convergence for the queueing regenerative processes.

Keywords: Regenerative process · Queuing theory · Rate of convergence · Total variation metrics · Coupling method

1 Introduction

We study the rate of convergence of distribution of regenerative process to the stationary distribution in the total variation metric.

Many queueing processes are regenerative, and establishing bounds for the rate of their convergence is a very important problem for the practical applications of the queueing theory. Recall the definition of regenerative process.

Definition 1. *The process $(X_t, t \geqslant 0)$ adapted to the filtration $\mathcal{F}_{t \geqslant 0}$ on a probability space $(\Omega, \mathcal{F}, \mathbf{P})$, with a measurable state space $(\mathcal{X}, \mathcal{F}(\mathcal{X}))$ is regenerative, if there exists an increasing sequence $\{\theta_n\}$ $(n \in \mathbb{Z}_+)$ of Markov moments with respect to the filtration $\mathcal{F}_{t \geqslant 0}$ such that $X_{\theta_i} = X_{\theta_j}$ for all $i, j \in \mathbb{Z}_+$ and the sequence*

$$\{\Theta_n\} = \left\{ X_{t+\theta_{n-1}} - X_{\theta_{n-1}}, \theta_n - \theta_{n-1}, t \in [\theta_{n-1}, \theta_n) \right\} \qquad (n \in \mathbb{N})$$

consists of independent identically distributed (i.i.d.) random elements on $(\Omega, \mathcal{F}, \mathbf{P})$. If $\theta_0 \neq 0$, then the process $(X_t, t \geqslant 0)$ is called delayed.

Denote $\zeta_n \stackrel{\text{def}}{=} \theta_n - \theta_{n-1}$, and let $F(s) = \mathbf{P}\{\zeta_n \leqslant s\} = \mathbf{P}\{\zeta_1 \leqslant s\}$ $(n \in \mathbb{N})$ be the distribution function of the length of the regeneration period; we assume that the distribution F is not lattice. Also denote $\zeta_0 \stackrel{\text{def}}{=} \theta_0$, $\mathbb{F}(s) \stackrel{\text{def}}{=} \mathbf{P}\{\zeta_0 \leqslant s\}$.

G. Zverkina—The author is supported by the RFBR, project No. 14-01-00319 A.

© Springer International Publishing AG 2016
V.M. Vishnevskiy et al. (Eds.): DCCN 2016, CCIS 678, pp. 381–393, 2016.
DOI: 10.1007/978-3-319-51917-3_34

Denote $\mathcal{P}_t^{X_0}(A) = \mathbf{P}\{X_t \in A\}$ for the process $(X_t, t \geqslant 0)$ with the initial state X_0. If $\mathbf{E}\,\zeta_i < \infty$, then for all X_0 we have $\mathcal{P}_t^{X_0} \Longrightarrow \mathcal{P}$ where $\mathcal{P}$ is the stationary distribution of the process $(X_t, t \geqslant 0)$.

It is known that if $\mathbf{E}\,\zeta_n^K < \infty$ for some $K > 1$ then for every $k \leqslant K - 1$ and $X_0 \in \mathcal{X}$

$$\lim_{t \to \infty} t^k \left\| \mathcal{P}_t^{X_0} - \mathcal{P} \right\|_{TV} = 0,$$

and if $\mathbf{E}\,e^{\alpha \zeta_n} < \infty$ for some $\alpha > 0$, then for every $a < \alpha$ and $X_0 \in \mathcal{X}$

$$\lim_{t \to \infty} e^{at} \left\| \mathcal{P}_t^{X_0} - \mathcal{P} \right\|_{TV} = 0$$

(see, e.g., [2,4,5,7,10–12] et al.). So, we know two statements:

1. If $\mathbf{E}\,\zeta_n^K < \infty$ for some $K > 1$, then for all $k \leqslant K - 1$ and $X_0 \in \mathcal{X}$ there exists $C(X_0, k)$ such that

$$\left\| \mathcal{P}_t^{X_0} - \mathcal{P} \right\|_{TV} \leqslant (1 + t)^{-k} C(X_0, k); \tag{1}$$

2. If $\mathbf{E}\,e^{\alpha \zeta_n} < \infty$ for some $\alpha > 1$, then for all $a < \alpha$ and $X_0 \in \mathcal{X}$ there exists $\mathfrak{C}(X_0, \alpha)$ such that

$$\left\| \mathcal{P}_t^{X_0} - \mathcal{P} \right\|_{TV} \leqslant e^{-at} \mathfrak{C}(X_0, a). \tag{2}$$

Our goal is to find the bounds of the constants $C(X_0, k)$ and $\mathfrak{C}(X_0, a)$ with sufficiently wide conditions; note that the behavior of the constants $C(X_0, k)$ and $\mathfrak{C}(X_0, a)$ has been studied in [8,9,13–16] for some special cases of regenerative processes. To achieve this goal, we will use a modified coupling method.

In the sequel, we suppose that

$$\int_{\{s:\,\exists\, F'(s)\}} F'(s)\,\mathrm{d}s > 0; \quad \mathbf{E}\,\zeta_i < \infty \tag{$*$}$$

2   Denotations and the Main Results

1. The times θ_i (see Definition 1) form the renewal process $N_t \stackrel{\text{def}}{=} \sum_{i=0}^{\infty} \mathbf{1}(\theta_i \leqslant t)$.

Denote $B_t \stackrel{\text{def}}{=} t - \theta_{N_t - 1}$. B_t is the backward renewal time of the process N_t. $\triangleright$

Remark 1. $(B_t, t \geqslant 0)$ is the Markov regenerative process. We call it *the backward renewal process*. $\triangleright$

2. For nondecreasing function $F(s)$ we put $F^{-1}(y) \stackrel{\text{def}}{=} \inf\{x : F(x) \geqslant y\}$. $\triangleright$

3. $\widetilde{F}(s) \overset{\text{def}}{=} \mu^{-1} \int_0^s (1 - F(u)) \, \mathrm{d}\,u$, where $\mu \overset{\text{def}}{=} \int_0^\infty u \, \mathrm{d}\, F(u) = \mathbf{E}\,\zeta$.   ▷

4. $F_a(s) \overset{\text{def}}{=} \dfrac{F(s+a) - F(a)}{1 - F(a)}$; $\mu_0 \overset{\text{def}}{=} \mathbf{E}\,\zeta_0$.   ▷

5. Let $\mathcal{U}, \mathcal{U}', \mathcal{U}'', \mathcal{U}_i, \mathcal{U}_i', \mathcal{U}_i'', \mathcal{U}_i'''$ be independent uniformly distributed on $[0,1)$ random variables on some probability space $\left(\widetilde{\Omega}, \widetilde{\mathcal{F}}, \widetilde{\mathbf{P}}\right)$.   ▷

6. Denote $\varphi(s) \overset{\text{def}}{=} \mathbf{1}(\exists\, F'(s)) \times (F'(s) \wedge \widetilde{F}'(s))$ where $\mathbf{1}(\cdot)$ is indicator, and we put $\Phi(s) \overset{\text{def}}{=} \int_0^s \varphi(u) \, \mathrm{d}\, u$.

The condition $(*)$ implies $\varkappa \overset{\text{def}}{=} \int_0^\infty \varphi(s) \, \mathrm{d}\, s = \Phi(+\infty) > 0$.

Denote $\overline{\varkappa} \overset{\text{def}}{=} 1 - \varkappa$.   ▷

7. Denote $\Psi(s) \overset{\text{def}}{=} F(s) - \Phi(s)$, $\widetilde{\Psi}(s) \overset{\text{def}}{=} \widetilde{F}(s) - \Phi(s)$; $\Psi(+\infty) = \widetilde{\Psi}(+\infty) = \overline{\varkappa}$.

Also denote $\mathfrak{P}_a(\zeta_1) \overset{\text{def}}{=} \int_0^\infty e^{as} \, \mathrm{d}\, \Psi(s)$.   ▷

Remark 2. Note that $\varkappa^{-1}\Phi(s)$ is the distribution function, and if $\varkappa < 1$ then $\overline{\varkappa}^{-1}\Psi(s)$ and $\overline{\varkappa}^{-1}\widetilde{\Psi}(s)$ are the distribution functions.

If $\varkappa = 1$ then $\Phi(s) \equiv F(s) \equiv \widetilde{F}(s) = 1 - e^{-\lambda s}$, and $\Psi(s) \equiv \widetilde{\Psi}(s) \equiv 0$; in this case we put $\overline{\varkappa}^{-1}\Psi(s) \equiv \overline{\varkappa}^{-1}\widetilde{\Psi}(s) \equiv 0$, and $\Psi^{-1}(u) = \widetilde{\Psi}^{-1}(u) = 0$.   ▷

8. Put $\Xi(\mathcal{U}, \mathcal{U}', \mathcal{U}'') \overset{\text{def}}{=} \mathbf{1}(\mathcal{U} < \varkappa)\Phi^{-1}(\varkappa\mathcal{U}') + \mathbf{1}(\mathcal{U} \geqslant \varkappa)\Psi^{-1}(\overline{\varkappa}\mathcal{U}'')$;

$\widetilde{\Xi}(\mathcal{U}, \mathcal{U}', \mathcal{U}'') \overset{\text{def}}{=} \mathbf{1}(\mathcal{U} < \varkappa)\Phi^{-1}(\varkappa\mathcal{U}') + \mathbf{1}(\mathcal{U} \geqslant \varkappa)\widetilde{\Psi}^{-1}(\overline{\varkappa}\mathcal{U}'')$.   ▷

Remark 3. Clearly,
$$F(s) = \varkappa\left(\varkappa^{-1}\Phi(s)\right) + \overline{\varkappa}\left(\overline{\varkappa}^{-1}\Psi(s)\right) \text{ and } \widetilde{F}(s) = \varkappa\left(\varkappa^{-1}\Phi(s)\right) + \overline{\varkappa}\left(\overline{\varkappa}^{-1}\widetilde{\Psi}(s)\right).$$
Hence,
$$\mathbf{P}\{\Xi(\mathcal{U}, \mathcal{U}', \mathcal{U}'') \leqslant s\} = F(s), \qquad \mathbf{P}\{\widetilde{\Xi}(\mathcal{U}, \mathcal{U}', \mathcal{U}'') \leqslant s\} = \widetilde{F}(s),$$
and $\mathbf{P}\{\Xi(\mathcal{U}, \mathcal{U}', \mathcal{U}'') = \widetilde{\Xi}(\mathcal{U}, \mathcal{U}', \mathcal{U}'')\} = \varkappa$.   ▷

9. Denote $C(X_0, k) \overset{\text{def}}{=} \mathbf{E}\,\zeta_0^k \varkappa \sum_{n=1}^\infty \left((n+1)^{k-1}\overline{\varkappa}^{\,n-1}\right)$
$$+ \mathbf{E}\,\zeta_1^k \sum_{n=1}^\infty \left(\left(\varkappa n(n+2)^{k-1} + (n+1)^{k-1}\right)\overline{\varkappa}^{\,n-1}\right).$$ ▷

10. Denote $\mathfrak{C}(X_0, a) \overset{\text{def}}{=} \dfrac{\mathbf{E}\,e^{a\zeta_1}\mathbf{E}\,e^{a\zeta_0}}{1 - \mathfrak{P}_a(\zeta_1)}.$ ▷

Theorem 1. *Suppose that the process* $(X_t, t \geqslant 0)$ *satisfies* $(*)$.

1. *If* $\mathbf{E}(\zeta_1)^K < \infty$ *and* $\mathbf{E}(\zeta_0)^K < \infty$ *for some* $K \geqslant 1$, *then for all* $k \in [1, K]$

$$\left\| \mathcal{P}_t^{X_0} - \mathcal{P} \right\|_{TV} \leqslant 2t^{-k} C(X_0, k).$$

2. *If* $\mathfrak{P}_a(\zeta_1) < 1$ *for some* $a > 0$ *and* $\mathbf{E}\, e^{a\zeta_0} < \infty$, *then*

$$\left\| \mathcal{P}_t^{X_0} - \mathcal{P} \right\|_{TV} \leqslant 2\mathfrak{C}(X_0, a)\, e^{-at}.$$

Remark 4. The bounds in Theorem 1 can be improved considering the properties of the distribution F.   ▷

Remark 5. Statement 1 of Theorem 1 improves the classic result (1). Statement 2 of Theorem 1 is weaker than the classical result (2), but here we have the bounds for the constant $\mathfrak{C}(\cdot)$.   ▷

3   Idea of the Proof of Theorem 1

Coupling Method (see [6]). To prove Theorem 1, we will use the *stationary coupling method*. The coupling method invented by Doeblin in [1] is used to obtain the bounds of the rate of convergence of a Markov process to the stationary regime.

Let $(X_t', t \geqslant 0)$ and $(X_t'', t \geqslant 0)$ be two versions of Markov process $(X_t, t \geqslant 0)$ with different initial states,

$$\mathcal{P}_t^{X_0'}(A) \stackrel{\text{def}}{=} \mathbf{P}\{X_t' \in A\}, \qquad \mathcal{P}_t^{X_0''}(A) \stackrel{\text{def}}{=} \mathbf{P}\{X_t'' \in A\},$$

and

$$\tau(X_0', X_0'') \stackrel{\text{def}}{=} \inf\{t > 0 : X_t' = X_t''\}.$$

We suppose that $\mathbf{E}\,\varphi(\tau(X_0', X_0'')) = C(X_0', X_0'') < \infty$ for some positive increasing function $\varphi(t)$. Then

$$\left| \mathcal{P}_t^{X_0'}(A) - \mathcal{P}_t^{X_0''}(A) \right| \leqslant \mathbf{P}\{\tau(X_0', X_0'') > t\}$$

$$= \mathbf{P}\{\varphi(\tau(X_0', X_0'')) > \varphi(t)\} \leqslant \frac{\mathbf{E}\,\varphi(\tau(X_0', X_0''))}{\varphi(t)}$$

by the coupling inequality and Markov's inequality. By integration of this inequality with respect to the stationary measure $\mathcal{P}$ we have

$$\left| \mathcal{P}_t^{X_0'}(A) - \mathcal{P}(A) \right| \leqslant \frac{\int_{\mathcal{X}} \varphi(\tau(X_0', X_0''))\, \mathrm{d}\mathcal{P}(X_0'')}{\varphi(t)} = \frac{\widehat{C}(X_0')}{\varphi(t)}, \qquad (3)$$

and $\left\|\mathcal{P}_t^{X_0'} - \mathcal{P}\right\|_{TV} \leqslant 2\dfrac{\widehat{C}(X_0')}{\varphi(t)}.$

Emphasize that the application of the coupling method is possible only for the Markov processes. However, in queuing theory, usually the regenerative queueing processes are not Markov. Therefore, the state space of considered regenerative process must be extended so that the regenerative process with this state space would become Markov.

So, for the use of the coupling method for the arbitrary regenerative process $(X_t,\, t \geqslant 0)$ we must extend the state space $\mathcal{X}$ of this process by such a way that the extended process $(\overline{X}_t,\, t \geqslant 0)$ with the extended state space $\overline{\mathcal{X}}$ is Markov. For markovization of non-Markov regenerative process we can (for example) for $t \in [\theta_{n-1}, \theta_n)$ include in the state X_t full history of the process $(X_t,\, t \geqslant 0)$ on the time interval $[\theta_{n-1}, t]$: the process $\overline{X}_t \stackrel{\text{def}}{=} \{X_s, s \in [\theta_{n-1}, t] \mid t \in [\theta_{n-1}, \theta_n)\}$ is Markov and regenerative with the extended state space $\overline{\mathcal{X}}$. Denote $\overline{\mathcal{P}}_t^{\overline{X}_0}(A) \stackrel{\text{def}}{=} \mathbf{P}\{\overline{X}_t \in A\}$ for the process $\overline{X}_t$ with the initial state $\overline{X}_0$ and $A \in \mathcal{B}(\overline{\mathcal{X}})$. If $\mathbf{E}\,\zeta_i < \infty$, then $\overline{\mathcal{P}}_t^{\overline{X}_0} \Longrightarrow \overline{\mathcal{P}}$.

If we can prove that $\left\|\overline{\mathcal{P}}_t^{\overline{X}_0} - \overline{\mathcal{P}}\right\|_{TV} \leqslant \varphi(t, \overline{X}_0)$ for all $t \geqslant 0$, then this inequality is true for the original non-Markov regenerative process $(X_t,\, t \geqslant 0)$.

For simplicity, we assume that the process $(\overline{X}_t,\, t \geqslant 0)$ is homogeneous Markov process, i.e. the transition function of this process in the period $[0, \theta_0]$ is the same as in the periods $[\theta_i, \theta_{i+1}],\ i \geqslant 1$.

Thus, in the sequel we suppose that the regenerative process $(X_t,\, t \geqslant 0)$ is homogeneous Markov process.

Notice, that in general case $\mathbf{P}\,\{\tau\,(X_0', X_0'') < \infty\} < 1$ (for the Markov processes in continuous time), and the "direct" use of coupling method is impossible.

Successful Coupling (see [3]) and Strong Successful Coupling. So, we will construct (in a special probability space) the paired stochastic process $(\mathcal{Z}_t,\, t \geqslant 0) = ((Z_t', Z_t''),\, t \geqslant 0)$ such that:

1. For all $t \geqslant 0$ $X_t' \stackrel{\mathcal{D}}{=} Z_t'$ and $X_t'' \stackrel{\mathcal{D}}{=} Z_t''$.
2. $\mathbf{P}\{\tau(Z_0', Z_0'') < \infty\} = 1$, where $\tau\,(Z_0', Z_0'') = \tau(\mathcal{Z}_0) \stackrel{\text{def}}{=} \inf\{t \geqslant 0 : Z_t' = Z_t''\}$.
3. $Z_t' = Z_t''$ for all $t \geqslant \tau\,(Z_0', Z_0'')$.

The paired stochastic process $(\mathcal{Z}_t,\, t \geqslant 0) = ((Z_t', Z_t''),\, t \geqslant 0)$ satisfying the conditions 1–3 is called *successful coupling*. If we replace the condition 2 by the condition
$2'$. $\mathbf{E}\,\tau(Z_0', Z_0'') < \infty$, where $\tau\,(Z_0', Z_0'') = \tau(\mathcal{Z}_0) \stackrel{\text{def}}{=} \inf\{t \geqslant 0 : Z_t' = Z_t''\}$, then the paired stochastic process $\mathcal{Z}_t = ((Z_t', Z_t''),\, t \geqslant 0)$ satisfying the conditions *1, 2′, 3* is called *strong successful coupling*.

The processes $(Z_t',\, t \geqslant 0)$ and $(Z_t'',\, t \geqslant 0)$ can be non-Markov, and its finite-dimensional distributions may differ from the finite-dimensional distributions of $(X_t',\, t \geqslant 0)$ and $(X_t'',\, t \geqslant 0)$ respectively; furthermore, the processes $(Z_t',\, t \geqslant 0)$ and $(Z_t'',\, t \geqslant 0)$ may be dependent.

For all $A \in \mathcal{B}(\mathcal{X})$ we can use the coupling inequality in the form:

$$
\begin{aligned}
\left|\mathcal{P}_t^{X_0'}(A) - \mathcal{P}_t^{X_0''}(A)\right| &= |\mathbf{P}\{X_t' \in A\} - \mathbf{P}\{X_t'' \in A\}| \\
&= |\mathbf{P}\{Z_t' \in A\} - \mathbf{P}\{Z_t'' \in A\}| \leqslant \mathbf{P}\{\tau(Z_0', Z_0'') \geqslant t\} \\
&\leqslant \frac{\mathbf{E}\,\varphi(\tau(Z_0', Z_0''))}{\varphi(t)} \leqslant \frac{C(Z_0', Z_0'')}{\varphi(t)},
\end{aligned}
\tag{4}
$$

where $C(Z_0', Z_0'') \geqslant \mathbf{E}\tau(Z_0', Z_0'')$. As $Z_0^{(i)} = X_0^{(i)}$, the right-hand side of the inequality depends only on $X_0^{(i)}$. Then we can integrate the inequality (4) with respect to the measure $\mathcal{P}$ as in (3):

$$
\left|\mathcal{P}_t^{X_0}(A) - \mathcal{P}(A)\right| \leqslant (\varphi(t))^{-1} \int\limits_{\mathcal{X}} C\left(Z_0', Z_0''\right) \mathcal{P}\left(\mathrm{d}\,Z_0''\right) = \frac{\widehat{C}\left(Z_0'\right)}{\varphi(t)}.
$$

However, this integration gives some trouble.

Stationary Coupling Method. We will construct a strong successful coupling $(\mathcal{Z}_t, t \geqslant 0) = ((Z_t, \widetilde{Z}_t), t \geqslant 0)$ for the process $(X_t, t \geqslant 0)$ and its stationary version $(\widetilde{X}_t, t \geqslant 0)$, so we will estimate the random variable $\widetilde{\tau}(X_0) = \widetilde{\tau}(Z_0) \stackrel{\text{def}}{=} \inf\left\{t > 0 : Z_t = \widetilde{Z}_t\right\}$. Then analogously to the inequality (4), we have

$$
\left\|\mathcal{P}_t^{X_0}(A) - \mathcal{P}(A)\right\|_{TV} \leqslant 2\,\mathbf{P}\left\{\widetilde{\tau}(X_0) > t\right\} \leqslant 2\,\frac{\mathbf{E}\,\varphi(\widetilde{\tau}(X_0))}{\varphi(t)}.
$$

4   Implementation of Idea

Theorem 2. *If the process $(X_t, t \geqslant 0)$ satisfies* $(*)$, *then there exists a strong successful coupling $(\mathcal{Z}_t, t \geqslant 0) = ((Z_t, \widetilde{Z}_t), t \geqslant 0)$ for the process $(X_t, t \geqslant 0)$ and its stationary version $(\widetilde{X}_t, t \geqslant 0)$.*

Proof. The proof of Theorem 2 consists of 4 steps.

1. Let us prove Theorem 2 for the process $(B_t, t \geqslant 0)$ (Denotation 1). We will give the construction of the strong successful coupling for the process $(B_t, t \geqslant 0)$ and its stationary version $(\widetilde{B}_t, t \geqslant 0)$ (on the some probability space $(\widetilde{\Omega}, \widetilde{\mathcal{F}}, \widetilde{\mathbf{P}})$ – Denotation 5).

Recall that the versions of the processes B_t and $\widetilde{B}_t$ can be constructed as follows: $\zeta_0 \stackrel{\text{def}}{=} \mathbb{F}^{-1}(\mathcal{U}_0)$, and $\zeta_i \stackrel{\text{def}}{=} F^{-1}(\mathcal{U}_i)$ for $i > 0$; $\theta_i \stackrel{\text{def}}{=} \sum\limits_{j=0}^{i} \zeta_j$;

$Z_t \stackrel{\text{def}}{=} t - \max\{\theta_i : \theta_i \leqslant t\} \stackrel{\mathcal{D}}{=} B_t$; $\widetilde{\theta}_0 = \widetilde{\zeta}_0 \stackrel{\text{def}}{=} \widetilde{F}^{-1}(\mathcal{U}_1')$, and

$\widetilde{\zeta}_i \overset{\text{def}}{=} F^{-1}(\mathcal{U}_i')$ for $i > 0$ $\widetilde{\theta}_i \overset{\text{def}}{=} \sum_{j=0}^{i} \widetilde{\zeta}_j$; $\widetilde{Z}_0 \overset{\text{def}}{=} F_{\widetilde{\theta}_0}^{-1}(\mathcal{U}_1'')$ (see Denotation 4);

$\widetilde{Z}_t \overset{\text{def}}{=} \mathbf{1}(t < \widetilde{\theta}_0)(t + \widetilde{Z}_0) + \mathbf{1}(t \geqslant \widetilde{\theta}_0)(t - \max\{\widetilde{\theta}_n : \widetilde{\theta}_n \leqslant t\}) \overset{\mathcal{D}}{=} \widetilde{B}_t.$

Remark 6. $\mathbf{P}\{\widetilde{Z}_0 \leqslant s\} = \int\limits_0^\infty F_u(s)\,\mathrm{d}\widetilde{F}(u) = \int\limits_0^\infty \dfrac{F(s+u) - F(u)}{1 - F(u)} \times \dfrac{1 - F(u)}{\mu}$

$\mathrm{d}u = \dfrac{1}{\mu}\int\limits_0^s (1 - F(u))\,\mathrm{d}u = \widetilde{F}(s).$ $\triangleright$

   This construction is the construction of independent versions of the processes $(B_t, t \geqslant 0)$ and $(\widetilde{B}_t, t \geqslant 0)$. Now we will transform this construction.

2. To construct a pair of the (dependent) backward renewal processes, it is enough to construct all renewal times of both processes (times ϑ_i in Fig. 1). We will construct this pair $(\mathcal{Z}_t, t \geqslant 0) = ((Z_t, \widetilde{Z}_t), t \geqslant 0)$ by induction.

Basis of Induction. Put

$$\theta_0 \overset{\text{def}}{=} \mathbb{F}^{-1}(\mathcal{U}_0), \qquad \widetilde{\theta}_0 \overset{\text{def}}{=} \widetilde{F}^{-1}(\mathcal{U}_0'), \qquad \widetilde{Z}_0 \overset{\text{def}}{=} F_{\widetilde{\theta}_0}^{-1}(\mathcal{U}_0'');$$

and we put $Z_t \overset{\text{def}}{=} t$, $\widetilde{Z}_t \overset{\text{def}}{=} t + \widetilde{Z}_0$ for $t \in [0, \vartheta_0)$, where $\vartheta_0 \overset{\text{def}}{=} t_0 \wedge \widetilde{t}_0$ (in Fig. 1 $\vartheta_0 = \widetilde{\theta}_0$).

Inductive Step. Suppose that we have constructed the process $(\mathcal{Z}_t, t \in [0, \vartheta_n))$, $\vartheta_n = \theta_i \wedge \widetilde{\theta}_j$. There are three alternatives.

1. $\vartheta_n = \theta_i = \widetilde{\theta}_j$ – in Fig. 1 this situation occurs for the first time at the point ϑ_5. In this case we put

$$Z_{\vartheta_n} = \widetilde{Z}_{\vartheta_n} = 0, \qquad \theta_{i+1} = \widetilde{\theta}_{j+1} = \vartheta_{n+1} = F^{-1}(\mathcal{U}_{n+1}) + \vartheta_n;$$

and $Z_t = \widetilde{Z}_t \overset{\text{def}}{=} t - \vartheta_n$ for $t \in [\vartheta_n, \vartheta_{n+1})$. After the first coincidence (time $\widetilde{\tau} = \vartheta_5$ in Fig. 1) the processes $(Z_t, t \geqslant \widetilde{\tau})$ and $(\widetilde{Z}_t, t \geqslant \widetilde{\tau})$ are identical.

2. $\vartheta_n = \widetilde{\theta}_j < \theta_i$ (the times $\widetilde{\theta}_0$ and $\widetilde{\theta}_3$ in Fig. 1). In this case we put

$$\widetilde{Z}_{\vartheta_n} = 0, \ Z_{\vartheta_n} = Z_{\vartheta_n - 0}, \qquad \widetilde{\theta}_{j+1} \overset{\text{def}}{=} \widetilde{\theta}_j + F^{-1}(\mathcal{U}_{n+1});$$

and $\widetilde{Z}_t \overset{\text{def}}{=} t - \vartheta_n$, $Z_t \overset{\text{def}}{=} t - \vartheta_n + Z_{\vartheta_n}$ for $t \in [\vartheta_n, \vartheta_{n+1})$ where $\vartheta_{n+1} \overset{\text{def}}{=} \theta_i \wedge \widetilde{\theta}_{j+1}$.

3. $\vartheta_n = \theta_i < \widetilde{\theta}_j$ (the times θ_0, θ_1 and θ_2 in Fig. 1). In this case we put

$$\theta_{i+1} \stackrel{\text{def}}{=} \theta_i + \Xi(\mathcal{U}_{n+1}, \mathcal{U}'_{n+1}, \mathcal{U}''_{n+1}); \qquad \widetilde{\theta}_j \stackrel{\text{def}}{=} \theta_i + \widetilde{A},$$

where $\widetilde{A} = \widetilde{\Xi}(\mathcal{U}_{n+1}, \mathcal{U}'_{n+1}, \mathcal{U}''_{n+1})$; and $Z_t \stackrel{\text{def}}{=} t - \vartheta_n$, $\widetilde{Z}_t \stackrel{\text{def}}{=} t - \vartheta_n + F_{\widetilde{A}}^{-1}(\mathcal{U}'''_{n+1})$ for $t \in [\vartheta_n, \vartheta_{n+1})$, where $\vartheta_{n+1} \stackrel{\text{def}}{=} \theta_i \wedge \widetilde{\theta}_{j+1}$.

3. Let us prove that *the process* $(\mathcal{Z}_t, t \geqslant 0) = ((Z_t, \widetilde{Z}_t), t \geqslant 0)$ *is a strong successful coupling for the processes* $(B_t, t \geqslant 0)$ *and* $(\widetilde{B}_t, t \geqslant 0)$, *and* $\mathbf{E}\widetilde{\tau}(B_0) \leqslant \mathbf{E}\zeta_0 + 2\varkappa^{-1}\mathbf{E}\zeta_1 < \infty$.

Denote $\mathcal{E}_n \stackrel{\text{def}}{=} \{Z_{\theta_n} = \widetilde{Z}_{\theta_n}\}$,

$$\mathfrak{E}_n \stackrel{\text{def}}{=} \left(\mathcal{E}_n \cap \bigcap_{i=0}^{n-1} \overline{\mathcal{E}}_i\right) = \{Z_{\theta_{n+1}} = \widetilde{Z}_{\theta_{n+1}} \& Z_{\theta_i} \neq \widetilde{Z}_{\theta_i}, i \leqslant n\}.$$

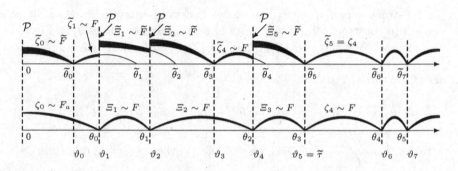

Fig. 1. Construction of the strong successful coupling $\mathcal{Z}_t$.

According to our construction of the pair $(\mathcal{Z}_t, t \geqslant 0)$, we have $\mathbf{P}\{Z_{\theta_0} \neq \widetilde{Z}_{\theta_0}\} = 1$ because the distribution $\widetilde{F}(s)$ is absolutely continuous, and $\mathbf{P}\{\widetilde{\tau} = \theta_{n+1}\} = \mathbf{P}(\mathfrak{E}_n) = \varkappa\overline{\varkappa}^n$, where $\overline{\varkappa} = 1 - \varkappa$. Now, using the inequality

$$\mathbf{E}(\xi \times \mathbf{1}(\mathcal{E}))\mathbf{P}(\mathcal{E}) \leqslant \mathbf{E}\xi \tag{5}$$

for non-negative random variable ξ, and considering that $\mathbf{E}\left(\zeta_n \mathbf{1}\left(\bigcap_{i=1}^{n-1}\overline{\mathcal{E}}_i\right)\right) = (\overline{\varkappa})^{n-1}\mathbf{E}\zeta_n$ for $n > 0$, we have

$$
\begin{aligned}
\mathbf{E}\widetilde{\tau} &= \mathbf{E}\zeta_0 + \mathbf{E}(\mathbf{1}(\mathcal{E}_1)\zeta_1) + \mathbf{E}(\mathbf{1}(\mathfrak{E}_2)(\zeta_1 + \zeta_2)) + \mathbf{E}(\mathbf{1}(\mathfrak{E}_3)(\zeta_1 + \zeta_2 + \zeta_3)) \\
&\quad + \ldots + \mathbf{E}\left(\mathbf{1}(\mathfrak{E}_n)\sum_{i=1}^{n}\zeta_i\right) + \ldots \leqslant \mathbf{E}\zeta_0 + \mathbf{E}\zeta_1\left(1 + \varkappa\sum_{i=0}^{\infty}\overline{\varkappa}^i\right) \\
&\quad + \overline{\varkappa}\mathbf{E}\zeta_2\left(1 + \varkappa\sum_{i=0}^{\infty}\overline{\varkappa}^i\right) + \overline{\varkappa}^2\mathbf{E}\zeta_3\left(1 + \varkappa\sum_{i=0}^{\infty}\overline{\varkappa}^i\right) + \ldots \\
&\quad + \overline{\varkappa}^n\mathbf{E}\zeta_{n+1}\left(1 + \varkappa\sum_{i=0}^{\infty}\overline{\varkappa}^i\right) + \ldots = \mathbf{E}\zeta_0 + 2\varkappa^{-1}\mathbf{E}\zeta_1.
\end{aligned}
\tag{6}
$$

4. Now we return to the Markov regenerative process $(X_t, t \geqslant 0)$. The regeneration times θ_0, θ_1, ... of X_t form an embedded process $Y_t \overset{\text{def}}{=} t - \max\{\theta_i : \theta_i \leqslant t\}$ (the backward renewal time of the embedded renewal process) with the distribution of the renewal time $\mathbf{P}\{\zeta \leqslant s\} = F(s)$; the length of i-th regeneration period is $\theta_i - \theta_{i-1} = \zeta_i \overset{\mathcal{D}}{=} \zeta$ $(i > 1)$.

We will apply the coupling method for extended process $(\mathbb{X}_t, t \geqslant 0) = ((X_t, Y_t), t \geqslant 0)$ *on the extended state space* $\widehat{\mathcal{X}} = \mathcal{X} \times \mathbb{R}_+$.

The random element $\mathfrak{X}_i = \{X_t, t \in [\theta_{i-1}, \theta_i)\}$ depends on the random variable $\zeta_i = \theta_i - \theta_{i-1}$; for $A \in \mathcal{B}(\mathcal{X})$ and $s \in [0, a)$ denote $\mathcal{G}_a(s, A) \overset{\text{def}}{=} \mathbf{P}\{X_{\theta_{i-1}+s} \in A | \zeta_i = a\}$; $\mathcal{G}_a(s, A)$ specify a conditional distribution of X_t on the time period $[\theta_{i-1}, \theta_i)$ given $\{\theta_i - \theta_{i-1} = a\}$.

Therefore if we know all regeneration times of the process $(X_t, t \geqslant 0)$, then we know the conditional distribution of process X_t (given realization of Y_t) in every time after the first regeneration time t_0: this distribution is determined by the conditional distribution $\mathcal{G}_{\zeta_i}(s, A)$ of the random elements $\mathfrak{X}_i$.

Also the first regeneration time depends on the initial state; denote $\mathcal{H}_a(A) \overset{\text{def}}{=} \mathbf{P}\{X_0 \in A | \theta_0 = a\} = \mathbf{P}\{X_t \in A | (\min\{\theta_i : \theta_i \geqslant t\} - t) = a\}$; $\mathcal{H}_a(A)$ specify a conditional distribution of X_t given $\{\zeta_t^* = a\}$, where $\zeta_t^* \overset{\text{def}}{=} (\min\{\theta_i : \theta_i \geqslant t\} - t)$ is a residual time of regeneration period at the time t.

Now we will construct the strong successful coupling for the extended process $(\mathbb{X}_t, t \geqslant 0) = ((X_t, Y_t), t \geqslant 0)$ and its stationary version $(\widetilde{\mathbb{X}}_t, t \geqslant 0) = ((\widetilde{X}_t, \widetilde{Y}_t), t \geqslant 0)$.

For this aim we construct the strong successful coupling $(\mathcal{W}_t, t \geqslant 0) = ((W_t, \widetilde{W}_t), t \geqslant 0)$ for the backward renewal process $(Y_t, t \geqslant 0)$ and its stationary version $(\widetilde{Y}_t, t \geqslant 0)$ considering that the first renewal time θ_0 of the process $(Y_t, t \geqslant 0)$ has a distribution $\mathbb{F}(s)$.

After construction of renewal points $\{\theta_i\}$ of the process $(W_t, t \geqslant 0)$ and renewal points $\{\widetilde{\theta}_i\}$ of the process $(\widetilde{W}_t, t \geqslant 0)$ we can complete them to the pairs $(Z_t, W_t) \overset{\mathcal{D}}{=} (X_t, Y_t)$ and $\left(\widetilde{Z}_t, \widetilde{W}_t\right) \overset{\mathcal{D}}{=} \left(\widetilde{X}_t, \widetilde{Y}_t\right)$ by using $\mathcal{G}_a(s, A)$ and $\mathcal{H}_a(A)$.

In the construction of the processes $(W_t, t \geqslant 0)$ and $(\widetilde{W}_t, t \geqslant 0)$ we can apply the technics used in the proof of Theorem 2 in such a way that

$$\widetilde{\tau}(X_0) = \inf\{t : W_t = \widetilde{W}_t\} \leqslant t_0 + \sum_{i=1}^{\nu} \zeta_i, \tag{7}$$

where $\mathbf{P}\{\nu > n\} = \overline{\varkappa}^n$.

So, for $t \geqslant \widetilde{\tau}(X_0)$ we have $W_t = \widetilde{W}_t$, by construction of renewal processes $(W_t, t \geqslant 0)$ and $(\widetilde{W}_t, t \geqslant 0)$. Then for $t \geqslant \widetilde{\tau}(X_0)$ we have

$$\mathcal{P}_t^{X_0}(A) = \mathbf{P}\{X_t \in A\} = \mathbf{P}\{\widetilde{X}_t \in A\} = \mathcal{P}(A)$$

as the distribution of the processes X_t and $\widetilde{X}_t$ (after the first renewal point) is determined only by renewal points.

Theorem 2 is proved. ∎

Proof (of Theorem 1).

1. Using the inequality (5) and Jensen's inequality for $k \geqslant 1$ in the form
$$\left(\sum_{i=1}^{n} a_i\right)^k \leqslant n^{k-1} \sum_{i=1}^{n} a_i^k \quad (a_i \geqslant 0)$$ we have from (6) and (7):

$$\mathbf{E}(\widetilde{\tau}(X_0))^k \leqslant \mathbf{E}\left(\sum_{n=1}^{\infty} \left((n+1)^{k-1}\left(\zeta_0^k + \sum_{i=1}^{n} \zeta_i^k\right) \mathbf{1}(\mathfrak{E}_n)\right)\right)$$

$$\leqslant \mathbf{E}\, \zeta_0^k \varkappa \sum_{n=1}^{\infty} \left((n+1)^{k-1} \overline{\varkappa}^{\,n-1}\right)$$

$$+ \mathbf{E}\, \zeta_1^k \sum_{n=1}^{\infty} \left((\varkappa n(n+2)^{k-1} + (n+1)^{k-1})\, \overline{\varkappa}^{\,n-1}\right) = C(X_0, k),$$

this inequality implies Statement 1 of Theorem 1.

2. Using the inequality (5) and considering that $\mathbf{E}(e^{a\zeta_i}\mathbf{1}(\overline{\mathcal{E}}_i)) = \mathfrak{P}_a(\zeta_1)$ for $i \geqslant 1$, we have from (6):

$$\mathbf{E}\, e^{a\widetilde{\tau}(X_0)} \leqslant \mathbf{E}\left(\sum_{n=1}^{\infty}\left(\exp\left(a\left(\zeta_0 + \sum_{i=1}^{n} \zeta_i\right)\right)\mathbf{1}(\mathfrak{E}_n)\right)\right)$$

$$\leqslant \mathbf{E}\, e^{a\zeta_0}\, \mathbf{E}\, e^{a\zeta_1}\left(1 + \sum_{n=1}^{\infty}(\mathfrak{P}_a(\zeta_1))^n\right) = \frac{\mathbf{E}\, e^{a\zeta_0}\, \mathbf{E}\, e^{a\zeta_1}}{1 - \mathfrak{P}_a(\zeta_1)} = \mathfrak{C}(X_0, a),$$

that implies Statement 2 of Theorem 1. Theorem 1 is proved. ∎

5   Applying to the Queueing Theory

In the queuing theory the distribution of the period of the regenerative process is often unknown. But often the regeneration period can be split into two parts, usually this is a busy period and an idle period. And as a rule the idle period has a known non-discrete distribution. So, in this situation the queueing process has an embedded *alternating renewal process.*

If the bounds of moments of a busy period are also known, then we can apply our construction for embedded alternating renewal process by some modification.

This modification is a construction of a strong successful coupling for alternating renewal process $(X_t, t \geqslant 0)$ and its stationary version, namely.

1. Let $(X_t, t \geqslant 0)$ be an alternating renewal process having two states, 1 and 2, say. The time of the stay of the process $(X_t, t \geqslant 0)$ in a state i has the distribution function $F_i(s) = \mathbf{P}\left\{\zeta^{(i)} \leqslant s\right\}$, and the periods of stay of the process $(X_t, t \geqslant 0)$ in the states 1 and 2 alternate. This process is non-Markov.

We complement the state of the process by the time, during which the process located continually in this state: if completed process is, say, $(Y_t, t \geqslant 0) = ((n_t, x_t), t \geqslant 0)$ (denote $n(Y_t) \overset{\text{def}}{=} n_t$, $x(Y_t) \overset{\text{def}}{=} x_t$), then at the time t the process is in the state n_t, and $x_t \overset{\text{def}}{=} t - \sup\{s < t : n(Y_s) \neq n_t\}$ (for definiteness, we assume $\mathbb{F}(s) = F_a(s)$, $x_0 = a$, and $x_t = a + t$ for $t \in [0, \inf\{s > 0 : n_s \neq n_0\})$). The Markov regenerative process $(Y_t, t \geqslant 0)$ has a state space $\{1, 2\} \times \mathbb{R}_+$.

Put $c(1) \overset{\text{def}}{=} 2$ and $c(2) \overset{\text{def}}{=} 1$.

If $Y_0 = (i, a)$, then the process $(Y_t, t \geqslant 0)$ changes its first component $n(Y_t)$ at the times $\zeta^{(a,i)} = \theta_{0,i}, \theta_{0,c(i)}, \theta_{1,i}, \theta_{1,c(i)}, \ldots$.

Denote $\zeta_j^{(i)} = \theta_{j,c(i)} - \theta_{j,i} \overset{\mathcal{D}}{=} \zeta^{(i)}$; $\zeta_j^{(c(i))} = \theta_{j,i} - \theta_{j-1,c(i)} \overset{\mathcal{D}}{=} \zeta^{(c(i))}$;

$$\mathbf{P}\{\zeta^{(a,i)} \leqslant s\} \overset{\text{def}}{=} \frac{F_i(a+s) - F_i(s)}{1 - F_i(s)}, \text{ and } \widetilde{F}_i(s) \overset{\text{def}}{=} \frac{1}{\mu_i} \int\limits_0^s (1 - F_i(u))\,\mathrm{d}u.$$

We assume that the distribution function $F_1(s)$ satisfies the condition $(*)$, and the random variables $\zeta^{(a,i)}$, $\zeta_j^{(1)}$, $\zeta_j^{(2)}$ are mutually independent.

Suppose that $\mathbf{E}\varphi\left(\zeta^{(i)}\right) < \infty$ for some increasing positive function $\varphi(t)$.

If $\mathbf{E}\zeta^{(i)} = \mu_i < \infty$, then distribution $\mathcal{P}_t^{Y_0}$ of the process $(Y_t, t \geqslant 0)$ with every initial state Y_0 weakly converges to the stationary distribution $\mathcal{P}$; for the stationary version $(\widetilde{Y}_t, t \geqslant 0)$ of the process $(Y_t, t \geqslant 0)$ we have $\mathbf{P}\{n(\widetilde{Y}_t) = 1\} = \frac{\mu_1}{\mu_1 + \mu_2} \overset{\text{def}}{=} p$. $\left(\text{If we know only an estimate } \mu_2 \leqslant m_2, \text{ then } p \geqslant \rho \overset{\text{def}}{=} \frac{\mu_1}{\mu_1 + m_2}\right)$.

For construction of the strong successful coupling $((Z_t, \widetilde{Z}_t), t \geqslant 0)$ of the processes $(Y_t, t \geqslant 0)$ and $(\widetilde{Y}_t, t \geqslant 0)$ we will again construct the times when at least one of them changes its first component.

At the times t_i' such that $n(Y_{t_i'-0}) = 2$ and $n(Y_{t_i'+0}) = 1$ we use (with probability p) the random variables Ξ_i for $(Y_t, t \geqslant 0)$ and $\widetilde{\Xi}_i$ for $(\widetilde{Y}_t, t \geqslant 0)$;

$$\mathbf{P}\{\Xi_i \leqslant s\} = F_1(s), \ \mathbf{P}\{\widetilde{\Xi}_i \leqslant s\} = \widetilde{F}_1(s) = (\mu_1)^{-1} \int\limits_0^s (1 - F_1(u))\,\mathrm{d}u, \text{ and}$$

$$\mathbf{P}\{\Xi_i = \widetilde{\Xi}_i\} = \varkappa = \int\limits_{\{s:\, \exists (F_1(s))'\}} (F_1(s))' \wedge (\widetilde{F}_1(s))'\,\mathrm{d}s.$$

And with probability $q = 1 - p$ at the time θ_i' we use a procedure of the prolongation of alternating renewal process $\widetilde{Y}_t$ by using the distribution $\widetilde{F}_2(s)$.

So, $\widetilde{\tau}(Y_0) \overset{\text{def}}{=} \inf\{t > 0 : Z_t = \widetilde{Z}_t\} \leqslant \theta_1' + \zeta_\nu^{(1)} + \sum\limits_{i=1}^{\nu-1}\left(\zeta_i^{(1)} + \zeta_i^{(2)}\right)$, where

$$\mathbf{P}\{\nu > n\} = (1 - p\varkappa)^n \left(\leqslant (1 - \rho\varkappa)^n\right).$$

Hence, if $(1 - \rho\varkappa)\mathbf{E}e^{a(\zeta^{(1)} + \zeta^{(2)})} < 1$ and $\mathbf{E}e^{at_0'} < \infty$, then we can find a bound $\mathfrak{C}(Y_0, a)$ for $\mathbf{E}e^{a\widetilde{\tau}(Y_0)}$: $\mathbf{E}e^{a\widetilde{\tau}(Y_0)} \leqslant \mathfrak{C}(Y_0, a)$.

Therefore we have $\left\| \mathcal{P}_t^{Y_0} - \mathcal{P} \right\|_{TV} \leqslant 2e^{-at}\mathfrak{C}(Y_0, a)$. The conditions for obtaining of such estimation may be relaxed.

And if $\mathbf{E}\left(\zeta^{(i)}\right)^K < \infty$ for some $K \geqslant 1$, then we can estimate $\mathbf{E}(\widetilde{\tau}(Y_0))^k$ for $k \in [1, K]$: $\mathbf{E}(\tau(Y_0))^k \leqslant C(Y_0, k)$.

Again we have $\left\| \mathcal{P}_t^{Y_0} - \mathcal{P} \right\|_{TV} \leqslant 2 \dfrac{C(Y_0, k)}{t^k}$.

2. Now back to the queueing process $(Q_t, t \geqslant 0)$. If the regeneration period of queueing process $(Q_t, t \geqslant 0)$ can be split into two independent parts, then this process has an embedded alternating renewal process.

Firstly we will extend this queueing process $(Q_t, t \geqslant 0)$ to the Markov process $(X_t, t \geqslant 0)$.

Then we will complete the process $(X_t, t \geqslant 0)$ by the embedded alternating renewal process $(Y_t, t \geqslant 0)$ competed by the time from the last change of its state (as in previous part).

Using the technique of the proof of Theorem 1 we can find the bounds for the convergence rate for the embedded alternating renewal process $(Y_t, t \geqslant 0)$; this bounds estimate the rate of convergence of the extended queueing process $(X_t, t \geqslant 0)$; also this bounds are useful for the original queueing process $(Q_t, t \geqslant 0)$.

Acknowledgments. The author is grateful to L.G. Afanasyeva, S.V. Anulova and A.Yu. Veretennikov for fruitful discussions and comments.

References

1. Doeblin, W.: Exposée de la théorie des chaînes simple constantes de Markov à un nombre fini d'états. Rev. Math. Union Interbalcan **2**, 77–105 (1938)
2. Down, D., Meyn, S.P., Tweedie, R.L.: Exponential and uniform ergodicity of Markov processes. Ann. Probab. **23**(4), 1671–1691 (1995)
3. Griffeath, D.: A maximal coupling for Markov chains. Zeitschrift für Wahrscheinlichkeitstheorie und Verwandte Gebiete. **31**(2), 95–106 (1975)
4. Kalashnikov, V.V.: Uniform estimation of the convergence rate in a renewal theorem for the case of discrete time. Theory Probab. Appl. **22**(2), 390–394 (1978)
5. Kalashnikov, V.V.: Topics on Regenerative Processes. CRC Press, Boca Raton (1994)
6. Lindvall, T.: Lectures on the Coupling Method. Wiley, New York (1992)
7. Lindvall, T.: On coupling of continuous-time renewal processes. J. Appl. Probab. **19**(1), 82–89 (1982)
8. Lund, R.B., Meyn, S.P., Tweedie, R.L.: Computable exponential convergence rates for stochastically ordered Markov processes. Ann. Appl. Probab. **6**(1), 218–237 (1996)
9. Roberts, G., Rosenthal, J.: Quantitative bounds for convergence rates of continuous time Markov processes. Electron. J. Probab. **1**(9), 1–21 (1996). https://projecteuclid.org/download/pdf_1/euclid.ejp/1453756472

10. Silvestrov, D.S.: Coupling for Markov renewal processes and the rate of convergence in ergodic theorems for processes with semi-Markov switchings. Acta Applicandae Mathematica **34**(1), 109–124 (1994)
11. Thorisson, H.: Coupling, Stationarity, and Regeneration. Springer, Heidelberg (2000)
12. Thorisson, H.: On maximal and distributional coupling. Ann. Probab. **14**(3), 873–876 (1986)
13. Veretennikov, A.Y.: On the rate of convergence to the stationary distribution in the single-server queuing system. Autom. Remote Control **74**(10), 1620–1629 (2013)
14. Veretennikov, A.Y.: On the rate of convergence for infinite server Erlang-Sevastyanov's problem. Queueing Syst. **76**(2), 181–203 (2014)
15. Veretennikov, A., Zverkina, G.A.: Simple proof of Dynkin's formula for single-server systems and polynomial convergence rates. Markov Process. Relat. Fields **20**(3), 479–504 (2014)
16. Veretennikov, A., Zverkina, G.: On polynomial bounds of convergence for the availability factor. In: Vishnevsky, V., Kozyrev, D. (eds.) DCCN 2015. CCIS, vol. 601, pp. 358–369. Springer, Heidelberg (2016). doi:10.1007/978-3-319-30843-2_37

Convergence Evaluation of Adaptation to Losses: The Case of Subscription Notification Delivery to Mobile Users in Smart Spaces

Dmitry Korzun$^{(\boxtimes)}$, Andrey Vdovenko, and Olga Bogoiavlenskaia

Petrozavodsk State University, Petrozavodsk, Russia
{dkorzun,vdovenko,olbgvl}@cs.karelia.ru

Abstract. A smart space provides a shared view on information, which is cooperatively produced, processed, and consumed by participants themselves in a computing environment. One of the most advanced networked operations on this information is the subscription operation. It supports the information-driven programming style: a notification is delivered to all interested participants when an appropriate information fact is formed in the smart space. In this work, we continue our study of the notification delivery when the latter is subject to losses. Notifications assigned to the mobile user in smart space are undelivered though several information updates have been made by other participants. The notification delivery performance can be improved by using active control: The client of the mobile user side proactively tracks information updates according to individually defined time points. These points can be selected rationally to adapt the notification delivery to observed losses. We analytically and experimentally evaluate the convergence of two active control strategies with adaptation to losses for different loss distributions.

Keywords: Smart spaces · Internet of Things · Publish/subscribe · Mobile user · Notification delivery · Active control · Adaptation to losses · Collaborative work environment · Convergence

1 Introduction

The emerging technologies of Internet of Things (IoT) lead now to the new type of ubiquitous computing environments (IoT environments) where the role of distributed processing of the information from multiple available sources by multiple participants is essential [3,6]. A promising paradigm for programming such an IoT environment, which consists of various information devices, is smart spaces [1,9]. A smart space is deployed in a given computing environment and provides a shared view on information, which is cooperatively produced, processed, and consumed by participants themselves. Service consumption by end-users is often performed using mobile personal devices (e.g., smartphones, tablets), which is more suitable for ubiquitous computing (anywhere, anytime).

The publish/subscribe (pub/sub) model is widely used for organizing multi-party interactions in distributed systems [4]. In smart spaces, the subscription

© Springer International Publishing AG 2016
V.M. Vishnevskiy et al. (Eds.): DCCN 2016, CCIS 678, pp. 394–405, 2016.
DOI: 10.1007/978-3-319-51917-3_35

operation is one of the most advanced networked operations on the shared information [12,16]. The operation supports the information-driven programming style: a notification is delivered to all interested participants when an appropriate information fact is formed in the smart space.

In existing solutions for smart spaces, the major role in information sharing is played by a semantic information broker (SIB) [5,9]. In particular, SIB is responsible for detection of information changes and for subsequent delivery of notifications to those clients that subscribed to the information. Both change detection and notification delivery are subject to losses in network environments.

We continue our study on the active control of notification delivery for subscription operation in the case of notification losses [7]. The client follows an adaptive strategy controlling the check interval based on the number of notifications lost in the latest round. This adaptive strategy is a generalization of the TCP algorithm of additive–increase/multiplicative–decrease (AIMD). This paper presents our extended convergence study of active control from [8]. We analyze adaptive and multiplicative–decrease strategies for different fixed distributions of notification losses.

Our previous analytical results characterize quantities of the steady state behavior for the check interval [7]. The question of the convergence speed to the steady state is topical especially for the IoT case with unstable networking environments, which suffer from random fluctuations, e.g., workload or capacity oscillations. Convergence properties characterize applicability of the analytical evaluations. Meanwhile these issues have paid little attention in the literature. A lot of studies, including [2,18], considered TCP congestion control convergence properties. Nevertheless, those works studied the convergence speed focusing on the aggressiveness and responsiveness indices. It is measured by the speed with which the system approaches the goal state. The latter reflects such properties of the congestion control as fairness and performance. In contrast in this work, we apply the method presented in [15] to obtain upper bounds of the difference between transient and steady state distributions during the discrete time evolution of the system adaptation and control.

The rest of the paper is organized as follows. Section 2 formulates the notification loss problem for the subscription operation in smart spaces. Section 3 introduces two control strategies for convergence analysis: the adaptive strategy, which is a generalization of the AIMD algorithm of TCP, and the multiplicative–decrease strategy. Section 4 discusses the applicability issues of the adaptive strategy for such a particular domain as collaborative work environments. Section 5 presents results of our convergence analysis, both theoretical and simulation estimates, for the two notification loss distributions (uniform and Poisson losses). Section 6 concludes the paper.

2   Notification Delivery Problem

Let us consider a smart space forming a sparse-connected multi-agent system deployed in a given IoT environment [9]. Such an environment consists of various digital devices, which act as IoT smart objects [6]. Software agents run on

the devices and interact over the shared information content. This type of interaction involves, in parallel and asynchronously, a lot of informational sources
and destinations. Information sharing makes the interaction indirect, based on a
semantic information broker (SIB) [11,16]. The latter implements a shared information storage, serving requests from agents on read/write operations. SIB acts
as an information hub maintaining knowledge of the whole environment and this
way enabling the agents to construct information services over this cooperative
knowledge generated in the smart space.

The subscription operation specifies a persistent query from each subscribed
agent (a subscriber or subscription client) to the SIB (a subscription server)
for a particular part of the shared content [12]. Whenever the specified part is
changed, the agent should receive the subscription notification. Changes are due
to parallel activity of other agents, which act as publishers in this interaction
(note that an agent may combine the roles of publisher and subscriber). SIB
monitors subscriptions of all clients and maps all incoming content changes to
the specified interests. Therefore, changes are controlled on the SIB side, and
corresponding notifications are sent to the clients. SIB acts as a passive receiver,
and we call such subscription notifications passive [7].

We employ Smart-M3 as a reference software platform for creating smart
spaces [5,11]. For each subscription, the SIB maintains a network connection
(e.g., a TCP connection) established by the client's request [12,13]. Knowing the
set of all subscriptions, the SIB regularly checks that they are alive, removing
the subscription if its network connection is lost. Smart-M3 follows the best
effort style in subscription notification delivery. A notification *should* be sent to
a client if a related change in the content has happened. Some notifications can
be unsent by SIB due to its overload or internal operability faults. SIB does not
check delivery for already sent notifications, and a new notification can be sent
although the underlying network connection is broken on the client side.

The above properties do not ensure the dependable notification delivery in
Smart-M3 even if reliable network protocols are used, such as TCP. For a client
a possible solution is to have an additional mechanism reducing the number of
undelivered notifications. The obvious way is augmenting the passive notification
delivery with an active control strategy that the client performs individually on
its own [7]. We focus on the case when clients are associated with mobile end-
users for whom the mobile personal device (e.g., smartphone or tablet) is the
primary tool to access the smart space and consume its services [17]. A particular
application domain here is collaborative work environments, such as SmartRoom
system, see [10,11].

Consider the following model to formalize the key properties of the subscription notification loss problem in smart spaces. Let $i = 1, 2, \ldots$ be the event-based
time evolution on the client side, where i is the index of notification events. An
event i is either a passive notification (i.e., received from SIB) or an explicit
check of the notification delivery (made by the client within its active control).
Denote by t_i and k_i the time elapsed and the number of losses occurred between
i and $i + 1$, respectively. Assume that some initial value t_0 is always defined.

The values for k_i are non-negative integers. The active control is interested in making t_i large while keeping k_i small (or even $k_i = 0$ for all intervals).

3   Strategies of Active Control with Adaptation to Losses

In accordance with our previous work [7,8], we consider the following two strategies of active control that implements "adaptation to losses" for the client. This kind of adaptation means that the client reduces its check interval t_i when losses are observed, and increases t_i in the case of no losses.

Adaptive Strategy. Let the client has observed no losses during t_{i-1}, i.e., $k_{i-1} = 0$. The observation indicates the system state. The client increases additively the check interval, i.e., $t_i = t_{i-1} + \delta$ for a fixed parameter $\delta > 0$. On the contrary, if the client has observed losses, i.e., $k_{i-1} > 0$, it reduces t_i to decrease the number of losses in the nearest future. The reduction applies the multiplicative average

$$t_i = \alpha t_{i-1} + (1 - \alpha) \frac{t_{i-1}}{k_{i-1} + 1}$$

for a fixed parameter $0 \le \alpha < 1$. In a result, we yield the recurrent system by which the check interval t_i is reduced (multiplicative decrease) in case of losses and incremented (additive increase) otherwise:

$$t_i = \begin{cases} t_{i-1} + \delta & \text{if } k_{i-1} = 0, \\ \dfrac{1 + \alpha k_{i-1}}{k_{i-1} + 1} t_{i-1} & \text{if } k_{i-1} > 0. \end{cases} \tag{1}$$

Note that (1) is valid only for active control of subscription notifications. When a passive notification i is delivered, then the value of t_i is not set by the client.

Multiplicative–Decrease Strategy. It is a semi-adaptive approach, with halving the check interval in case of losses and setting the check interval to some initial (reference) value t_0 otherwise:

$$t_i = \begin{cases} t_{i-1}/2 & \text{if } k_{i-1} > 0, \\ t_0 & \text{if } k_{i-1} = 0. \end{cases} \tag{2}$$

Evolution (2) can be described by a discrete-time Markov chain. Since the check interval accepts one of the values $t_0/2^i$ for $i = 0, 1, \ldots$, we consider it as a random variable. Assume that the sequence $\{i_n\}_{n \ge 0}$ of the indices of 2 forms a Markov chain which is a random walk with transition probabilities $p_{i,i+1} = q_i$ and $p_{i0} = p_i$. Denote $q_i = 1 - p_i$ for $i = 0, 1, \ldots$. According to [15] the chain has a steady state distribution $\pi_i = \lim\limits_{n \to \infty} p_{ij}$, if the following condition holds

$$\prod_{k=0}^{\infty} q_k = \lim_{n \to \infty} q_0 q_1 \cdots q_n = 0.$$

The expectation of the first recurrence time in the state with the maximal check interval $(i = 0)$ is

$$1 + \sum_{n=1}^{\infty} q_0 q_1 \cdots q_{n-1}.$$

4   Application to Collaborative Work Environments

A particular application domain for adaptive control strategy is collaborative work environments [14]. The smart spaces approach becomes popular for creating such environments [11]. Our case study is SmartRoom system [10], where human participants are mobile users. They consume services primarily from such mobile personal devices as smartphones and tablets. From the one hand, subscription is widely used by the mobile SmartRoom clients to detect appropriate events during the collaborative work. On the other hand, the subscription is clear subject to losses due to the mobility and wireless operation.

The SmartRoom system provides a set of services to assist such collaborative work as conferences or meetings. Services are deployed by the environment infrastructure and client applications are run by participants mobile devices. The presentation service displays the current presentation slide. The current speaker controls the slide show from his/her device. When the current slide is changing, other participants are notified, and the slide on their mobile clients is updated.

Figure 1 shows the role of subscribtion and its active control in smart space based collaborative work environment. Each publisher (service or participant from his device) makes updates with a certain rate. In sum, M publishers makes a flow of updates. Each of N subscribers perceives this flow as discrete process where k_i notification losses happen on time interval t_i. These parameters can be calculated on the mobile client side.

When a client uses the adaptive control strategy, an additional thread is started. It implements for the client an algorithm to check for changes on a subscription in parallel to client's main thread. In particular, an unique notification

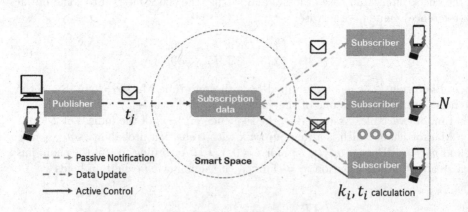

Fig. 1. Adaptive control strategy in smart space based collaborative work environment

number can be assigned to each notification. If the number is different from the last observed, then the value of k_i can be calculated, not only the fact that $k_i > 0$. After that the check interval t_i can be adjusted.

Parameter t_0 should reflect the time when one notification is lost on average— a tradeoff point of a control strategy. In practice, this average time is essentially varied. For instance, in the SmartRoom system, a speaker can switch most of slides more or less uniformly during the presentation. During the discussion (immediately after the presentation) the behavior of slide changes is radically different, e.g., a fast scrolling to reach a given slide and to answer to a question. Therefore a periodic recalculation of t_0 can be implemented. One way is to estimate t_0 as the average time between two subsequent passive notifications observed by the client.

Another important issue is the strategy parameters selection, see α and δ in (1). One of the main influencing factor is the total number of active agents, including N subscribers and M publishers. Even if some active agents are not involved into subscription they can create additional workload to the SIB and the network. In the simplest case, one can set $\alpha = 0.3$, i.e., the history has higher priority over the latest observation. Taking $\delta = t_0$ reflects a kind of doubling since the active control strategy tries to balance between $k_i = 0$ (no loss) and $k_i = 1$ (one loss).

In the simulation experiments below we applied the case when t_0 is fixed based on known characteristics of the notification loss distribution. Then $\alpha = 0.3$ and $\delta = t_0$. Study on effective methods for estimation and recalculation of t_0 as well as for selection of α and δ is subject of our further research.

5   Convergence Evaluation

In the subsequent analysis we consider the following distributions to model the notification losses, in accordance with [7].

1. Let the time elapsed between consecutive losses follow a uniform distribution $\mathcal{U}\{0, \xi t_0\}$. Hence, the average number of losses in any check interval is proportional to its length t_i
2. Let k_i follow a Poisson process of parameter λ. Hence, the number of losses during t_i has the probability mass function

$$\mathbb{P}(k_i = k) = \frac{(\lambda t_i)^k}{k!} e^{-\lambda t_i} \tag{3}$$

Convergence of the Multiplicative–Decrease Strategy to a Steady State Distribution. To estimate the convergence speed to the steady state distribution we apply the following criterion described in [15]. Let the ergodicity coefficient be defined as

$$k(n_0) = 1 - \frac{1}{2} \sup_{i,i+1} \sum_{m=0}^{\infty} |p_{im}(n_0) - p_{jm}(n_0)|, \tag{4}$$

where $p_{ij}(n_0)$ is the transition probability from the state i to the state j in n_0 time steps of the Markov chain.

Now we assume that $q_i \leq 1 - \Delta < 1$, i.e., the transition probability from any state $i \neq 0$ into the zero state satisfies the inequality $p_i \geq \Delta$. Therefore one can obtain the following estimation for the ergodicity coefficient $k(1)$:

$$k(1) \geq \inf_i p_i \geq \Delta.$$

Thus the convergence rate of the Markov chain under consideration to the steady state distribution $\{\pi_j\}_{j=0}^{\infty}$ could be estimated as

$$\sup_j |\pi_j(n) - \pi_j| \leq (1 - k(1))^{n-1}, \tag{5}$$

where $\pi_j(n)$ is the probability that the chain is in the state j at the step n.

For the Poisson notification losses we have $p_0 = e^{-\lambda t_0}$ and $p_n = p_0^{2^{-n}}$. Then it yields

$$\inf_i p_i = \inf_i \left(e^{-\lambda t_0}\right)^{2^{-i}} = \inf_i e^{\frac{-\lambda t_0}{2^i}} \geq e^{-\lambda t_0}.$$

Consequently,

$$\sup_j |\pi_j(n) - \pi_j| \leq \left(1 - e^{-\lambda t_0}\right)^{n-1}.$$

As a result, the Markov chain convergence rate is slower for higher initial values t_0.

For the uniform distribution one obtains $q_n = q_0 2^{-n}$ and correspondingly $p_n = 1 - q_0 2^{-n}$. Consequently, $\inf_i(1 - q_0 2^{-i}) \geq 1 - q_0$, and

$$\sup_j |\pi_j(n) - \pi_j| \leq (1 - 1 - q_0)^{n-1} = q_0^{n-1}.$$

As a result, the Markov chain convergence rate is slower for higher values of the probability q_0.

We can conclude this analytical evaluation that the convergence is exponential in dependence on the number of steps n.

Simulation Comparison of the Convergence for the Adaptive and Multiplicative–Decrease Strategies. Denote $l_k = \mathbb{P}[k_i = k]$ the probability that the number of losses is k on some interval t_i. We performed simulation experiments for the two active control strategies and for the two notification loss distributions. We estimate l_k as a frequency converging to theoretical values for the probabilities $\mathbb{P}[k_i = k]$. The results are visualized in Figs. 2, 3, 4 and 5.

For uniform losses we set $\xi = 0.1$. For Poisson losses we set $\lambda = 0.05$. In both control strategies, we take $t_0 = 20$ s. According to the selected parameters of the loss distributions, it means that every 20 s one notification is lost on average. For strategy (1) we use $\alpha = 0.3$ and $\delta = t_0$.

The multiplicative–decrease strategy for uniform losses (Fig. 2) has low values of losses (i.e., $k_i \leq 2$). In contrast, for Poisson losses (Fig. 3) the number of

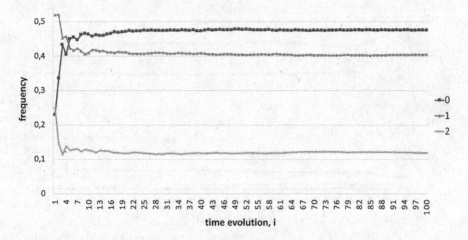

Fig. 2. Experimental convergence of l_k for the multiplicative–decrease strategy and uniform notification losses ($k = 0, 1, 2$).

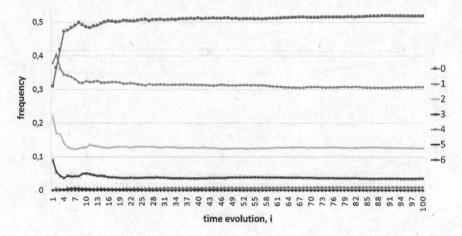

Fig. 3. Experimental convergence of l_k for the multiplicative–decrease strategy and Poisson notification losses ($k = 0, 1, \ldots, 6$).

simultaneous losses per interval can be high (i.e., $k_i = 6$ in some cases). The adaptive strategies has higher number of losses since it can set the check interval t_i more than t_0.

The experiments show that the probabilities l_k are decreasing in dependence on $k = 0, 1, \ldots$. As a result, many simultaneous losses happen rare. The probability of no loss is highest and lies in $0.4 \leq l_0 \leq 0.5$, which means that the strategies keep about half of check intervals in the "no loss state". Clearly, l_0 can be made higher by setting lower t_0.

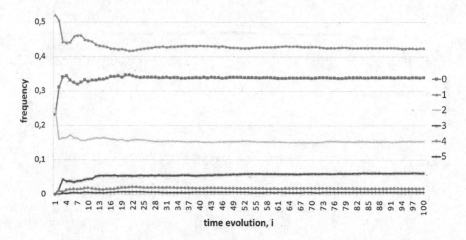

Fig. 4. Experimental convergence of l_k for the adaptive strategy and uniform notification losses ($k = 0, 1, \ldots, 5$).

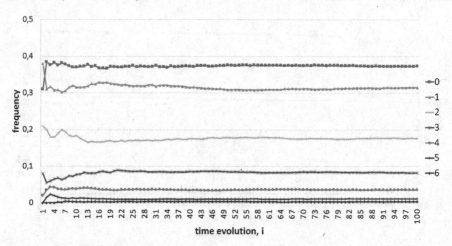

Fig. 5. Experimental convergence of l_k for the multiplicative–decrease strategy and Poisson notification losses ($k = 0, 1, \ldots, 6$).

The observed convergence of l_k is fast for both strategies and all k. Figures 2, 3, 4 and 5 show that l_k needs from 100 to 200 s (about 10 iterations) to reach very close to the steady state. Table 1 shows quantitative comparison based on efficiency metrics from [7]. Let $k_{\text{avg}} \approx \mathbb{E}[K]$ be the number of losses per interval on average and $t_{\text{avg}} \approx \mathbb{E}[T]$ be the average length of check interval:

$$k_{\text{avg}} = \frac{1}{n} \sum_{i=1}^{n} k_i, \quad t_{\text{avg}} = \frac{1}{n} \sum_{i=1}^{n} t_i.$$

The ratio $k_{\text{avg}}/t_{\text{avg}}$ shows the strategy performance: the lower the metric value the higher the performance.

Table 1. Performance comparison of the experimented strategies

Strategy	Metric	Distribution	
		Poisson	Uniform
Multiplicative-decrease strategy	k_{avg}	0.729	0.646
	t_{avg}	14.382	13.959
	k_{avg}/t_{avg}	0.051	0.046
Adaptive strategy	k_{avg}	1,138	0,996
	t_{avg}	22,989	20,838
	k_{avg}/t_{avg}	0,049	0,048

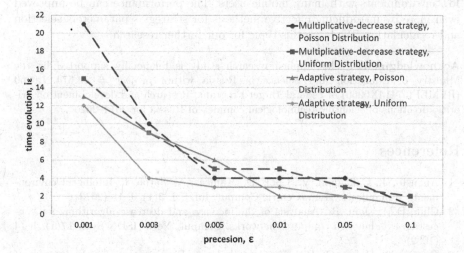

Fig. 6. Experimental convergence comparison: the number of iterations i_ϵ for a strategy to reach precision ϵ.

To estimate and compare the convergence we compute the metric i_ϵ (for each check interval evolution) for given small $\epsilon > 0$ as follows. Given evolution of $l_k(i)$ over the discrete time evolution with iterations i. Let us find the minimal i_k such that for all $i \geq i_k$ the difference $|l_k(i+1) - l_k(i)| < \epsilon$. Then $i_\epsilon = \max_k i_k$, i.e., after which step all the probabilities l_k become very close to their empirical steady state values. That is, i_ϵ is the number of iterations i_ϵ to reach the given precision ϵ. We experimented with $10^{-3} \leq \epsilon \leq 10^{-1}$; the measured behavior of i_ϵ is shown in Fig. 6.

We can conclude this experimental evaluation that the strategies have similar performance in the terms of metric k_{avg}/t_{avg}. Note that this result is mainly due to the static distribution of losses and apriory selection of t_0. Nevertheless, the adaptive strategy provides less number of iterations to converge to the steady state from the beginning.

6   Conclusion

This paper continued our study of active control of notification delivery for a mobile client subscribed to certain information in the smart space. We considered adaptive strategies of active control since they can achieve better performance of subscription notification delivery in the terms of ratio "the number of losses vs. the check interval". We analyzed on the convergence property: the speed to reach the steady state when the notification loss distribution is fixed. Our analytical and experimental evaluation shows that the convergence speed is reasonable for such an IoT-enabled application domain as collaborative work environments. The evaluation indicates that adaptive strategies are promising for use in modern IoT environments with many mobile users. The performance can be improved even more by introducing effective methods for strategy's parameters selection and recalculation. We leave this topic for our further research.

Acknowledgments. This applied research study is financially supported by the Ministry of Education and Science of Russia within project # 14.574.21.0060 (RFMEFI57414X0060) of Federal Target Program "Research and development on priority directions of scientific-technological complex of Russia for 2014–2020".

References

1. Augusto, J., Callaghan, V., Cook, D., Kameas, A., Satoh, I.: Intelligent environments: a manifesto. Hum.-Centric Comput. Inf. Sci. **3**(1), 1–18 (2013)
2. Chiu, D.M., Jain, R.: Analysis of the increase and decrease algorithms for congestion avoidance in computer networks. Comput. Netw. ISDN Syst. **17**(1), 1–14 (1989)
3. Cristea, V., Dobre, C., Pop, F.: Context-Aware Environments for the Internet of Things. In: Bessis, N., Xhafa, F., Varvarigou, D., Hill, R., Li, M. (eds.) Internet of Things and Inter-cooperative Computational Technologies for Collective Intelligence, vol. 460, pp. 25–49. Springer, Heidelberg (2013)
4. Eugster, P.T., Felber, P.A., Guerraoui, R., Kermarrec, A.M.: The many faces of publish/subscribe. ACM Comput. Surv. **35**, 114–131 (2003)
5. Honkola, J., Laine, H., Brown, R., Tyrkkö, O.: Smart-M3 information sharing platform. In: Proceedings of the IEEE Symposium on Computers and Communications (ISCC 2010), pp. 1041–1046. IEEE Computer Society, June 2010
6. Kortuem, G., Kawsar, F., Sundramoorthy, V., Fitton, D.: Smart objects as building blocks for the internet of things. IEEE Internet Comput. **14**(1), 44–51 (2010)
7. Korzun, D.G., Pagano, M., Vdovenko, A.: Control strategies of subscription notification delivery in smart spaces. In: Vishnevsky, V., Kozyrev, D. (eds.) DCCN 2015. CCIS, vol. 601, pp. 40–51. Springer, Heidelberg (2016). doi:10.1007/978-3-319-30843-2_5
8. Korzun, D., Vdovenko, A., Bogoiavlenskaia, O.: On convergence of active control strategies for subscription notification delivery in smart spaces. In: Proceedings of the 19-th International Scientific Conference on Distributed Computer and Communication Networks: Control, Computation, Communications (DCCN-2016), November 2016, in press

9. Korzun, D.G., Balandin, S.I., Gurtov, A.V.: Deployment of smart spaces in internet of things: overview of the design challenges. In: Balandin, S., Andreev, S., Koucheryavy, Y. (eds.) NEW2AN/ruSMART -2013. LNCS, vol. 8121, pp. 48–59. Springer, Heidelberg (2013). doi:10.1007/978-3-642-40316-3_5

10. Korzun, D., Galov, I., Kashevnik, A., Balandin, S.: Virtual shared workspace for smart spaces and M3-based case study. In: Balandin, S., Trifonova, U. (eds.)' Proceedings of the 15th Conference of Open Innovations Association FRUCT, pp. 60–68. ITMO University, April 2014

11. Korzun, D.G., Kashevnik, A.M., Balandin, S.I., Smirnov, A.V.: The smart-M3 platform: experience of smart space application development for internet of things. In: Balandin, S., Andreev, S., Koucheryavy, Y. (eds.) ruSMART 2015. LNCS, vol. 9247, pp. 56–67. Springer, Heidelberg (2015). doi:10.1007/978-3-319-23126-6_6

12. Lomov, A.A., Korzun, D.G.: Subscription operation in Smart-M3. In: Balandin, S., Ovchinnikov, A. (eds.) Proceedings of the 10th Conference of Open Innovations Association FRUCT and 2nd Finnish-Russian Mobile Linux Summit, pp. 83–94. SUAI, November 2011

13. Morandi, F., Roffia, L., D'Elia, A., Vergari, F., Cinotti, T.S.: RedSib: a Smart-M3 semantic information broker implementation. In: Balandin, S., Ovchinnikov, A. (eds.) Proceedings of the 12th Conference of Open Innovations Association FRUCT and Seminar on e-Tourism, pp. 86–98. SUAI, November 2012

14. Muntean, M.I.: Some collaborative systems approaches in knowledge-based environments. In: Hou, H.T. (ed.) New Research on Knowledge Management Models and Methods, pp. 379–394. InTech (2012)

15. Rozanov, Y.: Sluchaynyi protsessyi. Kratkiy kurs, 2nd edn. Nauka, (1979). (in Russian)

16. Smirnov, A., Kashevnik, A., Shilov, N., Oliver, I., Balandin, S., Boldyrev, S.: Anonymous agent coordination in smart spaces: state-of-the-art. In: Balandin, S., Moltchanov, D., Koucheryavy, Y. (eds.) NEW2AN/ruSMART -2009. LNCS, vol. 5764, pp. 42–51. Springer, Heidelberg (2009). doi:10.1007/978-3-642-04190-7_5

17. Vdovenko, A., Marchenkov, S., Korzun, D.: Mobile multi-service smart room client: initial study for multi-platform development. In: Balandin, S., Trifonova, U. (eds.) Proceedings of the 13th Conference of Open Innovations Association FRUCT and 2nd Seminar on e-Tourism for Karelia and Oulu Region, pp. 143–152. SUAI, April 2013

18. Yang, Y.R., Kim, M.S., Lam, S.S.: Transient behaviors of TCP-friendly congestion control protocols. Comput. Netw. **41**(2), 193–210 (2003)

Sojourn Time Analysis for Processor Sharing Loss Queuing System with Service Interruptions and MAP Arrivals

Konstantin Samouylov[1,2], Eduard Sopin[1,2($\boxtimes$)], and Irina Gudkova[1,2]

[1] Department of Applied Probability and Informatics, RUDN University,
Miklukho-Maklaya str. 6, 117198 Moscow, Russia
{samouylov_ke,sopin_es,gudkova_ia}@pfur.ru
[2] Institute of Informatics Problems, FRC CSC RAS,
Vavilova str. 44, 119333 Moscow, Russia

Abstract. Processor sharing (PS) queuing systems are widely investigated by research community and applied for the analysis of wire and wireless communication systems and networks. Nevertheless, only few works focus on finite queues with both PS discipline and service interruptions. In the paper, compared with the previous results we analyze a finite capacity PS queuing system with Markovian arrival process, unreliable server, service interruptions, and an upper limit of the number of customers it serves simultaneously. For calculating the mean sojourn time, unlike a popular but computational complex technique of inverse Laplace transform we use an effective method based on embedded Markov chain. A practical example concludes the paper.

Keywords: Queuing system · Processor sharing · Egalitarian processor sharing · Unreliable server · Interruption · Probability distribution · Recursive algorithm · Sojourn time · Absorbing Markov chain

1 Introduction

Processor sharing (PS) queueing systems has been widely adopted as a convenient models for bandwidth sharing in computer and communication systems [1]. Kleinrock [2] introduced the simplest and the best known class of egalitarian processor sharing (EPS) discipline [3]. There are two variants of such systems: without customer waiting and with customer waiting. Both groups can be further divided into subgroups of finite capacity queues and infinite capacity queues. In [4], an exact expression for the Laplace transform of the distribution of a customers response time was obtained for infinite capacity systems. In [5], the conditional moments of the sojourn time and in [6]. Laplace transform for the conditional sojourn time density was found for finite capacity mode. For

The reported study was funded by the Russian Science Foundation according to the research project No. 16-11-10227.

© Springer International Publishing AG 2016
V.M. Vishnevskiy et al. (Eds.): DCCN 2016, CCIS 678, pp. 406–417, 2016.
DOI: 10.1007/978-3-319-51917-3_36

infinite capacity system with exponentially distributed interarrival and service times, a method for calculating the moments and the distribution of the response time was presented in [7]. A recursive formula to compute the stationary sojourn time distribution in a infinite capacity system with Markovian arrival process was provided in [8]. General QBD model was investigated in [9], in which server unavailability and more settings regarding arrival and service time were incorporated. Note that the most papers supposes to calculate characteristics via inverse Laplace transform that is enough complicated process. For computational reasons, a more effective procedure is the use of algorithms, e.g. see [8,10,11]. In this paper, we study finite capacity EPS system with waiting, Markovian arrival process, unreliable server, and apply the similar with [10,16] method for derivation of an algorithm for calculating the mean sojourn time.

The paper is organized as follows. In Sect. 2, we propose a queuing system with service interruptions, finite buffer, PS discipline, and threshold on the number of customers as well as a method for calculating the mean sojourn time based on an embedded Markov chain. In Sect. 3, we give a numerical example. Section 4 concludes the paper.

2 Processor Sharing System with Markovian Arrival Process

2.1 Queuing Model

Consider a single-server queuing system with finite capacity r. The server is unreliable, on- and off-period durations are exponentially distributed with rates $\alpha > 0$ and $\beta > 0$ correspondingly. Customers arrive according to a Markovian Arrival Process (MAP) governed by a continuous-time finite-state Markov chain with M states, which is called the underlying Markov chain hereafter. Arrival process is described by matrices $\mathbf{A}_0$ and $\mathbf{A}_1$, $\mathbf{A}_0 + \mathbf{A}_1 = \mathbf{A}$, where $\mathbf{A}$ is the generator matrix of the underlying Markov chain [12,13]. We assume that matrix is irreducible and denote θ the stationary probability vector of the underlying Markov chain and $\mathbf{1}$ a vector of ones with appropriate size. Then customer arrival rate is given by $\lambda = \theta\mathbf{a}$, where $\mathbf{a} = \mathbf{A}_1\mathbf{1}$. Service times of customers are assumed to be i.i.d. having an exponential distribution with parameter $\mu > 0$. Upon arrival of a customer, it is placed in the queue. Customers are served according to PS discipline, but no more than N customers at once $0 < N \leq r$. It means, if there are k customers in the system, only $k^* = min\{k, N\}$ of them are served, and the rest $max\{k - k^*, 0\}$ customers wait in the queue. If upon arrival of a customer, the system is full already, then customer is lost.

Behaviour of the queue with unreliable server can be described by Markov process with states (k, j) and (i, k, j), $i = 0, 1, \ldots, N$, $k = 0, 1, \ldots, r$, $j = 1, 2, \ldots, M$. During the on-periods the process is in a state (k, j) and during the off-periods it is in a state (i, k, j). Here k is number of customers in the system, j is the state of the underlying Markov chain, and i indicates the number of servicing customers in the system at the moment of the last server failure. Total number of the states is equal to $M[(N + 2)(r + 1) - N(N + 1)/2]$.

Let us put the states of process in the following order:

$$
\begin{aligned}
&(0,1),...,(0,M),(1,1),...,(1,M),...,(r,1),...,(r,M),\\
&(0,0,1),...,(0,0,M),(0,1,1),...,(0,1,M),...,(0,r,1),...,(0,r,M),\\
&(1,1,1),...,(1,1,M),(1,2,1),...,(1,2,M),...,(1,r,1),...,(1,r,M),...,\\
&(N,N,1),...,(N,N,M),(N,N+1,1),...,(N,N+1,M),...,\\
&(N,r,1),(N,r,2),...,(N,r,M).
\end{aligned}
\tag{1}
$$

With this order, the generator matrix of $\xi(t)$ takes the following form

$$
\mathbf{Q} =
\begin{bmatrix}
\mathbf{U} & \mathbf{V}_0 & \mathbf{V}_1 & \cdots & \mathbf{V}_N\\
\mathbf{W}_0 & \mathbf{D}_0 & 0 & \cdots & 0\\
\mathbf{W}_1 & 0 & \mathbf{D}_1 & \ddots & \vdots\\
\vdots & \vdots & \ddots & \ddots & 0\\
\mathbf{W}_N & 0 & \cdots & 0 & \mathbf{D}_N
\end{bmatrix}.
\tag{2}
$$

All blocks in matrix $\mathbf{Q}$ in turn are also block matrices with size of blocks $M \times M$. Block size of matrix $\mathbf{U}$ is $(r+1) \times (r+1)$,

$$
\mathbf{U} =
\begin{bmatrix}
\mathbf{A}_0 - \alpha\mathbf{I} & \mathbf{A}_1 & & & \\
\mu\mathbf{I} & \mathbf{A}_0 - (\alpha+\mu)\mathbf{I} & \mathbf{A}_1 & & \\
& \ddots & \ddots & \ddots & \\
& & \mu\mathbf{I} & \mathbf{A}_0 - (\alpha+\mu)\mathbf{I} & \mathbf{A}_1\\
& & & \mu\mathbf{I} & \mathbf{A} - (\alpha+\mu)\mathbf{I}
\end{bmatrix}.
\tag{3}
$$

Matrices $\mathbf{W}_i$ with block size $(r+1-i) \times (r+1)$ have nonzero blocks $\beta\mathbf{I}$ only on the diagonal, which starts at the right bottom corner and continues up to the first row,

$$
\mathbf{W}_i =
\begin{bmatrix}
\beta\mathbf{I} & & & \\
& \beta\mathbf{I} & & \\
& & \ddots & \\
& & & \beta\mathbf{I}
\end{bmatrix}, i = 0,1,...,N.
\tag{4}
$$

Matrices $\mathbf{D}_i$ have block size $(r+1-i) \times (r+1-i)$ and have the following block two-diagonal structure:

$$
\mathbf{D}_i =
\begin{bmatrix}
\mathbf{A}_0 - \beta\mathbf{I} & \mathbf{A}_1 & & & \\
& \mathbf{A}_0 - \beta\mathbf{I} & \mathbf{A}_1 & & \\
& & \ddots & \ddots & \\
& & & \mathbf{A}_0 - \beta\mathbf{I} & \mathbf{A}_1\\
& & & & \mathbf{A} - \beta\mathbf{I}
\end{bmatrix}, i = 0,1,...,N.
\tag{5}
$$

And finally, matrices $\mathbf{V}_i$ have block size $(r+1) \times (r+1-i)$, all of them, except $\mathbf{V}_N$, have the only nonzero block $\alpha\mathbf{I}$ at the first column of the $(i+1)$−th block row. Blocks $\alpha\mathbf{I}$ in matrix $\mathbf{V}_N$ are placed at the diagonal that starts at the right bottom corner and continues up to the left edge:

$$\mathbf{V}_N = \begin{bmatrix} & & \alpha\mathbf{I} \\ & \alpha\mathbf{I} & \\ & \ddots & \\ \alpha\mathbf{I} & & \end{bmatrix}, \mathbf{V}_i = \begin{bmatrix} & & \alpha\mathbf{I} \\ & & \end{bmatrix}, i = 0, 1, ..., N-1. \tag{6}$$

2.2 Stationary Probability Distribution of the Process

Denote $\mathbf{q} = (\mathbf{q}_0, \mathbf{q}_1, ..., \mathbf{q}_r, \mathbf{q}_{00}, ..., \mathbf{q}_{0r}, \mathbf{q}_{11}, ..., \mathbf{q}_{1r}, ..., \mathbf{q}_{NN}, ..., \mathbf{q}_{Nr})$ stationary probabilities vector of process $\xi(t)$. All subvectors $\mathbf{q}_k$ and $\mathbf{q}_{ik}$ of vector $\mathbf{q}$ are M-dimensional and meet the system of equilibrium equations $\mathbf{qQ} = \mathbf{0}, \mathbf{q1} = 1$, that can be written in the following form:

$$\mathbf{q}_0(\mathbf{A}_0 - \alpha\mathbf{I}) + \mu\mathbf{q}_1 + \beta\mathbf{q}_{00} = \mathbf{0},$$

$$\mathbf{q}_k(\mathbf{A}_0 - (\alpha+\mu)\mathbf{I}) + \mathbf{q}_{k-1}\mathbf{A}_1 + \mu\mathbf{q}_{k+1} + \beta\sum_{i=0}^{k}\mathbf{q}_{ik} = \mathbf{0}, \quad 0 < k < N,$$

$$\mathbf{q}_k(\mathbf{A}_0 - (\alpha+\mu)\mathbf{I}) + \mathbf{q}_{k-1}\mathbf{A}_1 + \mu\mathbf{q}_{k+1} + \beta\sum_{i=0}^{N}\mathbf{q}_{ik} = \mathbf{0}, \quad N \leq k < r, \tag{7}$$

$$\mathbf{q}_r(\mathbf{A} - (\alpha+\mu)\mathbf{I}) + \mathbf{q}_{r-1}\mathbf{A}_1 + \beta\sum_{i=0}^{N}\mathbf{q}_{ir} = \mathbf{0};$$

$$\mathbf{q}_{ii}(\mathbf{A}_0 - \beta\mathbf{I}) + \alpha\mathbf{q}_i = \mathbf{0}, \quad 0 \leq i < N,$$

$$\mathbf{q}_{ik}(\mathbf{A}_0 - \beta\mathbf{I}) + \mathbf{q}_{ik-1}\mathbf{A}_1 = \mathbf{0}, \quad i < k < r, 0 \leq i < N, \tag{8}$$

$$\mathbf{q}_{ir}(\mathbf{A} - \beta\mathbf{I}) + \mathbf{q}_{ir-1}\mathbf{A}_1 = \mathbf{0}, \quad 0 \leq i < N;$$

$$\mathbf{q}_{NN}(\mathbf{A}_0 - \beta\mathbf{I}) + \alpha\mathbf{q}_N = \mathbf{0},$$

$$\mathbf{q}_{Nk}(\mathbf{A}_0 - \beta\mathbf{I}) + \mathbf{q}_{Nk-1}\mathbf{A}_1 + \alpha\mathbf{q}_k = \mathbf{0}, N < k < r, \tag{9}$$

$$\mathbf{q}_{Nr}(\mathbf{A} - \beta\mathbf{I}) + \mathbf{q}_{Nr-1}\mathbf{A}_1 + \alpha\mathbf{q}_r = \mathbf{0}.$$

Proposition 1. *Let matrices* $\mathbf{P}, \mathbf{R}, \mathbf{S}$ *be defined as* $\mathbf{P} = -(\mathbf{A}_0 - \beta\mathbf{I})^{-1}\mathbf{A}_1$, $\mathbf{R} = -(\mathbf{A}_0 - \beta\mathbf{I})^{-1}$, $\mathbf{S} = -(\mathbf{A} - \beta\mathbf{I})^{-1}$. *Then*

1. Vectors $\mathbf{q}_0, \mathbf{q}_1, ..., \mathbf{q}_r$ *are solutions of the following equilibrium equations with irreducible generator matrix*

$$\mathbf{q}_0(\mathbf{A}_0 - \alpha\mathbf{I}) + \mu\mathbf{q}_1 + \alpha\beta\mathbf{q}_0\mathbf{R} = \mathbf{0},$$

$$\mathbf{q}_k(\mathbf{A}_0 - (\alpha+\mu)\mathbf{I}) + \mathbf{q}_{k-1}\mathbf{A}_1 + \mu\mathbf{q}_{k+1} + \alpha\beta\sum_{i=0}^{k}\mathbf{q}_i\mathbf{P}^{k-i}\mathbf{R} = \mathbf{0},$$
$$0 < k < r, \tag{10}$$

$$\mathbf{q}_r(\mathbf{A} - (\alpha+\mu)\mathbf{I}) + \mathbf{q}_{r-1}\mathbf{A}_1 + \alpha\beta\sum_{i=0}^{r}\mathbf{q}_i\mathbf{P}^{r-i}\mathbf{S} = \mathbf{0}.$$

2. *Vectors* $\mathbf{q}_{ik}$, $0 \le i < N$, $i \le k < r$, *can be calculated using the following formulas:*

$$\mathbf{q}_{ik} = \alpha \mathbf{q}_i \mathbf{P}^{k-i} \mathbf{R}, \quad i \le k < r,$$
$$\mathbf{q}_{ir} = \alpha \mathbf{q}_i \mathbf{P}^{r-i} \mathbf{S}, \quad 0 \le i < N; \tag{11}$$

$$\mathbf{q}_{Nk} = \alpha \sum_{i=N}^{k} \mathbf{q}_i \mathbf{P}^{k-i} \mathbf{R}, \quad N \le k < r,$$
$$\mathbf{q}_{Nr} = \alpha \sum_{i=N}^{r} \mathbf{q}_i \mathbf{P}^{r-i} \mathbf{S}. \tag{12}$$

Proof. Since the generator matrix $\mathbf{A}$ of the underlying Markov chain is irreducible, matrices $\mathbf{A}_0 - \beta \mathbf{I}$ and $\mathbf{A} - \beta \mathbf{I}$ are non-singular with $(\mathbf{A}_0 - \beta \mathbf{I})^{-1} \le \mathbf{0}$ and $(\mathbf{A} - \beta \mathbf{I})^{-1} \le \mathbf{0}$. After the series of simplifications Eqs. (8) and (9) can be rewritten in form of expressions (11) and (12). Substitution of formulas (11–12) into (7) leads to system of Eq. (10) in vectors $\mathbf{q}_0, \mathbf{q}_1, ..., \mathbf{q}_r$.

Note that matrix $\mathbf{B}$ in system of Eq. (12) is block upper nearly triangular indecomposable generator matrix, which can be obtained from matrix $\mathbf{Q}$ by $\mathbf{B} = \mathbf{U} - \mathbf{V}\mathbf{D}^{-1}\mathbf{W}$ [15], where

$$\mathbf{V} = \begin{bmatrix} \mathbf{V}_0 \cdots \mathbf{V}_N \end{bmatrix}, \mathbf{D} = \begin{bmatrix} \mathbf{D}_0 & & \\ & \ddots & \\ & & \mathbf{D}_N \end{bmatrix}, \mathbf{W} = \begin{bmatrix} \mathbf{W}_0 \\ \vdots \\ \mathbf{W}_N \end{bmatrix}. \tag{13}$$

Thus, formulas (10–12) allow to calculate stationary probabilities up to a constant multiplier. Then, applying normalizing condition, vector $\mathbf{q}$ is obtained. One can note that the sum of all subvectors of vector $\mathbf{q}$ is the stationary probability distribution of the underlying Markov chain, i.e.

$$\sum_{k=0}^{r} \mathbf{q}_k + \sum_{n=0}^{N} \sum_{k=n}^{r} \mathbf{q}_{nk} = \theta. \tag{14}$$

2.3   Customer Sojourn Time in the System

Let us consider absorbing Markov process $\tilde{\xi}(t)$, that begins immediately after the arrival of a particular customer and falls to absorbing state ω at the end of service of the customer. Consequently, the customer sojourn time in the system is equal to the time before absorption in process $\tilde{\xi}(t)$. State space of $\tilde{\xi}(t)$ can be obtained from the state space (1) of the initial process $\xi(t)$ by adding absorbing state ω and by eliminating the states, that cannot be achieved immediately after an arrival of a customer:

$$(1,1), ..., (1, M), (2,1), ..., (2, M), ..., (r, 1), ..., (r, M),$$
$$(0,1,1), ..., (0,1, M), (0,2,1), ..., (0,2, M), ..., (0, r, 1), ..., (0, r, M),$$
$$(1,1,1), ..., (1,1, M), (1,2,1), ..., (1,2, M), ..., (1, r, 1), ..., (1, r, M), ...,$$
$$(N, N+1, 1), ..., (N, N+1, M), (N, N+2, 1), .., (N, N+2, M), ...,$$
$$(N, r, 1), ..., (N, r, M), \omega.$$

With the introduced states order, generator matrix of $\tilde{\xi}(t)$ takes the following form:

$$\tilde{C} = \begin{bmatrix} C & c \\ 0 & 0 \end{bmatrix}, C = \begin{bmatrix} \tilde{U} & \tilde{V}_0 & \tilde{V}_1 & \cdots & \tilde{V}_N \\ \tilde{W}_0 & \tilde{D}_0 & 0 & \cdots & 0 \\ \tilde{W}_1 & 0 & \tilde{D}_1 & \ddots & \vdots \\ \vdots & \vdots & \ddots & \ddots & 0 \\ \tilde{W}_N & 0 & \cdots & 0 & \tilde{D}_N \end{bmatrix}, c = \begin{bmatrix} \tilde{u} \\ 0 \\ 0 \\ \vdots \\ 0 \end{bmatrix}. \qquad (15)$$

All blocks of matric C are also block matrices with size of blocks $M \times M$. Block size of matrix $\tilde{U}$ is $r \times r$, and size of vector $\tilde{u}$ is rM. Matrix $\tilde{U}$ has block-tridiagonal structure and can written in the following form

$$\tilde{U} = \begin{bmatrix} A_0 - (\alpha + \mu)I & A_1 & & & & & \\ \frac{1}{2}\mu I & A_0 - (\alpha + \mu)I & \ddots & & & & \\ & \frac{2}{3}\mu I & \ddots & & & & \\ & & \ddots & & A_1 & & \\ & & & A_0 - (\alpha + \mu)I & \ddots & & \\ & & & \frac{N-1}{N}\mu I & \ddots & & \\ & & & & \ddots & A_1 & \\ & & & & & A - (\alpha + \mu)I \end{bmatrix},$$

$$\tilde{u} = \begin{bmatrix} \mu 1 \\ \frac{1}{2}\mu 1 \\ \vdots \\ \frac{1}{N-1}\mu 1 \\ \frac{1}{N}\mu 1 \\ \vdots \\ \frac{1}{N}\mu 1 \end{bmatrix}. \qquad (16)$$

Matrices $\tilde{V}_i$, $i \neq 0$, can be obtained from matrices V_i by eliminating of the first block row, matrices $\tilde{W}_i$, $i \neq 0$, are formed from matrices W_i by eliminating of the first block column, while matrices $\tilde{V}_0, \tilde{W}_0, \tilde{D}_0$ are obtained by eliminating of the first block row and the first block column from V_0, W_0, D_0.

The initial distribution of $\tilde{\xi}(t)$ is stationary probability distribution $p = (p_1, ..., p_r, p_{01}, ..., p_{0r}, p_{11}, ..., p_{1r}, ..., p_{NN}, ..., p_{Nr})$ of discrete time Markov chain $\xi(\tau_n)$, embedded at the moments just after an arrival of a customer. According to the properties of Markovian arrival process, probability distribution p can be calculated as it is shown below:

$$\begin{aligned}
\mathbf{p}_{ii} &= \mathbf{0}, \quad i = 1, 2, ..., N, \\
\mathbf{p}_k &= \tfrac{1}{\lambda(1-\pi)} \mathbf{q}_{k-1} \mathbf{A}_1, \quad k = 1, 2, ..., r, \\
\mathbf{p}_{ik} &= \tfrac{1}{\lambda(1-\pi)} \mathbf{q}_{ik-1} \mathbf{A}_1, \quad i = 0, 1, ..., N, k = i+1, i+2, ..., r,
\end{aligned} \tag{17}$$

where π is the blocking probability of a customer:

$$\pi = \lambda^{-1}(\mathbf{q}_r + \mathbf{q}_{0r} + \mathbf{q}_{1r} + ... + \mathbf{q}_{Nr})\mathbf{a}. \tag{18}$$

Probability density function $g(x)$ of the time before absorption of the $\tilde{\xi}(t)$ and its Laplace-Stieltjes transformation $f(s)$ can be found using well-known formulas for phase-type distribution functions [14]:

$$g(x) = \mathbf{p} \exp(\mathbf{C}x)\mathbf{c}, \; f(s) = \mathbf{p}\,(s\mathbf{I} - \mathbf{C})^{-1}\mathbf{c}. \tag{19}$$

Formula (19) leads to the following expression for the k-th moments of the customer sojourn time:

$$m_k = \mathbf{m}_k \mathbf{1}, \tag{20}$$

where

$$\mathbf{m}_k = k! \mathbf{p} \mathbf{M}^k, \quad \mathbf{M} = -\mathbf{C}^{-1}. \tag{21}$$

2.4   Computational Algorithms for the K-th Moment of the Sojourn Time

The correlation of vectors $\mathbf{m}_k$ that are needed for calculation of sojourn time moments by formula (21) can be described by a simple recurrent relation

$$\mathbf{m}_0 = \mathbf{p}, \quad \mathbf{m}_k = k\mathbf{m}_{k-1}\mathbf{M}, k = 1, 2, ... \tag{22}$$

Consequently, these vectors are easy to find having matrix $\mathbf{M} = -\mathbf{C}^{-1}$. Exploiting special structure (15) of matrix $\mathbf{C}$, we derive an efficient numerical algorithm for their calculation.

Multiplying both sides of the Eq. (22) by matrix $\mathbf{C}$, we obtain the following expression:

$$\mathbf{m}_k \mathbf{C} = -k\mathbf{m}_{k-1}, \quad k = 1, 2, ... \tag{23}$$

Thus, vectors $\mathbf{m}_k$ can be found from system of linear equations $\mathbf{x}\mathbf{C} = \mathbf{b}$ by sequential substitution the right sides of Eq. (23) to $\mathbf{b}$.

Proposition 2. *Solution* $\mathbf{x} = (\mathbf{x}_1, ..., \mathbf{x}_r, \mathbf{x}_{01}, ..., \mathbf{x}_{0r}, \mathbf{x}_{11}, ..., \mathbf{x}_{1r}, ..., \mathbf{x}_{NN}, ...,$ $\mathbf{x}_{Nr})$ *of system of linear equations* $\mathbf{x}\mathbf{C} = \mathbf{b}$ *with matrix* $\mathbf{C}$ *defined by formula (15) and any vector* $\mathbf{b} = (\mathbf{b}_1, ..., \mathbf{b}_r, \mathbf{b}_{01}, ..., \mathbf{b}_{0r}, \mathbf{b}_{11}, ..., \mathbf{b}_{1r}$ $, ..., \mathbf{b}_{NN}, \mathbf{b}_{NN+1}, ..., \mathbf{b}_{Nr})$ *can be found according to the following formulas:*

$$\mathbf{x}_{0k} = - \sum_{n=1}^{k} \mathbf{b}_{0n} \mathbf{P}^{k-n} \mathbf{R}, \quad 1 \leq k < r,$$

$$\mathbf{x}_{0r} = - \sum_{n=1}^{r} \mathbf{b}_{0n} \mathbf{P}^{r-n} \mathbf{S}; \tag{24}$$

$$\mathbf{x}_{ik} = \alpha \mathbf{x}_i \mathbf{P}^{k-i} \mathbf{R} - \sum_{n=i}^{k} \mathbf{b}_{in} \mathbf{P}^{k-n} \mathbf{R}, \quad i \leq k < r, 1 \leq i < N,$$

$$\mathbf{x}_{ir} = \alpha \mathbf{x}_i \mathbf{P}^{r-i} \mathbf{S} - \sum_{n=i}^{r} \mathbf{b}_{in} \mathbf{P}^{r-n} \mathbf{S}, \quad 1 \leq i < N; \tag{25}$$

$$\mathbf{x}_{Nk} = \alpha \sum_{n=N}^{k} \mathbf{x}_n \mathbf{P}^{k-n} \mathbf{R} - \sum_{n=N}^{k} \mathbf{b}_{Nn} \mathbf{P}^{k-n} \mathbf{R}, \quad N \leq k < r,$$

$$\mathbf{x}_{Nr} = \alpha \sum_{n=N}^{r} \mathbf{x}_n \mathbf{P}^{r-n} \mathbf{S} - \sum_{n=N}^{r} \mathbf{b}_{Nn} \mathbf{P}^{r-n} \mathbf{S}. \tag{26}$$

Here vectors $\mathbf{x}_1, ..., \mathbf{x}_r$ *give a unique solution of the following linear system:*

$$\mathbf{x}_1 (\mathbf{A}_0 - (\alpha + \mu)\mathbf{I}) + \tfrac{1}{2}\mu \mathbf{x}_2 + \alpha\beta \mathbf{x}_1 \mathbf{R} = \mathbf{b}_1 + \beta(\mathbf{b}_{01} + \mathbf{b}_{11})\mathbf{R},$$

$$\mathbf{x}_k (\mathbf{A}_0 - (\alpha + \mu)\mathbf{I}) + \mathbf{x}_{k-1} \mathbf{A}_1 + \tfrac{k}{k+1}\mu \mathbf{x}_{k+1} + \alpha\beta \sum_{i=1}^{k} \mathbf{x}_i \mathbf{P}^{k-i} \mathbf{R}$$

$$= \mathbf{b}_k + \beta \sum_{i=0}^{k} \sum_{n=i}^{k} \mathbf{b}_{in} \mathbf{P}^{k-n} \mathbf{R}, \quad 1 < k < N,$$

$$\mathbf{x}_k (\mathbf{A}_0 - (\alpha + \mu)\mathbf{I}) + \mathbf{x}_{k-1} \mathbf{A}_1 + \tfrac{N-1}{N}\mu \mathbf{x}_{k+1} + \alpha\beta \sum_{i=1}^{k} \mathbf{x}_i \mathbf{P}^{k-i} \mathbf{R} \tag{27}$$

$$= \mathbf{b}_k + \beta \sum_{i=0}^{N} \sum_{n=i}^{k} \mathbf{b}_{in} \mathbf{P}^{k-n} \mathbf{R}, \quad N \leq k < r,$$

$$\mathbf{x}_r (\mathbf{A} - (\alpha + \mu)\mathbf{I}) + \mathbf{x}_{r-1} \mathbf{A}_1 + \alpha\beta \sum_{i=1}^{r} \mathbf{x}_i \mathbf{P}^{r-i} \mathbf{S} =$$

$$= \mathbf{b}_r + \beta \sum_{i=0}^{N} \sum_{n=i}^{r} \mathbf{b}_{in} \mathbf{P}^{k-n} \mathbf{S},$$

with $\mathbf{b}_{00} = \mathbf{0}$.

Proof. Let us rewrite system of linear equations $\mathbf{x}\mathbf{C} = \mathbf{b}$ in vector $\mathbf{x}$, as the system of linear equations in its subvectors $\mathbf{x}_k$ and $\mathbf{x}_{ik}$:

$$\mathbf{x}_1(\mathbf{A}_0 - (\alpha + \mu)\mathbf{I}) + \tfrac{1}{2}\mu \mathbf{x}_2 + \beta(\mathbf{x}_{01} + \mathbf{x}_{11}) = \mathbf{b}_1,$$

$$\mathbf{x}_k(\mathbf{A}_0 - (\alpha + \mu)\mathbf{I}) + \mathbf{x}_{k-1}\mathbf{A}_1 + \tfrac{k}{k+1}\mu \mathbf{x}_{k+1} + \beta \sum_{i=0}^{k} \mathbf{x}_{ik} = \mathbf{b}_k,$$
$$1 < k < N,$$

$$\mathbf{x}_k(\mathbf{A}_0 - (\alpha + \mu)\mathbf{I}) + \mathbf{x}_{k-1}\mathbf{A}_1 + \tfrac{N-1}{N}\mu \mathbf{x}_{k+1} + \beta \sum_{i=0}^{N} \mathbf{x}_{ik} = \mathbf{b}_k, \tag{28}$$
$$N \leq k < r,$$

$$\mathbf{x}_r(\mathbf{A} - (\alpha + \mu)\mathbf{I}) + \mathbf{x}_{r-1}\mathbf{A}_1 + \beta \sum_{i=0}^{N} \mathbf{x}_{ir} = \mathbf{b}_r;$$

$$x_{01}(\mathbf{A}_0 - \beta\mathbf{I}) = \mathbf{b}_{01},$$
$$x_{0k}(\mathbf{A}_0 - \beta\mathbf{I}) + x_{0k-1}\mathbf{A}_1 = \mathbf{b}_{0k}, \quad 1 < k < r, \tag{29}$$
$$x_{0r}(\mathbf{A} - \beta\mathbf{I}) + x_{0r-1}\mathbf{A}_1 = \mathbf{b}_{0r};$$

$$x_{ii}(\mathbf{A}_0 - \beta\mathbf{I}) + \alpha x_i = \mathbf{b}_{ii}, \quad 1 \le i < N,$$
$$x_{ik}(\mathbf{A}_0 - \beta\mathbf{I}) + x_{ik-1}\mathbf{A}_1 = \mathbf{b}_{ik}, \quad i < k < r, 1 \le i < N, \tag{30}$$
$$x_{ir}(\mathbf{A} - \beta\mathbf{I}) + x_{ir-1}\mathbf{A}_1 = \mathbf{b}_{ir}, \quad 0 \le i < N;$$

$$x_{NN}(\mathbf{A}_0 - \beta\mathbf{I}) + \alpha x_N = \mathbf{b}_{NN},$$
$$x_{Nk}(\mathbf{A}_0 - \beta\mathbf{I}) + x_{Nk-1}\mathbf{A}_1 + \alpha x_k = \mathbf{b}_{Nk}, \quad N < k < r, \tag{31}$$
$$x_{Nr}(\mathbf{A} - \beta\mathbf{I}) + x_{Nr-1}\mathbf{A}_1 + \alpha x_r = \mathbf{b}_{Nr}.$$

Since matrices $\mathbf{A}_0 - \beta\mathbf{I}$ and $\mathbf{A} - \beta\mathbf{I}$ are non-singular, Eqs. (30) and (31) lead to expressions (24–26). By substitution of them into (28) system of linear Eq. (27) in vectors $x_1, ..., x_r$ is derived. Matrix $\tilde{\mathbf{B}}$ of system of Eq. (27) is non-singular generator matrix, since it can be obtained from blocks of non-singular generator matrix $\mathbf{C}$ in the following way: $\tilde{\mathbf{B}} = \tilde{\mathbf{U}} - \tilde{\mathbf{V}}\tilde{\mathbf{D}}^{-1}\tilde{\mathbf{W}}$ [15], where

$$\tilde{\mathbf{V}} = \begin{bmatrix} \tilde{\mathbf{V}}_1 \cdots \tilde{\mathbf{V}}_N \end{bmatrix}, \tilde{\mathbf{D}} = \begin{bmatrix} \mathbf{D}_1 & & \\ & \ddots & \\ & & \mathbf{D}_N \end{bmatrix}, \tilde{\mathbf{W}} = \begin{bmatrix} \tilde{\mathbf{W}}_1 \\ \vdots \\ \tilde{\mathbf{W}}_N \end{bmatrix}. \tag{32}$$

3   Numerical Analysis

In order to illustrate the application of developed recursive algorithms, plots of the blocking probability and the mean sojourn time for different values of α and β are presented in Figs. 1 and 2 accordingly. The calculations were performed for the service rate $\mu = 5$, $N = 15$ and $r = 25$ as it was done in [16] for analysis of web browsing performance metrics in an unreliable wireless infrastructure. Matrices defining the underlying Markov chain of the Markovian arrival process are

$$A = \begin{pmatrix} -1 & 1 & 0 \\ 1 & -3 & 2 \\ 0 & 2 & -2 \end{pmatrix}, A_1 = \begin{pmatrix} 2 & 0 & 0 \\ 0 & 3 & 0 \\ 0 & 0 & 4 \end{pmatrix}.$$

As it is seen on figures, the shorter the on-period in the system, the bigger blocking probability and longer the mean sojourn time.

Moreover, Fig. 2 shows that for nearly all of considered system parameters the mean sojourn time does not exceed 5 s, which is defined as a maximum permissible delay for web page download and rendering [16].

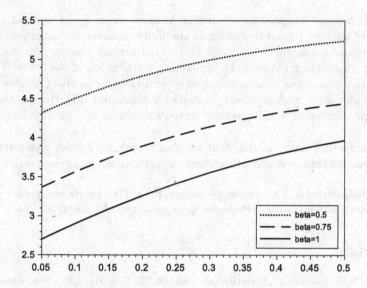

Fig. 1. Mean sojourn time of a customer depending on the on-period duration rate α

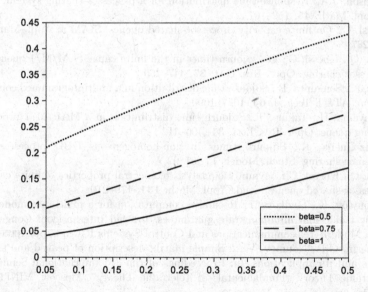

Fig. 2. Blocking probability of a customer depending on the on-period duration rate α

4 Conclusion

In this paper, we propose an absorbing Markov chain based method for the analysis of sojourn time distribution in the finite capacity queuing system with PS discipline, unreliable server and Markovian arrival process. We derived a recurrent algorithm for stationary probability distribution of the queue. For the analysis of mean sojourn time, we adjusted well-known absorbing Markov chain based technique to address some peculiarities connected with the threshold on maximum number of simultaneously served customers and unreliability of the server.

In our further study in this field we plan to calculate other characteristics, such as waiting time or waiting time due to interruptions, using similar technique.

Acknowledgements. The authors are grateful to the Director Research of the Service Innovation Research Institute, Professor Valeriy Naumov, for useful advices.

References

1. Fredj, S.B., Bonald, T., Proutiere, A., Regnie, G., Roberts, J.W.: Statistical bandwidth sharing: a study of congestion at flow level. In: ACM SIGCOMM 2001, pp. 111–122
2. Kleinrock, L.: Time-shared systems: a theoretical treatment. J. ACM **14**, 242–261
3. Yashkov, S.F., Yashkova, A.S.: Processor sharing: a survey of the mathematical theory. Autom. Remote Control **68**, 1662–1731
4. Morrison, J.A.: Response-time distribution for a processor-sharing system. SIAM J. Appl. Math. **45**, 152–167
5. Knessl, C.: On finite capacity processor-shared queues. SIAM J. Appl. Math. **50**, 264–287
6. Zhen, Q., Knessl, C.: On sojourn times in the finite capacity M/M/1 queue with processor sharing. Oper. Res. Lett. **37**, 447–450
7. Rege, K., Sengupta, B.: Sojourn time distribution in a multiprogrammed computer system. AT&T Tech. J. **64**, 1077–1090
8. Masuyama, H., Takine, T.: Sojourn time distribution in a MAP/M/1 processor-sharing queue. Oper. Res. Lett. **31**, 406–412
9. Nunez-Queija, R.: Sojourn times in non-homogeneous QBD processes with processor-sharing. Stoch. Models **17**, 61–92
10. Zhen, Q., Knessl, C.: Asymptotic analysis of spectral properties of finite capacity processor shared queues. Stud. Appl. Math. **131**, 179–210
11. Samouylov, K., Gudkova, I.: Recursive computation for a multi-rate model with elastic traffic and minimum rate guarantees. In: 2nd International Congress on Ultra Modern Telecommunications and Control Systems ICUMT, pp. 1065–1072
12. Basharin, G.P., Naumov, V.A.: Simple matrix description of peaked and smooth traffic and its applications. In: Proceedings of the 3rd International Seminar on Teletraffic Theory "Fundamentals of Teletraffic Theory", Moscow, VINITI, pp. 38–44
13. Lucantoni, D.M., Meier-Hellstern, K.S., Neuts, M.F.: A single server queue with server vacations and a class of non-renewal arrival processes. Adv. Appl. Probab. **22**, 676–705

14. Asmussen, S.: Applied Probability and Queues, vol. 51. Springer, New York (2003)
15. Naumov, V.A., Samouylov, K.E., Gaidamaka, Y.V.: Multiplicative solutions for the finite Markov chains. RUDN
16. Samouylov, K., Naumov, V., Sopin, E., Gudkova, I., Shorgin, S.: Sojourn time analysis for processor sharing loss system with unreliable server. In: Wittevrongel, S., Phung-Duc, T. (eds.) ASMTA 2016. LNCS, vol. 9845, pp. 284–297. Springer, Heidelberg (2016). doi:10.1007/978-3-319-43904-4_20

The Estimation of Probability Characteristics of Cloud Computing Systems with Splitting of Requests

Anastasia Gorbunova[1,2], Ivan Zaryadov[1,2], Sergey Matyushenko[1], and Eduard Sopin[1,2(✉)]

[1] Department of Applied Probability and Informatics, RUDN University, Miklukho-Maklaya Str. 6, 117198 Moscow, Russia
{gorbunova_av,zaryadov_is,matyushenko_si,sopin_es}@pfur.ru
[2] FRC CSC RAS, Institute of Informatics Problems, Vavilova Str. 44, 119333 Moscow, Russia

Abstract. Growing popularity of cloud services is explained by many advantages of them. The accessibility, flexibility, scalability, ease of management, the relatively low cost of implementation can be listed among the main advantages. The demand for cloud services with the ability to change one cloud service provider to another one without any significant cost for a user result in a high competition between cloud providers. Due to this reason, it became important to find the optimal performance measures of cloud systems. These measures, on the one hand, must meet all the requirements of Service Level Agreement (SLA), on the other hand, do not lead to excessive costs for provider. The paper presents the evaluation of the main service quality characteristics of cloud systems, including formulas for variance of residence time in the synchronization buffer. For the analysis of a cloud system, fork-join queues with corresponding methods of its approximation were used.

Keywords: Cloud computing · Fork-join queuing system · Response time · Synchronization buffer · Synchronization time

1 Introduction

The interest in cloud systems, in addition to the main advantages of their use, is defined by the need of commercial and research organizations in powerful computing infrastructure to perform resource-intensive tasks at relatively low cost [9]. The one of the most important indicators of quality of service for the cloud system is the response time. In the context of the resolution of cluster problems, that are characterized by a high level of parallelism, the request processing is completed at the end of processing of all of its constituent tasks, so the response time is the maximum of all tasks sojourn times. In addition, the one of the most effective ways to reduce the response time is to send the same request to several virtual machines, and wait for the earliest response. In this case, the response

© Springer International Publishing AG 2016
V.M. Vishnevskiy et al. (Eds.): DCCN 2016, CCIS 678, pp. 418–429, 2016.
DOI: 10.1007/978-3-319-51917-3_37

time is the minimum of the sojourn times of all tasks in the system [8]. Along with such an important quality of service metrics as the system response time, in recent years the time spent by sub-queries in the synchronization buffer (i.e. the time between the end of the first and last tasks of the same request), became an essential parameter for the parallel computing [13]. This trend is explained by the fact that the longer the tasks assembly time, the greater should be the size of the buffer. Moreover, as the number of virtual machines or task types increase, the buffer size is also should be increased. In this context, it is important to select the optimal values of system parameters to minimize the synchronization time without sacrificing the quality of service.

1.1   The Model of a Cloud Center

The functioning of a cloud computing center can be described in terms of a fork-join queueing system with K $M/M/1$-type subsystems with homogeneous servers [12] (Fig. 1). Let us denote ξ_k - the sojourn time of the k-th task of a request before it is collected at the synchronization buffer. Assume that joining up of all tasks in the synchronization buffer occurs immediately.

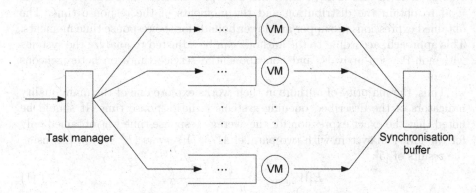

Fig. 1. The queueing system model of the cloud center

We should note that the task sojourn times ξ_k are dependent random variables due to the joint moment of arrival. Because the exact solution for fork-join system with $K > 2$ is unknown and the possibility of obtaining this solution is under question [10,15], we will approximate our system by K parallel independent $M_\lambda/M_\mu/1$ type queuing systems. We can use such approximation because the marginal probability distributions of k-th subsystem, $k = \overline{1,K}$, of a fork-join system are equal to probability distributions of the $M/M/1$ queuing system [6].

1.2   Probability Characteristics of the System

In this section, we consider the probability characteristics of random variables that determine performance measures of a cloud system. As mentioned above, for the cloud center analysis a fork-join queuing system was selected. It should be noted that the first results of its research have been obtained for the case of two servers with Poisson arrivals and exponential service time [10]. However, despite the Markovian assumptions, the stationary probabilities describing the system state explicitly have not been received even for such nominally simple version of this model.

A good approximation of the stationary distribution for infinite capacity queues is stationary distribution of a system of finite but comparatively large capacity, so that the blocking probability is negligible, and in this case iterative numerical methods are applied for solving the systems of linear equations of large dimension [12].

Another option for calculation of stationary probabilities, which were subsequently used to analyze the system response time was proposed in [1–3]. The authors consider two models approximating the original system: a model, in which all buffers, except the buffer of the first server, have finite capacity and the model with predefined threshold on maximum difference between each pair of queue lengths. Analysis of these systems is provided using matrix-geometric methods [11], since the transition probability matrices are reduced to block-triangular form and lead to the stationary probabilities distribution, which is used to obtain the distribution and the moments of the response time. The obtained expressions are upper and lower bounds for the response time moments. This approach, according to the authors, can be adjusted to analyze the systems with non-Poisson arrivals, and nonexponential service times on heterogeneous servers.

Thus, the majority of authors in their works explore one of the main quality indicators of the described queuing systems - the response time. It should be noted that the exact expression for the average response time was obtained only for the fork-join system with two parallel $M/M/1$ servers $(K = 2)$ in [10] using the results of [5]:

$$E[W_{2,\max}] = \frac{1}{\mu - \lambda} \cdot \frac{12 - \rho}{8}, \tag{1}$$

where $\rho = \lambda/\mu < 1$. For the case $K > 2$, the approximations of the average response time were obtained using different methods.

One way for the response time approximation of the fork-join system with K parallel $M/M/1$ servers with the same service intensities has also been proposed in [10]. The idea of proposed estimation method emerged from observing the behavior of the response time in numerical experiments. The upper bound for an average response time can be obtained assuming random variables of task sojourn time to be independent from each other, since it was proved in [10] that random variables of a task sojourn time $\xi_i, i = \overline{1, K}$ are associated random variables with the following property:

$$P(\max_{1 \leq i \leq K} \xi_i > x) \leq 1 - \prod_{i=1}^{K} P(\xi_i \leq x).$$

Lower bound for the average response time is obtained if we discard the buffer. In this case, the response time will be a maximum of K i.i.d. random variables with mean $1/\mu$.

Thus, the upper and lower bounds increase with the same rate $H_K = \sum_{i=1}^{K} 1/i$ (a partial sum of the harmonic series), in other words, for large values of K the limits have the order of $O(\ln K)$. Then, we determine the value of unknown constant using simulations and finally get

$$E[W_{K,\max}] \approx \left[\frac{H_K}{H_2} + \frac{4}{11}\left(1 - \frac{H_K}{H_2}\right)\rho \right] \frac{12 - \rho}{8} \frac{1}{\mu - \lambda}, \quad K \geq 2. \qquad (2)$$

This approach, as it is shown in the numerical analysis section, has an approximation error not exceeding 5% for $2 \leq K \leq 32$.

Authors of paper [15] also analyze a K-way exponential, infinite capacity fork-join system. The resulting approximation is arithmetic mean of the upper and lower bounds. For an upper bound, a modified expression of the upper bound from of the above-mentioned paper [10] is used. The lower bound is derived by analyzing the response time in the similar system with non-parallel queuing [14]. Thus,

$$\frac{1}{\mu}\left(H_K + \rho \sum_{k=1}^{K} \frac{1}{k(k - \rho)} \right) \leq E[W_{K,\max}] \leq \frac{1}{\mu}\left(H_K + \rho \sum_{k=1}^{K} \frac{1}{k(1 - \rho)} \right).$$

Next, one may calculate the arithmetic mean of the obtained bound and finally:

$$E[W_{K,\max}] \approx \frac{1}{\mu}\left[H_K + \frac{\rho}{2(1 - \rho)}\left(\sum_{k=1}^{K} \frac{1}{k - \rho} + (1 - 2\rho)\sum_{k=1}^{K} \frac{1}{k(k - \rho)} \right) \right]. \qquad (3)$$

In [16], a different method for analyzing response times is used. The basic idea is that, since an analytical solution is very difficult to find, but it is possible to get asymptotic formulas for the desired characteristics. In this connection, a combination of heavy and light traffic interpolation approximations is used. These techniques, unlike the above-described method, do not use simulations, but their use can be extended to the analysis of fork-join systems not only with the exponential service time or with Poisson arrivals.

Light traffic interpolation approximations is the result of system functioning in the low-load mode, i.e., when the intensity of the arrival flow λ is very small. In this case, it is advisable to refer to the expansion in a Taylor series of performance measures of the system: distribution function for the response time as a function of λ in the neighborhood of zero. In that way, we can determine the unknown values in the function representation as a polynomial of λ with degree n. We consider only polynomials of zero and first degree. In the case of the heavy traffic interpolation the fork-join system behavior is analysed in such a regime that the value λ is very close to the value μ. The key parameter here is β, two extreme values of which are interpreted as two bound cases: if $\beta = 0$, then it means that the incoming flow is deterministic, if $\beta = 1$, the service time is deterministic and hence the ways of fork-join system are K independent $D/GI/1$ or $GI/D/1$ queue, correspondingly.

Thus, thanks to the analysis of response time function behavior in the boundary values of system load, one is able - without carrying out of numerical experiments to estimate the constants in the interpolation formulas - to determine their exact expressions in closed form.

In the case of fork-join system with K $M/M/1$-ways, the approximation of response time is as follows:

$$E[W_{K,\max}] \approx \left[H_K + \left(V_K - H_K \right) \frac{\lambda}{\mu} \right] \frac{1}{\mu - \lambda}, \quad 0 \le \lambda < \mu, \quad K \ge 2, \quad (4)$$

where

$$V_K = \sum_{i=1}^{K} \binom{K}{i} (-1)^{i-1} \sum_{m=1}^{i} \binom{i}{m} \frac{(m-1)!}{i^{m+1}}.$$

Also in [16], in order to obtain an estimation of the average response time in cases with different arrival flows and distributions of service time, the heavy traffic interpolation method was modified, and analyzed for three values of key constant $\beta = 0, 1/2, 1$. As a result, we obtain a quadratic interpolation, and combining it with light traffic method, the approximation formulas for average response time were derived for the following four non-Markov cases:

- Erlang distribution with two stages for incoming flow and exponential service time;
- Poisson incoming flow (arrival) and hyperexponential service time;
- Poisson incoming flow and Erlang distribution with two stages of service time;
- hyperexponential incoming flow and exponential service time.

Since the response time of fork-join system is classically defined as a maximum, and in some cases, also as a minimum of K random variables of tasks sojourn times, it is natural that one of the alternative ways to measure response time is the use of order statistics theory. By definition, if $\xi_1, ..., \xi_K$ - final sample defined on a probability space (Ω, F, P) and $\omega \in \Omega : x_i = \xi_i(\omega), i = 1, ..., K$, enumerate the sequence $\{x_i\}_{i=1}^{K}$ in decreasing order so that $x_{(1)} \le x_{(2)} \le ... \le x_{(K-1)} \le x_{(K)}$, then this sequence is called the variational series and its members - order statistics. The random variable $\xi_{(k)} : \xi_{(k)}(\omega) = x_{(k)}$ is called the k-th order statistic of the original sample. From the definition it is obvious that

$$\xi_{(1)} = \min(\xi_1, ..., \xi_K), \quad \xi_{(K)} = \max(\xi_1, ..., \xi_K). \quad (5)$$

Thus, the response time is $W_{K,\max} = \xi_{(K)}$ or $W_{K,\min} = \xi_{(1)}$ in terms of the order statistics theory, where $\xi_k, k = \overline{1, K}$ - positive random variables of task sojourn time. Consequently, the mathematical expectation of the response time can be calculated from the distribution of extreme values $\xi_{(1)}$ and $\xi_{(K)}$, the function of the latter is actually the joint distribution function of random variable $\xi_1, ..., \xi_K$:

$$F_{\xi_{(K)}}(x) = P(\max(\xi_1, ..., \xi_K) < x) = P(\xi_1 < x, ..., \xi_K < x), \quad (6)$$

$$F_{\xi_{(1)}}(x) = P(\min(\xi_1, ..., \xi_K) < x) = 1 - P(\xi_1 > x, ..., \xi_K > x). \tag{7}$$

Denote $F_k(x)$ the cumulative distribution function (CDF), and $f_k(x)$ - probability density function (PDF) of random variables ξ_k, $k = \overline{1, K}$. If we assume that $\xi_1, ..., \xi_K$ - independent random variables, which in our case, as mentioned above, are simplifying assumptions, then

$$F_{\xi_{(K)}}(x) = P(\xi_1 < x, ..., \xi_K < x) = \prod_{i=1}^{K} F_i(x), \tag{8}$$

$$F_{\xi_{(1)}}(x) = 1 - P(\xi_1 > x, ..., \xi_K > x) = 1 - \prod_{i=1}^{K} [1 - F_i(x)], \tag{9}$$

a n-th moment of random variable of time response can be obtained by calculating the integral:

$$E[W_{K,\max}^n] \approx E[\xi_{(K)}^n] = \int_0^\infty x^n \sum_{j=1}^{K} \frac{f_j(x)}{F_j(x)} \prod_{i=1}^{K} F_i(x) dx, \tag{10}$$

$$E[W_{K,\min}^n] \approx E[\xi_{(1)}^n] = \int_0^\infty x^n \sum_{j=1}^{K} f_j(x) \prod_{i=1}^{K} \frac{1 - F_i(x)}{1 - F_j(x)} dx. \tag{11}$$

Further, if it is assumed that the servers are homogeneous, i.e., random variables $\xi_1, ..., \xi_K$ not only independent but also identically distributed, and, therefore, their CDFs and PDFs are equal $F_k(x) = F(x), f_k(x) = f(x), \forall k$, then

$$P(\xi_1 < x, ..., \xi_K < x) = F^K(x),$$
$$1 - P(\xi_1 > x, ..., \xi_K > x) = 1 - (1 - F(x))^K,$$

and accordingly

$$E[\xi_{(K)}^n] = K \int_0^\infty x^n f(x) F^{K-1}(x) dx. \tag{12}$$

$$E[\xi_{(1)}^n] = K \int_0^\infty x^n f(x)(1 - F(x))^{K-1} dx. \tag{13}$$

Note that the mathematical expectation for maximum of K i.i.d. random variables can be represented in the following form:

$$E[\xi_{(K)}] = \int_0^\infty x dF^K(x) = \int_0^\infty [1 - F^K(x)] dx. \tag{14}$$

Thus, to analyze the response time it is necessary to calculate the above integrals. To estimate the integrals, the numerical methods can be applied. For several types of distributions, such as exponential, hyperexponential, Cox and Erlang distribution with two stages, the result can be obtained in closed analytical form. Note that the computational complexity increases with the growth of number of parallel servers (K) and with the increase in number of stages of the Erlang distribution. To reduce computational complexity the characteristic maximum [12] can be used. In particular, if the task sojourn times are independent exponentially distributed random variables with the PDFs $f_k(x) = \lambda_k e^{-\lambda_k x}, x > 0, k = \overline{1, K}$, then the CDF of the response time is

$$F_{W_{K,\max}}(x) \approx F_{\xi_{(K)}}(x) = \prod_{i=1}^{K}(1 - e^{-\lambda_i x}). \tag{15}$$

After that we obtain explicit approximation of the mathematical expectation of the response time using formula (12):

$$E[W_{K,\max}] \approx \sum_{l=1}^{K}\frac{1}{\lambda_l} - \sum_{l \neq m}\frac{1}{\lambda_l + \lambda_m} + \sum_{l \neq m \neq k}\frac{1}{\lambda_l + \lambda_m + \lambda_k} + ...+$$
$$+ (-1)^{2K-1}\frac{1}{\lambda_1 + \lambda_2 + ... + \lambda_K}$$

and variance of the response time [6]:

$$D[W_{K,\max}] \approx \sum_{l=1}^{K}\frac{2}{\lambda_l^2} - \sum_{l \neq m}\frac{2}{(\lambda_l + \lambda_m)^2} + \sum_{l \neq m \neq k}\frac{2}{(\lambda_l + \lambda_m + \lambda_k)^2} + ...+$$
$$+ (-1)^{2K-1}\frac{2}{(\lambda_1 + \lambda_2 + ... + \lambda_K)^2} - \left(\sum_{l=1}^{K}\frac{1}{\lambda_l} - \sum_{l \neq m}\frac{1}{\lambda_l + \lambda_m} + \right.$$
$$\left. \sum_{l \neq m \neq k}\frac{1}{\lambda_l + +\lambda_m + \lambda_k} + ... + (-1)^{2K-1}\frac{1}{\lambda_1 + \lambda_2 + ... + \lambda_K}\right)^2.$$

Note that for the fork-join system with K $M_\lambda/M_{\mu_k}/1, k = \overline{1, K}$ ways, the last two formulas will be valid only if the following condition holds true: $\lambda_k = \mu_k - \lambda$, i.e., the task sojourn time in k-th $M_\lambda/M_{\mu_k}/1$ way will have an exponential distribution with parameter $(\mu_k - \lambda)$ [6]. If $\mu_k = \mu, k = \overline{1, K}$, then the following expression holds true:

$$E[W_{K,\max}] \approx \frac{1}{\mu - \lambda}H_K. \tag{16}$$

In order to determine the higher order moments of maximum of K independent exponential random variables, it is more effective from the computational point of view to differentiate the appropriate number of times the Laplace-Stieltjes transformation (LST) $\pi_{\xi_{(K)}}(s)$ of the CDF $F_{W_{K,\max}}(x)$ with respect to s and then equate s with zero. The general expression for LST $\pi_{\xi_{(K)}}(s)$ can be written as follows [12]:

$$\pi_{\xi_{(K)}}(s) = \sum_{i=1}^{K}(-1)^{i-1}\sum_{n=1}^{\binom{K}{i}}\sum_{k=1}^{n}\frac{\sum_{l=1}^{i}\lambda_{(k+l-1)K}}{s + \sum_{l=1}^{i}\lambda_{(k+l-1)K}} \tag{17}$$

where $(k + l - 1)_K$ denotes modulo K addition. Besides, LST of the CDF of the maximum of K for independent exponentially distributed random variables with parameters $\boldsymbol{\lambda} = (\lambda_1, ..., \lambda_K)$ and distribution density $f_K(\boldsymbol{\lambda}, x)$ with LST $\pi_K(\boldsymbol{\lambda}, s)$ can be represented by the recurrent formula [7]:

$$\left(s + \sum_{j=1}^{m} \lambda_j\right) \pi_{\xi_{(m)}}(\boldsymbol{\lambda}, s) = \sum_{j=1}^{m} \lambda_j \pi_{\xi_{(m-1)}}(\boldsymbol{\lambda}_{/j}, s), 1 \leq m \leq K, \qquad (18)$$

where $/j$ denotes exclusion of λ_j, i.e. $\boldsymbol{\lambda}_{/j} = (\lambda_1, ..., \lambda_{j-1}, \lambda_{j+1}, ..., \lambda_m)$, and $\pi_0(\boldsymbol{0}, s) = 1$. Then the n-th order moment of the maximum of independent exponentially distributed random variables is

$$M_K(\boldsymbol{\lambda}, n) = \frac{n}{\sum_{j=1}^{K} \lambda_j} M_K(\boldsymbol{\lambda}, n - 1) + \frac{\sum_{j=1}^{K} M_{K-1}(\boldsymbol{\lambda}_{/j}, n)}{\sum_{j=1}^{K} \lambda_j}, \qquad (19)$$

$K \geq 1$, $M_0(\boldsymbol{0}, n) = 0$ for all $n \geq 1$ and $M_K(\boldsymbol{\lambda}, 0) = 1$ for all $K \geq 0$.

Table 1. The random variables that determine performance measures of a cloud system

Performance metrics	Definition	Random variable	Notation in the order statistics theory
The response time	The sojourn time of the last served task.	$W_{K,max} = \max(\xi_1, ...\xi_K)$	$\xi_{(K)}$
	The sojourn time of the first served task	$W_{K,min} = \min(\xi_1, ...\xi_K)$	$\xi_{(1)}$
The synchronization time	The time between the end of the service of the first and the last tasks	$W_K = W_{K,max} - W_{K,min}$	$W_K = \xi_{(K)} - \xi_{(1)}$ (range)

The Table 1 shows the random variables that characterize the main performance metrics. In order to evaluate these metrics, we need to obtain expressions for the basic probability characteristics of random variables, i.e. the expressions for their mean and variance.

The upper bound of an approximation for the maximum of i.i.d. exponential random variables with parameter $\mu - \lambda$, as shown in [4], is

$$E[\xi_{(K)}] \leq \frac{1}{(\mu - \lambda)} \left(1 + \frac{K - 1}{\sqrt{2K - 1}}\right). \qquad (20)$$

Here after, we will consider the case of fork-join system with K $M_\lambda/M_\mu/1$ ways, or rather its approximation, i.e., we assume that the task sojourn times are i.i.d. RVs with the CDF $F(x) = 1 - e^{(\mu-\lambda)x}, x > 0$. The expression for the variance of the response time is given in [6,12]:

$$D[W_{K,\max}] \approx \frac{1}{(\mu-\lambda)^2}\left(2\sum_{k=1}^{K}\binom{K}{k}(-1)^{k-1}\frac{1}{k^2} - H_K^2\right). \tag{21}$$

The random variable $W_{K,\min}$ has the following characteristics:

$$E[W_{K,\min}] \approx \frac{1}{(\mu-\lambda)K}, \quad D[W_{K,\min}] \approx \frac{1}{(\mu-\lambda)K^2}. \tag{22}$$

Now let's analyze the synchronization time W_K. By the definition of mathematical expectation, the difference between two random variables $W_{K,\max}$ and $W_{K,\min}$ is not affected by their dependencies, it is sufficient to know only the first moments of them:

$$E[W_K] = E[W_{K,\max}] - E[W_{K,\min}] = \frac{1}{(\mu-\lambda)}\sum_{i=1}^{K-1}\frac{1}{i} = \frac{1}{(\mu-\lambda)}H_{K-1}. \tag{23}$$

But in order to obtain an expression for the dispersion of W_K, it is not enough to know the value of dispersions $W_{K,\max}$ and $W_{K,\min}$. Thus, it is possible to act in two ways. One of them is to use the CDF of the range of positive i.i.d. random variables [13]:

$$F_{W_K}(x) = \sum_{i=1}^{K}\int_0^{\infty} f_i(y)\prod_{j=1,j\neq i}^{K}[F_j(y+x) - F_j(y)]dy. \tag{24}$$

The second way is to use directly the formula for the dispersion of range of K i.i.d. random variables with the CDF $F(x)$ [4]:

$$D[W_K] = 2\int_{-\infty}^{\infty}(1 - F^K(y) - [1 - F(x)]^K + [F(y) - F(x)]^K)dxdy - (E[W_K])^2. \tag{25}$$

In this case, $F(x) = 1 - e^{-(\mu-\lambda)x}, x > 0$, $F(y) = 1 - e^{-(\mu-\lambda)y}, y > 0$, because the sojourn time in the $M_\lambda/M_\mu/1$ subsystem has an exponential distribution with parameter $(\mu-\lambda)$, and the expression under integral sign has the following form:

$$(1 - F^K(y) - [1 - F(x)]^K + [F(y) - F(x)]^K) = -\sum_{i=1}^{K}\binom{K}{i}(-1)^i e^{-(\mu-\lambda)iy} +$$

$$+ \sum_{i=1}^{K-1}\binom{K}{i}(-1)^{K-i}e^{-(\mu-\lambda)(K-i)y}e^{-(\mu-\lambda)ix} + (-1)^K e^{-(\mu-\lambda)Ky}.$$

Now we substitute this expression into (25), perform the necessary calculations and obtain the following expression:

$$D[W_K] = \frac{2}{(\mu - \lambda)}\left(\sum_{i=1}^{K}\frac{\binom{K}{i}(-1)^{i-1}}{i^2} - \sum_{i=1}^{K-1}\frac{\binom{K}{i}(-1)^{K-i}}{Ki} + \right.$$

$$\left. + \sum_{i=1}^{K-1}\frac{\binom{K}{i}(-1)^{K-i}}{i(K-i)} + \frac{(-1)^K}{K^2} - \frac{1}{2}\left(\sum_{i=1}^{K-1}\frac{1}{i}\right)^2\right).$$

Finally, after simplifications, we get

$$D[W_K] = \frac{1}{(\mu - \lambda)^2}\sum_{i=1}^{K-1}\frac{1}{i^2}. \tag{26}$$

2 Numerical Analysis

In order to illustrate the approximation accuracy of formulas (2, 3, 4, 16), the diagrams of the mean response time in dependence on the number of tasks K are presented in Fig. 2. The calculations were performed for the arrival rate $\lambda = 2\ s^{-1}$ and service rate $\mu = 2.2\ s^{-1}$. Since for $K = 2$ there is an exact expression for the mean response time (21), we consider only $K > 2$ case.

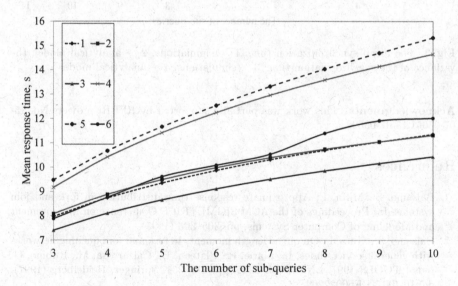

Fig. 2. The mean response time: 1 – formula (2), 2 – formula (3), 3 – (4), 4 – (16), 5 – (20), 6 – simulations

As it can be seen from the graphs, the formulas (2, 3, 4, 16) are presented in descending order of approximation accuracy, and the upper bound for the mean value of the maximum of order statistics is the upper bound for the mean response time.

The mean value and variance of the synchronization time are presented in Fig. 3. One can see that the estimation of the mean synchronization time has a low accuracy contrary to the estimation of the variance (the relative error is 7% on average).

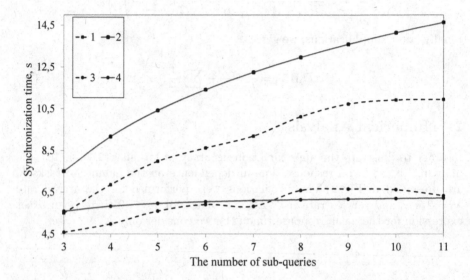

Fig. 3. The mean synchronization time: 1 – simulations, 2 – analytical model; the variance of the synchronization time: 3 – simulations, 4 – analytical model

Acknowledgments. This work was partially supported by RFBR, projects No. 15-07-03051,15-07-03608.

References

1. Balsamo, S., Mura, I.: Approximate response time distribution in fork and join systems. In: Proceedings of the ACM SIGMETRICS Conference on Measurement and Modeling of Computer Systems, pp. 305–306 (1995)
2. Balsamo, S., Mura, I.: On queue length moments in fork and join queuing networks with general service times. In: Marie, R., Plateau, B., Calzarossa, M., Rubino, G. (eds.) TOOLS 1997. LNCS, vol. 1245, pp. 218–231. Springer, Heidelberg (1997). doi:10.1007/BFb0022209
3. Balsamo, S., Donatiello, L., van Dijk, N.M.: Bound performance models of heterogeneous parallel processing systems. IEEE Trans. Parallel Distrib. Syst. 9(10), 1041–1056 (1998)
4. David, H.A.: Order Statistics. Wiley, New York (1981)
5. Flatto, L., Hahn, S.: Two parallel queues created by arrivals with two demands I. SIAM J. Appl. Math. 44(5), 1041–1053 (1984)

6. Gorbunova, A.V., Zaryadov, I.S., Matyushenko, S.I., Samouylov, K.E., Shorgin, S.Ya.: The approximation of response time of a cloud computing system. **9**, 32–38 (2015)
7. Harrison, P., Zertal, S.: Queueing models with maxima of service times. In: Kemper, P., Sanders, W.H. (eds.) TOOLS 2003. LNCS, vol. 2794, pp. 152–168. Springer, Heidelberg (2003). doi:10.1007/978-3-540-45232-4_10
8. Joshi, G., Soljanin, E., Wornell, G.: Efficient redundancy techniques for latency reduction in cloud systems (unpublished)
9. Khazae, H., Misic, J., Misic, V.B.: A fine-grained performance model of cloud computing centers. IEEE Trans. Parallel Distrib. Syst. **24**, 2138–2147 (2012)
10. Nelson, R., Tantawi, A.N.: Approximate analysis of fork/join synchronization in parallel queues. IEEE Trans. Comput. **37**, 739–743 (1988)
11. Neuts, M.F.: Matrix-geometric solutions in stochastic models. Johns Hopkins University Press, Baltimore (1981)
12. Thomasian, A.: Analysis of fork/join and related queueing systems. ACM Comput. Surv. (CSUR) **47**(17), 1–71 (2014)
13. Tsimashenka, I., Knottenbelt, W.J.: Reduction of subtask dispersion in fork-join systems. In: Balsamo, M.S., Knottenbelt, W.J., Marin, A. (eds.) EPEW 2013. LNCS, vol. 8168, pp. 325–336. Springer, Heidelberg (2013). doi:10.1007/978-3-642-40725-3_25
14. Varki, E.: Response time analysis of parallel computer and storage systems. IEEE Trans. Parallel Distrib. Syst. **12**(11), 1146–1161 (2001)
15. Varki, E., Merchant, A., Chen, H.: The M/M/1 fork-join queue with variable subtasks (unpublished)
16. Varma, S., Makowski, A.M.: Interpolation approximations for symmetric fork-join queues. Perform. Eval. **20**, 245–265 (1994)

Simulation of Medical Sensor Nanonetwork Applications Traffic

Rustam Pirmagomedov$^{(\boxtimes)}$, Ivan Hudoev, and Daria Shangina

Department of Telecommunication Networks and Data Transmision,
Saint-Petersburgs State University of Telecommunications,
Bolshevikov Pr. 22/1, Saint-Petersburg 193232, Russia
prya.spb@gmail.com, khvanches@gmail.com, shandaryna@gmail.com

Abstract. Nanonetworks is one of the most dynamically developing areas in the field of telecommunications. Nanonetworking promises new opportunities in different fields of science and technology. However along with obvious advantages, the implementation of nanonetwork applications can cause a number of problems for the functioning of modern telecommunication networks. One of them is a large number of data packets generated by nanonetwork applications. In this regard, the actual problem is the study of traffic nanoceramic applications and its impact on traditional telecommunication networks. The article deals with simulation of traffic from sensor nanetwork applications. The paper presents results of nanonetwork applications traffic simulation. Simulation based on the traffic models developed for M2M. In the simulation considered the possibility of gateway working in two modes: without processing messages received from nanonetwork and with it. Typical architecture of nanonetwork medical applications involving the use of remote Internet servers, was described. The results of traffic flow simulation were analysed on the self-similarity properties.

Keywords: Internet of nano-things · Nanonetworks · Internet of things · Traffic modeling

1 Introduction

The progress at Internet of Nano-things (IoNT) [1,2], supposes development of many applications, using nanonetwork structures in different spheres of human life whether it be industrial sphere, military sphere or everyday human life [3].

Medicine is one of the promising fields of nanonetwork structures use [4]. It is expected that nanonetwork applications will supplement the Internet of Things technologies, which have already existed in medical sphere [5,6] and will open new opportunities for diagnosing and diseases treating [7,8], environmental monitoring, performing of surgical operation for tissue reformation at molecular and DNA levels [9], making of smart medicines [10,11] and so on.

The development of medical applications of IoNT, promises to us personalized medicine. Such novel approach based on detailed, on-line, individual analysis

© Springer International Publishing AG 2016
V.M. Vishnevskiy et al. (Eds.): DCCN 2016, CCIS 678, pp. 430–441, 2016.
DOI: 10.1007/978-3-319-51917-3_38

of a wide range of health parameters of each patient and preventative actions. Preventive orientation will provide complete genetic information with subsequent determination of risk factors of the most significant diseases, the identification of pharmacogenetic features, followed by creating an individual program of primary (eliminate risk factors) and secondary (early detection of diseases prevention) treatment.

Development of such applications is supposed in three contexts: inside of human body, on the surface of human body and in human inhabitations (apartments, car, hospital room). It is planned that all above-listed applications will have the opportunity to transfer and receive the information from remote servers using the Internet resources [12].

Today nanonetwork structures can't make full convergence with traditional communication networks and now exist in the form of autonomous structures which were created to solve the narrow-purpose problems [10,11]. However, active researches carried out at this theme, led to the conclusion that in the future the question of nano- and traditional networks union will be decided [8]. The development and putting in operation of many nanoapplications, generating additional traffic flows in the Internet, will require reconsideration of established paradigms in the sphere of telecommunication. To forecast necessary changes of traditional networks, we need to research the behavior of nanonetwork applications traffic flows. First of all, it is necessary to answer the following questions:

1. What type of traffic flows will be generated by these nanonetwork applications?
2. What characteristics these traffic flows will have?

In view of the fact that the Internet of Nano-things is a logical continuation of the Internet of Things conception [12], it would be logically to suppose that characteristics of traffic generated by nanonetwork applications will be quite similar with the Internet of Things traffic. According to it, patterns have already been suggested and investigated in the context of the Internet of Things [13–15] may be also applied for Internet of Nano-things. Communication of nanonetwork structures with computer center (remote server) is suitable for the context of M2M conception [16,17], being its new manifestation.

In contradistinction from the Internet of Things traffic patterns, in nanonetwork applications sensor readouts are sent to server not immediately, but firstly get through the nanonetwork, which certainly has an influence on the transfer characteristics. One more factor having influence on traffic parameters is availability of intermediated node (gateway) between nanonetwork and the Internet. Gateway working algorithms, its opportunities of changing traffic parameters and message preprocessing should be taken into account during development and investigation of nanonetwork applications traffic patterns.

In this work we develop the pattern for nanosensor network with its connection with the Internet and make a comparison with the pattern of monitoring system and supervisory control traffic flow. The article has the following structure. In Sect. 2 different aspects of nanonetwork applications traffic simulation depending on their functionality and field of application are considered. In the third part peculiarities of traffic simulation for sensor nanonetwork are described. In the fourth part traffic pattern for nanosensor network is developing. In the fifths section, with the help of simulation modelling system the investigation of this pattern and comparison of results with the pattern of monitoring system and supervisory control traffic flow are described.

2   Architectures of Nanonetwork Medical Applications

Nanonetwork medical applications should solve a wide range of problems from sensor readouts capture, situated as well as in environment, and directly inside the human body, to tissue reformation and microsurgical operations. Depending on solving problems nanonetwork applications will have different functionality and therefore traffic flows of such applications will also be different. According to their functional peculiarities nanonetwork applications can be divided into a few basic classes:

(1) Autonomous application. It means autonomous work of nanonetwork developed in the environment, inside the human body or on the surface of human body. Data interchange with remote server is not supposed for nanoapplications of this class or data interchange will be restricted (technical message interchange in cases of critical condition, about workability of the network and so on). So, traffic between nanonetwork and remote server either is absent at all or relatively small volume of traffic is generated because application responds to the critical event. The events, caused generation of traffic are relatively rare.

(2) Sensor nanonetwork. It is the application based on nanosensor network use, in which nanosensors are activated on demand of server or local computer center. Such applications are used for periodical capture of information about the state of human health or about the state of environment. Traffic of such systems will represent determinated packets flow consists of server queries and nanonetwork application answers.

(3) Sensor-actuatory application. It is nanonetwork application which aims not only for capture of the information about state of organism with the help of nanosensors, but also at influence on processes in body with the help of nanoactuators (for example, remote microscopic surgical operations in human body). Applications of this type are very sensitive to delay and signal distortions and can't work off-line. Traffic of such systems can't be classified as M2M traffic, because of the presence of the operator (experienced surgeon), who performs the operation.

Final application can have mixed functionality and characteristics of different classes. Taking into account that sensor-actuatory applications are very demanding to QoS parameters, such systems should be developed locally, because it is very difficult to provide necessary QoS parameters using connection via the Internet. In view of this fact we can consider traffic of such applications only at local area networks. Hereinafter we will consider only sensor nanonetworks traffic.

3 The Peculiarities of Sensor Nanonetwork Traffic Simulation

The work of sensor nanonetwork builds on the master-slave principle (see Fig. 1), while data capture realized according to predetermined rules. Traffic generated by this application will be pseudodetermined.

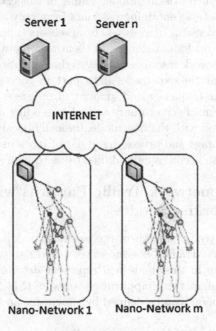

Fig. 1. General structure of sensor nanonetwork application

There is probability of message loss in nanonetwork applications - p because of realization difficulties at physical level and imperfection of routing algorithms. "Normal" probability of message loss for some nanonetwork technologies can account for 95 percents. Message loss can happen as well as during scanning of sensors by gateway, and during data transfer from sensors to gateway after receiving of query. Therefore, integrated probability of sensor readouts transfer to gateway Pdata will be expressed by (1).

$$P_{data} = (1-p)(1-p) = 1 - 2p + p^2 \tag{1}$$

p - is probability of message loss in nanonetwork.

In spite of probability of message loss, there are different working algorithms of gateway during scanning of sensors and during their data transfer to remote server:

1. Without processing of messages from sensors
2. With processing of messages from sensors

Working mode without processing of messages from sensors supposes all the messages transferred from sensor to be sent to the remote server directly. In this mode, sensors readouts are not processed and transferred to remote server from each sensor separately. Gateway, working in this mode, has high response speed concerning the second working mode and small buffer capacity. High traffic on communication channel between gateway and remote server can be referred to the disadvantages of such working mode. Traffic in this case is characterized by a great amount of packets containing readouts from only one sensor.

The mode with processing of messages from sensors supposes storage and pre-processing of sensors readouts before send them to remote server. In this mode efficiency of use of network resources increases during data interchange between gateway and server at the expense of data part in packets becomes considerably bigger, moreover frequency and amount of messages decreases. However, delay from the moment of server query arrival to sending of data from gateway increases in comparison with the first mode, basically because of additional time which requires for storage and processing of data from sensors. The architecture of gateway in this case becomes more difficult due to high system requirements.

4   Sensor Nanonetwork Traffic Patterns with Different Gateway Working Modes

Sensor nanonetwork traffic pattern in the working mode without processing of messages from sensors. Traffic between server i and nanonetwork application j with gateway working in this mode will represent determined flow consists of server queries and nanonetwork application answers (Fig. 2). Time parameters of system flow, in general, can be defined by time-table and have definite repeat period T_i.

We can observe burst of intensity of sending messages from the moment of server query arrival till the last sensor readout would be sent. Time between server query arrival to sending of the last sensor readout we shall designate t which will represent determinated flow. The amount of messages Npaceges sent by the application during t time will depend on amount of sensors in nanonetwork, which have delivered their messages.

Sensor network traffic pattern for gateway with processing of messages from sensors. Traffic between server i and nanonetwork application j with gateway working in this mode will be determined flow consists of server queries and nanonetwork application answers (Fig. 3). Time parameters of system flow, in general, can be defined by time-table and have definite repeat period T_i.

Time from the moment of server query arrival to sending readouts to server we shall designate Δt. This time gateway inquires sensors located in nanonetwork and processes received information. Sensors readouts are sent to server. The amount of messages sent by gateway to server will depend on the volume of information received from sensors. Limited quantity of information can be transmitted in one message D_{info}. Quantity of useful information can be calculated by (2).

$$D_{info} = V_{max} - D_{service} \qquad (2)$$

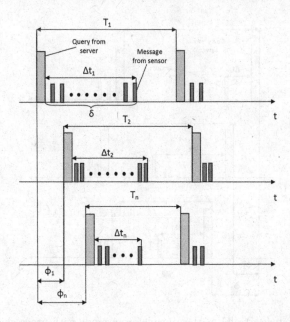

Fig. 2. Sensor network traffic pattern for gateway without processing of messages from sensors

where V_{max} is maximal packet size for this application, $D_{service}$ is a volume of ordering information.

In our work we will suppose that $V_{max} = 1500$ bytes (Ethernet frame size), $D_{service} = 62$ bytes (the sum of headers sizes Ethernet, IP, TCP). Using (2), 1438 bytes of useful information can be transferred. The volume of information, received from one sensor we will consider as $v_0 = 16$ bytes (4 bytes for sensors readouts + 4 bytes for sensor address in this nanonetwork + 4 bytes for sensor type). Using above-listed considerations we can transfer readouts of 89 sensors in one packet. If it is necessary to transfer readouts of greater amount of sensors, we shall send one more message. In general case, general amount of packets which would be necessary to transfer readouts of all the sensors $N_{paceges}$ is calculated by formula (3)

$$N_{paceges} = ceiling(N_{v0}/D_{info}) \qquad (3)$$

where N is the amount of sensors in an application. (ceiling is rounding upward), v_0 is the data volume of one sensor.

5   Results of Simulation

In this section results of traffic simulation presented for both modes. If there are n nanonetwork applications, so for each working mode of gateway, at unchanged phase shifts between moments of queries (answers) arrival φ_i for $i = 1..n$ general traffic will also represent determinated periodical process with a period equal

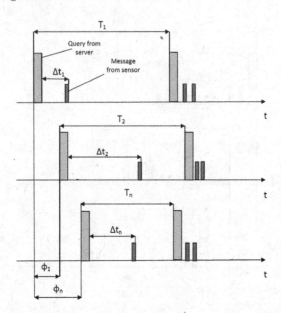

Fig. 3. Sensor network traffic pattern for gateway mode with processing of messages from sensors

to least common multiple of all scanning periods. Also at the beginning of each period of time T_i, from i server to j application a query for receiving sensors readouts arrives. Δt_i coefficient for each of systems is selected equal to $0,2*T_i$; $0,4*T_i$; $0,6*T_i$ and $0,8*T_i$. The amount of sensors located in one nanonetwork application is $N = 100$. For application pattern with gateway without preprocessing of messages selection of time for sending readouts of one sensor to server is drawing at random in the limits from 0 to Δt_i (selection of time for sending in a pattern is given by uniform law). Results of simulation are given at the Figs. 4 and 5.

For application pattern with gateway without preprocessing of messages selection of time for sending readouts of one sensor to server is drawing at random in the limits from 0 to Δt_i (selection of time for sending in a pattern is given by uniform law). Results of simulation are given at the Figs. 4 and 5.

Flows received during the simulation were checked for self-similarity features. Figures 6 and 7 show evaluation of Hurst coefficient with graphs of dispersion changing.

For gateway mode with preprocessing of messages from sensors estimated Hurst coefficient $H=0.12$ at $\Delta t_i= 0.2*T_i$, $H=0.11$ at $\Delta t_i= 0.4*T_i$, $H=0.19$ at $\Delta t_i= 0.6*T_i$, $H= 0.19$ at $\Delta t_i=0.8*T_i$. For gateway mode without preprocessing of messages from sensors Hurst coefficient made up $H=0.21$ at $\Delta t_i= 0.2*T_i$, $H=0.24$ at $\Delta t_i= 0.4*T_i$, $H=0.18$ at $\Delta t_i= 0.6*T_i$, $H=0.27$ at $\Delta t_i=0.8*T_i$.

From received results we can see that in all cases Hurst coefficient is considerably less than 0.5, therefore traffic in each experiment is self-similar stochastic

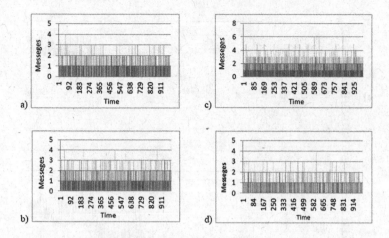

Fig. 4. Traffic flow for gateway mode with preprocessing of messages (a) $\Delta t_i = 0{,}2{*}T_i$ (b) $\Delta t_i = 0{,}4{*}T_i$ (c) $\Delta t_i = 0{,}6{*}T_i$ (d) $\Delta t_i = 0{,}8{*}T_i$

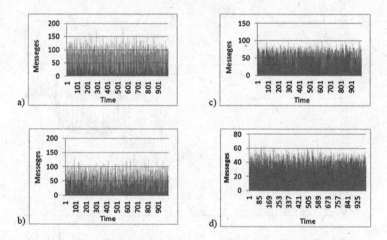

Fig. 5. Traffic flow for gateway mode without preprocessing of messages (a) $\Delta t_i = 0{,}2{*}T_i$ (b) $\Delta t_i = 0{,}4{*}T_i$ (c) $\Delta t_i = 0{,}6{*}T_i$ (d) $\Delta t_i = 0{,}8{*}T_i$

process, corresponding to the class of antipersistant processes. This result was received for all calculated data series for different correlations of periods between T queries and time of nanonetwork application reaction Δt_i.

6   Conclusions

At present, medical applications of the IoT are rapidly developing and can be useful for daily tasks (helps to keep healthy lifestyle) as well as for special tasks in the treatment and diagnosis of diseases. Medical applications of the IoNT combines the latest medical, technical and communicative knowledge, which leads

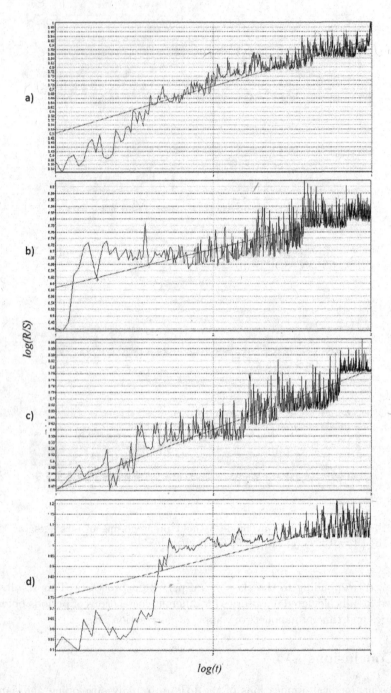

Fig. 6. Evaluation of Hurst coefficient (for gateway mode with preprocessing of messages from sensors) (a) $\Delta t_i = 0,2*T_i$ (b) $\Delta t_i = 0,4*T_i$ (c) $\Delta t_i = 0,6*T_i$ (d) $\Delta t_i = 0,8*T_i$

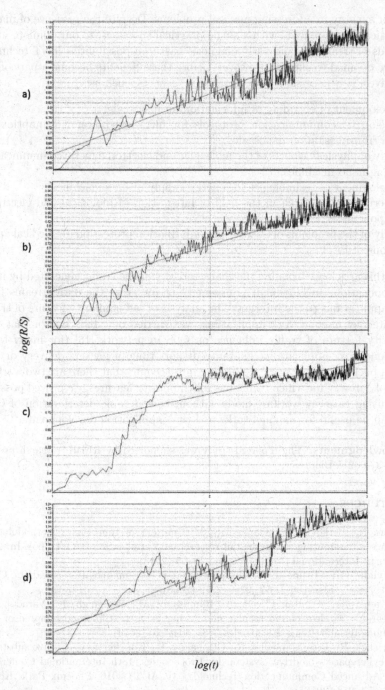

Fig. 7. Evaluation of Hurst coefficient (for gateway mode without preprocessing of messages from sensors) (a) $\Delta t_i = 0{,}2 * T_i$ (b) $\Delta t_i = 0{,}4 * T_i$ (c) $\Delta t_i = 0{,}6 * T_i$ (d) $\Delta t_i = 0{,}8 * T_i$

to their synergy in addressing medical problems. Despite the presence of mainly theoretical developments with a single practical successes, to our opinion, many methods of treatment and early diagnosis based on possibilities IoNT technologies are doomed to success. However, before these benefits become a part of our daily lives, it is necessary to develop many aspects, such as:

- Development of encoding of information on nanoscale;
- Develop telecommunication protocols for ultra-low power consumption for data transmission on nanoscale;
- Study of physical aspects of the molecular and electromagnetic communication among nano-machines;
- Development of technologies for energy supply of nanomachines;
- Experimental research in the field of integration of biological and electronic components;
- Study of the problems of security and information protection for medical applications of IoNT.

In this article we examine the characteristics of the traffic generated by medical applications IoNT that it is necessary to integrate the data streams from these applications to the Internet. The analysis of self-similarity feature of traffic generated by nanonetwork applications was carried out. It is known that self-similarity feature of traffic influence on QoS in network [18] (for many types of nanonetwork applications, guaranteed short time of packets delivery or/and level of packets loss are critical). Taking into account that there are two working mode of gateway for which traffic has self-similarity features, we proved possibility of using gateway working mode without considerable deterioration of QoS. It is can be useful for nanonetworks real-time application development.

Acknowledgments. The reported study was supported by RFBR, research project No. 16-37-00215 Biodriver.

References

1. Bari, N., Berkovich, S., Ganapathy, M.: Internet of things as a methodological concept, computing for geospatial research and application (COM.Geo). In: 2013 Fourth International Conference, pp. 48–55 (2013)
2. Akyildiz, A.F., Brunetti, F., Blazquez, C.: A new communication paradigm. Comput. Netw. (Elsevier) J. **52**, 2260–2279 (2008)
3. Shyamkumar, P., Rai, P., Oh, S., Ramasamy, M., Harbaugh, R., Varadan, V.: Wearable wireless cardiovascular monitoring using textile-based nanosensor and nanomaterial systems. Electronics **3**(3), 504–520 (2014)
4. Kirichek, R., Pirmagomedov, R., Glushakov, R., Koucheryavy, A.: Live substance in cyberspace - biodriver system. In: Proceedings, 18th International Conference on Advanced Communication Technology (ICACT) 2016, Phoenix Park, Korea, pp. 274–278 (2016)
5. Seyedi, M., Kibret, B., Lai, D., Faulkner, M.: A survey on intrabody communications for body area network applications. IEEE Trans. Biomed. Eng. **60**(8), 2067–2079 (2013)

6. Kumar, S., Nilsen, W., Abernethy, A., Atienza, A., Patrick, K., Pavel, M., Riley, W., Shar, A., Spring, B., Spruijt-Metz, D., Hedeker, D., Honavar, V., Kravitz, R., Lefebvre, R., Mohr, D., Murphy, S., Quinn, C., Shusterman, V., Swendem, D.: Mobile health technology evaluation: the mhealth evidence workshop. Am. J. Prev. Med. **45**(2), 228–236 (2013)

7. Yang, K., Chopra, N., Upton, J., Hao, Y., Philpott, M., Alomainy, A., Abbasi, Q.H., Qaraqe, K.: Characterising skin-based nano-networks for healthcare monitoring applications at THz. In: 2015 IEEE International Symposium on Antennas and Propagation & USNC/URSI National Radio Science Meeting, pp. 199–200 (2015)

8. Gopinath, S., Tang, T., Chen, Y., Citartan, M., Lakshmipriya, T.: Bacterial detection: from microscope to smartphone. Biosens. Bioelectron. **60**, 332–342 (2014)

9. Brufau, J., Puig-Vidal, M., Lopez-Sanchez, J., Samitie, J., Driesen, W., Breguet, J.: Small autonomous robot for cell manipulation applications. In: Proceedings of the 2005 IEEE International Conference on Robotics and Automation, pp. 844–849 (2005)

10. Gu, Z., Aimetti, A., Wang, Q., Dang, T., Zhang, Y., Veiseh, O., Cheng, H., Langer, R., Anderson, D.: Biomedical robotics and biomechatronics injectable nano-network for glucose- mediated insulin delivery. ACS Nano **7**(5), 4194–4201 (2013)

11. Cho, S., Park, S.J., Choi, Y.J., Jung, H., Zheng, S., Ko, S.Y., Park, J., Park, S.: Biomedical robotics and biomechatronics. In: 5th IEEE RAS EMBS International Conference, pp. 856–860 (2014)

12. Najah, A.A., Mervat, A.: Internet of NanoThings healthcare applications: requirements, opportunities, and challenges. In: Wireless and Mobile Computing, Networking and Communications (WiMob) pp. 9–14 (2015)

13. Chornaya, D., Paramonov, A., Koucheryavy, A.: Investigation of machine-to-machine traffic generated by mobile terminals. In: Special Session on Recent Advances in Broadband Access Networks, pp. 210–213 (2014)

14. Kirichek, R., Golubeva, M., Kulik, V., Koucheryavy, A.: The home network traffic models investigation. In: 2016 18th International Conference on Advanced Communication Technology (ICACT) (2016)

15. Rupp, M., Laner, M., Svoboda, P.: Detecting M2M traffic in mobile cellular networks. In: IWSSIP 2014 Proceedings, pp. 159–162 (2014)

16. Al-Khatib, O., Hardjawana, W., Vucetic, B.: Traffic Modeling for Machine-to-Machine (M2M) Last Mile Wireless Access Networks, Globecom 2014 - Communications QoS, Reliability and Modelling Symposium. pp. 1199–1204

17. Paramonov, A., Koucheryavy, A.: M2M traffic models and flow types in case of mass event detection. In: Balandin, S., Andreev, S., Koucheryavy, Y. (eds.) NEW2AN 2014. LNCS, vol. 8638, pp. 294–300. Springer, Heidelberg (2014). doi:10. 1007/978-3-319-10353-2_25

18. Parka, K., Kimb, G., Crovellab, M.: On the effect of traffic self-similarity on network performance. In: Proceedings of the SPIE International Conference on Performance and Control of Netork Systems, pp. 296–310 (1997)

Long-Range Data Transmission on Flying Ubiquitous Sensor Networks (FUSN) by Using LPWAN Protocols

Ruslan Kirichek[✉] and Vyacheslav Kulik

The Department of Communication Networks and Data Transmission,
State University of Telecommunications, Bolshevikov St., 22,
St. Petersburg, Russia 193232
kirichek@sut.ru, vslav.kulik@gmail.com
https://www.sut.ru/eng

Abstract. Flying Ubiquitous Sensor Networks (FUSN) are one of the new Internet of Things applications. In such networks, Unmanned Aerial Vehicles (UAVs) are used to collect data from the wireless sensor nodes, located in hard to reach remote areas, and the subsequent data delivery to the Internet. For the stable operation, it is required to solve complex scientific problems, one of which is the choice of connectivity technologies for the data delivery from the sensor nodes to the gateway with the IP-network as well as the calculation of quality of service parameters. The paper considers the problem of data delivery with the terrestrial segment of the flying ubiquitous sensor network over long distances using repeater chain. As technology of interaction nodes terrestrial sensor network considered IEEE 802.15.4 (6LoWPAN protocol), and technology IEEE 802.15.4g is considered for UAV interaction (LoRaWAN protocol). In the study, analytical and simulation models of the flying ubiquitous sensor network have been developed. As a result of experiments with the simulation model, delay and packet loss were investigated, occurring in the all stations of transmission network at different data rates, and conclusions on the optimal data rate of the network were made.

Keywords: FUSN · UAV · WSN · LoRa · Sensor node · Long-range · LPWAN · Transmission

1 Introduction

Currently, due to the active development of the concept of the Internet of Things, the questions of networks convergence of different technologies and standards are more relevant. We can confidently assert that the next generation networks will be of heterogeneous structure [1–3]. Modern communication systems evolve towards the introduction of a variety of Internet of Things devices in all spheres of human life, such as medicine, education, transport logistics, and other vehicular traffic [4]. Along with other technologies, FUSN concept has been actively developed [5]. Currently unmanned aerial vehicles are used for the delivery of small

© Springer International Publishing AG 2016
V.M. Vishnevskiy et al. (Eds.): DCCN 2016, CCIS 678, pp. 442–453, 2016.
DOI: 10.1007/978-3-319-51917-3_39

freights, traffic monitoring and environmental situation, prevention of emergency situations, data collection from the sensor fields and others [5,6]. Given that UAVs are mainly used for data collection from remote terrain, it is necessary to consider the problem of data delivery to the nearest gateway to the IP-network for onward data transmission to the Internet over distances of 100 km or more. Currently, there is a variety of data transmission protocols over long distances, which are mainly used for telemetry data collection. The paper describes the option to adapt long-range Internet of Things technology for use in flying ubiquitous sensor networks. As a basic LPWAN technology was considered by IEEE 802.15.4g [7] and LoRa communication protocol [8]. This protocol is designed for the data transmission on distance over a 10 km in an open space and more than 3 km in dense urban areas. However, some applications require data transmission over distances exceeding the maximum range of these technologies, and therefore, it is advisable to use a repeater nodes [9,10]. In turn, each node-repeater makes a certain delay in data transmission, which may be critical for certain applications. Also, an additional delay is introduced receiving and transmitting device located on the UAV that collect data from terrestrial segment (gateway WSN-LPWAN). LoRaWAN technology is widespread in the housing and utilities sector for data collection from meters and has a large number of hardware implementations. In view of the construction of the typical architecture of flying ubiquitous sensor networks, the architecture FUSN with a chain of repeating units was proposed. On the basis of this architecture, an analytical model and simulation model in AnyLogic software have been developed. While developing the model, it was taken into account that the hardware modules LoRaWAN function using the topology of "point to point", "star". Thus, for retransmitting the data, it is necessary to use an intermediate buffer, which introduces a delay network. As a result of research, loss and packet delay of all network site were revealed (from sensor nodes to the gateway with the Internet) depending on the data rate.

2 Architecture of Flying Ubiquitous Sensor Network

Taking into account the analysis of typical structures of building FUSN [5,11, 12], the architecture of data network over long distances was designed (Fig. 1), consisting of:

1. Terrestrial Segment FUSN, consisting of a plurality of sensor nodes, combined in a wireless sensor network (WSN) [13];
2. Flying FUSN segment, which includes:
   - Unmanned aerial vehicle (UAV-gateway WSN-LPWAN), which collects data from a terrestrial segment of the network and the subsequent data transmission, based on LoRa technology. In fact, the equipment, installed on the UAV, performs the role of Gateway 6LoWPAN - LoRa.
   - Group of unmanned aerial vehicles (UAV- repeater), performing retransmission of data on the basis of LoRaWAN technology for subsequent delivery to the gateway with the IP-network.

3. The base station network LPWAN - Internet (LoRa-IP base station). In this architecture, an unmanned aerial vehicle, mounted with a device acting as a gateway, is a node that collects data from terrestrial segment of the network.

Due to the Public UAV small flight time of the and a long distance to LoRa-IP base station, it is required to organize the data transmission channel in real time [9,10,14]. To organize such a channel, the interim UAV, located at various points between the UAV-gateway WSN-LPWAN and base station LoRa-IP base station, is used. UAV-gateway WSN-LPWAN, acting as a gateway, converts the data from the WSN and sends them to the next repeating node UAV- repeater. Next, there is a data transmission to LoRa-IP base station through relaying UAV network (UAV- repeater), connected to each other over the radio channel via LPWAN modules.

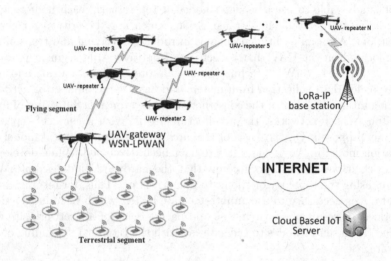

Fig. 1. Architecture FUSN for data transmission over long distances

As data transmission technology for the terrestrial segment of FUSN, IEEE 802.15.4 is considered as the most common in data collecting from the sensory fields [5,6,14]. As an example, there are protocols that use IEEE 802.15.4 standard: 6LoWPAN, RPL, ZigBee and others. As a protocol for data retransmission, LPWAN based on IEEE 802.15.4g technology were selected. As an example of protocols based on IEEE 802.15.4g standard, there are LoRa [8], SigFox, "Strij".

3   Selection of the Analytical Model

The analytical model should describe the main indicators of quality network performance, as we choose a probability of packet loss and latency data delivery. These rates depend on data delivery processes, running on the network layers from the Physical to Network. Therefore, the model must describe these layers sufficient degree for practical use.

3.1 Network Layer

Delivery route of data packets in the network between the data source S and the destination node D contains several sections (transits) k, formed by pairs of transceivers in the communications nodes (UAVs), Fig. 2.

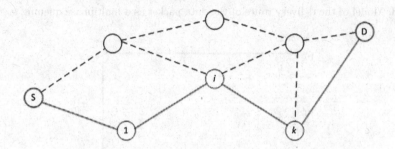

Fig. 2. Model of the delivery route of the data packet

At this layer, the model parameter is only the number of transits k. Later routing logic in the model is ignored.

3.2 Data Link Layer

Each of the transit nodes comprises a buffer for storing a certain (limited) amount of the received data that are transmitted to the next node of the route. Node model (route segment) may be represented as a queuing system with combined service discipline (delay and loss). The route is generally a multiphase queuing system, formed by a sequence of individual sections of the models [12,15], Fig. 3.

The probability of packet loss on the route can be defined as Eq. 1.

$$p = 1 - \prod_{i=1}^{k}(1 - p_i) \tag{1}$$

Where p_i may be determined as Eq. 2.

$$p_i = \frac{1 - \rho}{1 - \rho^{\frac{2n_b + C_a^2 + C_s^2}{C_a^2 + C_s^2}}} * \rho^{\frac{2n_b}{C_a^2 + C_s^2}} \tag{2}$$

where C_a^2 and C_s^2 - quadratic variation coefficients respectively distributions of the incoming flow and service time for the i-th node, n_b - the buffer size, ρ - the download of i-node.

To obtain an approximate estimate of the average time of package delivery various streams, we will use the expression [16] Eq. 3.

$$T_i = \frac{\rho * \bar{t}}{2(1 - \rho)} * (\frac{\bar{t}^2 + \sigma_a^2 + \sigma_s^2}{\bar{t}^2}) * (\frac{\bar{t}^2 + \sigma_s^2}{\bar{a}^2 + \sigma_s^2}) + \bar{t} \tag{3}$$

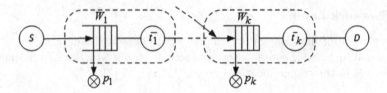

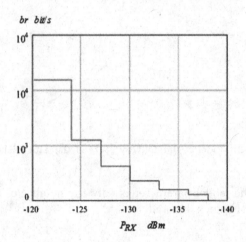

Fig. 3. Model of the delivery route of the data packet as a multiphase queuing system

Fig. 4. Dependence of the data rate from level of signal at the receiver input

Where σ_a^2, σ_s^2 - dispersion time intervals between packets and the service time i-th node, respectively, $\bar{a}$ - the average value of the interval between the i-th packet assembly, $\bar{t}$ - the average time to i-th node.

Then, the delivery time on the route will be the same Eq. 4.

$$T = \sum_{i=1}^{k} T_i \qquad (4)$$

Models Eqs. 1, 2, 3 and 4 provide an approximate estimate of the probability of packet loss and average the time of arrival of the data source node S to destination D. This model assumes that the known average valuation packet transmission time at each of the route segments $\bar{t}_i$. This time depends on the data rate at the site, and packet size. Data transmission rate, in turn, depends on the radio propagation conditions and can be described by one of the known models Eq. 5.

$$\bar{t}_i = \frac{\bar{L}}{br_i} \qquad (5)$$

where L - the average packet length (bits); br_i - data rate for i-th area (bit/s).

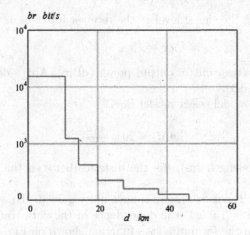

Fig. 5. Dependence of the data rate of the distance between transmitter and receiver (with antenna height h1 = h2 = 100 m)

3.3 Physical Layer

In general, at this level the processes, affecting the signal propagation conditions, should be regarded, such as attenuation in the propagation medium, signal fading in the channel and different kinds of interference to propagation conditions. To simplify the model, we will associate at a given layer the data rate in the channel only with the distance between nodes.

3.4 Distance Factor

According LoRaWAN specification [8], the data rate depends on the level at the receiver input signal as shown in Table 1.

Table 1. The dependence of the rate of signal level at the input of the receiver

Signal level (dBm)	Modulation type	Transmission speed (bit/s)
−122	GFSK	50000
−120	LoRa	10937
−123	LoRa	5468
−126	LoRa	3125
−129	LoRa	1757
−132	LoRa	976
−135	LoRa	537
−137	LoRa	292

We will describe the signal level at the receiver input as 6.

$$P_{RX} = P_{TX} - A(d) \tag{6}$$

where P_{TX} - at the transmitter output power (dBm); A(d) - dependence of the damping distance (dB).

As attenuation model select model Eq. 7.

$$A(d) = 20 \lg \frac{4\pi d}{\lambda} \tag{7}$$

where λ - the wavelength (m); d - the distance between the transmitter and receiver (m).

We believe that the used antennas have unity gain (0 dB).

Then, as shown in Table 1, the dependence of the data transmission rate of the signal will be in the form of a step function shown on Fig. 4.

As the sensitivity of LoRa radio receiving module according to specifications, can receive the signal with a low signal level (−137 dBm), and signal attenuation, as measured according to Eq. 7, the potential transmission distance are large enough to be considered that the curvature and height of the earth lifting antenna devices (UAV altitude).

The distance line of sight can be defined as Eq. 8.

$$d_{max} = 3.57(\sqrt{h_1} + \sqrt{h_2}) \tag{8}$$

We take into account the fast fading channel by entering the empirical coefficient increase damping $\gamma = 1.1$. Then, in view of Eqs. 7, 8 and the entered of the coefficient of dependency data rate of the distance will be of the form shown in Fig. 5.

Thus, the proposed model enables us to establish a relationship of packet service time in the nodes (transmission time on channels), the waiting delay and the probability of packet loss on the route parameters (distance between nodes and the number of transits).

4   Simulation Model of the FUSN

With the above-described architecture, simulation model of data transmission from wireless sensor network to the Internet has been developed. On the basis of research on a model network, the baseline data for the development of a simulation model were obtained. Thus, the volume of data transmitted from the wireless sensor network is 10 MB. These data must be transmitted over the FUSN a distance of 120 km. Flying ubiquitous sensor network model is represented as a queuing system [12, 15]. The model was performed with the help of simulation package AnyLogic. The model describes the dependence of the number of delivered packets and delays to the base station and the data transmission rate. Restrictions on the time of the UAV are not considered. In the simulation model, the parameters for the functioning of the modules LPWAN Semtech SX1272 were used [17].

Module Semtech SX1272 has the following characteristics:

1. Bandwidth (in LoRa modulation mode): 240–32600 (bit/s);
2. Range: up to 20 km (in an open space, without delivery confirmation);
3. Sensitivity transceiver (RSSI): (−117)–(−137) (dBm);
4. Transceiver power: 10–20 (dBm);
5. Data packet size: 30–256 (bytes);
6. Operating frequency: 860–1020 (MHz).

The above characteristics were used to construct LPWAN relay model. In particular, the RSSI was obtained on the basis of the distance and depending on the data rate of RSSI.

The developed simulation model consists of the following elements (Fig. 6):

(1) the source - an entity's source, simulating the operation of the terrestrial segment of the FUSN;
(2) the gateway - an agent, simulating the operation of WSN-LPWAN gateway;
(3) the *module*802_15_4*g* - an agent, simulating the operation of repeating LPWAN devices;
(4) the sink - an entity's point of destination, which performs the role of base station.

The model also uses elements selectOutput (selectOutput1 - selectOutput4), which is used to simulate the choice of communication channel LPWAN, according to the criteria of the buffer load and parameter of indicator RSSI of the adjacent unit [11]. The calculation of these criteria occurs in the agent *module*802_15_4*g* (Fig. 6).

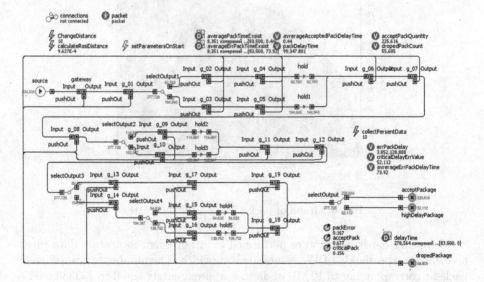

Fig. 6. Simulation model of FUSN

In this model, the *module*802_15_4*g* devices (Fig. 7), which simulate the work of the repeater LPWAN devices, are combined into a chain of 19 units. The data packet, from the moment of admission to the queuing system until reaching the destination, overcomes 12 nodes such as *module*802_15_4*g*. Package size of 30 bytes is selected. The distance between the devices is a dynamic value varies within 1 km from the initially set coordinates. As previously mentioned, the distance by which to transmit data is at 120 km. Data rate is constant throughout the transmission time. Data transfer mode: no acknowledgments and retries.

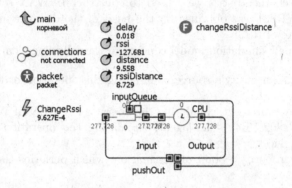

Fig. 7. Simulation model of agent *module*802_15_4*g*

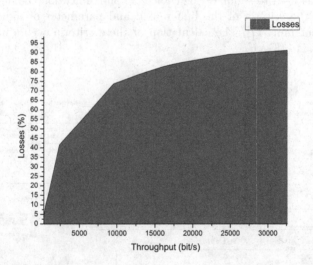

Fig. 8. Graph of packet loss on the data rate

Computer experiments were performed for 10 different indicators data rate, according to thresholds LPWAN module bandwidth. The number of transmitted packets, corresponding to 10 MB of data, is approximately equal to 333333 packets. The results of a series of computer experiments are given below (Table 2).

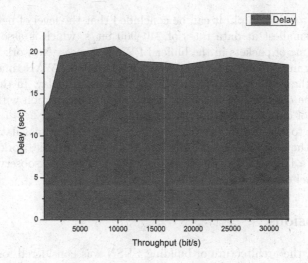

Fig. 9. Dependency network delay of data rate

Figure 8 shows the data transmission rate of loss. According to this graph, it can be concluded that the optimal data rate for a given network architecture is 240–480 bit/s, which allows to transmit data with an acceptable level of losses (for IEEE 802.15.4g devices). The losses are associated with a large number of nodes and constantly varying capacity for LPWAN devices, which in turn is connected with the distance between the devices and the values of RSSI parameter.

Figure 9 shows the network delay in sending packets on the data rate.

Table 2. The results of computer simulation

No	Throughput (bit/s)	Accepted packages (percent)	Lose packages (percent)	Delay (sec)	Model work time (sec)
1	32600	8.78	91.22	18.44	2491
2	24800	10.95	89.05	19.31	3332
3	18400	15.27	84.73	18.43	4412
4	16200	17.12	82.88	18.72	5010
5	12800	21.54	78.46	18.89	6280
6	9600	26.29	73.71	20.68	8431
7	2400	58.50	41.50	19.56	33380
8	960	58.50	16.68	14.23	83417
9	480	90.41	9.59	13.74	166720
10	240	96.06	3.94	12.67	333354

According to this graph, it can be concluded that the level of network delay becomes the smallest at data rates of 240–960 bit/s, which is associated with less waiting time of packets in the buffer LPWAN devices. Network latency can be reduced by introducing a simulation model module LPWAN timeout, which will increase the level of losses, but will reduce average delay. In this case, the network can be considered tolerant to delays [18], in connection with uncertain requirements of data transmission through the network.

According to the results, we can conclude that the optimal data transmission rate for the network architecture and the above scenario work is 240–480 bit/s. For data transmission rates, smallest delay (11–14 s) isobserved with an acceptable level of packet loss (3–10%).

5 Conclusions

In this paper, the architecture of building FUSN was considered for the transmission of data over long distances. In LPWAN, technology based on the IEEE 802.15.4g standard was considered as a technological basis for the organization of data repeat between the unmanned aerial vehicles. The characteristics of these devices are the most suitable for solving the transfer a small amount of data over long distances. Based on this architecture, analytical and simulation models were developed. Results were obtained on the basis of a series of computer experiments, which have revealed the delay and packet loss, occurring in the transmission network on all stations at different data rates, and conclusions were drawn on the optimal data rate in a given network architecture.

Acknowledgements. The reported study was supported by RFBR, research project No. 15 07-09431a "Development of the principles of construction and methods of self-organization for Flying Ubiquitous Sensor Networks".

References

1. Kirichek, R., Vladyko, A., Zakharov, M., Koucheryavy, A.: Model networks for Internet of Things and SDN. In: Proceedings of the 18th International Conference on Advanced Communication Technology (ICACT), pp. 76–79 (2016). doi:10.1109/ICACT.2016.7423280
2. Kirichek, R., Golubeva, M., Kulik, V., Koucheryavy, A.: The home network traffic models investigation. In: Proceedings of the 18th International Conference on Advanced Communication Technology (ICACT), pp. 97–100 (2016). doi:10.1109/ICACT.2016.7423288
3. Futahi, A., Koucheryavy, A., Paramonov, A., Prokopiev, A.: Ubiquitous sensor networks in the heterogeneous LTE network. In: Proceedings of the 17th International Conference on Advanced Communication Technology (ICACT), pp. 28–32 (2015). doi:10.1109/ICACT.2015.7224752
4. Y.2069. Terms and definitions for the Internet of things. Recommendation ITU-T, July 2012

5. Koucheryavy, A., Vladyko, A., Kirichek, R.: State of the art and research challenges for public flying ubiquitous sensor networks. In: Balandin, S., Andreev, S., Koucheryavy, Y. (eds.) ruSMART 2015. LNCS, vol. 9247, pp. 299–308. Springer, Heidelberg (2015). doi:10.1007/978-3-319-23126-6_27

6. Costa, F.G., Ueyama, J., Braun, T., Pessin, G., Osorio, F.S., Vargas, P.A.: The use of unmanned aerial vehicles and wireless sensor network in agricultural applications. In: 2012 IEEE International Geoscience and Remote Sensing Symposium, Munich, Germany, pp. 5045–5048 (2012)

7. IEEE Standard 802.15.4g. Part 15.4: Low-Rate Wireless Personal area Networks (LR-WPANs) Amendment 3: Physical Layer (PHY) Specifications for Low-Data-Rate, Wireless, Smart Metering Utility Networks, April 2012

8. Sornin, N., Luis, M., Eirich, T., Kramp, T.: LoRaWAN Specification V1.0, LoRa Alliance, January 2015

9. de Freitas, E.P., Heimfarth, T., Netto, I.F., Lino, C.E., Pereira, C.E., Ferreira, A.M., Wagner, F.R., Larsson, T.: UAV relay network to support WSN connectivity. In: International Congress on Ultra Modern Telecommunications and Control Systems, Moscow, Russia, pp. 309–314 (2010)

10. Marinho, M.A.M., de Freitas, E.P., Lustosa da Costa, J.P.C., de Almeida, A.L.F., da Sousa, R.T.: Using cooperative MIMO techniques and UAV relay networks to support connectivity in sparse wireless sensor networks. In: 2013 International Conference on Computing, Management and Telecommunications (ComManTel), Ho Chi Minh City, Vietnam, pp. 49–54 (2013)

11. Kirichek, R., Kulik, V.: Methods of test flying ubiquitous sensor networks. In: Proceedings of the 19-th International Conference on Distributed Computer and Communication Networks: Control, Computation, Communications (DCCN-2015), pp. 489–499 (2015)

12. Kirichek, R., Paramonov, A., Koucheryavy, A.: Swarm of public unmanned aerial vehicles as a queuing network. In: Vishnevsky, V., Kozyrev, D. (eds.) DCCN 2015. CCIS, vol. 601, pp. 111–120. Springer, Heidelberg (2016). doi:10.1007/978-3-319-30843-2_12

13. Kirichek, R., Koucheryavy, A.: Internet of things laboratory test bed. In: Zeng, Q.-A. (ed.) Wireless Communications, Networking and Applications. LNEE, vol. 348, pp. 485–494. Springer, Heidelberg (2016). doi:10.1007/978-81-322-2580-5_44

14. Kirichek, R., Paramonov, A., Vareldzhyan, K.: Optimization of the UAV-P's motion trajectory in public flying ubiquitous sensor networks (FUSN-P). In: Balandin, S., Andreev, S., Koucheryavy, Y. (eds.) ruSMART 2015. LNCS, vol. 9247, pp. 352–366. Springer, Heidelberg (2015). doi:10.1007/978-3-319-23126-6_32

15. Kirichek, R., Paramonov, A., Koucheryavy, A.: Flying ubiquitous sensor networks as a queueing system. In: Proceedings of the 17th International Conference on Advanced Communications Technology (ICACT), pp. 127–132 (2015). doi:10.1109/ICACT.2015.7224771

16. Villy, B.: Iversen Teletraffic Engineering Handbook. COM Center Technical University of Denmark Building 343, DK-2800 Lyngby Tlf.: 4525 3648. www.tele.dtu.dk/teletra

17. SX1272/73 - 860 MHz to 1020 MHz Low Power Long Range Transceiver Datasheet rev. 3, Semtech Corporation (2015)

18. Cerf, V., Burleigh, S., Hooke, A., Torgerson, L., Durst, R., Scott, K., Fall, K., Weiss, H.: RFC 4838. Delay-Tolerant Networking Architecture, April 2007

Hardware-Software Simulation Complex for FPGA-Prototyping of Fault-Tolerant Computing Systems

Oleg Brekhov[✉] and Alexander Klimenko

Department of Control Systems, Informatics and Electrical Engineering,
Moscow Aviation Institute (National Research University),
. 4 Volokolamskoe Shosse, Moscow 125993, Russia
obrekhov@mail.ru, a.v.klimenko@mai.ru

Abstract. The article presents the concept of building hardware-software complex based on field programmable gate array (FPGA) prototyping, which implements the methodology of fault-tolerant gate array (GA) or FPGA-based aerospace-born systems-on-chip (FTS) modeling, based on extended concept of fault injection. This complex allows to define functional FTS design for the given parameters of the negative impact of the external environment through the estimation of the FTS project fault tolerance level (FTL).

Keywords: FPGA-prototyping · Fault injection · Fault tolerance · Fault simulation

1 Introduction

The gate array (GA) or field programmable gate array (FPGA) based aerospace - born fault-tolerant systems-on-chip (FTS) development methods widely use means of electronic design automation (EDA). Herewith, the FTS design flow, involves FTS design project specification via hardware description language (Verilog/VHDL/SystemVerilog), passing it through a number of verification stages and eventual generation of the chip topology. It is well-known approach to use software simulators for verification, which, however, are characterized by a low simulation speed. This limitation can become an obstacle for the FTS complex projects optimization. Therefore, an alternative way is to use hardware software complexes.

This article presents the concept of building hardwaresoftware complex (HSC), used for the FTS projects simulation with verification, based on FPGA prototyping and fault injection method. The FTS verification incorporates both functional verification and the definition of the current fault tolerance level (FTL) of the project.

The rest of the paper is organized as follows. In Sect. 2 we briefly describe the proposed FTS design methodology. An extended concept of fault injection method used for FTS projects verification is proposed in Sect. 3. In

ⓒ Springer International Publishing AG 2016
V.M. Vishnevskiy et al. (Eds.): DCCN 2016, CCIS 678, pp. 454–467, 2016.
DOI: 10.1007/978-3-319-51917-3_40

Sect. 4 the comprehensive multi-phase FTS simulation methodology is proposed. Section 5 describes the proposed HSC implementation methodology. Finally, Sect. 6 presents our conclusions.

2   The FTS Design Methodology

We define three design stages in the FTS project development process based on hardware description languages usage: (1) The development of the FTS design, that provides a given functionality. (2) The FTS fault tolerance ensuring means (FTEM) development. (3) The development of the FTS design, having required FTL.

At the first stage, the FTS design project is created, after that the creation of the target FPGA/GA topology description file in the form of netlist is carried out.

The means ensuring various FTL (in other words fault tolerance ensuring means, FTEM) of the FTS functional units of various hierarchical levels (from individual registers and counters to large computational blocks) are developed at the second stage.

At the third stage the FTS design, having required FTL is determined by integrating the FTS design with given functionality, and FTEM.

The fault injection method is used for the FTS design project verification.

3   The Fault Injection Concept

To specify the environmental impact on the aerospaceborn FTS, we propose an extended concept of fault injection method [1] which is based on the joint use of the three models: the external influences model (EIM), the threats occurrence model (TOM) and the fault localization model (FLM).

For the aerospace-born FTS operating in harsh environmental conditions, the major factor limiting their lifetime is space radiation [2,3].

The EIM is used to determine the environmental parameters that have a negative impact on the FTS elements. The launch date of the spacecraft, where FTS is placed, the parameters of the orbit, and the characteristics of the taken radiation protection are used as input data for EIM. The output of the EIM is the environmental parameters (in particular, types, intensity and motion direction of the charged particles) that determine impact of the environment on physical environment of the chip.

The output of EIM is used as the input of TOM, which also uses the system-on-chip technological parameters and the topology of the chip as input data. The parameters of chip's physical environment changing form the output of the TOM.

The output of the TOM is in its turn the input for the FLM which also uses the topology of the chip as the input data. Based on the input data, the FLM generates the impact of the chip's physical environment changing parameters on the FTS logic elements: target elements and their type (combinational logic gates, memory elements), the moments of faults occurrence in this elements, types of logical faults that occur as a result of the influence of the external environment (bitflip, stuck-at, etc.).

Thus, when environmental influences on FTS are set, the joint usage of the three models (EIM, TOM and FLM) leads to the generation of fault list, which is used in the comprehensive multi-phase FTS simulation.

4   The Comprehensive Multi-phase FTS Simulation Methodology

A comprehensive multi-phase FTS simulation methodology includes the following phases: phase 1 - FTS model development; phase 2 - FTS model functional testing; phase 3 - FTS model with fault injection capability (FIFTS) development by adding fault injection means to FTS model; phase 4 - FIFTS functional testing in the absence of faults; phase 5 - FIFTS functional testing in the presence of faults; phase 6 - FIFTS FTL definition; phase 7 - FTEM model development; phase 8 - FTEM functional testing; phase 9 - FTS model with fault injection capability and integrated means of fault tolerance (FIFTMS) development through the FTEM to FIFTS inclusion; phase 10 - FIFTMS functional testing in the absence of faults; phase 11 - FIFTMS functional testing in the presence of faults; phase 12 - FIFTMS FTL definition.

The algorithm of implementing comprehensive multiphase FTS simulation, is as follows (see Fig. 1).

Phase 1: based on the original FTS specifications the initial FTS model that describes the logic circuit and the hierarchy of its functional modules, that implement FTS core functionality, is developed.

Phase 2: FTS functional testing.

Here and further under functional testing we understand the standard procedure of input actions submission to the inputs of the FTS model (FIFTS/FIFTMS) and the comparison of its utput values (hereinafter referred to as responses) with the reference values. The developer of the FTS model must directly specify the array of the reference values on this phase.

In the case of a successful test the transition to phase 3 is carried out, otherwise the return to phase 1 for performing FTS model redesign is implemented.

Phase 3: FIFTS model development. Fault injection means by defining the absence or the presence of faults in individual (figure here) FTS elements allow to imitate the impact of the external environment on FTS during simulation.

Phase 4: FIFTS functional testing in the absence of faults. If the test is successfully passed, the FIFTS model is equal to FTS model if we speak about FTS core functionality implementation. In this case, the transition to phase 5 is performed. Otherwise the transition to phase 3 for the FIFTS model refinement is performed.

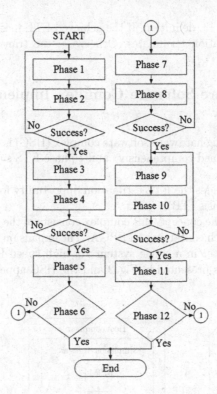

Fig. 1. The algorithm of implementing comprehensive multi-phase FTS simulation

Phase 5: FIFTS functional testing in the presence of faults. The fault types, fault injection time instants, as well as FTS specific elements, for which fault injection is performed, are determined via joint EIM, TOM and FLM models usage.

Phase 6: FIFTS FTL definition. If the achieved FTL meets the FTS requirements, the simulation terminates. Otherwise, the transition to phase 7 is performed.

Phase 7: FTEM model development. On each iteration of the simulation cycle a new FTEM model is developed on this phase. It has higher FTL related to the FTL of the FTEM developed on previous iterations.

Phase 8: FTEM functional testing. In case of successful test the transition to phase 9 is performed, otherwise the transition to phase 7 for FTEM model refinement is implemented.

Phase 9: FIFTMS model development via FIFTS and FTEM models integration.

Phase 10: FIFTMS functional testing in the absence of faults. In case of successful test the transition to phase 11 is carried out. Otherwise, the transition to phase 9 is performed.

Phase 11: FIFTMS functional testing in the presence of faults.

Phase 12: FIFTMS FTL definition. If the achieved FTL meets the FTS requirements, the simulation terminates. Otherwise, the transition to phase 7 is
performed.

5 The Hardware-Software Complex Implementation Methodology

The concept of building hardware-software complex (HSC) is developed to implement the aforementioned comprehensive multi-phase FTS simulation methodology.

The complex provides the FTL estimation opportunity for the FTS with GA
or FPGA as target basis (TB).

In accordance to the concept, this complex consists of the host computer, and
4 expansion cards, each containing an FPGA. The expansion cards are combined
with the host computer in a single system via high speed PCIe interface. The
HSC block diagram is presented in Fig. 2 and the HSC appearance is presented
in Fig. 3.

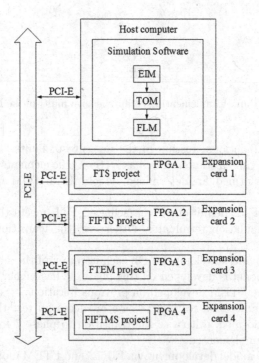

Fig. 2. The HSC block diagram

The FTS FTEM, FIFTS and FIFTMS models are implemented in FPGA 1,
FPGA 2, FPGA 3 and FPGA 4 correspondingly as same-name projects, developed using hardware description languages (phases 1, 3, 7, 9).

Fig. 3. The HSC appearance

These FPGAs are used for verification purposes (phases 2, 4, 5, 8, 10, 11).

Herewith, during the processes of the FIFTS and FIFTMS projects' verification (which are obtained via the sequential modification of the original FTS project), the responses received during the successful verification of the previous modification of the FTS project are used as reference responses. This means that when implementing phases 4–5 the responses of FTS project, obtained on phase 2, are used as reference responses and when implementing phases 10–11 the responses of FIFTS project, obtained on phase 4 are used in this capacity.

This solution allows to maintain continuity of the results during the simulation process.

During projects verification process the HSC software generates input actions, specifies the sequence of faults for FIFTS and FIFTMS projects (phases 5 and 11, respectively), in accordance with the fault list produced by the FLM. Herewith, FTL is defined (phases 6 and 12), and the conclusion about the necessity of project revision is made based on the achieved FTL.

The data exchange between the HSC software and the FTS FTEM, FIFTS and FIFTMS projects implemented in the corresponding FPGAs is performed via the PCIe interface.

The HSC contains the following functional components (see Fig. 4):

- The input actions formation system, consisting of software Input Action Generator (IAG) and hardware Input Actions Provider (IAP) modules.
- The fault injection system, consisting of software Fault Pattern Generator (FPG) (implementing the EIM, TOM and FLM models), and hardware Fault Injection Unit (FIU).
- The simulation management system, consisting of hardware (Local Control Unit (LCU) and RAM controller, implemented in FPGA) and software

(Simulation Control Unit (SCU), Project Processing Unit (PPU), interface module) parts.
- Data collection and modeling results analysis system, implemented in the form of software Results Analysis Unit (RAU) and hardware Response Collector (RC) modules.
- EDA tool for TB (marked on Fig. 4 as "EDA tool") that provides the synthesis and place & route processes implementation for FTS elements.

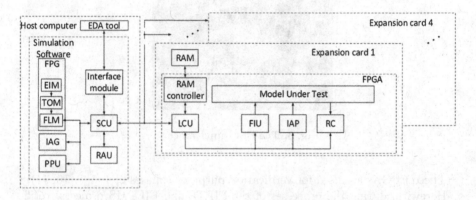

Fig. 4. The HSC functional diagram

The input data required for the HSC must contain: (1) The TB identification data (type, family); (2) The arrays of input actions and a reference responses; (3) The flight parameters of the spacecraft, where FTS is placed.

The flight parameters (orbit, launch time) are used to determine the characteristics of external influences while implementing functional testing of FIFTS and FIFTMS projects.

The FTS project that has an appropriate FTL for the specified external influences is the result of each comprehensive multi-phase simulation methodology implementation, provided by the HSC.

Further let us describe the operation of the HSC functional components.

A The input actions formation system
  1. Input Action Generator (IAG)
     The IAG unit provides the following functions:
     - Obtaining information about reference input data for the FTS model (hereinafter referred to as reference data) in a specified file containing information about the time-ordered changes in input and output reference signals of FTS model.
     - Processing of reference data: conversion reference data to the required format and generation of input actions (input actions vectors) for the simulation.
     - Transfer of input actions vectors to the simulation control unit (SCU).

- Convert and record the generated input actions in the "value change dump" file.
- Interaction with external test systems via the application programming interface (API) provided to the FTS developer.

There are two ways to generate input actions for the simulation: by using the reference data and by using API. In accordance with the first method the generation of input actions is based on the using of reference data provided by the FTS developer and performed before executing the simulation. After analyzing reference data, provided by the developer, IAG generates an array of input actions, a set of binary vectors, each of which corresponds to the value of the input signals of the FTS/FIFTS/FIFTMS model at a certain time of simulation. During the simulation input actions generated by the host computer are transferred to the expansion card one by one and supplied to the inputs of the FTS/FIFTS/FIFTMS model. In accordance with the second method, the input actions are generated by the FTS developer tools dynamically during the simulation. The HSC only provide an opportunities to transfer input actions vectors to the FTS/FIFTS/FIFTMS models and to get responses from them via API. API can be used in an external test system, developed by the FTS developer via SystemVerilog language. Using the API external test system receives the response of the model simulated by the means of HSC, analyzes it, generates the next input actions vector and transmits it to the HSC. This process is repeated until the completion of the simulation process. The advantages of this approach are greater flexibility and the ability to simulate wide range of FTS interface protocols. However, the FTS simulation time is increased due to the necessity, to transmit each response from the expansion card to the host computer for being analyzed by external test subsystem, as well as the necessity to transfer each input actions vector to the expansion card of HSC separately.

2. Input Actions Provider (IAP)

IAP unit is a hardware unit designed for the supplying input actions vectors to the inputs of the FTS/FIFTS/FIFTMS model at each cycle of simulation. IAP unit has three modes of operation: Idle, Wait load and Load. In the Load mode IAP unit asserts the ready to receive data signal for the DDR controller. After that DDR controller transfers words containing values of input actions vector to the IAP unit that loads data into the input actions register. When the data received from DDR controller have loaded into register the IAP unit deasserts the ready to receive data signal, generates the finish signal to control unit and switches to the Idle mode. If the control unit deasserts the active signal the IAP unit will switch from the Idle mode to the Wait load mode.

B The fault injection system

1. Fault Pattern Generator (FPG)

When modeling FIFTS and FIFTMS projects in addition to the input actions and reference responses arrays, formed by IAG and RAU respectively, it is also nessesary to have the fault injection program which is generated by FPG. This program is implemented in the form of a fault list (FL), obtained as the output of the FLM. There are two variants of internal organization of this list: horizontal and vertical organization. Horizontal organization of the list is used when the number of elements of the FTS project, for which there is a necessity to inject faults during simulation (denote them as "target elements") is negligible, and does not exceed some threshold value, which is due to hardware limitations. Otherwise, vertical organization is used. In both cases, the list is a set of q data packets, containing identification data of target elements, fault injection time points, and types of inserted faults. The number of packets (q) coincides with the number of cycles of the simulation on which fault injection is implemented. The data packets are listed in the order of use during simulation. The format of each data packet depends on the type of FL internal organization. When horizontal organization is used each data packet has format which is explained further. The packet size is fixed at $N*k+32$ bits in this case, where N is the number of target elements, k - is the number of bits required to uniquely identify the fault type. It is assumed that for all elements of the FTS project the same types of faults are used. The data packet contains a 32-bit field named "offset" which is located in the most significant bits of the packet. This field describes the fault injection time point (TP) for all target elements. Its value is determined as an offset relative to previous TP, measured in cycles of the simulation clock. For the first packet this field contains the offset from the beginning of the simulation. If all bits of the field "offset" is equal to "1", it means that this package is the latest in FL and simulation is completed after its processing. Behind the field "offset" the packet contains N k-bit fields, named as "type of fault for element i", (where i could take values from 1 to N) each of which corresponds to the i-th target element. The data packet format for vertical organization is explained further. Each data packet from FL has size $(Ni)*24+64$, where Ni is the number of target elements for which TP corresponds to i-th offset. In other words, the packet size can be different for different TPs (for different offsets). The most significant 32 bits of the data packet similar to the horizontal organization form field "offset" having the same sense. Next 32-bit word of the packet is reserved for the "target elements quantity". The value of this field indicates the number of 24-bit fields, named "data of target element i". The format of the field "data of target element i" has two subfields: "Type of fault" and "The element's number". The field, named "Type of fault" is intended to characterize the type of faults injected in the element, and contains the corresponding code. Used encoding allows to simulate either basic accepted fault models (such as bit-flip, stuck-at-0, stuck-at-1) or more complex fault models based on various combinations of basic models. The field "The element's number" is intended to uniquely identify the FTS target element. The sizes of the fields composing field "data of target element i" could be reviewed in the context of a specific

FTS project and in the subsequent implementations of the HSC. In the first case, the scope of the project determines the number of x binary digits required to uniquely identify each FTS element: x - is a minimal integer, that is greater or equal to log2 (number of FTS elements). As a result, x bits will be allocated to the field "item number". The size of the field "type of fault", y, is defined similarly: y - is a minimal integer, that is greater or equal to log2 (number of fault types). In the current HSC implementation the size of field "type of fault" is fixed at 3 bits. To implement functional testing of the FTS project in the presence of faults both input actions and faults arrays are required for each cycle of simulation clock. In other words, each input vector from the input actions array should be put in correspondence to the appropriate vector of faults from faults array. If on some simulation cycle there is no need to inject faults, then the vector of input actions corresponds to the zero vector of faults. A set of two vectors is an exchange data packet used for data transmission between software modules of the modeling system and hardware ones, implemented in FPGAs. To reduce the amount of data transmitted between hardware and software modules, the decision was made to introduce two types of exchange data packet instead of forming packages with a zero faults vectors. Both packages have the most significant bit, used as a marker. Its value determines whether the current packet have faults vector. When it is 0 - the packet contains only input actions vector, which size is equal to the number of the primary inputs of the simulated FTS project. When it is 1 - then, the packet contains the sequence input actions vector and faults vector. The size of the faults vector corresponds to the number of target elements of the simulated FTS project. The generation of packets is performed by the SCU module.

2. Fault Injection Unit (FIU)

   The fault injection unit is intended to provide fault codes to the target elements and could operate in three modes: the "Waiting download", "Uploading" and "Waiting". At power-up, the FIU switches to the "Waiting download" mode. In the "Waiting download" mode values from faults codes register are set to the control inputs of target elements of the FTS/FIFTS/FIFTMS project. If the activation signal from the LCU is received FIU switches to the "Uploading" mode. In the "Uploading" mode the code of "no faults" is set to the control inputs of target elements of the FTS/FIFTS/FIFTMS project. After that, FIU informs DDR controller that it is ready to get data. The DDR controller transfers data words with new faults codes which are loaded in the register of faults codes. After download is complete, FIU informs LCU about the end of operation and turns into "Waiting" mode. In the "Waiting" mode values from the faults codes register are set to the control inputs of target elements of the FTS/FIFTS/FIFTMS project. If the deactivation signal from the LCU is received, FIU turns into "Waiting to download" mode.

C The simulation management system
1. Local Control Unit (LCU)

HSC hardware units, FTS, FIFTS, FIFTMS and FTEM models integrated in the same FPGA form so-called LCU+FIU+IAP+RC+RAM controller. Control of LCU+FIU+IAP+RC+RAM controller is carried out by LCU that can operate in five modes:

- Get responses
- Wait input data
- Set input actions
- Fault injection
- Generate clock signal

LCU controls the clock signal for the model (FTS, FIFTS, FIFTMS or FTEM) simulated by means of the HSC. After turning on the power LCU sets the clock signal to 0 and enters the Get responses mode. In the Get responses mode the LCU generates the active signal for the RC. The LCU remains in the Get responses mode until the RC unit operation is completed and then enters the Wait input data mode. The LCU remains in the Wait input data mode until the RAM unit generates the data ready signal. After that LCU enters the Set input actions mode. In the Set input actions mode the LCU generates the active signal for the IAG unit. The LCU remains in the Get responses mode until the IAG unit operation is completed. After that LCU enters the Fault injection mode if input data received from RC contain fault vector. Otherwise the LCU enters the Generate clock signal mode. In the Fault injection mode the LCU generates the active signal for the FIU. The LCU remains in the Fault injection mode until the FIU operation is completed and then enters the Generate clock signal mode. In the Generate clock signal mode the LCU inverts the value of clock signal and then enters the Get responses mode.

2. RAM controller

RAM controller is intended for data exchange between HSC hardware or software units and RAM installed at the expansion card of HSC. RAM controller consists of unit that realizes the logic necessary to read/write data from/to RAM and units that serve the requests from such units as SCU, IAP, FIU and others. RAM controller uses RAM as two FIFO-buffers: one for storing data, coming from RC unit to the host-computer and another for storing data coming from the host computer to IAP and FIU units.

3. Simulation Control Unit (SCU)

SCU coordinates the operation of FPG, IAG, PPU, RAU, LCU modules, as well as the initialization of interaction with the EDA. The data exchange with the EDA is carried out through the description files in Verilog format. When implementing the FIFTS and FIFTMS models the SCU initiates operation of PPU and provides its interaction with EDA. While implementing the FTS/FIFTS/FIFTMS/FTEM simulation stages SCU at first initiates the FPG, IAG, PPU, and RAU modules and then initiates the download of fault injection microkernel and

FTS/FIFTS/FIFTMS/FTEM to the appropriate FPGA. Then the SCU initiates FPG and IAG operation and when they finished their work, generates data packets that contain information about the faults to be injected and input actions for each step of simulation time. After the generation of data packets SCU initiates their transmission to the LCU. When the transmission of the packets finished SCU initiates the simulation. During the simulation SCU provides the reception of the tested model's responses from LCU via the PCIe interface. SCU transmits these responses to RAU, which compares them with the reference responses. Upon completion of simulation and reception of the simulation results from RAU, SCU module provides the HSC user with the simulation results.

4. Project Processing Unit (PPU)

All modifications of FTS model, as well as its implementation in FPGA with embedded fault-injection microkernel are provided by the project processing unit - PPU. The input data for the PPU is:

- FTS model in the form of netlist of the FTS elements, described at the level of digital functional cells of TB (hereinafter, the cells) via Verilog language subset (structural Verilog netlist [4]), which contains the module, endmodule, input, output, inout, wire, assign keywords as well as identifiers, comments, and punctuation (hereinafter the structural Verilog language);
- cells library in the basis of FPGA used for simulation (Virtex 6 LX240T FPGA) in a structural Verilog language;
- Instances of the FTS model cells (optional), to which the faults can be injected during the simulation;
- library of the cells with embedded fault injection means in a basis of simulation FPGA in a structural Verilog language;
- Pins of the FTS cells, from which the responses will be read.

The results of PPU include:

- FTS model with fault injection capability and means to transmit input actions and read responses in a basis of simulation FPGA in a structural Verilog language;
- interface to the input actions transmission module;
- interface to fault activation unit;
- interface to response collection unit (RC).

FTS model processing consists of 7 stages.

Stage "Initial analysis of the FTS model" provides a lexical analysis, parsing and semantic analysis of the FTS model. Stage "Search unsupported cells" project provides the test of the FTL model for the presence of cells and difficult-functional units that are not supported by HSC simulation means. Stage "Analysis of cell connections" provides the generation of the netlist using the following fields: output of the given cell, as well as the type, name and the output of another cell, which is connected to it Stage "Analysis of the IC project" provides the determination and output of the cells that are connected to the outputs of each cell on the base of data obtained at the "Analysis of cell connections" stage. Stage "Implementation of the input actions transmission interface and responses collection interface" provides the generation of an interface to collect the responses from the given GA cell (triggers) outputs. The result of "Embedding of simulation tools without fault injection means" stage is the FTS with embedded simulation tools in a basis of simulation FPGA. The result of "Embedding of simulation tools and fault injection tools" stage is the FIFTS or FIFTMS depending on the phase of the multi-phase simulation algorithm.

D Data collection and modeling results analysis system

   1. Response Collector (RS)

RC unit is intended for collecting responses of the FTS/FIFTS/FIFTMS model elements during the simulation. After turning on the power RC enters the Wait responses mode. After LCU generates the active signal RC enters the Get responses mode. In the Get responses mode RC unit sends the responses from the simulated model to the RAM controller if it is ready to get data. After transfer is completed RC generates the complete signal to the LCU and enters the Idle mode. In the Idle mode RC remains until the LCU resets the active signal and then RC enters the Wait responses mode.

   2. Results Analysis Unit (RAU)

In accordance with the multi-phase FTS simulation methodology the HSC can control the phases of FTS, FIFTS and FIFTMS models functional testing. This control is provided by the Results Analysis Unit (RAU), which, in addition, evaluate the FTL level of FIFTS and FIFTMS models. In general the Results Analysis Unit (RAU) performs the following functions: Comparison and analysis of the output data generated by the FTS during the simulation (responses from FTS/FIFTS/FIFTMS/ FTEM), and reference output data. The reference responses during the FTS and FTEM simulation are the appropriate arrays of reference responses provided by the developers of related projects. The array of responses received in the result of a successful functional testing of the FTS is used as reference responses during the FIFTS simulation. The array of responses received in the result of successful functional testing of FIFTS is used as reference responses during the FIFTMS simulation. Identification of deviations from normal operation during the FTS simulation. Registration of discrepancy between obtained and reference data. Classification of deviations found, depending on the duration (failure/refusal) Storing all the settings and conditions of the HSC operation (model parameters, fault patterns, etc.), corresponding to the obtained simulation results. Classifying fault patterns depending on level of their influence on the performance of the simulated FTS model. Analysis and evaluation of the simulation results and the identification of the most dangerous (and most likely lead to unacceptable consequences) impacts, HCP settings, critical elements of FTS model. Determination of fault-tolerance level of the FTS model upon the simulation results. Generation of the report on simulation results.

E EDA for TB

When implementing a multi - phase complex simulation methodology the TSAB CAD is used at the first phase of simulation to generate the netlist (source FTS netlist) of FTS elements described using the primitives of TSAB technological library. At the same time the synthesizing compiler of the TSAB CAD provides the ability to restore the hierarchy of FTS functional modules using the original netlists.

6 Conclusions

This article presents an extended concept of fault injection method, implying the joint use of three models: the external influences model (EIM), the threats occurrence model (TOM) and the fault localization model (FLM). The comprehensive multi-phase aerospace-born FTS simulation methodology, based on the extended concept of fault injection method is then proposed. Eventually, the concept of hardware-software complex (HSC) is developed to implement the aforementioned comprehensive multi phase FTS simulation methodology. The proposed concept of HSC provides a continuous verification of the FTS project at various design stages and the continuity of the verification results during the whole simulation process.

Acknowledgments. This research has been supported by grant from the Ministry of Education and Science of the Russian Federation. It was conducted under Federal special purpose program Research and development on priority directions of scientific and technological complex of Russia in 2014–2020 under contract RFMEFI57715X0161.

References

1. Brekhov, O., Kordover, K., Klimenko, A., Ratnikov, M.: FPGA prototyping with advanced fault injection methodology for tolerant computing systems simulation. Distrib. Comput. Commun. Netw. **601**, 208–223 (2016)
2. Velazco, R., Fouillat, P., Reis, R.: Radiation Effects on Embedded Systems. Springer, Dordrecht (2007)
3. Mukherjee, S.: Architecture Design for Soft Errors. Elsevier, Burlington (2008)
4. Khosrow, G.: Physical Design Essentials: An ASIC Design Implementation Perspective. Springer, Heidelberg (2007). Conexant Systems Inc., Newport Beach

Mathematical Modeling
and Computation

Numerical and Analytical Modeling of Guided Modes of a Planar Gradient Waveguide

Edik Ayrjan[1,4], Migran Gevorkyan[2], Dmitry Kulyabov[1,2],
Konstantin Lovetskiy[2], Nikolai Nikolaev[2], Anton Sevastianov[2],
Leonid Sevastianov[2,3(✉)], and Eugeny Laneev[2]

[1] Laboratory of Information Technologies, Joint Institute for Nuclear Research,
6 Joliot-Curie, Dubna, Moscow Region 141980, Russia
ayrjan@jinr.ru, dharma@sci.pfu.edu.ru
[2] RUDN University (Peoples' Friendship University of Russia),
6 Miklukho-Maklaya Str., Moscow 117198, Russia
mngevorkyan@sci.pfu.edu.ru, lovetskiy@gmail.com,
alsevastyanov@gmail.com, nnikolaev@sci.pfu.edu.ru,
leonid.sevast@gmail.com
[3] Bogoliubov Laboratory of Theoretical Physics,
Joint Institute for Nuclear Research, 6 Joliot-Curie,
Dubna, Moscow Region 141980, Russia
[4] Yerevan Physics Institute, 2 Alikhanian Brothers St., 375036 Yerevan, Armenia

Abstract. The mathematical model of light propagation in a planar gradient optical waveguide consists of the Maxwell's equations supplemented by the matter equations and boundary conditions. In the coordinates adapted to the waveguide geometry, the Maxwell's equations are separated into two independent sets for the TE and TM polarizations. For each there are three types of waveguide modes in a regular planar optical waveguide: guided modes, substrate radiation modes, and cover radiation modes. We implemented in our work the numerical-analytical calculation of typical representatives of all the classes of waveguide modes.

In this paper we consider the case of a linear profile of planar gradient waveguide, which allows for the most complete analytical description of the solution for the electromagnetic field of the waveguide modes. Namely, in each layer we are looking for a solution by expansion in the fundamental system of solutions of the reduced equations for the particular polarizations and subsequent matching them at the boundaries of the waveguide layer.

The problem on eigenvalues (discrete spectrum) and eigenvectors is solved in the way that first we numerically calculate (approximately, with double precision) eigenvalues, then numerically and analytically— eigenvectors. Our modelling method for the radiation modes consists in

E. Ayrjan—The work was partially supported by RFBF grants No. 14-01-00628, No. 15-07-08795, No. 16-07-00556. The reported study was funded within the Agreement No. 02.a03.21.0008 dated 24.04.2016 between the Ministry of Education and Science of the Russian Federation and RUDN University.

reducing the initial potential scattering problem (in the case of the continuous spectrum) to the equivalent ones for the Jost functions: the Jost
solution from the left for the substrate radiation modes and the Jost
solution from the right for the cover radiation modes.

Keywords: Waveguide propagation of electromagnetic radiation ·
Equations of waveguide modes of regular waveguide · Numerical-
analytical modelling

1   Introduction

Waveguide propagation of polarized light is widely used in engineering, optoelectronics and nanophotonics [4,15,26]. Most of the integrated optical waveguide
structures are formed on the basis of thin-film planar waveguides [2,9,14] and
contain all sorts of waveguide transitions [3,7,8,10,11,20–22] with gradient planar waveguides [1,4,15,16,25–27]. In this connection the analysis of the propagation of guided and radiation waveguide modes in gradient waveguides is of
particular interest. Some works [5,6,17] are devoted to finding approximate solutions of the electromagnetic field of the waveguide modes under the assumption
of a given analytical behavior of the transverse distribution of refractive index
in the waveguide layer. In other works [12,13,18,19], this study is carried out
initially by approximate numerical methods.

In this paper we consider the case of a linear profile of planar gradient
waveguide, which allows for the most complete analytical description of the solution for the electromagnetic field of the waveguide modes. Namely, in each layer
we are looking for a solution by expansion in the fundamental system of solutions of the reduced equations for the particular polarizations and subsequent
matching them at the boundaries of the waveguide layer.

Propagation of monochromatic polarized electromagnetic radiation is
described by the system of vector homogeneous Maxwell's equations [2,9,14]:

$$\text{rot } \boldsymbol{H} = \frac{1}{c}\frac{\partial \boldsymbol{D}}{\partial t}, \quad \text{rot } \boldsymbol{E} = -\frac{1}{c}\frac{\partial \boldsymbol{D}}{\partial t}. \tag{1}$$

When there is a waveguide propagation of the radiation, at the interfaces of
the waveguide layer with the substrate and the cover (see Fig. 1) the tangential
boundary conditions are satisfied [2,9,14]:

$$\boldsymbol{E}^{\tau}\Big|_{1} = \boldsymbol{E}^{\tau}\Big|_{2}, \quad \boldsymbol{H}^{\tau}\Big|_{1} = \boldsymbol{H}^{\tau}\Big|_{2}. \tag{2}$$

And asymptotic conditions "at infinity" (at an infinite distance from the
waveguide layer):

$$\lim_{x\to\pm\infty}\big|\boldsymbol{E}(x,y,z)\big| \le C_{E}, \quad \lim_{x\to\pm\infty}\big|\boldsymbol{H}(x,y,z)\big| \le C_{H}. \tag{3}$$

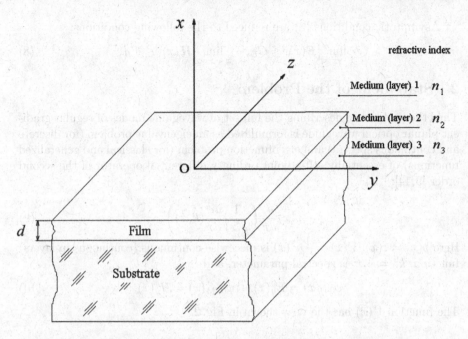

Fig. 1. Scheme of a flat three-layer dielectric waveguide. Waveguide is formed by media 1–3. The figure indications are: 1 is a framing medium or cover layer (air) with refractive index n_c; 2 is a waveguide layer (film) with a refractive index n_f; 3 is a substrate with refractive index n_s; d is the thickness of the waveguide layer. Film and substrate are homogeneous in the x and z directions, the substrate is usually much thicker than the film

In a Cartesian coordinate system associated with the geometry of the waveguide (see Fig. 1), the Maxwells equations, after the separation of variables, split into two linearly independent systems, which take the form [4,15]:

$$\frac{d^2 E_y}{dx^2} + k_0^2(\varepsilon\mu - \beta^2)E_y(x) = 0, \quad H_z = \frac{1}{ik_0\mu}\frac{dE_y}{dx}, \quad H_x = -\frac{\beta}{\mu}E_y, \quad (4)$$

$$\varepsilon\frac{d}{dx}\left(\frac{1}{\varepsilon}\frac{dH_y}{dx}\right) + k_0^2(\varepsilon\mu - \beta^2)H_y(x) = 0, \quad E_z = \frac{1}{ik_0\varepsilon}\frac{\partial H_y}{\partial x}, \quad E_x = \frac{\beta}{\varepsilon}H_y. \quad (5)$$

Here the invariance of the process in the direction Oy is taken into account: $\frac{\partial}{\partial y} = 0$.

The boundary conditions (2) are reduced to the following conditions: conditions for TE modes

$$E_y\Big|_1 = E_y\Big|_2, \quad H_z\Big|_1 = H_z\Big|_2, \quad (6)$$

and the boundary conditions for TM modes

$$H_y\Big|_1 = H_y\Big|_2, \quad E_z\Big|_1 = E_z\Big|_2. \quad (7)$$

Asymptotic conditions (3) are reduced to the following conditions:

$$\lim_{x \to \pm\infty} |E(x)| \le C_E, \quad \lim_{x \to \pm\infty} |H(x)| \le C_H. \tag{8}$$

2   Statement of the Problem

Thus the problem of describing the full set of waveguide modes of regular gradient planar optical waveguide is formulated as an eigenvalue problem (for discrete and continuous spectra) and eigenfunction problem (for classical and generalized functions) of essentially self-adjoint ordinary differential operator of the second order [9,14]:

$$- p(x) \frac{d}{dx} \left(\frac{1}{p(x)} \frac{d\psi}{dx}(k, x) \right). \tag{9}$$

Here $p(x) = \varepsilon(x)$, $V(x) = -n^2(x)$ is piecewise-continuous (continuous in layers) function, $k^2 = -\beta^2$ is spectral parameter, and

$$\psi_{\text{TE}}(x) = E_y(x), \quad \psi_{(\text{TM})}(x) = H_y(x). \tag{10}$$

The function $V(x)$ has the view shown in Fig. 2.

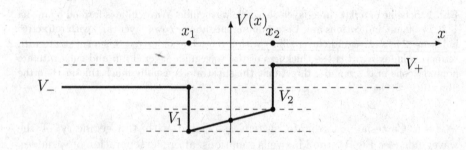

Fig. 2. The potential $V(x)$ graph

Lets introduce the auxiliary functions

$$\varphi_{\text{TE}}(x) = \frac{d\varphi_{\text{TE}}}{dx}(x), \quad \varphi_{\text{TM}}(x) = \frac{1}{p(x)} \frac{d\varphi_{\text{TM}}}{dx}(x). \tag{11}$$

Using these functions we can write down reduced boundary conditions at points of discontinuity of the potential, and therefore of the second derivative of the solution:

$$\psi\big|_{x_1-0} = \psi\big|_{x_1+0}, \quad \psi\big|_{x_2-0} = \psi\big|_{x_2+0}, \tag{12}$$

$$\varphi\big|_{x_1-0} = \varphi\big|_{x_1+0}, \quad \varphi\big|_{x_2-0} = \varphi\big|_{x_2+0}. \tag{13}$$

Besides, the asymptotic conditions are satisfied

$$|\psi(x)|_{x \to \pm\infty} \le C^\pm. \tag{14}$$

The spectrum of operator (9)–(14) consists of [23,24]:

- a finite number of discrete eigenvalues $k_j = i\kappa_j$: $k_j^2 \in \left(\min V(x), \min(V_-, V_+)\right)$ and the corresponding classical eigenfunctions (of guided waveguide modes);
- a single continuous spectrum k_-: $k_-^2 \in (V_-, \infty)$ and corresponding generalized eigenfunctions (substrate radiation modes);
- a single continuous spectrum k_+: $k_+^2 \in (V_+, \infty)$ and corresponding generalized eigenfunctions (cover radiation modes).

For a constructive description of the problem solutions, i.e. eigenfunctions of three types, we shall restrict our consideration to piecewise-linear potential:

$$V(x) = \begin{cases} V_-, & \text{when } x < x_1, \\ ax + b, & \text{when } x_1 < x < x_2, \quad \text{where } a = \frac{V_2 - V_1}{x_2 - x_1}, \ b = \frac{V_1 x_2 - V_2 x_1}{x_2 - x_1}, \\ V_+, & \text{when } x > x_2. \end{cases} \quad (15)$$

3   The Solution to the Problem on Eigenvalues (of the Discrete Spectrum) and Eigenfunctions (Classical)

The method of solution is the expansion on the sub-intervals of the general solution in terms of the fundamental system of solutions. To the left and to the right there are decreasing exponential functions in the case of real ε_s, ε_c (due to the asymptotic conditions):

$$\psi_s(k, x) = C_s \exp\{\gamma_s(x - x_1)\}, \quad (16)$$

$$\psi_c(k, x) = C_c \exp\{-\gamma_c(x - x_2)\}. \quad (17)$$

In the waveguide layer (with a linear potential in the subdomain) the fundamental system of solutions consists of the functions $Ai(x)$ and $Bi(x)$, such that

$$\psi_f(k, x) = C_1 Ai\left(\frac{a(x - x_2) + b}{(-a)^{2/3}}\right) + C_2 Bi\left(\frac{a(x - x_2) + b}{(-a)^{2/3}}\right). \quad (18)$$

These common solutions in the subdomains form a single particular solution of the problem (9)–(14), therefore, the equalities must be satisfied:

$$\psi_s(k, x_1) = \psi_f(k, x_1), \qquad \Phi_s(k, x_1) = \Phi_f(k, x_1), \quad (19)$$

$$\psi_f(k, x_2) = \psi_c(k, x_2), \qquad \Phi_f(k, x_2) = \Phi_c(k, x_2). \quad (20)$$

Thus we obtain a homogeneous system of linear algebraic equations for the indefinite coefficients of the expansion of common solutions in terms of the fundamental systems of solutions, which for the TE modes has the view:

$$C_s = C_1 Ai\left(\frac{-ad + b}{(-a)^{2/3}}\right) + C_2 Bi\left(\frac{-ad + b}{(-a)^{2/3}}\right), \quad (21)$$

$$\gamma_s C_s = -C_1(-a)^{1/3}\frac{dAi}{dx}\left(\frac{-ad + b}{(-a)^{2/3}}\right) - C_2(-a)^{1/3}\frac{dBi}{dx}\left(\frac{-ad + b}{(-a)^{2/3}}\right), \quad (22)$$

$$C_1 Ai(0) + C_2 Bi(0) = C_c, \tag{23}$$

$$- C_1(-a)^{1/3}\frac{dAi}{dx}(0) - C_2(-a)^{1/3}\frac{dBi}{dx}(0) = -\gamma_c C_c. \tag{24}$$

The resulting homogeneous system of linear algebraic equations

$$\hat{M}_{TE}(k)\,\boldsymbol{C}(k) = \boldsymbol{0} \tag{25}$$

has a non-trivial solution provided that

$$\det \hat{M}_{TE}(k) = 0. \tag{26}$$

Solutions of nonlinear transcendental algebraic equation k_j are substituted in SLAE(x) and then this system is solved with respect to $\boldsymbol{C}_j = \boldsymbol{C}(k_j)$. The obtained coefficients are substituted in the expressions for the fields. The results of calculations are presented in Figs. 3, 4, 5 and 6.

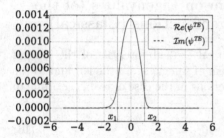

Fig. 3. Waveguide mode TE_0, $n_c = 1.0$, $n_f = 2.15$, $n_s = 1.515$, $\beta^{TE} = 1.6752$

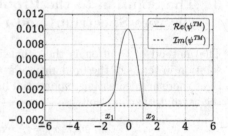

Fig. 4. Waveguide mode TM_0, $n_c = 1.0$, $n_f = 2.15$, $n_s = 1.515$, $\beta^{TE} = 1.5955$

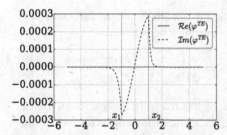

Fig. 5. Waveguide mode TE_0, $n_c = 1.0$, $n_f = 2.15$, $n_s = 1.515$, $\beta^{TE} = 1.6752$

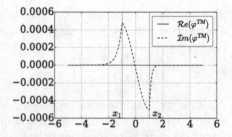

Fig. 6. Waveguide mode TM_0, $n_c = 1.0$, $n_f = 2.15$, $n_s = 1.515$, $\beta^{TE} = 1.5955$

4 Calculation of Cover Radiation Modes

Similarly to what was done in [15, 26] for piecewise-constant potentials, let's move from solutions of the problem (9)–(14) satisfying the asymptotic Jost conditions, to the solutions satisfying the "scattering problem" conditions. A one-to-one correspondence between them is set in [15, 26] for the potentials of a more general kind.

In particular, the asymptotics of the cover radiation modes $\psi_c(k, x)$ correspond to the problem of scattering of a plane Jost wave incident on the potential $V(x)$ from the right, from the region $x \sim +\infty$, which is reflected to the right with reflection coefficient $R_-(k)$, and is transmitted (through the potential $V(x)$) to the left with transmittance coefficient $T_-(k)$ in the form of a plane Jost wave propagating from right to left, in the region $x \sim -\infty$. All solutions $\psi_c(k, x)$ satisfy these asymptotics, when $k^2 \in (V_c, \infty)$. A sought solution, as in the case of guided modes, is constructed by matching at the boundaries of the general solutions of Eq. (9) in the regions of the argument $(-\infty, x_1)$, (x_1, x_2) and (x_2, ∞).

So, in the region $(-\infty, x_1)$ the general solutions of Eq. (9) with constant coefficient V_s are of the form (for TE modes):

$$\psi_c^{\text{TE}}(k, x) = T_-^{\text{TE}}(k) \exp\{-ip_s(x - x_1)\}. \tag{27}$$

In the region (x_2, ∞) the general solutions of Eq. (9) have the form

$$\psi_c^{\text{TE}}(k, x) = \exp\{-ip_c(x - x_2)\} + R_-^{\text{TE}}(k) \exp\{ip_c(x - x_2)\}. \tag{28}$$

In the region (a, b) the general solutions of Eq. (9) have the form (for TE and TM modes, respectively):

$$\psi_f(k, x) = C_f^1 Ai\left(\frac{a(x - x_2) + b}{(-a)^{2/3}}\right) + C_f^2 Bi\left(\frac{a(x - x_2) + b}{(-a)^{2/3}}\right). \tag{29}$$

Thus, the solutions (for TE modes) are given by sets of amplitude coefficients $(T_-^{\text{TE}}, C_f^1, C_f^2, R_-^{\text{TE}})^T$, satisfying the system of linear algebraic equations:

$$T_-^{\text{TE}}(k) = C_f^1 Ai\left(\frac{-ad + b}{(-a)^{2/3}}\right) + C_f^2 Bi\left(\frac{-ad + b}{(-a)^{2/3}}\right), \tag{30}$$

$$-\frac{p_s}{k_0 \mu_s} T_-^{\text{TE}}(k) = -C_f^1 (-a)^{1/3} \frac{dAi}{dx}\left(\frac{-ad + b}{(-a)^{2/3}}\right) - C_f^2 (-a)^{1/3} \frac{dAi}{dx}\left(\frac{-ad + b}{(-a)^{2/3}}\right), \tag{31}$$

$$C_f^1 Ai(0) + C_f^2 Bi(0) = 1 + R_-^{\text{TE}}(k), \tag{32}$$

$$-C_f^1 (-a)^{1/3} \frac{dAi}{dx}(0) - C_f^2 (-a)^{1/3} \frac{dBi}{dx}(0) = -\frac{p_c}{k_0 \mu_c}\left[1 - R_-^{\text{TE}}(k)\right]. \tag{33}$$

The resulting SLAE can be rewritten as:

$$\hat{M}^{\text{TE}}(k)(T_-^{\text{TE}}, C_f^1, C_f^2, R_-^{\text{TE}})^T = \left(0, 0, 1, -\frac{p_c}{k_0 \mu_c}\right)^T, \tag{34}$$

so that the solution exists for any $k^2 \in (V_c, \infty)$ and is unique up to a complex factor (Figs. 7 and 8).

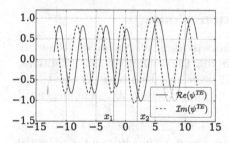

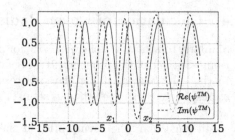

Fig. 7. Cover radiation TE mode $n_c = 1.0$, $n_f = 1.59$, $n_s = 1.515$, $k^2 = 0.250$

Fig. 8. Cover radiation TM mode $n_c = 1.0$, $n_f = 1.59$, $n_s = 1.515$, $k^2 = 0.250$

5   Calculation of Substrate Radiation Modes

The asymptotics of substrate radiation modes $\psi_s(k, x)$ correspond to the scattering problem of plane Jost wave, the potential $V(x)$ from the left, from the region $x \sim -\infty$, which is reflected to the left with reflection coefficient $R_+(k)$. At the same time, the Jost wave coming from the left, passing through the potential $V(x)$, propagates to the right as the plane Jost wave with the transmittance coefficient $T_+(k)$ when $k^2 \in (V_c, \infty)$, and as an evanescent wave decaying to the right with a weighting factor $A_c(k)$ when $k^2 \in (V_s, V_c)$.

Solutions have different view for different values of spectral parameter k from the two spectral subregions $k^2 \in (V_s, V_c)$ and $k^2 \in (V_c, \infty)$. But for both regions the solution, as in the case of guided modes, is constructed by matching at the boundaries of the general solutions of Eq. (9) in the regions of the argument $(-\infty, x_1)$, (x_1, x_2) and (x_2, ∞).

In the region $(-\infty, x_1)$ the general solutions of Eq. (9) with a spectral parameter $k^2 \in (V_s, V_c)$ have the form:

$$\psi_s^{\text{TE}}(k, x) = \exp\{i p_s(k)(x - x_1)\} + R_+^{\text{TE}}(k) \exp\{i p_s(k)(x - x_1)\}. \tag{35}$$

In the region (x_1, x_2) the general solutions of Eq. (9) with a spectral parameter $k^2 \in (V_s, V_c)$ have the form:

$$\psi_f(k, x) = C_f^1 Ai\left(\frac{a(x - x_2) + b}{(-a)^{2/3}}\right) + C_f^2 Bi\left(\frac{a(x - x_2) + b}{(-a)^{2/3}}\right). \tag{36}$$

In the region (x_2, ∞) the general solutions of Eq. (9) with a spectral parameter $k^2 \in (V_s, V_c)$ have the form (by virtue of the asymptotic decay at infinity):

$$\psi_s^{\text{TE}}(k, x) = A_c \exp\{-\gamma_c(x - x_2)\}. \tag{37}$$

Thus, the solutions (for TE modes) are given by sets of amplitude coefficients $(R_+^{\text{TE}}, C_f^1, C_f^2, A_c)^T$ satisfying the system of linear algebraic equations:

$$1 + R_+^{\text{TE}}(k) = C_f^1 Ai\left(\frac{-ad + b}{(-a)^{2/3}}\right) + C_f^2 Bi\left(\frac{-ad + b}{(-a)^{2/3}}\right), \tag{38}$$

$$\frac{p_s}{k_0\mu_s}\left[1-R_+^{\text{TE}}(k)\right] = -C_f^1(-a)^{1/3}\frac{dAi}{dx}\left(\frac{-ad+b}{(-a)^{2/3}}\right) - C_f^2(-a)^{1/3}\frac{dAi}{dx}\left(\frac{-ad+b}{(-a)^{2/3}}\right),$$
(39)

$$C_f^1 Ai(0) + C_f^2 Bi(0) = A_c,$$
(40)

$$-C_f^1(-a)^{1/3}\frac{dAi}{dx}(0) - C_f^2(-a)^{1/3}\frac{dBi}{dx}(0) = -\frac{\gamma_c}{ik_0\mu_c}A_c.$$
(41)

The resulting SLAE can be rewritten as:

$$\hat{M}^{\text{TE}}(k)(R_+^{\text{TE}}, C_f^1, C_f^2, A_c)^T = \left(-1, -\frac{p_c}{k_0\mu_s}, 0, 0\right)^T,$$
(42)

so that there exists a solution for any $k^2 \in (V_s, V_c)$ and it is unique up to a complex multiplier (Figs. 9 and 10).

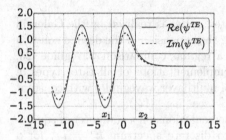

Fig. 9. Substrate radiation TE mode $n_c = 1.0$, $n_f = 1.59$, $n_s = 1.515$, $k^2 = -1.648 \in (V_s, V_c)$

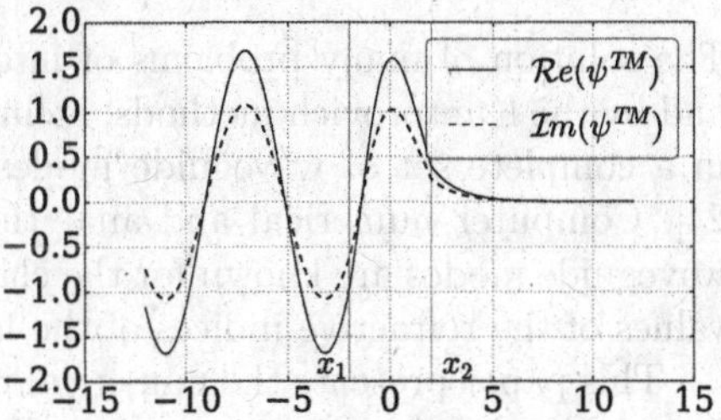

Fig. 10. Substrate radiation TM mode $n_c = 1.0$, $n_f = 1.59$, $n_s = 1.515$, $k^2 = -1.648 \in (V_s, V_c)$

For the spectral parameter k from the region $k^2 \in (V_c, \infty)$, in the coordinate regions $(-\infty, x_1)$ and (x_1, x_2) common solutions have the same form as in the case $k^2 \in (V_s, V_c)$, and in the region (x_2, ∞), they take the form:

$$\psi_s^{\text{TE}}(k, x) = T_+^{\text{TE}}(k)\exp\{ip_c(k)(x - x_2)\}.$$
(43)

Consequently, the second pair of boundary equations at the point $x = x_2$ for TE modes take the form:

$$C_f^1 Ai(0) + C_f^2 Bi(0) = T_+^{\text{TE}}(k),$$
(44)

$$-C_f^1(-a)^{1/3}\frac{dAi}{dx}(0) - C_f^2(-a)^{1/3}\frac{dBi}{dx}(0) = \frac{p_c(k)}{k_0\mu_c}T_+^{\text{TE}}(k).$$
(45)

The resulting SLAE can be rewritten as:

$$\hat{M}^{\text{TE}}(k)(R_+^{\text{TE}}, C_f^1, C_f^2, T_+^{\text{TE}})^T = \left(-1, -\frac{p_s}{k_0\mu_s}, 0, 0\right)^T,$$
(46)

so that there exists a solution for any $k^2 \in (V_c, \infty)$ and it is unique up to a complex multiplier (Figs. 11 and 12).

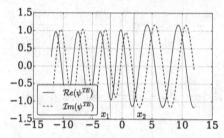
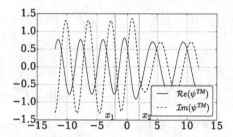

Fig. 11. Substrate radiation TE mode $n_c = 1.0$, $n_f = 1.59$, $n_s = 1.515$, $k^2 = 0.500 \in (V_c, \infty)$

Fig. 12. Substrate radiation TM mode $n_c = 1.0$, $n_f = 1.59$, $n_s = 1.515$, $k^2 = 0.500 \in (V_c, \infty)$

6   Conclusion

The solution of many problems of integrated optics is realized by the Galerkin and by the Kantorovich methods, including the expansion of the desired solution in a complete set of waveguide modes of a regular comparison waveguide [23, 24]. Computer numerical and analytical implementations of all three types of waveguide modes are known for the class of multilayer waveguides with constant values of the refractive indices of the layers (see., e.g., [14]).

This paper presents the numerical implementations on a computer of square-integrable eigenfunctions corresponding to discrete spectrum $k_j = i\kappa_j$ for a piecewise-linear potential $V(x)$ (for the gradient waveguide). The present study also shows the numerical implementations on a computer of the cover radiation modes and substrate radiation modes. For modeling these modes, the problems of scattering on the potential $V(x)$ of Jost functions equivalent to the original problem in the case of the continuous spectrum were used: the problems of scattering on the left for the substrate radiation modes and the problems of scattering on the right for the cover radiation modes.

References

1. Adams, M.J.: An Introduction to Optical Waveguides. Wiley, Chichester (1981)
2. Ayrjan, E.A., Egorov, A.A., Michuk, E.N., Sevastyanov, A.L., Sevastianov, L.A., Stavtsev, A.V.: Representations of guided modes of integrated-optical multilayer thin-film waveguides, p. 52. Dubna, preprint JINR E11-2011-31 (2011)
3. Ayryan, E.A., Egorov, A.A., Sevastyanov, L.A., Lovetskiy, K.P., Sevastyanov, A.L.: Mathematical modeling of irregular integrated optical waveguides. In: Adam, G., Buša, J., Hnatič, M. (eds.) MMCP 2011. LNCS, vol. 7125, pp. 136–147. Springer, Heidelberg (2012)
4. Barnoski, M.: Introduction to Integrated Optics. Plenunm, New York (1974)
5. Conwell, E.: Modes in optical waveguides formed by diffusion. Appl. Phys. Lett. **23**, 328–329 (1973)
6. Conwell, E.: WKB approximation for optical guide modes in a medium with exponentially varying index. J. Appl. Phys. **47**, 1407 (1975)

7. Divakov, D.V., Sevastianov, L.A.: Application of incomplete Galerkin method to irregular transition in open planar waveguides. Matematicheskoe Modelirovanie **27**(7), 44–50 (2015)
8. Egorov, A.A., Sevastyanov, A.L., Airyan, E.A., Lovetskiy, K.P., Sevastianov, L.A.: Adiabatic modes of smoothly irregular optical wavegide: zero-order vector theory. Matematicheskoe Modelirovanie **22**(8), 42–54 (2010)
9. Egorov, A.A., Lovetskii, K.P., Sevastianov, A.L., Sevastianov, L.A.: Integrated Optics: Theory and Computer Modelling. RUDN Publisher, Moscow (2015)
10. Egorov, A.A., Lovetskiy, K.P., Sevastianov, A.L., Sevastianov, L.A.: Simulation of guided modes (eigenmodes) and synthesis of a thin-film generalised waveguide luneburg lens in the zero-order vector approximation. Quantum Electron. **40**(9), 830–836 (2010)
11. Egorov, A.A., Sevastyanov, L.A.: Structure of modes of a smoothly irregular integrated-optical four-layer three-dimensional waveguide. Quantum Electron. **39**(6), 566–574 (2009)
12. Fitio, V.M., Romakh, V.V., Bobitski, Y.V.: Numerical method for analysis of waveguide modes in planar gradient waveguides. Mater. Sci. **20**(3), 256–261 (2014)
13. Fitio, V.M., Romakh, V.V., Bobitski, Y.V.: Search of mode wavelengths in planar waveguides by using Fourier transform of wave equation. Semicond. Phys. Quantum Electron. Optoelectron. **19**(1), 28–33 (2016)
14. Gevorkyan, M.N., Kulyabov, D.S., Lovetskiy, K.P., Sevastyanov, A.L., Sevastyanov, L.A.: Waveguide modes of a planar optical waveguide. Math. Modell. Geom. **3**(1), 43–63 (2015)
15. Hunsperger, R.G.: Integrated Optics: Theory and Technology. Springer, Heidelberg (1995)
16. Marcuse, D.: Light Transmission Optics. Van Nostrand Reinhold Company, New York (1972)
17. Nikolaev, N., Shevchenko, V.V.: Inverse method for the reconstruction of refractive index profile and power management in gradient index optical waveguides. Opt. Quantum Electron. **39**(10), 891–902 (2007)
18. Rganov, A.G., Grigas, S.E.: Defining the parameters of multilayer waveguide modes of dielectric waveguides. Numer. Methods Program. **10**, 258–262 (2009)
19. Rganov, A.G., Grigas, S.E.: Numerical algorithm for waveguide and leaky modes determination in multilayer optical waveguides. Tech. Phys. **55**(11), 1614–1618 (2010)
20. Sevastianov, L., Divakov, D., Nikolaev, N.: Modelling of an open transition of the "horn" type between open planar waveguides. In: EPJ Web of Conferences, vol. 108, p. 02020 (2016)
21. Sevastianov, L.A., Egorov, A.A.: The theoretical analysis of waveguide propagation of electromagnetic waves in dielectric smoothly-irregular integrated structures. Math. Modell. Geom. **105**(4), 576–584 (2008)
22. Sevastianov, L.A., Egorov, A.A., Sevastyanov, A.L.: Method of adiabatic modes in studying problems of smoothly irregular open waveguide structures. Phys. At. Nucl. **76**(2), 224–239 (2013)
23. Sevastyanov, L.A.: The complete system of modes of open planar waveguide. In: Proceedings of the VI International Scientific Conference Lasers in Science, Technology, and Medicine, pp. 72–76. Publishing House of IRE, Suzdal (1995)
24. Shevchenko, V.V.: On the spectral expansion in eigenfunctions and associated functions of a non self-adjoint problem of sturm-liouville type on the entire axis. Differ. Equ. **15**, 2004–2020 (1979)

25. Snyder, A.W., Love, J.D.: Optical Waveguide Theory. Chapman and Hall, New York (1983)
26. Tamir, T.: Integrated Optics. Springer-Verlag, Berlin (1979)
27. Unger, H.G.: Planar Optical Waveguides and Fibres. Clarendon Press, Oxford (1977)

Diagram Representation for the Stochastization of Single-Step Processes

Ekaterina G. Eferina[1], Michal Hnatich[3,4,5], Anna V. Korolkova[1],
Dmitry S. Kulyabov[1,2(✉)], Leonid A. Sevastianov[1,3], and Tatiana R. Velieva[1]

[1] Department of Applied Probability and Informatics,
RUDN University (Peoples' Friendship University of Russia),
6 Miklukho-Maklaya Street, Moscow 117198, Russia
eg.eferina@gmail.com, {akorolkova,dharma,sevast}@sci.pfu.edu.ru,
trvelieva@gmail.com
[2] Laboratory of Information Technologies, Joint Institute for Nuclear Research,
6 Joliot-Curie, Dubna, Moscow Region 141980, Russia
[3] Bogoliubov Laboratory of Theoretical Physics,
Joint Institute for Nuclear Research,
6 Joliot-Curie, Dubna, Moscow Region 141980, Russia
hnatic@saske.sk
[4] Department of Theoretical Physics, SAS, Institute of Experimental Physics,
Watsonova 47, 040 01 Košice, Slovakia
[5] Faculty of Science, Pavol Jozef Šafárik University in Košice (UPJŠ),
Šrobárova 2, 041 80 Košice, Slovakia

Abstract. Background. By the means of the method of stochastization of one-step processes we get the simplified mathematical model of the original stochastic system. We can explore these models by standard methods, as opposed to the original system. The process of stochastization depends on the type of the system under study. **Purpose.** We want to get a unified abstract formalism for stochastization of one-step processes. This formalism should be equivalent to the previously introduced. **Methods.** To unify the methods of construction of the master equation, we propose to use the diagram technique. **Results.** We get a diagram technique, which allows to unify getting master equation for the system under study. We demonstrate the equivalence of the occupation number representation and the state vectors representation by using a Verhulst model. **Conclusions.** We have suggested a convenient diagram formalism for unified construction of stochastic systems.

Keywords: Occupation numbers representation · Fock space · Dirac notation · One-step processes · Master equation · Diagram technique

1 Introduction

When modeling various physical and technical systems, we often can model them in the form of a one-step processes (see [1,3,4,24]). Then there is the problem

© Springer International Publishing AG 2016
V.M. Vishnevskiy et al. (Eds.): DCCN 2016, CCIS 678, pp. 483–497, 2016.
DOI: 10.1007/978-3-319-51917-3_42

of adequate representation and study of the resulting model. For the statistical systems in addition to representation of the state vectors (combinatorial approach) [3,4] the representation of the occupation numbers (operator approach) (see [12,13,17,19,20,23]) is also used. This representation is especially well suited for the system with a variable number of elements description.

However, technique for obtaining models for the combinatorial approach is quite different from the technique for the operator approach. In this paper, we want to propose a unified methodology for both approaches on the basis of the diagram technique.

The structure of the article is as follows. In the Sect. 2 basic notations and conventions are introduced. The ideology of the method of stochastization of one-step process and its components are described in the Sect. 3. Then the interaction schemes and master equation overview are presented in the next Sects. 5 and 4. The combinatorial method of modelling is discussed in the following Sect. 6. The operator model approach is presented in the Sect. 7. In fact diagram technique introduced in Sect. 8. Application of of this technique is described in Sect. 9 on the example of Verhulst model.

2   Notations and Conventions

1. The abstract indices notation (see [22]) is used in this work. Under this notation a tensor as a whole object is denoted just as an index (e.g., x^i), components are denoted by underlined index (e.g., $x^{\underline{i}}$).
2. We will adhere to the following agreements. Latin indices from the middle of the alphabet (i, j, k) will be applied to the space of the system state vectors. Latin indices from the beginning of the alphabet (a) will be related to the Wiener process space. Greek indices (α) will set a number of different interactions in kinetic equations.

3   General Review of the Methodology

Our methodology is completely formalized in such a way that it is sufficient when the original problem is formulated accordingly. It should be noted that the most of the models under our study can be formalized as a one-step process (see [10,18]). In fact, for this type of models we developed this methodology, but it may be expanded for other processes.

First we transform our model to the one-step process (see Fig. 1). Next, we need to formalize this process in the form of interaction schemes[1] (see [3,4,13]).

Each scheme has its own interaction semantics. Semantics leads directly to the master equation (see [10,18]). However, the master equation has usually rather complex structure that makes it difficult for direct study and solution. Our technique involves two possibilities (see Fig. 2):

---

[1] The analogs of the interaction schemes are the equations of chemical kinetics, reaction particles and etc.

– computational approach—the solution of the master equation with help of perturbation theory;
– modeling approach—the approximate models are obtained in the form of Fokker–Planck and Langevin equations.

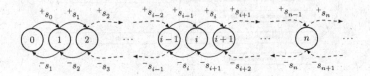

Fig. 1. One-step process

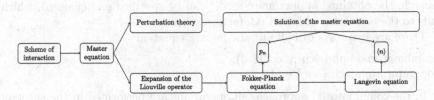

Fig. 2. The general structure of the methodology

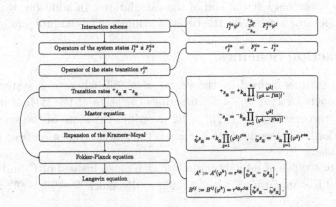

Fig. 3. Combinatorial modeling approach

The computational approach allows to obtain a concrete solution for the studied model. In our methodology, this approach is associated with perturbation theory (see [14–16]).

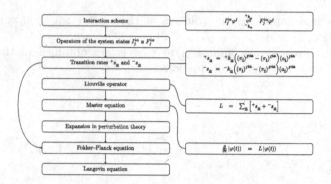

Fig. 4. Operator modeling approach

The model approach provides a model that is convenient to study numerically and qualitatively. In addition, this approach assumes the iterative process of research: the obtained approximate model can be specified and changed, which leads to the correction of initial interaction schemes.

There are two ways of building the master equation[2]

– combinatorial approach (see Fig. 3);
– operator approach (see Fig. 4).

In the combinatorial approach, all operations are performed in the space of states of the system, so we deal with a particular system throughout manipulations with the model.

For the operator approach we can abstract from the specific implementation of the system under study. We are working with abstract operators. We return to the state space only at the end of the calculations. In addition, we choose a particular operator algebra on the basis of symmetry of the problem.

4   Interaction Schemes

The system state is defined by the vector $\varphi^i \in \mathfrak{R}^n$, where n is system dimension[3]. The operator $I_j^i \in \mathfrak{N}_0^n \times \mathfrak{N}_0^n$ describes the state of the system before the interaction, the operator $F_j^i \in \mathfrak{N}_0^n \times \mathfrak{N}_0^n$ describes the state of the system after the interaction[4]. The result of interaction is the system transition from one state to another one.

There are s types of interaction in our system, so instead of I_j^i and F_j^i operators we will use operators $I_j^{i\alpha} \in \mathfrak{N}_0^n \times \mathfrak{N}_0^n \times \mathfrak{N}_+^s$ and $F_j^{i\alpha} \in \mathfrak{N}_0^n \times \mathfrak{N}_0^n \times \mathfrak{N}_+^s$[5].

---

[2] In quantum field theory the path integrals approach can be considered as an analogue of the combinatorial approach and the method of second quantization as analog of the operator approach.

[3] We denote the module over the field $\mathbb{R}$ as $\mathfrak{R}$. Accordingly, $\mathfrak{N}$, $\mathfrak{N}_0$, $\mathfrak{N}_+$ are modules over rings N, N_0 (cardinal numbers with 0), N_+ (cardinal numbers without 0).

[4] The component dimension indices take on values $\underline{i}, j = \overline{1, n}$.

[5] The component indices of number of interactions take on values $\underline{\alpha} = \overline{1, s}$.

The interaction of the system elements will be described by interaction schemes, which are similar to schemes of chemical kinetics [11,26]:

$$I_j^{i\alpha}\varphi^j \underset{-k_\alpha}{\overset{+k_\alpha}{\rightleftharpoons}} F_j^{i\alpha}\varphi^j, \qquad \alpha = \overline{1,s}, \tag{1}$$

the Greek indices specify the number of interactions and Latin are the system order. The coefficients $^+k_\alpha$ and $^-k_\alpha$ have meaning intensity (speed) of interaction.

We can also write (1) not in the form of vector equations but in the form of sums:

$$I_j^{i\alpha}\varphi^j\delta_i \underset{-k_\alpha}{\overset{+k_\alpha}{\rightleftharpoons}} F_j^{i\alpha}\varphi^j\delta_i, \tag{2}$$

where $\delta_i = (1,\ldots,1)$.

Also the following notation will be used:

$$I^{i\alpha} := I_j^{i\alpha}\delta^j, \quad F^{i\alpha} := F_j^{i\alpha}\delta^j, \quad r^{i\alpha} := r_j^{i\alpha}\delta^j.$$

The state transition is given by the operator:

$$r_j^{i\alpha} = F_j^{i\alpha} - I_j^{i\alpha}. \tag{3}$$

5   The Master Equation

For the system description we will use the master equation,[6] which describes the transition probabilities for Markov process (see [10,18]):

$$\frac{\partial p(\varphi_2,t_2|\varphi_1,t_1)}{\partial t} = \int \left[w(\varphi_2|\psi,t_2)p(\psi,t_2|\varphi_1,t_1) - w(\psi|\varphi_2,t_2)p(\varphi_2,t_2|\varphi_1,t_1) \right]\mathrm{d}\psi,$$

where $w(\varphi|\psi,t)$ is the probability of transition from the state ψ to the state φ for unit time.

By fixing the initial values of φ_1,t_1, we can write the equation for subensemble:

$$\frac{\partial p(\varphi,t)}{\partial t} = \int [w(\varphi|\psi,t)p(\psi,t) - w(\psi|\varphi,t)p(\varphi,t)]\mathrm{d}\psi. \tag{4}$$

For the discrete domain of φ, the (4) can be written as follows (the states are numbered by n and m):

$$\frac{\partial p_n(t)}{\partial t} = \sum_m [w_{nm}p_m(t) - w_{mn}p_n(t)], \tag{5}$$

where the p_n is the probability of the system to be in a state n at time t, w_{nm} is the probability of transition from the state m into the state n per unit time.

---

[6] Master equation can be considered as an implementation of the Kolmogorov equation. However, the master equation is more convenient and has an immediate physical interpretation (see [18]).

6   Combinatorial Approach

There are two types of system transition from one state to another (based on one–step processes) as a result of system elements interaction: in the forward direction $(\varphi^i + r^{i\underline{\alpha}})$ with the probability $^+s_{\underline{\alpha}}(\varphi^k)$ and in the opposite direction $(\varphi^i - r^{i\underline{\alpha}})$ with the probability $^-s_{\underline{\alpha}}(\varphi^k)$ (Fig. 1). The matrix of transition probabilities has the form:

$$w_{\underline{\alpha}}(\varphi^i|\psi^i, t) = {}^+s_{\underline{\alpha}}\delta_{\varphi^i,\psi^i+1} + {}^-s_{\underline{\alpha}}\delta_{\varphi^i,\psi^i-1}, \qquad \alpha = \overline{1, s},$$

where $\delta_{i,j}$ is Kronecker delta.

Thus, the general form of the master equation for the state vector φ^i, changing by steps with length $r^{i\underline{\alpha}}$, is:

$$\frac{\partial p(\varphi^i, t)}{\partial t} = \sum_{\underline{\alpha}=1}^{s} \left\{ {}^-s_{\underline{\alpha}}(\varphi^i + r^{i\underline{\alpha}}, t)p(\varphi^i + r^{i\underline{\alpha}}, t) \right.$$
$$\left. + {}^+s_{\underline{\alpha}}(\varphi^i - r^{i\underline{\alpha}}, t)p(\varphi^i - r^{i\underline{\alpha}}, t) - \left[{}^+s_{\underline{\alpha}}(\varphi^i) + {}^-s_{\underline{\alpha}}(\varphi^i) \right] p(\varphi^i, t) \right\}. \quad (6)$$

7   Operator Approach

7.1   Occupation Numbers Representation

Occupation number representation is the main language in the description of many-body physics. The main elements of the language are the wave functions of the system with information about how many particles are in each single-particle state. The creation and annihilation operators are used for system states change.

The method of application of the formalism of second quantization for the non-quantum systems (statistical, deterministic systems) was studied in a series of articles (see [6, 7, 12, 21]).

The Dirac notation is commonly used for occupation numbers representation recording.

7.2   Dirac Notation

This notation is proposed by Dirac[7] (see [5]). Under this notation, state of the system is described by an element of the projective Hilbert space $\mathcal{H}$. The vector $\varphi^i \in \mathcal{H}$ is defined as $|i\rangle$, and covariant vector (covector) $\varphi_i \in \mathcal{H}^* := \mathcal{H}_{\bullet}$ is defined as $\langle i|$. Conjunction operation is used for raising and lowering of indices[8]:

$$\varphi_i^* := \varphi_i = (\varphi^i)^\dagger \equiv \langle i| = |i\rangle^\dagger.$$

---

[7] The notation is based on the notation, proposed by G. Grassmann in 1862 (see [2, p. 134]).

[8] In this case, we use Hermitian conjugation $\bullet^\dagger$. The sign of the complex conjugate $\bullet^*$ in this case is superfluous.

The scalar product is as follows:

$$\varphi_i \varphi^i \equiv \langle i|i\rangle.$$

The tensor product is:

$$\varphi_j \varphi^i \equiv |i\rangle\langle j|.$$

7.3 Creation and Annihilation Operators

The transition to the space of occupation numbers is not a unitary transformation. However, the algorithm of transition (specific to each task) can be constructed.

Let's write the master equation (5) in the occupation number representation. We will consider a system that does not depend on the spatial variables. For simplicity, we consider the one-dimensional version.

Let's denote in (5) the probability that there are n particles in our system as φ_n:

$$\varphi_n := p_n(\varphi, t).$$

The vector space $\mathcal{H}$ consists of states of φ.

Depending on the structure of the model, we can introduce the probability-based or the moment-based inner products [12]. We introduce a scalar product, exclusive ($\langle |\rangle_{ex}$) and inclusive ($\langle |\rangle_{in}$). Let $|n\rangle$ are basis vectors.

$$\langle \varphi|\psi\rangle_{ex} = \sum_n n! p_n^*(\varphi) p^n(\psi); \tag{7}$$

$$\langle \varphi|\psi\rangle_{in} = \sum_n \frac{1}{k!} n_k^*(\varphi) n^k(\psi).$$

There n_k are factorial moments:

$$n_k(\varphi) = \langle n(n-1)\cdots(n-k+1)\rangle = \frac{\partial k}{\partial z^k} G(z,\varphi)|_{z=1},$$

G is generating function:

$$G(z,\varphi) = \sum_n z^n p_n(\varphi).$$

Let's use creation and annihilation operators:

$$\pi|n\rangle = |n+1\rangle,$$
$$a|n\rangle = n|n-1\rangle$$

and commutation rule[9]:

$$[a,\pi] = 1. \tag{8}$$

If the form of scalar product is (7) then from (8) follows that our system is described by Bose–Einstein statistics.

---

[9] In fact, $a\pi|n\rangle - \pi a|n\rangle = (n+1)|n\rangle - n|n\rangle = |n\rangle$.

7.4   The Liouville Operator

In the occupation numbers formalism the master equation becomes the Liouville equation:

$$\frac{\partial}{\partial t}|\varphi(t)\rangle = L|\varphi(t)\rangle.$$

The Liouville operator L satisfies the relation:

$$\langle 0|L = 0. \tag{9}$$

8   Diagram Representation

We describe our proposed diagram technique for the stochastization of one-step processes.

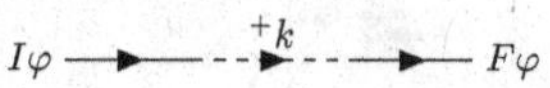

Fig. 5. Forward interaction 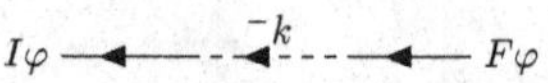   **Fig. 6.** Backward interaction

We will write the the scheme of interaction in the form of diagrams. Each scheme (1) or (2) corresponds to a pair of diagrams (see Figs. 5 and 6) for forward and backward interaction respectively.[10] The diagram consists of the following elements.

- Incoming lines (in the Fig. 5 are denoted by the solid line). These lines are directed to the line of interaction. These lines are marked with the number and type of interacting entities. You can write a single entity per a line or group them.
- Outgoing lines (in the Fig. 5 are denoted by the solid line). These lines are directed from the line of interaction. These lines are marked with the number and type of interacting entities. You can write a single entity per a line or group them.
- Line of interaction (in the Fig. 5 is denoted by the dotted line). The direction of time is denoted by the arrow. This line is marked by the coefficient of intensity of the interaction.

Each line is attributed to a certain factor (depending on the the approach chosen). The resulting expression is obtained by multiplying these factors.

---

[10] In order not to clutter the diagram, we have only one type of interacting entities left in these schemes.

8.1   Operator Approach

We obtain the Liouville operator when using interaction diagrams in the operator approach. Let us assign the corresponding factor for each line. The resulting term is obtained as the normal ordered product of factors.[11]

Fig. 7. Forward interaction (operator approach)

Fig. 8. Backward interaction (operator approach)

We use the following factors for each type of line (Fig. 7).

- Incoming line. This line corresponds to the disappearance of one entity from the system. Therefore, it corresponds to the annihilation operator a. It is clear that the line with combined capacity I corresponds to the operator a^I.
- Outgoing line. This line corresponds to the emergence of one entity in the system. Therefore, it corresponds to the creation operator π. It is clear that the line with combined capacity F corresponds to the operator π^F.
- Line of interaction. This line corresponds to the ratio of the interaction intensity.

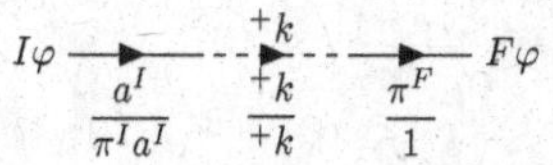

Fig. 9. Forward interaction (operator approach), extended notation

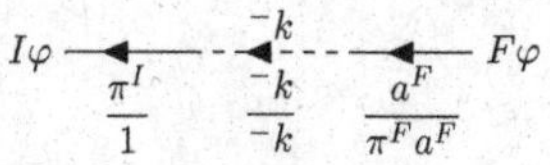

Fig. 10. Backward interaction (operator approach), extended notation

That is, for the Fig. 7 we obtain a factor $^{+}k\pi^F a^I$. However, this violates the Eq. (9). Redressing this, we have to subtract the number of entities that have entered into interaction, multiplied by the intensity of the interaction. Then we get a following term of the Liouville operator:

$$^{+}k\pi^F a^I - {}^{+}k\pi^I a^I = {}^{+}k\left(\pi^F - \pi^I\right)a^I. \tag{10}$$

To backward interactions (Fig. 8), we use the same rules.

---

[11] In normal ordering product all creation operators are moved so as to be always to the left of all the annihilation operators.

To account for the additional factor of (10) we will use the extended diagrams (Figs. 9 and 10). Here, from the the normal ordered product of the numerators the normal product of the denominators is subtracted.

Thus, the following Liouville operator corresponds to the scheme (1):

$$L = \sum_{\underline{\alpha},\underline{i}} \left[{}^+k_{\underline{\alpha}} \left((\pi_{\underline{i}})^{F^{i\alpha}} - (\pi_{\underline{i}})^{I^{i\alpha}} \right) (a_{\underline{i}})^{I^{i\alpha}} + {}^-k_{\underline{\alpha}} \left((\pi_{\underline{i}})^{I^{i\alpha}} - (\pi_{\underline{i}})^{F^{i\alpha}} \right) (a_{\underline{i}})^{F^{i\alpha}} \right].$$

(11)

8.2   Combinatorial Approach

For the combinatorial approach we get the master equation in the representation of the state vectors. In this approach, with the help of diagrams, we obtain the transition probability ${}^+s_{\underline{\alpha}}$ and ${}^-s_{\underline{\alpha}}$. They are, as in the case of operator approach, obtained by multiplying the diagrams factors.

However, the structure of the right-hand side of the Eq. (6) more complicated than Liouville operator. In the representation the state vectors the additive terms are presented in the functions arguments (dependency of the arguments from the operator r, see (3).). Therefore, we can not use only the factors multiplication.

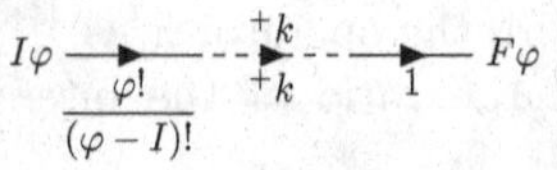

Fig. 11. Forward interaction (combinatorial approach)

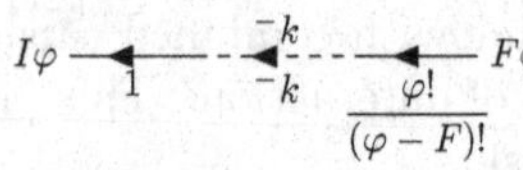

Fig. 12. Backward interaction (combinatorial approach)

We use the following factors for each type of line (Fig. 11).

- Incoming line. If all lines correspond to different state vectors, the factor of each line is the corresponding state vector. If there are several lines corresponding to the same state vector, the first line corresponds to the actual state vector (φ), the second line corresponds to the value of $\varphi - 1$ (as the first line has reduced the number of entities of this type in the system by one), and so further. That is, for a combined line factor can be written as follows:

$$\frac{\varphi!}{(\varphi - I)!}.$$

- Outgoing line do not give multiplicative contribution. It serves to obtain the step coefficient r:

$$r = F - I.$$

- Line of interaction. This line corresponds to the ratio of the interaction intensity.

In addition, we need the transition probabilities for the previous and the next steps:

$$\varphi^i + r^{i\underline{\alpha}},$$
$$\varphi^i - r^{i\underline{\alpha}}.$$

Thus, for the Fig. 11 transition probability will be as follows:

$$^+s(\varphi) = {}^+k \frac{\varphi!}{(\varphi - I)!},$$
$$^-s(\varphi) = {}^-k \frac{\varphi!}{(\varphi - F)!}.$$

For the general case:

$$^+s_{\underline{\alpha}}(\varphi^i) = {}^+k_{\underline{\alpha}} \prod_{i=1}^{n} \frac{\varphi^i!}{(\varphi^i - I^{i\underline{\alpha}})!},$$
$$^-s_{\underline{\alpha}}(\varphi^i) = {}^-k_{\underline{\alpha}} \prod_{i=1}^{n} \frac{\varphi^i!}{(\varphi^i - F^{i\underline{\alpha}})!}.$$

To backward interactions (Fig. 12), we use the same rules.

The general form of the master equation for the state vector φ^i we obtain on the basis of formula (6).

9   Verhulst Model

As a demonstration of the method, we consider the Verhulst model [8,9,25], which describes the limited growth[12]. Initially, this model was written down as the differential equation:

$$\frac{d\varphi}{dt} = \lambda\varphi - \beta\varphi - \gamma\varphi^2,$$

where λ denotes the breeding intensity factor, β—the extinction intensity factor, γ—the factor of population reduction rate (usually the rivalry of individuals is considered)[13].

The interaction scheme for the stochastic version of the model is:

$$\varphi \underset{\gamma}{\overset{\lambda}{\rightleftharpoons}} 2\varphi,$$
$$\varphi \overset{\beta}{\rightarrow} 0. \tag{12}$$

The interaction schemes (12) match Figs. 13, 14, and 15.

The first relation means that an individual who eats one unit of meal is immediately reproduced, and in the opposite direction is the rivalry between individuals. The second relation describes the death of an individual.

---

[12] The attractiveness of this model is that it is one-dimensional and non-linear.

[13] The same notation as in the original model [25] is used.

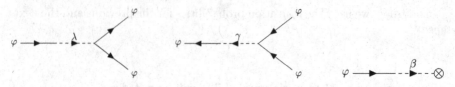

Fig. 13. First forward interaction

Fig. 14. First backward interaction

Fig. 15. Second forward interaction

9.1 Combinatorial Approach

The interaction schemes (12) in combinatorial approach match Figs. 16, 17, and 18.

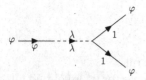

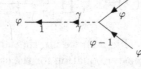

Fig. 16. First forward interaction (combinatorial approach)

Fig. 17. First backward interaction (combinatorial approach)

Fig. 18. Second forward interaction (combinatorial approach)

Let's define transition rates within the Verhults model as follows:

$$^+s_1(\varphi) = \lambda\varphi, \qquad ^+s_1(\varphi - 1) = \lambda(\varphi - 1), \qquad ^+s_1(\varphi + 1) = \lambda(\varphi + 1),$$
$$^-s_1(\varphi) = \gamma\varphi(\varphi - 1), \quad ^-s_1(\varphi - 1) = \gamma(\varphi - 1)(\varphi - 2), \quad ^-s_1(\varphi + 1) = \gamma(\varphi + 1)\varphi,$$
$$^+s_2(\varphi) = \beta\varphi. \qquad ^+s_2(\varphi - 1) = \beta(\varphi - 1). \qquad ^+s_2(\varphi + 1) = \beta(\varphi + 1).$$

$$r^1 = 1, \qquad r^2 = -1.$$

Then, based on (6), the form of the master equation is:

$$\frac{\partial p(\varphi, t)}{\partial t} = - \left[\lambda\varphi + \beta\varphi + \gamma\varphi(\varphi - 1) \right] p(\varphi, t)$$
$$+ \left[\beta(\varphi + 1) + \gamma(\varphi + 1)\varphi \right] p(\varphi + 1, t) + \lambda(\varphi - 1)p(\varphi - 1, t).$$

For particular values of φ (as in (5)):

$$\frac{\partial p_n(t)}{\partial t} := \frac{\partial p(\varphi, t)}{\partial t} \bigg|_{\varphi = n} = - \left[\lambda n + \beta n + \gamma n(n - 1) \right] p_n(t)$$
$$+ \left[\beta(n + 1) + \gamma(n + 1)n \right] p_{n+1}(t) + \lambda(n - 1)p_{n-1}(t). \quad (13)$$

9.2 Operator Approach

The interaction schemes (12) in operator approach match Figs. 19, 20, and 21.
From (12) and (11) the Liouville operator is:

$$
L = \lambda(\pi^2 - \pi)a + \gamma(\pi - \pi^2)a^2 + \beta(1 - \pi)a
$$
$$
= \lambda\left((a^\dagger)^2 - a^\dagger\right)a + \gamma\left(a^\dagger - (a^\dagger)^2\right)a^2 + \beta\left(1 - a^\dagger\right)a
$$
$$
= \lambda\left(a^\dagger - 1\right)a^\dagger a + \beta\left(1 - a^\dagger\right)a + \gamma\left(1 - a^\dagger\right)a^\dagger a^2.
$$

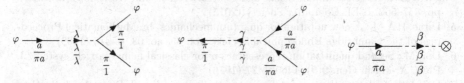

Fig. 19. First forward interaction (operator approach)

Fig. 20. First backward interaction (operator approach)

Fig. 21. Second forward interaction (operator approach)

The master equation by Liouville operator:

$$
\frac{\partial p_n(t)}{\partial t} = \frac{1}{n!}\langle n|L|\varphi\rangle
$$
$$
= \frac{1}{n!}\langle n| - [\lambda a^\dagger a + \beta a^\dagger a + \gamma a^\dagger a^\dagger aa] + [\beta a + \gamma a^\dagger aa] + \lambda a^\dagger a^\dagger a|\varphi\rangle
$$
$$
= -[\lambda n + \beta n + \gamma n(n-1)]\langle n|\varphi\rangle
$$
$$
+ [\beta(n+1) + \gamma(n+1)n]\langle n+1|\varphi\rangle + \lambda(n-1)\langle n-1|\varphi\rangle
$$
$$
= -[\lambda n + \beta n + \gamma n(n-1)]p_n(t) +
$$
$$
+ [\beta(n+1) + \gamma(n+1)n]p_{n+1}(t) + \lambda(n-1)p_{n-1}(t). \quad (14)
$$

The result (14) coincides with the formula (13), which was obtained by combinatorial method.

10 Conclusions

The authors proposed a diagram technique for the stochastization of one-step processes. At the moment, this technique allows to get main master equation. Also, this technique makes it possible to unify different approaches to the stochastization of one-step processes.

Acknowledgments. The work is partially supported by RFBR grants No's. 14-01-00628, 15-07-08795, and 16-07-00556. Also the publication was supported by the Ministry of Education and Science of the Russian Federation (the Agreement No. 02.a03.21.0008).

References

1. Basharin, G.P., Samouylov, K.E., Yarkina, N.V., Gudkova, I.A.: A new stage in mathematical teletraffic theory. Autom. Remote Control **70**(12), 1954–1964 (2009)
2. Cajori, F.: A History of Mathematical Notations, vol. 2 (1929)
3. Demidova, A.V., Korolkova, A.V., Kulyabov, D.S., Sevastianov, L.A.: The method of stochastization of one-step processes. In: Mathematical Modeling and Computational Physics, p. 67, JINR, Dubna (2013)
4. Demidova, A.V., Korolkova, A.V., Kulyabov, D.S., Sevastyanov, L.A.: The method of constructing models of peer to peer protocols. In: 6th International Congress on Ultra Modern Telecommunications and Control Systems and Workshops (ICUMT), pp. 557–562. IEEE Computer Society (2015)
5. Dirac, P.A.M.: A new notation for quantum mechanics. In: Mathematical Proceedings of the Cambridge Philosophical Society, vol. 35, no. 03, p. 416 (1939)
6. Doi, M.: Second quantization representation for classical many-particle system. J. Phys. A: Math. Gen. **9**(9), 1465–1477 (1976)
7. Doi, M.: Stochastic theory of diffusion-controlled reaction. J. Phys. A: Math. Gen. **9**(9), 1479–1495 (1976)
8. Feller, W.: Die Grundlagen der Volterraschen Theorie des Kampfes ums Dasein in wahrscheinlichkeitstheoretischer Behandlung. Acta Biotheor. **5**(1), 11–40 (1939)
9. Feller, W.: On the theory of stochastic processes, with particular reference to applications. In: Proceedings of the [First] Berkeley Symposium on Mathematical Statistics and Probability, pp. 403–432 (1949)
10. Gardiner, C.W.: Handbook of Stochastic Methods: for Physics, Chemistry and the Natural Sciences. Springer Series in Synergetics, Springer, Heidelberg (1985)
11. Gorban, A.N., Yablonsky, G.S.: Three waves of chemical dynamics. Math. Model. Nat. Phenom. **10**(5), 1–5 (2015)
12. Grassberger, P., Scheunert, M.: Fock-space methods for identical classical objects. Fortschr. Phys. **28**(10), 547–578 (1980)
13. Hnatič, M., Eferina, E.G., Korolkova, A.V., Kulyabov, D.S., Sevastyanov, L.A.: Operator approach to the master equation for the one-step process. In: EPJ Web of Conferences, vol. 108, p. 02027 (2016)
14. Hnatič, M., Honkonen, J., Lučivjanský, T.: Field-theoretic technique for irreversible reaction processes. Phys. Part. Nuclei **44**(2), 316–348 (2013)
15. Hnatich, M., Honkonen, J.: Velocity-fluctuation-induced anomalous kinetics of the $A + A \rightarrow$ reaction. Phys. Rev. E, 3904–3911 (2000). Statistical physics, plasmas, fluids, and related interdisciplinary topics 61(4 Pt A)
16. Hnatich, M., Honkonen, J., Lučivjanský, T.: Field theory approach in kinetic reaction: role of random sources and sinks. Theoret. Math. Phys. **169**(1), 1489–1498 (2011)
17. Janssen, H.K., Täuber, U.C.: The field theory approach to percolation processes. Ann. Phys. **315**(1), 147–192 (2005)
18. van Kampen, N.G.: Stochastic Processes in Physics and Chemistry. Elsevier Science, North-Holland Personal Library, Amsterdam (2011)
19. Korolkova, A.V., Eferina, E.G., Laneev, E.B., Gudkova, I.A., Sevastianov, L.A., Kulyabov, D.S.: Stochastization of one-step processes in the occupations number representation. In: Proceedings - 30th European Conference on Modelling and Simulation, ECMS 2016, pp. 698–704 (2016)
20. Mobilia, M., Georgiev, I.T., Täuber, U.C.: Fluctuations and correlations in lattice models for predator-prey interaction. Phys. Rev. E **73**(4), 040903 (2006)

21. Peliti, L.: Path integral approach to birth-death processes on a lattice. J. de Phys. **46**(9), 1469–1483 (1985)
22. Penrose, R., Rindler, W.: Spinors and Space-Time: Volume 1, Two-Spinor Calculus and Relativistic Fields, vol. 1. Cambridge University Press, Cambridge (1987)
23. Täuber, U.C.: Field-theory approaches to nonequilibrium dynamics. In: Ageing and the Glass Transition, vol. 716, pp. 295–348. Springer, Heidelberg (2005)
24. Velieva, T.R., Korolkova, A.V., Kulyabov, D.S.: Designing installations for verification of the model of active queue management discipline RED in the GNS3. In: 6th International Congress on Ultra Modern Telecommunications and Control Systems and Workshops (ICUMT), pp. 570–577. IEEE Computer Society (2015)
25. Verhulst, P.F.: Notice sur la loi que la population suit dans son accroissement, vol. 10 (1838)
26. Waage, P., Gulberg, C.M.: Studies concerning affinity. J. Chem. Educ. **63**(12), 1044 (1986)

Construction and Analysis of Nondeterministic Models of Population Dynamics

A.V. Demidova[1]([✉]), Olga Druzhinina[2], Milojica Jacimovic[3], and Olga Masina[4]

[1] Department of Applied Probability and Informatics, RUDN University
(Peoples' Friendship University of Russia), Miklukho-Maklaya Street 6,
Moscow, Russia 117198
avdemidova@sci.pfu.edu.ru
[2] Federal Research Center "Computer Science and Control" of Russian Academy of
Sciences, Vavilov Street 44, Building 2,
Moscow, Russia 119333
ovdruzh@mail.ru
[3] Department of Mathematics, University of Montenegro,
Džordž Washington Street, Podgorica, Montenegro 81000
milojica@jacimovic.me
[4] Department of Mathematical Modeling and Computer Technologies,
Bunin Yelets State University, Communards Street 28,
Yelets, Russia 399770
olga121@inbox.ru

Abstract. Three-dimensional mathematical models of population dynamics are considered in the paper. Qualitative analysis is performed for the model which takes into account the competition and diffusion of species and for the model which takes into account mutual interaction between the species. Nondeterministic models are constructed by means of transition from ordinary differential equations to differential inclusions, fuzzy and stochastic differential equations. Using the principle of reduction, which allows us to study stability properties of one type of equations, using stability properties of other types of equations, as a basis, sufficient conditions of stability are obtained. The synthesis of the corresponding stochastic models on the basis of application of the method of construction of stochastic self-consistent models is performed. The structure of these stochastic models is described and computer modelling is carried out. The obtained results are aimed at the development of methods of analysis of nondeterministic nonlinear models.

Keywords: Stochastic model · Single-step processes · Population dynamics · Differential equations · Stability · Principle of a reduction · Computer modelling

1 Introduction

In the study of mathematical biology models one of the most pressing problems is the problem of stability of population dynamics models [1–8]. An effective

© Springer International Publishing AG 2016
V.M. Vishnevskiy et al. (Eds.): DCCN 2016, CCIS 678, pp. 498–510, 2016.
DOI: 10.1007/978-3-319-51917-3_43

method for the analysis of stability is the method of Lyapunov functions [6–10]. A systematic approach to stability research, which considers the stability properties of the models defined by the differential equations of various types from the uniform point of view with application of Lyapunov functions is described in [6,9,10]. The specified approach is based on the transition from deterministic to stochastic models and on the principle of reduction of the problem of stability of solutions of differential inclusions to the problem of stability of other types of equations. This approach allows performing a comparative analysis of the qualitative properties of mathematical models in the transition from deterministic description to non-deterministic and justifying the construction and using of the models of one type or another.

Two models of dynamics of populations are studied in this paper: the model which takes into account the competition and diffusion of species and the model considering the competition and a mutualism of populations. The deterministic description of each of the models is given by a system of three ordinary nonlinear differential equations. Qualitative research of the specified models is conducted. The analysis of stability is made on the basis of the principle of reduction.

The two-dimensional models which take into account the symbiosis were researched in [4,11–13]. The model of the population dynamics of prey and two predators, in the presence of interactions prey-mutualist and predator-mutualist, is considered in [14]. The models of interactions "predator–prey–mutualist" are studied in [15]. The stability of the four-dimensional model of interaction of two competing species with two symbionts in the deterministic and stochastic cases are investigated in [16–18].

We use the method of constructing a self-consistent stochastic model developed in [19–21], which is based on the idea of combinatorial methodology described in [22,23]. The synthesis of stochastic models which take into account the competition and diffusion of species, as well as models incorporating competition and mutualism of populations is carried out. We describe the structure of stochastic models by means of schemes of interaction of elements and the operator of change the state of the system. With the aid of Fokker–Planck equation we formulate the transition rule to a stochastic differential equation in the form of is carried out. We use the method of Lyapunov functions and the theory of stochastic calculation for comparative analysis of deterministic and stochastic models.

2   Qualitative Analysis of Deterministic Models

We consider a model described by a system of three ordinary differential equations of the form

$$
\begin{aligned}
x_1' &= x_1(1 - x_1 - qy_1) + \beta x_2 - \gamma x_1, \\
x_2' &= x_2(1 - x_2) + \gamma x_1 - \beta x_2, \\
y_1' &= y_1(1 - rx_1 - y_1),
\end{aligned}
\tag{1}
$$

where x_1 and y_1 are densities of populations of the competing species in the first area of the species x and y, x_2 is density of population of the species x in the second area, $q > 0$ and $r > 0$ are coefficients of the competition in the first area, β and γ are coefficients of the diffusion of the species between the two areas, while the second area is a refuge and $\beta \neq \gamma$. Model (1) is a generalization of the model considered in [8] in the case when diffusion velocities don't coincide.

Besides, we consider a model described by a system of three ordinary differential equations of the form

$$x' = \delta x(1 - \frac{x}{K}) - \frac{\eta xy}{1 + mu},$$
$$y' = y(-s + \frac{c\eta x}{1 + mu}),$$
$$u' = \tau u(1 - \frac{u}{L_0 + lx}),$$
(2)

where x, y, u represent the prey population, the predator population and the mutualist population, respectively, τ, L_0, δ, K, η, s, c are positive parameters, l and m are mutualistic constants. Model (2) as a special case of the model considered in [15] is characterized by logistic type growth of prey populations in the absence of predator and by logistic growth of a population of mutualist. In the absence of the mutualism model (2) represents a classical Lotka–Volterra model.

Four equilibrium states of the model (1) are found: $O(0,0,0)$, $A_1(0,0,1)$, $A_2(\bar{x}_1, \bar{x}_2, 0)$ and $A_3(\hat{x}_1, \hat{x}_2, \hat{y}_1)$, as a result of solving of the corresponding algebraic equations. Coordinates $\bar{x}_1$, $\bar{x}_2$, $\hat{x}_1$, $\hat{x}_2$, $\hat{y}_1$ are found by means of Mathematica computing system.

The conditions of existence of nonnegative equilibrium state A_2 nd positive equilibrium state A_3 are obtained. In particular, it was discovered that if one of the conditions: (C_1) $0 < \beta < 1$, $\gamma > 1 - \beta$, (C_2) $\beta > 1$, $\gamma > 0$ holds, then model (1) has the nonnegative equilibrium state A_2. The conditions of existence of positive equilibrium state A_3 are formulated analogically. Estimations of the model parameters carried out and local phase portraits are constructed. Conditions of asymptotic stability of equilibrium state A_3 are obtained on the basis of the method of Lyapunov functions.

The following states of equilibrium are found for the model (2):

$$E_0(0,0,0), E_1(0, K, 0), E_2\left(0, \frac{s}{c\eta}, \frac{\delta}{\eta}\left(1 - \frac{s}{c\eta K}\right)\right),$$
$$E_3(L_0, 0, 0), E_4(L_0 + lK, K, 0), E_5\left(L_0 + \frac{ls\lambda}{c\eta}, \frac{s\lambda}{c\eta}, \frac{\delta\lambda}{\eta}\left(1 - \frac{s\lambda}{c\eta K}\right)\right),$$

where $\lambda = c\eta(1 + mL_0)/(c\eta - lms)$. Let us note that all equilibrium states $E_0 - E_5$ exist under condition

$$c\eta K > s(1 + m(L_0 + lK)).$$
(3)

If condition (3) and condition $s \geq \tau$ are satisfied, then there is a unique positive equilibrium of the model (2) and this equilibrium is asymptotically stable [15].

3   Construction and Stability Analysis of the Models of Population Dynamics on the Basis of the Principle of Reduction

From properties of differential inclusion it is possible a transition to the properties of fuzzy differential equations and stochastic differential equations. This transition is based on the principle of reduction of the stability problem for differential inclusions to the stability problem for fuzzy differential equations. The fuzzy equation for each α-level, where $\alpha \in (0, 1]$, is given by the corresponding differential inclusion. The set of all motions of the inclusion generates a multivalued mapping which takes into account the α-level of fuzzy function while this function is the solution of the corresponding fuzzy differential equation.

Model (1) is presented in the form of the vector equation

$$dx/dt = f(x), \tag{4}$$

where $x = (x_1, x_2, y_1)$, $f(x) = (f_1, f_2, f_3) = (x_1(1 - x_1 - qy_1) + \beta x_2 - \gamma x_1,$ $x_2(1-x_2)+\gamma x_1 - \beta x_2, y_1(1-rx_1-y_1))$, $x \in R_+^3 = R_+ \times R_+ \times R_+$, $R_+ = [0, \infty)$, $f : R_+^3 \to R_+^3$.

The differential inclusion which corresponds to deterministic equation (4) takes the form

$$dx/dt \in F(x), \tag{5}$$

where $F(x) = \{f(x) | \beta \in B, \gamma \in C, q \in Q, r \in R\}$, $B ::= [\beta_1, \beta_2]$, $C ::= [\gamma_1, \gamma_2]$, $Q ::= [q_1, q_2]$, $R ::= [r_1, r_2]$, $F : R_+^3 \to 2^{R_+^3}$.

Let Φ be a set of all motions of the inclusion (5). By

$$B(M, r) = \{x \in R_+^3 | e(x, M) \le r\}$$

denote the r-neighborhood of the set M.

Let us formulate the definitions of stability and the definition of Lyapunov function for the differential inclusion (5).

The closed set $M \subset R_+^3$ regarding the inclusion (5) is called:

(1) *stable in small*, if

$$\forall t_0 \ge s, \ \forall \varepsilon > 0 \ \exists \delta ::= \delta(\varepsilon) > 0, \ \forall \psi \in \Phi$$
$$e(\psi(t_0), M) < \delta \Rightarrow \text{Dom}\psi \supset [t_0, \infty), \ e(\psi(t), M) < \varepsilon \ \forall t \in [t_0, \infty] \cap \text{Dom}\psi;$$

(2) *attracting in small*, if

$$\forall t_0 \ge s, \ \exists h > 0, \ \forall \eta > 0 \ \exists T ::= T(h, \eta) > s \ \forall \psi \in \Phi \ e(\psi(t_0), M) \le h \Rightarrow$$
$$\Rightarrow \text{Dom}\psi \supset [t_0, \infty), \ e(\psi(t), M) < \eta \ \text{at} \ t \in [t_0 + T, \infty] \cap \text{Dom}\psi;$$

(3) *asymptotically stable in small* if it is stable in small and attracting in small.

Let us formulate the notion of Lyapunov function regarding the differential inclusion (5).

Continuous function $V : B(M, r) \to R$ is called Lyapunov function for the closed set $M \subset R_+^3$ regarding the inclusion (5), if there is number $r > 0$ and the non-negative non-decreasing functions $w_1, w_2 : (0, r] \to R$ such that $w_1(\rho) > 0$ at $\rho \in (0, r]$, $w_2(0) = 0$ and $w_1(e(x, M)) \leq V(x) \leq w_2(e(x, M)) \ \forall x \in B(M, r)$.

The derivative of Lyapunov function V along the motions of inclusion (5) is called the multivalued function $DV : B(M, r) \to 2^R$ defined by equality

$$DV(x) ::= \lim_{\Delta t \to 0} \{[V(\varphi(t + \Delta t)) - V(x)]/\Delta t : \varphi \in \Phi, \ \varphi(t) = x\}.$$

Functions D_+V and D_-V, for which $D_+V(x) :: = \sup DV(x)$ and $D_-V ::= \inf DV(x)$, are called the upper and lower derivative of Lyapunov function,
respectively.

The principle of reduction of the stability problem of the differential inclusion to the stability problem of fuzzy differential equation is considered in [6,9,10]. The following stability conditions of differential inclusion are obtained by means of this principle and by means of transition from model (1) to models (4) and (5):

(1) if there is a Lyapunov function V for the closed set $M \subset R_+^3$ regarding the inclusion (5), such that the inequality $D_+V(x) \leq 0 \ \forall x \in B(M, r)$ is satisfied, then the set M is stable in small regarding this inclusion. If the inequality $D_+V(x) \leq -w_3(e(x, M)) \ \forall x \in B(M, r)$ is satisfied, where $w_3 : B(M, r) \to R$ is the continuous and positive function in $R_+^3 \ M$, then the set M is asymptotically stable in small regarding the inclusion (5);
(2) if there is a Lyapunov function V for the closed set $M \subset R_+^3$ regarding the inclusion (5), such that the inequality $D_-V(x) \leq 0 \ \forall x \in B(M, r)$ is satisfied, then the set M is stable in small regarding this inclusion. If there is $h > 0$ and positive continuous function $w_3 : (0, r) \to R$ such that $D_-V(x) < -w_3(e(x, M)) \ \forall x \in B(M, h)$, then the set M is asymptotically stable in small regarding the inclusion (5).

By means of subsets of α-level $B_\alpha = \{\beta | \mu_B(\beta) \geq \alpha\}$, $C_\alpha = \{\gamma | \mu(\gamma) \geq \alpha\}$, $Q_\alpha = \{q | \mu_Q(q) \geq \alpha\}$ and $R_\alpha = \{r | \mu_R(r) \geq \alpha\}$, where $\alpha \in (0, 1]$, we make the transition from Eq. (4) to fuzzy differential equation

$$dX/dt = F(x), \tag{6}$$

where $F : z_+^3 \to P(R_+^3)$, $P(R_+^3)$ is the set of all fuzzy subsets of R_+^3. In terms of subsets of α-level the corresponding to Eq. (6) the differential inclusion takes the form $d\varphi/dt \in F_\alpha(\varphi)$, where $\alpha \in (0, 1]$, $F_\alpha(\varphi) = \{f(\varphi(t)) | \beta \in B_\alpha, \gamma \in C_\alpha, q \in Q_\alpha, r \in R_\alpha\}$.

Definitions of α-stability, α-attraction and asymptotic α-stability for the model of the form (6) are given. The theorems of stability on the basis of Lyapunov functions are proven.

Let us formulate the definition of Lyapunov function regarding the fuzzy differential equation (6).

Continuous function $V : B(M, r) \to R$ is called Lyapunov function for the closed set $M \subset (R^3_+)$ regarding the Eq. (6), if for each $\alpha \in (0, 1]$ there is number $r ::= r(\alpha) > 0$ and the non-negative non-decreasing functions $w_{1\alpha}, w_{2\alpha} : (0, r] \to R$ such that

$$w_{1\alpha}(\rho) > 0 \text{ at } \rho \in (0, r], \; w_{2\alpha}(0) = 0,$$
$$w_{1\alpha}(e(x, M_\alpha)) \leq V(x) \leq w_{2\alpha}(e(x, M_\alpha)) \; \forall x \in B(M_\alpha, r).$$

The derivative of Lyapunov function V along the motions of the Eq. (6) is called multivalued function $DV(x) : B(M, r) \to P(R)$, for which α-levels values are defined by the equality

$$DV_\alpha ::= \left\{ \lim_{\Delta t \to 0} [V(\varphi(t + \Delta t)) - V(x)] / \Delta t : \; \varphi \in \Phi_\alpha, \; \varphi(t) = x \right\}.$$

Functions $D_+V_\alpha(t, x) ::= \sup DV_\alpha(t, x)$ and $D_-V_\alpha ::= \inf DV_\alpha(t, x)$ are called the upper and lower derivative of α-level of Lyapunov function, respectively.

The following stability conditions of fuzzy Eq. (6) are obtained by means of the principle of reduction and by means of transition from model (1) to models (4) and (6):

(1) if there is a Lyapunov function V regarding the equation (6) for the closed set $M \subset P(R^3_+)$, such that the inequality $D_+V_\alpha(x) \leq 0 \; \forall x \in B(M, r)$ is satisfied, then the set M is α-stable in small regarding this equation. If the inequality $D_+V_\alpha(x) \leq -w_{3\alpha}(e(x, M)) \; \forall x \in B(M, r)$ is satisfied, where $w_{3\alpha} : (0, r) \to R$ is the continuous and positive function, then the set M is asymptotically α-stable in small regarding the equation (6);

(2) if there is a Lyapunov function V regarding the Eq. (6) for the closed set $M \subset P(R^3_+)$ and for some $r ::= r(\alpha)$, such that $D_-V_\alpha(x) \leq 0 \; \forall x \in B(M_\alpha, r)$, then the set is α-stable in small regarding this equation. If there is $h > 0$ and positive continuous function $w_{3\alpha} : (0, r) \to R$, such that $D_-V_\alpha(x) \leq -w_{3\alpha}(e(x, M_\alpha)) \; \forall x \in B(M_\alpha, h)$, then the set is asymptotically α-stable in small regarding the Eq. (6).

By means of the principle of reduction we formulate the stability conditions of stochastic equations which correspond to model (6). It is shown that if the trivial solution of a fuzzy equation is α-stable for every $\alpha \in (0, 1]$, then the trivial solution of the corresponding stochastic equation is stable on probability. If the trivial solution of the fuzzy equation is asymptotically α-stable for any $\alpha \in (0, 1]$, then the trivial solution of the corresponding stochastic equation is asymptotically stable on probability.

Similar stability conditions can be formulate for model (2) on the basis of transition to the differential inclusion, to the fuzzy differential equation and to the stochastic equation.

4   Synthesis of Stochastic Models of Population Dynamics by Means of the Method for Construction of Self-consistent Stochastic Models

Let us a realize a transition to stochastic models corresponding models (1) and (2) using the method of construction of the self-consistent stochastic models. Application of this method allows to obtain the stochastic differential equation in Langevin form with coordination of stochastic and deterministic parts. Consistency is understood in the sense that the stochastics in the constructed stochastic model is associated with the structure of the system, but isn't the description of external perturbations.

According to the main idea of the method it is possible for the system under consideration to describe the scheme of interaction in the form of symbolic representation of all possible interaction between the system elements.

Then we give the intensities of transitions and master equation, for which we can obtain an approximate Fokker–Planck equation by the aids of formal series expansion. It is not difficult to transit from the Fokker–Planck equation to the equivalent stochastic differential equation in Langevin form. In the practice of the method the stochastic differential equation can be written immediately after the representation of the interaction scheme. It is connected with the fact that for obtained the coefficients of the Fokker–Planck equation is necessary to know only the intensities of transitions and operators of state changes.

We present the scheme of interaction elements and the operator of state change for the system (1) in the form:

$$
\begin{aligned}
X_1 &\to 2X_1, \\
X_2 &\to 2X_2, \\
Y &\to 2Y, \\
X_1 + X_1 &\to X_1, \\
X_2 + X_2 &\to X_2, \\
Y + Y &\to Y, \\
X_1 + Y &\xrightarrow{q} Y, \\
X_1 + Y &\xrightarrow{r} X_1, \\
X_1 &\xrightarrow{\gamma} X_2, \\
X_2 &\xrightarrow{\beta} X_1,
\end{aligned}
\qquad
R =
\begin{pmatrix}
1 & 0 & 0 \\
0 & 1 & 0 \\
0 & 0 & 1 \\
1 & 0 & 0 \\
0 & 1 & 0 \\
0 & 0 & 1 \\
-1 & 0 & 0 \\
0 & 0 & -1 \\
-1 & 1 & 0 \\
1 & -1 & 0
\end{pmatrix}.
$$

The first three rows of the scheme of interaction correspond to natural reproduction species in the absence of other factors, lines 4–6 and lines 7–8 symbolize intraspecific and interspecific competition, respectively, and the last two lines describe the migration of species x_1 and x_2 between the first and the second areas.

The state of the system can be described by means of vector $x = (x_1, x_2, y_1)$. The following relations are given for intensities of transitions from the state x to the state $x \pm R^A$ in the unit of the time:

$$s_1^+(x_1, x_2, y) = x_1, \qquad s_2^+(x_1, x_2, y) = x_2, \qquad s_3^+(x_1, x_2, y) = y,$$
$$s_4^+(x_1, x_2, y) = x_1{}^2, \qquad s_5^+(x_1, x_2, y) = x_2{}^2, \qquad s_6^+(x_1, x_2, y) = y^2,$$
$$s_7^+(x_1, x_2, y) = qx_1 y, \qquad s_8^+(x_1, x_2, y) = rx_1 y,$$
$$s_9^+(x_1, x_2, y) = \gamma x_1, \qquad s_{10}^+(x_1, x_2, y) = \beta x_2.$$

Let us present Fokker–Planck equation corresponding to the model in the form:

$$\frac{\partial P(x, t)}{\partial t} = -\sum_a \partial_a \left(A_a(x) P(x, t) \right) + \frac{1}{2} \sum_{a,b} \partial_a \partial_b \left(B_{ab}(x) P(x, t) \right), \qquad (7)$$

where

$$A(x) = \sum_{A=\overline{1,10}} \left[s_A^+(x) - s_A^-(x) \right] R^A = \begin{pmatrix} x_1(1 - x_1 - qy) + \beta x_2 - \gamma x_1, \\ x_2(1 - x_2) - \beta x_2 + \gamma x_1, \\ y(1 - rx_1 - y) \end{pmatrix},$$

$$B(x) = \sum_{A=\overline{1,10}} \left[s_A^+(x) - s_A^-(x) \right] R^A (R^A)^T =$$

$$= \begin{pmatrix} x_1(1 + x_1 + qy) + \beta x_2 + \gamma x_1 & -\beta x_2 - \gamma x_1 & 0 \\ -\beta x_2 - \gamma x_1 & x_2(1 + x_2) + \beta x_2 + \gamma x_1 & 0 \\ 0 & 0 & y(1 + rx_1 + y) \end{pmatrix}.$$

Then we obtain Langeven equations equivalent to Fokker–Planck equations in the form

$$dx = a(x, t)dt + b(x, t)dW, \qquad (8)$$

where $x \in R^3$ is the function of a state of the system, nd $W \in R^3$ is the standard three-dimensional Brownian motion.

We have the following relations for the coefficients:

$$A(x) = a(x), \quad B(x) = b(x)b^T(x).$$

It is easy to see that the equation in the moments for a stochastic differential equation in the form of Langeven completely coincides with model (1) and this fact can serve for study of deterministic behavior.

Investigation of the stochastic component of a stochastic differential equation in the form of Langeven allows us to studying the influence of introduction of stochastics on the behavior of the considered system. Transition from the vector differential equation corresponding to model (2) to the stochastic differential equation is similarly realized.

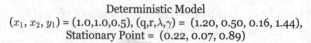

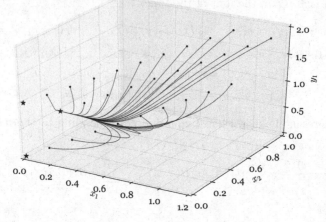

Fig. 1. The phase portrait for model (1)

Similar stochastic model can be construct for model (2). In this case, the coefficients of the Fokker–Planck equation will be:

$$A(z) = \begin{pmatrix} x\delta - x^2\frac{\delta}{K} - xy\nu + \nu mxyu, \\ -sy + c\nu xy - c\nu mxyu, \\ \tau u - \frac{\tau}{L}u^2 + \frac{\tau l}{L}xu^2 \end{pmatrix},$$

$$B(z) = \begin{pmatrix} x\delta + x^2\frac{\delta}{K} + xy\nu + \nu mxyu & 0 & 0 \\ 0 & sy + c\nu xy + c\nu mxyu & 0 \\ 0 & 0 & \tau u + \frac{\tau}{L}u^2 + \frac{\tau l}{L}xu^2 \end{pmatrix}.$$

where $z = (x, y, u)$ is the phase vector of the system.

Numerical experiments for models (1) and (2) we have made by means of the developed software package for the numerical solving of systems of differential equations by stochastic Runge–Kutta methods [24,25]. The library is prepared in the Python language with using of Numpy and Scipy modules. Algorithms for generation of trajectories of Wiener process and multipoint distributions, approximation of the multiple stochastic integrals, testing strong and weak convergence of numerical methods and directly numerical algorithms of a stochastic Runge–Kutta method are realized.

Numerical experiments showed that the developed software package gives results which are completely coordinated with analytical conclusions for the discussed deterministic models of population dynamics.

As a verification of numerical methods and the developed software package with the analytical results, we give the phase portraits for the first model (Fig. 1) and for the second model (Fig. 2). Here, asterisks indicate the stationary points, obtained by analytical formulas. The graphs show that for the given parameters the trajectories "roll" in the stationary points that correspond to the analytic results.

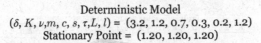

Deterministic Model
$(\delta, K, \nu, m, c, s, \tau, L, l) = (3.2, 1.2, 0.7, 0.3, 0.2, 1.2)$
Stationary Point = $(1.20, 1.20, 1.20)$

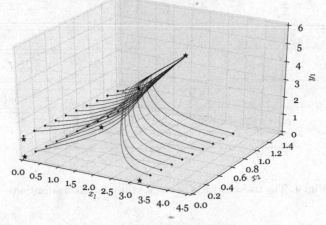

Fig. 2. The phase portrait for model (2)

Stochastic model
$(x_1, x_2, y_1) = (1.0, 1.0, 0.5), (q, r, \gamma, \beta) = (1.20, 0.50, 0.16, 1.44)$

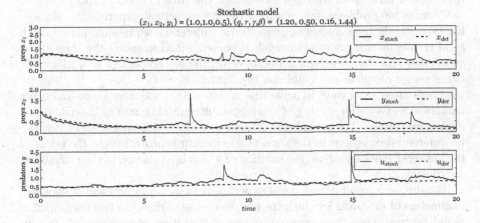

Fig. 3. The trajectories of average values for 100 realizations

The following parameters for the numerical experiment concerning the stochastic model which takes into account the competition and diffusion of species were selected: initial values of populations densities $(x_1(0), x_2(0), y_1(0)) ==$ $(1.0, 1.0, 0.5)$, parameter values $q = 1.2$, $r = 0.5$, $\gamma = 0.16$, $\beta = 1.44$. The graph (Fig. 3) shows the trajectory of the average values of 100 realizations with these initial values and parameter values for the time interval $[0, 20]$.

The following parameters for the numerical experiment concerning the stochastic model "predator–prey–mutualist" were selected: initial values of populations densities $(x(0), y(0), u(0)) = (1.0, 1.0, 0.5)$, parameter values $\delta = 1.2$, $K = 3.2$, $\eta = 0.2$, $m = 1.2$, $c = 2.2$, $s = 0.2$, $\tau = 2.2$, $L = 1.2$, $l = 1.2$. The

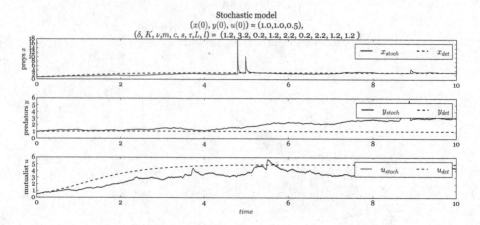

Fig. 4. The trajectories of average values for 100 realizations

graph (Fig. 4) shows the trajectory of the average values of 100 realizations with these initial values and parameter values for the time interval $[0, 10]$.

Number of problems was revealed by the numerical experiments. Runge–Kutta methods used for solving stochastic differential equations, give enough good results on a short time interval. However, with increasing the time of the experiment, due to the accumulation of error, the method ceases to work stable. Large bursts of values are visible on the figure.

There are several ways to solve this problem. One way may be to find other numerical methods for solving of stochastic differential equations describing the models we studied. The second way consists in transition from solving of stochastic differential equations in the form of specific implementations to the study of the Fokker–Planck equation, the solution of which is a function of the probability distribution.

Besides, as a result of numerical experiment there was a problem of choice of parameters of modeling in which the physical sense of the modelled phenomenon would remain. Thus there was a problem of detection of intervals which save physical sense, and also tracing of change of qualitative behavior depending on a choice of parameters not only by means of numerical modeling, but also by preliminary qualitative analysis of model.

5   Conclusions

The developed combined approach to analysis of models of nonlinear dynamics is based on the principle of reduction, and on the construction of self-consistent stochastic models. The method for construction of self-consistent stochastic model allowed us to synthesize the stochastic model which takes into account the competition and the diffusion of species, and the stochastic model which takes into account the mutualism. The principle of reduction allowed us to obtain the conditions of stability with the transition to differential inclusions, fuzzy differential equations and stochastic differential equations. Numerical experiments conducted using the developed software package show consistency with the analytical results for the studied deterministic models. A number of problems arising in the numerical study of corresponding stochastic models identified, and methods for their solving are indicated. The obtained results can be used for the study of nondeterministic nonlinear dynamic models.

Acknowledments. The work is partially supported by RFBR grant No. 15-07-08795.

References

1. Svirezhev, Y.M., Logofet, D.O.: Stability of Biological Communities. MIR, Moscow (1983)
2. Pykh, Y.A.: Equilibrium and Stability in the Models of Population Dynamics. Nauka, Moscow (1983)
3. Murray, J.D.: Mathematical Biology I: An Introduction. Interdisciplinary Applied Mathematics, vol. 7. Springer, Heidelberg (2003)
4. Bazykin, A.D.: Nonlinear Dynamics of Interacting Populations. Institute of Computer Science, Moscow-Izhevsk (2003)
5. Srinivasa Rao, A.S.R.: Population stability and momentum. Not. AMS. **61**(9), 1062–1065 (2014)
6. Druzhinina, O.V., Masina, O.N.: Methods of Stability Research and Controllability of Fuzzy and Stochastic Dynamic Systems. Dorodnicyn Computing Center of RAS, Moscow (2009)
7. Demidova, A.V., Druzhinina, O.V., Masina, O.N.: Stability research of models of population dynamics based on the construction of self-consistent stochastic models and principle reduction. Bull. Peoples' Friendship Univ. Russia Ser. Math. Inf. Sci. Phys. **3**, 18–29 (2015)
8. Xin-an, Z., Chen, L.: The linear and nonlinear diffusion of the competitive Lotka-Volterra model. Nonlinear Anal. **66**, 2767–2776 (2007)
9. Merenkov, Y.N.: Stability-like Properties of Differential Inclusions Fuzzy and Stochastic Differential Equations. PFUR, Moscow (2000)
10. Shestakov, A.A.: Generalized Direct Lyapunov Method for Systems with Distributed Parameters. URSS, Moscow (2007)
11. Dean, A.M.: A simple model of mutualism. Am. Nat. **121**, 409–417 (1983)
12. Kumar, R., Freedman, H.I.: A mathematical model of facultative mutualism with populations interacting in a food chain. Math. Biosci. **97**, 235–261 (1989)
13. Addicott, J.F.: Stability properties of 2-species models of mutualism: simulation studies. Oecologia **49**, 42–49 (1981)
14. Freedman, H.I., Rai, B.: Uniform persistence and global stability in models involving mutualism competitor-competitor-mutualist systems. Indian J. Math. **30**, 175–186 (1988)
15. Rai, B., Freedman, H.I., Addicott, J.F.: Analysis of three species models of mutualism in predator-prey and competitive systems. Math. Biosci. **63**, 13–50 (1983)
16. Freedman, H.I., Rai, B.: Can mutualism alter competitive outcome: a mathematical analysis. Rocky Mt. J. Math. **25**(1), 217–230 (1995)

17. Masina, O.N., Shcherbakov, A.V.: The model stability analysis of the interaction between two competing species taking into account the symbiosis. In: Proceedings of the International Conference "Systems of Control, Technical Systems: Stability, Stabilization, Ways and Methods of Research", Dedicated to the 95th Anniversary from Birthday of Professor A.A. Shestakov, Yelets, 2–3 April 2015, pp. 93–97. YSU Named After Ivan Bunin, Yelets (2015)
18. Demidova, A.V., Druzhinina, O.V., Masina, O.N., Shcherbakov, A.V.: Synthesis of nonlinear models of population dynamics based on the combination of deterministic and stochastic approaches. In: Abstracts of the Third International Scientific Conference Modeling Nonlinear Processes and Systems (MNPS - 2015), Moscow, STANKIN, 22–26 June 2015 , pp. 87–88. Janus-K, Moscow (2015)
19. Kulyabov, D.S., Demidova, A.V.: The introduction of an agreed term in the equation of stochastic population model. Bull. Peoples' Friendship Univ. Russia Ser. Math. Inf. Sci. Phys. **3**, 69–78 (2012)
20. Demidova, A.V.: The equations of population dynamics in the form of stochastic differential equations. Bull. Peoples' Friendship Univ. Russia Ser. Math. Inf. Sci. Phys. **1**, 67–76 (2013)
21. Demidova, A.V., Gevorkyan, M.N., Egorov, A.D., Korolkova, A.V., Kulyabov, D.S., Sevastyanov, L.A.: Influence of stochastization to one-step model. Bull. Peoples' Friendship Univ. Russia Ser. Math. Inf. Sci. Phys. **1**, 71–85 (2014)
22. Gardiner, C.W.: Handbook of Stochastic Methods: for Physics, Chemistry and the Natural Sciences. Springer Series in Synergetics. Springer, Heidelberg (1985)
23. Van Kampen, N.G.: Stochastic Processes in Physics and Chemistry. Elsevier Science, Amsterdam (1992)
24. Gevorkyan, M.N., Velieva, T.R., Korolkova, A.V., Kulyabov, D.S., Sevastyanov, L.A.: Stochastic Runge–kutta software package for stochastic differential equations. In: Zamojski, W., Mazurkiewicz, J., Sugier, J., Walkowiak, T., Kacprzyk, J. (eds.) Dependability Engineering and Complex Systems. AISC, vol. 470, pp. 169–179. Springer, Heidelberg (2016). doi:10.1007/978-3-319-39639-2_15
25. Eferina, E.G., Korolkova, A.V., Gevorkyan, M.N., Kulyabov, D.S., Sevastyanov, L.A.: One-step stochastic processes simulation software package. Bull. Peoples' Friendship Univ. Russia Ser. Math. Inf. Sci. Phys. **3**, 45–59 (2014)

Model of Diatomic Homonuclear Molecule Scattering by Atom or Barriers

A.A. Gusev[1]([✉]), O. Chuluunbaatar[1,5], S.I. Vinitsky[1,2], L.L. Hai[1,6],
V.L. Derbov[3], and P.M. Krassovitskiy[4]

[1] Joint Institute for Nuclear Research, Dubna, Russia
gooseff@jinr.ru
[2] RUDN University (Peoples' Friendship University of Russia),
6 Miklukho-Maklaya Str., Moscow 117198, Russia
[3] Saratov State University, Saratov, Russia
[4] Institute of Nuclear Physics, Almaty, Kazakhstan
[5] Institue of Mathematics, National University of Mongolia, University Street,
Sukhbaatar District, Ulaanbaatar, Mongolia
[6] Ho Chi Minh City University of Education, Ho Chi Minh City, Vietnam

Abstract. The mathematical model of quantum tunnelling of diatomic *homonuclear* molecules through repulsive barriers *or scattering by an atom* is formulated in the s-wave approximation. The 2D boundary-value problem (BVP) in polar coordinates is reduced to a 1D BVP for a set of second-order ODEs by means of Kantorovich expansion over the set of parametric basis functions. The algorithm for calculating the asymptotic form of the parametric basis functions and effective potentials of the ODEs at large values of the parameter (hyperradial variable) is presented. The solution is sought by matching the numerical solution in one of the subintervals with the analytical solution in the adjacent one. The efficiency of the algorithm is confirmed by comparing the calculated solutions with those of the parametric eigenvalue problem obtained by applying the finite element method in the entire domain of definition at large values of the parameter. The applicability of algorithms and software are demonstrated by the example of benchmark calculations of discrete energy spectrum of the trimer Be_3 in collinear configuration.

Keywords: Parametric boundary-value problems · Second-order ordinary differential equations · Finite element method

1 Introduction

The studies of tunnelling of bound particles through repulsive barriers revealed the effect of resonance quantum transparency of the barrier: when the cluster size is comparable with the spatial width of the barrier, one can observe enhanced barrier transparency, the mechanism of which is analogous to the mechanism of blooming of optical systems. At present this effect and its possible applications is a subject of extensive studies in different physical fields, e.g., the quantum diffusion of molecules [14,18].

© Springer International Publishing AG 2016
V.M. Vishnevskiy et al. (Eds.): DCCN 2016, CCIS 678, pp. 511–524, 2016.
DOI: 10.1007/978-3-319-51917-3_44

The formulation of the model of quantum tunnelling of a diatomic molecule through Gaussian barriers in the s-wave approximation is given in the form of 2D boundary-value problems in the Cartesian and polar coordinates [11,19]. Evidently, it corresponds to the scattering of a diatomic homonuclear molecule in a potential field of the third atom having infinite mass. Therefore, we can consider the tunneling of a diatomic molecule through the barriers as a limiting case of the molecule scattering by an atom, e.g., the resonance scattering of Be_2 dimer by the Be atom via the compound trimer Be_3: Be_2+Be states. Below we consider the statement of both problems and show both common and specific properties of the corresponding solutions.

In the present paper using different solutions of the auxiliary boundary-value problems with respect to the transverse variable, or the angular variable, with parametric dependence upon the *hyperradial* variable as basis functions, the 2D boundary-value problem is reduced to a system of coupled ODEs of the second order. In the Cartesian coordinates the effective potentials decrease exponentially (below the dissociation threshold) and in the polar coordinates they decrease as inverse powers of the independent variable. Therefore, in the latter case it is necessary to calculate the asymptotic expansions of matrix elements and fundamental solutions of the system of coupled ODEs. For their calculation it is necessary to develop symbolic-numeric algorithms, implemented in the Maple computer algebra system [19].

The paper is organised as follows. In Sect. 2 we give the setting of the 2D BVP. In Sect. 3 the reduction of the BVP using the Kantorovich method is executed. As an example, the eigenvalues and the hyperradial components of eigenfunctions are calculated for the model of Be_3 trimer in the collinear configuration. In Sect. 4 we present the algorithms for calculating the asymptotes of parametric basis functions in polar coordinates at large values of the parameter (hyperradial variable) and the effective potentials. In Conclusion the results and perspectives are discussed.

2   Setting of the Problem

Consider a 2D model of three identical particles with the mass M and the coordinates $x_i \in \mathbf{R}^1$, $i = 1, 2, 3$, coupled via the pair potential $\tilde{V}(|x_i - x_j|)$ $i, j = 1, 2, 3$. Performing the change of variables at cyclic permutation $(\alpha, \beta, \gamma) = (1, 2, 3)$:

$$x \equiv x_{(\alpha\beta)} = x_\alpha - x_\beta, \quad y \equiv y_{(\alpha\beta)\gamma} = \frac{x_\alpha + x_\beta - 2x_\gamma}{\sqrt{3}}, \quad x_0 = \frac{\sqrt{2}}{\sqrt{3}}(x_1 + x_2 + x_3),$$

we arrive at the Schrödinger equation for the wave function in a center-of-mass system $\{x_i \in \mathbf{R}^1 | x_1 + x_2 + x_3 = 0\}$

$$\left(-\frac{\partial^2}{\partial y^2} - \frac{\partial^2}{\partial x^2} + \frac{M}{\hbar^2}(\tilde{V}(x, y) - \tilde{E}) \right) \Psi(y, x) = 0. \tag{1}$$

In the case of a diatomic molecule with identical nuclei coupled via the pair potential $\tilde{V}(|x_1 - x_2|)$ and moving in the external potential field $\tilde{V}^b(|x_i - x_3|)$,

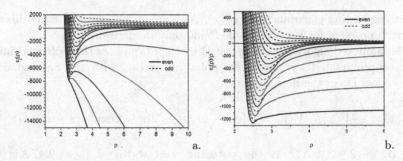

Fig. 1. The potential curves of Be$_2$ (in K, $1\,\mathrm{K} = 0.18\,\text{Å}^{-2}$), i.e., the energy eigenvalues depending upon the parameter ρ (in Å): a. $\varepsilon_j(\rho)$ and b. $\tilde{\varepsilon}_j = \varepsilon_j(\rho)/\rho^2$.

$i = 2, 1$ of the third atom having the infinite mass, the same Eq. (1) is valid for the variables

$$x = x_1 - x_2, \quad y = x_1 + x_2,$$

the origin of the coordinate frame being placed on the infinite-mass atom, $x_3 = 0$.

Here the potential function for a trimer with the pair potentials (below this case is referred to as Task 2),

$$\tilde{V}(x,y) = \tilde{V}(|x|) + \tilde{V}(|\frac{x - \sqrt{3}y}{2}|) + \tilde{V}(|\frac{x + \sqrt{3}y}{2}|), \tag{2}$$

or the potential function for a dimer in the field of barrier potentials (below this case is referred to as Task 3)

$$\tilde{V}(x,y) = \tilde{V}(|x|) + \tilde{V}^b(|\frac{x - y}{2}|) + \tilde{V}^b(|\frac{x+y}{2}|), \tag{3}$$

is symmetric with respect to the straight line $x = 0$ (i.e., $x_1 = x_2$), which allows one to consider the solutions of the problem in the half-plane $x \geq 0$. Using the Dirichlet or Neumann boundary condition at $x = 0$ allows one to obtain the solutions, symmetric and antisymmetric with respect to the permutation of two particles. If the pair potential possesses a high maximum in the vicinity of the pair collision point, then the solution of the problem in the vicinity of $x = 0$ is exponentially small and can be considered in the half-plane $x \geq x_{\min}$. In this case setting the Neumann or Dirichlet boundary condition at $x_{\min}$ gives only a minor contribution to the solution. The equation, describing the molecular subsystem, has the form

$$\left(-\frac{d^2}{dx^2} + \frac{M}{\hbar^2}(\tilde{V}(x) - \tilde{\varepsilon}) \right) \phi(x) = 0. \tag{4}$$

We assume that the molecular subsystem has the discrete spectrum, consisting of a finite number n_0 of bound states with the eigenfunctions $\phi_j(x)$, $j = 1, n$ and eigenvalues $\tilde{\varepsilon}_j = -|\tilde{\varepsilon}_j|$, and the continuous spectrum of eigenvalues $\tilde{\varepsilon} > 0$ with the corresponding eigenfunctions $\phi_{\tilde{\varepsilon}}(x)$. As a rule, the solution of the discrete spectrum problem for Eq. (4) can be found only numerically.

The proposed algorithm is illustrated by the example of the molecular interaction approximated by the Morse potential of Be$_2$ with the reduced mass $M/2 = 4.506$ Da of the nuclei [14,19]

$$V(x) = \frac{M}{\hbar^2}\tilde{V}(x), \quad \tilde{V}(x) = D\{\exp[-2(x - \hat{x}_{eq})\alpha] - 2\exp[-(x - \hat{x}_{eq})\alpha]\}. \quad (5)$$

Here $\alpha = 2.96812\,\text{Å}^{-1}$ is the potential well width, $\hat{x}_{eq} = 2.47\,\text{Å}$ is the average distance between the nuclei, and $D = 1280\,\text{K}$ ($1\,\text{K} = 0.184766$ $\text{Å}^{-2}, 1\,\text{Å}^{-2} = 5.412262\,\text{K}$) is the potential well depth. This potential supports five bound states [20] having the energies $\varepsilon_i = (M/\hbar^2)\tilde{\varepsilon}_i$, $i = 1,...,n_0 = 5$ presented in Table 1. The parameter values are determined from the condition $(\tilde{\varepsilon}_2 - \tilde{\varepsilon}_1)/(2\pi\hbar c) = 277.124\,\text{cm}^{-1}$, $1\,\text{K}/(2\pi\hbar\,c) = 0.69503476\,\text{cm}^{-1}$.

To solve the discrete spectrum problem we applied the finite element method of the seventh order using the Hermitian interpolation polynomials with double nodes [12]. The grid $\{x_0, ..., x_i, ..., x_n\}$ was used to calculate the values of both the function and its derivatives.

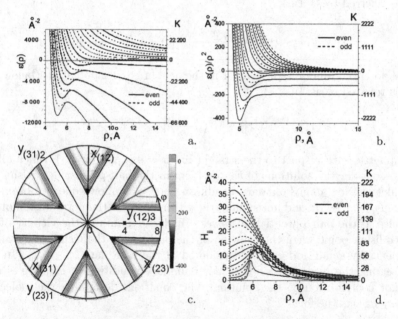

Fig. 2. Be+Be$_2$: The potential curves of Be$_3$ (in K), i.e., the energy eigenvalues depending upon the parameter ρ (in Å): a. $\varepsilon_j(\rho)$ and b. $\tilde{\varepsilon}_j = \varepsilon_j(\rho)/\rho^2$, c. the isolines of 2D potentials of Be$_3$ trimer, and d the diagonal effective potentials $H_{jj}(\rho)$.

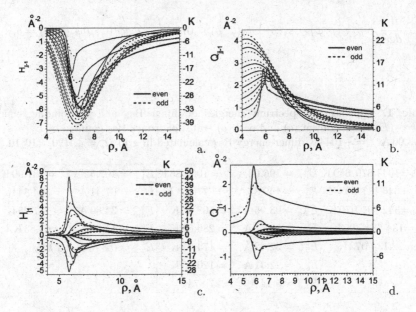

Fig. 3. The effective potentials (13) a. $H_{jj-1}(\rho)$, b. $Q_{jj-1}(\rho)$, c. $H_{j1}(\rho)$, d. $Q_{j1}(\rho)$.

3   Reduction of the BVP Using the Kantorovich Method

Using the change of variables $x = \rho \sin \varphi$, $y = \rho \cos \varphi$, we rewrite Eq. (1) in polar coordinates (ρ, φ), $\Omega_{\rho,\varphi} = (\rho \in (0, \infty), \varphi \in [0, 2\pi])$

$$\left(-\frac{1}{\rho} \frac{\partial}{\partial \rho} \rho \frac{\partial}{\partial \rho} - \frac{1}{\rho^2} \frac{\partial^2}{\partial \varphi^2} + V(\rho, \varphi) - E \right) \Psi(\rho, \varphi) = 0, \tag{6}$$

where for a trimer with pair potentials

$$V(\rho, \varphi) = V(\rho | \sin \varphi|) + V(\rho | \sin(\varphi - 2\pi/3)|) + V(\rho | \sin(\varphi - 4\pi/3)|), \tag{7}$$

and for a dimer with pair potential in the external field of barrier potentials:

$$V(\rho, \varphi) = V(\rho | \sin \varphi|) + V^b(\rho | \sin(\varphi - \pi/4)|) + V^b(\rho | \sin(\varphi + \pi/4)|). \tag{8}$$

The solution of Eq. (6) is sought in the form of Kantorovich expansion

$$\Psi_{i_o}(\rho, \varphi) = \sum_{j=1}^{j_{\max}} \phi_j(\varphi; \rho) \chi_{ji_o}(\rho). \tag{9}$$

Here $\chi_{ji_o}(\rho)$ are unknown functions and the orthogonal normalised basis functions $\phi_j(\varphi; \rho)$ in the interval $\varphi \in [0, \pi]$ are defined as eigenfunctions, corresponding to the eigenvalues of the Sturm-Liouville problem for the equation

$$\left(-\frac{d^2}{d\varphi^2} + \rho^2 V(\rho, \varphi) - \varepsilon_j(\rho)\right)\phi_j(\varphi; \rho) = 0, \qquad \int_0^{2\pi} d\varphi \phi_i(\varphi; \rho)\phi_j(\varphi; \rho) = \delta_{ij}.$$

$$(10)$$

Table 1. The discrete spectrum energies of dimer Be$_2$ and the binding energy $E^b = -(E - E^a)$ of even (e) and odd (o) states of trimer Be$_3$ counted of $E^a = \tilde{\varepsilon}_1 = -193.06\,\text{Å}^{-2} = -1044\,\text{K}$ dimer energy Be$_2$ calculated in grid $\Omega_\rho = 4.1(20)7(10)10$.

$-\tilde{\varepsilon}_1 = 1044.879\,649\,\text{K}$	$E^b_{1,e} = 196.02\,\text{Å}^{-2} = 1060.86\,\text{K}$	$E^b_{1,o} = 107.52\,\text{Å}^{-2} = 581.90\,\text{K}$
$-\tilde{\varepsilon}_2 = 646.157\,093\,\text{K}$	$E^b_{2,e} = 142.37\,\text{Å}^{-2} = 770.51\,\text{K}$	$E^b_{2,o} = 67.41\,\text{Å}^{-2} = 364.84\,\text{K}$
$-\tilde{\varepsilon}_3 = 342.791\,979\,\text{K}$	$E^b_{3,e} = 93.95\,\text{Å}^{-2} = 508.50\,\text{K}$	$E^b_{3,o} = 34.60\,\text{Å}^{-2} = 187.28\,\text{K}$
$-\tilde{\varepsilon}_4 = 134.784\,305\,\text{K}$	$E^b_{4,e} = 52.77\,\text{Å}^{-2} = 285.63\,\text{K}$	$E^b_{4,o} = 11.79\,\text{Å}^{-2} = 63.83\,\text{K}$
$-\tilde{\varepsilon}_5 = 22.134\,073\,\text{K}$	$E^b_{5,e} = 32.32\,\text{Å}^{-2} = 174.95\,\text{K}$	$E^b_{5,o} = 0.8\,\text{Å}^{-2} = 4.4\,\text{K}$
	$E^b_{6,e} = 22.31\,\text{Å}^{-2} = 120.75\,\text{K}$	
	$E^b_{7,e} = 5.14\,\text{Å}^{-2} = 27.87\,\text{K}$	

For the problems under consideration the potential function $V(\rho, \varphi)$ depending on the parameter ρ can be defined as follows.

Task 1. The case of one pair potential in the intervals $\varphi \in (0, 2\varphi_\alpha)$ $(\varphi_\alpha = \pi/3, \pi/4$ or $\pi/2)$ $V(\rho, \varphi) = V(\rho \sin \varphi)$.

Task 2. The case of three pair potentials (7), in the interval $\varphi \in (0, 2\varphi_\alpha = \pi/3)$.

Task 3. The case of one pair potential and two penetrable or almost impenetrable barrier potentials (8), in the interval $\varphi \in (0, \varphi_\alpha = \pi/2)$ or in the intervals $\varphi \in (0, \varphi_\alpha = \pi/4 - \epsilon)$ and $\varphi \in (\varphi_\alpha = \pi/4 - \epsilon, \pi/2)$, $0 < \epsilon \ll \pi/4$.

The solutions symmetric with respect to the permutation of two particles satisfy the Neumann boundary condition at $\varphi = 0$ and $\varphi = 2\varphi_\alpha$, while the antisymmetric ones satisfy the Dirichlet boundary condition. If the pair potential possesses a high peak in the vicinity of the pair collision point, then the solution of the problem (6) will be considered in the half-plane $\Omega_{\rho,\varphi} = (\rho \in (\rho_{\min}, \infty), \varphi \in [\varphi_{\min}(\rho), 2\varphi_\alpha - \varphi_{\min}(\rho)])$ with the Neumann or Dirichlet boundary condition. Since the potential of the boundary-value problem (10) is symmetric with respect to $\varphi = \varphi_\alpha$, the even $\phi_j(\varphi; \rho) = \phi_j(2\varphi_\alpha - \varphi; \rho)$ and odd $\phi_j(\varphi; \rho) = -\phi_j(2\varphi_\alpha - \varphi; \rho)$ solutions, satisfying the Neumann or the Dirichlet boundary condition respectively, will be considered separately in the interval $\varphi \in [\varphi_{\min}(\rho), \varphi_\alpha]$.

The system of coupled ODEs in the Kantorovich form has the form

$$\left[-\frac{1}{\rho}\frac{d}{d\rho}\rho\frac{d}{d\rho} + \frac{\varepsilon_i(\rho)}{\rho^2} - E\right]\chi_{ii_o}(\rho) + \sum_{j=1}^{j_{\max}} W_{ij}(\rho)\chi_{ji_o}(\rho) = 0, \qquad (11)$$

$$W_{ij}(\rho) = H_{ji}(\rho) + \frac{1}{\rho}\frac{d}{d\rho}\rho Q_{ji}(\rho) + Q_{ji}(\rho)\frac{d}{d\rho}. \qquad (12)$$

The effective potentials $Q_{ij}(\rho) = -Q_{ji}(\rho)$, $H_{ij}(\rho) = H_{ji}(\rho)$ are given by the integrals calculated using the above symmetry on reduced intervals $\varphi \in [0, \varphi_\alpha]$:

$$Q_{ij}(\rho) = -\int_0^{\varphi_\alpha} d\varphi \, \phi_i(\varphi; \rho) \frac{d\phi_j(\varphi; \rho)}{d\rho}, \quad H_{ij}(\rho) = \int_0^{\varphi_\alpha} d\varphi \frac{d\phi_i(\varphi; \rho)}{d\rho} \frac{d\phi_j(\varphi; \rho)}{d\rho}. \tag{13}$$

For Task 3 the effective potentials $\hat{W}_{ij}(\rho) = W_{ij}(\rho) + V_{ij}^b(\rho)$ are sums of $W_{ij}(\rho)$, calculated using the potential curves and the parametric basis functions of Task 1, and the integrals of barrier potentials $V_{ij}^b(\rho)$ mutiplied by the basis functions

$$V_{ij}^b(\rho) = \int_0^{\varphi_\alpha} d\varphi \, \phi_i(\varphi; \rho)(V^b(\rho \sin(\varphi - \pi/4)) + V^b(\rho \sin(\varphi + \pi/4)))\phi_j(\varphi; \rho).$$

As an example, we calculated the parametric basis functions of BVP (10) and the effective potentials (13) for the models of Be$_2$ dimer and Be$_3$ trimer in collinear configuration using the programme ODPEVP [1]. The results are shown in Figs. 1, 2, and 3. For this model the eigenvalues and the hyperradial components of 2D eigenfunctions of the BVP for the set of ODEs (11) were calculated using the program KANTBP [12]. The discrete energy spectrum of the dimer Be$_2$ and a set of the binding energies of the trimer Be$_3$ is shown in Table 1, and the components of the trimer eigenfunctions (9) are shown in Fig. 4.

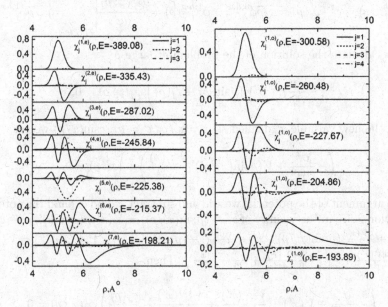

Fig. 4. Components $\chi_j^{i,\sigma=e,o}(\rho, E) \equiv \chi_j^{(i)}(\rho)$ of even (e) and odd (o) bound states with total energy E in Å^{-2}.

4   Asymptotes of the Parametric Basis Functions

In polar coordinates at large ρ the width of the potential well decreases with the growth of ρ. This fact allows the linearisation of the argument $\rho \sin \varphi - \hat{x}_{eq} \rightarrow \rho(\varphi - \arcsin(\hat{x}_{eq}/\rho))$ at $|x - \hat{x}_{eq}|/\rho \ll 1$ in the expression for the potential function $V(\rho \sin \varphi)$ and the reformulation of Eq. (10) in the interval $\varphi = (0, \varphi_\alpha)$ as

$$\left(-\frac{\partial^2}{\partial \varphi^2} + \rho^2 V(\rho \varphi) - \varepsilon_j(\rho) \right) \phi_j(\varphi; \rho) = 0. \tag{14}$$

By the change of variables $x = \rho \varphi$ this equation is reduced to Eq. (4).

The cluster eigenfunctions of the lower part of discrete spectrum $\varepsilon_j(\rho) < 0$ are known to be localised in the potential well and exponentially small beyond it. Therefore, the solutions of the BVP for Eq. (14) of Task 1 are determined from the solutions of the BVP for Eq. (4) $\varepsilon_j(\rho) = \rho^2 \tilde{\varepsilon}_j$, $\phi_j(\varphi = x/\rho; \rho) = \sqrt{\rho} \phi_j(x)$, $j = 1, ..., n_0$. Provided that the solution of the BVP was earlier calculated in Cartesian coordinates on the grid $\{x_0 = 1.7665, ..., x_i, ..., x_{7999} = 9.7655\}$ by means of the program KANTBP 4M using FEM with the interpolating Hermite polynomials of the fifth order and double nodes, the solution in polar coordinates on the grid $\varphi_i = (x_i + \frac{x_{eq}^3}{6\rho^2})/\rho$ is recalculated as

$$\phi_{j;i}^{0;h\varphi}(\rho) = \phi_j^h(\varphi_i; \rho) = \sqrt{\rho} \phi_{j;i}^{0;hx}, \quad \phi_{j;i}^{1;h\varphi}(\rho) = \left. \frac{\phi_j^h(\varphi; \rho)}{d\varphi} \right|_{\varphi = \varphi_i} = \rho \sqrt{\rho} \phi_{j;i}^{1;hx}. \tag{15}$$

Let us calculate the solution of the problem at large ρ

$$\left(-\frac{\partial^2}{\partial \varphi^2} + \rho^2 V(\rho \sin \varphi) - \varepsilon_j(\rho) \right) \phi_j(\varphi; \rho) = 0. \tag{16}$$

Using the new variable x' defined as $\varphi = x'/\rho$, $x' = \rho \arcsin(x/\rho)$ we get

$$\left(-\frac{\partial^2}{\partial x'^2} + V(\rho \sin(x'/\rho)) - \frac{\varepsilon_j(\rho)}{\rho^2} \right) \phi_j(x'; \rho) = 0. \tag{17}$$

In the argument of the potential we add and subtract x' and expand the potential in Taylor series in the vicinity of x', $V(\rho \sin(x'/\rho)) = V(x' + (\rho \sin(x'/\rho) - x')) = V(x') + \frac{dV(x')}{dx'}(\rho \sin(x'/\rho) - x') + \frac{1}{2}\frac{d^2V(x')}{dx'^2}(\rho \sin(x'/\rho) - x')^2 + O(\rho^{-6}) = V(x') - \frac{1}{\rho^2}\frac{x'^3}{6}\frac{dV(x')}{dx'} + \frac{1}{\rho^4}\left(\frac{x'^5}{120}\frac{dV(x')}{dx'} + \frac{x'^6}{36}\frac{d^2V(x')}{dx'^2} \right)$. Then

$$\left(-\frac{\partial^2}{\partial x'^2} + V(x') - \frac{V^{(1)}(x')}{\rho^2} + \frac{V^{(2)}(x')}{\rho^4} - \frac{\varepsilon_j(\rho)}{\rho^2} \right) \phi_j(x'; \rho) = 0, \tag{18}$$

$$\langle \phi_i(\rho) | \phi_j(\rho) \rangle \equiv \int_{x_0'}^{x'_{max}} dx'(\phi_i(x'; \rho)\phi_j(x'; \rho) = \delta_{ij}. \tag{19}$$

For the Morse potential (5) the corrections to the potential are expressed as

$$V^{(1)}(x') \equiv \frac{x'^3}{6} \frac{dV(x')}{dx'} = \frac{D\alpha x'^3}{3} \{\exp[-2(x' - \hat{x}_{eq})\alpha] - 2\exp[-(x' - \hat{x}_{eq})\alpha]\}$$

$$V^{(2)}(x') \equiv \left(\frac{x'^5}{120} \frac{dV(x')}{dx'} + \frac{x'^6}{72} \frac{d^2V(x')}{dx'^2} \right)$$

$$= \frac{D\alpha x'^5}{180} \{(10\alpha x' - 3)\exp[-2(x' - \hat{x}_{eq})\alpha] - (5\alpha x' - 3)\exp[-(x' - \hat{x}_{eq})\alpha]\}$$

We seek the solution in the form of the power series using the second-order perturbation theory

$$\phi_j(x'; \rho) = \phi_j^{(0)}(x') + \frac{\phi_j^{(1)}(x')}{\rho^2} + \frac{\phi_j^{(2)}(x')}{\rho^4}, \quad \frac{\varepsilon_j(\rho)}{\rho^2} = E_j^{(0)} + \frac{E_j^{(1)}}{\rho^2} + \frac{E_j^{(2)}}{\rho^4} \quad (20)$$

that yields the recurrence set of nonuniform ODEs

$$\left(L - E_j^{(0)}\right)\phi_j^{(0)}(x') = 0, \quad L = -\frac{\partial^2}{\partial x'^2} + V(x'),$$

$$\left(L - E_j^{(0)}\right)\phi_j^{(1)}(x') + \left(V^{(1)}(x') - E_j^{(1)}\right)\phi_j^{(0)}(x') = 0,$$

$$\left(L - E_j^{(0)}\right)\phi_j^{(2)}(x') + \left(V^{(1)}(x') - E_j^{(1)}\right)\phi_j^{(1)}(x') + \left(V^{(2)}(x') - E_j^{(2)}\right)\phi_j^{(0)}(x') = 0.$$

The first- and second-order corrections of the eigenfunctions satisfy the relations

$$\langle \phi_i^{(0)}(x')|\phi_j^{(0)}(x')\rangle \equiv \int_{x_0'}^{x'_{max}} dx' (\phi_i^{(0)}(x')\phi_j^{(0)}(x') = \delta_{ij},$$

$$\langle \phi_j^{(0)}(x')|\phi_j^{(1)}(x')\rangle = 0, \quad \langle \phi_j^{(1)}(x')|\phi_j^{(1)}(x')\rangle + 2\langle \phi_j^{(0)}(x')|\phi_j^{(2)}(x')\rangle = 0.$$

and corrections of the eigenvalues are determined by integrals

$$E_j^{(1)} = \langle \phi_j^{(0)}|V^{(1)}(x')|\phi_j^{(0)}\rangle, \quad E_j^{(2)} = \langle \phi_j^{(0)}|V^{(2)}(x')|\phi_j^{(0)}\rangle + \langle \phi_j^{(0)}|V^{(1)}(x')|\phi_j^{(1)}\rangle.$$

Substituting (20) into the effective potentials $Q_{ij}(\rho)$ and $H_{ij}(\rho)$ defined by

$$Q_{ij}(\rho) = -\langle \phi_i(\rho)|\frac{\partial}{\partial \rho}|\phi_j(\rho)\rangle + Q_{ij}^{(0)}, \quad Q_{ij}^{(0)} = -\langle \phi_i(\rho)|\frac{x}{\rho}\frac{\partial}{\partial x} + \frac{1}{2\rho}|\phi_j(\rho)\rangle, \quad (21)$$

$$H_{ij}(\rho) = K_{ij}(\rho) - \frac{\partial Q_{ij}(\rho)}{\partial \rho}, \quad K_{ij}(\rho) = -\langle \phi_i(\rho)|(\frac{\partial}{\partial \rho} + \frac{x}{\rho}\frac{\partial}{\partial x} + \frac{1}{2\rho})^2|\phi_j(\rho)\rangle,$$

$$H_{ij}(\rho) = \langle \frac{\phi_i(\rho)}{\partial \rho}|\frac{\partial \phi_j(\rho)}{\partial \rho}\rangle + \frac{2}{\rho}\langle(-\frac{1}{2} - x\frac{\partial}{\partial x})\phi_i(\rho)|\frac{\partial \phi_j(\rho)}{\partial \rho}\rangle + H_{ij}^{(0)}(\rho), \quad (22)$$

$$H_{ij}^{(0)}(\rho) = \frac{1}{\rho^2}\langle \frac{\partial \phi_i(\rho)}{\partial x}|x^2|\frac{\partial \phi_j(\rho)}{\partial x}\rangle - \frac{1}{4\rho^2}\langle \phi_i(\rho)|\phi_j(\rho)\rangle, \quad (23)$$

we arrive at the required asymptotic expansions for cluster states, $i, j = 1, ..., n_0$:

$$Q_{ij}(\rho) = \frac{Q_{ij}^{(1)}}{\rho^1} + \frac{Q_{ij}^{(3)}}{\rho^3} + O(\rho^{-5}), \quad H_{ij}(\rho) = \frac{H_{ij}^{(2)}}{\rho^2} + \frac{H_{ij}^{(4)}}{\rho^4} + O(\rho^{-6}). \quad (24)$$

Remark. The effective potentials $E_j^{(1)} + H_{jj}^{(0)} = (1/4)\rho^{-2}$ lead to asymptotic cluster fundamental solutions of the ODEs (11), defined by the Bessel functions $J_{1/2}(\sqrt{-E + \varepsilon_j}\rho)$, $j = 1, ..., n_0$, while for pseudostates the asymptotic fundamental solutions of the ODEs are $J_m(\sqrt{E}\rho)$ with integer $m = (j - n_0) = 1, 2, ...$

The eigenfunctions of pseudostates $\varepsilon_j(\rho) \geq 0$, $(j - n_0) = 1, 2, ...,$, are localized in out of the potential well. Then the $(n_0 - 1)$-th node is located at the boundary of the potential well. Here and below we consider the case of $\varphi_\alpha = \pi/2$ illustrated by Fig. 5. From this fact the estimate of the eigenvalues for pseudostates $\varepsilon_j(\rho) \approx (j - n_0)^2$ follows, namely, the eigenvalues of the corresponding BVPs in Cartesian coordinates, $\varepsilon_j = \varepsilon_j(\rho)/\rho^2$, will be a small quantity (see Fig. 1b). Then the numerical values of the function $B(\varphi_i; \rho) = B(x_i)$ and its derivative $B'(\varphi_i; \rho) = \rho B'(x_i)$ on the specified grid $\Omega_\varphi = \{\varphi_1 = \varphi_0, ..., \varphi_i = x_i/\rho, ..., \varphi_N = \varphi_\varepsilon\}$ in the polar system of coordinates are determined via the values of the function $B(x_i)$ and its derivative $B'(x_i)$ on the grid $\Omega_x = \{x_1 = x_0, ..., x_i, ..., x_N = x_\varepsilon\}$, found in the form of the power series of small parameter ε_n:

$$B_j(x_i) = B_i^{(0)} + B_i^{(1)}\varepsilon_n + B_i^{(2)}\varepsilon_n^2, \quad B_j'(x_i) = b_i^{(0)} + b_i^{(1)}\varepsilon_n + b_i^{(2)}\varepsilon_n^2, \quad (25)$$

using the Runge-Kutta method, in which the third power of unknown ε_n and the higher ones are neglected. The expansion coefficients $B_i^{(k)} \equiv B_i^{(k)}(x_i)$ and $b_i^{(k)} \equiv b_i^{(k)}(x_i)$, $k = 0, 1, 2$, calculated at the grid nodes x_i for the potential (5) are presented in Fig. 6. One can see that in the vicinity of the potential well the corrections to the eigenfunctions are small, and at $x > 6$ they become essential. The coefficient $b_i^{(0)}$, the derivative of the wave function with $\varepsilon_n = 0$, becomes constant for $x > 5.5$. From these observations the condition for choosing x_ε follows. The interval $\varphi_0 \leq \varphi \leq \pi/2$ is divided into two subintervals by the point $\varphi_\varepsilon = x_\varepsilon/\rho$: $\varphi_0 < \varphi \leq \varphi_\varepsilon$ and $\pi/2 > \varphi > \varphi_\varepsilon$. In the calculations the point x_ε was chosen from the condition $|V(x > x_\varepsilon)| < \varepsilon$, where ε is a preassigned number, and the left-hand boundary of the interval $\varphi_0 = 0$. In the case of a high barrier, at the pair collision point, when the eigenfunctions in its vicinity are close to zero, the left boundary of the interval changes, $\varphi_0 = x_0/\rho > 0$. The eigenfunctions $\phi_j(\varphi; \rho)$ are calculated in the form

$$\phi_j(\varphi; \rho) = \begin{cases} A_j(\rho)B_j(\varphi; \rho), & \varphi_0 \leq \varphi \leq \varphi_\varepsilon, \\ C_j(\rho)\sqrt{\frac{2}{\pi}} \begin{Bmatrix} \cos \\ \sin \end{Bmatrix} (\sqrt{\varepsilon_j(\rho)}(\varphi - \pi/2)), & \varphi_\varepsilon < \varphi \leq \pi/2, \end{cases} \quad (26)$$

$$2\int_{\varphi_0}^{\pi/2} d\varphi(\phi_n(\varphi; \rho))^2 = 1. \quad (27)$$

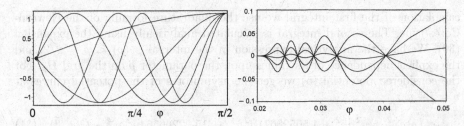

Fig. 5. Eigenfunctions $\phi_j(\varphi; \rho)$ corresponding to the eigenvalues $\varepsilon_j(\rho) \geq 0$ of the even pseudostates $j = n + 1 = 6, ..., 10$ at $\rho = 100$.

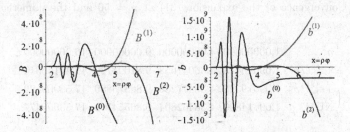

Fig. 6. Expansion coefficients $B_i^{(k)}$ and $b_i^{(k)}$, $k = 0, 1, 2$, calculated at the nodes x_i.

Here $A_j(\rho)$ and $C_j(\rho)$ are the normalisation factors, and $B(\varphi; \rho)$ is determined from the numerical solution $B(x)$ in Cartesian coordinates using the transformation $\varphi = x/\rho$. From the continuity of the eigenfunctions and their derivatives,

$$\phi_n(\varphi_\varepsilon - 0; \rho) = \phi_n(\varphi_\varepsilon + 0; \rho), \frac{d\phi_n}{d\varphi}(\varphi_\varepsilon - 0; \rho) = \frac{d\phi_n}{d\varphi}(\varphi_\varepsilon + 0; \rho), \qquad (28)$$

we get the equation for the eigenvalue $\varepsilon_n(\rho)$:

$$\begin{Bmatrix} \tan(\sqrt{\varepsilon_n(\rho)}(\varphi_\varepsilon - \frac{\pi}{2})) & \text{even } n \\ -\cot(\sqrt{\varepsilon_n(\rho)}(\varphi_\varepsilon - \frac{\pi}{2})) & \text{odd } n \end{Bmatrix} - \frac{\sqrt{\varepsilon_n(\rho)}}{R} = 0, \quad R = \frac{B_n'(\varphi_\varepsilon; \rho)}{B_n(\varphi_\varepsilon; \rho)} = \frac{\rho B_n'(x_\varepsilon)}{B_n(x_\varepsilon)}.$$
$$(29)$$

The solution $\varepsilon_n(\rho)$ ($\varepsilon_n = \varepsilon_n(\rho)/\rho^2$) of the derived equation is sought in the form of a power series

$$\varepsilon_n(\rho) = n^2 + \varepsilon_n^{(1)}\rho^{-1} + \varepsilon_n^{(2)}\rho^{-2} + O(\rho^{-3}). \qquad (30)$$

Substitute (25) into (29), and then substitute (30) into the resulting equation. Expanding both sides of the equation in inverse powers of ρ and neglecting the terms, containing the third and higher powers of $1/\rho$, we arrive at the system of linear equations, from which the expansion coefficients $\varepsilon_n^{(1)}$ and $\varepsilon_n^{(2)}$, and then the coefficients $A_n(\rho)$ and $C_n(\rho)$ are determined. Since the values of the function $B_n(\varphi; \rho)$ and its derivative $B_n'(\varphi; \rho)$ on the grid Ω_φ are known, for the calculation of the first integral we use the quadrature formula of the Newton-Cotes type. The second integral is calculated analytically using the expansion (30). We have the analytical expression in the interval $\varphi_\varepsilon(\rho) < \varphi \le \pi/2$, and the explicit dependence of its values upon the parameter ρ on the grid Ω_φ. For the considered potential (5) we get the asymptotes of the potential curves at $n = j - n_0$:

$$\varepsilon_n(\rho)\rho^{-2} = n^2\rho^{-2} + 4.50520671n^2\rho^{-3} + 15.22266564n^2\rho^{-4} + O(\rho^{-5}). \quad (31)$$

Table 2. Convergence of the expansion (31) at $\rho = 50$ and the numerical results (NUM).

n^2	1.00000000	4.00000000	9.00000000	16.00000000
$+\varepsilon_n^{(1)}/\rho$	1.09010413	4.36041653	9.81093720	17.44166614
$+\varepsilon_n^{(2)}/\rho^2$	1.09619320	4.38477280	9.86573880	17.53909120
NUM	1.09614800	4.38462804	9.86554769	17.53908477

The calculated eigenvalues in comparison with the numerical solution obtained by means of the program ODPEVP [1] are presented in Table 2. The described algorithm is implemented in the Maple system. The asymptotic expansions, obtained using it at $\rho = 50$, coincide with the numerical solution given by the finite element method to 4–5 significant digits for the eigenvalues and to 3–4 significant digits for the eigenfunctions. The asymptotes of the effective potentials (13) between the states $n_1 = i - n_0$ and $n_2 = j - n_0$ of the same parity at $n_0 = 5$, $i,j = n_0 + 1, \ldots$ have the form:

$$Q_{n_1 n_2}(\rho) = \frac{2.27}{\rho^2} \frac{n_2 n_1}{(n_1^2 - n_2^2)} + \frac{5.14}{\rho^3} \frac{n_2 n_1}{(n_1^2 - n_2^2)} + O\left(\frac{1}{\rho^4}\right),$$

$$H_{n_1 n_2}(\rho) = \frac{10.27}{\rho^4} \frac{n_2 n_1 (n_1^2 + n_2^2)}{(n_1^2 - n_2^2)^2} + \frac{68.20}{\rho^5} \frac{n_2 n_1 ((n_1^2 - n_2^2)^2 + 0.68(n_1^2 + n_2^2))}{(n_1^2 - n_2^2)^2} + O\left(\frac{1}{\rho^6}\right),$$

$$H_{n_1 n_1}(\rho) = (0.64 + 2.11n_1^2)\frac{1}{\rho^4} + (2.91 - 10.03n_1^2)\frac{1}{\rho^5} + O\left(\frac{1}{\rho^6}\right). \quad (32)$$

Using (15), (20) and (26) we get the asymptotic expansions for $Q_{ij}(\rho)$ and $H_{ij}(\rho)$ between the cluster states $i = 1, \ldots, n_0$ and pseudostates $(j - n_0) = 1, 2, \ldots,$

$$Q_{ij}(\rho) = \frac{Q_{ij}^{(5/2)}}{\rho^{5/2}} + \frac{Q_{ij}^{(7/2)}}{\rho^{7/2}} + O(\rho^{-9/2}), \quad H_{ij}(\rho) = \frac{H_{ij}^{(7/2)}}{\rho^{7/2}} + \frac{H_{ij}^{(9/2)}}{\rho^{9/2}} + O(\rho^{-11/2}).$$

$$(33)$$

5   Conclusion

The model for quantum tunneling of a diatomic molecule through repulsive barrier is formulated as a 2D boundary-value problem for the Schrödinger equation.

This problem is reduced using the Kantorovich expansions to the boundary-value problem for a set of second-order ordinary differential equations with the third-type boundary conditions. The symbolic-numeric algorithms are proposed and implemented in Maple to evaluate the asymptotic expansions (20), (32), (24) and (33) of the parametric BVP eigensolutions and the effective potentials $W_{ij}(\rho)$ in inverse powers of ρ, used for calculation of the asymptotes of the fundamental solutions of the system of second-order ODEs at large values of ρ [19].

The proposed approach can be applied to the analysis of quantum transparency effect, quantum diffusion of molecules Be_2 and the Efimov effect [5–8,21] in $Be+Be_2$ scattering using modern theoretical and experimental results [13,15–17] and algorithms and programs [1–4,9,10,12].

The work was supported by the Russian Foundation for Basic Research (grant No. 14-01-00420) and the Ministry of Education and Science of the Republic of Kazakhstan (Grant No. 0333/GF4). The reported study was funded within the Agreement N 02.03.21.0008 dated 24.04.2016 between the Ministry of Education and Science of the Russian Federation and RUDN University.

References

1. Chuluunbaatar, O., Gusev, A.A., Vinitsky, S.I., Abrashkevich, A.G.: ODPEVP: a program for computing eigenvalues and eigenfunctions and their first derivatives with respect to the parameter of the parametric self-adjoined Sturm-Liouville problem. Comput. Phys. Commun. **181**, 1358–1375 (2009)
2. Chuluunbaatar, O., Gusev, A.A., Abrashkevich, A.G., et al.: KANTBP: a program for computing energy levels, reaction matrix and radial wave functions in the coupled-channel hyperspherical adiabatic approach. Comput. Phys. Commun. **177**, 649–675 (2007)
3. Chuluunbaatar, O., Gusev, A.A., Vinitsky, S.I., Abrashkevich, A.G.: KANTBP 2.0: new version of a program for computing energy levels, reaction matrix and radial wave functions in the coupled-channel hyperspherical adiabatic approach. Comput. Phys. Commun. **179**, 685–693 (2008)
4. Chuluunbaatar, O., Gusev, A.A., Vinitsky, S.I., Abrashkevich, A.G.: KANTBP 3.0: new version of a program for computing energy levels, reflection and transmission matrices, and corresponding wave functions in the coupled-channel adiabatic approach. Comput. Phys. Commun. **185**, 3341–3343 (2014)
5. Efimov, V.N.: Weakly-bound states of three resonantly-interacting particles. Soviet J. Nucl. Phys. **12**, 589–595 (1971)
6. Efimov, V.: Energy levels of three resonantly interacting particles. Nucl. Phys. A **210**, 157–188 (1973)
7. Efimov, V.: Few-body physics: Giant trimers true to scale. Nat. Phys. **5**, 533–534 (2009)
8. Fonseca, A.C., Redish, E.F., Shanley, P.E.: Efimov effect in an analytically solvable model. Nucl. Phys. A **320**, 273–288 (1979)
9. Gusev, A.A., Chuluunbaatar, O., Vinitsky, S.I., Abrashkevich, A.G.: POTHEA: a program for computing eigenvalues and eigenfunctions and their first derivatives with respect to the parameter of the parametric self-adjoined 2D elliptic partial differential equation. Comput. Phys. Commun. **185**, 2636–2654 (2014)

10. Gusev, A.A., Chuluunbaatar, O., Vinitsky, S.I., Abrashkevich, A.G.: Description of a program for computing eigenvalues and eigenfunctions and their first derivatives with respect to the parameter of the coupled parametric self-adjoined elliptic differential equations. Bull. Peoples' Friendsh. Univ. Russ. Ser. Math. Inf. Sci. Phys. **2**, 336–341 (2014)

11. Gusev, A.A., Hai, L.L.: Algorithm for solving the two-dimensional boundary value problem for model of quantum tunneling of a diatomic molecule through repulsive barriers. Bull. Peoples' Friendship Univ. Russia. Ser. Math. Inf. Sci. Phys. **1**, 15–36 (2015)

12. Gusev, A.A., Hai, L.L., Chuluunbaatar, O., Vinitsky, S.I.: Programm KANTBP 4M for solving boundary problems for a system of ordinary differential equations of the second order (2015). http://wwwinfo.jinr.ru/programs/jinrlib/kantbp.4m

13. Koput, J.: The ground-state potential energy function of a beryllium dimer determined using the single-reference coupled-cluster approach. Chem. Phys. **13**, 20311–20317 (2011)

14. Krassovitskiy, P.M., Pen'kov, F.M.: Contribution of resonance tunneling of molecule to physical observables. J. Phys. B **47**, 225210 (2014)

15. Merritt, J.M., Bondybey, V.E., Heaven, M.C.: Beryllium dimercaught in the act of bonding. Science **324**, 1548–1551 (2009)

16. Mitin, A.V.: Ab initio calculations of weakly bonded He_2 and Be_2 molecules by MRCI method with pseudo-natural molecular orbitals. Int. J. Quantum Chem. **111**, 2560–2567 (2011)

17. Patkowski, K., Špirko, V., Szalewicz, K.: On the elusive twelfth vibrational state of beryllium dimer. Science **326**, 1382–1384 (2009)

18. Pijper, E., Fasolino, A.: Quantum surface diffusion of vibrationally excited molecular dimers. J. Chem. Phys. **126**, 014708 (2007)

19. Vinitsky, S., Gusev, A., Chuluunbaatar, O., Hai, L., Góźdź, A., Derbov, V., Krassovitskiy, P.: Symbolic-numeric algorithm for solving the problem of quantum tunneling of a diatomic molecule through repulsive barriers. In: Gerdt, V.P., Koepf, W., Seiler, W.M., Vorozhtsov, E.V. (eds.) CASC 2014. LNCS, vol. 8660, pp. 472–490. Springer, Heidelberg (2014). doi:10.1007/978-3-319-10515-4_34

20. Wang, J., Wang, G., Zhao, J.: Density functional study of beryllium clusters, with gradient correction. J. Phys.: Condens. Matter **13**, L753–L758 (2001)

21. Zaccanti, M., Deissler, B., D'Errico, C., et al.: Observation of an Efimov spectrum in an atomic system. Nat. Phys. **5**, 586–591 (2009)

The Coupled-Channel Method for Modelling Quantum Transmission of Composite Systems

S.I. Vinitsky[1,2(✉)], A.A. Gusev[1], O. Chuluunbaatar[1,5], A. Góźdź[3], and V.L. Derbov[4]

[1] Joint Institute for Nuclear Research, Dubna, Russia
vinitsky@theor.jinr.ru
[2] RUDN University (Peoples' Friendship University of Russia),
6 Miklukho-Maklaya Street, Moscow 117198, Russia
[3] Institute of Physics, University of Maria Curie-Sklodowska, Lublin, Poland
[4] Saratov State University, Saratov, Russia
[5] Institue of Mathematics, National University of Mongolia, University Street,
Sukhbaatar District, Ulaanbaatar, Mongolia

Abstract. The description of quantum transmission of composite systems of barriers or wells using the coupled-channel method is presented. In this approach the multichannel scattering problem for the Schrödinger equation is reduced to a set of coupled second-order ordinary differential equations with the boundary conditions of the third type and solved using the finite element method. The efficiency of the proposed approach is demonstrated by the example of analyzing metastable states that appear in composite quantum systems tunnelling through barriers and wells and give rise to the quantum transparency and total reflection effects.

Keywords: Coupled-channel method · Quantum tunnelling · Second-order ordinary differential equations · Finite element method software

1 Introduction

Quantum tunnelling of composite systems through barriers is one of the problems most often occurring in nuclear physics, physics of solid state and semiconductor nanostructures. Usually the theory is based on considering the penetration of a structureless particle through barriers within the effective mass approximation [19]. However, the majority of important applications deal with tunnelling of structured objects (clusters), e.g., atomic nuclei through Coulomb barrier, where the effects of structure (multiple particles) manifest themselves in anomalous behaviour of nuclear reaction cross-sections below the Coulomb barrier [20]. Indeed, when the cluster size is comparable with the spatial width of the barrier, the mechanisms arise that enhance the barrier transparency. The effect of quantum barrier transparency depending on the internal structure of the incident particles was revealed for a pair of coupled particles tunnelling through a repulsive barrier [9]. The effect was shown to be due to the barrier resonance formation under the condition that the potential energy of the compound system (cluster + barriers) possesses local minima, thus providing the

© Springer International Publishing AG 2016
V.M. Vishnevskiy et al. (Eds.): DCCN 2016, CCIS 678, pp. 525–537, 2016.
DOI: 10.1007/978-3-319-51917-3_45

appearance of metastable states of the moving cluster [8]. The manifestations
and the underlying mechanisms of the effect were extensively studied in multiple
quantum phenomena [14–18, 20–23], for example, near-surface quantum diffusion
of molecules [10], channelling and tunnelling of ions through multidimensional
barriers [2,5,11,22,24], and sub-barrier tunnelling of light nuclei [12], and the
collinear ternary fission [13]. A method and programs for solving the tunnelling
of a system of n identical particles coupled by oscillator-type potentials through
repulsive barriers has been presented in [1,3,4,6,7], while their application to
study of a transmission of composite systems of both barriers and wells is actual
problem in the field.

In present paper we consider the problem of a transmission of composite
systems of barriers or wells in the framework of the coupled-channel method
basing on the Galerkin-type and Kantorovich methods and discuss conditions of
their applicability. By the examples of particles with different coupling potentials, transmission of composite systems as of Gaussian barriers or wells, the
transmission coefficients, and the metastable states are analyzed. The energy
dependencies of these coefficients demonstrate the phenomena of quantum transparency and total reflection.

The structure of paper is following. In Sect. 2 the coupled-channel method
and the multichannel scattering problem are formulated. In Sect. 3 the transmission of clusters comprising several identical particles coupled by oscillator
and double-well polynomial potentials are studied separately: tunneling through
barrier, transmission above barriers and wells. In Conclusion the results and
perspectives are discussed.

2   Problem Statement

2.1   Coupled-Channel Method

Consider the boundary-value problem (BVP) for the equation

$$\left(\hat{H}_f(\mathbf{x}_f; x_s) + \hat{H}_s(x_s) + \check{V}_{fs}(\mathbf{x}_f, x_s) - \mathcal{E}_t \right) \Psi_t(\mathbf{x}_f, x_s) = 0 \tag{1}$$

with fast $\mathbf{x}_f$ and slow x_s variables. The operators $\hat{H}_f(\mathbf{x}_f; x_s)$ and $\hat{H}_s(x_s)$ describe
the fast and slow subsystem

$$\hat{H}_f(\mathbf{x}_f; x_s) = -\frac{1}{g_{1f}(\mathbf{x}_f)} \frac{\partial}{\partial \mathbf{x}_f} g_{2f}(\mathbf{x}_f) \frac{\partial}{\partial \mathbf{x}_f} + \check{V}_f(\mathbf{x}_f; x_s), \tag{2}$$

$$\hat{H}_s(x_s) = -\frac{1}{g_{1s}(x_s)} \frac{\partial}{\partial x_s} g_{2s}(x_s) \frac{\partial}{\partial x_s} + \check{V}_s(x_s), \tag{3}$$

$\check{V}_f(\mathbf{x}_f; x_s)$ and $\check{V}_s(x_s)$ are the potentials of the fast and slow subsystem, and
$\check{V}_{fs}(\mathbf{x}_f, x_s)$ is the interaction potential. The solution $\Psi_t(\mathbf{x}_f, x_s)$ of the problem (1)
with the appropriate boundary conditions is sought in the form of Kantorovich
expansion

$$\Psi_t(\mathbf{x}_f, x_s) = \sum_{j=1}^{j_{\max}} B_j(\mathbf{x}_f; x_s)\chi_{jt}(x_s). \tag{4}$$

The trial functions $B_j(\mathbf{x}_f; x_s)$ are chosen to be eigenfunctions of the Hamiltonian $\hat{H}_f(\mathbf{x}_f; x_s)$ with the eigenvalues $\hat{E}_j(x_s)$, parametrically depending on $x_s \in \Omega(x_s)$:

$$\hat{H}_f(\mathbf{x}_f; x_s)B_j(\mathbf{x}_f; x_s) = \hat{E}_j(x_s)B_j(\mathbf{x}_f; x_s). \tag{5}$$

These functions satisfy the orthonormality conditions with the weighting function $g_{1f}(\mathbf{x}_f)$ in the same interval $\mathbf{x}_f \in \Omega_{\mathbf{x}_f}(x_s)$:

$$\int_{\mathbf{x}_f^{\min}(x_s)}^{\mathbf{x}_f^{\max}(x_s)} B_i(\mathbf{x}_f; x_s)B_j(\mathbf{x}_f; x_s)g_{1f}(\mathbf{x}_f)d\mathbf{x}_f = \delta_{ij}. \tag{6}$$

Substitution of (4) into (1) yields a BVP for a set of ODEs with respect to the unknown vector functions $\chi_t(x_s) = (\chi_{1;t}(x_s), ..., \chi_{j_{\max};t}(x_s))^T$ of the slow subsystem, corresponding to the unknown eigenvalues $2E_t \equiv \mathcal{E}_t$,

$$\left(\mathbf{D} + \mathbf{E}(x_s) + \mathbf{W}(x_s) - \mathbf{I}\mathcal{E}_t\right)\chi_t(x_s) = 0,$$

$$\mathbf{D} = -\frac{1}{g_{1s}(x_s)}\mathbf{I}\frac{d}{dx_s}g_{2s}(x_s)\frac{d}{dx_s} + \mathbf{I}\check{V}_s(x_s), \tag{7}$$

$$\mathbf{W}(x_s) = \mathbf{V}(x_s) + \frac{g_{2s}(x_s)}{g_{1s}(x_s)}\mathbf{H}(x_s) + \frac{1}{g_{1s}(x_s)}\frac{dg_{2s}(x_s)\mathbf{Q}(x_s)}{dx_s} + \frac{g_{2s}(x_s)}{g_{1s}(x_s)}\mathbf{Q}(x_s)\frac{d}{dx_s}$$

with the effective potentials $H_{ij}(x_s)$ and $Q_{ij}(x_s)$ defined as

$$V_{ij}(x_s) = V_{ji}(x_s) = \int_{\mathbf{x}_f^{\min}(x_s)}^{\mathbf{x}_f^{\max}(x_s)} B_i(\mathbf{x}_f; x_s)\check{V}_{fs}(\mathbf{x}_f, x_s)B_j(\mathbf{x}_f; x_s)g_{1f}(\mathbf{x}_f)d\mathbf{x}_f,$$

$$H_{ij}(x_s) = H_{ji}(x_s) = \int_{\mathbf{x}_f^{\min}(x_s)}^{\mathbf{x}_f^{\max}(x_s)} \frac{\partial B_i(\mathbf{x}_f; x_s)}{\partial x_s}\frac{\partial B_j(\mathbf{x}_f; x_s)}{\partial x_s}g_{1f}(\mathbf{x}_f)d\mathbf{x}_f, \tag{8}$$

$$Q_{ij}(x_s) = -Q_{ji}(x_s) = -\int_{\mathbf{x}_f^{\min}(x_s)}^{\mathbf{x}_f^{\max}(x_s)} B_i(\mathbf{x}_f; x_s)\frac{\partial B_j(\mathbf{x}_f; x_s)}{\partial x_s}g_{1f}(\mathbf{x}_f)d\mathbf{x}_f.$$

If the potential of the fast subsystem $\check{V}_f(\mathbf{x}_f; x_s)$ is independent of the slow variable, then the expansion is referred to as Galerkin-type expansion. Its advantage is that the eigenvalue problem (5) should be solved only once. However, if the position of the potential well and, therefore, the localization of eigenfunctions changes, the convergence of Galerkin-type expansions becomes very slow [5]. The example of the effective potentials of double-well potential (from Fig. 1) for Galerkin-type and Kantorovich expansions are shown in Fig. 2. In considered case the Galerkin method is a more appropriate because effective potentials have a smooth behavior, while in Kantorovich method effective potentials have a sharp behavior with a large magnitude due to series of quasicrossing of the potential curves.

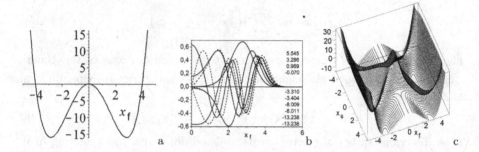

Fig. 1. Double-well interaction potential (a), the first even (solid lines) and odd (dashed lines) eigenfunctions (b), and the corresponding 2D potential $V(x_f) + V^b(x_f; x_s)$ (c).

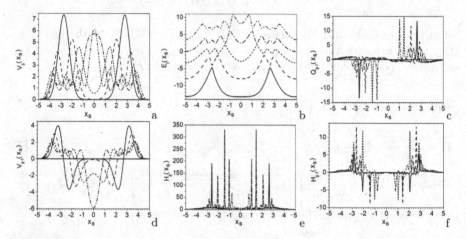

Fig. 2. Even effective potentials for Galerkin-type (a, d) and Kantorovich (b, c, e, f) expansions.

2.2 Scattering Problem

Consider the scattering problem with the homogeneous boundary conditions of the third kind at $x_s = x_s^{\min} \ll 0$ and $x_s = x_s^{\max} \gg 0$:

$$\frac{d\Phi(x_s)}{dx_s}\bigg|_{x_s=x_s^{\min}} = \mathcal{R}(x_s^{\min})\Phi(x_s^{\min}), \quad \frac{d\Phi(x_s)}{dx_s}\bigg|_{x_s=x_s^{\max}} = \mathcal{R}(x_s^{\max})\Phi(x_s^{\max}), \quad (9)$$

where $\mathcal{R}(x_s)$ is an unknown $N \times N$ matrix function, $\Phi(x_s) = \{\chi^{(j)}(x_s)\}_{j=1}^{N_o}$ is the desired $N \times N_o$ matrix solution and N_o is the number of open channels, $N_o = \max_{2E \geq \epsilon_j} j \leq N$.

The matrix solution $\Phi_v(x_s) = \Phi(x_s)$, describing the incidence of the particle and its scattering, with the asymptotic form "incident wave + outgoing waves" (see Fig. 4a) is written as

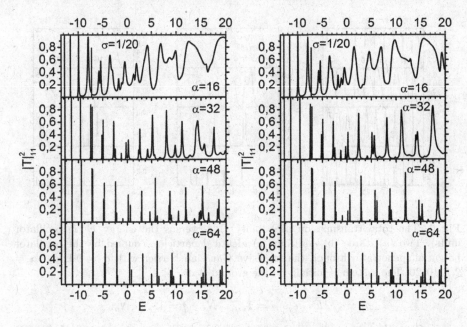

Fig. 3. The total probability of penetration through repulsive Gaussian potential barriers $|\mathbf{T}|^2_{11}$ versus the energy E with the ground and excited initial states.

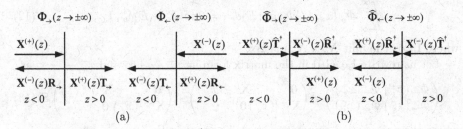

Fig. 4. Schematic diagrams of the wave functions $\Phi_v(z)$ at $z \equiv x_s$ having the asymptotic form: (a) "incident wave + outgoing waves", (b) "incident waves + ingoing wave".

$$\Phi_v(x_s \to \pm\infty) = \begin{cases} \begin{cases} \mathbf{X}^{(+)}(x_s)\mathbf{T}_v, & x_s > 0, \\ \mathbf{X}^{(+)}(x_s) + \mathbf{X}^{(-)}(x_s)\mathbf{R}_v, & x_s < 0, \end{cases} & v =\to, \\ \begin{cases} \mathbf{X}^{(-)}(x_s) + \mathbf{X}^{(+)}(x_s)\mathbf{R}_v, & x_s > 0, \\ \mathbf{X}^{(-)}(x_s)\mathbf{T}_v, & x_s < 0, \end{cases} & v =\leftarrow, \end{cases} \quad (10)$$

where $\mathbf{R}_v$ and $\mathbf{T}_v$ are the reflection and transmission $N_o \times N_o$ matrices, $v =\to$ and $v =\leftarrow$ denote the initial direction of the particle motion along the x_s axis. The leading term of the asymptotic rectangular matrix functions $\mathbf{X}^{(\pm)}(x_s)$ has the form [5]

$$X^{(\pm)}_{ij}(x_s) \to p_j^{-1/2} \exp\left(\pm\imath\left(p_j x_s - \frac{Z_j}{p_j} \ln(2p_j|x_s|)\right)\right) \delta_{ij}, \quad (11)$$

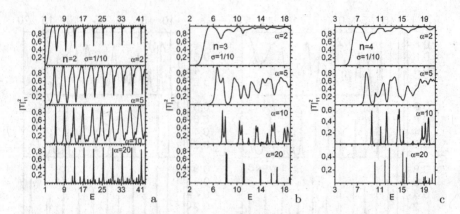

Fig. 5. The total transmission probability $|\mathbf{T}|_{11}^2$ versus the energy E (in oscillator units). Two (a), three (b) and four (c) identical particles, coupled by the oscillator potential, penetrate through the repulsive Gaussian barrier with $\sigma = 0.1$ and $\alpha = 2, 5, 10, 20$. The system is initially in the ground state.

$$p_j = \sqrt{2E - \epsilon_j} \quad i = 1, \ldots, N, \quad j = 1, \ldots, N_o,$$

where $Z_j = Z_j^+$ at $x_s > 0$ and $Z_j = Z_j^-$ at $x_s < 0$. The matrix solution $\Phi_v(x_s, E)$ is normalized by the condition

$$\int_{-\infty}^{\infty} \Phi_{v'}^{\dagger}(x_s, E')\Phi_v(x_s, E)dx_s = 2\pi\delta(E' - E)\delta_{v'v}\mathbf{I}_{oo}, \tag{12}$$

where $\mathbf{I}_{oo}$ is the unit $N_o \times N_o$ matrix.

Let us rewrite Eq. (10) in the matrix form at $x_s^+ \to +\infty$ and $x_s^- \to -\infty$ as

$$\begin{pmatrix} \Phi_\to(x_s^+) & \Phi_\leftarrow(x_s^+) \\ \Phi_\to(x_s^-) & \Phi_\leftarrow(x_s^-) \end{pmatrix} = \begin{pmatrix} \mathbf{0} & \mathbf{X}^{(-)}(x_s^+) \\ \mathbf{X}^{(+)}(x_s^-) & \mathbf{0} \end{pmatrix} + \begin{pmatrix} \mathbf{0} & \mathbf{X}^{(+)}(x_s^+) \\ \mathbf{X}^{(-)}(x_s^-) & \mathbf{0} \end{pmatrix}\mathbf{S}, \tag{13}$$

where the unitary and symmetric scattering matrix $\mathbf{S}$

$$\mathbf{S} = \begin{pmatrix} \mathbf{R}_\to & \mathbf{T}_\leftarrow \\ \mathbf{T}_\to & \mathbf{R}_\leftarrow \end{pmatrix}, \qquad \mathbf{S}^{\dagger}\mathbf{S} = \mathbf{S}\mathbf{S}^{\dagger} = \mathbf{I} \tag{14}$$

is composed of the reflection and transmission matrices. Detailed calculation of the matrix solution $\Phi_v(x_s)$ is presented in Reference [4].

3   Transmission of Clusters Comprised by Several Identical Particles

Consider a cluster of two or three identical particles with the masses m coupled via the pair potentials $\tilde{U}^{pair}(x_{tt'})$, $x_{tt'} = x_t - x_{t'}$ propagated the barrier or well $\tilde{V}(x_t)$. The wave function of this system satisfies the Schrödinger equation

$$\left[-\sum_{t=1}^{n} \frac{\partial^2}{\partial x_t^2} + \sum_{t,t'=1;t<t'}^{n} \frac{(x_{tt'})^2}{n} + \sum_{t,t'=1;t<t'}^{n} U^{pair}(x_{tt'}) + \sum_{t=1}^{n} V(x_t) - E \right] \Psi(\mathbf{x}) = 0. \tag{15}$$

Here E is the total energy of n particles, $V(x_t) = \tilde{V}(x_t x_{osc})/E_{osc}$ is the barrier or the well potential, $V^{hosc}(x_{tt'}) = \tilde{V}^{hosc}(x_{tt'} x_{osc})/E_{osc} = \frac{1}{n}(x_{tt'})^2$ is the harmonic oscillator potential, $V^{pair}(x_{tt'}) = \tilde{V}^{pair}(x_{tt'} x_{osc})/E_{osc}$ and $U^{pair}(x_{tt'}) = V^{pair}(x_{tt'}) - V^{hosc}(x_{tt'})$ is the effective pair potential given in the oscillator units. In the symmetric coordinates [4, 6]:

$$\xi_0 = \frac{1}{\sqrt{n}} \sum_{t=1}^{n} x_t, \quad \xi_{t'} = \frac{1}{\sqrt{n}} \left(x_1 + \sum_{t=2}^{n} a_0 x_t + \sqrt{n} x_{t'+1} \right), \quad t' = 1, ..., n-1, \quad (16)$$

where $a_0 = 1/(1 - \sqrt{n}) < 0$, $a_1 = a_0 + \sqrt{n}$, Eq. (15) takes the form

$$\left[-\frac{\partial^2}{\partial \xi_0^2} + \sum_{i=1}^{n-1} \left(-\frac{\partial^2}{\partial \xi_i^2} + (\xi_i)^2 \right) + U(\xi_0, ..., \xi_{n-1}) - E \right] \Psi(\xi_0, ..., \xi_{n-1}) = 0,$$

$$U(\xi_0, ..., \xi_{n-1}) = \sum_{i,j=1; i<j}^{n} U^{pair}(x_{ij}(\xi_1, ..., \xi_{n-1})) + \sum_{i=1}^{n} V(x_i(\xi_0, ..., \xi_{n-1})). \quad (17)$$

Here $x_s = \xi_0$ in the center-of-mass variable and $\mathbf{x}_f = \{\xi_1, ..., \xi_{n-1}\}$ is the set of relative variables, such that at $n = 2$ they correspond to the Jacobi coordinates (Fig. 8).

Double-Well Interaction Potential. Now consider a pair of particles, coupled by the double-well interaction potential $V(x_f) = x_f^4/4 - 4x_f^2$ (see Fig. 1a) tunnelling through the repulsive Gaussian barriers $V_i(x_i) = \alpha \exp(-x_i^2/2\sigma)$ with $\alpha = 16, 32, 48, 64$, $\sigma = 1/20$. In this case Eq. (15) takes the form

$$\left(-\frac{\partial^2}{\partial x_s^2} - \frac{\partial^2}{\partial x_f^2} + V(x_f) + V^b(x_f; x_s) - 2E \right) \Psi(x_f, x_s) = 0, \quad (18)$$

where $V^b(x_f; x_s) = V_1(x_1) + V_2(x_2)$.

The first even and odd eigenfunctions are presented in Fig. 1b. The typical behaviour of symmetric double-well potential eigenfunctions is seen, namely, for $E < 0$ there are pairs of even and odd eigenfunctions localized in the potential wells, with closely spaced energy levels. For $E > 0$ the energy levels of even and odd states alternate. The corresponding 2D potential is demonstrated in Fig. 1c.

In this case we have two possibilities to construct the fast, slow, and interaction potential, corresponding either to the Galerkin-type expansion

$$\check{V}_f(x_f; x_s) = V(x_f), \quad \check{V}_s(x_s) = 0, \quad \check{V}_{fs}(x_f, x_s) = V^b(x_f; x_s),$$

or the Kantorovich expansion

$$\check{V}_f(x_f; x_s) = V(x_f) + V^b(x_f; x_s), \quad \check{V}_s(x_s) = 0, \quad \check{V}_{fs}(x_f, x_s) = 0.$$

The effective potentials (8) are presented in Fig. 2. It is seen that the non-diagonal matrix elements in the case of Kantorovich expansion are small as compared to the case of Galerkin-type expansion, except some areas, corresponding to quasi-crossing of the energy levels in the problem (5) (see Fig. 2b).

Figure 3 shows the energy dependence of the total transmission probability $|\mathbf{T}|_{ii}^2 = \sum_{j=1}^{N_o} |T_{ji}(E)|^2$. This is the probability of a transition from a chosen state i into any of N_o states, found by solving the boundary-value problem in the Galerkin form. The behaviour of the probability versus the energy is non-monotonic, and the observed resonances are manifestations of the quantum transparency effect. This effect is caused by the existence of barrier metastable states, embedded in the continuum.

Parabolic Interaction Potential. Two, three or four identical particles (n=2,3,4) are coupled by the harmonic oscillator potential $V(x_t - x_{t'}) = (x_t - x_{t'})^2$, $t', t = 1, ..., n$ and the Gaussian barrier $(\alpha > 0)$ or well $(\alpha < 0)$: $V(x_t) = \alpha/(\sqrt{2\pi}\sigma) \exp(-x_t^2/\sigma^2)$.

Figure 5 shows the energy dependence of the total transmission probability $|\mathbf{T}|_{ii}^2 = \sum_{j=1}^{N_o} |T_{ji}(E)|^2$. This is the probability of a transition from the ground state i to any of N_o eigenstates of the BVP in the Galerkin form solved using the program KANTBP [1,3]. The dependence of the probability upon the energy is non-monotonic, and the observed resonance peaks are manifestations of the quantum transparency effect. The multiplet structure of the peaks for symmetric and antisymmetric states is similar. Due to the symmetry of the potential in the case of two identical particles, the position of the maxima for symmetric and antisymmetric states coincide. In the case of three particles peak positions for symmetric and antisymmetric states are different, but due to the symmetry with respect to the plane $\xi_0 = 0$, explain the presence of doublets.

Figure 6 shows the profiles of $|\Psi|^2 \equiv |\Psi_{Em\to}^{(-)}|^2$ with $\alpha = 20$, $\sigma = 1/10$ at the resonance energies of the first three maxima and the second maximum and the first minimum of the transmission coefficient, illustrating the resonance transmission. It is seen that in the case of resonance transmission the energy is transferred from the centre-of-mass degree of freedom, described by the coordinate ξ_0, to the internal (transverse) one, described by ξ_1 i.e., the transverse oscillator undergoes a transition from the ground state to the excited state. On the contrary, in the case of total reflection the energy transfer is extremely small, and the transverse oscillator returns to infinity in the initial state. In Fig. 7 the first three metastable states are presented. The wave function amplitudes for these states are seen to differ from the amplitudes of the states, corresponding to the first three maxima in the vicinity of wells.

Figure 9 shows the profiles of probability density $|\Psi(\xi_0, \xi_1)|^2$ for the symmetric states of $A = 2$ particles transmitting above Gaussian barrier $\alpha = 2$, $\sigma = 1/10$, revealing total reflection at resonance energies. In Table 1 the values of energies $E_m^M = \Re E_m^M + i\Im E_m^M$ of corresponding metastable states for a transmission of $A = 2$ particles above the Gaussian barrier $\alpha = 2$, $\sigma = 1/10$ are presented. One can see that the series of resonances in the transmission $|\mathbf{T}|_{11}^2$ from the ground state 1 are induced by metastable states from second, third, fourth and seventh closed channels, respectively from left to right panels.

In Fig. 10 the total transmission probability $|\mathbf{T}|_{11}^2$ versus the energy E (in oscillator units) for systems of the $A = 2, 3, 4$ particles, coupled by the oscillator potential, propagating above the Gaussian well with $\sigma = 0.1$ and $\alpha = -1, -2$

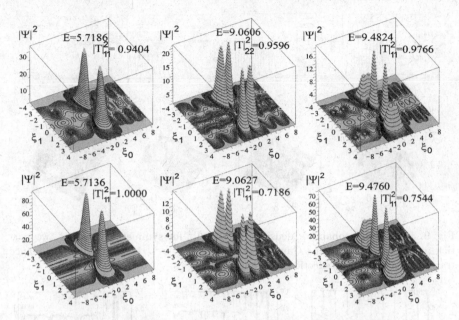

Fig. 6. Profiles of probability densities $|\Psi(\xi_0, \xi_1)|^2$ for symmetric (top panel) and anti-symmetric (bottom panel) states of two particles, revealing resonance transmission and total reflection at resonance energies, shown in Fig. 5.

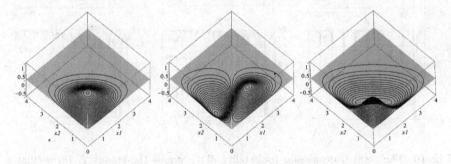

Fig. 7. The first three metastable states corresponding to $E_i^D = 5.76$, 9.12, 9.53.

E_i^S	$|\mathbf{T}|_{11}^2$	$|\mathbf{T}|_{33}^2$	E_m^M
5.8228	0.3794		
9.6479	0.3779		$9.614 - i0.217$
13.5548	0.4765		$13.505 - i0.144$
13.9648		0.8536	$14.018 - i0.286$
17.4512	0.4874		$17.445 - i0.103$

Fig. 8. The 2D potential for propagation of two particles ($n = 2$) above the Gaussian barrier $\alpha = 2$, $\sigma = 1/10$ and the values of energies $E_m^M = \Re E_m^M + i\Im E_m^M$ of metastable states corresponding to the peaks of $|\mathbf{T}|_{11}^2$ shown in Fig. 5a.

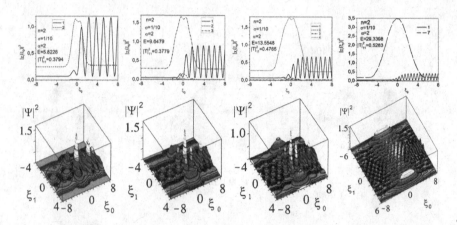

Fig. 9. Profiles of probability densities $|\Psi(\xi_0,\xi_1)|^2$ for symmetric states of two particles transmitted above the Gaussian barrier $\alpha = 2$, $\sigma = 1/10$, revealing resonance transmission and total reflection at resonance energies.

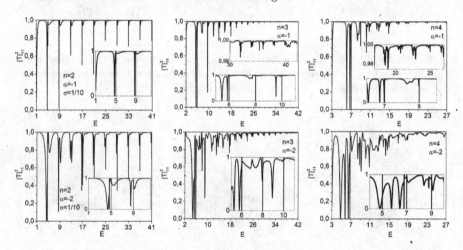

Fig. 10. The total transmission probability $|\mathbf{T}|^2_{11}$ versus the energy E (in oscillator units). The cluster of $n = 2, 3, 4$ particles, coupled by the oscillator potential, propagates above the Gaussian well with $\sigma = 0.1$ and $\alpha = -1, -2$. The system is initially in the ground state. The vertical lines on the epures denote the threshold energies.

are presented. In Table 1 the values of energies $E_m^M = \Re E_m^M + i\Im E_m^M$ of the corresponding metastable states for the transmission of $A = 2$, 3 and 4 particles above the Gaussian well $\alpha = -2$, $\sigma = 1/10$ are shown. The energies $E_m^B < E_1^{th}$ of bound states below first threshold E_1^{th} shown in last row. One can see that the resonance structure becomes enriched with increasing the number of transmitted particles. So, in the case of $A = 2$ we see double-resonance structures, similar to the double-well case. In the case of $A = 3$ and 4 the double structure can appear with increasing the depth of wells $|\alpha|$. Figure 11 presents the profiles of

Table 1. The values of energies $E_m^M = \Re E_m^M + i\Im E_m^M$ of metastable states for a transmission of a cluster of $A = 2$, 3 and 4 particles above the Gaussian well $\alpha = -2$, $\sigma = 1/10$ shown in Figs. 10 and 11. The energies $E_m^B < E_1^{th}$ of bound states below the first threshold E_1^{th} are shown in last row.

E_i^{th}	$E_m^M(A=2)$	E_i^{th}	$E_m^M(A=3)$	E_i^{th}	$E_m^M(A=4)$
1	$4.4348-i0.2572$	2	$5.3307-i0.0620$	3	$5.7747-i0.0742$
	$4.6764-i0.0058$		$5.7911-i0.0621$		$6.4441-i0.1050$
5	$8.5158-i0.0506$	6	$6.9922-i0.0751$		$6.7934-i0.0033$
	$8.7675-i0.1261$		$7.9457-i0.0565$	7	$8.3668-i0.0651$
9	$12.6009-i0.1215$	8	$8.9601-i0.0588$		$8.7797-i0.0080$
	$12.7330-i0.0142$		$9.4950-i0.2251$	9	$9.4050-i0.1995$
13	$16.6841-i0.0364$		$9.8617-i0.0852$		$9.9926-i0.1225$
	$16.7050-i0.0914$	10	$11.4173-i0.1678$		$10.0755-i0.0676$
Bound states: -0.3588		$\{-0.2605, 1.5082\}$		$\{-0.1938, 1.7084\ 2.7046\}$	

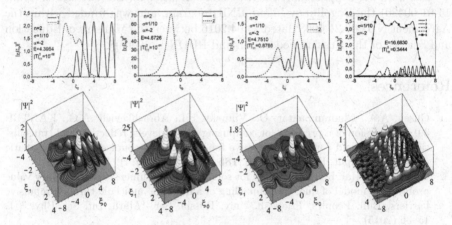

Fig. 11. Profiles of probability density $|\Psi(\xi_0,\xi_1)|^2$ for symmetric states of two particles transmitted above the Gaussian well $\alpha = 2$, $\sigma = 1/10$, revealing total reflection and resonance transmission at the resonance energies.

probability density $|\Psi(\xi_0,\xi_1)|^2$ for the symmetric states of two particles transmitted above the Gaussian well $\alpha = 2$, $\sigma = 1/10$, revealing the resonance transmission and total reflection at resonance energies. One can see that the series of resonances in the transmission $|\mathbf{T}|_{11}^2$ from the ground state 1 are induced by Feshbach metastable states from second and the fifth closed channels, respectively, from left to right panels. In contrast to the case of barrier in the vicinity of the well resonance, we see both the resonance reflection and the transmission (see two middle panels in Fig. 11).

4   Conclusion

We considered the application of the coupled-channel methods to the problem of quantum tunnelling of a cluster of particles coupled by the oscillator-type interactions, through Gaussian potential barriers and above wells. The initial boundary problem is reduced to that for a set ordinary differential equations of the second order. By a few examples we demonstrate the efficiency of the proposed approach for the cluster tunnelling problem and the capability of the method to provide correct description of the cluster tunnelling specific features, including the quantum transparency and total reflection phenomena induced by the shape and Feshbach metastable states. The Kantorovich method finds a more general application in solving multichannel scattering problems with long-range interactions [5, 11] and the break-up processes in few-body systems in hyperspherical adiabatic representation [25]. An important advantage of the approach is the possibility of efficient use of symbolic-numeric software packages that considerably simplify the calculations as compared to direct numerical approaches.

The work was supported by the Russian Foundation for Basic Research (grant 14-01-00420) the Bogoliubov-Infeld JINR program, and was funded within the Agreement N 02.03.21.0008 dated 24.04.2016 between the Ministry of Education and Science of the Russian Federation and RUDN University.

References

1. Gusev, A.A., Chuluunbaatar, O., Vinitsky, S.I., Abrashkevich, A.G.: KANTBP 3.0: new version of a program for computing energy levels, reflection and transmission matrices, and corresponding wave functions in the coupled-channel adiabatic approach. Comput. Phys. Commun. **185**, 3341–3343 (2014)
2. Gusev, A.A., Hai, L.L.: Algorithm for solving the two-dimensional boundary value problem for model of quantum tunneling of a diatomic molecule through repulsive barriers. Bull. Peoples' Friendsh. Univ. Russia. Ser. "Math. Inf. Sci. Phys." **1**, 15–36 (2015)
3. Gusev, A.A., Hai, L.L., Chuluunbaatar, O., Vinitsky, S.I.: Programm KANTBP 4M for solving boundary-value problems for systems of ordinary differential equations of the second order (2015). http://wwwinfo.jinr.ru/programs/jinrlib/kantbp4m
4. Gusev, A.A., Vinitsky, S.I., Chuluunbaatar, O., Derbov, V.L., Góźdź, A., Krassovitskiy, P.M.: Metastable states of a composite system tunneling through repulsive barriers. Theor. Math. Phys. **186**, 21–40 (2016)
5. Gusev, A.A., Vinitsky, S.I., Chuluunbaatar, O., Gerdt, V.P., Rostovtsev, V.A.: Symbolic-numerical algorithms to solve the quantum tunneling problem for a coupled pair of ions. In: Gerdt, V.P., Koepf, W., Mayr, E.W., Vorozhtsov, E.V. (eds.) CASC 2011. LNCS, vol. 6885, pp. 175–191. Springer, Heidelberg (2011). doi:10.1007/978-3-642-23568-9_14
6. Gusev, A., Vinitsky, S., Chuluunbaatar, O., Rostovtsev, V., Hai, L., Derbov, V., Góźdź, A., Klimov, E.: Symbolic-numerical algorithm for generating cluster eigenfunctions: identical particles with pair oscillator interactions. In: Gerdt, V.P., Koepf, W., Mayr, E.W., Vorozhtsov, E.V. (eds.) CASC 2013. LNCS, vol. 8136, pp. 155–168. Springer, Heidelberg (2013). doi:10.1007/978-3-319-02297-0_14

7. Vinitsky, S., Gusev, A., Chuluunbaatar, O., Rostovtsev, V., Hai, L., Derbov, V., Krassovitskiy, P.: Symbolic-numerical algorithm for generating cluster eigenfunctions: tunneling of clusters through repulsive barriers. In: Gerdt, V.P., Koepf, W., Mayr, E.W., Vorozhtsov, E.V. (eds.) CASC 2013. LNCS, vol. 8136, pp. 427–442. Springer, Heidelberg (2013). doi:10.1007/978-3-319-02297-0_35

8. Pen'kov, F.M.: Quantum transmittance of barriers for composite particles. J. Exp. Theor. Phys. **91**, 698–705 (2000)

9. Pen'kov, F.M.: Metastable states of a coupled pair on a repulsive barrier. Phys. Rev. A **62**, 044701-1-4 (2000)

10. Pijper, E., Fasolino, A.: Quantum surface diffusion of vibrationally excited molecular dimers. J. Chem. Phys. **126**, 014708-1-10 (2007)

11. Chuluunbaatar, O., Gusev, A.A., Derbov, V.L., Krassovitskiy, P.M., Vinitsky, S.I.: Channeling problem for charged particles produced by confining environment. Phys. Atom. Nucl. **72**, 768–778 (2009)

12. Shotter, A.C., Shotter, M.D.: Quantum mechanical tunneling of composite particle systems: linkage to sub-barrier nuclear reactions. Phys. Rev. C **83**, 054621 (2011)

13. Tashkhodjaev, R.B., Muminov, A.I., Nasirov, A.K., von Oertzen, W., Yongseok, O.: Theoretical study of the almost sequential mechanism of true ternary fission. Phys. Rev. C **91**, 054612 (2015)

14. Gusev, A.A., Vinitsky, S.I., Chuluunbaatar, O., Hai, L.L., Derbov, V.L., Gozdz, A., Krassovitskiy, P.M.: Resonant tunneling of a few-body cluster through repulsive barriers. Phys. At. Nucl. **77**, 389–413 (2014)

15. Sato, T., Kayanuma, Y.: Quantum inelasticity in reflection of a composite particle. Europhys. Lett. **60**, 331–336 (2002)

16. Bertulani, C.A., Flambaum, V.V., Zelevinsky, V.G.: Tunneling of a composite particle: effects of intrinsic structure. J. Phys. G: Nucl. Part. Phys. **34**, 2289–2295 (2007)

17. Flambaum, V.V., Zelevinsky, V.G.: Quantum tunnelling of a complex system: effects of a finite size and intrinsic structure. J. Phys. G: Nucl. Part. Phys. **31**, 355–360 (2005)

18. Bertulani, C.A.: Tunneling of atoms, nuclei and molecules. Few-Body Syst. **56**, 727–736 (2015)

19. Esaki, L.: Long journey into tunneling. Rev. Mod. Phys. **46**, 237–244 (1973)

20. Lemasson, A., et al.: Modern rutherford experiment: tunneling of the most neutron-rich nucleus. Phys. Rev. Lett. **103**, 232701 (2009)

21. Lugovskoy, A.V., Bray, I.: Pseudostate description of diatomic-molecule scattering from a hard-wall potential. Phys. Rev. A **87**, 012904 (2013)

22. Bulatov, V.L., Kornilovitch, P.E.: Anomalous tunneling of bound pairs in crystal lattices. Europhys. Lett. **71**, 352–358 (2005)

23. Muradyan, G., Hakobyan, H., Muradyan, A.Z.: Tunneling dynamics of a two-atom system. J. Phys: Conf. Ser. **672**, 012013 (2016)

24. Bondar, D.I., Liu, W.-K., Ivanov, M.Y.: Enhancement and suppression of tunneling by controlling symmetries of a potential barrier. Phys. Rev. A **82**, 052112 (2010)

25. Chuluunbaatar, O., Gusev, A.A., Derbov, V.L., Kaschiev, M.S., Melnikov, L.A., Serov, V.V., Vinitsky, S.I.: Calculation of a hydrogen atom photoionization in a strong magnetic field by using the angular oblate spheroidal functions. J. Phys. A **40**, 11485–11524 (2007)

The Stochastic Processes Generation in OpenModelica

Migran Gevorkyan[1], Michal Hnatich[3,4,5], Ivan M. Gostev[6], A.V. Demidova[1], Anna V. Korolkova[1], Dmitry S. Kulyabov[1,2], and Leonid A. Sevastianov[1,3(✉)]

[1] Department of Applied Probability and Informatics,
RUDN University (Peoples' Friendship University of Russia),
6 Miklukho-Maklaya str., Moscow 117198, Russia
{mngevorkyan,avdemidova,akorolkova,dharma,sevast}@sci.pfu.edu.ru
[2] Laboratory of Information Technologies, Joint Institute for Nuclear Research,
6 Joliot-Curie, Dubna, Moscow Region 141980, Russia
hnatic@saske.sk
[3] Bogoliubov Laboratory of Theoretical Physics,
Joint Institute for Nuclear Research,
6 Joliot-Curie, Dubna, Moscow Region 141980, Russia
[4] Department of Theoretical Physics, SAS, Institute of Experimental Physics,
Watsonova 47, 040 01 Košice, Slovakia
[5] Faculty of Science, Pavol Jozef Šafárik University in Košice (UPJŠ),
Šrobárova 2, 041 80 Košice, Slovakia
[6] National Research University Higher School of Economics,
20 Myasnitskaya Ulitsa, Moscow 101000, Russia
igostev@hse.ru

Abstract. This paper studies program implementation problem of pseudo-random number generators in OpenModelica. We give an overview of generators of pseudo-random uniform distributed numbers. They are used as a basis for construction of generators of normal and Poisson distributions. The last step is the creation of Wiener and Poisson stochastic processes generators. We also describe the algorithm to call external C-functions from programs written in Modelica. This allows us to use random number generators implemented in the C language.

Keywords: Modelica · OpenModelica · Random generator · Wiener process · Poisson process · SDE

1 Introduction

In this article we study the problem of generation of uniformly distributed pseudo-random numbers, stochastic Wiener and Poisson processes in OpenModelica framework [7]. OpenModelica is one of the open source implementations of Modelica [5] modeling language (for other implementations see [1,3,4,6,8,10]). This language is designed for modeling various systems and processes that can be represented as a system of algebraic or differential equations. For the numerical solution of the equations OpenModelica uses a number of open source

© Springer International Publishing AG 2016
V.M. Vishnevskiy et al. (Eds.): DCCN 2016, CCIS 678, pp. 538–552, 2016.
DOI: 10.1007/978-3-319-51917-3_46

libraries [2,9,18,31]. However, in OpenModelica standard library there is no any function even for generating uniformly distributed pseudo-random numbers.

The first part of the article provides an overview of some algorithms for pseudo-random numbers generation, including description of pseudo-device /dev/random of Unix OS. For most of them we provide the algorithm written in pseudocode. We implement all described algorithms in the C language and partly in OpenModelica. Also we tested them with dieharder—a random number generator testing suite [17].

In the second part of the paper we describe algorithms for normal and Poisson distributions generation. These algorithms are based on the generators of uniformly distributed pseudo-random numbers. Then we study the problem of computer generation of stochastic Wiener and Poisson processes.

The third part of the article has a practical focus and is devoted to the description of external functions (written in C language) calling directly from OpenModelica programs code.

2 Algorithms for Uniformly Distributed Pseudo-random Numbers Generating

In this section we will describe some of the most common generators of uniformly distributed pseudo-random numbers. These generators are the basis for obtaining a sequence of pseudo-random numbers of other distributions.

2.1 Linear Congruential Generator

A linear congruential generator (LCG) was first proposed in 1949 by Lehmer [24]. The algorithm is given by the formula:

$$x_{n+1} = (ax_n + c) \mod m, \quad n \geqslant 0,$$

where m is *the mask* or *the modulus* $m > 1$, a is *the multiplier* $(0 \leqslant a < m)$, c is *the increment* $(0 \leqslant c < m)$, x_0 is *the seed* or initial value. The result of the repeated application of this recurrence formula is *linear congruential sequence* $x_1, \ldots, x_n$. A special case $c = 0$ is called *multiplicative* congruential method.

The numbers m, a, c are called "magic" because their values are specified in the code of the program and are selected based on the experience of the use of the generator. The quality of the generated sequence depends essentially on the correct choice of these parameters. The sequence $\{x\}_1^n$ is periodic and its period depends on the number m, which must therefore be large. In practice, one chooses m equal to the machine word size (for 32-bit architecture—2^{32}, for 64-bit architecture—2^{64}). Knuth [24] recommends to choose

$$a = 6364136223846793005, \quad c = 1442695040888963407,$$
$$m = 2^{64} = 18446744073709551616.$$

In the article [26], you can find large tables with optimal values a, b m.

Also there are generalisations of LCG, such as quadratic congruential method $x_n = (ax_{n-1}^2 + bx_{n-1} + d) \mod m$ cubic congruential method $x_n = (ax_{n-1}^3 + bx_{n-1}^2 + cx_{n-1} + d) \mod 2^e$.

Currently, the linear congruential method has mostly a historical value, as it generates relatively low-quality pseudo-random sequence compared to other, equally simple generators.

2.2 Lagged Fibonacci Generator

The lagged Fibonacci generation can be considered as the generalization of the linear congruential generator. The main idea of this generalisation is to use multiple previous elements to generate current one. Knuth [24] claims that the first such generator was proposed in the early 50-ies and based on the formula:

$$x_{n+1} = (x_n + x_{n-1}) \mod m.$$

In practice, however, it showed itself not the best way. In 1958 George. J. Mitchell and D. Ph. Moore invented a much better generator

$$x_n = (x_{n-n_a} + x_{n-n_b}) \mod m, \ n \geqslant \max(n_a, n_b).$$

It was the generator that we now call LFG—lagged Fibonacci Generator.

As in the case of LCG generator the "magical numbers" n_a and n_b greatly affect the quality of the generated sequence. The authors proposed to use the following magic numbers n_a and n_b

$$n_a = 24, n_b = 55.$$

Knuth [24] gives a number of other values, starting from $(37, 100)$ and finishing with $(9739, 23209)$. Period length of this generator is exactly equal to $2^{e-1}(2^{55} - 1)$ when choosing $m = 2^e$.

As can be seen from the algorithm an initial value and a sequence of $\max(n_a, n_b)$ random numbers must be used for the initialization of this generator.

In open source GNU Scientific Library (GSL) [20] the *composite multirecursive* generator is used. It was proposed in paper [25]. This generator is a generalisation of LFG and may be expressed by the following formulas:

$$x_n = (a_1 x_{n-1} + a_2 x_{n-2} + a_3 x_{n-3}) \mod m_1,$$
$$y_n = (b_1 y_{n-1} + b_2 y_{n-2} + b_3 y_{n-3}) \mod m_2,$$
$$z_n = (x_n - y_n) \mod m_1.$$

The composite nature of this algorithm allows to obtain a large period equal to $10^{56} \approx 2^{185}$. The GSL uses the following parameter values of a_i, b_i, m_1, m_2:

$$
\begin{aligned}
a_1 &= 0, & b_1 &= 86098, & m_1 &= 2^{32} - 1 = 2147483647, \\
a_2 &= 63308, & b_2 &= 0, & m_2 &= 2145483479, \\
a_3 &= -183326, & b_3 &= -539608.
\end{aligned}
$$

Another method suggested in the paper [27] is also a kind of Fibonacci generator and is determined by the formula:

$$x_n = (a_1 x_{n-1} + a_5 x_{n-5}) \mod 5,$$

The GSL used the following values: $a_1 = 107374182$, $a_2 = 0$, $a_3 = 0$, $a_4 = 0$, $a_5 = 104480$, $m = 2^{31} - 1 = 2147483647$. The period of this generator is equal to 10^{46}.

2.3   Inverse Congruential Generator

Inverse congruential method based on the use of inverse modulo of a number.

$$x_{i+1} = (a x_i^{-1} + b) \mod m$$

where a is a *multiplier* $(0 \leqslant a < n)$, b is an *increment* $(0 \leqslant b < n)$, x_0 is an initial value (seed). In addition $GCD(x_0, m) = 1$ and $HCF(a, m) = 1$ are required.

This generator is superior to the usual linear method, however, it is more complicated algorithmically, since it is necessary to find the inverse modulo integers which leads to performance reduction. The extended Euclidean algorithm [24, §4.3.2] is usually applied for compution of the inverse of the number.

2.4   Generators with Bitwise Operations

Most generators that produce high quality pseudo-random numbers sequence use bitwise operations, such as conjunction, disjunction, negation, exclusive disjunction (xor) and bitwise right/left shifting.

Mersenne Twister. Mersenne twister is considered to be one of the best pseudo-random generators. It was developed in 1997 by Matsumoto and Nishimura [30]. There are 32-,64-,128-bit versions of the Mersenne twister. The name of the algorithm derives from the use of Mersenne primes $2^{19937} - 1$. Depending on the implementation the period of this generator can be up to $2^{216091} - 1$.

The main disadvantage of the algorithm is the relative complexity and, consequently, relatively slow performance. Otherwise, this generator provides high-quality pseudo-random sequence. An important advantage is the requirement of only one initiating number (seed). Mersenne twister is used in many standard libraries, for example in the Python 3 module **random** [12].

Due to the complexity of the algorithm, we do not give its pseudocode in this article, however, the standard implementation of the algorithm created by Matsumoto and Nishimura freely available at the link http://www.math.sci.hiroshima-u.ac.jp/~m-mat/MT/emt64.html.

XorShift Generator. Some simple generators giving a high quality pseudo-random sequence were developed in 2003 by George. Marsala (Marsaglia) [29,33].

KISS Generator. Another group of generators, giving a high quality sequence of pseudo-random numbers, is KISS generators family [35] (Keep It Simple Stupid). They are used in the procedure `random_number()` of Frotran language (`gfortran` compiler [11]).

2.5   Pseudo Devices `/dev/random` and `/dev/urandom`

To create a truly random sequence of numbers using a computer some Unix systems (in particular GNU/Linux) use the collection of "background noise" from the operating system environment and hardware. Source of this random noise are moments of time between keystrokes (inter-keyboard timings), various system interrupts and other events that meet two requirements: to be non-deterministic and be difficult for access and for measurement by external observer.

Randomness from these sources is added to an "entropy pool", which is mixed using a CRC-like function. When random bytes are requested by the system call, they are retrieved from the entropy pool by taking the SHA hash from its content. Taking the hash allows not to show the internal state of the pool. Thus the content restoration by hash computing is considered to be an impossible task. Additionally, the extraction procedure reduces the content pool size to prevent hash calculation for the entire pool and to minimize the theoretical possibility of determining its content.

External interface for the entropy pool is available as symbolic pseudo-device `/dev/random`, as well as the system function:

```
void get_random_bytes(void *buf, int nbytes);
```

The device `/dev/random` can be used to obtain high-quality random number sequences, however, it returns the number of bytes equal to the size of the accumulated entropy pool, so if one needs an unlimited number of random numbers, one should use a character pseudo-device `/dev/urandom` which does not have this restriction, but it also generates good pseudo-random numbers, sufficient for the most non-cryptographic tasks.

2.6   Algorithms Testing

A review of quality criterias for a sequence of pseudo-random numbers can be found in the third chapter of the book [24], as well as in paper [28]. All the algorithms, which we described in this articles, have been implemented in C-language and tested with Dieharder test suite, available on the official website [17].

Dieharder Overview. Dieharder is a tests suite, which is implemented as a command-line utility that allows one to test a quality of sequence of uniformly distributed pseudorandom numbers. Also Dieharder can use any generator from GSL library [20] to generate numbers or for direct testing.

- `dieharder -l`—show the list of available tests,
- `dieharder -g -1`—show the list of available random number generators; each generator has an ordinal number, which must be specified after `-g` option to activate the desired generator.
  - 200 `stdin_input_raw`—to read from standard input binary stream,
  - 201 `file_input_raw`—to read the file in binary format,
  - 202 `file_input`—to read the file in text format,
  - 500 `/dev/random`—to use a pseudo-device `/dev/random`,
  - 501 `/dev/urandom`—to use a pseudo-device `/dev/urandom`.

Each pseudo-random number should be on a new line, also in the first lines of the file one must specify: type of number (d—integer double-precision), the number of integers in the file and the length of numbers (32 or 64 - bit). An example of such a file:

```
type: d
count: 5
numbit: 64
1343742658553450546
16329942027498366702
3111285719358198731
2966160837142136004
17179712607770735227
```

When such a file is created, you can pass it to `dieharder`

```
dieharder -a -g 202 -f file.in > file.out
```

where the flag `-a` denotes all built-in tests, and the flag `-f` specifies the file for analysis. The test results will be stored in `file.out` file.

Table 1. Test results

The generator	Fail	Weak	Pass
LCG	52	6	55
LCG2	51	8	54
LFG	0	2	111
ICG	0	6	107
KISS	0	3	110
jKISS	0	4	109
XorShift	0	4	109
XorShift+	0	2	111
XorShift*	0	2	111
Mersenne Twister	0	2	111
dev/urandom	0	2	111

Test Results and Conclusions. The best generators with bitwise operations are `xorshift*`, `xorshift+` and Mersenne Twister (see Table 1). They all give the sequence of the same quality. The algorithm of the Mersenne Twister, however, is far more cumbersome than `xorshift*` or `xorshift+`, thus, to generate large sequences is preferable to use `xorshift*` or `xorshift+`.

Among the generators which use bitwise operations the best result was showed by Lagged Fibonacci generator. The test gives results at the level of XorShift+ and Mersenne Twister. However, one has to set minimum 55 initial values to initialize this generator, thus its usefulness is reduced to a minimum. Inverse congruential generator shows slightly worse results, but requires only one number to initiate the algorithm.

3   Generation of Wiener and Poisson Processes

Let us consider the generation of normal and Poisson distributions. The choice of these two distributions is motivated by their key role in the theory of stochastic differential equations. The most general form of these equations uses two random processes: Wiener and Poisson [34]. Wiener process allows to take into account the implicit stochasticity of the simulated system, and the Poisson process— external influence.

3.1   Generation of the Uniformly Distributed Pseudo-random Numbers from the Unit Interval

Generators of pseudo-random uniformly distributed numbers are the basis for other generators. However, most of the algorithms require a random number from the unit interval $[0, 1]$, while the vast majority of generators of uniformly distributed pseudo-random numbers give a sequence from the interval $[0, m]$ where the number m depends on the algorithm and the bitness of the operating system and processor.

To obtain the numbers from the unit interval one can proceed in two ways. First, one can normalize existing pseudo-random sequence by dividing each its element on the maximum element. This approach is guaranteed to give 1 as a random number. However, this method is bad when a sequence of pseudo-random numbers is too large to fit into memory. In this case it is better to use the second method, namely, to divide each of the generated number by m.

3.2   Normal Distribution Generation

An algorithm for normal distributed numbers generation has been proposed in 1958 by Bux and Mueller [16] and named in their honor *Box-Muller transformation*. The method is based on a simple transformation. This transformation is usually written in two formats:

- standard form (was introduce in the paper [16]),
- polar form (suggested by Bell [15] and Knop [23]).

Standard Form. Let x and y are two independent, uniformly distributed pseudo-random numbers from the interval $(0, 1)$, then numbers z_1 and z_2 are calculated according to the formula

$$z_1 = \cos(2\pi y)\sqrt{-2\ln x}, \; z_2 = \sin(2\pi y)\sqrt{-2\ln x}$$

and they are independent pseudo-random numbers distributed according to a standard normal law $\mathcal{N}(0, 1)$ with expectation $\mu = 0$ and the standard deviation $\sigma = 1$.

Polar Form. Let x and y—two independent, uniformly distributed pseudo-random numbers from the interval $[-1, 1]$. Let us compute additional value $s = x^2 + y^2$. If $s > 1$ or $s = 0$ then existing x and y values should be rejected and the next pair should be generated and checked. If $0 < s \geqslant 1$ then the numbers z_1 and z_2 are calculated according to the formula

$$z_1 = x\sqrt{\frac{-2\ln s}{s}}, \; z_2 = y\sqrt{\frac{-2\ln s}{s}}$$

and they are independent random numbers distributed according to a standard normal law $\mathcal{N}(0, 1)$.

For computer implementation is preferable to use a polar form, because in this case one has to calculate only single transcendental function ln, while in standard case three transcendental functions (ln, sin cos) have to be calculated. An example of the algorithm shown in Fig. 1.

To obtain a general normal distribution from the standard normal distribution, one can use the formula $Z = \sigma \cdot z + \mu$ where $z \sim \mathcal{N}(0, 1)$, and $Z \sim \mathcal{N}(\mu, \sigma)$.

3.3   The Generation of a Poisson Distribution

To generate a Poisson distribution there is a wide variety of algorithms [13, 14, 19]. The easiest was proposed by Knut [24]. This Algorithm 2.1 uses uniform pseudo-random number from the interval $[0, 1]$ for it's work. The algorithm's output example is depicted on Fig. 2.

Algorithm 2.1. The generator of the Poisson distribution

Require: *seed*, λ
   $\Lambda \leftarrow \exp(-\lambda)$, $k \leftarrow 0$, $p \leftarrow 1$, $u \leftarrow seed$
   repeat
      $k \leftarrow k + 1$
      $u \leftarrow rand(u)$ ▷ generation of uniformly distributed random number
      $p = p \cdot u$
   until $p > \Lambda$
   return $k - 1$

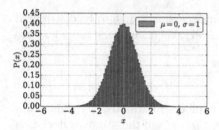

Fig. 1. Normal distribution   **Fig. 2.** Poisson distribution

3.4   Generation of Poisson and Wiener Processes

Now we going to use generators of normal and Poisson distributions to generate Wiener and Poisson stochastic processes. For definitions of Poisson and Wiener processes see, for example, [22, 32, 34].

The Generation of the Wiener Process. To simulate one-dimensional Wiener process, one should generate the N normally distributed random numbers $\varepsilon_1, \ldots, \varepsilon_N$ and build their cumulative sums of ε_1, $\varepsilon_1 + \varepsilon_2$, $\varepsilon_1 + \varepsilon_2 + \varepsilon_3$. As result we will get *a trajectory* of the Wiener process $W(t)$ see Fig. 3.

In the case of multivariate random process, one needs to generate m sequences of N normally distributed random variables.

The Generation of a Poisson Process. A simulation of the Poisson process is much like Wiener one, but now we need to generate a sequence of numbers distributed according to the Poisson law and then calculate their cumulative sum. The plot of Poisson process is shown in Fig. 4. The figure shows that the Poisson process represents an abrupt change in numbers that has occurred over time events. The intensity λ depends on the average number of events over a period of time.

Because of this characteristic of behavior the Poisson process is also called as a process with jumps, and stochastic differential equations, with Poisson process as second driving process, are called equations with jumps [34]

4   Simulation of Stochastic Processes in OpenModelica

As already mentioned in the introduction, there are no any pseudorandom numbers generators in OpenModelica. Thus that makes this system unusable for stochastic processes modeling. However `Noise` library build.openmodelica.org/Documentation/Noise.html developed by Klockner (Klockner) [21] should be mentioned. The basis of this library are `xorshift` generators, written in C. However, an inexperienced user may face a problem, because one needs compile C-source files first to use that library.

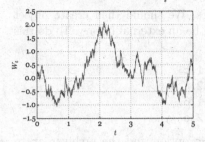

Fig. 3. Wiener process

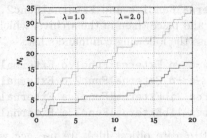

Fig. 4. Poisson process

In this article we will describe the procedure required for connection of external C functions to OpenModelica programme. That will allow the user to install the Noise library and to connect their own random number generators. We also provide a minimal working example of stochastic Wiener process generator and the example of ordinary differential equation with additive stochastic part.

4.1 Connection of External C-Functions to OpenModelica Program

Let us consider the process of connection of external functions to modelica program. The relevant section in the official documentation misses some essential steps that's why it will lead to an error. All steps we described, had been performed on a computer with Linux Ubuntu 16.04 LTS and OpenModelica 1.11.0-dev-15.

When the code is compiled the OpenModelica program is translated to C code that then is processed by C-compiler. Therefore, OpenModelica has built-in support of C-functions. In addition to the C language OpenModelica also supports Fortran (F77 only) and Python functions. However, both languages are supported indirectly, namely via wrapping them in the appropriate C-function.

The usage of external C-functions may be required for various reasons, for example, implementations of performance requiring components of the program, the usage of a fullscale imperative programming language, or the use of existing sourcecode in C.

We give a simple example of calling C-functions from Modelica program. Let's create two source files: ExternalFunc1.c and ExternalFunc2.c. These files will contain simple functions that we want to use in our Modelica program.

```
// File ExternalFunc1.c
double ExternalFunc1_ext(double x) { return x+2.0*x*x;}
```

```
// File ExternalFunc2.c
double ExternalFunc2(double x){return (x-1.0)*(x+2.0);}
```

In the directory, where the source code of Modelica program is placed, we must create two directories: Resources and Library, which will contain

`ExternalFunc1.c` and `ExternalFunc2.c` files. We should then create object files and place them in the archive, which will be an external library. To do this we use the following command's list:

```
gcc -c -o ExternalFunc1.o ExternalFunc1.c
gcc -c -o ExternalFunc2.o ExternalFunc2.c
ar rcs libExternalFunc1.a ExternalFunc1.o
ar rcs libExternalFunc2.a ExternalFunc2.o
```

To create object files, we use gcc with `-c` option and the archiver **ar** to place generated object files in the archive. As a result, we get two of the file `libExternalFunc1.a` and `libExternalFunc2.a`. There is also the possibility to put all the needed object files in a single archive.

To call external functions, we must use the keyword **external**. The name of the wrapper function in Modelica language can be differ from the name of the external function. In this case, we must explicitly specify which external functions should be wrapped.

```
model ExternalLibraries
  function ExternalFunc1 // Function name differs
    input Real x;
    output Real y;
    external y=ExternalFunc1_ext(x); // Explicitly specifying C-function name
    annotation(Library="ExternalFunc1");
  end ExternalFunc1;
  function ExternalFunc2
    input Real x;
    output Real y;
    // The functions names are the same
    external "C" annotation(Library="ExternalFunc2");
  end ExternalFunc2;
  Real x(start=1.0, fixed=true), y(start=2.0, fixed=true);
equation
  der(x)=-ExternalFunc1(x);
  der(y)=-ExternalFunc2(y);
end ExternalLibraries;
```

Note that in the annotation the name of the external library is specified as `ExternalFunc1`, while the file itself is called `libExternalFunc1.a`. This is not a mistake and the prefix `lib` must be added to all library's files.

The example shows that the type **Real** corresponds to the C type **double**. Additionally, the types of **Integer** and **Boolean** match the C-type **int**. Arrays of type **Real** and **Integer** transferred in arrays of type **double** and **int**.

It should be noted that consistently works only call -functions with arguments of **int** and **double** types, as well as arrays of these types. The attempt to use specific c-type, for example, **long long int** or an unsigned type such as **unsigned int**, causes the error.

4.2 Modeling Stochastic Wiener Process

Let us describe the implementation of a generator of the normal distribution and Wiener process. We assume that the generator of uniformly-distributed random numbers is already implemented in the functions **urand**. To generate the normal distribution we will use Box-Muller transformation and Wiener process can be calculated as cumulative sums of normally-distributed numbers.

The minimum working version of the code is shown below. The key point is the use of an operator **sample(t_0, h)**, which generates events using **h** seconds starting from the time **t_0**. For every event the operator **sample** calls the function **urand** that returns a new random number.

```
model generator
  Integer x1, x2;
  Port rnd; "Random number generator's port"
  Port normal; "Normal numbers generator's port"
  Port wiener; "Wiener process values port"
  Integer m = 429496729; "Generator modulo"
  Real u1, u2;
initial equation
  x1 = 114561;
  x2 = 148166;
algorithm
  when sample(0, 0.1) then
    x1 := urand(x1);
    x2 := urand(x2);
  end when;
  // normalisation of random sequence
  rnd.data[1] := x1 / m;
  rnd.data[2] := x2 / m;
  u1 := rnd.data[1];
  u2 := rnd.data[2];
  // normal generator
  normal.data[1] := sqrt(-2 * log(u1)) * sin(6.28 * u2);
  normal.data[2] := sqrt(-2 * log(u1)) * cos(6.28 * u2);
  // Wiener process
  wiener.data[1] := wiener.data[1] + normal.data[1];
  wiener.data[2] := wiener.data[2] + normal.data[2];
end generator;
```

Note also the use of a special variable of type Port which serves to connect the various models together. In our example we have created three such variables: **1g**, **normal**, **wiener**. Because of this, other models can access the result of our generator.

```
connector Port
  Real data[2];
end Port;
```

A minimal working code below illustrates the connection example between two models. A system of two ordinary differential equations describes van der PolDuffing oscillator with additive stochastic part in the form of a Wiener process (see 5).

$$\begin{cases} \dot{x} = y, \\ \dot{y} = x(1.0 - x^2) - y + x \cdot W_t. \end{cases}$$

It is important to mention that this equation is not stochastic. Built-in Open-Modelica numerical methods do not allow to solve stochastic differential equations.

```
// the model specifies a system of ODE
  model ODE
    Real x, y;
    Port IN;
  initial equation
    x = 2.0;
    y = 0.0;
  equation
    der(x) = y ;
    der(y) = x*(1-x*x) - y + x*IN.data[1];
  end ODE;
  model sim
    generator gen;
    ODE eq;
  equation
    connect(gen.wiener, eq.IN);
  end sim;
```

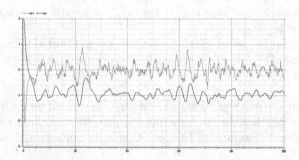

Fig. 5. Results of van der Pol–Duffing oscillator simulation. The graphs are created using the functionality OMEditor'a

5   Conclusion

We reviewed the basic algorithms for generating uniformly distributed pseudo-random numbers. All algorithms were implemented by the authors in C language and tested using DieHarder utility. The test results revealed that the most effective algorithms are xorshift and Mersenne Twister algorithms.

Due to the fact that OpenModelica does not implement bitwise logical and shifting operators, generators of uniformly distributed pseudo-random numbers have to be implemented in C language and connected to the program as external functions. We gave a rather detailed description of this process, that, as we hope, will fill a gap in the official documentation.

Acknowledgments. The work is partially supported by RFBR grants No's 14-01-00628, 15-07-08795, and 16-07-00556. Also the publication was supported by the Ministry of Education and Science of the Russian Federation (the Agreement No. 02.a03.21.0008).

References

1. Jmodelica.org. http://www.jmodelica.org/
2. LAPACKLinear Algebra PACKage. http://www.netlib.org/lapack/
3. LMS Imagine.Lab Amesim. http://www.plm.automation.siemens.com/en_us/products/lms/imagine-lab/amesim/index.shtml
4. MapleSim - High Performance Physical Modeling and Simulation - Technical Computing Software. http://www.maplesoft.com/products/maplesim/index.aspx
5. Modelica and the Modelica Association Official Site. https://www.modelica.org/
6. Multi-Engineering Modeling and Simulation - Dymola - CATIA. http://www.3ds.com/products-services/catia/products/dymola
7. OpenModelica Official Site. https://www.openmodelica.org/
8. SciLab Official Site. http://www.scilab.org/
9. SuiteSparse: A Suite of Sparse Matrix Software. http://faculty.cse.tamu.edu/davis/suitesparse.html
10. Wolfram SystemModeler. http://www.wolfram.com/system-modeler/index.html
11. Using GNU Fortran (2015). https://gcc.gnu.org/onlinedocs/
12. Python 3.5.1 Documentation, March 2016. https://docs.python.org/3/
13. Ahrens, J.H., Dieter, U.: Computer methods for sampling from gamma, beta, poisson and bionomial distributions. Computing **12**(3), 223–246 (1974)
14. Ahrens, J.H., Dieter, U.: Computer generation of poisson deviates from modified normal distributions. ACM Trans. Math. Softw. **8**(2), 163–179 (1982)
15. Bell, J.R.: Algorithm 334: normal random deviates. Commun. ACM **11**(7), 498 (1968)
16. Box, G.E.P., Muller, M.E.: A note on the generation of random normal deviates. Ann. Math. Stat. **29**(2), 610–611 (1958)
17. Brown, R.G., Eddelbuettel, D., Bauer, D.: Dieharder: A Random Number Test Suite (2013). http://www.phy.duke.edu/~rgb/General/rand_rate.php
18. Collier, A.M., Hindmarsh, A.C., Serban, R., Dward, C.S.W.: User Documentation for KINSOL v2.8.2 (2015). http://computation.llnl.gov/sites/default/files/public/kin_guide.pdf

19. Devroye, L.: Non-Uniform Random Variate Generation. Springer-Verlag, New York (1986)
20. Galassi, M., Gough, B., Jungman, G., Theiler, J., Davies, J., Booth, M., Rossi, F.: The GNU Scientific Library Reference Manual (2015). https://www.gnu.org/software/gsl/manual/gsl-ref.pdf
21. Klckner, A., van der Linden, F.L.J., Zimmer, D.: Noise generation for continuous system simulation. In: Proceedings of the 10th International Modelica Conference, Lund, Sweden, pp. 837–846 (2014)
22. Kloeden, P.E., Platen, E.: Numerical Solution of Stochastic Differential Equations, 2nd edn. Springer, Heidelberg (1995)
23. Knop, R.: Remark on algorithm 334 [g5]: normal random deviates. Commun. ACM **12**(5), 281 (1969)
24. Knuth, D.E.: The Art of Computer Programming, Volume 2 (3rd Ed.): Seminumerical Algorithms, vol. 2. Addison-Wesley Longman Publishing Co. Inc., Boston (1997)
25. L'Ecuyer, P.: Combined multiple recursive random number generators. Oper. Res. **44**(5), 816–822 (1996)
26. L'Ecuyer, P.: Tables of linear congruential generators of different sizes and good lattice structure. Math. Comput. **68**(225), 249–260 (1999)
27. L'Ecuyer, P., Blouin, F., Couture, R.: A search for good multiple recursive random number generators. ACM Trans. Modeling Comput. Simul. (TOMACS) **3**(2), 87–98 (1993)
28. L'Ecuyer, P., Simard, R.: Testu01: AC library for empirical testing of random number generators. ACM Trans. Mathe. Softw. (TOMS) **33**(4), 22 (2007)
29. Marsaglia, G.: Xorshift RNGs. J. Stat. Softw. **8**(1), 1–6 (2003)
30. Matsumoto, M., Nishimura, T.: Mersenne twister: A 623-dimensionally equidistributed uniform pseudo-random number generator. ACM Trans. Model. Comput. Simul. **8**(1), 3–30 (1998)
31. Nishida, A., Fujii, A., Oyanagi, Y.: Lis: Library of Iterative Solvers for Linear Systems. http://www.phy.duke.edu/~rgb/General/rand_rate.php
32. Øksendal, B.: Stochastic Differential Equations: An Introduction with Applications, 6th edn. Springer, Heidelberg (2003)
33. Panneton, F., L'Ecuyer, P.: On the xorshift random number generators. ACM Trans. Model. Comput. Simul. **15**(4), 346–361 (2005)
34. Platen, E., Bruti-Liberati, N.: Numerical Solution of Stochastic Differential Equations with Jumps in Finance. Springer, Heidelberg (2010)
35. Rose, G.: Kiss: A Bit Too Simple (2011). https://eprint.iacr.org/2011/007.pdf

Metric Analysis as a Tool for Interpolating Multivariate Functions in the Case of an Information Lack

Alexander Kryanev[1]([✉]), Gleb Lukin[1], and David Udumyan[1,2,3]

[1] National Research Nuclear University "MEPhI",
Kashirskoe shosse 31, 115409 Moscow, Russia
avkryanev@mephi.ru
[2] Department of Applied Probability and Informatics,
RUDN University (Peoples' Friendship University of Russia),
6 Miklukho-Maklaya str., Moscow 117198, Russia
[3] Department of Mathematics, The College of Arts and Sciences at the University of
Miami, 1320 South Dixie Highway 1320, Coral Gables, FL 33146, USA

Abstract. In report the ill-posed problems in view arising at the solution of applied problems by means of the metric analysis are considered. In the report new schemes and algorithms for smoothing and restoration based on the metric analysis were presented. These schemes and algorithms have demonstrated a high accuracy of smoothing and retrieving the values of functions of one or many variables. Examples of such problems are problems of interpolation, filtration and forecasting of values of functions of one and many variables claimed at the solution of applied problems physicists, technicians, economy and other areas of researches.

Keywords: Function of many variables · Interpolation · Ill-posed problems · Metric analysis

1 Introduction

The problem of interpolation is one of primary objectives in mathematics, not to mention its broad-ranging applications in practical problem solving. Approaches to solving interpolation problems of one-variable functions have been proposed since the times of Lagrange and Newton. To date, sufficiently complete results are obtained for different interpolation methods, including an analysis of interpolation errors and the convergence of interpolation values to the exact values [1,2]. The classical scheme involves representing the interpolated function in a form of an expansion to basis functions. Thus, in Lagrange's scheme the basic functions are monomials. However, the Lagrange interpolation provides the uniform convergence of interpolation polynomials to the function only for a certain class of smooth functions, for example, for the class of entire functions. The reason for this divergence is the discontinuity of derivatives. As an alternative to Lagrange's scheme, the scheme of spline-interpolation has been developed.

© Springer International Publishing AG 2016
V.M. Vishnevskiy et al. (Eds.): DCCN 2016, CCIS 678, pp. 553–564, 2016.
DOI: 10.1007/978-3-319-51917-3_47

Spline-interpolation allows to localize the contagion of angular points and provides the uniform convergence for any continuous function [2]. The scheme of representation of functions in the form of linear combinations of basic functions, including polynomials and spline-approximations, simply, can be generalized on multivariate functions, but such schemes are efficient only for functions of two, at maximum three variables. For functions of a variables more than three there are no effective general schemes of interpolation. Only crude approximation schemes of local linear interpolation are in use. These schemes require a large number of data, and even in a presence of a large quantity of data often do not provide the necessary accuracy. An example of such schemes is the neural networks, which allow to interpolate functions of one or many variables [3]. In this report a universal approach for solving problems of multidimensional interpolation without holding fixed the species of functional connection of the function to its arguments is presented. This approach takes into account only the information on the mutual arrangement of the point at which we retrieve the value of function and the interpolation nodes $X_1, ..., X_n$ as well as the values of function $Y_i, i = 1, ..., n$ at interpolation nodes and can be put to use even in the case of an insufficiency of initial information.

2   The Metric Analysis Interpolation Scheme

We consider the problems associated with functionality:

$$Y = F(X), \tag{1}$$

where the function $F(X)$ is unknown, and we aim to retrieve its value at point X^* using the function values $Y_i, i = 1, ..., n$ at interpolation nodes $X_i = (X_{i1}, ..., X_{im})^T, i = 1, ..., n.$

In accordance with the scheme of the metrical analysis we form a matrix W of metric uncertainty for the point X^* on a set of points X $X_i = (X_{i1}, ..., X_{im})^T, i = 1, ..., n.$ [4]

$$W = \begin{pmatrix} \rho^2(X_1, X^*) & (X_1, X_2) & \cdots & (X_1, X_n) \\ (X_2, X_1) & \rho^2(X_2, X^*) & & (X_2, X_n) \\ \cdots & \cdots & \cdots & \cdots \\ (X_n, X_1) & (X_n, X_2) & \cdots & \rho^2(X_n, X^*) \end{pmatrix}, \tag{2}$$

where $\rho^2(X_i, X^*) = \sum_{k=1}^m \nu_k \cdot (X_{ik} - X_k^*)^2$,

$(X_i, X_j) = \sum_{k=1}^m \nu_k \cdot (X_{ik} - X_k^*) \cdot (X_{jk} - X_k^*)$, ν_k determine the function's degree of tolerance with respect to its arguments. Remark. When solving applied problems it is necessary to normalize all argument values X to an identical interval before generating a matrix of metric uncertainty. This can be achieved, for instance, by means of linear replacement of each variable $X_i, i = 1, ..., n$ by its increment in an interval $[0, 1]$. According to (2) W is a symmetric non-negative matrix. It is further assumed that the matrix is positive definite. The interpolation formula for the retrieved value Y^* at the point $\boldsymbol{X}^*$ has the form:

$$Y^* = \sum_{i=1}^{n} z_i \cdot Y_i, \tag{3}$$

where the interpolation weights $z_i, i = 1, ..., n$ as expected for the interpolated formulas satisfy to normalization condition $\sum_{i=1}^{n} z_i = 1$ We define the numerical characteristic $\sigma_{mu}^2(Y^*)$ of metrical uncertainty of the retrieved value Y^* at the point $\boldsymbol{X}^*$ by the equality:

$$\sigma_{mu}^2(Y^*) = (W\boldsymbol{z}, \boldsymbol{z}), \tag{4}$$

where $\boldsymbol{z} = (z_1, ..., z_n)^T$. We pose the problem of selecting interpolation weight values $z_i, i = 1, ..., n$, that satisfy the normalization condition $\sum_{i=1}^{n} z_i = 1$, for which the numerical value of the uncertainty is minimal:

$$\begin{cases} (W\boldsymbol{z}, \boldsymbol{z}) - \min \boldsymbol{z} \\ (\boldsymbol{z}, \boldsymbol{1}) = 1, \boldsymbol{1} = (1, ..., 1)^T. \end{cases} \tag{5}$$

The problem (5) is solved by Lagrange method. The required vector $\boldsymbol{z}^*$ and interpolated value Y^* are given by equalities:

$$\boldsymbol{z}^* = \frac{(W^{-1}\boldsymbol{1})}{(W^{-1}\boldsymbol{1})}, \quad Y^* = \frac{(W^{-1}\boldsymbol{1}, \boldsymbol{Y})}{(W^{-1}\boldsymbol{1}, \boldsymbol{1})}, \quad \boldsymbol{Y} = (Y_1, ..., Y_n)^T, \tag{6}$$

where W^{-1} is the inverse matrix. The approach to interpolating multivariate functions named metric analysis allows to consider various levels of tolerance of function with respect to changes of its arguments' values by means of the coefficients $\nu_j, j = 1, ..., m$. To find the coefficients ν_j j we consecutively exclude function arguments and trace the values of changes of function at exclusions, as have been proposed earlier [4–9]. In the present work two new approaches to calculating the coefficients $\nu_j, j = 1, ..., m$, based on comparing the values of "correlation rates" to the rates of function variance with regard to its arguments are presented. The first scheme of definition of metric scales $\nu_j, j = 1, ..., m$ is based on calculating the "correlation rates" from the realized values of required function (1) and then normalizing according to formulas

$$\nu_j = \frac{|r_j|}{\sum_{j=1}^{m} |r_j|}, j = 1, ..., m, \tag{7}$$

where

$$r_j = \frac{cov(Y, X_j)}{\sigma(Y) \cdot \sigma(X_j)}, j = 1, ..., m,$$

$$cov(Y, X_j) = \frac{1}{n-1} \sum_{k=1}^{n} (Y_k - \overline{Y})(X_{kj} - \overline{X_j}), j = 1, ..., m,$$

$$\sigma^2(Y) = \frac{1}{n-1} \sum_{k=1}^{n} (Y_k - \overline{Y})^2, \tag{8}$$

$$\sigma^2(X_j) = \frac{1}{n-1} \sum_{k=1}^{n} (X_{kj} - X_k)^2, j = 1, ..., m,$$

$$\overline{Y} = \frac{1}{n} \sum_{k=1}^{n} Y_k, \overline{X_j} = \frac{1}{n} \sum_{k=1}^{n} X_{kj}, j = 1, ..., m.$$

The second scheme of defining the metric coefficients $\nu_j, j = 1, ..., m$ is based on calculating weight multipliers $\nu_j, j = 1, ..., m$ of the linear regression model based on the realized values of the relevant function (1):

$$Y = u_0 + \sum_{j=1}^{m} u_j \cdot X_j + \epsilon. \tag{9}$$

According to the method of the least squares (MLS), estimations of parameters $u = (u_1, ..., u_m)^T$ of model (9) are given by equality:

$$u = K_X^{-1} cov(Y, X), \tag{10}$$

where the elements of covariance matrix K_X are defined by equalities:

$$cov(X_i, X_j) = \frac{1}{n-1} \sum_{k=1}^{n} (X_{ki} - \overline{X_i}) \cdot (X_{ki} - \overline{X_i}), \tag{11}$$

and the vector components $cov(Y, X) = (cov(Y, X_1), ..., cov(Y, X_m))^T$ are calculated from equalities:

$$cov(Y, X_j) = \frac{1}{n-1} \sum_{k=1}^{n} (Y_k - \overline{Y})(X_{kj} - \overline{X_j}), j = 1, ..., m, \tag{12}$$

Then values of metric coefficients on the realized values of required function (1) are calculated under formulas:

$$w_j = \frac{|u_j|}{\sum_{j=1}^{m} |u_j|} \cdot m, j = 1, ..., m. \tag{13}$$

In the case of strong correlation within a share of arguments $X_i = (X_{i1}, ..., X_{im})^T, i = 1, ..., n$ and, thereby, a case of matrix singularity or ill-conditioning of matrix K_X it is necessary to conduct a regularization, replacing matrix K_X by a regularized matrix, for example, by matrix $K_{X,\alpha} = K_X + \alpha \cdot diag(K_{11}, ..., K_{mm})$, $\alpha \geq 0$, where $K_{ij}, i, j = 1, ..., m$ are the elements of matrix K_X. Let's consider the problem of smoothing and restoration of functional dependence $Y = F(X_1, ..., X_n) = F(\boldsymbol{X})$ in the presence of chaotic deviations from exact values. The values of function $Y_i = F(\boldsymbol{X}_i), i = 1, ..., n$ are given with errors at the nodes $\boldsymbol{X}_i = (X_{i1}, ..., X_{im})^T, i = 1, ..., n$. We assume that the matrix of metric uncertainty W is singular. For every node $\boldsymbol{X}^*$ we aim to find the smoothing value Y^* in the form of

$$Y^* = \sum_{i=1}^{n} z_i \cdot Y_i = (\boldsymbol{z}, \boldsymbol{Y}), \tag{14}$$

where the vector of weights $\boldsymbol{z}$ is the solution of the following problem of minimization of the total uncertainty:

$$(W\boldsymbol{z}, \boldsymbol{z}) + \alpha \cdot (K_Y \boldsymbol{z}, \boldsymbol{z}) - \min \boldsymbol{z},$$
$$(\boldsymbol{z}, \boldsymbol{1}) = 1, \tag{15}$$

where α is the smoothing parameter.

While the expression $(W\boldsymbol{z}, \boldsymbol{z})$ is responsible for metric uncertainty, the expression $(K_Y \boldsymbol{z}, \boldsymbol{z})$ is responsible for stochastic uncertainty. The problem (15) can be solved using the Lagrange's function. The smoothed value at the point $\boldsymbol{X}^*$ is given by the equality:

$$Y_{sm}^* = ((W + \alpha \cdot K_Y)^{-1}\boldsymbol{1}, \boldsymbol{Y})/((W + \alpha \cdot K_Y)^{-1}\boldsymbol{1}, \boldsymbol{1}). \tag{16}$$

When $\alpha \to +\infty$, the smoothed value Y_{sm}^* at any node $\boldsymbol{X}^*$ is given by

$$Y_{sm}^* = (K_Y^{-1}\boldsymbol{1}, \boldsymbol{Y})/(K_Y^{-1}\boldsymbol{1}, \boldsymbol{1}), \tag{17}$$

Theorem. At points $\boldsymbol{X}^* = \boldsymbol{X}_l$ when $\alpha \to +0$ the solution of problem (15) converges to the value of function $Y_i = F(\boldsymbol{X}_l)$.

The proof can be realized with the help of the Lagrange function and the analysis of the limits at $\alpha \to +0$ components of the vector $z = (z_1, ..., z_n)^T$ and a Lagrange multiplier.

The problem (15) can be solved also by means of eigen values and eigen vectors of matrix W. Let $\lambda_1, ..., \lambda_n$ be the eigen values, and $\phi_1, ..., \phi_n$ the corresponding orthonormal system of eigen vectors of the matrix W. We decompose the vector of weights $\boldsymbol{z}$ in system $\phi_1, ..., \phi_n$

$$\boldsymbol{z} = \sum_{i=1}^{n} a_i \cdot \phi_i. \tag{18}$$

Let's denote by $\Phi = (\phi_1, ..., \phi_n)$ the matrix columns of which are the eigen vectors of matrix W. Then

$$\boldsymbol{z} = \Phi a, a = (a_1, ..., a_n)^T. \tag{19}$$

We have:

$$(W z, z) + \alpha \cdot (K_Y z, z) = \sum_{i=1}^{n} \lambda_i \cdot a_i^2 + \alpha \cdot \sum_{i,j=1}^{n} a_i \cdot a_j \cdot (K_y \phi_i, \phi_j) \tag{20}$$
$$= (D a, a) + \alpha \cdot (B a, a)$$

where $D = diag(\lambda_1, ..., \lambda_n)$, $B_{ij} = (K_Y \phi_i, \phi_j)$, $i, j = 1, ..., n$,

$(z, 1) = \sum_{i=1}^{n} c_i \cdot (\phi_i, 1) = (c, \phi)$, $\phi = ((\phi_1, 1), ..., (\phi_n, 1))^T$

From problem (15) for vector z we turn to the following problem for vector a

$$(D a, a) + \alpha \cdot (B a, a) - \min a \tag{21}$$
$$(a, \phi) = 1.$$

We have:

$$L(a, u) = \frac{1}{2} \cdot ((D + \alpha \cdot B) a, a) + u \cdot (1 - (a, \phi)),$$
$$\nabla_c L = (D + \alpha \cdot B) a - u \cdot 1 = 0 \Rightarrow a = u \cdot (D + \alpha \cdot B)^{-1} \phi, \tag{22}$$
$$(a, \phi) = 1 \Rightarrow u = \frac{1}{((D + \alpha \cdot B)^{-1} \phi, \phi)}.$$

From here we deduce the vector a

$$a = \frac{(D + \alpha \cdot B)^{-1} \phi}{((D + \alpha \cdot B)^{-1} \phi, \phi)} \tag{23}$$

From (24) we deduce the vector of weights z

$$z = \Phi \frac{(D + \alpha \cdot B)^{-1} \phi}{((D + \alpha \cdot B)^{-1} \phi, \phi)} \tag{24}$$

and we deduce the smoothed value

$$Y_{smo}^* = \frac{(\Phi(D + \alpha \cdot B)^{-1} \phi, Y)}{((D + \alpha \cdot B)^{-1} \phi, \phi)} \tag{25}$$

If a matrix of metric uncertainty W is singular, the required vector z^* and the required interpolated value Y^*, except for specially stipulated cases, are defined by equalities [4]:

$$z^* = \frac{W^+ 1}{(W^+ 1, 1)}, \tag{26}$$

$$Y^* = \frac{(W^+ 1, Y)}{(W^+ 1, 1)}, \tag{27}$$

where W^+ is the pseudoinverse or regularized matrix.

Remark. At application of Formulas (26 and 27) it is supposed that $(W^+1, 1) > 0$. Let's define size of metric uncertainty of value Y^*, given by the Formula (27). We have:

$$\sigma_{ND}^2(Y^*) = (Wz^*, z^*) = \frac{(WW^+1, W^+1)}{(W^+1, 1)^2} \qquad (28)$$

$$= \frac{(W^+WW^+1, 1)}{(W^+1, 1)^2} = \frac{(W^+1, 1)}{(W^+1, 1)^2} = \frac{1}{(W^+1, 1)}$$

Here we have considered symmetry of a pseudo-return matrix W^+ and equality $W^+W^*W^+ = W^+$. Quantity

$$\sigma_{ND}^2(X^*/X_1, ..., X_n) = \frac{1}{(W^+1, 1)} > 0 \qquad (29)$$

we name a measure of metric uncertainty of restoration of function at a point X^* using values of function at aggregate points $X_i, i = 1, ..., n$.

The inverse quantity

$$I(X^*/X_1, ..., X_n) = (W^+1, 1) = \sum_{i=1}^{n} \sum_{j=1}^{n} W_{ij}^+ > 0 \qquad (30)$$

we name the metric information at a point X^*, concerning the set of points $X_1, ..., X_n$.

From properties of pseudoinverce matrices it follows that at addition of each new point X_{n+1} to the set $X_1, ..., X_n$ the metric information in any point X^* concerning the set of points $X_1, ..., X_n, X_{n+1}$ will be not less than the metric information at a point X^* concerning the set of points $X_1, ..., X_n$:

$$I(X^*/X_1, ..., X_n, X_{n+1}) \geq I(X^*/X_1, ..., X_n). \qquad (31)$$

For a measure of metric uncertainty we arrive at equality:

$$\sigma_{ND}^2(X^*/X_1, ..., X_n, X_{n+1}) \leq \sigma_{ND}^2(X^*/X_1, ..., X_n) \qquad (32)$$

Numerical results of interpolation of functions of many variables by means of schemes of the metric analysis are presented in Figs. 1, 2, 3, 4, 5, 6, 7, 8 and Table 1. Figures 1 and 2 are presented in the form of graphs of recovery functions of one variable using the methods of the metric analysis presented in this paper. The initial function, which is unknown before the implementation of recovery methods is depicted in red; green line shows the given noisy realization of the function; the dashed line represents the function recovered by the methods of the metric analysis presented in this paper. As can be seen from the figures, the average error of the recovered values of the function equals to less than 2 percent. Function presented in Table 1 is $Y = f(x) = (Vx, x) + (c, x), x = (x_1, ..., x_m)^T, 0 \leq x_i \leq 1, i = 1, ..., m$, where V - $m \times m$ constant matrix, $c = (c_1, ..., c_m)^T$ - constant vector, $m = 12$ - dimension of

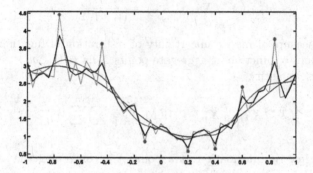

Fig. 1. Recovery function 1 (Color figure online)

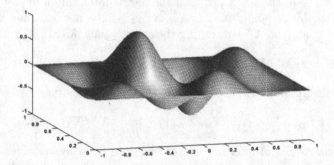

Fig. 2. Recovery function 2 (Color figure online)

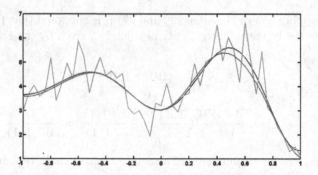

Fig. 3. Exact function of two variables

space. $Y(\boldsymbol{X}_k) = Y_k, k = 1, ..., n$, where $\boldsymbol{X}_k$ - nodes of interpolation, Y_k - values of the function at nodal points, $n = 25$ - number of nodal points. The results of interpolating are presented in Table 1. In Fig. 3, 4, 5 the exact function of two variables, the measured function, and the interpolated function are presented, correspondingly.

Figures 6 and 7 exemplify the results of applying the recovery function of two variables, the mean error of the recovered values equals to less than 1 percent.

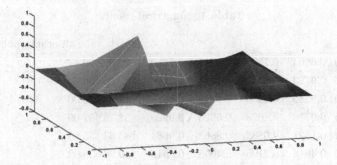

Fig. 4. Measured function of two variables

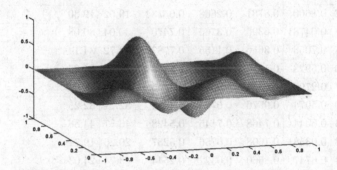

Fig. 5. Recovery function of two variables

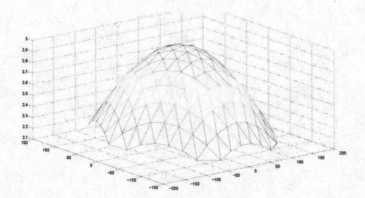

Fig. 6. The surface of exact values

Table 1. Numerical results

X					Y	Result of interpolation
0.3210	0.6604	0.1332	0.2217	0.7252	20.99	20.67
0.5536	0.9723	0.8545	0.1759	0.5116	37.00	36.34
0.7910	0.9280	0.1469	0.1180	0.007696	11.48	13.00
0.8445	0.9716	0.1673	0.5628	0.9920	52.78	52.09
0.5111	0.1521	0.0883	0.8863	0.9684	33.33	35.65
0.5060	0.4818	0.7105	0.4053	0.8512	39.96	39.02
0.1462	0.4970	0.5390	0.8821	0.9541	43.71	43.74
0.2640	0.0875	0.6904	0.1256	0.6399	17.26	15.52
0.3095	0.1707	0.7593	0.9395	0.7266	39.98	40.53
0.7409	0.02982	0.4092	0.01672	0.8803	20.66	21.50
0.2377	0.8600	0.8781	0.2668	0.05134	19.52	19.30
0.7078	0.04781	0.6390	0.3765	0.7470	30.04	30.08
0.8857	0.7058	0.3656	0.1155	0.7787	33.72	34.14
0.5480	0.5923	0.5340	0.9722	0.2310	34.40	34.69
0.7977	0.2799	0.01847	0.9505	0.4345	27.43	28.66
0.8280	0.5974	0.9576	0.07910	0.4444	32.21	32.25
0.7173	0.6044	0.7468	0.7145	0.5428	46.57	44.86
0.5938	0.03581	0.1608	0.6756	0.5257	20.26	19.71
0.7413	0.6747	0.8660	0.03228	0.2544	23.74	24.03

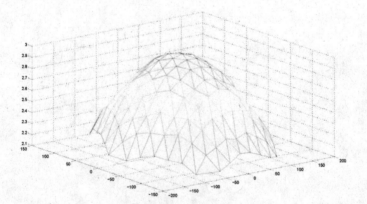

Fig. 7. Resurfacing

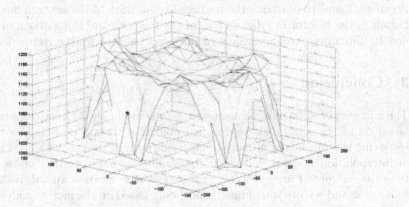

Fig. 8. Exact surface

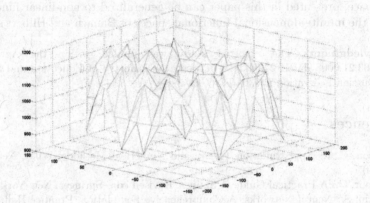

Fig. 9. Noisy surface

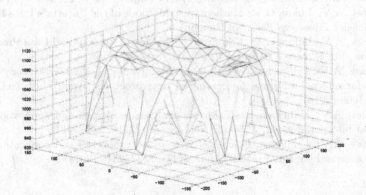

Fig. 10. Recovery surface

Figures 8, 9 and 10 illustrate the results of restoration of the neutron flux distribution in the reactor in cross-section of the reactor core. Comparison of Figs. 8 and 10 illustrates good agreement of the exact values with the restored values.

3 Conclusion

The numerical results of implementing the presented approaches to interpolation based on metric analysis have demonstrated that metric analysis is a tool for retrieving the values of multivariate functions even in the case of a little number of interpolation nodes (even when the number of interpolation nodes is less than the number of arguments). In the report new schemes and algorithms for smoothing and restoration of functional values based on the metric analysis were presented. These schemes and algorithms have demonstrated a high accuracy of smoothing and retrieving the values of functions of one or many variables.

Methods of interpolation and smoothing of functions of several variables in a finite space, presented in this paper can be generalized to non-linear functional such as the infinite-dimensional functional spaces of Banach and Hilbert spaces.

Acknowledgments. The reported study was funded within the Agreement No 02.a03.21.0008 dated 24.04.2016 between the Ministry of Education and Science of the Russian Federation and RUDN University.

References

1. Watson, G.A.: Approximation Theory and Numerical Methods. Wiley, New York (1980)
2. de Boor, C.: A Practical Guide to Splines, Revised edn. Springer, New York (2001)
3. Haykin, S.: Neural Networks: A Comprehensive Foundation. Prentice Hall, Upper Saddle River (1999)
4. Kryanev, A.V., Lukin, G.V., Udumyan, D.K.: Metric analysis and applications, numerical methods and programming. Adv. Comput. Sci. J. **10**, 408–414 (2009)
5. Kryanev, A.V., Lukin, G.V.: Mathematical Methods of the Uncertain Data Processing. Science, Moscow (2006) (in Russian)
6. Kryanev, A.V., Lukin, G.V., Udumyan, D.K.: Metric Analysis and Data Processing. Moscow, Ed. Science (2012). (in Russian)
7. Kryanev, A.V., Udumyan, D.K., Lukin, G.V., Ivanov, V.V.: Metric analysis approach for interpolation and forecasting of time processes. Appl. Math. Sci. **8**(22), 1053–1060 (2014)
8. Kryanev, A.V., Udumyan, D.K.: Metric analysis, properties and applications as a tool for interpolation Int. J. Math. Anal. **8**(45), 2221–2228 (2014)
9. Kryanev, A.V., Udumyan, D.K.: Metric analysis, properties and applications as a tool for smoothing Int. J. Math. Anal. **8**(47), 2337–2346 (2014)

Systems of Differential Equations of Infinite Order with Small Parameter and Countable Markov Chains

Galina Bolotova[1], S.A. Vasilyev[1(✉)], and Dmitry N. Udin[2]

[1] Department of Applied Probability and Informatics, RUDN University, Miklukho-Maklaya Street 6, Moscow, Russia 117198
galinabolotova@gmail.com, svasilyev@sci.pfu.edu.ru
[2] IBM Österreich Internationale Büromaschinen Gesellschaft m.b.H., Obere Donaustrasse 95, 1020 Wien, Austria
udin.pfur@gmail.com

Abstract. Tikhonov-type Cauchy problems are investigated for systems of ordinary differential equations of infinite order with a small parameter μ and initial conditions. It is studying the singular perturbated systems of ordinary differential equations of infinite order of Tikhonov-type $\mu\dot{x} = F(x(t, g_x), y(t, g_y), t), \dot{y} = f(x(t, g_x), y(t, g_y), t)$ with the initial conditions $x(t_0) = g_x$, $y(t_0) = g_y$, where x, $g_x \in X$, $X \subset l_1$ and y, $g_y \in Y$, $Y \in \mathbf{R}^\mathbf{n}$, $t \in [t_0, t_1]$ $(t_0 < t_1)$, $t_0, t_1 \in T$, $T \in \mathbf{R}$, g_x and g_y are given vectors, $\mu > 0$ is a small real parameter. The results may be applied to the queueing networks, which arise from the modern telecommunications.

Keywords: Systems of differential equations of infinite order · Singular perturbated systems of differential equations · Small parameter · Markov chains

1 Introduction

The recent research of service networks with complex routing discipline in [13,20–22] transport networks [1,5,6] and the asymptotic behavior of Jackson networks [16] faced with the problem of proving the global convergence of the solutions of certain infinite systems of ordinary differential equations to a time-independent solution. Scattered results of these studies, however, allow a common approach to their justification. This approach will be expounded here. In work [14] the countable systems of differential equations with bounded Jacobi operators are studied and the sufficient conditions of global stability and global asymptotic stability are obtained. In [12] it was considered finite closed Jackson networks with N first come, first serve nodes and M customers. In the limit $M \to \infty$, $N \to \infty$, $M/N \to \lambda > 0$, it was got conditions when mean queue lengths are uniformly bounded and when there exists a node where the mean queue length tends to ∞ under the above limit (condensation phenomena, traffic jams), in terms of the

© Springer International Publishing AG 2016
V.M. Vishnevskiy et al. (Eds.): DCCN 2016, CCIS 678, pp. 565–576, 2016.
DOI: 10.1007/978-3-319-51917-3_48

limit distribution of the relative utilizations of the nodes. It was deriven asymptotics of the partition function and of correlation functions. Cauchy problems for the systems of ordinary differential equations of infinite order was investigated Tihonov [17], Persidsky [15], Zhautykov [23,24], Korobeinik [8] other researchers. For example, Kreer, Ayse and Thomas [10] investigated fractional Poisson processes, a rapidly growing area of non-Markovian stochastic processes, that are useful in statistics to describe data from counting processes when waiting times are not exponentially distributed. They showed that the fractional Kolmogorov-Feller equations for the probabilities at time t could be represented by an infinite linear system of ordinary differential equations of first order in a transformed time variable. These new equations resemble a linear version of the discrete coagulation-fragmentation equations, well-known from the non-equilibrium theory of gelation, cluster-dynamics and phase transitions in physics and chemistry.

It was studied the singular perturbated systems of ordinary differential equations by Tihonov [18], Vasil'eva [19], Lomov [11] other researchers.

A particular our interest is the synthesis all these methods and its applications in telecommunications. In this paper we apply methods from [14] for the singular perturbated systems of ordinary differential equations of infinite order of Tikhonov-type.

1.1 Auxiliary Notations and Statements

Denote by $\mathbf{R}$ the set of real numbers and denote by $\mathbf{R}^n$ n-dimensional Euclidean space where $n \in \mathbf{N} = \{1, 2, ...\}$. The time parameter is denoted by t where $t \in [t_0, t_1]$ $(t_0 < t_1)$, $t \in T$, $T \in \mathbf{R}$. Let $x(t) = (x_1(t), x_2(t), ..., x_n(t))^T$ be a n-dimensional vector function, $x(t) \in X$, $X \in \mathbf{R}^n$.

Let the sum

$$\|x(t)\|_n = |x_1(t)| + |x_2(t)| + ... + |x_n(t)| = \sum_{i=0}^{n} |x_i(t)| \qquad (1)$$

exists for any $t \in T$. This sum induces the norm and the topology on the set of bounded operators $A: x_1 \to x_2$, $\|A\|_n = \sup_{x=0} \|Ax\|_n / \|x\|_n$. Let l_1 be a sequence space $y(t) = (y_1(t), y_2(t), ...)^T$, $y(t) \in Y$, $Y \subset l_1$ where $y(t)$ is an infinite-dimensional vector function.

Let the sum

$$\|y(t)\| = |y_1(t)| + |y_2(t)| + ... = \sum_{i=0}^{\infty} |y_i(t)| \qquad (2)$$

exists for any $t \in T$. This sum induces the norm and the topology on the set of bounded operators $B: l_1 \to l_2$,

$$\|B\| = \sup_{y=0} \|By\| / \|y\|. \qquad (3)$$

Let

$$z(t) = (x_1(t), x_2(t), ..., x_n(t), y_1(t), y_2(t), ...)^T = (z_1(t), z_2(t), ...)^T \qquad (4)$$

be a infinite-dimensional vector $z(t) \in Z$, $Z \in X \times Y$ where Z is a direct products of the spaces X, Y. Denote by $S = Z \times T = X \times Y \times T$ the set where S is a direct products of the spaces X, Y and T.

Let exists the sum

$$\begin{cases} \rho(z', z'') = \sum_{i=0}^{\infty} |z_i'(t) - z_i''(t)|, \\ z', z'' \in Z, \end{cases} \qquad (5)$$

where $\rho(z', z'')$ defines the distance in the space Z. All vectors and matrices inequalities should be understood as component-wise.

Denote by

$$U_\varepsilon(G) = \{z \in l_1 \mid \|G - z\| < \varepsilon\} \qquad (6)$$

the ε-neighborhood of the point $G = (G_1, G_2, ...)^T$ where $G \in Z$. Let's remind that a set is called compact if any open cover of it has a finite subcover. Subset l_1 has a compact closure in l_1, and called precompact. It is known that a set is precompact if and only if for any $\varepsilon > 0$ there is a finite ε-grid. Let $T_k^+ z = z_k^+$ be a vector $(0, ..., 0, z_k(t), z_{k+1}(t), ...)$ $(k \in \mathbf{N})$, which has all co-ordinates starting with $k + 1$-th congruent with the corresponding coordinates of vector $z(t)$, and all the previous co-ordinates are equal to zero, and let $T_k^- z = z_k^-$ be a vector $(z_1(t), ..., z_k(t), 0, ...)$, which has all co-ordinates starting with k-th congruent with the corresponding coordinates of vector $z(t)$, and all the next co-ordinates are equal to zero.

Statement 1 (The criterion of precompactness) *[9]. A set $Z \subset l_1$ is precompact if and only if when Z is finite and*

$$\forall \varepsilon > 0 \exists k \in N \forall z \in Z \, \|z_k^+\| < \varepsilon. \qquad (7)$$

Statement 2 (The generalized fixed point principle) *[9]. Let $F \colon Z \to Z$ is a mapping of a complete metric space (Z, ρ) has the property that for any*

$$z', z'' \in Z, \rho(Fz', Fz'') \le q(\alpha, \beta)\rho(z', z''), \, (\alpha \le \rho(z', z'') \le \beta) \qquad (8)$$

and $q(\alpha, \beta) < 1$ when $0 < \alpha \le \beta < \infty$ than F has the only fixed point z^* and $\lim_{n \to \infty} \rho(F^n z', z^*) = 0$ with $\forall z' \in Z$.

Statement 3. *Let $F \colon Z \to Z$ is a mapping of compact metric domain (Z, p) to itself decrease the distance, i.e. for any*

$$z', z'' \in Z, \, z' \neq z'', \, \rho(Fz', Fz'') < \rho(z', z''). \qquad (9)$$

So function F has the only fixed point

$$z^* = F(z^*), \, \lim_{m \to \infty} \rho(F^m z', z*) = 0, \, \forall z' \in Z. \qquad (10)$$

Proof. Lets remove all pairs of points in Cartesian product $Z \times Z$ which have the distance between each over strictly less than ε. For any pair of points (z', z''), which is belong to a remaining set R, it is possible to define the compression ratio

$$k(z', z'') = \rho(Fz', Fz'')/\rho(z', z'') < 1 \tag{11}$$

correctly. Inasmuch as R is compact, there is such a $k_0 < 1$ that

$$\forall (z', z'') \in R: k(z', z'') \leq k_0. \tag{12}$$

Thus

$$\forall z', z'' \in Z, \forall \varepsilon > 0, \exists N = N(z', z'', \varepsilon) > 0: \forall m > N(z', z'', \varepsilon), \tag{13}$$

$$\rho(F^m z', F^m z'') < \varepsilon,$$

i.e. $\rho(F^m z', F^m z'') \to 0$ and $m \to \infty$.

Lets examine the sequence $z', Fz', F^2 z', \dots$ ($z' \in Z$). Because of Z is compact this sequence has a limit point z^*, i.e.

$$\exists k_1 < k_2 < \dots: F^{k_i} z' \to z^*, (i \in \mathbf{N}). \tag{14}$$

Lets show that point z^* could be transferred to itself by function F. The sequence $F^{k_i+1} z' = F(F^{k_i})z' \to Fz^*$. Hence $\rho(z^*, Fz^*) \leftarrow \rho(F^{k_i} z', F(F^{k_i} z'))$. But we have earlier proved that $\rho(F^{k_i+1} z', F^{k_i} Fz') \to 0$, in particular for z' and $z'' = Fz'$. Hence we get $z^* = Fz^*$, i.e. fixity of point z^*. The unicity is obvious, and we have already proved the convergence.

1.2   Tikhonov-Type Cauchy Problems for Systems of Ordinary Differential Equations of Infinite Order with a Small Parameter

Let's consider Tikhonov-type Cauchy problems for systems of ordinary differential equations of infinite order with a small parameter μ and initial conditions:

$$\begin{cases} \dot{x} = f(x(t, g_x), y(t, g_y), t), \\ \mu\dot{y} = F(x(t, g_x), y(t, g_y), t); \\ x(t_0, g_x) = g_x, \\ y(t_0, g_y) = g_y, \end{cases} \tag{15}$$

where $x, f \in X$, $X \in \mathbf{R}^n$ are n-dimensional functions; $y, F \in Y$, $Y \subset l_1$ are infinite-dimensional functions and $t \in [t_0, t_1]$ ($t_0 < t_1 \leq \infty$), $t \in T$, $T \in \mathbf{R}$; $g_x \in X$ and $g_y \in Y$ are given vectors, $\mu > 0$ is a small real parameter; $x(t, g_x)$ and $y(t, g_y)$ are solutions of (15). Given functions $f(x(t, g_x), y(t, g_y), t)$ and $F(x(t, g_x), y(t, g_y), t)$ are continuous functions for all variables. Let S is an integral manifold of the system (15) in $X \times Y \times T$. If any point $t^* \in [t_0, t_1]$ $(x(t^*), y(t^*), t^*) \in S$ of trajectory of this system has at least one common point on S this trajectory $(x(t, G), y(t, g), t) \in S$ belongs the integral manifold S

totally. If we assume in (15) that $\mu = 0$ than we have a degenerate system of the ordinary differential equations and a problem of singular perturbations

$$\begin{cases} \dot{x} = f(x(t, g_x), y(t), t), \\ 0 = F(x(t, g_x), y(t), t); \\ x(t_0, g_x) = g_x, \end{cases} \tag{16}$$

where the dimension of this system is less than the dimension of the system (15), since the relations $F(x(t), y(t), t) = 0$ in the system (16) are the algebraic equations (not differential equations). Thus for the system (16) we can use limited number of the initial conditions then for system (15). Most natural for this case we can use the initial conditions $x(t_0, g_x) = g_x$ for the system (16) and the initial conditions $y(t_0, g_y) = g_y$ disregard otherwise we get the overdefined system. We can solve the system (16) if the equation $F(x(t), y(t), t) = 0$ could be solved. If it is possible to solve we can find a finite set or countable set of the roots $y_q(t, g_x) = u_q(x(t, g_x), t)$ where $q \in \mathbf{N}$.

If the implicit function $F(x(t), y(t), t) = 0$ has not simple structure we must investigate the question about the choice of roots. Hence we can use the roots $y_q(t, g_x) = u_q(x(t, g_x), t)$ $(q \in \mathbf{N})$ in (16) and solve the degenerate system

$$\begin{cases} \dot{x}_d = f(x_d(t, g_x), u_q(x_d(t, g_x), t), t); \\ y_d(t_0, g_x) = g_x. \end{cases} \tag{17}$$

Since it is not assumed that the roots $y_q(t, g_x) = u_q(x(t, g_x), t)$ satisfy the initial conditions of the Cauchy problem (15) $(y_q(t_0) \neq g_x, q \in \mathbf{N})$, the solutions $y(t, g_y)$ (15) and $y_q(t, g_x)$ do not close to each other at the initial moments of time $t > 0$. Also there is a very interesting question about behaviors of the solutions $x(t, g_x)$ of the singular perturbated problem (15) and the solutions $x_d(t, g_x)$ of the degenerate problem (17). When $t = 0$ we have $x(t_0, g_x) = x_d(t_0, g_x)$. Do these solutions close to each other when $t \in (t_0, t_1]$? The answer to this question depends on using roots $y_q(t, g_x) = u_q(x(t, g_x), t)$ and the initial conditions which we apply for the systems (15) and (16).

1.3   Local Existence Theorem for Cauchy Problems for Systems of Ordinary Differential Equations of Infinite Order

Let Tikhonov-type Cauchy problems for systems of ordinary differential equations of infinite order with a small parameter $\mu > 0$ and initial conditions (15) has a form:

$$\begin{cases} \dot{z} = P(z(t, G, \mu), t, \mu), \\ z(t_0, G, \mu) = G, \end{cases} \tag{18}$$

where

$$z = (x_1, x_2, ..., x_n, y_1, y_2, ...)^T,$$

$$P(z(t, G, \mu), t, \mu) = (f_1, f_2, ..., f_n, \mu^{-1} F_1, \mu^{-1} F_2, ...)^T$$

are the infinite-dimensional function;

$$G = (g_{x1}, g_{x2}, ..., g_{xn}, g_{y1}, g_{y2}, ...)^T$$

is the given vector; $t \in [t_0, t_1]$ $(t_0 < t_1 \leq \infty)$.

Let $z(t, G, \mu)$ be a continuously differentiable solution of the Cauchy problems (18) then there are

$$\Phi(t, G, \mu) = \partial z(t, G, \mu) / \partial G, \tag{19}$$

$$\Psi(t, G, \mu) = \partial z(t, G, \mu) / \partial \mu, \tag{20}$$

where $\Phi(t, G, \mu)$ and $\Psi(t, G, \mu)$ satisfy of the system of ordinary differential equations in variations:

$$\begin{cases} \dot{z} = P(z(t, G, \mu), t, \mu), \\ \dot{\Phi}(t, G, \mu) = J_z(t, G, \mu)\Phi(t, G, \mu), \\ \dot{\Psi}(t, G, \mu) = J_z(t, G, \mu)\Psi(t, G, \mu) + \Lambda_\mu(t, G, \mu), \\ z(t_0, G, \mu) = G, \ \Phi(t_0, G, \mu) = I, \ \Psi(t_0, G, \mu) = 0, \\ t_0 \in T, \end{cases} \tag{21}$$

where

$$J_z(t, G, \mu) = (\partial P_i / \partial z_j)_{i,j=1}^\infty \tag{22}$$

is Jacobis matrix, I is an identity operator and $\Lambda_\mu(t, G, \mu) = (\partial P_i / \partial \mu)_{i=1}^\infty$ is a vector.

Theorem 1 (local existence theorem). *Let* $P(z(t, G, \mu), t, \mu)$, $J_z(t, G, \mu)$, $\Lambda_\mu(t, G, \mu)$ *be continuous and meet Gelder's local condition with* $z \in U_\epsilon(G)$ *then the system (21) has only one solution, which meet the conditions* $z(t_0, G, \mu) = G$, $z(t, G, \mu) \in U_\epsilon(G)$. *Thus* $z(t, G, \mu)$ *continuously differentiable with respect to the initial condition, and its derivative meet the Eq. (21).*

Proof. This statement is following from [4] (theorem 3.4.4) when the unlimited operator be $A = 0$.

The behavior of the solution $z(t, G, \mu)$ (18) and the nonnegative condition for the off-diagonal elements of the matrix $J_z(t, G, \mu)$ is demonstrated by the following theorem.

Theorem 2. *Let the solution* z *(18) be* $z(t, G, \mu) \in l_1$ *for any* $t \geq 0$, $G \in l_1$ *and* μ. *The following claims are equal: (i) the off-diagonal elements* $J_z(t, G, \mu)$ *are non-negative for any* G; *(ii) for any* G *and any vector* $h \in l_1, h \geq 0$,

$$z(t, G + h, \mu) \geq z(t, G, \mu). \tag{23}$$

Proof. Lets examine a convex set Z, and $z(t, G, \mu) \in Z$ for any $G \in Z$, derivative $\varPhi(t, G, \mu)$ of function $z(t, G, \mu)$ can be specify by simultaneous equations (21). In that case the following formula is fair for any $G^0, G^1 \in Z$:

$$z(t, G^1, \mu) - z(t, G^0, \mu) = \int_0^1 \varPhi(t, \gamma(s), \mu)(G^1 - G^0)ds \tag{24}$$

where

$$\gamma(s) = (1 - s)G^0 + sG^1, 0 \le s \le 1. \tag{25}$$

In fact the function $z(t, G, \mu)$ transfer the segment $\gamma(s)$ into the curve $z(t, \gamma(s), \mu)$. The following formula is fair because of the continuous differentiability of function $z(t, G, \mu)$

$$z(t, \gamma(\tau), \mu) = z(t, G^0, \mu) + \int_0^\tau \frac{\partial z(t, \gamma(s), \mu)}{\partial s} ds.$$

By the formula of complex derivative

$$\frac{\partial z(t, \gamma(s), \mu)}{\partial s} = \frac{\partial z}{\partial G}(\gamma(s))\gamma'(s)$$

Recalling that $\partial z / \partial G = \varPhi$ and $\gamma'(s) = G^1 - G^0$, with $\tau = 1$ we get (26). Lets suppose that statement (i) is fair. So because of (26)

$$z(t, G + h, \mu) - z(t, G, \mu) = \int_0^1 \varPhi(t, \gamma(s), \mu)hds$$

where $\gamma(s) = G + sh, 0 \le s \le 1$. Because of non-negativeness of function $J_z(t, G, \mu)$ outside of diagonal from (21) we get $\varPhi(t, \gamma(s), \mu) \ge 0$, so

$$\varPhi(t, \gamma(s), \mu)h \ge 0$$

whence we get statement (ii).

Lets suppose that (ii) is fair. Under the conditions of Theorem 1 P, J_z with $z \in U_\epsilon(G)$ be continuous and meet Gelder's local condition. Let Gelder's local condition be $\|P\| < M_0$, $\|J\| < M_1$, and there are numbers

$$\delta = min(\epsilon/M_0, 1/M_1), \ \delta > 0.$$

Let $z(t, G, \mu) = G + z^*(t, G, \mu)$ be a solution of (21), where $z^*(t, G, \mu)$ is a fixed point of Picard's mapping

$$\left(\prod \theta\right)(t) = \int_{t_0}^t P(G + \theta(\tau))d\tau$$

under conditions $t \in [t_0 - \delta_1, t_0 + \delta_1], \delta_1 < \delta$. Mapping $\prod$ is contraction with coefficient $\lambda = \delta_1 M_1 < 1$. Consider the approximation to solution

$$\tilde{z}(t, G, \mu) = G + \tilde{z}^*(t, G, \mu) = G + (t - t_0) P(z(t, G, \mu), t, \mu).$$

Using Statement 2 we can see that

$$\|\tilde{z}(t, G, \mu) - z(t, G, \mu)\|$$

$$= \|\tilde{z}^*(t, G, \mu) - z^*(t, G, \mu)\|$$

$$\leq \frac{1}{1 - \lambda} \| \prod \tilde{z}(t, G, \mu) - \tilde{z}(t, G, \mu)\|,$$

$$\prod \tilde{z}(t, G, \mu) - \tilde{z}(t, G, \mu)$$

$$= \int_{t_0}^{t} P(G + (\tau - t_0)P)d\tau - \int_{t_0}^{t} P d\tau$$

$$= \int_{t_0}^{t} (P(G + (\tau - t_0)P) - P)d\tau = D.$$

Because of the derivative of the function P is limited and P meet Gelder's local condition with the constant M_1, where

$$\|P(G + (\tau - t_0)P(G)) - P(G)\| \leq M_1 \|(\tau - t_0)P(G)\| \leq M_0 M_1 |\tau - t_0|,$$

so

$$\|D\| \leq M_0 M_1 (t - t_0)^2 / 2(1 - \lambda),$$

$$\|\tilde{z}(t, G, \mu) - z(t, G, \mu)\| \leq M_0 M_1 (t - t_0)^2 / 2(1 - \lambda).$$

Using this estimation and for all small $\zeta > 0$ we have that

$$0 \leq z(t, G + \zeta e_j, \mu) - z(t, G, \mu) = \zeta e_j + (t - t_0)[P(G + \zeta e_j) - P(G)] + \gamma(G, t),$$

where

$$\|\gamma(G, t)\| \leq M_0 M_1 (t - t_0)^2 / 2(1 - \lambda)$$

and e_j is a vector, which has all coordinates equal to 0 but j-th coordinate equal to 1. Component $i \neq j$ of this inequality is given by

$$0 \leq (t - t_0)[P_i(G + \zeta e_j) - P_i(G)] + \gamma_i(G, t).$$

Dividing by $t - t_0 > 0$ and directing $t \to t_0$ on the right, considering $\gamma_i(G, t)/(t - t_0) \to 0$ we get $0 \leq P(G + \zeta e_j) - P(G)$. Let's divide last expression by ζ and direct $\zeta \to 0$

$$0 \leq \lim_{\zeta \to 0+} \frac{P(G + \zeta e_j) - P(G)}{\zeta} = \frac{\partial P_i}{\partial G_i} = J_{ij}$$

what is mean the fairing of statement (i).

Theorem 3. *Let Φ be Markovian mapping and $G^0, G^1 \in X$, $t \geq 0$, $\mu > 0$ than*
$$\|z(t, G^1, \mu) - z(t, G^0, \mu)\| \leq \|G^1 - G^0\|.$$

Proof. Using (26) from the proofing of Theorem 4 we have

$$\|z(t, G^1, \mu) - z(t, G^0, \mu)\| \leq \int_0^1 \|\Phi(t, \gamma(s))(G^1 - G^0)\| ds \qquad (26)$$

Let function $\Phi(t, \gamma(s))$ is Markovian mapping for any

$$\|t \geq 0, \ s \in [0, 1], \ \|\Phi(t, \gamma(s))(G^1 - G^0)\| \leq \|G^1 - G^0\|.$$

Estimating the integral, considering this inequality, we get required.

This theorem shows us the following sufficient condition for the boundedness of the norm-solution $z(t, G, \mu)$.

Corollary fact from Theorem 3. *Let* $\exists G^* \in X: z(t, G^*, \mu) = G^*$. *Then* $\|z(t, G, \mu) - G^* \leq \|G - G^*\|$ *with* $t \geq 0, G \in X$.

This fact we can use for solutions analysis of the systems (18).

1.4 Using Cutting Method for Systems of Ordinary Differential Equations of Infinite Order

There is a cutting method of solving of systems of ordinary differential equations of infinite order where for (18) we can get Tikhonov-type Cauchy problems

$$\dot{z}^{(M)} = P^{(m)}(z^{(M)}(t, G, \mu), t, \mu); \ z^{(M)}(t_0, G, \mu) = G, \qquad (27)$$

where $M = n + m$ and

$$z^{(M)} = (x_1, x_2, ..., x_n, y_1, y_2, ..., y_m)^T,$$

$$P^{(M)}(z^{(M)}(t, G, \mu), t, \mu) = (f_1, f_2, ..., f_n, \mu^{-1}F_1, \mu^{-1}F_2, ..., \mu^{-1}F_m)^T$$

are the M-dimensional function;

$$G^M = (g_{x1}, g_{x2}, ..., g_{xn}, g_{y1}, g_{y2}, ..., g_{ym})^T$$

is the given vector; $t \in [t_0, t_1]$ $(t_0 < t_1 \leq \infty)$ and $\mu > 0$.

Next theorem give a sufficient conditions using of this method.

Theorem 4. *Let* $\|P(z^{(M)}(s, G^{(M)}, \mu)) - P^{(M)}(z^{(M)}(s, G^{(M)}, \mu))\| < C(M, t^*)$ *and* $P(z^{(M)}(s, G^{(M)}, \mu)) - P(z(s, G, \mu)) \leq K\|z^{(M)}(s, G^{(n)}, \mu) - z(s, G, \mu)\|$ *be for all of* $s \leq t^*$ *amd* $\mu > 0$ *then*

$$\|z^{(M)}(t, G^{(M)}, \mu) - z(t, G, \mu)\| \leq [C(M, t^*) + \|G^{(M)} - G\|e^{Kt} - C(M, t*)]$$

when $0 < t < t*$.

Proof. Transforming the differential equations to integral, we get

$$Z(t, G, \mu) = G + \int_0^t P(z(s, G))ds, z^{(M)}(t, G^{(M)})$$

$$= G^{(M)} + \int_0^t P^{(M)}(z^{(M)}(s, G^{(M)}))ds.$$

Considering the Gelder's condition on function P at $t < t^*$

$$\|z^{(M)}(t, G^{(M)}) - z(t, G)\| \leq \|G^{(M)} - G\|$$

$$+ \int_0^t \|P^{(M)}(z^{(M)}(s, G^{(M)})) - P(z, s, G))\|ds$$

$$\leq \|G^{(M)} - G\| + \int_0^t \|P^{(M)}(z^{(M)}(s, G^{(M)})) - P(z^{(M)}(s, G^{(M)}))ds$$

$$+ \int_0^t \|P(z^{(M)}(s, G^{(M)})) - P(z(s, G^{(M)}))\|ds$$

$$\leq \|G^{(M)} - G\| + C(M, t^*)t + K \int_0^t \|z^{(M)}(s, G^{(M)}) - z(s, G)\|ds.$$

Let's note $\varphi(t) = \|z^{(M)}(t, G^{(M)}) - z(t, G)\|ds$, so we get the integral inequality

$$\varphi(t) \leq \|G^{(M)} - G\| + C(M, t^*)t + K \int_0^t \varphi(s)ds.$$

Let $\varphi(t) \leq \psi(t)$ where

$$\psi(t) = \|G^{(M)} - G\| + C(M, t^*)t + K \int_0^t \varphi(s)ds,$$

where

$$\psi(t) = (\|G^{(M)} - G\| + C(M, t^*))e^{Kt} - C(M, t^*).$$

Corollary fact from Theorem 4. *If* $\|G^{(M)} - G\| \to 0$ *and* $C(M, t^*) \to 0$ *and any* $\mu > 0$, *with fixed* t^* *in conditions of Theorem 4, then* $\|z^{(M)}(t, G^{(M)} - z(t, M)\| \to 0$ *uniformly on any subset of the segment* $[0, t^*]$.

2   Conclusions

The boundaries of applications and possible generalizations. Some works in the routing disciplines. All systems can be analyzed for the global stability but with some condition that the convergence to the steady-state solution will not coordinate-wise, but the norm. We have seen that the most serious constraints of our methods are non-negativity of the Jacobi matrix off-diagonal elements and the availability of the first integral, which equal to the sum of the components.

It would be interesting to understand the physical meaning of these conditions (mean-field conditions). It is necessary to remember that such systems describes the behavior of the queue lengths on the devices. Roughly speaking, z_k is the proportion of units in the queue for a service, to which there is at least k requests (including requests, which are serviced at the moment). Non-negative elements of the Jacobi matrix indicate that the rate of change of z_k (i.e., the time derivative of z_k) can only grow at the expense of z_j with $j \neq k$. It can be reduced (or decrease) only due to u_k. Thus, with the increase of the portion of queues with a minimum number of requests j in the system, the percentage change in intensity with the minimum number of queues requests $k \neq j$ can only increase.

For example, mean-field conditions could be used in active queue management schemes like RED (random early detection) that had been suggested when multiple TCP sessions are multiplexed through a bottleneck buffer [2]. The idea was to detect congestion before the buffer overflows and packets are lost. When the queue length reacheed a certain threshold RED schemes drop/mark incoming packets with a probability that increases as the queue size increases. The objectives was an equitable distribution of packet loss, reduced delay and delay variation and improved network utilization. Here we could modeling multiple connections maintained in the congestion avoidance regime by the RED mechanism. The window sizes of each TCP session evolve like independent dynamical systems coupled by the queue length at the buffer. We could introduce a mean-field approximation to one such RED system as the number of flows tends to infinity. The deterministic limiting system was described by a transport equation.

Acknowledgments. The reported study was funded within the Agreement 02.03.21.0008 dated 24.11.2016 between the Ministry of Education and Science of the Russian Federation and RUDN University.

References

1. Afanassieva, L.G., Fayolle, G., Popov, S.Y.: Models for transportation networks. J. Math. Science. **84**(3), 1092–1103 (1997)
2. McDonald, D.R., Reynier, J.: A mean-field model for multiple TCP connections through a buer implementing RED. Perform. Eval. **49**(14), 77–97 (2002)
3. Daletsky, Y.L., Krein, M.G.: Stability of solutions of differential equations in Banach space. OSCOW, Science Pub. (1970)
4. Henry, D.: Geometric theory of semilinear parabolic equations. Lecture Notes in Mathematics. Springer-Verlag, Berlin (1981)
5. Khmelev, D.V., Oseledets, V.I.: Mean-field approximation for stochastic transportation network, stability of dynamical system: Preprint No. 434 of University of Bremen (1999)
6. Khmelev, D.V.: Limit theorems for nonsymmetric transportation networks. Fundamentalnaya i Priklladnaya Matematika **7**(4), 1259–1266 (2001)
7. Kirstein, B.M., Franken, D.E., Stoian, D.: Comparability and monotonicity of Markov processes. Theory Probab. Appl. **22**(1), 43–54 (1977)

8. Korobeinik, J.: Differential equations of infinite order and infinite systems of differential equations. Izv. Akad. Nauk SSSR Ser. Mat. **34**, 881–922 (1970)
9. Krasnoselsky, M.A., Zabreyko, P.P.: Geometrical Methods of Nonlinear Analysis. Springer-Verlag, Berlin (1984)
10. Kreer, M., Ayseand, K., Thomas, A.W.: Fractional Poisson processes and their representation by infinite systems of ordinary differential equations. Stat. Probab. Lett. **84**, 27–32 (2014)
11. Lomov, S.A.: The construction of asymptotic solutions of certain problems with parameters. Izv. Akad. Nauk SSSR Ser. Mat. **32**, 884–913 (1968)
12. Malyshev, V., Yakovlev, A.: Condensation in large closed Jackson networks. Ann. Appl. Probab. **6**(1), 92–115 (1996)
13. Mitzenmacher, M.: The Power of Two Choices in Randomized Load Balancing. Ph.D. thesis, University of California at Berkley (1996)
14. Oseledets, V.I., Khmelev, D.V.: Global stability of infinite systems of nonlinear differential equations, and nonhomogeneous countable Markov chains. Problemy Peredachi Informatsii **36**(1), 60–76 (2000). (Russian)
15. Persidsky, K.P.: Izv. AN KazSSR. Ser. Mat. Mach. (2), 3–34 (1948)
16. Scherbakov, V.V.: Time scales hierarchy in large closed Jackson networks: Preprint No. 4. French-Russian A.M. Liapunov Institute of Moscow State University, Moscow (1997)
17. Tihonov, A.N.: Ber unendliche Systeme von Dierentialgleichungen. Rec. Math. **41**(4), 551–555 (1934)
18. Tihonov, A.N.: Systems of differential equations containing small parameters in the derivatives. Mat. Sbornik N. S. **31**(73), 575–586 (1952)
19. Vasil'eva, A.B.: Asymptotic behaviour of solutions of certain problems for ordinary non-linear differential equations with a small parameter multiplying the highest derivatives. Uspehi Mat. Nauk. **18**(111), 15–86 (1963). no. 3
20. Vvedenskaya, N.D., Dobrushin, R.L., Kharpelevich, F.I.: Queueing system with a choice of the lesser of two queues the asymptotic approach. Probl. Inform. **32**(1), 15–27 (1996)
21. Vvedenskaya, N.D., Suhov, Y.M.: Dobrushin's Mean-Field Approximation for a Queue with Dynamic Routing. Markov Processes and Related Fields. (3), 493–526 (1997)
22. Vvedenskaya, N.D.: A large queueing system with message transmission along several routes. Problemy Peredachi Informatsii **34**(2), 98–108 (1998)
23. Zhautykov, O.A.: On a countable system of differential equations with variable parameters. Mat. Sb. (N.S.) **49**(91), 317–330 (1959)
24. Zhautykov, O.A.: Extension of the Hamilton-Jacobi theorems to an infinite canonical system of equations. Mat. Sb. (N.S.) **53**(95), 313–328 (1961)

Applying OpenCL Technology for Modelling Seismic Processes Using Grid-Characteristic Methods

Nikolay Khokhlov[1], Andrey Ivanov[1(✉)], Michael Zhdanov[1,2], Igor Petrov[1], and Evgeniy Ryabinkin[1,3]

[1] Moscow Institute of Physics and Technology (State University), Dolgoprudny, Russia
ip-e@mail.ru
[2] The University of Utah, Salt Lake City, USA
[3] NRC, Kurchatov Institute, Moscow, Russia

Abstract. This paper is concerned with CUDA and OpenCL technologies used to solve seismic problems in elastic media. We solve the problem of the dynamic wave disturbances spreading in geologic environment in an elastic approach in the two-dimensional case. A grid-characteristic method is used for numerical solution. Performance of problem solving algorithm with the GPU is compared with the performance of solving it on a single core CPU. We also study the influence of various optimizations on the performance of the algorithm. We measured the effectiveness of parallelization on multiple graphics processors.

Keywords: Seismic · Grid-characteristic · CUDA · OpenCL

1   Introduction

The technology of high-performance computing on graphics processors is being more and more intensively used in recent years. These technologies well suit to the tasks of seismology in elastic media, since they require a large number of computing resources. A numerical solution of hyperbolic equations is required to solve these problems. Some papers include solutions of various problems of seismology, which come to the solution of hyperbolic systems on GPU. The paper [1] describes the implementation of a numerical method ADER-DG using CUDA technology. The implementation of WENO schemes in GPUs is examined in other works [2]. They study the acceleration at the solution of hyperbolic systems of equations on structured grids in GPU, compared to CPU [3]. Other problems also come to the solution of hyperbolic systems of equations. The paper [4] contains calculations in GPU with improved accuracy - up to 60 decimal places. Authors [8,9] obtained performance improvement in solving shallow-water equations. Magnetohydrodynamic phenomena processes are simulated on GPUs [5]. The authors of [10] implemented a discontinued Galerkin method on GPUs and its profiling is described in detail in this paper. The same method has been

© Springer International Publishing AG 2016
V.M. Vishnevskiy et al. (Eds.): DCCN 2016, CCIS 678, pp. 577–588, 2016.
DOI: 10.1007/978-3-319-51917-3_49

implemented on a multi-GPU in work [11], they obtained acceleration in 28.3 times on the GPU cluster, compared with CPU cluster. The comparative study of acceleration on GPU cluster compared with a CPU cluster is presented in the paper [12], calculated with the help of the method of the spectral components of seismic wave propagation. Additionally CUDA technology is used to solve other problems. The influence of a large number of I/O operations on the performance of the algorithm is shown in [13]. GPU and CPU performances are compared in [14–17]. The attempts are made to reduce power consumption on GPU during computing at the same time [18] In this paper we use the grid-characteristic method, which has proved itself well in the solution of seismic problems [20], requiring numerical solution of hyperbolic systems [19]. This method lends itself well to parallelization, since it uses explicit method and large computational grids. This algorithm has been previously parallelized using MPI and OpenMP. In this paper, the algorithm was implemented using CUDA and OpenCL technology. The impact of various performance optimizations on the implementation of algorithm using CUDA technology was considered as well. After that, the most effective implementation has been rewritten using OpenCL technology. Besides, this algorithm involves multiple GPUs, so the parallelization efficiency was measured in multiple GPUs. NVIDIA and AMD graphic cards were used for testing. The results of OpenCL implementation on the NVIDIA GPU were compared with the same implementation on CUDA. NVIDIA GeForce and Tesla cards were used as well, including the latest models: Tesla k80 and Tesla k40m. AMD GPUs of Radeon HD and Radeon R9 series were tested. We considered differences in implementation efficiency with single and double precision. Intel Xeon E5-2697 CPU was chosen to test consistent implementation.

2   Mathematical Model

Environment behavior is described by the model of an ideal isotropic linear-elastic material. We consider the two-dimensional problem. The following system of partial differential equations describes the state of the elementary volume of elastic material in the approximation of small deformations:

$$\rho \frac{\partial v_x}{\partial t} = \frac{\partial \sigma_{xx}}{\partial x} + \frac{\partial \sigma_{xy}}{\partial y}, \rho \frac{\partial v_y}{\partial t} = \frac{\partial \sigma_{xy}}{\partial x} + \frac{\partial \sigma_{yy}}{\partial y},$$

$$\frac{\partial \sigma_{xx}}{\partial t} = (\lambda + 2\mu)\frac{\partial v_x}{\partial x} + \lambda \frac{\partial v_y}{\partial y}, \frac{\partial \sigma_{yy}}{\partial t} = \lambda \frac{\partial v_x}{\partial x} + (\lambda + 2\mu)\frac{\partial v_y}{\partial y},$$

$$\frac{\partial \sigma_{xy}}{\partial t} = \mu \left(\frac{\partial v_x}{\partial x} + \frac{\partial v_y}{\partial y} \right),$$

where ρ is the density of the medium; Λ, μ Lame parameters; V_x and V_y are the horizontal and vertical components of the velocity of the particles of the medium; $\sigma_{xx}, \sigma_{yy}, \sigma_{xy}$ are the components of the stress tensor. This system can be presented in the matrix form:

$$\frac{\partial \mathbf{u}_p}{\partial t} + A_{pq} \frac{\partial u_q}{\partial x} + B_{pq} \frac{\partial u_q}{\partial y} = 0, \tag{1}$$

where $\mathbf{u}$ is the vector of 5 independent variables $\mathbf{u} = (\sigma_{xx}, \sigma_{yy}, \sigma_{xy}, v_x, v_y)^T$. The explicit form of the matrices A_{pq}, B_{pq} is presented in [6]. Here in after we mean summation over repeated indices. The eigenvalues of matrices A_{pq} and B_{pq} are as follows: $s_1 = -c_p, s_2 = -c_s, s_3 = 0, s_4 = c_s, s_5 = c_p$, where c_p and c_s are the propagation speeds of longitudinal and transverse waves in the medium.

3   Numerical Method

Using coordinate-wise splitting we can reduce the problem of constructing a difference scheme for the system of equations (1) to the problem of constructing a difference scheme for systems of the form:

$$\frac{\partial u_p}{\partial t} + A_{pq} \frac{\partial u_q}{\partial x} = 0 \qquad (2)$$

For hyperbolic system of equations (2) matrix $\mathbf{A}$ can be represented as $\mathbf{A} = \mathbf{R}\mathbf{\Lambda}\mathbf{R}^{-1}$, where $\mathbf{\Lambda}$ is a diagonal, the elements of which are the eigenvalues of $\mathbf{A}$, and $\mathbf{R}$ is the matrix consisting of right eigenvectors of $\mathbf{A}$. We introduce new variables: $\mathbf{w} = \mathbf{R}^{-1}\mathbf{u}$ (The so-called Riemann invariants). Then the system of equations (2) will be reduced to a system of 5 independent scalar transport equations.

Let's reduce a third-order accuracy scheme to the numerical solution of one-dimensional linear transfer equation $u_t + au_x = 0, a > 0$, $sigma = a\tau/h$, τ is a time step, h is step on coordinate:

$$u_m^{n+1} = u_m^n + \sigma(\Delta_0 + \Delta_2)/2 + \sigma^2(\Delta_0 - \Delta_2)/2 + \frac{\sigma(\sigma^2 - 1)}{6}(\Delta_1 - 2\Delta_0 + \Delta_2), \quad (3)$$

$$\Delta_0 = u_{m-1}^n - u_m^n,$$
$$\Delta_1 = u_{m-2}^n - u_{m-1}^n,$$
$$\Delta_2 = u_m^n - u_{m+1}^n.$$

Scheme (3) is tolerant to Courant numbers not bigger than 1. We used a grid-characteristic criterion of monotony, it is based on the characteristic property of the accurate solutions:

$$min(u_{m-1}^n, u_m^n) \le u_m^{n+1} \le max(u_{m-1}^n, u_m^n).$$

In places where this criterion is met, the order of scheme falls to the second one.

Once the values of the Riemann invariants on the next time step are found, the solution: $\mathbf{u}^{n+1} = \mathbf{R}\mathbf{w}^{n+1}$ is recovered.

4   Statement of Problem

Test model is shown in Fig. 1. The dimensions are given in kilometers. A non-reflecting boundary condition is set for bottom and side borders and free boundary is set for the top. The source of perturbations is a vertical force applied to the site from 925.7 m to 974.1 m on the day surface; its amplitude is set by Ricker pulse frequency of 40 Hz. The calculation results are presented in Fig. 2.

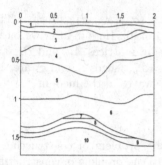

Fig. 1. Geological model of anticlinal trap [7]

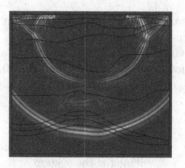

Fig. 2. The calculation result is the wave pattern in time moment $t = 0.38$ s

5   Test Conditions

The two-dimensional test problem with the number of nodes 4096×4096 is considered. 6500 time steps were carried out. Each grid point stored 5 floating point variables. All calculations were performed both with a single (SP), and a double (DP) accuracy. Grid size in memory is 320 MB for computing with single precision and 640 MB for the calculations with double accuracy. However, various optimizations required more memory. If the amount of stream processors (CUDA cores) is C, frequency - F, the number GFLOPS of single accuracy - $2CF$, where "2" is used due to the fact that 2 FMA operations can be carried out per cycle (fused multiply-add). It is known how many units are contained in different processor architectures for single-precision and double-precision. SP:DP column contains their ratio. The amount of GFLOPS can calculated from this for double accuracy (Table 1).

Table 1. Features of tested graphics cards

GPU	Cores	Clock rate, MHz	GFlops (SP)	SP:DP	GFlops (DP)
GeForce GT 640	384	900	691	24	29
GeForce GTX 480	480	1401	1345	8	168
GeForce GTX 680	1536	1006	3090	24	129
GeForce GTX 760	1152	980	2258	24	94
GeForce GTX 780	2304	863	3977	24	166
GeForce GTX 780 Ti	2880	876	5046	24	210
GeForce GTX 980	2048	1126	4612	32	144
Tesla M2070	448	1150	1030	2	515
Tesla K40m	2880	745	4291	3	1430
Tesla K80	2496	562	2806	1.5	1870
Radeon HD 7950	1792	800	2867	4	717
Radeon R9 290	2560	947	4849	8	606

6   Description of Algorithm

The CPU optimized version of this program was taken as the basis for the algorithm implementation on GPUs. The most computationally expensive parts of the algorithm were optimized. As the spatial coordinate splitting was used, two steps were required to transfer the entire grid: on the X axis and Y axis. At that the number of arithmetic floating point operation was calculated, required for the conversion of one grid point in two steps - 190 Flops. Therefore, by knowing the number of grid nodes, the number of time steps, the theoretical amount of GFlops consumed by algorithm, can be determined. Next, knowing the number of stream processors in GPU, its clock frequency and the number of FMA (fused multiply-add) processors in a single processor, we calculated a peak performance for each GPU. The real algorithm tests on GPUs have shown lower values of performance.

Figure 4 shows results for double accuracy. The percentage of the peak performance of algorithms was estimated as the ratio between two values - theoretically required amount of Flops for grid converting and real consumed amount of Flops.

6.1   Transferring Implementation from CPU to GPU - CUDA1

In the original version, the algorithm was redesigned for execution on GPUs using CUDA technology, but it is not optimized for execution on GPUs. In this technology, the graphic processor was assigned 2 times more memory than required to store the computational grid. It is a standard practice when working with the algorithm designing technologies for GPUs. As a result, only the synchronization is performed between function calls that run on GPU (CUDA kernels). This was

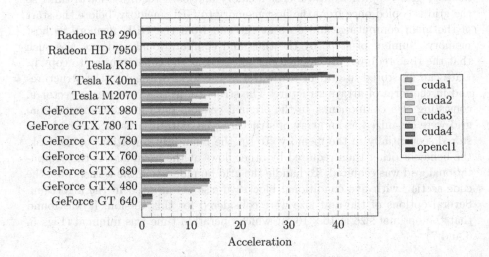

Fig. 3. Acceleration on GPU compared to single core of CPU, double precision

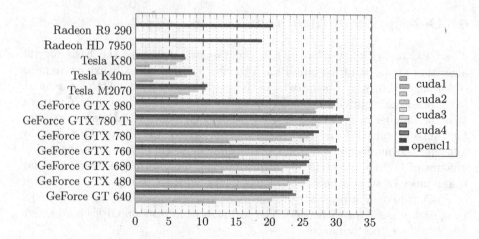

Fig. 4. Percentage of peak performance, double precision

done due to the fact that global synchronization of all graphics processor causes large time delays, so these delays were inserted between calls of kernels. Thus, it became possible to reduce the number of global synchronization up to two times by one time step. Architecture of CUDA stream processors means that in one CUDA unit, flows that perform the same code on different areas of memory are executed simultaneously, if there are no branches in the program code. Therefore, a situation when all flows wait for completion of one occurs only if this flow executes any operations that differ from the rest.

All operations on the memory assigning on GPU and calls of functions running on the graphic processor, are produced by the host - CPU. Operations of memory grid copping from the host memory and back require a lot of time, so the grid is copied once from the host memory to GPU memory, before the start of the main computing, and once at the end from the GPU memory to host memory. Number of steps by the time and computational grid size was such that the time required for calculations far exceeds the time required to copy it. Data can be stored in grid in two ways: in the form of the array of structures and in the form of arrays structure. The structure in this case means vector u, consisting of 5 components. In the original version, data in the computational grid were organized as an array of structures, i.e. data in a particular node are stored sequentially in the memory. To set the boundary conditions of the task, the behavior other than required by other flows at the boundary of the computational grid was required. To handle the grid boundaries in the kernel code, the code section with five conditional blocks "if-else" is inserted in the algorithm. Several options of the unit size were considered for this version. It was found that the optimal size is 16×16, at which operation time was minimal (Figs. 5, 6 and 7).

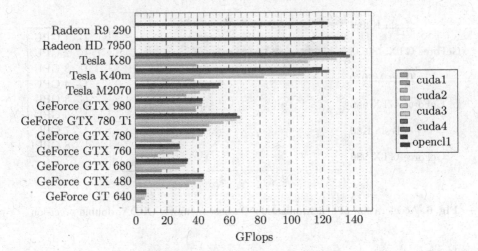

Fig. 5. Performance, double precision

6.2 Structure of Arrays and Sequential Memory Access - CUDA2

Shared memory was used in the next optimization of CUDA block. This is due to the fact that the latency during the interaction with this memory is less than the interaction with the global memory. Optimization means there is only one reading from the global memory at each step along X and Y at the beginning of a function call that is running on the GPU and the data is copied in total memory. Immediately after that all the flows are synchronized in the block; after that, all calculations are performed on the data in the shared memory block. AT the end of kernel, the result is written to the second copy of the computational grid in the global memory.

Another optimization was the selection of a different storing method of the calculated grid in GPU memory - array structure. After that, calls to the global memory become coalesced which led to the increase in performance.

The above mentioned optimizations reduced the number of if-else blocks for grid boundaries processing up to 2.

6.3 Options of Kernel Call - CUDA3

Before that, additional information, such as grid size, material, variables values, resulting from intermediate calculations valid along all time steps were passed to the structure in the GPU global memory as a pointer. All flows constantly addressed the same memory section. Such data were calculated on the CPU in CUDA3 algorithm version once before the execution of code on GPU and were transferred through the parameters of kernel call. It was expected that each flow would create a local copy of these values.

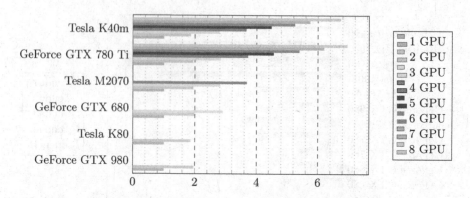

Fig. 6. Acceleration of synchronization via host memory, CUDA, double precision

6.4   Block Sizes - CUDA4

It was important to choose the dimensions of the blocks so that the graphics processor would be constantly loaded, that is, there were no part-time loaded flow processors. It was also important to choose the block sizes multiple of the warp sizes, as the sequential access to memory is carried out within a warp. Another reason is that the flows of a warp can execute code simultaneously on different memory sections, if there are no branches. In version CUDA4, block sizes are selected so as to satisfy the above mentioned requirements and minimize the number of nodes which require memory exchanging between the memory blocks. At the step on X axis to convert each node of the computational grid, values in two adjacent nodes on the X axis are required, and at step on Y-axis, two adjacent nodes on the Y-axis are required. Therefore, one should choose the block sizes so that the number of nodes that require values in the adjacent blocks would be smaller. It was necessary, since the adjacent blocks of nodes require more memory in a general memory block. If, for example, we take a block of size $M \times N$, then in step X for storage of nodes of adjacent blocks $4N$, additional nodes are required, and in step Y - $4M$. Therefore, in step X, the size equal to 256×1 has been selected, as a result, block required only 4 additional nodes in memory. In step Y, the block size was set as 16×16 to reach a compromise between the number of additional memory (64 grid nodes) and the requirements for the serial memory access.

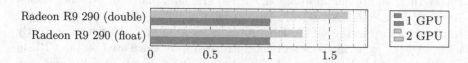

Fig. 7. Acceleration of synchronization via host memory, OpenCL

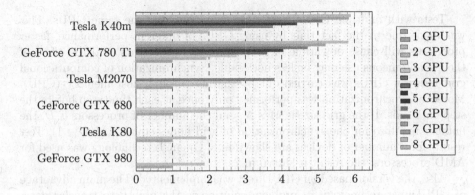

Fig. 8. Acceleration using GPUDirect, single precision

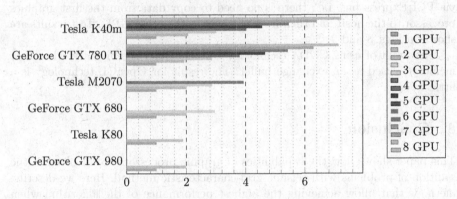

Fig. 9. Acceleration using GPUDirect, double precision

6.5 OpenCL1

The next step was to create the OpenCL implementation of this algorithm. Optimized version of the CUDA4 algorithm was taken as the basis for this. Acceleration test results for all implementations are shown in Fig. 3. The maximum obtained acceleration compared with a single CPU on one graphic CPU - 55 times on the GeForce GTX 780 Ti in the computations of single precision and 44 times at Tesla K80 in double precision computations. It good results for AMD devices should be noted, inspite of the fact that cheap desktop card were used, they showed good results on implementations of single- and double-precision.

7 Multiple GPUs

The algorithm has been parallelized to run on multiple graphics processors. The most optimized version was used to perform several unused GPU. At that the computational grid was divided into several equal-sized rectangular areas. The division was made on the Y axis, due to the selected size of the block on X axis.

Tests with multiple GPUs were carried out only for the same GPUs. This was made due to the fact that by using GPUs of different performance, faster processors will cause downtime of slower graphics processors, and the effect of their simultaneous use may be less noticeable. Synchronization of computational grid between GPUs was carried out by sharing through host memory (CPU). Moreover synchronization was performed only once at each time step before the step on Y axis. If the grid size is $M \times N$, and the number of processors is D, the number required for the synchronization of grid nodes equals $4M(D - 1)$. Test results for a number of devices are shown in 6. OpenCL technology was used for AMD processors; the result is shown in 7.

Also, the version based on GPUDirect was implemented. The main advantage of GPUDirect technology in problem solution is the ability to transfer data, which is located in GPU memory directly without the involvement of the host via PCI Express bus, i.e., there is no need to copy data from the first graphics processor to the host, and then from the host to the other GPU. Test results are shown in Figs. 8 and 9.

The result of using CUDA technology does not differ significantly, when host memory is used as an exchange buffer. The result for OpenCL technology is a little worse.

8   Conclusion

This paper shows how the capabilities of graphic processors are involved in the solution of problems with seismic grid-characteristic method. Here we describe methods that allow achieving the highest performance of the algorithm when computing on the GPU. The problem of effective GPU memory use both in case of using one GPU, and in case of multiple GPUs is also studied. The influence of different optimizations on the performance of the algorithm was considered. The maximum acceleration obtained on graphic CPU compared with a single CPU - 55 times on GeForce GTX 780 Ti when computing with a single precision and 44 times at Tesla K80 at double precision computations. We managed to achieve the performance of 460 GFlops for single accuracy, and the maximum performance of 138 GFlops for double accuracy was obtained. Maximum achieved acceleration of the graphics processors is 7.1 times for 8 GPUs for double precision. GPUDirect technology raised acceleration to 10% of what has been achieved without the calculations with a single precision.

Based on these results we can draw come to a conclusion that GPUs can be used for the solution of such problems. The results are similar for other tasks, which use similar numerical methods (finite volume method, finite difference methods). It is also worth noting that there are good results for the AMD processors and OpenCL technology, while CUDA technology for NVidia GPUs is used in most works.

Acknowledgments. The scientific research was sponsored by RFBR (Russian Foundation for Basic Research) as a part of the research project 16-29-02018 ofi_m.

References

1. Castro, C.E., Behrens, J., Pelties, C.: CUDA-C implementation of the ADER-DG method for linear hyperbolic PDEs. Geoscientific Model Dev. Discuss. **80**(3), 3743–3786 (2013)
2. Esfahanian, V., Darian, H.M., Gohari, S.I.: Assessment of WENO schemes for numerical simulation of some hyperbolic equations using GPU. Comput. Fluids **6**, 260–268 (2013)
3. Rostrup, S., De Sterck, H.: Parallel hyperbolic PDE simulation on clusters: cell versus GPU. Comput. Phys. Commun. **181**(12), 2164–2179 (2010)
4. Khanna, G.: High-precision numerical simulations on a CUDA GPU: Kerr black hole tails. J. Sci. Comput. **56**(2), 366–380 (2013)
5. Wong, H.C., Wong, U.H., Feng, X., Tang, Z.: Efficient magnetohydrodynamic simulations on graphics processing units with CUDA. Comput. Phys. Commun. **182**(10), 2132–2160 (2011)
6. LeVeque, R.J.: Finite Volume Methods for Hyperbolic Problems, p. 588. Cambridge University Press, Cambridge (2002)
7. Jose, M.: Review article: seismic modeling. Geophysics **67**(4), 1304–1325 (2002)
8. Gallardo, J.M., Ortega, S., de la Asuncion, M., Mantas, J.M.: Two-dimensional compact third-order polynomial reconstructions. Solving nonconservative hyperbolic systems using GPUs. J. Sci. Comput. **48**(1–3), 141–163 (2011)
9. Castro, M.J., Ortega, S., de la Asuncion, M., Mantas, J.M., Gallardo, J.M.: GPU computing for shallow water flow simulation based on finite volume schemes. Comptes Rendus Mecanique **339**(2–3), 165–184 (2011)
10. Fuhry, M., Giuliani, A., Krivodonova, L.: Discontinuous Galerkin methods on graphics processing units for nonlinear hyperbolic conservation laws. Int. J. Numer. Methods Fluids **76**(12), 982–1003 (2014)
11. Mu, D., Chen, P., Wang, L.: Accelerating the discontinuous Galerkin method for seismic wave propagation simulations using multiple GPUs with CUDA and MPI earthquake. Science **26**(6), 377–393 (2013)
12. Komatitsch, D.: Fluid-solid coupling on a cluster of GPU graphics cards for seismic wave propagation. Comptes Rendus Mecanique **339**(2–3), 125–135 (2011)
13. Mielikainen, J., Huang, B., Huang, H., Goldberg, M.: Improved GPU/CUDA based parallel weather and research forecast (WRF) single moment 5-class (WSM5) cloud microphysics. Sel. Top. Appl. Earth Obs. Remote Sens. **5**, 1256–1265 (2012)
14. Nickolls, J., Buck, I., Garland, M., Skadron, K.: Scalable parallel programming with CUDA. Queue **6**, 40–53 (2008)
15. Priimak, D.: Finite difference numerical method for the superlattice Boltzmann transport equation and case comparison of CPU(C) and GPU(CUDA) implementations. J. Comput. Phys. **278**, 182–192 (2014)
16. Tuttafesta, M., Dangola, A., Laricchiuta, A., Minelli, pp, Capitelli, M., Colonna, G.: GPU and multi-core based reaction ensemble Monte Carlo method for non-ideal thermodynamic systems. Comput. Phys. Commun. **185**(2), 540–549 (2014)
17. Ferroni, F., Tarleton, E., Fitzgerald, S.: GPU accelerated dislocation dynamics. J. Comput. Phys. **272**, 619–628 (2014)
18. Tamascelli, D., Dambrosio, F.S., Conte, R., Ceotto, M.: GPU accelerated semiclassical initial value representation molecular dynamics. ArXiv e-prints (2013)

19. Karavaev, D.A., Glinsky, B.M., Kovalevsky, V.V.: A technology of 3D elastic wave propagation simulation using hybrid supercomputers. In: Proceedings of the 1st Russian Conference on Supercomputing - Supercomputing Days 2015, Moscow, Russia, 28–29 September, pp. 26–33 (2015)
20. Khokhlov, N.I., Petrov, I.B.: Application of modern high-performance techniques for solving local and global seismic problems. In: Proceedings of the 1st Russian Conference on Supercomputing - Supercomputing Days 2015, Moscow, Russia, 28–29 September, pp. 380–391 (2015)

Linear Approach for Mathematical Modelling as a Tool for Efficient Portfolio Selection

Alexander Kryanev[✉], Darya Sliva, and Andrey Sinitsin

National Research Nuclear University "MEPhI",
Kashirskoe shosse 31, 115409 Moscow, Russia
avkryanev@mephi.ru

Abstract. This report introduces two approaches to the efficient portfolio selection problem, wherein the criteria and the constraints are linear with respect to control variables. The first approach consists of unconditioned optimization of the average expected efficiency value of a portfolio without imposing any additional constraints on the structure of selected portfolio. For this scheme the problem of effective portfolio formation is reduced to two linear programming problems, solving these for an efficient frontier may be effectively accomplished in closed form. The second scheme considers an additional set of group constraints, which can also be reduced to the problem of finding the Pareto fronts of two linear programming problems.

Keywords: Efficient portfolio · Linear formulation · Pareto frontier · Linear programming problem

1 Introduction

One of the primary goals of economical science is the distribution of resources under conditions when the future efficiency of use of resources after they have been allocated is uncertain. Currently there are a number of approaches to setting up and solving problems of effective portfolio selection. The most widely used are the setting up proposed by Markowitz and the problem statement implementing the "Value at Risk" technique. The present work also addresses the statement of effective portfolio selection problem that uses fuzzy numbers to determine the uncertainty in efficiency values.

2 Mathematical Models of Effective Portfolio Formation

The main objective of portfolio optimization is to maximize the aggregated efficiency (return) of investing or allocating the available resources. However, under the conditions of uncertainty in true efficiency values, the portfolio optimization problem should take into account the risk of realization of unacceptably small values of efficiency. Currently the approach to statement and the solution of

© Springer International Publishing AG 2016
V.M. Vishnevskiy et al. (Eds.): DCCN 2016, CCIS 678, pp. 589–600, 2016.
DOI: 10.1007/978-3-319-51917-3_50

problems of formation of investment portfolios, proposed by Markowitz, in which the efficiency values are treated as random variables, and the variance values are considered to account for risk measures for both individual assets and a portfolio as a whole, is generally used. The other criterion characterizing the investment portfolio selected via the Markowitz's scheme is the average expected value m_p (mathematical expectation) of return of portfolio R_p. Thus, the Markowitz's scheme belongs to a class of double criteria problems in which one of criteria (m_p) is aimed at maximizing the average expected value of efficiency of a portfolio, and the second (an efficiency variance σ_p^2) is aimed at minimizing. The mathematical model of the Markowitz's scheme without additional constraints can be defined as follows [1–3]:

$$\sigma_p^2 = (W\boldsymbol{x}, \boldsymbol{x}) - min$$
$$m_p = (\boldsymbol{m}, \boldsymbol{x}) - max$$
$$\sum_{i=1}^{n} x_i = 1, x_i \geq 0, i = 1, ..., n. \tag{1}$$

where $\boldsymbol{x}$ is a vector of shares of an investment portfolio; W is the covariance matrix of efficiency values R_i; $\boldsymbol{m} = (m_1, ..., m_n)^T$ is a vector of average expected values of asset efficiency. Let's consider statement with both of Markowitz's criteria whilst short-selling or selling assets on credit is allowed. Mathematically the borrowed resources correspond to negative values of shares $x_i, i = 1, ..., n$.

Hence, the mathematical model of effective portfolio selection problem with both Markowitz's criteria and without additional restrictions that includes short-selling opportunities becomes:

$$\sigma_p^2 = (W\boldsymbol{x}, \boldsymbol{x}) - min$$
$$m_p = (\boldsymbol{m}, \boldsymbol{x}) - max$$
$$\sum_{i=1}^{n} x_i = 1. \tag{2}$$

Unlike Markowitz's problem, the short-selling statement does not require conditions $x_i \geq 0, i = 1, ..., n$ to be satisfied. To find the Pareto set of solutions to the double criterion problem (2) that corresponds to the set of effective portfolios in short-selling statement, as a preliminary we seek to solve the single criterion problem of minimizing σ_p^2 :

$$\sigma_p^2 = (W\boldsymbol{x}, \boldsymbol{x}) - min$$
$$\sum_{i=1}^{n} x_i = 1. \tag{3}$$

The solution of the problem (3) is given by the equality:

$$\boldsymbol{x}^* = \frac{W^{-1}\mathbf{1}}{(W^{-1}\mathbf{1}, \mathbf{1})}, \tag{4}$$

where $\mathbf{1} = (1, ..., 1)^T$.

The problem of finding all of Pareto solutions of the problem (2) is further reduced to the parametric class of single criterion problems concerning parameter m_p:

$$\sigma_p^2 = (W\boldsymbol{x}, \boldsymbol{x}) - min$$
$$(\boldsymbol{m}, \boldsymbol{x}) = m_p \geq m_p^*,$$
$$\sum_{i=1}^{n} x_i = 1, \tag{5}$$

where $m_p^* = (\boldsymbol{m}, \boldsymbol{x}^*)$. The entire set of Pareto solutions of problem (2) as the solutions of a problem (5) is given by equality:

$$\boldsymbol{x} = W^{-1} \cdot A^T \cdot (A \cdot W^{-1} \cdot A^T)^{-1} \cdot \boldsymbol{f}, \tag{6}$$

where $A = \begin{bmatrix} 1 & \cdots & 1 \\ m_1 & \cdots & m_n \end{bmatrix}$ is a $(2 \times n)$ matrix, $\boldsymbol{f} = (1, m_p)^T, m_p \geq m_p^* = (\boldsymbol{m}, \boldsymbol{x}^*)$.

Vector equality (6) with regard to the components of vector $\boldsymbol{x}$ will become:

$$x_i = a_i + b_i \cdot m_p, i = 1, ..., n, \tag{7}$$

where $\sum_{i=1}^{n} a_i = 1, \sum_{i=1}^{n} b_i = 0$.

Equalities (6) and (7) define the apparent functional dependence of the vector $\boldsymbol{x}$ components with regard to parameter m_p and, thereby, allow to calculate the structure of the effective portfolio set for any values m_p using the formula (7) without solving anew problem (5). Figure 1 illustrates an exemplificative result of numerical computation of the effective portfolios presented on a plane of the criteria m_p and σ_p. The values used to calculate the Pareto set in Fig. 1 are listed in Table 1.

Table 1. Input data

i	m	Covariance matrix									
1	1.35	1.07	−0.00	−0.06	−0.02	−0.04	−0.05	−0.03	0.10	−0.04	0.09
2	1.45	−0.00	2.81	0.13	−0.07	0.54	0.00	0.08	−0.04	−0.04	−0.11
3	1.51	−0.06	0.13	2.76	−0.04	−0.07	−0.04	0.12	0.01	−0.11	−0.13
4	1.23	−0.02	−0.07	−0.04	0.61	−0.03	−0.01	−0.01	−0.01	−0.01	−0.01
5	1.26	−0.04	0.54	−0.07	−0.03	0.51	0.04	0.01	−0.01	0.01	−0.01
6	1.28	−0.05	0.00	−0.04	−0.01	0.04	0.35	−0.01	−0.00	−0.01	−0.06
7	1.27	−0.03	0.08	0.12	−0.01	0.01	−0.01	0.43	−0.02	−0.03	−0.06
8	1.23	0.10	−0.04	0.01	−0.01	−0.01	−0.00	−0.02	0.43	−0.03	0.04
9	1.30	−0.04	−0.04	−0.11	−0.01	0.01	−0.01	−0.03	−0.03	0.60	−0.02
10	1.42	0.09	−0.11	−0.13	−0.01	−0.01	−0.06	−0.06	0.04	−0.02	0.93

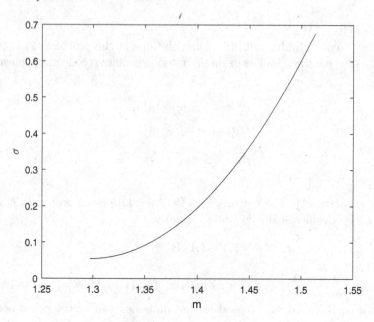

Fig. 1. Set of Pareto solutions.

In the statements of the effective portfolio selection under conditions of uncertainty problems presented above the values of return of developing the distributed resources were treated as random variables. Recently to characterise the uncertainty in random variables they have been treated as fuzzy numbers with invoking the fuzzy set framework [4].

In the present report proposes the mathematical model and the solution on its basis of problems of optimum distribution of resources, wherein the uncertainty in efficiency of resource development is characterised by fuzzy numbers rather than by random variables. One of the problem statements considered allows to reduce the effective portfolio selection problem to a linear programming problem. This gives the opportunity to apply an effective algorithm of finding numerically the set of Pareto solutions corresponding to solutions of the optimum distribution of resources problems. Below one of the possible portfolio optimization problem statements that implements the fuzzy set framework is considered [5]. In this problem statement the mathematical model of formation of effective portfolios, which takes into account a set of constraints can be specified as follows:

$$R_p = \sum_{i=1}^{n} x_i \cdot R_i - max$$

$$r_p = \sum_{i=1}^{n} x_i \cdot r_i - min$$

$$\alpha_j \leq x_{j_1} + ... + x_{j_{n_j}} \leq \beta_j, j = 1, ..., m,$$

$$0 \leq x_i \leq 1, i = 1, ..., n,$$

$$\sum_{i=1}^{n} x_i \cdot V_i = S, \tag{8}$$

where R_p is the cumulative portfolio efficiency; $R_i, i = 1, ..., n$ is the expected value of return for i-th asset; r_p is the risk value of the decrease in actualized value of the cumulative portfolio; $r_i, i = 1, ..., n$ is the risk value for the decrease in realized values of efficiency for i-th asset; $x_i, i = 1, ..., n$ are the shares of resources invested in i-th asset included in portfolios under consideration; $V_i, i = 1, ..., n$ is the maximal possible volume that can be invested in i-th object; S - total volume of resources of a portfolio $(0 < S < V = \sum_{i=1}^{n} V_i)$; m - number of groups.

Problem (8), also as well as problems (1), (2) is a double criterion problem with regard to control variables $x_i, i = 1, ..., n$ with two linear criteria R_p and r_p, aimed at maximization and minimization accordingly.

Values $R_i, i = 1, ..., n$ of the expected values of return and $r_i, i = 1, ..., n$ - risk measures of decrease in actualized values of return can be ascertained on the basis of expert estimations by means of fuzzy numbers.

To find the set of Pareto solutions it is necessary to find the solutions of two single criterion problems (9) and (10):

$$R_p = \sum_{i=1}^{n} x_i \cdot R_i - max$$

$$\alpha_j \leq x_{j_1} + ... + x_{j_{n_j}} \leq \beta_j, j = 1, ..., m,$$

$$0 \leq x_i \leq 1, i = 1, ..., n,$$

$$\sum_{i=1}^{n} x_i \cdot V_i = S, \tag{9}$$

and

$$r_p = \sum_{i=1}^{n} x_i \cdot r_i - min$$

$$\alpha_j \leq x_{j_1} + ... + x_{j_{n_j}} \leq \beta_j, j = 1, ..., m,$$

$$0 \leq x_i \leq 1, i = 1, ..., n,$$

$$\sum_{i=1}^{n} x_i \cdot V_i = S \tag{10}$$

(9) and (10) are linear programming problems, both have the unique solutions $\boldsymbol{x}_1^*$ and $\boldsymbol{x}_2^*$, accordingly $\boldsymbol{x}_1^* = (x_{11}^*, ..., x_{1n}^*)^T, \boldsymbol{x}_2^* = (x_{21}^*, ..., x_{2n}^*)^T$.

The entire set of Pareto solutions of problem (8) is defined by equality:

$$\boldsymbol{x}^*(\alpha) = \alpha \cdot \boldsymbol{x}_1^* + (1 - \alpha) \cdot \boldsymbol{x}_2^*, \tag{11}$$

where $\alpha \in [0, 1]$.

By varying the numerical values of the priority parameter α, the individual effective portfolio structures can be computed. In particular, at $\alpha = 1$ we attain the structure of the effective portfolio of a company that corresponds to the maximum expected value of efficiency, and at $\alpha = 0$ we attain the structure of the effective portfolio of the company that corresponds to the minimum value of risk.

The numerical solutions of the linear programming problems (9), (10) can be carried out by means of an effective method of solving the linear programming problems of the considered class [6]. In the report numerical results of effective portfolio formation under the problem statement (8) are adduced.

In the case when the group constraints are simplified to exclusive constraints, problem (8) takes on the following form:

$$R_p = \sum_{i=1}^{n} x_i \cdot R_i - max$$

$$r_p = \sum_{i=1}^{n} x_i \cdot r_i - min$$

$$\alpha_j \le x_j \le \beta_j, j = 1, ..., m,$$

$$0 \le x_i \le 1, i = 1, ..., n,$$

$$\sum_{i=1}^{n} x_i \cdot V_i = S \tag{12}$$

This setting allows to illustrate the constraints, the solutions to single criterion problems, and the calculated set of Pareto optimal solutions all on one graph. Figures 2 and 3 represents some numerical results of solving problem (12). The solutions to the partial optimization problems are depicted in red and blue, the limitations imposed on values of x_i are shaded in grey, while the set of Pareto optimal solutions to problem (12) that correspond to various values of α ($\alpha = 0, 0.1, ..., 1$) are represented by dotted lines. It is obvious that whenever the solutions to the single criterion problems coincide, the set of Pareto optimal solutions of the multiple criteria problem is limited to the singular portfolio.

The group constraints in problem (8) do not allow for such a clear way of illustrating the principle of finding the Pareto set of optimal solutions to the multiple criteria optimization problems. For the purpose of demonstrating this method numerical results are also presented below. Figure 4 presents the solutions to the following problem:

$$R_p = 1.77x_1 + 1.66x_2 + 1.32x_3 + 1.22x_4 + 1.35x_5$$
$$+ 1.79x_6 + 1.99x_7 + 1.85x_8 + 1.05x_9 + 1.05x_{10} - max$$
$$r_p = 4.76x_1 + 3.33x_2 + 3.58x_3 + 3.79x_4 + 4.14x_5$$
$$+ 4.54x_6 + 4.31x_7 + 4.04x_8 + 3.94x_9 + 3.81x_{10} - min$$
$$0.02 \le x_3 + x_4 + x_5 + x_6 + x_9 \le 1,$$

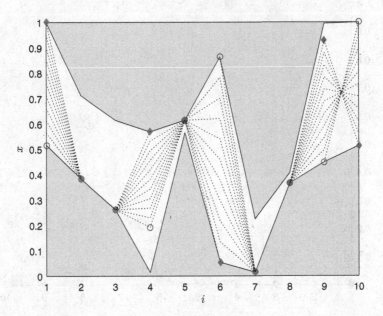

Fig. 2. Numerical results of solving problem (12) (Color figure online)

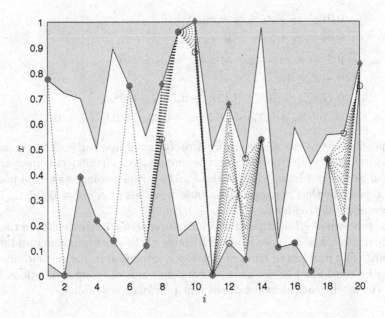

Fig. 3. Numerical results of solving problem (12) (Color figure online)

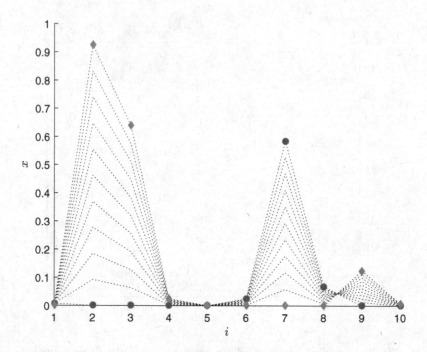

Fig. 4. Numerical results of solving problem (13) (Color figure online)

$$0.01 \leq x_1 + x_4 + x_9 + x_{10} \leq 1,$$
$$0.04 \leq x_1 + x_2 + x_3 + x_4 + x_5 + x_7 + x_9 \leq 1,$$
$$0 \leq x_1 + x_3 + x_4 + x_5 + x_6 + x_{10} \leq 0.67,$$
$$0 \leq x_i \leq 1, i = 1, ..., 10,$$
$$0.77x_1 + 0.19x_2 + 0.42x_3 + 0.76x_4 + 0.98x_5$$
$$+ \ 0.50x_6 + 0.77x_7 + 0.49x_8 + 0.30x_9 + 0.06x_{10} = 0.5 \tag{13}$$

The solutions to the single criterion problems of maximizing the cumulative portfolio efficiency and minimizing the cumulative portfolio risk measure are depicted in red and blue, whereas the Pareto optimal solutions of the multiple criteria problem that correspond to various values of α ($\alpha = 0, 0.1, ..., 1$) are represented by dotted lines.

The problem of efficient portfolios formation arises, in particular, in the reinsurance market, and in these markets it is one of the most important and urgent problems. The problem of efficient portfolios formation in the reinsurance market can be considered in the more severe conditions of group restrictions under which the mathematical model of efficient portfolios is as follows:

$$\sum_{i=1}^{n} x_i \cdot w_i \cdot R_i - max$$

$$\sum_{i=1}^{n} x_i \cdot w_i \cdot r_i - min$$

$$0 \le \alpha_j \le x_{j_1} + x_{j_2} + ... + x_{j_{n_j}} \le \beta_j \le 1, j = 1, ..., m,$$

$$\sum_{j=1}^{m} \beta_j \ge 1, \sum_{j=1}^{m} \alpha_j \le 1, \sum_{j=1}^{m} n_j = n,$$

$$\sum_{i=1}^{n} x_i \cdot V_i = S, \sum_{i=1}^{n} \beta_i \cdot V_i > S, w_i = \frac{V_i}{V}, V = \sum_{i=1}^{n} V_i, \tag{14}$$

where $x_i, i = 1, ..., n$ - the share of the insurance sum of the ith reinsurance contract included in the portfolio; $R_i, i = 1, ..., n$ - the expected value of return for the ith reinsurance contract (for example, profitability and normalized return); $r_i, i = 1, ..., n$ - the value of the insured event risk for the ith reinsurance contract (for example, the value of the likelihood lower than expected yield); $V_i, i = 1, ..., n$ - the total insurance sum of the ith reinsurance contract; S - the total insurance sum of the portfolio $(0 < S < V)$, n - the number of potential reinsurance contracts, which are tested for their possible inclusion in the portfolio of the insurance company contracts. Then problems (9), (10) transform, respectively, to problems (15), (16):

$$\sum_{i=1}^{n} x_i \cdot w_i \cdot R_i - max$$

$$0 \le \alpha_j \le x_{j_1} + x_{j_2} + ... + x_{j_{n_j}} \le \beta_j \le 1, j = 1, ..., m,$$

$$\sum_{j=1}^{m} \beta_j \ge 1, \sum_{j=1}^{m} \alpha_j \le 1, \sum_{j=1}^{m} n_j = n,$$

$$\sum_{i=1}^{n} x_i \cdot V_i = S, \sum_{i=1}^{n} \beta_i \cdot V_i > S, w_i = \frac{V_i}{V}, V = \sum_{i=1}^{n} V_i, \tag{15}$$

$$\sum_{i=1}^{n} x_i \cdot w_i \cdot r_i - min$$

$$0 \le \alpha_j \le x_{j_1} + x_{j_2} + ... + x_{j_{n_j}} \le \beta_j \le 1, j = 1, ..., m,$$

$$\sum_{j=1}^{m} \beta_j \ge 1, \sum_{j=1}^{m} \alpha_j \le 1, \sum_{j=1}^{m} n_j = n,$$

$$\sum_{i=1}^{n} x_i \cdot V_i = S, \sum_{i=1}^{n} \beta_i \cdot V_i > S, w_i = \frac{V_i}{V}, V = \sum_{i=1}^{n} V_i, \tag{16}$$

problem (12) transforms to problem (17):

$$\sum_{i=1}^{n} x_i \cdot w_i \cdot R_i - max$$

$$\sum_{i=1}^{n} x_i \cdot w_i \cdot r_i - min$$

$$0 \leq \alpha_i \leq x_i \leq \beta_i \leq 1, i = 1, ..., n,$$

$$\sum_{i=1}^{n} x_i \cdot V_i = S, \sum_{i=1}^{n} \beta_i \cdot V_i > S, w_i = \frac{V_i}{V}, V = \sum_{i=1}^{n} V_i, \qquad (17)$$

Then, problems (9), (10) transform, respectively, to problems (18), (19):

$$\sum_{i=1}^{n} x_i \cdot w_i \cdot R_i - max$$

$$0 \leq \alpha_i \leq x_i \leq \beta_i \leq 1, i = 1, ..., n,$$

$$\sum_{i=1}^{n} x_i \cdot V_i = S, \sum_{i=1}^{n} \beta_i \cdot V_i > S, w_i = \frac{V_i}{V}, V = \sum_{i=1}^{n} V_i, \qquad (18)$$

$$\sum_{i=1}^{n} x_i \cdot w_i \cdot r_i - min$$

$$0 \leq \alpha_i \leq x_i \leq \beta_i \leq 1, i = 1, ..., n,$$

$$\sum_{i=1}^{n} x_i \cdot V_i = S, \sum_{i=1}^{n} \beta_i \cdot V_i > S, w_i = \frac{V_i}{V}, V = \sum_{i=1}^{n} V_i, \qquad (19)$$

Problems (15), (16) are linear programming problems, each of which has a unique solution x_1^*, x_2^*, respectively $x_1^* = (x_{11}^*, ..., x_{1n}^*)^T$, $x_2^* = (x_{21}^*, ..., x_{2n}^*)^T$. The entire set of Pareto's solutions of (14) is defined by the equality

$$x^*(\alpha) = \alpha \cdot x_1^* + (1 - \alpha) \cdot x_2^*, \qquad (20)$$

where $\alpha \in [0, 1]$.

By varying the numerical value of priority parameter α, we obtain the specific formulation of effective reinsurance portfolios. In particular, at $\alpha = 1$ we obtain the effective portfolio of reinsurance companies that corresponds to the maximum expected value of return, and at $\alpha = 0$, we obtain the effective portfolio that corresponds to the minimum value of risk.

Similarly problems (18), (19) are linear programming problems, each of which has a unique solution x_1^{**}, x_2^{**}, respectively, $x_1^{**} = (x_{11}^{**}, ..., x_{1n}^{**})^T$, $x_2^{**} = (x_{21}^{**}, ..., x_{2n}^{**})^T$. The entire set of Pareto's solutions of problem (17) is defined by the equality

$$x^{**}(\alpha) = \alpha \cdot x_1^{**} + (1 - \alpha) \cdot x_2^{**}, \qquad (21)$$

where $\alpha \in [0, 1]$.

The linear relationship between the risk value r and the R is illustrated in Fig. 5.

Numerical solutions of specific problems of linear programming type (15)–(19) are also attained by means of the effective two-step method [6].

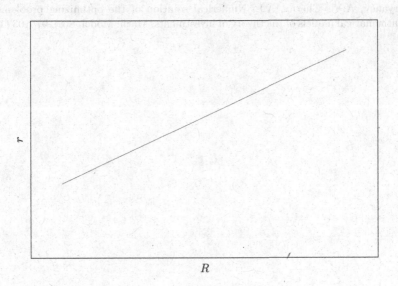

Fig. 5. Linear relationship between the risk value r and the R

3   Conclusion

In the report statements and solutions of effective portfolio problems are presented. For statements that allow short-selling the solutions are attained in closed form in the analytical kind. This allows to calculate the structure of effective portfolios without solving anew the parent criterion problem. Statement with double linear criteria, wherein the criteria can be defined by means of fuzzy numbers, is presented. The proposed problem statement allows to account for the relation between the possible uncertain future efficiency values of resource development in the case of absence of preliminary statistics on return actualization, based entirely on the use of expert estimations.

References

1. Markowitz, H.M.: Mean Variance Analysis in Portfolio Choice and Capital Markets. Basil Blackwell, New York (1990)
2. Sharpe, W.F., Alexander, G.J., Bailey, J.W.: Investments. Infra-M, Moscow (2001)
3. Kryanev, A.V., Lukin, G.V.: On formulation and solutions of investment portflio optimization problems. Preprint MEPhI 006-2001. MEPhI (2001)

4. Orlov, A.I.: Non-numeric Statistics. MZ-Press, Moscow (2004)
5. Klimanov, S.G., Rostovsky, N.S., Sliva, D.E., Smirnov, D.S., Udumyan, D.K., Balashov, R.B.: Statement and solution of the problems of effective investment portfolios formation for the companies in SWU markets. Appl. Math. Sci. 8(107), 5329–5335 (2014)
6. Kryanev, A.V., Cherny, A.I.: Numerical solution of the optimizing problems for mathematical models of the theory of investments. Math. Model. 8(8), 97–103 (1996)

Mathematical Modeling of Smoothly-Irregular Integrated-Optical Waveguide and Mathematical Synthesis of Waveguide Luneburg Lens

Edik Ayrjan[1,3], Genin Dashitsyrenov[2], Konstantin Lovetskiy[2],
Nikolai Nikolaev[2], Anton Sevastianov[2], Leonid Sevastianov[1,2(✉)],
and Eugeny Laneev[2]

[1] Joint Institute for Nuclear Research (JINR), Joliot-Curie 6, 141980 Dubna,
Moscow Region, Russia
edik.hayryan@jinr.ru, leonid.sevast@gmail.com
[2] RUDN University (Peoples' Friendship University of Russia),
6 Miklukho-Maklaya Street, 117198 Moscow, Russia
genin_d@mail.ru, lovetskiy@gmail.com, nnikolaev@sci.pfu.edu.ru,
alsevastyanov@gmail.com, elaneev@yandex.ru
[3] Yerevan Physics Institute, 2 Alikhanian Brothers street, 375036 Yerevan, Armenia
http://www.jinr.ru, http://www.rudn.ru

Abstract. In this paper we consider a class of multilayer integrated-optical waveguides consisting of homogeneous dielectric layers of constant or variable thickness, which are being systematically numerically studied using the cross-section method. The method is based on the adiabatic approximation of the asymptotic expansion on the one hand and the expansion in the complete system of modes of regular comparison waveguide. The paper discusses the problems of numerical implementation of the cross-section method to the transformation of particular mode in a smooth transition from one planar regular open waveguide to another.

Luneburg proposed a model of the ideal optical instrument (in the framework of geometrical optics), afterwards called Luneburg lens. Later classical Luneburg lens was included in the family of the ideal optical instruments - generalized Luneburg lenses. Zernike in his work showed that a local increase in thickness of the waveguiding layer leads to a local deceleration of phase velocity of the propagating waveguide mode. This effect has led to the idea of manufacturing the waveguide (two-dimensional) Luneburg lenses instead of volume (three-dimensional) lenses. In this work we synthesized mathematically the thickness profiles of the additional (irregular in thickness) waveguide layer forming the thin-film generalized waveguide Luneburg lens.

E. Ayrjan—The work was partially supported by RFBF grants No 14-01-00628, No 15-07- 08795, No 16-07-00556. The reported study was funded within the Agreement No 02.a03.21.0008 dated 24.04.2016 between the Ministry of Education and Science of the Russian Federation and RUDN University.

V.M. Vishnevskiy et al. (Eds.): DCCN 2016, CCIS 678, pp. 601–611, 2016.
DOI: 10.1007/978-3-319-51917-3_51

Keywords: Multilayer integrated-optical waveguides · Generalized luneburg lenses · The cross-section method

1   Introduction

We are interested in the class of so-called multilayer integrated optical waveguides consisting of homogeneous dielectric layers of constant or variable thickness. If the layers thicknesses are constant and the interfaces between them are parallel planes, planar waveguides are called regular planar open waveguides. Theoretical (analytical) description of such waveguides is made nearly half a century ago and described in many reviews and books (see, e.g., [1–6]). A systematic description of the numerical (computer) models of regular planar open waveguides is less common (see, e.g., [7,8]).

If one or more layers are of variable thickness, the waveguide is called irregular. Among such waveguides we distinguish the class of waveguides with smooth change of thickness, the so-called smoothly-irregular integrated-optical waveguides. The theory of such waveguides, the so-called cross-section method is created by Shevchenko V.V. and published in [9]. Subsequently relatively successful attempts to generalize this method have been made (see, e.g., [10–16]). At the same time, to date, the authors are unknown of systematic numerical studies of smoothly-irregular integrated-optical waveguides based on the cross-section method. We use a very particular embodiment of the method for the design of thin-film generalized waveguide Luneburg lens.

2   The Cross-Section Method

The basis of the cross-section method is the adiabatic approximation of the asymptotic expansion of locally plane waves, simplified by two additions:

(1) in the derivatives of adiabatic waveguide modes only the zero-order contributions are taken into account;
(2) instead of the tangent planes at irregular boundaries for the formulation of boundary conditions their approximations by "horizontal projections" are used.

Using the second assumption reduces the complete system of modes used in the method to the system of modes of the regular comparison waveguide. Using the first assumption leads to a system of ordinary differential equations for the coefficients of the expansion of the general solution in the cross-section method.

In the cross-section method the propagation of polarized monochromatic light in a smooth dielectric transition (irregular along the axis Oz) between two regular planar dielectric waveguides is described by a complete set of modes of the regular comparison waveguide [17,18]:

In the left regular part of the waveguide (see Fig. 1) the electromagnetic radiation (monochromatic, linearly polarized) propagates in two forms: TE modes, obeying the Helmholtz equation

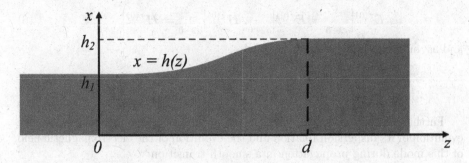

Fig. 1. A homogeneous smooth dielectric transition between two planar regular dielectric waveguides.

$$\left(\frac{\partial^2}{\partial x^2} + \frac{\partial^2}{\partial z^2}\right) E_y + k_0^2 n^2(x) E_y = 0 \tag{1}$$

and TM modes, obeying the Helmholtz equation

$$n^2(x) \frac{\partial}{\partial x} \left(\frac{1}{n^2(x)} \frac{\partial}{\partial x}\right) H_y + \frac{\partial^2 H_y}{\partial z^2} + k_0^2 n^2(x) H_y = 0 \tag{2}$$

There can be modes, running from left to right, and modes, running from right to left. Solutions of (1) and (2) (i.e., propagating modes) in the case of regular waveguide are described in [7,17].

When the waveguide mode, which has reached during propagation from left to right the beginning of the waveguide transition, propagates further through the waveguide with a variable (continuously variable) section, the mode undergoes (transverse, geometrical) transformation, its velocity of propagation changes, and in general it no longer strictly satisfies the Helmholtz equation (1) or (2) (for the regular part of the waveguide). Propagation in smooth transitions of open waveguides is often described by the cross-section method. Namely, the field in each cross-section z of this transition is represented as an expansion in a complete system of modes of the (regular) comparison waveguide (in the section z).

We limit our consideration by the waveguide modes corresponding to the discrete spectrum $E_y^j(x; z)$ and $H_y^j(x; z)$. Each of them satisfies the equations

$$\frac{d^2}{dx^2} \begin{pmatrix} E_y^j \\ H_y^j \end{pmatrix} + \chi^2(\beta_j(z)) \begin{pmatrix} E_y^j \\ H_y^j \end{pmatrix} = \begin{pmatrix} 0 \\ 0 \end{pmatrix}. \tag{3}$$

The remaining components of the electromagnetic field satisfy the relations

$$H_z = \frac{1}{ik_0\mu} \frac{dE_y}{dx}, H_x = -\frac{\beta}{\mu} E_y, E_z = -\frac{1}{ik_0\varepsilon} \frac{\partial H_y}{\partial x}, E_x = \frac{\beta}{\varepsilon} H_y. \tag{4}$$

At the same time the tangential boundary conditions

$$\boldsymbol{E}^{\tau(0)}\Big|_{a1-0} = \boldsymbol{E}^{\tau(0)}\Big|_{a1+0}, \boldsymbol{H}^{\tau(0)}\Big|_{a1-0} = \boldsymbol{H}^{\tau(0)}\Big|_{a1+0}, \tag{5}$$

$$E^{\tau(0)}\Big|_{a2-0} = E^{\tau(0)}\Big|_{a2+0}, \quad H^{\tau(0)}\Big|_{a2-0} = H^{\tau(0)}\Big|_{a2+0}, \tag{6}$$

and asymptotic conditions are satisfied

$$\left|E^{\tau(0)}\Big|_{x\to\pm\infty}\right| < +\infty, \left|H^{\tau(0)}\Big|_{x\to\pm\infty}\right| < +\infty. \tag{7}$$

Further in this paper we consider the evolution of a particular mode, the evolution of its dispersion relation and the evolution of the electromagnetic field of this mode during propagation in a smooth transition:

$$C_j^+(z)\,\psi_j^+(x;z)\exp\left\{-ik_0\int\limits^z \beta_j(z)\,dz\right\} \tag{8}$$

We do not consider the transfer of modes (this question remains outside our consideration), but this does not mean that it does not exist.

3 Dispersion Relations for Homogeneous Smooth Transitions

Consider a very smooth transition between the two dielectric waveguides with thicknesses d_{in} of the left and d_{fin} of the right waveguides. The optical refractive indices are the same throughout the structure: n_s is the refractive index of the substrate, n_f is the refractive index of the waveguide layer, n_c is the refractive index of the coating layer (air). Thickness of the waveguide transition over the interval $[z_{in}, z_{fin}]$ changes from d_{in} to d_{fin} so that for any current $z \in [z_{in}, z_{fin}]$ the thickness $d \in [d_{in}, d_{fin}]$.

The problem of describing guided TE mode in a smooth transition (by the cross-section method [19]) is formulated in the form

$$\begin{aligned}
&A_s^+ = A_f^+ + A_f^- \\
&\tfrac{\gamma_s^j}{ik_0}A_s^+ = \tfrac{\chi_1^j}{k_0}\left(A_f^+ - A_f^-\right) \\
&A_f^+\exp\left\{i\chi_f^j d(z)\right\} + A_f^-\exp\left\{-i\chi_f^j d(z)\right\} = A_c^-\exp\left\{-\gamma_c^j d(z)\right\} \\
&\tfrac{\chi_f^j}{k_0}\left(A_f^+\exp\left\{i\chi_f^j d(z)\right\} - A_f^-\exp\left\{-i\chi_f^j d(z)\right\}\right) \\
&= -\tfrac{\gamma_c^j}{ik_0}A_c^-\exp\left\{-\gamma_c^j d(z)\right\}
\end{aligned} \tag{9}$$

So the solution on the entire axis Ox is given (under the described matrix model [7,17]) by the homogeneous system of linear algebraic equations

$$\mathbf{M}_4^{TE}\left(d(z),\beta^j(z)\right)\boldsymbol{A} = \boldsymbol{0} \tag{10}$$

This system has a non-trivial solution, in particular, provided that

$$\det\left(\mathbf{M}_4^{TE}\left(d(z),\beta^j(z)\right)\right) = 0 \tag{11}$$

It is well known that the waveguide dispersion relation $\beta^j \left(d\left(z \right) \right)$ takes values in the range $\beta^j \in \left(n_s, n_f \right)$. Moreover, for each root β^j there exist its critical thickness (see, e.g., [7])

$$d_{cr}^j = \frac{arctg\left(\gamma_c\left(n_s \right) / \chi_f\left(n_s \right) \right)}{\chi_f\left(n_s \right)} + j\pi \tag{12}$$

from which $\beta^j > n_s$. Note also that $\beta^j(d(z))_{d(z) \to \infty} \to n_f$.

Non-linear transcendental algebraic equation (11) for each fixed $d > d_{cr}^m > \ldots > d_{cr}^1 > d_{cr}^0$ has a finite number of roots $\beta^0 > \beta^1 > \beta^2 > \ldots > \beta^m$. In order to calculate in a sustainable way the dispersion curves $\beta^j\left(d \right)$ at $d > d_{cr}^j$, we apply a stable procedure of numerical solution of the equation (11) [16].

4   Dispersion Relations for the Non-uniform (Two-Layer) Smooth Transitions

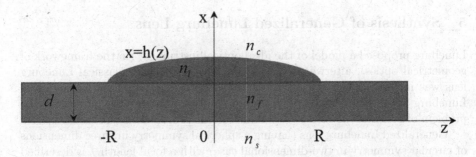

Fig. 2. A non-homogeneous smooth dielectric transition between planar regular dielectric waveguides.

The problem of describing guided TE mode in a non-homogeneous (see Fig. 2) smooth transition (by the cross-section method [19]) is formulated as:

$$
\begin{aligned}
& A_s^+ \exp\left\{ -\gamma_s^j d \right\} = A_f^+ \exp\left\{ -i\chi_f^j d \right\} + A_f^- \exp\left\{ i\chi_f^j d \right\} \\
& \tfrac{\gamma_s^j}{ik_0} A_s^+ \exp\left\{ -\gamma_s^j d \right\} = \tfrac{\chi_f^j}{k_0}\left(A_f^+ \exp\left\{ -i\chi_f^j d \right\} - A_f^- \exp\left\{ i\chi_f^j d \right\} \right) \\
& A_f^+ + A_f^- = A_l^+ + A_l^- \\
& \tfrac{\chi_f^j}{k_0}\left(A_f^+ - A_f^- \right) = \tfrac{\chi_l^j}{k_0}\left(A_l^+ - A_l^- \right) \\
& A_l^+ \exp\left\{ i\chi_l^j h\left(z \right) \right\} + A_l^- \exp\left\{ -i\chi_l^j h\left(z \right) \right\} = A_c^- \exp\left\{ -\gamma_c^j h\left(z \right) \right\} \\
& \tfrac{\chi_l^j}{k_0}\left(A_l^+ \exp\left\{ i\chi_l^j h\left(z \right) \right\} - A_l^- \exp\left\{ -i\chi_l^j h\left(z \right) \right\} \right) \\
& = -\tfrac{\gamma_c^j}{ik_0} A_c^- \exp\left\{ -\gamma_c^j h\left(z \right) \right\}
\end{aligned}
\tag{13}
$$

So the solution for the TE mode in the whole range $x \in (-\infty, \infty)$ is defined by a set A_k satisfying the homogeneous system of linear algebraic equations

$$\mathbf{M}_6^{TE}\left(d, h\left(z\right), \beta^j\left(z\right)\right) \mathbf{A} = 0 \qquad (14)$$

This system has a non-trivial solution, in particular, provided that

$$\det\left(\mathbf{M}_6^{TE}\left(d, h\left(z\right), \beta^j\left(z\right)\right)\right) = 0 \qquad (15)$$

Similar reasoning for the TM mode leads to the homogeneous system of linear algebraic equations

$$\mathbf{M}_6^{TM}\left(d, h\left(z\right), \beta^j\left(z\right)\right) \mathbf{B} = 0 \qquad (16)$$

which has a non-trivial solution, in particular, provided that

$$\det\left(\mathbf{M}_6^{TM}\left(d, h\left(z\right), \beta^j\left(z\right)\right)\right) = 0 \qquad (17)$$

Dispersion relations $\beta^j\left(d, h\left(z\right)\right)$ are calculated using stable numerical methods [16] for solving non-linear transcendental equations (15) and (17).

5   Synthesis of Generalized Luneburg Lens

Luneburg proposed a model of the ideal optical instrument (in the framework of geometrical optics), afterwards called Luneburg lens. Later classical Luneburg lens was included in the family of the ideal optical instruments - generalized Luneburg lenses. Kepler and Morgan proposed alternative ways of solving the problem of synthesis of such lenses, and Kotlyar proved their equivalence.

Generalized Luneburg lens (having a spherical symmetry in three dimensions or circular symmetry in two-dimensional case) with a focal length f is described by the relations:

$$\rho\left(r\right) = r n\left(r, f\right), \quad n\left(r, f\right) = \exp\left\{\omega\left(r, f\right)\right\} \qquad (18)$$

where

$$\omega\left(r, f\right) = \frac{1}{\pi} \int_{\rho}^{1} \frac{\arcsin\left(x/f\right)}{\left(\rho^2 + x^2\right)^{1/2}} dx \qquad (19)$$

The parallel light beam (with a plane wave front) incident on the Luneburg lens is focused on the axis of the lens at a distance $F = Rf$, where R is the radius of the lens. If $f = 1$ the beam is focused at a point on the lens surface, and it is called classical.

The refractive index profile of classical Luneburg lens (see [24]) has the form

$$n\left(r, 1\right) = \sqrt{2 - \left(r/R\right)^2} \qquad (20)$$

In the case $f > 1$, the analytical solution $n\left(r, f\right)$ of relations (19) does not exist. It was calculated using different approximate methods in [22,23]. Each of the found solutions has limitations, and only under them these solutions have acceptable precision. In [15,16] non-linear relations (19) are solved by the deformed polyhedron method of Nelder-Mead of the minimization of the residual, with "relative accuracy" for all values of the parameters. The value of the integral is calculated using the adaptive QUANC8 program based on the Newton-Cotes formula of the 8th order.

6 Thin Film Generalized Waveguide Luneburg Lens

In the work by Zernike [21] it was shown that a local increase in thickness of the waveguiding layer leads to a local deceleration of phase velocity of the propagating waveguide mode. This effect has led to the idea of manufacturing the waveguide (two-dimensional) Luneburg lenses instead of volume (three-dimensional) lenses. Manufacturing of the volume lens with a variable refractive index is difficult materials science problem. At the same time the formation of a complex profile of the thickness of a homogeneous waveguide layer is a completely solvable problem in the dielectric thin-film technology.

Suppose that the TE (or TM) mode propagates through the regular three-layer planar dielectric waveguide with refractive indices n_s, n_f, n_c with a thickness d/λ of the waveguide layer and the phase retardation coefficient β_{TE}^j (β_{TM}^j), and this mode is incident from the left on a local thickening of the waveguide layer. Generalized Luneburg lens (volume lens) with a focal length $f = F/R$ has a refractive index distribution $n(r, f), r \leq R$. We design thin-film generalized waveguide Luneburg lens in the form of additional cylindrically symmetric waveguide layer with a thickness profile along the radius $h(r)$, ensuring deformation of incident waveguide mode. That means focusing of the flat (on the regular region) wave front and of the family of rays locally orthogonal to it by means of the formed by this thickening distribution of the effective refractive index $n_{eff}(r, f) = n(r, f), r \leq R$ of propagating deformed waveguide mode. In this case, the distribution of the effective coefficient of phase retardation of the mode is equal $\beta_{TE}^j(r, f) = \beta_{TE}^j n_{eff}(r, f), r \leq R$ ($\beta_{TM}^j(r, f) = \beta_{TM}^j n_{eff}(r, f), r \leq R$). Additional waveguide layer may be formed from the material of the main waveguide layer, as well as from a different material with a refractive index n_l.

7 Designing the Thickness Profile of the Thin-Film Generalized Waveguide Luneburg Lens on the Base of Dispersion Relations of the Cross-Section Method

We now pass on to the solution of the problem of design (mathematical synthesis) of the thickness profile of additional waveguide layer $h(r)$, which ensures "focusing" of the TE (TM) mode having "flat" (rectilinear in the plane yOz) wave front, incident on the thin-film generalized waveguide (TFGW) Luneburg lens. By focusing, we mean the formation "in plane" (on the line in the plane yOz), spaced at a distance F from the center of the lens with radius R (in the plane yOz), of diffraction pattern of the "infinitely thin comparison lens" with the angular aperture $\Omega = 2arctg\,(R/F\,)$.

Consider the cross-section of TFGW Luneburg lens along the diameter of the additional layer, so that when $z \in (-1,0)$ the radius changes $r \in (1,0)$, and further by changing $z \in (0,1)$ the radius changes $r \in (0,1)$. The guided mode of the regular three-layer waveguide with the thickness of waveguide layer equal to d/λ , has the phase retardation coefficient β^j_{TE} (β^j_{TM}). When the waveguide mode is incident on the lens, it is deformed by additional waveguide layer, which changes the thickness of the lens from zero at the edge of the lens to $h\,(0)$ at its center, and then again to zero on the opposite side of the lens. With this $\beta^j_{TE}\,(r,f) = \beta^j_{TE}n_{eff}\,(r,f)\,,r \leq R$ changes from $\beta^j_{TE}\,(d,0)$ to $\beta^j_{TE}\,(d,h\,(0)) = \beta^j_{TE}n_{eff}\,(0,f)$.

For each incident TE_j (TM_j) waveguide mode of the regular three-layer waveguide with the parameters n_s, n_f, n_c, d, we calculate the phase retardation coefficient β^j_{TE} (β^j_{TM}). For each value $r \in (0,R)$ we have the calculated value $n_{eff}\,(r,f) = n\,(r,f)\,,r \leq R$ with a given focal length. As a result we get $\beta^j_{TE}\,(r,f) = \beta^j_{TE}n_{eff}\,(r,f)\,,r \leq R$ $(\beta^j_{TM}\,(r,f) = \beta^j_{TM}n_{eff}\,(r,f)\,,r \leq R)$. The relations (15) and (17) are the theoretical basis (a mathematical model) of the design (synthesis) of the thickness profile $h\,(r)$ of additional waveguide layer with refractive index n_l.

Namely, using the Nelder-Mead method of constrained minimization of zero order [20] we minimize the functional at each subsequent point r_k

$$\det \mathbf{M}(\beta\,(d,h\,(r_k))\,,d,h\,(r_k))_{h(r_k)} \to \min \qquad (21)$$

with the penalty function $\Omega\,[h\,(r_k)] = |h\,(r_k) - h\,(r_{k-1})|^2$ with the initial vector $h\,(R) = 0$, and $\beta^j_{TE}\,(R,f) = \beta^j_{TE}$ $(\beta^j_{TM}\,(R,f) = \beta^j_{TM})$.

Naturally, for each TE_j (TM_j) mode of the regular three-layer waveguide as a result of minimization we obtain the particular thickness profile $h^{TE}_j\,(r)$ $(h^{TM}_j\,(r))$. In [22,23] the solution to the synthesis problem of TFGW Luneburg lens was proposed only for the TE_0 mode. Moreover, the method used in these studies has not allowed to generalize the results (with different focal lengths) to the other modes.

Our results coincide with the results in [22,23] within their accuracy, with our results outperforming the compared results in accuracy by eight orders of magnitude. Besides, the proposed method allows us to further improve the accuracy of calculations and to obtain double-precision results at any given in advance focal length ($f \to 1$), while the method of Southwell does not allow it. The thickness profiles $h^j_{TE}\,(r)$ and $h^j_{TM}\,(r)$ of additional waveguide layer for different waveguide modes are shown in Figs. 3 and 4.

8   Conclusion

The paper discusses the problems of numerical implementation of the cross-section method to calculate an evolution of the particular mode in a smooth

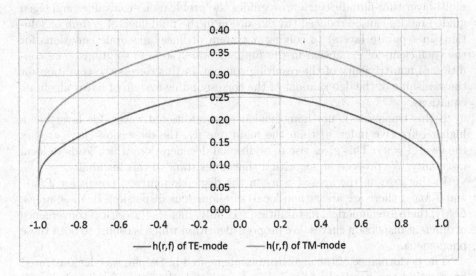

Fig. 3. Graphs of the thickness profiles $h_0^{TE}(r)$ and $h_0^{TM}(r)$.

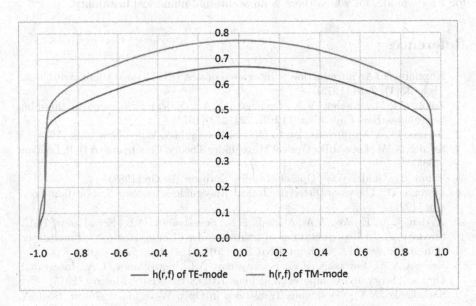

Fig. 4. Graphs of the thickness profiles $h_1^{TE}(r)$ and $h_1^{TM}(r)$.

transition from one planar regular open waveguide to another. In the class of multi-layer thin-film dielectric waveguides the problem on eigenvalues and eigen guided modes can be reduced (by expansion in the fundamental system of solutions in separate layers) to solving a system of linear algebraic equations for the coefficients of expansion in the fundamental system of solutions. The condition of non-triviality of the resulting solution in this case is the condition for the vanishing of the determinant of the corresponding system of linear algebraic equations.

If this transition is not homogeneous and additional waveguide layer has a higher refractive index n_l than the main one n_f, the dispersion curve crosses the level $\beta = n_f$. This gives rise to additional degeneration, which leads to local instability. We propose an algorithm that is resistant to this instability.

In this case, in the area $\beta > n_f$ in the dispersion curves, except for TE_0- and TM_0-, there are areas (non-local) of anomalous dispersion. In these areas, firstly, there are numerical instabilities, secondly, there is the critical convergence of different dispersion curves. We propose algorithm which is robust to both these phenomena.

The performance of the algorithm implemented in Delphi, was demonstrated by plotting the particular dispersion curves and plotting a family of dispersion curves, demonstrating a critical convergence. As an additional result, the thickness profiles of additional (irregular in thickness) waveguide layer, forming a thin-film generalized waveguide Luneburg lens were synthesized. This result generalizes results by Southwell [22,23], considering only the thickness profiles for TE_0- mode, for which there is no additional numerical instability.

References

1. Kogelnik, H.: An introduction to integrated optics. IEEE Trans. Microwave Theor. Tech. **23**(1), 2–16 (1975)
2. Zolotov, E.M., Kiselev, V.A., Sychugov, V.A.: Optical phenomena in thin-film waveguides. Sov. Phys. Usp. **112**(2), 231–273 (1974)
3. Adams, M.J.: An Introduction to Optical Waveguides. Wiley, New York (1981)
4. Snyder, A.W., Love, J.D.: Optical Waveguides Theory. Chapman and Hall, London (1983)
5. Tamir, T.: Guides-Wave Optoelectronics. Springer, Berlin (1988)
6. Marcuse, D.: Theory of Dielectric Optical Waveguides. Academic Publication, London (1991)
7. Ayrjan, E.A., Egorov, A.A., Michuk, E.N., Sevastyanov, A.L., Sevastianov, L.A., Stavtsev, A.V.: Representations of guided modes of integrated-optical multilayer thin-film waveguides. Preprint JINR E11-2011-31, Dubna, 52 p. (2011)
8. Egorov, A.A., Lovetskiy, K.P., Sevastyanov, A.L., Sevastianov, L.A.: Integrated Optics: Theory and Computer Modelling. RUDN University, Moscow (2015)
9. Shevchenko, V.V.: Continuous Transitions in Open Waveguides. Golem, Boulder (1971)
10. Sveshnikov, A.G.: The basis for a method of calculating irregular waveguides. Comput. Math. Math. Phys. **3**(1), 170–179 (1963)

11. Sveshnikov, A.G.: A substantiation of a method for computing the propagation of electromagnetic oscillations in irregular waveguides. Comput. Math. Math. Phys. **3**(2), 314–326 (1963)

12. Fedoryuk, M.V.: A justification of the method of transverse sections for an acoustic wave guide with nonhomogeneous content. Zh. Vychisl. Mat. Mat. Fiz. **13**(1), 127–135 (1973)

13. Ayryan, E.A., Egorov, A.A., Sevastyanov, L.A., Lovetskiy, K.P., Sevastyanov, A.L.: Mathematical modeling of irregular integrated optical waveguides. In: Adam, G., Buša, J., Hnatič, M. (eds.) MMCP 2011. LNCS, vol. 7125, pp. 136–147. Springer, Heidelberg (2012). doi:10.1007/978-3-642-28212-6_12

14. Sevastianov, L.A., Egorov, A.A., Sevastyanov, A.L.: Method of adiabatic modes in studying problems of smoothly irregular open waveguide structures. Phys. At. Nucl. **76**(2), 224–239 (2013)

15. Egorov, A.A., Sevast'yanov, L.A., Sevast'yanov, A.L.: Method of adiabatic modes in research of smoothly irregular integrated optical waveguides: zero approximation. Qant. Electron. **44**(2), 167–173 (2014)

16. Egorov, A.A., Sevastyanov, A.L., Ayryan, E.A., Sevastianov, L.A.: Stable computer modeling of thin-film generalized waveguide Luneburg lens. Matem. Mod. **26**(11), 37–44 (2014)

17. Gevorkyan, M.N., Kulyabov, D.S., Lovetskiy, K.P., Sevastyanov, A.L., Sevastyanov, L.A.: Waveguide modes of a planar optical waveguide. Math. Model. Geom. **3**(1), 43–63 (2015)

18. Shevchenko, V.V.: On the spectral decomposition in eigenfunctions and associated functions of a nonself-adjoint Sturm-Liouville problem on the whole axis. Diff. Eqn. **15**, 2004–2020 (1979). (in Russian)

19. Tyutyunnik, A.: Obtaining ODE system for coefficients functions in the cross-section method using computer algebra system. In: Book of Thes. V International Conference on Problems of Mathematical and Physics and Mathematical Modeling, pp. 105–107. MEPHI Publication, Moscow (2016)

20. Nelder, J.A., Mead, R.A.: Simplex method for function minimization. Comput. J. **7**(4), 308–313 (1965)

21. Zernike, F.: Luneburg lens for optical waveguide use. Opt. Comm. **12**(4), 379–381 (1974)

22. Southwell, W.H.: Inhomogeneous optical waveguide lens analysis. JOSA **67**(8), 1004–1009 (1977)

23. Southwell, W.H.: Index profiles for generalized Luneburg lenses and their use in planar optical waveguides. JOSA **67**(8), 1010–1014 (1977)

24. Born, M., Wolf, E.: Principles of Optics, 7th edn. Cambridge University Press, Cambridge (1999)

Damping Problem for Multidimensional Control System with Delays

A.S. Adkhamova$^{(\boxtimes)}$ and A.L. Skubachevskii

Department of Applied Mathematics, People's Friendship University of Russia,
Miklukho-Maklaya Street 6, Moscow 117198, Russia
ami_adhamova@mail.ru, skub@lector.ru

Abstract. We consider damping problem for control system with delay described by the system of differential-difference equations of neutral type, and establish the relationship of the variational problem for the nonlocal functionals and the corresponding boundary value problem for differential-difference equations. We prove the existence and uniqueness of generalized solution to the boundary value problem for this system of differential-difference equations.

Keywords: Control system with delay · Boundary value problem · Differential-difference equations · Generalized solution

1 Introduction

In recent years an interest to differential-difference equations is associated with applications to the theory of control systems with delay [1–7], the theory of multilayer plates and shells in the aircraft technology [8–10], to the theory of multidimensional diffusion processes [10–13], to plasma theory [10,14,15], to the theory of nonlinear laser systems with feedback loop [16–18] and others.

In [1], Krasovskii considered damping problem for control system with after-effect described by differential-difference equations of delay type. He has brought this problem to the boundary value problem for systems of differential-difference equations with deviating argument in lower order terms. Boundary value problems for differential-difference equations with shifts of argument in highest derivatives have been studied in [19,20]. In [10,21] the Krasovskii problem was generalized to the case when the equation describing the control system has neutral type. In [22], this problem was considered in multidimensional case with one delay.

We consider a damping problem for multidimensional control system with several delays. The paper consists of four sections. The second section contains the statement of problem and some auxiliary results. In the third section we establish the relationship between a variational problem for a nonlocal functional

---

A.S. Adkhamova and A.L. Skubachevskii—This research was carried out with the support of the State's Programme of the Russian Ministry for Education and Science (No. 1.1974.2014/K).

© Springer International Publishing AG 2016
V.M. Vishnevskiy et al. (Eds.): DCCN 2016, CCIS 678, pp. 612–623, 2016.
DOI: 10.1007/978-3-319-51917-3_52

and boundary value problem for a system of differential-difference equations. The fourth section deals with a solvability of the boundary value problem for the above system of differential-difference equations.

2 Statement of Problem and Auxiliary Results

We consider a linear control system with delay described by the system of differential equations with delays

$$\sum_{m=0}^{M} A_m y'(t - m\tau) + \sum_{m=0}^{M} B_m y(t - m\tau) = u(t), \quad 0 < t, \tag{1}$$

$$y(t) = \begin{pmatrix} y_1(t) \\ \vdots \\ y_n(t) \end{pmatrix}, u(t) = \begin{pmatrix} u_1(t) \\ \vdots \\ u_n(t) \end{pmatrix},$$

where A_m, B_m are $n \times n$ matrixes with constant elements, A_0 is nonsingular matrix, the delay $\tau > 0$ is constant, and $u(t)$ is a control vector-function.

A previous history of the system is defined by the initial condition

$$y(t) = \varphi(t), \ t \in [-M\tau, 0], \tag{2}$$

where $\varphi(t) = \begin{pmatrix} \varphi_1(t) \\ \vdots \\ \varphi_n(t) \end{pmatrix}$ is a given vector-function.

We shall study the problem of how to reduce the system (1)–(2) to equilibrium. Let us find a control vector-function $u(t), \ 0 < t < T$, such that

$$y(t) = 0, \ t \in [T - M\tau, T], \tag{3}$$

where $T \geq (M + 1)\tau$.

Vector-function $y(t)$ satisfying the conditions (1)–(3) is not unique. So we also assume that

$$\int_0^T |u(t)|^2 dt \to min,$$

where $|\cdot|$ is the Euclidean norm. We obtain the variational problem for functional

$$J(y) = \int_0^T \left| \sum_{m=0}^{M} A_m y'(t - m\tau) + \sum_{m=0}^{M} B y(t - m\tau) \right|^2 dt \to min \tag{4}$$

with boundary conditions (2)–(3).

We denote by $W_2^k(a,b)$ the space of continuous real-valued functions having a derivative of k-order from the space $L_2(a,b)$, with the scalar product

$$(v,w)_{W_2^k(a,b)} = \sum_{i=0}^{k} \int_a^b v^{(i)}(t)w^{(i)}(t)dt.$$

Let $\dot{W}_2^k(a,b) = \{w \in W_2^k(a,b) : w^{(i)}(a) = w^{(i)}(b) = 0, i = 0, ..., k-1\}$.

We introduce the real spaces of vector-functions

$$W_2^{k,n}(a,b) = \prod_{i=1}^{n} W_2^k(a,b),$$

$$\dot{W}_2^{k,n}(a,b) = \prod_{i=1}^{n} \dot{W}_2^k(a,b),$$

$$L_2^n(a,b) = \prod_{i=1}^{n} L_2(a,b),$$

where $L_2^n(a,b)$ is a Hilbert space with the scalar product

$$(v,w)_{L_2^n(a,b)} = \sum_{i=1}^{n}(v_i,w_i)_{L_2(a,b)} = \sum_{i=1}^{n} \int_a^b v_i(t)w_i(t)dt.$$

We consider the matrix operator $R : L_2^n(\mathbb{R}) \to L_2^n(\mathbb{R})$

$$R = \begin{pmatrix} R_{11} & \cdots & R_{1n} \\ \vdots & \ddots & \vdots \\ R_{n1} & \cdots & R_{nn} \end{pmatrix}.$$

Difference operators $R_{ik} : L_2(\mathbb{R}) \to L_2(\mathbb{R})$ are given by the formula

$$R_{ik}y = \sum_{j=-N}^{N} b_{ik}^j y(t - j\tau), \tag{5}$$

where b_{ik}^j are real numbers $(i, k = 1, ..., n)$.

Let $d = (N + \theta)\tau$ $(\theta \leq 1)$.

We introduce the operators

$$I_Q : L_2(0,d) \to L_2(\mathbb{R}), P_Q : L_2(\mathbb{R}) \to L_2(0,d), R_Q : L_2(0,d) \to L_2(0,d),$$

by the formulas

$$(I_Q v)(t) = v(t) \ (t \in (0,d)), v(t) = 0 \ (t \in (-\infty,0) \cup (d,+\infty)),$$
$$(P_Q v)(t) = v(t) \ (t \in (0,d)), \tag{6}$$
$$R_{ikQ} = P_Q R_{ik} I_Q : L_2(0,d) \to L_2(0,d).$$

We define the matrix operator $R_Q : L_2^n(0, d) \to L_2^n(0, d)$ as following:

$$R_Q = \begin{pmatrix} R_{11Q} & \cdots & R_{1nQ} \\ \vdots & \ddots & \vdots \\ R_{n1Q} & \cdots & R_{nnQ} \end{pmatrix}.$$

Below we shall formulate some properties of difference operators. The proofs follow from S Sect. 2 in [10].

Lemma 1. $I_Q^* = P_Q, P_Q^* = I_Q$, i.e. for any $u \in L_2(0, d)$, $v \in L_2(\mathbb{R})$ we have

$$(I_Q u, v)_{L_2(R)} = (u, P_Q v)_{L_2(0,d)}. \tag{7}$$

Lemma 2. The operators $R_{ik} : L_2(\mathbb{R}) \to L_2(\mathbb{R})$, $R_{ikQ} : L_2(0, d) \to L_2(0, d)$ are bounded, and

$$R_{ik}^* y(t) = \sum_{j=-N}^{N} b_{ik}^j y(t + j\tau); \ R_{ikQ}^* = P_Q R_{ik}^* I_Q. \tag{8}$$

Lemma 3. The matrix operators R, R_Q are bounded, and

$$R^* = \begin{pmatrix} R_{11}^* & \cdots & R_{n1}^* \\ \vdots & \ddots & \vdots \\ R_{1n}^* & \cdots & R_{nn}^* \end{pmatrix}^T, \ R_Q^* = \begin{pmatrix} R_{11Q}^* & \cdots & R_{n1Q}^* \\ \vdots & \ddots & \vdots \\ R_{1nQ}^* & \cdots & R_{nnQ}^* \end{pmatrix}^T.$$

Lemma 4. Matrix operator R (R_Q) is self-adjoint if and only if

$$R_{ik} = R_{ki}^* \ (R_{ikQ} = R_{kiQ}^*).$$

Definition 1. Bounded self-adjoint operator A in a Hilbert space H is called positive, if for all nonzero $y \in H$

$$(Ay, y)_H > 0.$$

Definition 2. Bounded self-adjoint operator A in a Hilbert space H is called positive definite if there is a constant $c > 0$ such that for all $y \in H$

$$(Ay, y)_H \geq c(y, y)_H.$$

Lemma 5. Let R be positive operator. Then R_Q is also positive operator.

To study the properties of the operator R_Q we introduce some additional notation. If $0 < \theta < 1$, we denote $Q_{1s} = ((s-1)\tau, (s-1+\theta)\tau)$, $s = 1, ..., N+1$, and $Q_{2s} = ((s-1+\theta)\tau, s\tau)$, $s = 1, ..., N$. If $\theta = 1$, we denote $Q_{1s} = ((s-1)\tau, s\tau)$, $s = 1, ..., N + 1$. Thus, there are two classes of disjoint intervals, if $0 < \theta < 1$, and there is only one class of intervals, if $\theta = 1$. Every two intervals of the same class can be obtained one from another by a shift $j\tau$.

Let $P_\alpha : L_2^n(0,d) \to L_2^n(\bigcup_s Q_{\alpha s})$ be the operator of orthogonal projection onto $L_2^n(\bigcup_s Q_{\alpha s})$, where $L_2^n(\bigcup_s Q_{\alpha s}) = \{y \in L_2^n(0,d) : y(t) = 0, t \in (0,d) \setminus \bigcup_s Q_{\alpha s}\}$; if $\theta < 1$, then $\alpha = 1,2$; if $\theta = 1$, then $\alpha = 1$ and P_α is the identity operator.

Lemma 6. $L_2^n(\bigcup_s Q_{\alpha s})$ is an invariant subspace of the operator R_Q.

We introduce an isomorphism

$$U_\alpha : L_2(\bigcup_s Q_{\alpha s}) \to L_2^K(Q_{\alpha 1})$$

by the formula

$$(U_\alpha u)_k(t) = u(t + (k-1)\tau), \quad t \in Q_{\alpha 1},$$

$k = 1, ..., K; K = N + 1$, if $\alpha = 1; K = N$, if $\alpha = 2$.

We also introduce an isomorphism of the Hilbert spaces

$$\tilde{U}_\alpha : L_2^n(\bigcup_s Q_{\alpha s}) \to L_2^{nM}(Q_{\alpha 1})$$

by the formula

$$(\tilde{U}_\alpha y)(t) = ((U_\alpha y_1)^T, ..., (U_\alpha y_n)^T)^T(t),$$

where $y(t) = \begin{pmatrix} y_1(t) \\ \vdots \\ y_n(t) \end{pmatrix} \in L_2^n(0,d)$, $(U_\alpha y_j)(t) = ((U_\alpha y_j)_1(t), ..., (U_\alpha y_j)_M(t))^T$.

For every $\alpha = 1,2$ we consider a block matrix $R_\alpha = \{R_{ik\alpha}\}_{i,k=1}^n$. Here R_{ik1} is the matrix of order $(N+1) \times (N+1)$ with the elements $r_{lp} = b_{ik}^{p-l}$, and R_{ik2} is the matrix of order $N \times N$, obtained from R_{ik1} by deleting the last column and the last row.

Lemma 7. The operator

$$R_{Q\alpha} = \tilde{U}_\alpha R_Q \tilde{U}_\alpha^{-1} : L_2^{nM}(Q_{\alpha 1}) \to L_2^{nM}(Q_{\alpha 1})$$

is the operator of multiplication by the matrix R_α.

From Lemma 7 we obtain the following statement:

Lemma 8.

$$\sigma(R_Q) = \begin{cases} \sigma(R_1) \cup \sigma(R_2), & \text{if } \theta < 1; \\ \sigma(R_1), & \text{if } \theta = 1, \end{cases}$$

where $\sigma(\cdot)$ denotes the spectrum of the operator.

From Lemma 8 we derive the following result:

Lemma 9. If operator R_Q is positive, then operator R_Q is positive definite.

3   Relation Between Variational Problem and Boundary Value Problem

We show that the variational problem (4), (2), (3) is equivalent to the boundary value problem for system of differential-difference equations of second order.

Let $y \in W_2^{1,n}(-M\tau, T)$ be a solution of the variational problem (2)–(4), where $\varphi \in W_2^{1,n}(-M\tau, 0)$. We denote

$$\widetilde{W} = \{v \in W_2^{1,n}(-M\tau, T) : v(t) = 0, t \in [-M\tau, 0] \cup [T - M\tau, T]\}.$$

Let $v(t) \in \widetilde{W}$ be an arbitrary fixed function. Then a function $y + sv \in W_2^{1,n}(-M\tau, T)$ and satisfies the boundary conditions (2), (3) for every $s \in \mathbb{R}$.

Denote $J(y + sv) = F(s)$. Since $J(y + sv) \geq J(y)$ $(s \in \mathbb{R})$, we have

$$\left. \frac{dF}{ds} \right|_{s=0} = 0.$$

It follows that

$$B(y,v) = \int_0^T \left(\sum_{l=0}^M A_l y'(t - l\tau) + \sum_{l=0}^M B_l y(t - l\tau) \right) \\ \times \left(\sum_{m=0}^M A_m v'(t - m\tau) + \sum_{m=0}^M B_m v(t - m\tau) \right) dt = 0. \tag{9}$$

Let $d = (N + \theta)\tau = T - M\tau$, i.e. $T = (N + M + \theta)\tau$.

In the terms containing $v(t - m\tau)$ or $v'(t - m\tau)$, we change the variable $\xi = t - m\tau$ and return to the old variable $t = \xi$. Then, since $v(t) = 0, t \in [-M\tau, 0] \cup [T - M\tau, T]$, and integrating by the parts, we have

$$B(y,v) = \int_0^T \sum_{l,m=0}^M \left[y'^T(t - (l - m)\tau) A_l^T A_m v'(t) \\ + y'^T(t - (l - m)\tau) A_l^T B_m v(t) - y'^T(t - (l - m)\tau) B_l^T A_m v(t) \\ + y^T(t - (l - m)\tau) B_l^T B_m v(t) \right] dt = 0. \tag{10}$$

From (10) we obtain

$$(\sum_{l,m=0}^M y'^T(t - (l - m)\tau) A_l^T A_m)^T \in W_2^{1,n}(0, T - N\tau). \tag{11}$$

Integrating by parts, we obtain

$$-(\sum_{l,m=0}^M A_m^T A_l y'(t - (l - m)\tau))' + \sum_{l,m=0}^M [B_m^T A_l y'(t - (l - m)\tau) \\ -A_m^T B_l y'(t - (l - m)\tau) + B_m^T B_l y(t - (l - m)\tau)] = 0 \ (t \in (0, T - N\tau). \tag{12}$$

Definition 3. A function $y \in W_2^{1,n}(-M\tau, T)$ is called a generalized solution of problem (12), (2), (3), if condition (11) holds, and the function $y(t)$ satisfies system of Eq. (12) and the boundary conditions (2), (3).

Clearly, a function $y \in W_2^{1,n}(-M\tau, T)$ is a generalized solution of problem (12), (2), (3) if and only if it satisfies integral identity (10) for all $v \in \mathring{W}_2^{1,n}(0, T - M\tau)$ and the boundary conditions (2), (3).

We have proved, that if a function $y \in W_2^{1,n}(-M\tau, T)$ gives a minimum to variational problem (4) with boundary conditions (2), (3), then y is a generalized solution of the boundary value problem (12), (2), (3).

Let $y \in W_2^{1,n}(-M\tau, T)$ be a generalized solution of the boundary value problem (12), (2), (3). Then we obtain for all $v \in \widetilde{W}$

$$J(y + v) = J(y) + J(v) + 2B(y, v),$$

where $J(v)$ is non-negative quadratic functional. Since y is a generalized solution of problem (12), (2), (3), then $B(y, v) = 0$. Therefore,

$$J(y + v) \geq J(y)$$

for all $v \in \widetilde{W}$. Hence we have proved the following statement.

Theorem 1. A function $y \in W_2^{1,n}(-M\tau, T)$ gives a minimum to functional (4) with boundary conditions (2), (3) if and only if it is a generalized solution of the boundary problem (12), (2), (3).

4   Solvability of Boundary Value Problem

In order to prove the existence and uniqueness of generalized solution of boundary value problem (12), (2), (3), we obtain some auxiliary results.

We denote

$$J_1(w) = \int_0^T \left| \sum_{l=0}^M A_l w'(t - l\tau) \right|^2 dt,$$

where $w \in \widetilde{W}$.

Lemma 10. There exists a constant $c_1 > 0$ such that, for all $w \in \widetilde{W}$

$$J_1(w) \geq c_1 \|w\|^2_{W_2^{1,n}(0, T - M\tau)}. \tag{13}$$

Proof. 1. First we prove that for all $0 \neq y \in L_2(\mathbb{R})$

$$J_0(y) = \int_{-\infty}^{\infty} \left| \sum_{m=0}^M A_m y(t - m\tau) \right|^2 dt > 0. \tag{14}$$

Using the Fourier transform, from the Plancherel theorem we obtain

$$J_0(y) = \int_{-\infty}^{\infty} \left| \sum_{m=0}^{M} (A_m e^{-im\xi}) \hat{y}(\xi) \right|^2 d\xi,$$

where

$$\hat{y}(\xi) = \frac{1}{\sqrt{2\pi}} \int_{-\infty}^{\infty} e^{-it\xi} y(t) dt$$

is the Fourier transform of the function $y(t)$. Since

$$\Phi(\xi) := det(\sum_{m=0}^{M} A_m e^{-im\xi})$$

is an analytic function and $\hat{y}(\xi) \neq 0$ almost everywhere on $\mathbb{R}$, then

$$J_0(y) = 0$$

if and only if

$$det(\sum_{m=0}^{M} A_m e^{-im\xi}) = 0 \tag{15}$$

for all $\xi \in \mathbb{R}$.

On the other hand, $det A_0 \neq 0$. Therefore the polynomial $det(\sum_{m=0}^{M} A_m \lambda^m) = 0$ has at most nM roots $\lambda_1, ..., \lambda_{nM}$. Hence Eq. (15) can have only countable number of real roots $\xi_1, \xi_2, ...$ Thus $J_0(y) > 0$.

2. Now we can prove inequality (13).

Let $R_Q = P'_Q R I'_Q$, where $R : L_2^n(\mathbb{R}) \rightarrow L_2^n(\mathbb{R}), I'_Q : L_2^n(0, T - M\tau) \rightarrow L_2^n(\mathbb{R}), P'_Q : L_2^n(\mathbb{R}) \rightarrow L_2^n(0, T - M\tau)$ are bounded operators given by

$$Ry(t) = \sum_{l,m=0}^{M} A_m^T A_l y(t - (l - m)\tau),$$

I' is the operator of extension by zero outside of $(0, T - M\tau)$, P'_Q is the restriction operator of functions onto $(0, T - M\tau)$. Then similarly to (9) from (14) we derive

$$J_1(w) = J_0(w') = (R_Q w', w')_{L_2^n(0, T-M\tau)} > 0. \tag{16}$$

for all $0 \neq w \in \widetilde{W}$. Here we assume that $w(t) = 0$ for $t \in \mathbb{R}\backslash(0, T - M\tau)$. By virtue of (16) the self-adjoint operator $R_Q : L_2^n(0, T - M\tau) \rightarrow L_2^n(0, T - M\tau)$ is positive. From Lemmas 5 and 9 it follows that it is positive definite. Finally, using theorem on the equivalent norms, we obtain

$$J_1(w) \geq c_0 ||w'||^2_{L_2^n(0,T-M\tau)} \geq c_1 ||w||_{W_2^{1,n}(0,T-M\tau)},$$

where $c_0, c_1 > 0$ do not depend on w. $\square$

Lemma 11. There exist constants $c_2, c_3 > 0$ such that, for all $w \in \widetilde{W}$

$$c_2 \|w\|_{W_2^{1,n}(0,T-M\tau)} \leq J(w) \leq c_3 \|w\|^2_{W_2^{1,n}(0,T-M\tau)}. \tag{17}$$

Proof. First we prove the left side of (17).

Assume to the contrary that inequality (17) does not hold for any $c_2 > 0$. Then, for any $K = 1, 2, \ldots$ there exist a function $w_K \in \widetilde{W}$ such that

$$J(w_K) \leq \frac{1}{K} \|w_K\|^2_{W_2^{1,n}(0,T-M\tau)}.$$

Without loss the generality, we assume that

$$\|w_K\|^2_{W_2^{1,n}(0,T-M\tau)} = 1. \tag{18}$$

In opposite case we consider $w_K / \|w_K\|_{W_2^{1,n}(0,T-M\tau)}$. Then

$$J(w_K) \leq \frac{1}{K}. \tag{19}$$

On the other hand, from the inequality $(\alpha + \beta)^2 \geq \alpha^2/2 - \beta^2 (\alpha, \beta \in \mathbb{R})$ and Lemma 2, for every $v \in \widetilde{W}$, we obtain

$$J(v) \geq k_1 \|v\|^2_{W_2^{1,n}(0,T-M\tau)} - k_2 \|v\|^2_{L_2^n(0,T-M\tau)}. \tag{20}$$

By virtue of the compactness of the imbedding operator from W into $L_2^n(-M\tau, T)$, the unit ball $\widetilde{W}$ is a compact set in $L_2^n(-M\tau, T)$. It means that there exists a subsequence w_{K_l}, which converges to w_0 in the space $L_2^n(-\tau, T)$, i.e.

$$\|w_{K_l} - w_{K_m}\|_{L_2^n(0,T-M\tau)} \to 0, \ l, m \to \infty.$$

Thus, from (19), (20) it follows that

$$k_1 \|w_{K_l} - w_{K_m}\|^2_{W_2^{1,n}(0,T-M\tau)} \leq k_2 \|w_{K_l} - w_{K_m}\|^2_{L_2(0,T-M\tau)}$$
$$+ J(w_{K_l} - w_{K_m}) \leq k_2 \|w_{K_l} - w_{K_m}\|^2_{L_2(0,T-M\tau)}$$
$$+ 2/K_l + 2/K_m \to 0, k, m \to \infty.$$

Hence, $w_{K_l} \to w_0$ in the space $\widetilde{W}$.

Passing to the limit in (18), we obtain

$$\|w_0\|_{W_2^{1,n}(0,T-M\tau)} = 1.$$

Convergence to the limit in (19) gives

$$J(w_0) = \int_0^T \left| \sum_{m=0}^M A_m w_0'(t - m\tau) + \sum_{m=0}^M \dot{B}_m w_0(t - m\tau) \right|^2 dt = 0, \tag{21}$$

i.e.

$$\sum_{m=0}^{M} A_m w_0'(t - m\tau) + \sum_{m=0}^{M} B_m w_0(t - m\tau) = 0 \qquad (22)$$

for almost all $t \in (0, T)$.

Since $w_0 \in \widetilde{W}$, a function w_0 satisfies the initial condition

$$w_0(t) = 0 \ (t \in [-M\tau, 0]). \qquad (23)$$

Then, if $0 < t \le \tau$, Eq. (22) takes the form

$$A_0 w_0'(t) + B_0 w_0(t) = 0, \ (t \in (0, \tau)), \qquad (24)$$

and $w_0(0) = 0$. Hence,

$$w_0(t) = 0 \ (t \in [0, \tau]). \qquad (25)$$

Then Eq. (22) on the interval $t \in (\tau, 2\tau)$ takes form (24) and $w_0(\tau) = 0$. Hence $w_0(t) = 0$ ($t \in [\tau, 2\tau]$). For a finite number of steps we have $w_0(t) = 0$ ($t \in [0, T - M\tau]$). But this is impossible, since

$$\|w_0\|_{W_2^{1,n}(0, T - M\tau)} = 1.$$

It remains to prove the right part of (17).

From the Cauchy-Bunyakovskii inequality it follows that

$$|J(w)| = |B(w, w)| \le k_3 \|w'\|_{L_2^n(0, T-M\tau)}^2 + k_4 \|w\|_{L_2^n(0, T-M\tau)}^2 \le c_3 \|w\|_{W_2^{1,n}(0, T-M\tau)}^2.$$

$\square$

Theorem 2. For every $\varphi \in W_2^{1,n}(-M\tau, 0)$, there exists a unique generalized solution $y \in W_2^{1,n}(-M\tau, T)$ of the boundary value problem (12), (2), (3), and

$$\|y\|_{W_2^{1,n}(-M\tau, T)} \le c \|\varphi\|_{W_2^{1,n}(-M\tau, 0)}, \qquad (26)$$

where $c > 0$ does not depend on φ.

Proof. We denote

$$\Phi(t) = \begin{cases} \varphi(t), & \text{if} \ -M\tau \le t \le 0; \\ 0, & \text{if} \ T - M\tau \le t \le T; \\ \varphi(0) - \varphi(0)t/(T - M\tau), & \text{if} \ 0 < t < T - M\tau. \end{cases}$$

It is clear that $\Phi \in W_2^{1,n}(-M\tau, T)$ and

$$\|\Phi\|_{W_2^{1,n}(-M\tau, T)} \le k_1 \|\varphi\|_{W_2^{1,n}(-M\tau, 0)}. \qquad (27)$$

Let $x = y - \Phi$, then $x \in \widetilde{W}$. Integral identity (11) takes the form

$$B(\Phi, v) + B(x, v) = 0, \ v \in \widetilde{W}. \qquad (28)$$

By Lemma 11 in the space $\dot{W}_2^{1,n}(0, T - M\tau)$ we can introduce an equivalent scalar product by the formula

$$(x, v)'_{\dot{W}_2^{1,n}(0,T-M\tau)} = B(x, v). \tag{29}$$

Therefore, the Eq. (28) can be rewritten as

$$B(\Phi, v) + (x, v)'_{\dot{W}_2^{1,n}(0,T-M\tau)} = 0. \tag{30}$$

For a fixed $\Phi \in W_2^{1,n}(-M\tau, T)$ functional $B(\Phi, v)$ is linear in $\widetilde{W}, v \in \widetilde{W}$. By the Cauchy-Bunyakovskii inequality and inequalities (27), (17) we have

$$\begin{aligned}|B(\Phi, v)| &\leq k_2\|\Phi\|_{W_2^{1,n}(-M\tau,T)}\|v\|_{W_2^{1,n}(0,T-M\tau)} \\ &\leq k_3\|\varphi\|_{W_2^{1,n}(-M\tau,0)}\|v\|_{W_2^{1,n}(0,T-M\tau)} \\ &\leq k_4\|\varphi\|_{W_2^{1,n}(-M\tau,0)}\|v\|'_{W_2^{1,n}(0,T-M\tau)}.\end{aligned} \tag{31}$$

Thus the functional $B(\Phi, v)$ is bounded on $\widetilde{W}$. By virtue of (31), the norm of the functional $B(\Phi, v)$ in $\dot{W}_2^{1,n}(0, T - M\tau)$ does not exceed $k_4\|\varphi\|_{W_2^{1,n}(-M\tau,0)}$.

According to the Riesz theorem, there exists a function $F \in \widetilde{W}$, such that

$$B(\Phi, v) = (F, v)'_{\dot{W}_2^{1,n}(0,T-M\tau)}$$

and

$$\|F\|'_{\dot{W}_2^{1,n}(0,T-M\tau)} \leq k_4\|\varphi\|_{W_2^{1,n}(-M\tau,0)}.$$

This function is unique. Thus, identity (30) can be rewritten as

$$(x, v)'_{\dot{W}_2^{1,n}(0,T-M\tau)} + (F, v)'_{\dot{W}_2^{1,n}(0,T-M\tau)} = 0.$$

Consequently, the problem (12), (2), (3) has a unique generalized solution $y = \Phi - F$ and inequality (26) holds. This proves the theorem. $\qquad\square$

References

1. Krasovskii, N.N.: Control Theory of Motion. Nauka, Moscow (1968)
2. Halanay, A.: Optimal controls for systems with time lag. SIAM J. Control **6**, 213–234 (1968)
3. Gabasov, R., Kirillova, F.M.: Qualitative Theory of Optimal Processes. Nauka, Moscow (1971)
4. Weiss, L.: On the controllability of delay-differential systems. SIAM J. Control **5**, 575–587 (1967)
5. Kent, G.A.: A maximum principle for optimal control problems with neutral functional differential systems. Bull. Am. Math. Soc. **77**, 565–570 (1971)
6. Banks, H.T., Kent, G.A.: Control of functional differential equations of retarded and neutral type to target sets in function space. SIAM J. Control **10**, 567–593 (1972)

7. Kharatishvili, G.L., Tadumadze, T.A.: Nonlinear optimal control systems with variable delays. Mat. Sb. **107**(149), 613–628 (1978). English transl. in Mathem. USSR Sb. 35 (1979)
8. Onanov, G.G., Skubachevskii, A.L.: Differential equations with displaced arguments in stationary problems in the mechanics of a deformed body. Prikladnaya Mekh. **15**, 39–47 (1979). English transl. in Soviet Applied Mech. 15
9. Onanov, G.G., Tsvetkov, E.L.: On the minimum of the energy with respect to the functions with deviating argument in a stationary problem of elasticity theory. Russ. J. Math. Phys. **3**(4), 491–500 (1996)
10. Skubachevskii, A.L.: Elliptic functional differential equations and applications. Birkhauser, Basel, Boston, Berlin (1997)
11. Feller, W.: Diffusion processes in one dimension. Trans. Am. Math. Soc. **77**, 1–30 (1954)
12. Ventsel', A.D.: On boundary conditions for multidimensional diffusion processes. Theoriya Veroyatn. i ee Primen. **4**, 172–185 (1959). English transl. in Theory Prob. and its Appl. 4
13. Skubachevskii, A.L.: On some problems for multidimensional diffusion processes. Dokl. Akad. Nauk SSSR **307**, 287–292 (1989)
14. Bitsadze, A.V., Samarskii, A.A.: On some simple generalizations of linear elliptic boundary value problems. Dokl. Akad. Nauk SSSR **185**, 739–740 (1969). English transl. in Soviet Math. Dokl. 10
15. Samarskii, A.A.: Some problems of the theory of differential equations. Differentsial'nye Uravneniya **16**, 1925–1935 (1980). English transl. in Differential Equations 16
16. Vorontsov, M.A., Ricklin, J.C., Carhart, G.W.: Optical simulation of phase-distorted imaging systems: nonlinear and adaptive optics approach. Opt. Eng. **34**, 3229–3238 (1995)
17. Razgulin, A.V.: Rotational multi-petal waves in optical system with 2D feedback, In: Roy, R. (ed.) Proceedings SPIE 2039, Chaos in Optics, pp. 342–352 (1993)
18. Skubachevskii, A.L.: Bifurcation of periodic solutions for nonlinear parabolic functional differential equations arising in optoėlectronics. Nonlinear Anal. **32**(2), 267–278 (1998)
19. Kamenskii, G.A., Myshkis, A.D.: Formulation of boundary-value problems for differential equations with deviating arguments containing highest-order terms. Differentsial'nye Uravneniya **10**, 409–418 (1974). English transl. in Differential Equations 10 (1975)
20. Kamenskii, A.G.: Boundary value problems for equations with formally symmetric differential-difference operators. Differentsial'nye Uravneniya **12**, 815–824 (1976). English transl. in Differential Equations 12 (1977)
21. Skubachevskii, A.L.: On the damping problem for control system with delay. Dokl. Akad. Nauk Ross. **335**, 157–160 (1994)
22. Leonov, D.D.: On damping problem for control system with delay. Contemp. Math. Fundam. Dir. **37**, 28–37 (2010)

Nonclassical Hamilton's Actions and the Numerical Performance of Variational Methods for Some Dissipative Problems

Vladimir Savchin[✉] and Svetlana Budochkina

Peoples' Friendship University of Russia,
Miklukho-Maklaya str. 6, 117198 Moscow, Russia
{vsavchin,sbudotchkina}@yandex.ru
http://www.rudn.ru

Abstract. The use of variational methods for the construction of sufficiently accurate approximate solutions of a given system requires the existence of the corresponding variational principle - a solution of the inverse problems of the calculus of variations. In the frame of the Euler's functionals there may not exist variational principles. But if we extend the class of functionals then it could allow to get the variational formulations of the given problems. There naturally arises the problem of the constructive determination of the corresponding functionals - nonclassical Hamilton's actions - and their application for the search of approximate solutions of the given boundary value problems. The main goal of the paper is to present a scheme for the construction of indirect variational formulations for given evolutionary problems and to demonstrate the effective use of the nonclassical Hamilton's action for the construction of approximate solutions with the high accuracy for the given dissipative problem.

Keywords: Nonpotential operators · Non-Eulerian functionals · Approximate solutions · Dissipative problems · Variational methods

1 Introduction

An important problem in applications of variational methods is a representation of the given equations in the form of the Euler-Lagrange equations. It means the construction of the functional F_N such that its extremals are solutions of the given equations. This is known as the classical inverse problem of the calculus of variations.

In spite of the remarkable number of papers on the subject different approaches for constructing of integral variational principles for equations with nonpotential operators should be developed. They will allow to obtain so-called indirect variational formulations of given problems.

First let us introduce the following concepts of bilinear forms, Gâteaux derivatives and B_u-potential operators.

© Springer International Publishing AG 2016
V.M. Vishnevskiy et al. (Eds.): DCCN 2016, CCIS 678, pp. 624–634, 2016.
DOI: 10.1007/978-3-319-51917-3_53

Let N be an operator such that its domain of definition $D(N) \subseteq U$ and the range of values $R(N) \subseteq V$, where U and V are real linear normed spaces, i.e.

$$N(u) = v, \qquad u \in U, \qquad v \in V.$$

If there exists a limit

$$\delta N(u,h) = \lim_{\varepsilon \to 0} \frac{1}{\varepsilon}\{N(u+\varepsilon h) - N(u)\}, \qquad u \in D(N), \quad (u+\varepsilon h) \in D(N), \quad (1)$$

then it is called the Gâteaux variation of the operator N at the point u or the first variation of the operator N at the point u.

$\delta N(u,h)$ is homogeneous relative to h : $\delta N(u, \lambda h) = \lambda \delta N(u,h)$, but the operator $\delta N(u, \cdot) : U \to V$ is not always additive relative to h.

If $\delta N(u,h)$ is a linear operator relative to h, when u is a fixed element of $D(N)$, then we say that the operator N is Gâteaux differentiable at the point u. The expression $\delta N(u,h)$ is called the Gâteaux differential and denoted by $DN(u,h)$. In this case we shall also write $DN(u,h) = N'_u h$ and say that N'_u is the Gâteaux derivative of operator N at the point u.

If N is a linear operator then $N'_u h = Nh$, i.e. the Gâteaux derivative of the linear operator coincides with it.

Further assume that for any given operator $N : D(N) \subset U \to V$ there exists its Gâteaux derivative at any point $u \in D(N)$. The domain of definition $D(N'_u)$ consisits of elements $h \in U$ such that $(u+\varepsilon h) \in D(N)$ for all ε sufficiently small. In this case $h \in D(N'_u)$ is called an admissible element.

Note that for any linear operator $\tilde{N}_u$ which may depend on u in a nonlinear way, the Gâteaux derivative is defined by

$$\tilde{N}'_u(g;h) = \lim_{\varepsilon \to 0} \frac{\tilde{N}_{u+\varepsilon h} g - \tilde{N}_u g}{\varepsilon}. \qquad (2)$$

The second Gâteaux derivative N''_u of the operator N is given by

$$N''_u(h_1, h_2) = \frac{\partial^2}{\partial \varepsilon^1 \partial \varepsilon^2} N(u + \varepsilon^1 h_1 + \varepsilon^2 h_2)\Big|_{\varepsilon^1 = \varepsilon^2 = 0}. \qquad (3)$$

In the most general applications N''_u satisfies the symmetry condition

$$N''_u(h_1, h_2) = N''_u(h_2, h_1).$$

Definition 1. A mapping $\Phi : V \times U \to \mathbb{R}$ is said to be a bilinear form if it is linear relative to every argument.

Definition 2. A bilinear form $\Phi : V \times V \to \mathbb{R}$ is called symmetric if

$$\Phi(v, g) = \Phi(g, v) \quad \forall g, v \in V.$$

Consider a bilinear form

$$\Phi(\cdot,\cdot) \equiv \int_{t_0}^{t_1} \langle\cdot,\cdot\rangle \, dt : V \times U \to \mathbb{R} \qquad (4)$$

such that the bilinear mapping $\Phi_1(\cdot,\cdot) \equiv \langle\cdot,\cdot\rangle$ satisfies the following conditions:

$$\langle v_1(t), v_2(t)\rangle = \langle v_2(t), v_1(t)\rangle \qquad \forall v_1(t), v_2(t) \in V_1, \qquad (5)$$

$$D_t \langle v(t), g(t)\rangle = \langle D_t v(t), g(t)\rangle + \langle v(t), D_t g(t)\rangle \qquad \forall v, g \in C^1([t_0, t_1]; U_1). \quad (6)$$

If $v = v(x,t), x \in \Omega \subset \mathbb{R}^n, t \in (t_0, t_1), U_1 = V_1 = C(\overline{\Omega})$, then we can take for example

$$\langle v, g\rangle = \int_{\Omega} v(x,t)g(x,t) \, dx. \qquad (7)$$

Definition 3. The operator $N : D(N) \subset U \to V$ is said to be B_u-potential on the set $D(N)$ relative to the bilinear form $\Phi : V \times U \to \mathbb{R}$, if there exist a functional $F_N : D(F_N) = D(N) \to \mathbb{R}$ and a linear operator $B_u : D(B_u) \subset V \to V$ such that

$$\delta F_N[u, h] = \Phi(N(u), B_u h) \qquad \forall u \in D(N), \quad \forall h \in D(N_u', B_u),$$

where $D(N_u', B_u) = D(N_u') \cap D(B_u)$.

If $B_u \equiv I$ is the identical operator then the operator N is called potential on $D(N)$ relative to bilinear form Φ.

The following theorem is needed for the sequel.

Theorem 1 [1]. Consider the operator $N : D(N) \subset U \to V$ and the bilinear form $\Phi : V \times V \to \mathbb{R}$ such that for any fixed elements $u \in D(N), g, h \in D(N_u', B_u)$ the function $\psi(\varepsilon) = \Phi(N(u + \varepsilon h), B_{u+\varepsilon h}g)$ belongs to class $C^1[0, 1]$. For N to be B_u-potential on the convex set $D(N)$ relative to Φ it is necessary and sufficient to have

$$\Phi(N_u' h, B_u g) + \Phi(N(u), B_u'(g; h)) = \Phi(N_u' g, B_u h) + \Phi(N(u), B_u'(h; g)) \quad (8)$$
$$\forall u \in D(N), \quad \forall g, h \in D(N_u', B_u).$$

Under this condition the functional F_N is given by

$$F_N[u] = \int_0^1 \Phi\left(N(\tilde{u}(\lambda)), B_{\tilde{u}(\lambda)}\frac{\partial \tilde{u}(\lambda)}{\partial \lambda}\right) d\lambda + F_N[u_0], \qquad (9)$$

where $\tilde{u}(\lambda) = u_0 + \lambda(u - u_0), u_0$ is a fixed element of $D(N)$.

In the paper we shall use notations and notions of [1–5].

2   An Operator Equation with the Second Time Derivative and Variational Principles

Consider the following operator equation

$$N(u) \equiv P_{2u,t}u_{tt} + P_{1u,t}u_t + P_{3u,t}u_t^2 + Q(t,u) = 0, \qquad (10)$$

$$u \in D(N) \subseteq U \subseteq V, \quad t \in [t_0, t_1] \subset \mathbb{R},$$

$$u_t \equiv D_t u \equiv \frac{d}{dt}u, \qquad u_{tt} \equiv \frac{d^2}{dt^2}u.$$

Here $\forall t \in [t_0, t_1]$, $\forall u \in U_1$ $P_{iu,t} : U_1 \to V_1$ $(i = \overline{1,3})$ are linear operators; $Q : [t_0, t_1] \times U_1 \to V_1$ is an arbitrary operator; $D(N)$ is the domain of definition of the operator N,

$$D(N) = \{u \in U : u(t) \in W \, \forall t \in [t_0, t_1], \, u|_{t=t_0} = \varphi_1, \, u|_{t=t_1} = \varphi_2,$$
$$u_t|_{t=t_0} = \varphi_3, \, u_t|_{t=t_1} = \varphi_4, \, \varphi_i \in U_1 \, (i = \overline{1,4})\}; \quad (11)$$

$U = C^2([t_0, t_1]; U_1)$, $V = C([t_0, t_1]; V_1)$, U_1, V_1 are real linear normed spaces, $U_1 \subseteq V_1$. The set $W \subseteq U_1$ is defined by the external constraints imposed on the system.

Assume that for every $t \in [t_0, t_1]$ and $g(t), u(t) \in U_1$ the functions $P_{1u,t}g(t)$, $P_{3u,t}g(t)$ are continuously differentiable and $P_{2u,t}g(t)$ is twice continuously differentiable on (t_0, t_1).

Any function $u \in D(N)$ is called a solution of problem (10) if it satisfies Eq. (10).

For notational simplicity, Eq. (10) hereafter written as

$$N(u) \equiv P_{2u}u_{tt} + P_{1u}u_t + P_{3u}u_t^2 + Q(u) = 0,$$

assuming that the operators P_{iu} $(i = \overline{1,3})$ and Q additionally depend on t.

Theorem 2. Let $D_t^* = -D_t$ on $D(N_u', B_u)$. Operator N (10) is B_u-potential on the set $D(N)$ (11) relative to bilinear form (4) $\iff$ $\forall u \in D(N)$, $\forall t \in [t_0, t_1]$, $\forall h \in D(N_u', B_u)$ the following conditions are fulfilled on $D(N_u', B_u)$:

$$B_u^* P_{2u} - P_{2u}^* B_u = 0, \qquad (12)$$

$$u_t P_{3u}^* B_u - P_{2u}^{*\prime}(B_u(\cdot); u_t) - P_{2u}^* B_u'(\cdot; u_t) + B_u^* P_{3u}(u_t(\cdot)) = 0, \qquad (13)$$

$$-2\frac{\partial}{\partial t}(P_{2u}^* B_u) + P_{1u}^* B_u + B_u^* P_{1u} = 0, \qquad (14)$$

$$-\frac{\partial^2}{\partial t^2}(P_{2u}^* B_u)h + [B_u'(\cdot; h)]^* Q(u) - [B_u'(h; \cdot)]^* Q(u) + \frac{\partial}{\partial t}(P_{1u}^* B_u)h$$
$$+ B_u^* Q_u' h - Q_u'^* B_u h = 0, \qquad (15)$$

$$P_{1u}^{*'}(B_u h; u_t) + B_u^* P_{1u}'(u_t; h) - [P_{1u}'(u_t; \cdot)]^* B_u h + 2u_t \frac{\partial}{\partial t}(P_{3u}^* B_u)h$$

$$+P_{1u}^* B_u'(h; u_t) - 2\frac{\partial}{\partial t}P_{2u}^{*'}(B_u h; u_t) + [B_u'(\cdot; h)]^* P_{1u} u_t$$

$$-2\frac{\partial}{\partial t}(P_{2u}^* B_u'(h; u_t)) - [B_u'(h; \cdot)]^* P_{1u} u_t = 0, \tag{16}$$

$$B_u^* P_{2u}'(u_{tt}; h) - P_{2u}^{*'}(B_u h; u_{tt}) - [P_{2u}'(u_{tt}; \cdot)]^* B_u h + 2u_{tt} P_{3u}^* B_u h$$

$$+[B_u'(\cdot; h)]^* P_{2u} u_{tt} - P_{2u}^* B_u'(h; u_{tt}) - [B_u'(h; \cdot)]^* P_{2u} u_{tt} = 0, \tag{17}$$

$$-P_{2u}^{*''}(B_u h; u_t; u_t) + B_u^* P_{3u}'(u_t^2; h) - [P_{3u}'(u_t^2; \cdot)]^* B_u h$$

$$+2u_t P_{3u}^{*'}(B_u h; u_t) + [B_u'(\cdot; h)]^* P_{3u} u_t^2 - 2P_{2u}^{*'}(B_u'(h; u_t); u_t)$$

$$-P_{2u}^* B_u''(h; u_t; u_t) + 2u_t P_{3u}^* B_u'(h; u_t) - [B_u'(h; \cdot)]^* P_{3u} u_t^2 = 0. \tag{18}$$

Proof. By using Eq. (10), we get

$$N_u' h = 2P_{3u}(u_t h_t) + P_{3u}'(u_t^2; h) + P_{2u} h_{tt} + P_{2u}'(u_{tt}; h) + P_{1u} h_t + P_{1u}'(u_t; h) + Q_u' h.$$

In this case, criterion (8) acquires the form

$$\int_{t_0}^{t_1} \left(\langle 2P_{3u}(u_t h_t) + P_{3u}'(u_t^2; h) + P_{2u} h_{tt} + P_{2u}'(u_{tt}; h) + P_{1u} h_t \right.$$

$$+P_{1u}'(u_t; h) + Q_u' h, B_u g \rangle + \langle P_{2u} u_{tt} + P_{1u} u_t + P_{3u} u_t^2$$

$$+Q(u), B_u'(g; h) \rangle \big) dt = \int_{t_0}^{t_1} \left(\langle 2P_{3u}(u_t g_t) + P_{3u}'(u_t^2; g) + P_{2u} g_{tt} \right.$$

$$+P_{2u}'(u_{tt}; g) + P_{1u} g_t + Q_u' g + P_{1u}'(u_t; g), B_u h \rangle$$

$$+ \langle P_{2u} u_{tt} + P_{1u} u_t + P_{3u} u_t^2 + Q(u), B_u'(h; g) \rangle \big) dt,$$

or

$$\int_{t_0}^{t_1} \left\{ \langle 2B_u^* P_{3u}(u_t h_t) + B_u^* P_{3u}'(u_t^2; h) + B_u^* P_{2u} h_{tt} + B_u^* P_{2u}'(u_{tt}; h) \right.$$

$$+B_u^* P_{1u} h_t + B_u^* P_{1u}'(u_t; h) + B_u^* Q_u' h, g \rangle$$

$$+ \langle [B_u'(\cdot; h)]^*(P_{2u} u_{tt} + P_{1u} u_t + P_{3u} u_t^2 + Q(u)), g \rangle$$

$$- \langle -2D_t(u_t P_{3u}^* B_u h) + [P_{3u}'(u_t^2; \cdot)]^* B_u h + D_t^2(P_{2u}^* B_u h)$$

$$+[P_{2u}'(u_{tt}; \cdot)]^* B_u h - D_t(P_{1u}^* B_u h) + [P_{1u}'(u_t; \cdot)]^* B_u h + Q_u'^* B_u h, g \rangle$$

$$- \langle [B_u'(h; \cdot)]^*(P_{2u} u_{tt} + P_{1u} u_t + P_{3u} u_t^2 + Q(u)), g \rangle \right\} dt = 0$$

$$\forall u \in D(N), \quad \forall g, h \in D(N_u', B_u). \tag{19}$$

Taking into account the second Gâteaux derivative we obtain

$$D_t^2(P_{2u}^* B_u h) = D_t[D_t(P_{2u}^* B_u h)]$$

$$= D_t\left[P_{2u}^* B_u h_t + \frac{\partial}{\partial t}(P_{2u}^* B_u)h + P_{2u}^{*\prime}(B_u h; u_t) + P_{2u}^* B_u'(h; u_t)\right]$$

$$= \frac{\partial^2}{\partial t^2}(P_{2u}^* B_u)h + 2\frac{\partial}{\partial t}P_{2u}^{*\prime}(B_u h; u_t) + 2\frac{\partial}{\partial t}(P_{2u}^* B_u'(h; u_t))$$

$$+2\frac{\partial}{\partial t}(P_{2u}^* B_u)h_t + P_{2u}^{*\prime\prime}(B_u h; u_t; u_t) + 2P_{2u}^{*\prime}(B_u'(h; u_t); u_t)$$

$$+2P_{2u}^{*\prime}(B_u h_t; u_t) + P_{2u}^{*\prime}(B_u h; u_{tt}) + P_{2u}^* B_u''(h; u_t; u_t)$$

$$+2P_{2u}^* B_u'(h_t; u_t) + P_{2u}^* B_u'(h; u_{tt}) + P_{2u}^* B_u h_{tt}. \tag{20}$$

Next, we have

$$D_t[P_{1u}^* B_u h] = \frac{\partial}{\partial t}(P_{1u}^* B_u)h + P_{1u}^{*\prime}(B_u h; u_t) + P_{1u}^* B_u'(h; u_t) + P_{1u}^* B_u h_t, \tag{21}$$

$$D_t[u_t P_{3u}^* B_u h] = u_{tt} P_{3u}^* B_u h + u_t \frac{\partial}{\partial t}(P_{3u}^* B_u)h + u_t P_{3u}^{*\prime}(B_u h; u_t)$$

$$+u_t P_{3u}^* B_u'(h; u_t) + u_t P_{3u}^* B_u h_t. \tag{22}$$

From $(19) - (22)$ it follows that

$$\int_{t_0}^{t_1} \langle 2B_u^* P_{3u}(u_t h_t) + B_u^* P_{3u}'(u_t^2; h) + B_u^* P_{2u} h_{tt} + B_u^* P_{2u}'(u_{tt}; h)$$

$$+B_u^* P_{1u} h_t + B_u^* P_{1u}'(u_t; h) + B_u^* Q_u' h + [B_u'(\cdot; h)]^*(P_{2u} u_{tt} + P_{1u} u_t + P_{3u} u_t^2$$

$$+Q(u)) - \frac{\partial^2}{\partial t^2}(P_{2u}^* B_u)h - 2\frac{\partial}{\partial t}P_{2u}^{*\prime}(B_u h; u_t) - 2\frac{\partial}{\partial t}(P_{2u}^* B_u'(h; u_t))$$

$$-2\frac{\partial}{\partial t}(P_{2u}^* B_u)h_t - P_{2u}^{*\prime\prime}(B_u h; u_t; u_t) - 2P_{2u}^{*\prime}(B_u'(h; u_t); u_t)$$

$$-2P_{2u}^{*\prime}(B_u h_t; u_t) - P_{2u}^{*\prime}(B_u h; u_{tt}) - P_{2u}^* B_u''(h; u_t; u_t) - 2P_{2u}^* B_u'(h_t; u_t)$$

$$-P_{2u}^* B_u'(h; u_{tt}) - P_{2u}^* B_u h_{tt} - [P_{2u}'(u_{tt}; \cdot)]^* B_u h + 2u_{tt} P_{3u}^* B_u h$$

$$+2u_t \frac{\partial}{\partial t}(P_{3u}^* B_u)h + 2u_t P_{3u}^{*\prime}(B_u h; u_t) + 2u_t P_{3u}^* B_u'(h; u_t) + 2u_t P_{3u}^* B_u h_t$$

$$-[P_{3u}'(u_t^2; \cdot)]^* B_u h + P_{1u}^* B_u h_t + \frac{\partial}{\partial t}(P_{1u}^* B_u)h + P_{1u}^{*\prime}(B_u h; u_t)$$

$$+P_{1u}^* B_u'(h; u_t) - [P_{1u}'(u_t; \cdot)]^* B_u h - Q_u'^* B_u h$$

$$-[B_u'(h; \cdot)]^*(P_{2u} u_{tt} + P_{1u} u_t + P_{3u} u_t^2 + Q(u)), g\rangle\, dt = 0.$$

Therefore, condition (19) can be reduced to the form

$$
\int_{t_0}^{t_1} \langle (B_u^* P_{2u} - P_{2u}^* B_u)\, h_{tt} + (2B_u^* P_{3u}(u_t(\cdot)) + B_u^* P_{1u} + 2u_t P_{3u}^* B_u
$$

$$
-2\frac{\partial}{\partial t}(P_{2u}^* B_u) - 2P_{2u}^{*\prime}(B_u(\cdot); u_t) - 2P_{2u}^* B_u'(\cdot; u_t) + P_{1u}^* B_u)\, h_t
$$

$$
+ B_u^* P_{3u}'(u_t^2; h) + B_u^* P_{2u}'(u_{tt}; h) + B_u^* P_{1u}'(u_t; h) + B_u^* Q_u' h
$$

$$
+ [B_u'(\cdot; h)]^* P_{2u} u_{tt} + [B_u'(\cdot; h)]^* P_{1u} u_t + [B_u'(\cdot; h)]^* P_{3u} u_t^2
$$

$$
+ [B_u'(\cdot; h)]^* Q(u) + 2u_{tt} P_{3u}^* B_u h + 2u_t \frac{\partial}{\partial t}(P_{3u}^* B_u) h + 2u_t P_{3u}^{*\prime}(B_u h; u_t)
$$

$$
+ 2u_t P_{3u}^* B_u'(h; u_t) - [P_{3u}'(u_t^2; \cdot)]^* B_u h - \frac{\partial^2}{\partial t^2}(P_{2u}^* B_u) h
$$

$$
-2\frac{\partial}{\partial t} P_{2u}^{*\prime}(B_u h; u_t) - 2\frac{\partial}{\partial t}(P_{2u}^* B_u'(h; u_t)) - P_{2u}^{*\prime\prime}(B_u h; u_t; u_t)
$$

$$
-2P_{2u}^{*\prime}(B_u'(h; u_t); u_t) - P_{2u}^{*\prime}(B_u h; u_{tt}) - P_{2u}^* B_u''(h; u_t; u_t)
$$

$$
- P_{2u}^* B_u'(h; u_{tt}) - [P_{2u}'(u_{tt}; \cdot)]^* B_u h + \frac{\partial}{\partial t}(P_{1u}^* B_u) h + P_{1u}^{*\prime}(B_u h; u_t)
$$

$$
+ P_{1u}^* B_u'(h; u_t) - [P_{1u}'(u_t; \cdot)]^* B_u h - Q_u'^* B_u h - [B_u'(h; \cdot)]^* P_{2u} u_{tt}
$$

$$
- [B_u'(h; \cdot)]^* P_{1u} u_t - [B_u'(h; \cdot)]^* P_{3u} u_t^2 - [B_u'(h; \cdot)]^* Q(u), g \rangle \, dt = 0
$$

$$
\forall u \in D(N), \quad \forall g, h \in D(N_u', B_u).
$$

This condition is satisfied identically if and only if

$$
(B_u^* P_{2u} - P_{2u}^* B_u)\, h_{tt} + (2B_u^* P_{3u}(u_t(\cdot)) + B_u^* P_{1u} + 2u_t P_{3u}^* B_u
$$

$$
-2\frac{\partial}{\partial t}(P_{2u}^* B_u) - 2P_{2u}^{*\prime}(B_u(\cdot); u_t) - 2P_{2u}^* B_u'(\cdot; u_t) + P_{1u}^* B_u)\, h_t
$$

$$
+ B_u^* P_{3u}'(u_t^2; h) + B_u^* P_{2u}'(u_{tt}; h) + B_u^* P_{1u}'(u_t; h) + B_u^* Q_u' h
$$

$$
+ [B_u'(\cdot; h)]^* P_{2u} u_{tt} + [B_u'(\cdot; h)]^* P_{1u} u_t + [B_u'(\cdot; h)]^* P_{3u} u_t^2
$$

$$
+ [B_u'(\cdot; h)]^* Q(u) + 2u_{tt} P_{3u}^* B_u h + 2u_t \frac{\partial}{\partial t}(P_{3u}^* B_u) h + 2u_t P_{3u}^{*\prime}(B_u h; u_t)
$$

$$
+ 2u_t P_{3u}^* B_u'(h; u_t) - [P_{3u}'(u_t^2; \cdot)]^* B_u h - \frac{\partial^2}{\partial t^2}(P_{2u}^* B_u) h
$$

$$
-2\frac{\partial}{\partial t} P_{2u}^{*\prime}(B_u h; u_t) - 2\frac{\partial}{\partial t}(P_{2u}^* B_u'(h; u_t)) - P_{2u}^{*\prime\prime}(B_u h; u_t; u_t)
$$

$$
-2P_{2u}^{*\prime}(B_u'(h; u_t); u_t) - P_{2u}^{*\prime}(B_u h; u_{tt}) - P_{2u}^* B_u''(h; u_t; u_t)
$$

$$
- P_{2u}^* B_u'(h; u_{tt}) - [P_{2u}'(u_{tt}; \cdot)]^* B_u h + \frac{\partial}{\partial t}(P_{1u}^* B_u) h + P_{1u}^{*\prime}(B_u h; u_t)
$$

$$
+ P_{1u}^* B_u'(h; u_t) - [P_{1u}'(u_t; \cdot)]^* B_u h - Q_u'^* B_u h - [B_u'(h; \cdot)]^* P_{2u} u_{tt}
$$

$$
- [B_u'(h; \cdot)]^* P_{1u} u_t - [B_u'(h; \cdot)]^* P_{3u} u_t^2 - [B_u'(h; \cdot)]^* Q(u) = 0
$$

$$
\forall u \in D(N), \quad \forall h \in D(N_u', B_u),
$$

and for the last relation to hold, in turn, it is necessary and sufficient that conditions (12)–(18) be satisfied. The proof of the theorem is complete.

Note that, the corresponding functional F_N is represented in the form

$$F_N[u] = \int\limits_{t_0}^{t_1} \{\langle R_{3u}(u_t u), B_u u_t \rangle + \langle R_{2u} u_t, B_u u_t \rangle$$

$$+ \langle R_1(u), B_u u_t \rangle - \left\langle \frac{\partial}{\partial t}(B_u^* R_{2u})u, u_t \right\rangle + \tilde{B}[u] \} \, dt + F_N[u_0],$$

where

$$\Phi(R_1(u), B_u u_t) = \int\limits_{t_0}^{t_1} \int\limits_0^1 \left\langle -P_{1\tilde{u}(\lambda)}(u - u_0), B_{\tilde{u}(\lambda)} \frac{\partial \tilde{u}(\lambda)}{\partial t} \right\rangle \, d\lambda dt,$$

$$\Phi(R_{2u} u_t, B_u u_t) = \int\limits_{t_0}^{t_1} \int\limits_0^1 \left\langle -P_{2\tilde{u}(\lambda)}(u_t - u_{0_t}), B_{\tilde{u}(\lambda)} \frac{\partial \tilde{u}(\lambda)}{\partial t} \right\rangle \, d\lambda dt,$$

$$\Phi(R_{3u}(u_t u), B_u u_t) = \int\limits_{t_0}^{t_1} \int\limits_0^1 \left\langle -P_{3\tilde{u}(\lambda)} \left(\frac{\partial \tilde{u}(\lambda)}{\partial t}(u - u_0) \right), B_{\tilde{u}(\lambda)} \frac{\partial \tilde{u}(\lambda)}{\partial t} \right\rangle \, d\lambda dt,$$

$$\tilde{B}[u] = \int\limits_0^1 [\langle Q(\tilde{u}(\lambda)), B_{\tilde{u}(\lambda)}(u - u_0) \rangle$$

$$+ \lambda \left\langle \frac{\partial}{\partial t}(B_{\tilde{u}(\lambda)}^* P_{1\tilde{u}(\lambda)})(u - u_0), u - u_0 \right\rangle$$

$$- \lambda \left\langle \frac{\partial^2}{\partial t^2}(B_{\tilde{u}(\lambda)}^* P_{2\tilde{u}(\lambda)})(u - u_0), u - u_0 \right\rangle] \, d\lambda,$$

$\tilde{u}(\lambda) = u_0 + \lambda(u - u_0)$, u_0 is a fixed element of $D(N)$.

3   On Performance of a Variational Method for Some Dissipative Problems

Let us present the numerical performance of a variational method for a simple linear ordinary differential equation with nonpotential operator.

Let us consider the following problem

$$N(u) \equiv u'' - 5u' + x^4 = 0,$$
$$x \in [0,1], \quad u(0) = u(1) = 0. \tag{23}$$

By $D(N) = \{u \in C^2[0,1] : u(0) = u(1) = 0\}$ we denote the domain of definition of the operator N.

The exact solution of problem (23) has the form

$$u(x) := \frac{434}{3125\,(e^5 - 1)}\left(1 - e^{5x}\right) + \frac{125x^5 + 125x^4 + 100x^3 + 60x^2 + 24x}{3125}.$$

Operator N (23) is not potential relative to the classical bilinear form

$$\Phi(v, g) = \int_0^1 v(x)g(x)dx. \tag{24}$$

There exists the variational multiplyer $M(x) = e^{-5x}$ such that the operator $\tilde{N}(u) = M(x)N(u)$ is potential relative to bilinear form (24). It means that operator N (23) is B-potential on its domain of definition $D(N)$ relative to bilinear form (24), where $B = M(x)I$, I is the identical operator.

The corresponding functional of $\tilde{N}(u)$ has the form

$$F_{\tilde{N}}[u] = \int_0^1 e^{-5x}\left(-0.5\,(u')^2 + x^4 u\right) dx.$$

Applying the Ritz-process [6] we find out three approximate solutions of problem (23).

The first one is

$$u_0(x) := 7.491414512583465465 \cdot 10^{-3} \cdot x \cdot (1 - x),$$

the second one -

$$u_1(x) := x \cdot (1 - x) \cdot \left(5.5511435176018702506 \cdot 10^{-2} \cdot x\right.$$
$$\left. - 4.0678902532769005056 \cdot 10^{-3}\right),$$

the third one -

$$u_2(x) := x \cdot (1 - x) \cdot (-4.336202537698754952 \cdot 10^{-2} \cdot x$$
$$+ 0.1610369015630196335 \cdot x^2 + 7.230056822340569444 \cdot 10^{-3}).$$

Let us choose the auxiliary operator $\overline{B}$ of the kind $\overline{B}u(x) = u(1 - x)$ and consider the following convolution bilinear form

$$\Phi_1(v, g) = \int_0^1 v(1 - x)g(x)dx.$$

Operator N (23) is potential relative to that bilinear form and the corresponding functional has the form

$$F_N[u] = \int_0^1 \left(0.5u'(x)u'(1 - x) - 2.5u'(x)u(1 - x) + x^4 u(1 - x)\right) dx.$$

Applying the Ritz-process [6] we find out three approximate solutions of problem (23).

The first approximate solution is

$$u_0(x) := \frac{1}{14} \cdot x \cdot (1 - x),$$

the second one -

$$u_1(x) := x \cdot (1 - x) \cdot \left(\frac{20}{119} \cdot x - \frac{13}{238} \right),$$

and the third one -

$$u_2(x) := x \cdot (1 - x) \cdot \left(\frac{1495}{4824} \cdot x^2 - \frac{835}{4824} \cdot x + \frac{130}{4221} \right).$$

Let us evaluate the deviations between approximate and the exact solutions in the norm of space L_2.

First let us consider the approximate solutions, found out with the use of the classical bilinear form. Their deviations with the exact solution $u(x)$ are equal correspondingly to

$$R_0 = 6.7713069428403088645 \cdot 10^{-3},$$

$$R_1 = 3.73153634159812502980 \cdot 10^{-3},$$

$$R_2 = 1.138326186371896467 \cdot 10^{-3}.$$

The deviations of approximate solutions, found out with the use of the convolution bilinear form, with the exact solution $u(x)$ are equal correspondingly to

$$r_0 = 8.701 \cdot 10^{-3},$$

$$r_1 = 2.526 \cdot 10^{-3},$$

$$r_2 = 7.145 \cdot 10^{-4}.$$

4 Conclusions

In the paper we obtained the necessary and sufficient conditions for the given operator equation with the second time derivative to admit in general an indirect variational formulation and constructed the corresponding functional - variational principle. The results are applied for finding of approximate solutions of some boundary value problems.

Acknowledgments. This paper was financially supported by the Ministry of Education and Science of the Russian Federation on the program to improve the competitiveness of Peoples' Friendship University among the world's leading research and education centers in the 2016-2020.

The work was also supported by RFBR grant No. 16-01-00450.

References

1. Savchin, V.M.: Mathematical methods in the mechanics of infinite-dimensional non-potential systems. Univ. Druzhby Narodov, Moscow (1991)
2. Savchin, V.M., Budochkina, S.A.: On the structure of a variational equation of evolution type with the second t-derivative. Differ. Equ. **39**(1), 127–134 (2003)
3. Savchin, V.M., Budochkina, S.A.: On the existence of a variational principle for an operator equation with the second derivative with respect to "time". Math. Notes **80**(1), 83–90 (2006)
4. Budotchkina, S.A., Savchin, V.M.: On indirect variational formulations for operator equations. J. Funct. Spaces Appl. **5**(3), 231–242 (2007)
5. Budochkina, S.A., Savchin, V.M.: On direct variational formulations for second order evolutionary equations. Eurasian Math. J. **3**(4), 23–34 (2012)
6. Mikhlin, S.G.: Numerical Performance of Variational Methods. Nauka, Moscow (1965)

Modeling of Spinning Sphere Motion in Shear Flow of Viscous Fluid

Yuri P. Rybakov[✉]

Department of Theoretical Physics and Mechanics, RUDN University,
6, Miklukho-Maklaya Street, Moscow 117198, Russia
soliton4@mail.ru
http://www.rudn.ru

Abstract. Modeling the motion of a small rigid spinning spherical particle in viscous Navier—Stokes fluid, we generalize the Rubinow—Keller and Maxey—Riley method of estimating the force and the torque acting on the particle to the case of shear flow and arbitrary Reynolds number. We represent the velocity of the flow near the particle as solid body part and small perturbation. As for the velocity far from the particle, it includes a steady external shear flow part and again small perturbation. We use the simplest quadratic polynomial approximation for the small velocity parts and insert it in matching condition at some intermediate spherical surface. It appears that the force parallel to the angular velocity of the particle proves to contain the oscillatory part, with the frequency being proportional to the gradient of the external steady velocity.

Keywords: Viscous fluid · Spinning particle · Shear flow

1 Introduction

The equations of motion for small spinning grains in viscous fluid flow were investigated in numerous papers [1–20]. The main approach was based on the small Reynolds number approximation. We do not use this supposition and suggest the polynomial development of the flow velocity near the particle. As a result the linearization procedure appears to be effective both in the nearest domain and far from the particle. General formulae will be given for the force $\mathbf{F}$ and the torque $\mathbf{T}$ acting on the particle. In the sequel we use the following notations:

a – radius of the sphere particle,

$\mathbf{\Omega}(t)$ – angular velocity of the particle,

$\xi(t)$ – the radius-vector of the particle center,

$\mathbf{V}(t) = \dot{\xi}(t)$ – the velocity of the particle center,

$\mathbf{r} = (x_1, x_2, x_3)$ – the coordinate radius-vector,

Y.P. Rybakov—The author expresses his gratitude to Dr. Pavel Vlasak for fruitful discussion of the paper.

V.M. Vishnevskiy et al. (Eds.): DCCN 2016, CCIS 678, pp. 635–645, 2016.
DOI: 10.1007/978-3-319-51917-3_54

x_i – the Cartesian coordinates,

$\mathbf{u(r)}$ – the unperturbed velocity of the steady fluid flow,

$\mathbf{v(r,}t)$ – the perturbed velocity of the fluid flow,

S – the surface of the particle (sphere),

$\mathbf{a = r} - \xi(t)$ – the relative radius-vector,

$\mathbf{n}$ – the unit normal vector on the surface S,

ρ – the density of the fluid,

μ – the dynamical viscosity coefficient of the fluid,

$\nu = \mu/\rho$ – the kinematic viscosity coefficient of the fluid,

σ_{ik} – the stress tensor of the fluid,

p – the pressure in the fluid.

2   Main Equations

The Navier—Stokes equations for the viscous fluid read:

$$\partial_t \mathbf{v} + (\mathbf{v}\nabla)\mathbf{v} = -\nabla p/\rho + \nu \triangle \mathbf{v}, \tag{1}$$

with the condition of incompressibility

$$\operatorname{div} \mathbf{v} = 0. \tag{2}$$

We divide the space out of the sphere particle into two domains:

$$I = \{a \leq |\mathbf{r} - \xi(t)| \leq r_0\},$$
$$II = \{r_0 \leq |\mathbf{r} - \xi(t)| \leq \infty\},$$

with some parameter $r_0 \sim a$. In the domain I one can put

$$\mathbf{v} = \mathbf{U} + \mathbf{w_1}, \tag{3}$$

where $\mathbf{U}$ stands for

$$\mathbf{U} = \mathbf{V} + \mathbf{\Omega} \times \mathbf{a} \tag{4}$$

and it is supposed that

$$|\mathbf{w_1}| \ll |\mathbf{U}|. \tag{5}$$

It can be easily seen that (4) corresponds to solid body motion of the fluid.

Inserting (3) into (1), in view of the restriction (5) one can linearize the Eq. (1) in the domain I:

$$\partial_t \mathbf{w_1} + (\mathbf{w_1}\nabla)\mathbf{U} + (\mathbf{U}\nabla)\mathbf{w_1} - \nu \triangle \mathbf{w_1} = -\nabla p_1/\rho - \partial_t \mathbf{U} - (\mathbf{U}\nabla)\mathbf{U}, \tag{6}$$

with the trivial boundary condition on the surface of the particle:

$$\mathbf{w_1}|_S = 0. \tag{7}$$

As for the domain II, one can put

$$\mathbf{v} = \mathbf{u} + \mathbf{w}_2, \tag{8}$$

with the natural restriction

$$|\mathbf{w}_2| \ll |\mathbf{u}|. \tag{9}$$

Inserting (8) into (1), one can linearize that equation due to (9), thus implying

$$\partial_t \mathbf{w}_2 + (\mathbf{w}_2 \nabla)\mathbf{u} + (\mathbf{u}\nabla)\mathbf{w}_2 - \nu \triangle \mathbf{w}_2 = -\nabla p_2/\rho - (\mathbf{u}\nabla)\mathbf{u}. \tag{10}$$

Taking into account that

$$\mathrm{div}\mathbf{U} = \mathrm{div}\mathbf{u} = 0,$$

one also derives from (2) the similar equations:

$$\mathrm{div}\mathbf{w}_1 = \mathrm{div}\mathbf{w}_2 = 0, \tag{11}$$

with the natural boundary and matching conditions for the velocity, pressure and their normal derivatives:

$$\mathbf{w}_2|_{r=\infty} = 0, \ (1 + \varkappa\partial_n)(\mathbf{w}_1 + \mathbf{U} - \mathbf{w}_2 - \mathbf{u})|_{r_0} = 0, \ (1 + \varkappa\partial_n)(p_1 - p_2)|_{r_0}, \tag{12}$$

where $\varkappa$ is an arbitrary parameter.

In particular, one can search for the solution to (6) in the form of the polynomial decomposition:

$$w_{1i} \approx A_i + B_{ik}a_k + C_{ijk}a_ja_k, \tag{13}$$

with coefficients vanishing on S and being some functions of time t and $|a|$, and use (13) to calculate the stress tensor

$$\sigma_{ij} = -p\,\delta_{ij} + \mu\left(\partial_i v_j + \partial_j v_i\right). \tag{14}$$

First of all we take into account that due to the formulae

$$U_i = \dot{\xi}_i + \epsilon_{ijk}\Omega_j a_k, \quad \partial_j a_k = \delta_{jk},$$

the rotational part of the velocity does not give any contribution to the stress tensor σ_{ij} since

$$\partial_j U_i = \epsilon_{ilj}\Omega_l = -\partial_i U_j.$$

Inserting (13) into (14), one gets in the first approximation with respect to a for the viscous part of the stress tensor the following expression:

$$\sigma_{ij}^{\mathrm{visc}} = \mu\,\mathrm{Sym}\left(A'_j n_i + B_{ji} + B'_{jl}a_l n_i + 2C_{jil}a_l\right),$$

that permits one to calculate the force $\mathbf{F}$ and the torque $\mathbf{T}$ acting on the particle. To this end, it is convenient to use the formula for spherical averaging of the product of $2k$ components of the unit vector $\mathbf{n}$:

$$\langle n_{i(1)}n_{i(2)}\cdots n_{i(2k)}\rangle = \frac{1}{(2k+1)!!}\mathrm{cycle}\left(\delta_{i(1)i(2)}\cdots\delta_{i(2k-1)i(2k)}\right).$$

In particular case $k = 1$ one has $\langle n_i n_k \rangle = \delta_{ik}/3$ and for $k = 2$:

$$\langle n_i n_k n_l n_s \rangle = \frac{1}{15} \left(\delta_{ik}\delta_{ls} + \delta_{il}\delta_{ks} + \delta_{is}\delta_{kl} \right).$$

For the viscous part of the force one finds

$$
\begin{aligned}
F_i^{\text{visc}} = \oint_S n_j \sigma_{ij}^{\text{visc}} dS &= 4\pi a^2 \langle n_j \sigma_{ij}^{\text{visc}} \rangle \\
&\approx 8\pi a^2 \mu \left(A'_j \langle n_i n_j \rangle + 2C_{jil} a \langle n_l n_j \rangle + C'_{jkl} a^2 \langle n_k n_l n_i n_j \rangle \right) \\
&= \frac{8\pi}{3} a^2 \mu \left(A'_i + a S_i + \frac{3}{5} a^2 S'_i \right),
\end{aligned}
$$

where the denotation is used $S_i = C_{ikk}$ for the trace of the totally symmetric tensor C_{ijk}. Finally, one can obtain the total force $\mathbf{F}$ acting on the particle:

$$F_i = \oint_S n_j \sigma_{ij} dS = \frac{4\pi}{3} a^3 \left(-\partial_i p + \mu \triangle v_i \right)_0 \approx \frac{4\pi}{3} a^3 \left(-\partial_i p + \frac{2\mu}{a} A'_i \right)_0, \quad (15)$$

where the prime denotes the radial derivative and the subscript "0" stands for the mean value of the corresponding function in the particle domain. The latter value, due to small size of the particle, can be obtained by extrapolation.

Similar calculations can be performed for the torque $\mathbf{T}$:

$$T_i = a\epsilon_{ijk} \oint_S n_j \sigma_{kl} n_l dS \approx 8\pi a^3 \mu \left(B_{kl} \langle n_j n_l \rangle + B'_{ls} a \langle n_s n_k n_j n_l \rangle \right).$$

Finally, one gets for the torque $\mathbf{T}$ the following expression:

$$T_i \approx \frac{8\pi}{3} a^3 \mu \epsilon_{ijk} \left(B_{kj} + \frac{2}{5} a B'_{kj} \right), \quad (16)$$

where it was taken into account that $B_{kk} = 0$, the latter property being proven later [cf. (26)].

Denoting the mass of the particle by m_p, one deduces from (15) and (16) the following equations of motion for the spherical grain:

$$m_p \frac{dV_i}{dt} \approx \frac{4\pi}{3} a^2 \left(-a\partial_i p + 2\mu A'_i \right)_0,$$

$$m_p \frac{d\Omega_i}{dt} \approx \frac{8\pi}{3} a^3 \mu \epsilon_{ijk} \left(B_{kj} + \frac{2}{5} a B'_{kj} \right)_0.$$

3   Structure of Solution in the Domain I

In the domain I we use the first three terms in the decomposition (13) and the analogous one for the pressure:

$$p_1 \approx P_0 + P_k a_k + Q_{kl} a_k a_l, \quad (17)$$

where the coefficients are functions of $|\mathbf{a}|$ and t with the evident restrictions:

$$C_{ikl} = C_{ilk}, \quad Q_{ik} = Q_{ki}, \quad Q_{ii} = 0, \tag{18}$$

and the Einstein rule of summation over the repeating indices is used.

Now we choose the following Cartesian components for the vectors $\boldsymbol{\Omega}$ and $\mathbf{U}$:

$$\Omega_i = \Omega(t)\delta_{i3}, \quad U_i = \dot{\xi}_i - \Omega\epsilon_{ik3}a_k,$$

where the dot denotes the time derivative, and insert the decompositions (13) and (17) into (6). Using the well-known formulae:

$$\partial_t|\mathbf{a}| = -n_k\dot{\xi}_k, \quad \partial_j|\mathbf{a}| = n_j, \quad \partial_j a_k = \delta_{jk}, \quad n_k \equiv a_k/|\mathbf{a}|,$$

one deduces from (6) the following equation of motion:

$$\dot{A}_i - \Omega\epsilon_{js3}a_s B_{ij} + \dot{B}_{ik}a_k - 2\Omega\epsilon_{js3}C_{ijl}a_s a_l$$
$$+ \dot{C}_{ikl}a_k a_l - \Omega\epsilon_{ij3}\left(A_j + B_{jk}a_k + C_{jkl}a_k a_l\right)$$
$$-\nu\left(A_i'' + \frac{2}{|\mathbf{a}|}A_i' + B_{ik}''a_k + 4B_{is}'\frac{a_s}{|\mathbf{a}|} + C_{ikl}''a_k a_l + 2S_i + 6C_{isl}'\frac{a_s a_l}{|\mathbf{a}|}\right)$$
$$= \dot{\Omega}\epsilon_{ik3}a_k + \Omega^2 a_k\delta_{ik}^{\perp} - \ddot{x}i_i$$
$$-\frac{1}{\rho}\left(P_k'\frac{a_i a_k}{|\mathbf{a}|} + P_i + Q_{kl}'\frac{a_i a_k a_l}{|\mathbf{a}|} + 2Q_{il}a_l + P_0'\frac{a_i}{|\mathbf{a}|}\right). \tag{19}$$

Multiplying this expression consequently by 1, a_j, $a_j a_k$ and averaging over the sphere $n_k n_k = 1$, one gets the following relations:

$$\dot{A}_i + \frac{1}{3}\dot{S}_j|\mathbf{a}|^2 - \Omega\epsilon_{ij3}\left(A_j + \frac{1}{3}S_i|\mathbf{a}|^2\right)$$
$$-\nu\left(A_i'' + \frac{2}{|\mathbf{a}|}A_i' + \frac{1}{3}|\mathbf{a}|^2 S_i'' + 2S_i + 2S_i'|\mathbf{a}|\right)$$
$$= -\ddot{\xi}_i - \frac{1}{\rho}\left(P_i + \frac{1}{3}|\mathbf{a}|P_i'\right); \tag{20}$$

$$\dot{B}_{ik} - \Omega\left(\epsilon_{jk3}B_{ij} + \epsilon_{ij3}B_{jk}\right) - \nu\left(B_{ik}'' + \frac{4}{|\mathbf{a}|}B_{ik}'\right)$$
$$= \dot{\Omega}\epsilon_{ik3} + \Omega^2\delta_{ik}^{\perp} - \frac{1}{\rho}\left(\frac{1}{|\mathbf{a}|}P_0'\delta_{ik} + 2Q_{ik} + \frac{2}{15}|\mathbf{a}|Q_{ik}'\right); \tag{21}$$

$$\dot{A}_i\delta_{jk} - 2\Omega\frac{|\mathbf{a}|^2}{5}\left(\epsilon_{mj3}C_{imk} + \epsilon_{mk3}C_{imj}\right) + \frac{|\mathbf{a}|^2}{5}\left(\dot{S}_i\delta_{jk} + 2\dot{C}_{ijk} - \Omega\epsilon_{im3}S_m\delta_{jk}\right)$$
$$-\frac{|\mathbf{a}|^2}{5}\left[2\epsilon_{im3}\Omega C_{mjk} + \nu\left(S_i''\delta_{jk} + 2C_{ijk}''\right)\right] - 6\nu\frac{|\mathbf{a}|}{5}\left(S_i'\delta_{jk} + 2C_{ijk}'\right)$$
$$= -\frac{|\mathbf{a}|}{5\rho}\left(P_i'\delta_{jk} + P_k'\delta_{ji} + P_j'\delta_{ik}\right) - \ddot{\xi}\delta_{jk} \tag{22}$$

and for $j \neq k$ in (22)

$$\dot{C}_{isl} - \Omega \left(\epsilon_{ij3}C_{jsl} + 2\epsilon_{j(s3}C_{ijl)} \right) - \nu \left(C''_{isl} + \frac{6}{|\mathbf{a}|}C'_{isl} \right) = -\frac{1}{\rho|\mathbf{a}|}P'_{(l}\delta_{is)}, \quad (23)$$

where the special notations were used for the symbol $\delta^\perp_{ik} = \epsilon_{ij3}\epsilon_{kj3}$ and for the symmetrization operation: $f_{(ik)} \equiv 1/2\left[f_{ik} + f_{ki}\right]$.

Executing similar procedures with the incompressibility Eq. (11), one easily finds the following two new relations:

$$B_{ii} + \frac{|\mathbf{a}|}{3}B'_{ii} = 0; \quad (24)$$

$$\frac{1}{|\mathbf{a}|}A'_i = -2S_i - \frac{1}{5}|\mathbf{a}|S'_i. \quad (25)$$

Taking into account the vanishing of the radial functions on the surface of the particle, one deduces from (24) that

$$B_{ii} = \frac{B(t)}{|\mathbf{a}|^3},$$

the latter formula implying

$$B_{ii} = 0. \quad (26)$$

Then the Eqs. (26) and (21) imply, due to (18), the following relation:

$$P'_0 = \frac{\rho}{3}|\mathbf{a}|\Omega \left(2\Omega + \epsilon_{jk3}B_{kj} \right). \quad (27)$$

Now we search for the solutions to the Eqs. (19)–(26) in the form of the radial polynomial decomposition:

$$Q_{ik} = \sum_{n=0}^\infty Q_{ik}^{(n)}x^n, \quad B_{ik} = \sum_{n=1}^\infty B_{ik}^{(n)}x^n, \quad C_{ijk} = \sum_{n=1}^\infty C_{ijk}^{(n)}x^n, \quad (28)$$

where $x \equiv |\mathbf{a}| - a$.

In the cubic approximation one gets

$$B_{ik}^{(2)} = -\frac{2}{a}B_{ik}^{(1)} - \frac{1}{2\nu}\left[-\frac{2}{\rho}\left(Q_{ik}^{(0)} + \frac{a}{5}Q_{ik}^{(1)}\right) + \dot{\Omega}\epsilon_{ik3} + \Omega^2\left(\delta^\perp_{ik} - \frac{2}{3}\delta_{ik}\right)\right]; \quad (29)$$

$$6\nu B_{ik}^{(3)} = \dot{B}_{ik}^{(1)} - \Omega \left(\epsilon_{jk3}B_{ij}^{(1)} + \epsilon_{ij3}B_{jk}^{(1)} - \frac{2}{3}\delta_{ik}\epsilon_{js3}B_{sj}^{(1)} \right) + \frac{20\nu}{a^2}B_{ik}^{(1)}$$
$$+ \frac{4}{a}\left[-\frac{1}{5\rho}\left(10Q_{ik}^{(0)} - aQ_{ik}^{(1)} - a^2Q_{ik}^{(2)}\right) + \dot{\Omega}\epsilon_{ik3} + \Omega^2\left(\delta^\perp_{ik} - \frac{2}{3}\delta_{ik}\right)\right]. \quad (30)$$

From (25) one derives

$$A'_i = -(x+a)\left(2S_i + \frac{x+a}{5}S'_i\right), \quad (31)$$

whence

$$A_i = -\int\limits_0^x dx(x+a)\left(2S_i + \frac{x+a}{5}S_i'\right). \tag{32}$$

Thus, one can exclude A_i from equations of motion.

4   Structure of Solution in the Domain II

In the domain II we consider the simplest steady shear flow of the form $u_1 = \Lambda x_2 + u_0$, where Λ stands for the constant velocity gradient and u_0 is the constant velocity. In this case the Eqs. (10) take the form:

$$\partial_t w_{2i} + (\Lambda x_2 + u_0)\partial_1 w_{2i} + \Lambda\delta_{i1}w_{22} - \nu\triangle w_{2i} = -\frac{1}{\rho}\partial_i p_2. \tag{33}$$

In much the same way as in the previous section, we insert into (33) the polynomial decompositions of the form:

$$w_{2i} \approx A_i^* + B_{ik}^* a_k + C_{ijk}^* a_j a_k; \quad p_2 \approx P_0^* + P_k^* a_k + Q_{kl}^* a_k a_l. \tag{34}$$

As a result we obtain the following equations for the radial functions:

$$\dot{A}_i^* - B_{ik}^* \dot{\xi}_k - \frac{|a|}{3}B_{ik}^{*'}\dot{\xi}_k + \frac{\Lambda}{3}\overline{\xi}_2|a|B_{i1}^{*'} + (\Lambda\xi_2 + u_0)B_{i1}^* + \frac{2\Lambda}{15}|a|^3 C_{i12}^{*'}$$
$$+\frac{2\Lambda}{3}|a|^2 C_{i12}^* + \Lambda\delta_{i1}A_2^* - \nu\left(A_i^{*''} + \frac{2}{|a|}A_i^{*'}\right) = -\frac{1}{\rho}\left(P_i^* + \frac{|a|}{3}P_i^{*'}\right); \tag{35}$$

$$\dot{B}_{ik}^* - \frac{1}{|a|}A_i^{*'} - 2\dot{\xi}_s\left(C_{iks}^* + \frac{1}{5}|a|C_{iks}^{*'}\right)$$
$$+\Lambda\left[\overline{\xi}_2\left(\frac{1}{|a|}A_i^{*'}\delta_{k1} + C_{ik1}^* + \frac{|a|}{5}C_{ik1}^{*'}\right) + B_{i1}^*\delta_{k2} + B_{2k}^*\delta_{i1} + \frac{|a|}{5}\left(B_{i1}^{*'}\delta_{k2} + B_{i2}^{*'}\delta_{k1}\right)\right]$$
$$-\nu\left(B_{ik}^{*''} + \frac{4}{|a|}B_{ik}^{*'}\right) = -\frac{1}{\rho}\left(\frac{1}{|a|}P_0^{*'}\delta_{ik} + 2Q_{ik}^* + \frac{2}{5}|a|Q_{ik}^{*'}\right); \tag{36}$$

$$\dot{C}_{ikl}^* - \frac{1}{|a|}B_{i(k}^{*'}\dot{\xi}_{l)} + \Lambda\left[C_{2kl}^*\delta_{i1} + 2C_{i(k1}^*\delta_{l)2} + \frac{1}{|a|}\left(A_i^{*'}\delta_{(k1}\delta_{l)2} + \overline{\xi}_2 B_{i(k}^{*'}\delta_{l)1}\right)\right.$$
$$\left.+\frac{2}{7}|a|\left(C_{i2(k}^{*'}\delta_{l)1} + C_{i1(l}^{*'}\delta_{k)2}\right)\right] - \nu\left(C_{ikl}^{*''} + \frac{6}{|a|}C_{ikl}^{*'}\right) = -\frac{1}{\rho|a|}P_{(k}^{*'}\delta_{il)}, \tag{37}$$

where $\overline{\xi}_2 = \xi_2 + u_o/\Lambda$. Finally, the incompressibility Eq. (11) gives the relations similar to (24) and (25):

$$B_{ii}^* + \frac{|a|}{3}B_{ii}^{*'} = 0; \tag{38}$$

$$\frac{1}{|a|}A_i^{*'} = -2C_{kki}^* - \frac{2}{5}|a|C_{kki}^{*'}. \tag{39}$$

From (38) and (26) one deduces, in view of the matching condition (12), that

$$B_{ii}^* = 0. \tag{40}$$

Taking into account (39) and (40), one obtains from (35), (36) and (37) the following relations containing the radial gradient of pressure:

$$\Lambda\left[B_{12}^* + B_{21}^* + \frac{|\mathbf{a}|}{5}\left(B_{12}^{*'} + B_{21}^{*'}\right)\right] = -\frac{3}{\rho|\mathbf{a}|}P_0^{*'}; \tag{41}$$

$$\Lambda\left[2C_{i12}^* + \frac{4|\mathbf{a}|}{7}C_{i12}^{*'} + \frac{\overline{\xi_2}}{|\mathbf{a}|}B_{i1}^{*'}\right] - \frac{1}{|\mathbf{a}|}B_{ik}^{*'}\dot{\xi}_k = -\frac{1}{\rho|\mathbf{a}|}P_i^{*'}; \tag{42}$$

$$\dot{A}_i^* - B_{ik}^*\dot{\xi}_k + \Lambda\left(A_2^*\delta_{i1} + \overline{\xi_2}B_{i1}^* - \frac{2}{35}|\mathbf{a}|^3C_{i12}^{*'}\right)$$
$$-\nu\left(A_i^{*''} + \frac{2}{|\mathbf{a}|}A_i^{*'}\right) = -\frac{1}{\rho}P_i^*. \tag{43}$$

Now we search for the solutions to the Eqs. (35)–(43) in the form of decompositions in decreasing degrees of $|\mathbf{a}|$:

$$P_0^* = \sum_{n=0}^{\infty} P_0^{*(n)}|\mathbf{a}|^{-n}; \quad P_i^* = \sum_{n=2}^{\infty} P_i^{*(n)}|\mathbf{a}|^{-n}; \quad Q_{ik}^* = \sum_{n=3}^{\infty} Q_{ik}^{*(n)}|\mathbf{a}|^{-n};$$

$$A_i^* = \sum_{n=1}^{\infty} A_i^{*(n)}|\mathbf{a}|^{-n}; \quad B_{ik}^* = \sum_{n=2}^{\infty} B_{ik}^{*(n)}|\mathbf{a}|^{-n}; \quad C_{ikl}^* = \sum_{n=3}^{\infty} C_{ikl}^{*(n)}|\mathbf{a}|^{-n},$$

with the coefficients depending on time t. Thus, the following relations arise:

$$A^{*(n)} = \frac{2}{5n}(3-n)C_{kki}^{*(n+2)}; \tag{44}$$

$$\frac{1}{\rho}P_0^{*(n)} = \frac{\Lambda}{15n}\left[-(2+n)B_{12}^{*(n+2)} + (8-n)B_{21}^{*(n+2)}\right]; \quad n \geq 1; \quad \cdot$$
$$B_{12}^{*(2)} = 4B_{21}^{*(2)}; \tag{45}$$

$$\frac{1}{\rho}P_i^{*(n)} = B_{ik}^{*(n)}\dot{\xi}_k - (\Lambda\xi_2 + u_0)B_{i1}^{*(n)} + \frac{2\Lambda}{7n}(3-2n)C_{i12}^{*(n+2)}; \quad n \geq 2;$$
$$C_{i12}^{*(3)} = 0; \tag{46}$$

$$\dot{A}_i^{*(n)} + \Lambda\delta_{i1}A_2^{*(n)} - \nu(n-2)(n-3)A_i^{*(n-2)} + \frac{2\Lambda}{35n}(n-3)(n-5)C_{i12}^{*(n+2)} = 0; \tag{47}$$

$$\dot{C}_{ikl}^{*(n+2)} + nB_{i(k}^{*(n)\cdot}\xi_{l)} - nB_{(ks}^{*(n)}\dot{\xi}_s\delta_{il)} + \Lambda\Big[n\overline{\xi}_2\Big(B_{(k1}^{*(n)}\delta_{il)} - B_{i(k}^{*(n)}\delta_{l)1}\Big)$$

$$-nA_i^{*(n)}\delta_{(k1}\delta_{l)2} + 2C_{i(k1}^{*(n+2)}\delta_{l)2} + C_{2kl}^{*(n+2)}\delta_{i1}$$

$$-\frac{2}{7}\Big[(3-2n)C_{(k12}^{*(n+2)}\delta_{il)} + (n+2)\Big(C_{i2(k}^{*(n+2)}\delta_{l)1} + C_{i1(l}^{*(n+2)}\delta_{k)2}\Big)\Big]\Big]$$

$$= \nu n(n-5)C_{ikl}^{*(n)}; \quad (48)$$

$$\frac{2}{\rho}(n-5)Q_{ik}^{*(n)} = 5\dot{B}_{ik}^{*(n)} - 5\nu(n-2)(n-5)B_{ik}^{*(n-2)}$$

$$+5A^{*(n-2)}\Big(\dot{\xi}_k - (\Lambda\xi_2 + u_0)(n-2)\delta_{k1}\Big) + 2(n-5)C_{iks}^{*(n)}\dot{\xi}_s$$

$$+\Lambda\Big[(5-n)B_{i1}^{*(n)}\delta_{k2} + 5B_{2k}^{*(n)}\delta_{i1} - nB_{i2}^{*(n)}\delta_{k1} - 2(n-5)\overline{\xi}_2 C_{ik1}^{*(n)}$$

$$-\frac{1}{3}\delta_{ik}\Big[(10-n)B_{21}^{*(n)} - nB_{12}^{*(n)}\Big]\Big], \quad (49)$$

with the natural conditions being imposed:

$$B_{ii}^{*(n)} = Q_{ii}^{*(n)} = 0. \quad (50)$$

In particular, putting $n = 1$, $n = 2$ in (48) one finds in the first approximation the following nontrivial tensor components:

$$C_{111}^{*(3)} = \alpha = \text{const}, \quad C_{311}^{*(3)} = \beta = \text{const}, \quad B_{13}^{*(2)} = \gamma = \text{const}; \quad (51)$$

$$C_{122}^{*(3)} = \frac{2}{5}\alpha, \quad C_{133}^{*(3)} = -\frac{7}{5}\alpha, \quad C_{322}^{*(3)} = 34\beta, \quad C_{333}^{*(3)} = -35\beta; \quad (52)$$

$$A_1^{*(1)} = \frac{4}{5}\alpha, \quad A_3^{*(1)} = -28\beta; \quad (53)$$

$$B_{31}^{*(2)} = \epsilon\cos(\omega t) + \sqrt{\frac{2}{3}}\delta\sin(\omega t); \quad B_{32}^{*(2)} = \delta\cos(\omega t) - \sqrt{\frac{3}{2}}\epsilon\sin(\omega t), \quad (54)$$

with ϵ, δ being arbitrary integration constants and ω standing for the frequency

$$\omega = \frac{\sqrt{6}}{5}\Lambda. \quad (55)$$

5   Matching of Solutions in Domains I and II

Substituting the solutions found above into the matching conditions (12) one obtains the following relations for the decomposition coefficients considered as radial functions defined on the matching sphere $|\mathbf{a}| = r_0(t)$:

$$A_i - A_i^* + \dot{\xi}_i - \Lambda\delta_{i1}\overline{\xi_2} = 0, \quad A_i' - A_i^{*'} = 0; \tag{56}$$

$$B_{ik} - B_{ik}^* - \Omega\epsilon_{ik3} - \Lambda\delta_{i1}\delta_{k2} = 0, \quad B_{ik}' - B_{ik}^{*'} = 0; \tag{57}$$

$$C_{ikl} - C_{ikl}^* = 0, \quad C_{ikl}' - C_{ikl}^{*'} = 0; \tag{58}$$

$$P_0 - P_0^* = 0, \quad P_0' - P_0^{*'} = 0; \tag{59}$$

$$P_i - P_i^* = 0, \quad P_i' - P_i^{*'} = 0; \tag{60}$$

$$Q_{ik} - Q_{ik}^* = 0, \quad Q_{ik}' - Q_{ik}^{*'} = 0. \tag{61}$$

6   Conclusions

Using polynomial decompositions of the viscous fluid velocity in the vicinity of small rigid particle moving in the shear flow, we found the approximate expressions for the pressure and velocity and calculated the force and the torque acting on the particle. The force parallel to the angular velocity of the particle proves to contain the oscillatory part, with the frequency being proportional to the gradient of the external steady velocity.

It should be emphasized that this oscillatory effect can be very important for solving the problem of particle saltation in channels with a rough bed [21–27].

References

1. Rubinow, S.I., Keller, J.B.: The transverse force on a spinning sphere moving in a viscous fluid. J. Fluid Mech. **11**, 447–459 (1961)
2. Maxey, M.R., Riley, J.J.: Equation of motion for a small rigid sphere in a nonuniform flow. Phys. Fluids **28**, 883–889 (1983)
3. Fenton, J.D., Abbott, J.E.: Initial movement of grains on a stream bed: the effect of relative protrusion. Proc. Roy. Soc. London **A352**, 523–537 (1977)
4. Ikeda, S.: Incipient motion of sand particles on side slopes. J. Hydraul. Eng. Div. ASCE **108**, 95–114 (1982)
5. Ling, C.-H.: Criteria for incipient motion of spherical sediment particles. J. Hydraul. Eng. Div. ASCE **121**, 472–478 (1995)
6. Nishimura, K., Hunt, J.C.R.: Saltation and incipient suspension above a flat particle bed below a turbulent boundary layer. J. Fluid Mech. **417**, 72–102 (2000)
7. Van Rijn, L.C.: Sediment transport, part I: bed load transport. J. Hydraul. Eng. Div. ASCE **110**, 1431–1456 (1984)

8. Yalin, M.S., Karahan, E.: Inception of sediment transport. J. Hydraul. Eng. Div. ASCE **105**, 1433–1444 (1979)

9. Bagnold, R.A.: The nature of saltation and of "bed-load" transport in water. Proc. Roy. Soc. London **A332**, 473–504 (1973)

10. Barkla, H.M., Auchterlonie, L.J.: The Magnus or Robins effect on rotating spheres. J. Fluid Mech. **47**, 437–447 (1971)

11. Francis, J.R.D.: Experiments on the motion of solitary grains along the bed of a water stream. Proc. Roy. Soc. London **A332**, 443–471 (1973)

12. Lee, H.-Y., Hsu, I.-S.: Investigation of saltating patricle motions. J. Hydraul. Eng. Div. ASCE **120**, 831–845 (1994)

13. Lee, H.-Y., Chen, Y.-H., You, J.-Y., Lin, Y.-T.: Investigation of continuous bed load saltating process. J. Hydraul. Eng. Div. ASCE **126**, 691–700 (2000)

14. Lee, H.-Y., Lin, Y.-T., Chen, Y.-H., You, J.-Y., Wang, H.-W.: On three- dimensional continuous saltating process of sediment particles near the channel bed. J. Hydraul. Res. **44**, 374–389 (2006)

15. Mei, R., Adrian, R.J., Hanratty, T.J.: Particle dispersion in isotropic turbulence under Stokes drag and Basset force with gravitational settling. J. Fluid Mech. **225**, 481–495 (1991)

16. Michaelides, E.E.: Hydrodynamic force and heat/mass transfer from particles, bubbles, and drops - the Freeman scholar lecture. J. Fluids Eng. ASME **125**, 209–238 (2003)

17. Murphy, P.J., Hooshiari, H.: Saltation in water dynamics. J. Hydraul. Eng. Div. ASCE **108**, 1251–1267 (1982)

18. Oesterle, B., Bui Dinh, T.: Experiments on the lift of a spinning sphear in the range of intermediate Reynolds numbers. Exp. Fluids **25**, 16–22 (1998)

19. Sawatzki, O.: Das Strömungsfeld um eine Rotiendre Kugel. Acta Mechanica **9**, 159–214 (1970)

20. Tsuji, Y., Morikawa, Y., Mizuno, O.: Experimental measurements of the Magnus force on a rotating sphere at Low Reynolds numbers. ASME J. Fluid Eng. **107**, 484–488 (1985)

21. Ancey, C., Bigillon, F., Frey, P., Lanier, J., Ducret, R.: Saltating motion of a bead in a rapid water stream. Phys. Rev. E **66**, 036306 (2002)

22. Colombini, M.: Turbulence-driven secondary flows and formation of sand ridges. J. Fluid Mech. **254**, 701–719 (1993)

23. Lukerchenko, N., Platsevich, S., Chara, Z., Vlasak, P.: 3D numerical model of the spherical particle saltation in a channel with a rough fixed bed. J. Hydraul. Hydromech. **57**, 100–112 (2009)

24. Lukerchenko, N., Kvurt, Y., Kharlamov, A., Chara, Z., Vlasak, P.: Experimental evaluation of the drag force and drag torque acting on a rotating spherical particle moving in fluid. J. Hydraul. Hydromech. **56**, 88–94 (2008)

25. Nino, Y., Garcia, M.: Experiments on saltation of sand in water. J. Hydraul. Eng. ASCE **124**, 1014–1025 (1998)

26. Reizes, J.A.: Numerical study of continuous saltation. J. Hydraul. Hydromech. **104**(HY9), 1303–1321 (1978)

27. Wiberg, D.L., Smith, J.D.: A theoretical model for saltating grains in water. J. Geophys. Res. **90**(C4), 7341–7354 (1985)

Fast Two-Dimensional Smoothing
with Discrete Cosine Transform

Pavel Lyubin[✉] and Eugeny Shchetinin[✉]

Department of Applied Mathematics,
Moscow State Technology University "STANKIN",
3a, Vadkovsky lane, Moscow, Russia 119136
lyubin.p@gmail.com, riviera-molto@mail.ru

Abstract. Smoothing is the process of removing "noise" and "insignificant" fragments while preserving the most important properties of the data structure. We propose a fast spline method for two-dimensional smoothing. Data smoothing usually attained by parametric and nonparametric regression. The nonparametric regression requires a prior knowledge of the regression equation form. However, most of the investigated data can't be parameterized simply. From this point of view, our algorithm belongs to nonparametric regression. Our simulation study shows that smoothing with discrete cosine transform is orders of magnitude faster to compute than other two-dimensional spline smoothers.

Keywords: Nonparametric regression · Two-dimensional estimation · Penalized splines · Smoothing splines · Cross-validation · Discrete cosine transform

1 Problem Statement

Raw data of real processes are noisy and need "smoothing" before analyse. Smoothing is attempt to filter "noise" or "insignificant" fragments while preserving the most important properties of data structure. Consider the following model

$$y = \hat{y} + \varepsilon \tag{1}$$

where ε - Gaussian white noise. There are supposed that function $\hat{y}$ should be smooth, i.e. has continuous derivatives up to some order. Data smoothing is usually carried out by a parametric or nonparametric regression. In the case of parametric regression, it requires some a priori knowledge of regression equation form, which must well described original proccess. However, most of the observed data is impossible to parameterize and function $f(x)$ can't be determined analytically. From this point of view, nonparametric and semiparametric regression is the best approach to solving the problem (1). One of the classical methods for smoothing data is the use of various modifications least squares with

© Springer International Publishing AG 2016
V.M. Vishnevskiy et al. (Eds.): DCCN 2016, CCIS 678, pp. 646–656, 2016.
DOI: 10.1007/978-3-319-51917-3_55

penalty. It was first introduced in 1920 [1] and it has been extensively studied ever since 1990 [2]. This technique consists in minimize some functional that balances between "approximation" and "smoothness" of estimation and it has follow form

$$F(\hat{y}) = RSS + \lambda \cdot P(\hat{y}) = ||\hat{y} - y||^2 + \lambda \cdot P(\hat{y}), \qquad (2)$$

where $|| \cdot ||$ - Euclidean norm. The parameter λ is a real positive number controlling the smoothness of solutions: smoothness of $\hat{y}$ growing when parameter increases. The regression is called smoothing spline [1,3,4], when the penalty function written like square integral of p-order derivatives of $\hat{y}$. Apart from this, simple and effective approach to solving problem (1) is squared form of penalty function [5]:

$$P(\hat{y}) = ||D\hat{y}||^2 \qquad (3)$$

where D - tridiagonal matrix as

$$\begin{bmatrix} -1 & 1 & & & \\ 1 & -2 & 1 & & \\ & \ddots & \ddots & \ddots & \\ & & 1 & -2 & 1 \\ & & & 1 & -1 \end{bmatrix}$$

This paper is a continuation of the authors research reviewed in papers [6,7]. Additionally, some main ideas gleaned from the articles [8,9] (Fig. 2).

2 One-Dimensional Smoothing

Suppose $\{x_i\}_{1 \le i \le n}$ is equally spaced points and response function follows

$$y_i = f(x_i) + \varepsilon_i \qquad (4)$$

where $\varepsilon_i \sim N(0, \sigma^2)$. Let $\hat{y}$ is an estimate of $f(x_i)$. After minimization (2) we have

$$\hat{y} = H(\lambda) \cdot y, \qquad (5)$$

where $H(\lambda) = (I + \lambda \cdot D^T D)^{-1}$ is a projection matrix and λ is smoothing parameter. Smoothing parameter selecting by minimization of following equation

$$GCV(\lambda) = \frac{RSS(\lambda)/n}{(1 - Tr(H(\lambda))/n)^2}. \qquad (6)$$

This approach is called as method of cross-validation. Matrix D has some special properties if observations is equidistant. That's possible to simplify the calculation GCV, because matirx D can explain $U\Gamma U^T$, where matrix U is unitary and it is a discrete cosine transformation [9]. Then RSS can be rewritten as follows:

$$\begin{aligned} RSS &= ||\hat{y} - y||^2 = ||H(\lambda) \cdot y - y||^2 \\ &= ||((I + \lambda \cdot D^T D)^{-1} - I) \cdot y||^2 \\ &= ||(U \cdot (I + \lambda \cdot \Gamma^2)^{-1} - I) \cdot U^T \cdot y||^2 \\ &= \sum_i (\frac{1}{1 + \lambda \gamma_{n_i}^2} - 1)^2 \cdot DCT_i^2(y). \end{aligned}$$

In this case, (6) can be rewritten in more convenient for computing form

$$GCV(\lambda) = \frac{n \cdot \sum_i (\frac{1}{1+\lambda\gamma_i^2} - 1)^2 \cdot DCT_i^2(y)}{(n - \sum_i (\frac{1}{1+\lambda\gamma_i^2})^2}. \tag{7}$$

3 Two-Dimensional Smoothing

Suppose $\{(x_{1,i}, x_{2,j})\}_{1 \le i \le n_1, 1 \le j \le n_2}$ is uniform grid and response function follows

$$y_{i,j} = f(x_{1,i}, x_{2,j}) + \varepsilon_{i,j} \tag{8}$$

where $\varepsilon_{i,j} \sim N(0, \sigma^2)$. In this case, values of response function can represent like matrix Y, where element of i-th row and j-th column is value $y_{i,j}$. Then the smoothed values will be denoted by $\hat{Y}$. Introduce the operation vec, which represents matrix in column vector form. Then $vec(\hat{Y})$ can be written:

$$vec(\hat{Y}) = (H_{x_2} \otimes H_{x_1}) \cdot vec(Y) = H_{x_2, x_1} \cdot vec(Y), \tag{9}$$

where H_{x_1}, H_{x_2} - projection matrix for corresponding dimension. Obviously, the projection matrix has follow form

$$H_{x_i} = (I_{n_i} + \lambda_i D_{n_i}^T D_{n_i})^{-1}, \quad i = 1, 2. \tag{10}$$

Applying the approach and properties of the tensor product [10], expression (9) can be simplified as follows:

$$\begin{aligned} \hat{y} &= (H_{x_2} \otimes H_{x_1}) \cdot y \\ &= (I_{n_2} + \lambda_2 D_{n_2}^T D_{n_2})^{-1} \otimes (I_{n_1} + \lambda_1 D_{n_1}^T D_{n_1})^{-1} \cdot y \\ &= U_{x_2} \cdot (\frac{1}{1 + \lambda_2 \gamma_{x_2}^2}) \cdot U_{x_2}^T \otimes U_{x_1} \cdot (\frac{1}{1 + \lambda_1 \gamma_{x_1}^2}) \cdot U_{x_1}^T \cdot y \\ &= U_{x_2} \otimes U_{x_1} \cdot (\frac{1}{1 + \lambda_1 \gamma_{x_1}^2}) \otimes (\frac{1}{1 + \lambda_2 \gamma_{x_2}^2}) \cdot U_{x_2}^T \otimes U_{x_1}^T \cdot y \\ &= U_{x_2, x_1} \cdot \Gamma_{x_2, x_1} \cdot U_{x_2, x_1}^T \cdot y \end{aligned}$$

To automatically search for the best values λ_1 and λ_2, we use a cross-validation adapted for two-dimensional case:

$$GCV(\lambda_1, \lambda_2) = \frac{RSS/n}{(1 - Tr(H_{x_2, x_1})/n^2)}. \tag{11}$$

Properties of tensor product of matrices [10] denotes $Tr(H_{x_2,x_1}) = \sum \frac{1}{1+\lambda_1\gamma_{x_1}^2} \cdot \sum \frac{1}{1+\lambda_2\gamma_{x_2}^2}$. Obviously, main consuming place of the estiomation is a calculation RSS, because it requires evaluation of $\hat{y}$ for all combinations λ_1 and λ_2. This calculation can be simplified:

$$
\begin{aligned}
RSS &= ||\hat{y} - y||^2 = ||H_{x_2,x_1} \cdot y - y||^2 = ||(H_{x_2,x_1} - I_n) \cdot y||^2 \\
&= ||U_{x_2,x_1} \cdot (\Gamma_{x_2,x_1} - I_n) \cdot U_{x_2,x_1}^T \cdot y||^2 \\
&= (U_{x_2,x_1} \cdot (\Gamma_{x_2,x_1} - I_n) \cdot U_{x_2,x_1}^T \cdot y)^T \cdot (U_{x_2,x_1} \cdot (\Gamma_{x_2,x_1} - I_n) \cdot U_{x_2,x_1}^T \cdot y) \\
&= (DCT_2 \cdot y)^T \cdot (\Gamma_{x_2,x_1} - I_n)^2 \cdot DCT_2 \cdot y \\
&= \sum (\gamma_{x_2,x_1} - 1)^2 \cdot (DCT_2 \cdot y)^2,
\end{aligned}
$$

where DCT_2 - is a two-dimensional discrete cosine transform. From the simplified equation shows the transformation must evaluate one times and result change with values γ_{x_2,x_1} depending values λ_1 and λ_2. This approach implemented in R. To demonstrate the advantages of considered approach performed numerical experiments: with model and real data (Fig. 4).

4   Experiments

Model data: To illustrate the effectiveness of the algorithm, sample data have been modeled from function $sin(2\pi(x_1 - 0.5)^3) \cdot cos(4\pi x_2)$ with noise - random values from normal distribution of $N(0, 0.2^2)$ (Fig. 1). Smoothing was carried by presented approach and MGCV package [11], which implements smoothing with penalized splines, including multidimensional case with tensor product of basic functions. Below is a table contains result of smoothing with different methods (Table 1).

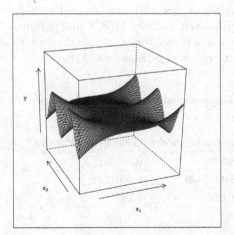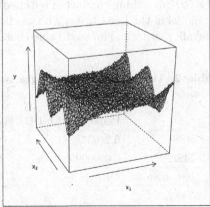

Fig. 1. Function $sin(2\pi(x - 0.5)^3) \cdot cos(4\pi y)$: raw (left) and with noise (right).

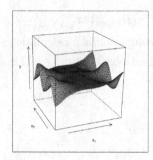 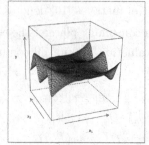 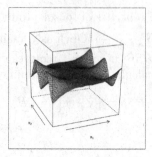

Fig. 2. Results of smoothing $sin(2\pi(x-0.5)^3) \cdot cos(4\pi y)$: GAM with 10^2 knots (top-left), GAM with 20^2 knots (top-right) and DCT (bottom).

Table 1. The results of smoothing model data with different methods.

	P-splines with DCT	GAM with 10^2 knots	GAM with 20^2 knots
RSS	9.488243	11.72485	9.87163
MSE	0.001483	0.001832	0.00154
Corr. with true values	0.9993394	0.996919	0.9991624
Est. time (s)	1.941	10.237	29.875

Real data: To demonstrate the practical application of the approach, real data of mortality in Russia have been smoothed and compared with results of another approaches. The data are taken from the open source [12] and contains observations for ages of 0 and 110 between years of 1959 and 2010. For experiment was taken part of data, which belongs to the older ages (50–101, Fig. 3). That part was chosen, because observations contain many errors and outliers. Thus, analyzed data are evenly spaced values of mortality rates on grid with size 52×52. Smoothing conducted outlined approach, package MGCV and parametric model of the Lee-Carter, who has become a classic for appraisals dimensional mortality surface. The next table contains result of estimations (Table 2).

Table 2. The results of smoothing a two-dimensional surface of Russian mortality rates.

	P-splines with DCT	Lee-Carter model	GAM with 12^2 knots
RSS	0.21637	18.5092	0.41395
MSE	0.0000905	0.0077379	0.0001731
Est. time (s)	0.49	1.194	4.185

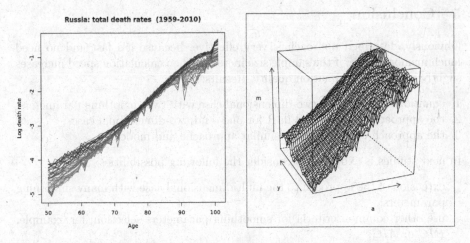

Fig. 3. The raw mortality rates in Russia for ages of 50 and 101 between years of 1959 and 2010.

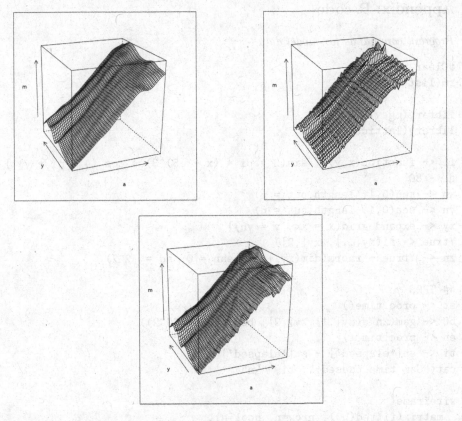

Fig. 4. Results of smoothing mortality data: GAM with 12^2 knots (top-left), Lee-Carter (top-right) and DCT (bottom).

5 Conclusion

Obviously, described approach is very effective, because it's fast and no need much memory. Note, if the sample size increases then calculation speed increases slightly with same estimation quality. Results:

1. equations obtained for two-dimensional case with two smoothing parameters;
2. the approach implemented in R for one- and two-dimensional cases;
3. the approach compared with similar approache and model.

In next studies is expected to consider the following possibilities:

– extension of the approach to the multidimensional case with many smoothing parameters;
– use other common criteria for smoothing parameters selection, for example, *BIC* or *AIC*;
– use a faster method for minimization *GCV* instead of grid search.

Appendix: R code

Program commands for model data

```
#Cléar workspace
rm(list=ls(all=T))

library(mgcv)
library(lattice)

f1 <- function(x,y) { sin(2 * pi * (x - .5)^3) * cos (4 * pi * y) }
n <- 80
xn <- seq(0,1, length.out = n)
yn <- seq(0,1, length.out = n)
xy <- expand.grid(x = xn, y = yn)]
Ytrue <- f1(xy[,1], xy[,2])
zn <- Ytrue + rnorm(dim(xy)[1], mean = 0, sd = .2^2)

## TPRS
st <- proc.time()
b0 <- gam(zn~s(xy[,1],xy[,2], bs='ts', k=20^2))
en <- proc.time()
ti <- en['elapsed'] - st['elapsed']
cat("Gam time passed:", ti, "\n")

wireframe(
  matrix(fitted(b0), nrow=n, ncol=n),
  zlim = c ( -2, 2),
  xlab = expression(x[1]),
```

```
  ylab = expression(x[2]),
  zlab = expression(y),
  screen = list(z = 20, x = -70, y = 3)
  )


#DCT
lr <- seq( 64, 66, by = .1 )
lc <- seq( 36, 38, by = .1 )
fit <- psdct2d(matrix(zn, nrow=n, ncol=n))
plot(fit, theta = -15, phi = 30, zlim = c ( -2, 2))
summary(fit)

wireframe(
  matrix(fitted(fit), nrow=n, ncol=n),
  zlim = c ( -2, 2),
  xlab = expression(x[1]),
  ylab = expression(x[2]),
  zlab = expression(y),
  screen = list(z = 20, x = -70, y = 3)
  #screen = list(z = -60, x = -60)
  )

cat("RSS DCT:", sum( (residuals(fit))^2 ), "\n")
cat("RSS GAM:", sum( (residuals(b0))^2 ), "\n")
cat("Corr DCT:", cor(fitted(fit), Ytrue), "\n")
cat("Corr GAM:", cor(fitted(b0), Ytrue))
```

Program commands for model data

```
#Clear workspace
rm(list=ls(all=T))

library(demography)

#Raw
ru.mort <- read.demogdata("data/Mx_1x1.txt",
"data/Exposures_1x1.txt", "mortality", "Russia")
plot(ru.mort, series="total")

ru.ext <- extract.ages(ru.mort, 50:101 , FALSE)
plot(ru.ext, series="total")

wireframe(
  matrix(log(ru.ext$rate$total), nrow=52, ncol=52),
  xlab = expression(a),
  ylab = expression(y),
```

```
      zlab = expression(m),
      screen = list(z = 20, x = -70, y = 3)
      )


#GAM
library(mgcv)
gamst <- proc.time()
z <- as.vector(log(ru.ext$rate$total))
x <- 1:nrow(ru.ext$rate$total)
y <- 1:ncol(ru.ext$rate$total)
xy <- expand.grid(x, y)
ru.gam <- gam(z~s(xy[,1],xy[,2], bs='ts', k=12^2))
gamen <- proc.time()
gamel <- gamen['elapsed'] - gamst['elapsed']
cat("Gam time passed:", gamel, "\n")
persp(matrix(fitted(ru.gam), nrow=length(x), ncol=length(y)))
persp(matrix(residuals(ru.gam), nrow=length(x), ncol=length(y)))
levelplot(matrix(residuals(ru.gam), nrow=length(x), ncol=length(y)))

wireframe(
  matrix(fitted(ru.gam), nrow=52, ncol=52),
  xlab = expression(a),
  ylab = expression(y),
  zlab = expression(m),
  screen = list(z = 20, x = -70, y = 3)
  )


#Lee-Carter
lcst <- proc.time()
ru.lc <- lca(ru.ext, adjust="e0")
plot(ru.lc)
persp(ru.lc$fitted$y)
persp(ru.lc$residuals$y)
levelplot(ru.lc$residuals$y)
lcen <- proc.time()
lcel <- lcen['elapsed'] - lcst['elapsed']
cat("LC time passed:", lcel, "\n")

wireframe(
  ru.lc$fitted$y,
  xlab = expression(a),
  ylab = expression(y),
  zlab = expression(m),
  screen = list(z = 20, x = -70, y = 3)
  )
```

```
#DCT
ru.dct <- psdct2d(log(ru.ext$rate$total))
persp(matrix(residuals(ru.dct2), nrow=length(x), ncol=length(y)))
levelplot(matrix(residuals(ru.dct2), nrow=length(x), ncol=length(y)))

wireframe(
  matrix(fitted(ru.dct), nrow=52, ncol=52),
  xlab = expression(a),
  ylab = expression(y),
  zlab = expression(m),
  screen = list(z = 20, x = -70, y = 3)
  )
```

References

1. Whittaker, E.T.: On a new method of graduation. Proc. Edinb. Math. Soc. **41**, 62–75 (1923)
2. Wahba, G.: Spline Models for Observational Data. Society for Industrial Mathematics, Philadelphia (1990)
3. Schoenberg, I.J.: Spline functions and the problem of graduation. Proc. Natl. Acad. Sci. USA **52**, 947–950 (1964)
4. Takezawa, K.: Introduction to Nonparametric Regression. Wiley, New York (2005)
5. Weinert, H.L.: Efficient computation for whittaker-henderson smoothing. Comput. Stat. Data Anal. **52**, 959–974 (2007)
6. Yu, S.E., Lyubin, P.G.: Robust smoothing with splines. Sci. Rev. **1**, 86–94 (2015)
7. Lyubin, P.G., Shetinin, E.Y.: Stochastic models of mortality estimation. Sci. Rev. **18**, 147–155 (2015)
8. Xiao, L., Li, Y., Ruppert, D.: Fast bivariate p-splines: the sandwich smoother. J. Roy. Stat. Soc. **75**, 577–599 (2013)
9. Garcia, D.: Robust smoothing of gridded data in one and higher dimensions with missing values. Comput. Stat. Data Anal. **54**, 1167–1178 (2010)
10. Seber, G.: A Matrix Handbook for Statisticians. Wiley-Interscience, Hoboken (2007)
11. Wood, S.: mgcv: mixed gam computation vehicle with gcv/aic/reml smoothness estimation. R package version 1.8.10
12. Department of Demography at the University of California. The human mortality database. Last visited on 25.02.2016
13. Wood, S.N.: Thin plate regression splines. R. Stat. Soc. B **65**, 95–114 (2003)
14. Ruppert, D.: Selecting the number of knots for penalized splines. Comput. Graph. Stat. **1**, 735–757 (2006)
15. Li, Y., Ruppert, D.: On the asymptotics of penalized splines. Biometrika **95**, 415–436 (2008)
16. Dierckx, P.: A fast algorithm for smoothing data on a rectangular grid while using spline functions. SIAM J. Numer. Anal. **19**(6), 1286–1304 (1982)
17. Wood, S.N., Smith, L., Hyndman, R.J.: Spline interpolation for demographic variables: the monotonicity problem. J. Popul. Res. **21**(1), 95–98 (2004)

18. Eilers, P.H.C., Marx, B.D.: Generalized linear additive smooth structures. J. Comput. Graph. Stat. **11**(4), 758–783 (2002)
19. Eilers, P.H.C., Marx, B.D.: Multidimensional penalized signal regression. Technometrics **47**(1), 13–22 (2005)
20. Hyndman, R.J., Booth, H., Tickle, L., Maindonald, J.: Demography: forecasting mortality, fertility, migration and population data, R package version 1.18 (2014)
21. Yueh, W.C.: Eigenvalues of several tridiagonal matrices. Appl. Math. E-Not. **5**, 66–74 (2005)
22. Hoaglin, D.C., Welsch, R.E.: The hat matrix in regression and ANOVA. Am. Stat. **32**, 17–22 (1978)
23. Eilers, P.H.C., Marx, B.D.: Splines, knots, and penalties. Wiley Interdisc. Rev. Comput. Stat. **2**, 637–653 (2010)

Cluster Method of Description of Information System Data Model Based on Multidimensional Approach

Maxim Fomin[✉]

Department of Information Technologies, RUDN University,
Miklukho-Maklaya st. 6, Moscow 117198, Russia
mfomin@sci.pfu.edu.ru

Abstract. Multidimensional data cube is a data model at the information systems based on the multidimensional approach. If one uses a large
set of aspects for the analysis of data domain the data cubes are characterized by substantial sparseness. It complicates the organization of data
storage. The proposed cluster method of description of multidimensional
data cube is based on the investigation of data domain semantics. The
dimensionalities of the multidimensional cube are the dimensions corresponding to the aspects of analysis. The basis of the cluster method is a
construction of the groups of members which are semantically related to
the groups of other members. Building of associations between the groups
of different members allows to reveal the clusters in the data cube – the
sets of cells with similar properties which may be described in a same
way. Clusters are used as the main element of information system data
model.

Keywords: Multidimensional information system · Multidimensional
data model · Sparse data cube · Set of possible member combinations ·
Cluster of member combinations

1 Introduction

Multidimensional information systems based on the principles of OLAP are used
for the operational analysis of large datasets. Analytical space in a system of
this type is a multidimensional data cube. The role of the cube dimensionalities
is played by the dimensions corresponding to various aspects of the observed
phenomenon for which description the system is developed. If we use a large
amount of semantically heterogeneous data for the description of the observed
phenomenon the multidimensional cube is characterized by high sparseness and
irregular filling [1–8]. As a result, there is a problem of developing an adequate
way to describe the structure of an analytical space which use would make it
possible to effectively organize the data analysis process [9–18]. Such a correct
way should provide the accounting of semantics of the observed phenomenon.

© Springer International Publishing AG 2016
V.M. Vishnevskiy et al. (Eds.): DCCN 2016, CCIS 678, pp. 657–668, 2016.
DOI: 10.1007/978-3-319-51917-3_56

2 Structure of Sparse Multidimensional Data Cube

The structure of analytical space of multidimensional information system should reflect the characteristics of those aspects of the observed phenomenon which are used in the data analysis process. Each aspect corresponds to one dimension of a multidimensional cube H. A full set of dimensions forms a set $D(H) = \{D^1, D^2, ..., D^n\}$, there D^i is i-dimension, and $n = dim(H)$ – dimensionality of multidimensional cube. Each dimension is characterized by a set of members $D(H) = \{d_1^i, d_2^i, ..., d_{k_i}^i\}$, there i is a number of dimension, k_i – the quantity of members. Members of D^i are drawn from a set of positions of the basic classifier which corresponds to an aspect of the observed phenomenon associated with D^i [19–24].

The multidimensional data cube is a structured set of cells. Each cell c is defined by a combination of members $c = \left(d_{i_1}^1, d_{i_2}^2, ..., d_{i_n}^n \right)$. The combination includes one member for each of the dimensions. If the analysis of the observed phenomenon is performed using a large set of diverse aspects, not all member combinations define the possible cells of multidimensional cube, i.e. the cells corresponding to a certain fact. This effect occurs due to semantic inconsistencies of some members from different dimensions to each other and generates a sparseness in the cube.

The complex structure of the compatibility of members may lead to a situation where a certain dimension becomes semantically uncertain if combined with a set of members from other dimensions. In this situation, while describing the possible cell of multidimensional cube we will use the special value "Not in use" to set the member of semantically unspecified dimension [25].

Thus, the structure of a multidimensional information system analytical space defines a set of possible member combinations comporting with a set of possible cells of multidimensional cube. To denote this set we will use the abbreviation "SPMC". To set the members during the process of SPMC combinations forming we will use the data taken from the classifiers which match the dimensions, and the special member "Not in use". The set of possible member combinations should meet the following requirements:

– if there is a combination in SPMC in which a special member "Not in use" is set for one or more dimensions in combination with a certain set of other members, the other combination with the same set of other members can not exist in SPMC. In other words, the dimension is either used or not in combination with a certain set of other members;
– there should be no combination in SPMC in which all dimensions are defined with a special member "Not in use".

The observed phenomenon is characterized by the values of measures specified in the possible cells of multidimensional cube. The full set of measures composes the set $V(H) = \{v_1, v_2, ..., v_m\}$, there v_j is a j measure, and m – the quantity of measures in a hypercube. Not all measures from $V(H)$ may be set in a possible cell. The possibility of such a situation arises in the case of semantic mismatch between the cell-defining members and some measures. Describing the analytical space for each possible cell requires to specify its own set of $V(c) = \{v_1, v_2, ..., v_{m_c}\}$, consisting of measures specified in this cell, $m_c \leq m$. To describe the measures in cell c outside the set $V(c)$ we introduce the special value "Not in use". The rule must be hold: a set of measures $V(c)$ defined in a possible cell c can not be empty. Description of measures in cells of multidimensional cube matching the combinations of members not included into the SPMC does not make sense.

The challenge here is to develop a formal approach to describing of SPMC, which allows to present the metadata of multidimensional information system in a compact form reflecting the semantics of the observed phenomenon.

3   Cluster Approach to the Description of the Analytical Space

To properly describe the structure of an analytical space one should perform a semantic analysis of the compatibility of members. There may be regularities in the compatibility of two, three or more members defining the structure of SPMC, but in most cases the rules of SPMC compatibility are specified by the pairwise associations between dimensions. Let us limit ourselves to such situation. As an illustrative example we consider the structure of an analytical space of information system that describes the observed phenomenon of "Granting of loans". The data of the system measures will be represented in six aspects corresponding to the following dimensions: "Time of loan granting", "Place of loan granting", "Debtor type", "Debtor gender", "Occupation" and "Type of loan". The first dimension is based on calendar data specified in the time range which is used in the analysis. The second dimension is based on the reference book of the territorial administrative division. The remaining dimensions are defined with the following members:

- Debtor type = {"Legal entity", "Natural person"};
- Debtor gender = {"Male", "Female"};
- Occupation = {"Construction engineering", "Trade", "Banking"};
- Type of loan = {"Operating", "Interbank", "Mortgage", "Consumer"}.

The source of information about the semantic relationships between the dimensions is the normative documentation relating to the observed phenomenon. The analyst should formalize this information in the form of rules of compatibility allowing to build SPMC. If pairwise associations are analyzed the rules should determine which pair of two members can occur in the SPMC combinations, and which members of one dimension are incompatible with all members of the other dimension. This approach allows to allocate the groups of members in a set of members. The group of members is a set including one or several members which combine with the members of some other dimension within SPMC in a similar way.

The method based on the allocation of groups in a set of members allows to describe the pairwise relations between dimensions. These pairwise relations are specified by the determination of conformity between the two groups of members from the different dimensions for which the "identity" of compatibility or consistency between the group in one dimension and "Not in use" member in the other one were revealed. For the pairwise relations the following conditions must be held:

1. If some member of the first dimension is included in the group that corresponds to the group in the second dimension, it can not be included in the group which corresponds to the "Not in use" member;
2. If "Not in use" member for the second dimension corresponds to a certain group of members of the first dimension, the members of this group can present in SPMC only in combination with the "Not in use" member for the second dimension;
3. If a certain member of the first dimension is included into the group that corresponds to the group in the second dimension, for the combination of SPMC including this member the second dimension must either take the member from the second group, or there must be the "Not in use" member set for it.

There may be several types of relations between the dimensions divided in "simple" and "complex". Simple types of relations are as follows:

1. Association. There is an association in a pair of dimensions D^1 and D^2 if n groups, $n \geq 2$, can be singled out of a set of members of each of them, and a bijection can be established between these groups which manifests as follows: if a combination of SPMC includes the members D^1 and D^2, they come in pairs, taken from the corresponding groups of members;
2. Full association. There is a full association in a pair of dimensions D^1 and D^2 if a bijection can be established between the members of these dimensions which manifests as follows: the members D^1 and D^2 can come in SPMC in pairs and in any combinations;
3. Dependence. There is a dependence between dimensions D^1 and D^2 (D^2 depends on D^1) if the members of D^1 can be divided in two groups of members in such way that if a certain combination from SPMC includes the member from the first group of members D^1, the member of D^2 in this combination is possible, and if the member of the second group of members D^1 is included into the combination, the D^2 in such combination is set to the "Not in use" member.

There may be complex relationships specified in a pair of dimensions which are the combinations of a few simple relationships:

1. Association and dependence. There is an association and dependence between D^1 and D^2 if n groups can be singled out of D^2, $n \geq 1$, and $(n+1)$ – out of D^1 in such way that there is an association between first n groups from D^1 and

D^2, and if the combination of SPMC includes the member from $(n+1)$ group of D^1 members, D^2 in this combination is set to the "Not in use" member. Besides, the members from $(n + 1)$ group of D^1 members can not be met in other groups of this dimension;

2. Association and two-sided dependence. There is an association and two-sided dependence between D^1 and D^2 if n groups can be singled out of a set of members of each of that dimensions, $n \geq 2$, in a such way that if the combination of SPMC includes the member from the first group D^1, the D^2 in this combination is set to the "Not in use" member, and if the combination of SPMC includes the member from the first group D^2, the D^1 in such combination is set to the "Not in use" member; herewith, the remaining $(n - 1)$ groups of members of D^1 and D^2 dimensions form an association;

3. Two-sided dependence. There is a two-sided dependence between D^1 and D^2 dimensions if the following rule holds: in case of SPMC combination includes the member from D^1, the D^2 in this combination is set to the "Not in use" member, and when the combination includes the member from D^2, the D^1 the in this combination is set to the "Not in use" member.

Figure 1 presents the diagrams containing the designations of the pairwise relations between the dimensions for the case of illustrative example described above.

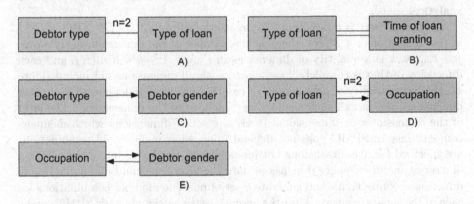

Fig. 1. The types of diagrams describing the pairwise relations between the dimensions: association (A), full association (B), dependence (C), association and dependence (D), two-sided dependence (E)

It is convenient to use the compliance charts of the groups of members for the description of the pairwise relations between dimensions. Figure 2 presents the pairwise compliance charts of the groups for the proposed illustrative example.

After building of the pairwise relations between dimensions of the multidimensional cube one can draw a diagram of dimensions connectivity. This diagram should present all dimensions with the indication of all relations between them. On the basis of this diagram the other diagram can be built – a compliance

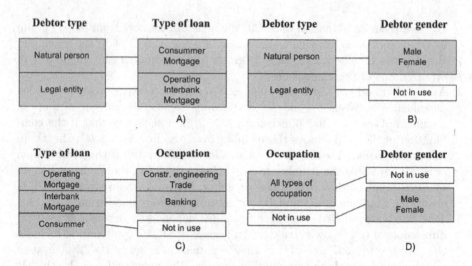

Fig. 2. Pairwise compliance charts of the groups of members: association (A), dependence (B), association and dependence (C), two-sided dependence (D)

chart for the groups of members which shows the all groups and specifies the relations between them. These diagrams may be used in the formation of SPMC analytical space.

In case of there is a possibility to isolate the subset $L^i = \left\{ D^{j_1}, D^{j_2}, ..., D^{j_k} \right\}$ in the dimensions set $D(H)$, there j_k is a number of the dimension in the layer, $j = 1, ..., k$, k is a quantity of dimensions in i layer, $1 \leq k < dim(H)$, and each dimension of that is completely associated with all dimensions not included into L^i, the compatibility of members in L^i can be considered independently of other dimensions. Let us call such a subset as "the layer of the dimensions". The layer of the dimensions, or dimensional layer, is a set of dimensions which members compatibility in SPMC does not depend upon what members in combinations are specified for the dimensions not included into the layer. In case of splitting of a set of analytical space dimensions onto the layers one can build a diagram of dimensions connectivity and generate a set of possible member combinations for each of the separate layers. After the analysis of dimensional layers SPMC can be obtained by the Cartesian product: $SPMC(H) = SPMC(L^1) \times SPMC(L^2) \times ... \times SPMC(L^m)$, there m is a quantity of layers. In the present example there are three layers: $L^1 = \{$Debtor type, Debtor gender, Occupation, Type of loan$\}$, $L^2 = \{$Time of loan granting$\}$ and $L^3 = \{$Place of loan granting$\}$.

Figure 3 presents the diagram of dimensions connectivity for the L^1 layer from the illustrative example.

If we analyze some dimension as an element of the diagram of layer connectivity and take into account the relations between the considered dimension and all the rest dimensions of the layer, the groups of members available in this dimension can be transformed so that they will comply with all relations of the considered dimension simultaneously. New groups must lie at the intersection of

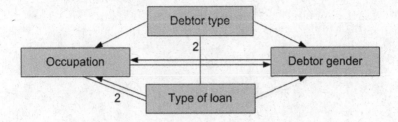

Fig. 3. The diagram of dimensions connectivity for the L^1 layer

the groups participating in the description of pairwise relations with different dimensions. Using such procedure one can describe the compatibility of the full set of dimensions in the layer. Let us call such procedure of groups formation as a subdivision of groups of members describing the pairwise relations. When the groups are subdivided the relations between the dimensions revealed at the stage of the pairwise analysis must be inherited.

Figure 4 presents a fragment of the chart of group compliance illustrating the procedure of subdivision of the groups for the dimension "Type of loan".

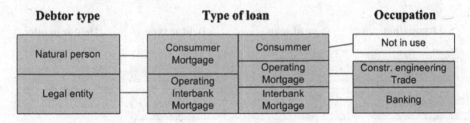

Fig. 4. Fragment of the compliance chart of the groups of members for the L^1 layer

All pairwise relations from the diagram of layer connectivity are used in the procedure of subdivision of the groups. This complete set of relations allows to distinguish the relations of "Full association" type and relations describing the compliance of the groups which have been already accounted in the remaining relations. These distinguished relations do not influence the result of the groups subdivision and can be removed from the connectivity graph. Thus, the graph can be reduced to a much more simple form without the loss of information about the compatibility of members (Fig. 5).

After subdividing the groups describing the pairwise relations between dimensions, one can bypass the compliance chart of the groups of members of the analytical space or the dimensional layer. While bypassing the compliance chart one can reveal the chains of groups of members longwise its relations, and for several dimensions – also the special member "Not in use" instead of the group which members are combined in the SPMC by the "all-to-all" rule. Such chains define a set of combinations included in the SPMC which can be obtained by the

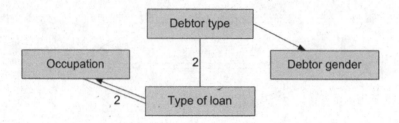

Fig. 5. Reduced dimensions connectivity diagram for the L^1 layer

Cartesian product of the groups of members and the special member "Not in use" if it is present in the chain. Let us call such set of combinations as the "cluster of member combinations". Cluster of member combinations is a set of combinations of members which can be obtained by means of the Cartesian product operation in which the operands are the groups of members or the special member "Not in use", one operand for each of the dimensions, assigned in the multidimensional cube or in the dimensional layer of the multidimensional cube. Figure 6 presents the clusters of member combinations corresponding to the dimensions connectivity diagram for the L^1 layer from the illustrative example.

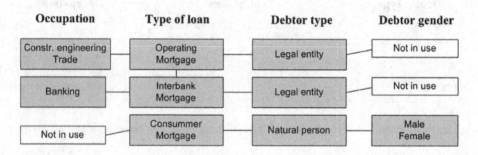

Fig. 6. Clusters of member combinations for the L^1 layer

In an absence of subdivision of the dimensions set $D(H)$ onto the layers, SPMC can be represented as the association of clusters corresponding to the compatibility diagram of the analytical space dimensions.

In case of the subdivision of the dimensions set $D(H)$ onto the layers, SPMC for each layer must be built as an association of clusters of member combinations of the layer, and SPMC of the members of multidimensional cube is obtained as a result of the Cartesian product of SPMC for the layers.

There may be a situation when the very different semantic components can be distinguished within the observed phenomenon. In this case it is possible to separately form the subsets of member combinations corresponding to different semantic components. For this purpose it is necessary to analyze the compatibility of members for each component and in accordance with this analysis to form the clusters of member combinations. The SPMC of the members of multidimensional cube can be computed using the set theory operations. Operands in these operations are subsets of member combinations for the components.

We can identify two cases where such approach can be successfully used. The first one takes place when different subdivisions of the dimensions onto the layers occur during the analysis of different semantic components, and the second one – when there is a simple way of building a subset describing the SPMC redundantly, and the efficient way to describe the combinations which are to be excluded from this subset to reduce it to the SPMC. Let's consider these cases in more detail.

In the first case, the decomposition of the observed phenomenon on l semantic components corresponds to the union of member combinations subsets:

$$SPMC(H) = Q_1 \cup Q_2 \cup .. \cup Q_l.$$

Set of analytical space dimensions can be divided into layers in different ways due to differences in semantics of the observed phenomenon components:

$$D(H) = L_i^1 \cup L_i^2 \cup .. \cup Q_i^{m_i},$$

there $i = 1, ..., l$ – number of component, and m_i – the quantity of layers in $i$component. Each subset Q_i is formed according to its split of set of the dimensions can into layers.

In the second case, set of possible member combinations is represented as the difference of two subsets:

$$SPMC(H) = R \backslash Q,$$

there R – set of member combinations, described with an excess (set to reduce), and Q – set of combinations to be excluded. Set to reduce may be formed using the following rules. It should include member combinations obtained by the Cartesian product of all members of all dimensions. It must be supplemented with a set of combinations that contain the special value "Not in use" for some dimensions, for which this value is acceptable. From this set it should be excluded those combinations which can be obtained by replacing the special value "Not in use" by the member. This approach can be used in case the set $SPMC(H)$ has a complex structure and it may be offered a simple algorithm of forming a subset Q.

4   Method of Construction of Set of Possible Member Combinations

We can propose the algorithm of SPMC description basing on the cluster approach and consisting of the following steps:

1. Allocate the n semantic components ($n \geq 1$) within the observed phenomenon and juxtapose these components with the subsets of combinations Q_i, $i = 1, ..., n$;

2. Construct a formula for $SPMC(H)$ using Q_i and operations of set theory according to the revealed relationships between the components of the observed phenomenon;
3. Form a subset of combinations for each Q_i:
   (a) perform the analysis of pairwise relations between the dimensions corresponding to Q_i semantics, and form the groups of members expressing these relations;
   (b) allocate the layers of dimensions in a set of dimensions and build the dimensions connectivity diagram for each layer;
   (c) make the subdivision of the groups of members specified in layers according to the relations available from the diagrams of layers connectivity;
   (d) realize the formation of clusters of member combinations and consolidation of these clusters in subsets of combinations for layers;
   (e) execute the formation of a subset of Q_i combinations by the Cartesian product of subsets of combinations for the dimensional layers;
4. Calculate the $SPMC(H)$ using the constructed formula.

5   Conclusion

In case of the development of large multiple-aspect multidimensional information system the use of the cluster approach for describing the set of possible member combinations allows to provide the compactness while specifying the metadata and to express the semantics of the analyzed phenomenon observed. The proposed approach is based on the identification of relations between the dimensions which reflect the properties of the observed phenomenon, and on the formation of the groups of members which elements are united by the similar behavior towards these relations.

Acknowledgments. The work is partially supported by the Ministry of Education and Science of the Russian Federation (the Agreement number 02.a03.21.0008).

References

1. Thomsen, E.: OLAP Solution: Building Multidimensional Information System. Willey Computer Publishing, New York (2002). ISBN 0-471-40030-0
2. Hirata, C.M., Lima, J.C.: Multidimensional cyclic graph approach: representing a data cube without common sub-graphs. Inf. Sci. **181**, 2626–2655 (2011)
3. Karayannidis, N., Sellis, T., Kouvaras, Y.: CUBE file: a file structure for hierarchically clustered OLAP cubes. In: Bertino, E., Christodoulakis, S., Plexousakis, D., Christophides, V., Koubarakis, M., Böhm, K., Ferrari, E. (eds.) EDBT 2004. LNCS, vol. 2992, pp. 621–638. Springer, Heidelberg (2004). doi:10.1007/978-3-540-24741-8_36. ISBN 978-3-540-21200-3
4. Chun, S.-J.: Partial prefix sum method for large data warehouses. In: Zhong, N., Raś, Z.W., Tsumoto, S., Suzuki, E. (eds.) ISMIS 2003. LNCS (LNAI), vol. 2871, pp. 473–477. Springer, Heidelberg (2003). doi:10.1007/978-3-540-39592-8_67. ISBN 978-3-540-39592-8

5. Messaoud, R.B., Boussaid, O., Rabaseda, S.L.: A multiple correspondence analysis to organize data cube. In: Databases and Information Systems IV DB & IS 2006, pp. 133–146. IOS Press, Vilnius (2007). ISBN 978-1-58603-715-4

6. Jin, R., Vaidyanathan, J.K., Yang, G., Agrawal, G.: Communication and memory optimal parallel data cube construction. IEEE Trans. Parallel Distrib. Syst. **16**, 1105–1119 (2005)

7. Luo, Z.W., Ling, T.W., Ang, C.H., Lee, S.Y., Cui, B.: Range top/bottom k queries in OLAP sparse data cubes. In: Mayr, H.C., Lazansky, J., Quirchmayr, G., Vogel, P. (eds.) Database and Expert Systems Applications - DEXA 2001, vol. 2113, pp. 678–687. Springer, Heidelberg (2001). ISBN 978-3-540-42527-4

8. Fu, L.: Efficient evaluation of sparse data cubes. In: Li, Q., Wang, G., Feng, L. (eds.) WAIM 2004. LNCS, vol. 3129, pp. 336–345. Springer, Heidelberg (2004). doi:10.1007/978-3-540-27772-9_34. ISBN 978-3-540-27772-9

9. Chen, C., Feng, J., Xiang, L.: Computation of sparse data cubes with constraints. In: Kambayashi, Y., Mohania, M., Wöß, W. (eds.) DaWaK 2003. LNCS, vol. 2737, pp. 14–23. Springer, Heidelberg (2003). doi:10.1007/978-3-540-45228-7_3. ISBN 978-3-540-40807-9

10. Salmam, F.Z., Fakir, M., Errattahi, R.: Prediction in OLAP data cubes. J. Inf. Knowl. Manag. **15**, 449–458 (2016)

11. Romero, O., Pedersen, T.B., Berlanga, R., Nebot, V., Aramburu, M.J., Simitsis, A.: Using semantic web technologies for exploratory OLAP: a survey. IEEE Trans. Knowl. Data Eng. **27**, 571–588 (2015)

12. Gomez, L.I., Gomez, S.A., Vaisman, A.: A generic data model and query language for spatiotemporal OLAP cube analysis. In: Proceedings of the 15-th International Conference on Extending Database Technology – EDBT 2012, pp. 300–311, Berlin (2012). ISBN: 978-1-4503-0790-1

13. Tsai, M.-F., Chu, W.: A multidimensional aggregation object (MAO) framework for computing distributive aggregations. In: Kambayashi, Y., Mohania, M., Wöß, W. (eds.) DaWaK 2003. LNCS, vol. 2737, pp. 45–54. Springer, Heidelberg (2003). doi:10.1007/978-3-540-45228-7_6. ISBN 978-3-540-40807-9

14. Vitter, J.S., Wang, M.: Approximate computation of multidimensional aggregates of sparse data using wavelets, In: Proceedings of the 1999 International Conference on Management of Data - SIGMOD 1999, pp. 193–204. ACM, New York (1999). ISBN 1-58113-084-8

15. Leonhardi, B., Mitschang, B., Pulido, R., Sieb, C., Wurst, M.: Augmenting OLAP exploration with dynamic advanced analytics. In: Proceedings of the 13th International Conference on Extending Database Technology - EDBT 2010, pp. 687–692. ACM, New York (2010). ISBN 978-1-60558-945-9

16. Wang, W., Lu, H., Feng, J., Yu, J.X.: Condensed cube: an effective approach to reducing data cube size. In: Proceedings of the 18th International Conference on Data Engineering - ICDE 2002, pp. 155–165. IEEE Computer Society, Washington (2002). ISBN 0-7695-1531-2

17. Goil, S., Choudhary, A.: Design and implementation of a scalable parallel system for multidimensional analysis and OLAP. In: Parallel and Distributed Processing - 11th IPPS/SPDP 1999, pp. 576–581. Springer, Heidelberg (1999). ISBN 978-3-540-65831-3

18. Cuzzocrea, A.: OLAP data cube compression techniques: a ten-year-long history. In: Kim, T., Lee, Y., Kang, B.-H., Ślęzak, D. (eds.) FGIT 2010. LNCS, vol. 6485, pp. 751–754. Springer, Heidelberg (2010). doi:10.1007/978-3-642-17569-5_74. ISBN 978-3-642-17568-8

19. Hirata, C.M., Lima, J.C., Silva, R.R.: A hybrid memory data cube approach for high dimension relations. In: Proceedings of the 17-th International Conference on Enterprise Information Systems - ICEIS 2015, vol. 1, pp. 139–149. SciTePress, Barselona (2015). ISBN 978-989-758-096-3

20. Le, P.D., Nguyen, T.B.: OWL-based data cube for conceptual multidimensional data model. In: Proceedings of the First International Conference on Theories and Applications of Computer Science - ICTACS 2006, pp. 247–260. World Scientific Publishing, Ho Chi Minh (2006). ISBN 978-981-270-063-6

21. Viswanathan, G., Schneider, M.: BigCube: a metamodel for managing multidimensional data. In: Proceedings of the 19-th Conference on Software Engineering and Data Engineering - SEDE 2010, pp. 237–242. World Scientific Publishing, Singapore (2010). ISBN 978-981-270-063-6

22. Loh, Z.X., Ling, T.W., Ang, C.H., Lee, S.Y.: Adaptive method for range top-k queries in OLAP data cubes. In: Proceedings of International Conference on Information and Knowledge Management - CIKM 2002, pp. 60–67. ACM, New York (2002). ISBN 1-58113-492-4

23. Simić, D., Kurbalija, V., Budimac, Z.: An application of case-based reasoning in multidimensional database architecture. In: Kambayashi, Y., Mohania, M., Wöß, W. (eds.) DaWaK 2003. LNCS, vol. 2737, pp. 66–75. Springer, Heidelberg (2003). doi:10.1007/978-3-540-45228-7_8. ISBN 978-3-540-40807-9

24. Thanisch, P., Niemi, T., Niinimaki, M., Nummenmaa, J.: Using the entity-attribute-value model for OLAP cube construction. In: Grabis, J., Kirikova, M. (eds.) BIR 2011. LNBIP, vol. 90, pp. 59–72. Springer, Heidelberg (2011). doi:10.1007/978-3-642-24511-4_5. ISBN 978-3-642-24510-7

25. Viskov, A.V., Fomin, M.B.: Methods of description of possible combinations of signs and details while using the multidimensional models in infocomm systems. T-Comm. - Telecommun. Transp. 7, 45–47 (2012)

Author Index

Printed in the United States
By Bookmasters

Printed in the United States
By Bookmasters